S*pringer* **M***onographs* in **M***athematics*

More information about this series at http://www.springer.com/series/3733

Thomas E. Cecil • Patrick J. Ryan

Geometry of Hypersurfaces

Springer

Thomas E. Cecil
Department of Mathematics
and Computer Science
College of the Holy Cross
Worcester, MA, USA

Patrick J. Ryan
Department of Mathematics
and Statistics
McMaster University
Hamilton, ON, Canada

ISSN 1439-7382 ISSN 2196-9922 (electronic)
Springer Monographs in Mathematics
ISBN 978-1-4939-3245-0 ISBN 978-1-4939-3246-7 (eBook)
DOI 10.1007/978-1-4939-3246-7

Library of Congress Control Number: 2015947495

Mathematics Subject Classification (2010): 53B25, 53C40, 53C42

Springer New York Heidelberg Dordrecht London

Printed on acid-free paper

Springer Science+Business Media LLC New York is part of Springer Science+Business Media (www.springer.com)

To our wives,
Patsy and Ellen

Preface

The purpose of this book is to provide a thorough, up-to-date treatment of the differential geometry of hypersurfaces in real, complex, and quaternionic space forms. Special emphasis is placed on isoparametric and Dupin hypersurfaces in real space forms and Hopf hypersurfaces in complex space forms. An in-depth discussion of these topics and the contents of each chapter is given in the introduction.

The presentation is aimed at a reader who has completed a one-year graduate course in differential geometry and manifold theory. This book could be used for a second graduate course in differential geometry, a research seminar or as a reference.

The material in Chapters 2 and 3 has substantial overlap with our book *Tight and Taut Immersions of Manifolds* [95], published in 1985. For many topics, the order of the presentation has been changed significantly from our earlier book, and the material has been updated to include results published after 1985. Chapter 4 contains a brief introduction to submanifold theory in the context of Lie sphere geometry. This is studied in more detail in the book *Lie Sphere Geometry* [77], published in 2008. Parts of Chapter 5 on Dupin hypersurfaces are also treated in [77] or [95].

The material in Chapters 6–9 on real hypersurfaces in complex and quaternionic space forms was only mentioned briefly in our previous book [95]. The treatment of these subjects follows the notation and terminology of the survey article [399] of Niebergall and Ryan published in 1997.

All of the figures in this book are adapted from figures in the book [77]. These figures were constructed by Andrew D. Hwang, College of the Holy Cross, using his ePiX program for constructing figures in the LaTeXpicture environment. We are grateful to Professor Hwang for the excellent quality of the figures, and for his time and effort in constructing them. See the project page: http://math.holycross.edu/~ahwang/software/ePiX.html for more information on the ePiX program.

This book grew out of lectures given in the Differential Geometry Seminar of the Clavius Group during the summers of 2009–2014 at the University of Notre Dame, the College of the Holy Cross, Boston College, Fairfield University, and Loyola University Maryland. We are grateful to our fellow members of the Clavius

Group for their support of these lectures and for many enlightening remarks. We also acknowledge with gratitude the hospitality of the institutions mentioned above.

We wish to acknowledge the personal contributions to our understanding of this subject of several mathematicians both living and deceased. Katsumi Nomizu was our doctoral thesis advisor at Brown University, and he introduced us to the theory of isoparametric hypersurfaces in the early 1970s. Thomas Banchoff introduced us to tight and taut immersions and the cyclides of Dupin, and he has given us many important insights over the years. We are also deeply indebted to S.-S. Chern, Quo-Shin Chi, Thomas Ivey, Gary Jensen, and Ross Niebergall for sustained collaborations over many years.

We also gratefully acknowledge helpful conversations and correspondence on various aspects of the subject with Jürgen Berndt, Sheila Carter, Lawrence Conlon, José Carlos Díaz-Ramos, Miguel Domínguez-Vázquez, Josef Dorfmeister, Hermann Karcher, Makoto Kimura, Mayuko Kon, Nicolaas Kuiper, Hiroyuki Kurihara, Sadahiro Maeda, Martin Magid, Reiko Miyaoka, Thomas Murphy, Tetsuya Ozawa, Richard Palais, Juan de Dios Pérez, Ulrich Pinkall, Helmut Reckziegel, Paul Schweitzer, S.J., Young Jin Suh, Z.-Z. Tang, Chuu-Lian Terng, Gudlaugur Thorbergsson, McKenzie Wang, Alan West, and Andrew Whitman, S.J.

While writing this book during the period 2009–2015, Professor Cecil was supported by the Anthony and Renee Marlon Professorship in the Sciences, the Mary Louise Marfuggi Award, and by a sabbatical leave in academic year 2013–2014 at the College of the Holy Cross. He was also supported by a grant from the National Science Foundation (DMS–0405529). This support was invaluable in completing the project, and it is gratefully acknowledged.

We are also grateful to Elizabeth Loew, Ann Kostant, and the staff at Springer for their excellent professional support of this project.

Finally, we are most grateful to our wives, Patsy and Ellen, and to our families for their support and encouragement of this lengthy project.

Worcester, MA, USA — Thomas E. Cecil
Hamilton, ON, Canada — Patrick J. Ryan
July 2015

Contents

Chapter 1
Introduction

A smooth real-valued function F defined on a Riemannian manifold $\tilde{M}$ is called an *isoparametric function* if both of its classical Beltrami differential parameters $\Delta_1 F = |\text{grad } F|^2$ and $\Delta_2 F = \Delta F$ (Laplacian of F) are smooth functions of F itself. That is, both of the differential parameters are constant on each level set of F. An *isoparametric family* of $\tilde{M}$ is the collection of level sets of a nonconstant isoparametric function F on $\tilde{M}$.

In the case where $\tilde{M}$ is a real space form $\mathbf{R}^n$, S^n or H^n (hyperbolic space), a necessary and sufficient condition for an oriented hypersurface $M \subset \tilde{M}$ to belong to an isoparametric family is that all of its principal curvatures are constant (see Section 3.1). Thus, an oriented hypersurface of a real space form $\tilde{M}$ is called an *isoparametric hypersurface* if it has constant principal curvatures.

For $\tilde{M}$ equal to $\mathbf{R}^n$ or H^n, the classification of isoparametric hypersurfaces is complete and relatively simple, but as Cartan [52–55] showed in a series of four papers in 1938–1940, the subject is much deeper and more complicated for hypersurfaces in the sphere S^n.

A hypersurface M^{n-1} in a real space form $\tilde{M}^n$ is *proper Dupin* if the number g of distinct principal curvatures is constant on M^{n-1}, and each principal curvature function is constant along each leaf of its corresponding principal foliation. This is an important generalization of the isoparametric property that traces back to the book of Dupin [143] published in 1822. Proper Dupin hypersurfaces have been studied effectively in the context of Lie sphere geometry.

The theories of isoparametric and Dupin hypersurfaces are beautiful and filled with well-known important examples, and they have been analyzed from several points of view: geometric, algebraic, analytic, and topological. In this book, we cover the fundamental framework of these theories, and we study the main examples in detail. We also give a comprehensive treatment of the extension of these theories to real hypersurfaces with special curvature properties in complex and quaternionic space forms. We now give a brief overview of the contents of the book.

T.E. Cecil, P.J. Ryan, *Geometry of Hypersurfaces*, Springer Monographs in Mathematics, DOI 10.1007/978-1-4939-3246-7_1

Chapter 2 contains important results from the theory of submanifolds of real space forms that are needed in our study of isoparametric and Dupin hypersurfaces. In Sections 2.1–2.4, we find formulas for the shape operators of parallel hypersurfaces and tubes over submanifolds, and we discuss the focal submanifolds of a given submanifold. This leads naturally to the notions of curvature surfaces and Dupin hypersurfaces in Section 2.5. There we prove Pinkall's [446] local result (Theorem 2.25) which states that given any positive integer g, and any positive integers $m_1, \ldots, m_g$ with $m_1 + \cdots + m_g = n - 1$, there exists a proper Dupin hypersurface M^{n-1} in $\mathbf{R}^n$ with g distinct principal curvatures having respective multiplicities $m_1, \ldots, m_g$.

In Sections 2.6 and 2.7, we cover some basic results concerning tight and taut immersions of manifolds into real space forms. These are fundamental ideas in themselves, and they are needed to develop certain important results in the theory of isoparametric and Dupin hypersurfaces. In Section 2.8, we study the close relationship between the concepts of taut and Dupin submanifolds. Finally, we close the chapter with a treatment of the standard embeddings of projective spaces into Euclidean spaces. These examples play a significant role in the theories of tight, taut, and isoparametric hypersurfaces.

Chapter 3 is devoted to the basic theory of isoparametric hypersurfaces in real space forms developed primarily by Cartan [53–56] and Münzner [381, 382] (first published as preprints in the early 1970s). In Section 3.1, we describe the aspects of the theory that are common to all three space forms, and then prove the classification of isoparametric hypersurfaces M^{n-1} in Euclidean space $\mathbf{R}^n$ and in hyperbolic space H^n using Cartan's formula involving the principal curvatures of M^{n-1}.

The rest of the chapter is devoted to the much more complicated theory of isoparametric hypersurfaces in the sphere S^n. In Sections 3.2–3.6, we present Münzner's theory, including the proof that an isoparametric hypersurface in $S^n \subset \mathbf{R}^{n+1}$ with g distinct principal curvatures is always contained in a level set of a homogeneous polynomial of degree g on $\mathbf{R}^{n+1}$ satisfying certain differential equations on the length of its gradient and its Laplacian. From this it can be shown that every connected isoparametric hypersurface in S^n is contained in a unique compact, connected isoparametric hypersurface in S^n.

Using Münzner's construction, it can also be shown that each compact, connected isoparametric hypersurface $M^{n-1} \subset S^n$ has two focal submanifolds of codimension greater than one. These codimensions are determined by the multiplicities of the principal curvatures of M^{n-1}. From this it follows that M^{n-1} separates S^n into two ball bundles over these two focal submanifolds. Münzner then used cohomology theory to show that this topological situation implies that the number g of distinct principal curvatures of M^{n-1} can only be $1, 2, 3, 4$, or 6. At approximately the same time as Münzner's work, Takagi and Takahashi [511] classified homogeneous isoparametric hypersurfaces and found examples having g distinct principal curvatures for each of the values $g = 1, 2, 3, 4$ or 6.

Thorbergsson [533] applied Münzner's theory to show that the number g of distinct principal curvatures of a compact proper Dupin hypersurface M^{n-1} embedded in S^n is always $1, 2, 3, 4$, or 6, since M^{n-1} separates S^n into two ball bundles over two focal submanifolds of M^{n-1}, as in the isoparametric case. Several authors then used this same topological information to find a complete list of possibilities for the multiplicities of the principal curvatures of a compact proper Dupin hypersurface in S^n. This is discussed in Section 3.7.

In Section 3.8, we describe many important examples of isoparametric hypersurfaces in S^n from various points of view, and we discuss many classification results that have been obtained. Then in Section 3.9, we give a thorough treatment of the important paper of Ferus, Karcher, and Münzner [160], who used representations of Clifford algebras to construct an infinite collection of isoparametric hypersurfaces with $g = 4$ principal curvatures, now known as isoparametric hypersurfaces of *FKM-type*. Many of the hypersurfaces of FKM-type are not homogeneous. At the end of that section (see Subsection 3.9.1), we discuss progress that has been made on the classification of isoparametric hypersurfaces with four principal curvatures.

Isoparametric hypersurfaces in spheres have also occurred in considerations of several concepts in Riemannian geometry, such as the spectrum of the Laplacian, constant scalar curvature, and Willmore submanifolds. These applications and others are discussed in Section 3.10.

Chapter 4 describes the method for studying submanifolds of Euclidean space $\mathbf{R}^n$ or the sphere S^n in the setting of Lie sphere geometry. For proper Dupin hypersurfaces this has proven to be a valuable approach, since Dupin hypersurfaces occur naturally as envelopes of families of spheres, which can be handled well in Lie sphere geometry. Since the proper Dupin condition is invariant under Lie sphere transformations, this is also a natural setting for classification theorems. In Section 4.6, we formulate the related notion of tautness in the setting of Lie sphere geometry and prove that it is also invariant under Lie sphere transformations. The material in this chapter is covered in more detail in Chapters 2–4 of the book [77].

In Chapter 5, we study proper Dupin hypersurfaces in a real space form $\tilde{M}^n$ in detail. As noted above, proper Dupin hypersurfaces can also be studied in the context of Lie sphere geometry, and many classification results have been obtained in that setting. In this chapter, we use the viewpoint of the metric geometry of $\tilde{M}^n$ and that of Lie sphere geometry to obtain results about proper Dupin hypersurfaces.

An important class of proper Dupin hypersurfaces consists of the isoparametric hypersurfaces in S^n, and those hypersurfaces in $\mathbf{R}^n$ obtained from isoparametric hypersurfaces in S^n via stereographic projection. For example, the well-known cyclides of Dupin in $\mathbf{R}^3$ are obtained from a standard product torus $S^1(r) \times S^1(s) \subset S^3$, $r^2 + s^2 = 1$, in this way. These examples are discussed in more detail in Section 5.5.

However, in contrast to the situation for isoparametric hypersurfaces, there are both local and global aspects to the theory of proper Dupin hypersurfaces with quite different results. As noted above, Thorbergsson [533] proved that the restriction $g = 1, 2, 3, 4$, or 6 on the number of distinct principal curvatures of an isoparametric hypersurface in S^n also holds for a compact proper Dupin hypersurface M^{n-1} embedded in S^n. On the other hand, Pinkall [446] (see Theorem 2.25) showed that it is possible to construct a non-compact proper Dupin hypersurface with any number

g of distinct principal curvatures having any prescribed multiplicities. The proof involves Pinkall's standard constructions of building tubes, cylinders, and surfaces of revolution over lower dimensional Dupin submanifolds. These constructions are described in Section 5.1.

Pinkall's constructions lead naturally to proper Dupin hypersurfaces with the property that each point has a neighborhood with a local principal coordinate system, i.e., one in which the coordinate curves are principal curves. We discuss this type of hypersurface in Section 5.2. In particular, we show that if $M \subset S^n$ is an isoparametric hypersurface with $g \geq 3$ principal curvatures, then there does not exist any local principal coordinate system on M (see Pinkall [442, p. 42] and Cecil–Ryan [95, pp. 180–184]). We then give necessary and sufficient conditions for a hypersurface in a real space form with a fixed number g of distinct principal curvatures to have a local principal coordinate system in a neighborhood of each of its points.

An important notion in the local classification of proper Dupin hypersurfaces is reducibility. A proper Dupin hypersurface is called *reducible* if it is locally Lie equivalent to a proper Dupin hypersurface in $\mathbf{R}^n$ obtained as the result of one of Pinkall's standard constructions. In Section 5.3, we discuss reducible proper Dupin hypersurfaces in detail and develop Lie geometric criteria for reducibility.

In Section 5.4, we introduce the method of moving frames in Lie sphere geometry, which has been used to obtain local classifications of proper Dupin hypersurfaces with 2, 3, or 4 distinct principal curvatures. In Section 5.5, we use this method to give a complete local classification of proper Dupin hypersurfaces with $g = 2$ distinct principal curvatures, i.e., the cyclides of Dupin. This is a nineteenth century result for $n = 3$, and it was obtained in dimensions $n > 3$ by Pinkall [446] in 1985. In Sections 5.6 and 5.7, we discuss local classification results for the cases $g = 3$ and $g = 4$, respectively, that have been obtained using the moving frames approach.

As demonstrated by Thorbergsson's restriction on the number of distinct principal curvatures, compact proper Dupin hypersurfaces in S^n are relatively rare, and several important classification results have been obtained for them. These results are discussed in detail in Section 5.8 together with the important counterexamples of Pinkall-Thorbergsson [448] and Miyaoka–Ozawa [377] to the conjecture of Cecil and Ryan [95, p. 184] that every compact proper Dupin hypersurface embedded in S^n is Lie equivalent to an isoparametric hypersurface.

As noted earlier, the Dupin and taut conditions for submanifolds of real space forms are very closely related (see Section 2.8). In Sections 5.9 and 5.10, we discuss important classification results that have been obtained for taut submanifolds in Euclidean space $\mathbf{R}^n$. Many of these have been proven by using classifications of compact proper Dupin hypersurfaces.

The study of real hypersurfaces in complex projective space $\mathbf{CP}^n$ and complex hyperbolic space $\mathbf{CH}^n$ began at approximately the same time as Münzner's work on isoparametric hypersurfaces in spheres. A key early work was Takagi's [507] classification in 1973 of homogeneous real hypersurfaces in $\mathbf{CP}^n$. These hypersurfaces necessarily have constant principal curvatures, and they serve as model spaces for many subsequent classification theorems. Later Montiel [378] provided a similar list of standard examples in complex hyperbolic space $\mathbf{CH}^n$. These examples of Takagi and Montiel are presented in detail in Sections 6.3–6.5.

Let M be an oriented real hypersurface in $\mathbf{CP}^n$ or $\mathbf{CH}^n$, $n \geq 2$, with field of unit normals ξ. The hypersurface M is said to be *Hopf* if the *structure vector* $W = -J\xi$ is a principal vector at every point of M, where J is the complex structure on the ambient space. In that case, if $AW = \alpha W$, then α is called the *Hopf principal curvature* on M. A fundamental result is that the Hopf principal curvature α is always constant on a Hopf hypersurface M.

The hypersurfaces on the lists of Takagi and Montiel are Hopf hypersurfaces with constant principal curvatures. Furthermore, all tubes over complex submanifolds of $\mathbf{CP}^n$ or $\mathbf{CH}^n$ are Hopf. Conversely, if the rank of the focal map determined by the Hopf principal curvature is constant, then the image of that focal map is a complex submanifold of the ambient space. This was first shown by Cecil–Ryan [94] in $\mathbf{CP}^n$, and by Montiel [378] in $\mathbf{CH}^n$. These basic results concerning Hopf hypersurfaces are covered in Sections 6.6–6.8.

In Section 6.7, we study parallel hypersurfaces, focal sets and tubes over submanifolds of complex space forms using techniques similar to those used in Sections 2.2–2.4 for submanifolds of real space forms. This yields formulas that can be used to compute the principal curvatures of the hypersurfaces on the lists of Takagi and Montiel. In Section 6.9, we present an alternative approach to the study of parallel hypersurfaces and tubes using the method of Jacobi fields. This method has been effective in proving some important results in the field, and we will use it extensively.

Most of the examples on the lists of Takagi and Montiel are tubes over complex submanifolds. In Chapter 7, we study the basic geometry of complex submanifolds in complex space forms, and we focus on certain important examples in $\mathbf{CP}^n$ that arise in the classification of Hopf hypersurfaces with constant principal curvatures. Specifically, in Sections 7.2–7.5, we determine the behavior of the principal curvatures of the Veronese embedding of $\mathbf{CP}^m$ in $\mathbf{CP}^n$, the Segre embedding of $\mathbf{CP}^h \times \mathbf{CP}^k$ in $\mathbf{CP}^n$, the Plücker embedding of complex Grassmannians in $\mathbf{CP}^n$, and the half-spin embedding of $SO(2d)/U(d)$ in $\mathbf{CP}^n$.

In Chapter 8, we present the classification of Hopf hypersurfaces with constant principal curvatures. This is due to Kimura [270] in $\mathbf{CP}^n$ and to Berndt [27] in $\mathbf{CH}^n$. Simply stated, these theorems say that any connected Hopf hypersurface in a complex space form is an open subset of a hypersurface on Takagi's list for $\mathbf{CP}^n$, and on Montiel's list for $\mathbf{CH}^n$. These classifications are major results in the field.

In the case of $\mathbf{CH}^n$, the classification follows from a generalization to complex space forms of Cartan's formula for isoparametric hypersurfaces in real space forms (see Section 8.1). The proof of the classification theorem in the case of $\mathbf{CP}^n$ is more involved. A Hopf hypersurface M in $\mathbf{CP}^n$ with constant principal curvatures gives rise to an isoparametric hypersurface $\tilde{M} = \pi^{-1}M$ in the sphere S^{2n+1}, where $\pi : S^{2n+1} \to \mathbf{CP}^n$ is the Hopf fibration. By Münzner's results, the number $\tilde{g}$ of principal curvatures of $\tilde{M}$ can only be $1, 2, 3, 4$, or 6. By a careful analysis of the relationship between the principal curvatures of M and those of $\tilde{M}$, one can show that the number

g of principal curvatures of M is either 2, 3 or 5, the same as for the hypersurfaces on Takagi's list. When $g = 2$ or 3, we prove that M is an open subset of a hypersurface on Takagi's list by an elementary argument involving the shape operators of the complex focal submanifold of M corresponding to the Hopf principal curvature α.

In the case $g = 5$, the classification is much more difficult. We first prove that the complex focal submanifold determined by α is a parallel submanifold, i.e., it has parallel second fundamental form. Such parallel complex submanifolds of $\mathbf{CP}^n$ were classified by Nakagawa and Takagi [391], and the list includes the special embeddings studied in Chapter 7. Using the analysis of the shape operators of these parallel submanifolds in Sections 7.2–7.5, we ultimately determine which parallel submanifolds have tubes with constant principal curvatures. From this we can deduce that every connected Hopf hypersurface with constant principal curvatures in $\mathbf{CP}^n$ is an open subset of a hypersurface on Takagi's list.

In Section 8.5, we study other characterizations of the hypersurfaces on the lists of Takagi and Montiel. In particular, a real hypersurface M in $\mathbf{CP}^n$ or $\mathbf{CH}^n$ is said to be *pseudo-Einstein* if there exist functions ρ and σ on M such that the Ricci tensor S of M satisfies the equation $SX = \rho X + \sigma \langle X, W\rangle W$, for all tangent vectors X to M, where W is the structure vector defined above.

Of course, if σ is identically zero, then M is Einstein, but there do not exist any Einstein real hypersurfaces in $\mathbf{CP}^n$ or $\mathbf{CH}^n$. For $n \geq 3$, Cecil and Ryan [94] proved in 1982 that a pseudo-Einstein hypersurface in $\mathbf{CP}^n$ is an open subset of a geodesic sphere, a tube of a certain radius over a totally geodesic $\mathbf{CP}^k$, $1 \leq k \leq n-2$, or a tube of a certain radius over a complex quadric $Q^{n-1} \subset \mathbf{CP}^n$. All of these hypersurfaces are on Takagi's list. M. Kon [289] obtained the same conclusion in 1979 under the assumption that the functions ρ and σ are constant.

In 1985, Montiel [378] showed that a pseudo-Einstein hypersurface in $\mathbf{CH}^n$, $n \geq 3$, is an open subset of a geodesic sphere, a tube over a complex hyperplane, or a horosphere. It is important to note that the classifications of pseudo-Einstein hypersurfaces in $\mathbf{CP}^2$ by H.S. Kim and Ryan [260], and in $\mathbf{CH}^2$ by Ivey and Ryan [222], are different than the classification theorems for $n \geq 3$. These classifications of pseudo-Einstein hypersurfaces will be discussed in detail in Section 8.5. There we also study several other related classifications of hypersurfaces based on conditions on the shape operator, the curvature tensor, or the Ricci tensor.

In Section 8.6, we study non-Hopf hypersurfaces in complex space forms. These include the *Berndt orbits*, which are a family of non-Hopf homogeneous hypersurfaces in $\mathbf{CH}^n$ for $n \geq 2$ having three distinct constant principal curvatures. In Section 8.7, we discuss various ways to extend the definition of "isoparametric" to hypersurfaces in complex forms. These formulations of the concept are equivalent for hypersurfaces in real space forms, but different in complex space forms. In Section 8.8, we discuss some open problems that remain in the theory of real hypersurfaces in complex space forms.

In 1986, Martinez and Pérez [353] began the study of real hypersurfaces in quaternionic space forms, and in 1991 Berndt [28] found a list of standard examples of real hypersurfaces in quaternionic space forms with constant principal curvatures, leading to further research in this area. These examples together with classification results and open problems are described in Chapter 9.

In this book, all manifolds and maps are taken to be smooth, i.e., C^∞, unless explicitly stated otherwise. Notation generally follows the book of Kobayashi and Nomizu [283].

Chapter 2
Submanifolds of Real Space Forms

In this chapter, we review the basic theory of submanifolds of real space forms needed for our in-depth treatment of isoparametric and Dupin hypersurfaces in later chapters. In Sections 2.1–2.4, we find the formulas for the shape operators of parallel hypersurfaces and tubes over submanifolds, and we discuss the focal submanifolds of a given submanifold.

We then define curvature surfaces and Dupin hypersurfaces in Section 2.5, and prove Pinkall's [446] result (Theorem 2.25) that given any positive integer g, and any positive integers $m_1, \ldots, m_g$ with $m_1 + \cdots + m_g = n - 1$, there exists a proper Dupin hypersurface M^{n-1} in $\mathbf{R}^n$ with g distinct principal curvatures having respective multiplicities $m_1, \ldots, m_g$.

In the next two sections, we define the notions of tight and taut immersions of manifolds into real space forms and develop the basic properties of these types of immersions. These concepts are important in themselves, and they are needed in the theory of isoparametric and Dupin hypersurfaces. In Section 2.8, we study the close relationship between the concepts of taut and Dupin submanifolds in detail.

Finally, in Section 2.9, we describe the standard embeddings of projective spaces into Euclidean spaces. These examples have many remarkable properties, and they are important in the theories of tight, taut, and isoparametric hypersurfaces.

2.1 Real Space Forms

We let $\mathbf{R}^n$ denote n-dimensional Euclidean space endowed with the standard Euclidean metric of constant sectional curvature zero. The theory of isoparametric and Dupin hypersurfaces in the sphere $S^n(c)$ of constant sectional curvature $c > 0$ is essentially the same for all values of $c > 0$, and so we restrict our attention to the sphere S^n of constant sectional curvature 1, that is, the unit sphere in $\mathbf{R}^{n+1}$ with the Riemannian metric induced from the Euclidean metric in $\mathbf{R}^{n+1}$.

T.E. Cecil, P.J. Ryan, *Geometry of Hypersurfaces*, Springer Monographs in Mathematics, DOI 10.1007/978-1-4939-3246-7_2

Similarly, for ambient spaces of constant negative sectional curvature, we restrict our attention to the hyperbolic space H^n of constant sectional curvature -1. To get a model for H^n, we consider the Lorentz space $\mathbf{R}_1^{n+1}$ endowed with the Lorentz metric of signature $(1, n)$,

$$\langle x, y\rangle = x_1y_1 + \cdots + x_ny_n - x_{n+1}y_{n+1}, \tag{2.1}$$

for $x = (x_1, \ldots, x_{n+1})$ and $y = (y_1, \ldots, y_{n+1})$ in $\mathbf{R}_1^{n+1}$. Then real hyperbolic space of constant sectional curvature -1 is the hypersurface in $\mathbf{R}_1^{n+1}$ given by

$$H^n = \{x \in \mathbf{R}_1^{n+1} \mid \langle x, x\rangle = -1,\ x_{n+1} \geq 1\}, \tag{2.2}$$

on which the Lorentz metric $\langle\ ,\ \rangle$ restricts to a Riemannian metric of constant sectional curvature -1 (see Kobayashi–Nomizu [283, Vol. II, pp. 268–271] for more detail).

By a real *space form* of dimension n, we mean a complete, connected, simply connected manifold $\tilde{M}^n$ with constant sectional curvature c. If $c = 0$, then $\tilde{M}^n = \mathbf{R}^n$; if $c = 1$, then $\tilde{M}^n = S^n$, and if $c = -1$, then $\tilde{M}^n = H^n$ (see, for example, [283, Vol. I, pp. 204–209]).

Let $f : M^n \to \tilde{M}^{n+k}$ for $k \geq 1$ be an immersion with codimension k of an n-dimensional manifold M into one of the three space forms $\tilde{M}^{n+k}$ mentioned above. For $x \in M$, let T_xM denote the tangent space to M at x, and let $T_x^{\perp}M$ denote the normal space to $f(M)$ at the point $f(x) \in \tilde{M}$. Let

$$NM = \{(x, \xi) \mid x \in M,\ \xi \in T_x^{\perp}M\}, \tag{2.3}$$

be the normal bundle of $f(M)$ with natural bundle projection $\pi : NM \to M$ defined by $\pi(x, \xi) = x$. Let η be a local cross section of NM. For any vector X in the tangent space T_xM, we have the fundamental equation

$$\tilde{\nabla}_{f_*(X)}\eta = -f_*(A_\eta X) + \nabla^{\perp}_{f_*(X)}\eta, \tag{2.4}$$

where $\tilde{\nabla}$ is the Levi-Civita connection in $\tilde{M}$, f_* is the differential of f, A_η is the *shape operator* determined by the normal vector $\eta(x)$, and $\nabla^{\perp}$ is the connection in the normal bundle.

The shape operator defines smooth map $(x, \xi) \mapsto A_\xi$ from the normal bundle NM into the space of symmetric tensors of type $(1, 1)$ on M. An eigenvalue λ of A_ξ is called a *principal curvature* of A_ξ, and its corresponding eigenvector is called a *principal vector*. Since $A_{t\xi} = tA_\xi$, for $t \in \mathbf{R}$, it is sufficient to know the principal curvatures on the bundle BM of unit normal vectors to M, i.e., the *unit normal bundle* of M.

2.2 Focal Points

Let $f : M \to \tilde{M}$ be an embedded submanifold of a real space form. Let $T\tilde{M}$ denote the tangent bundle of $\tilde{M}$, and let $\exp : T\tilde{M} \to \tilde{M}$ be the exponential map of $\tilde{M}$. The *normal exponential map* or *end-point map* $E : NM \to \tilde{M}$ is the restriction of the exponential map of $\tilde{M}$ to the normal bundle NM of the submanifold M. Thus, if ξ is a nonzero normal vector to $f(M)$ at $f(x)$, then $E(x, \xi)$ is the point of $\tilde{M}$ reached by traversing a distance $|\xi|$ along the geodesic in $\tilde{M}$ with initial point $f(x)$ and initial tangent vector ξ. If ξ is the zero vector in the tangent space to $\tilde{M}$ at $f(x)$, then $E(x, \xi)$ is the point $f(x)$. It is well known (see, for example, [283, Vol. I, p. 147]) that exp is smooth in a neighborhood of the 0-section in $T\tilde{M}$, and so E is also smooth in a neighborhood of the 0-section in NM. It is easy to show that the differential E_* is nonsingular at points on the zero section, so we restrict our attention to points in NM that are not in the 0-section in trying to locate the critical values of E.

The focal points of M are the critical values of the normal exponential map E. Specifically, a point $p \in \tilde{M}$ is called a *focal point of* (M, x) *of multiplicity* m if $p = E(x, \xi)$ and the differential E_* at the point (x, ξ) has nullity $m > 0$. The *focal set* of M is the set of all focal points of (M, x) for all $x \in M$. Since NM and $\tilde{M}$ have the same dimension, it follows from Sard's Theorem (see, for example, [359, p. 33]) that the focal set of M has measure zero in $\tilde{M}$.

We now assume that ξ is a unit length normal vector to $f(M)$ at a point $x \in M$. The following theorem gives the location of the focal points of (M, x) along the geodesic $E(x, t\xi)$, for $t \in \mathbf{R}$, in terms of the eigenvalues of the shape operator A_ξ at x. We will give a proof for part (a) of the theorem, the case $\tilde{M}^{n+k} = \mathbf{R}^{n+k}$. (See also Milnor [359, pp. 32–35] for a proof in the Euclidean case, and Cecil [70] for a proof in the hyperbolic case. The proof in the spherical case is similar to that in the hyperbolic case.)

Theorem 2.1. *Let $f : M^n \to \tilde{M}^{n+k}$ be a submanifold of a real space form $\tilde{M}^{n+k}$, and let ξ be a unit normal vector to $f(M^n)$ at $f(x)$. Then $p = E(x, t\xi)$ is a focal point of (M^n, x) of multiplicity $m > 0$ if and only if there is an eigenvalue λ of the shape operator A_ξ of multiplicity m such that*

(a) $\lambda = 1/t$, *if* $\tilde{M}^{n+k} = \mathbf{R}^{n+k}$,
(b) $\lambda = \cot t$, *if* $\tilde{M}^{n+k} = S^{n+k}$,
(c) $\lambda = \coth t$, *if* $\tilde{M}^{n+k} = H^{n+k}$.

Proof. (a) In the following local calculation, we consider $M^n \subset \mathbf{R}^{n+k}$ as an embedded submanifold and do not mention the embedding f explicitly. We also consider the tangent space T_xM to be a subspace of $T_x\mathbf{R}^{n+k}$. We first recall some standard terminology and equations of submanifold theory. We will denote the Levi-Civita connection on $\mathbf{R}^{n+k}$ by D rather than $\tilde{\nabla}$. For locally defined smooth vector fields X and Y defined on M, we have the decomposition of D_XY into tangential and normal components,

$$D_XY = \nabla_XY + \sigma(X, Y), \tag{2.5}$$

which defines the *Levi-Civita connection* ∇ of the induced Riemannian metric on M and the *second fundamental form* σ. For a local field of unit normal vectors ξ on M, we have the decomposition of $D_X\xi$ into tangential and normal components,

$$D_X\xi = -A_\xi X + \nabla^\perp_X \xi, \tag{2.6}$$

which defines the shape operator A_ξ and normal connection $\nabla^\perp$, as in equation (2.4) above.

Since we know that there are no focal points on the 0-section, we will compute E_* at a point of NM that is not on the 0-section. We can consider this point to have the form $(x, t\xi)$, where $|\xi| = 1$ and $t > 0$. Let $\xi_1, \ldots, \xi_k$ be an orthonormal frame of normal vectors to M at x with $\xi_1 = \xi$. Let U be a normal coordinate neighborhood of x in M as defined in [283, Vol. I, p. 148]. In order to simplify the calculations below, we extend $\xi_1, \ldots, \xi_k$ to orthonormal normal vector fields on U by parallel translation with respect to the normal connection $\nabla^\perp$ along geodesics in U through x.

Let $\{\epsilon_1, \ldots, \epsilon_k\}$ be the standard orthonormal basis of $\mathbf{R}^k$. Let S^{k-1} be the unit sphere in $\mathbf{R}^k$ given by

$$S^{k-1} = \{a = \sum_{j=1}^{k} a_j\epsilon_j \mid a_1^2 + \cdots + a_k^2 = 1\}. \tag{2.7}$$

We parametrize the normal bundle NM locally in a neighborhood of the point $(x, t\xi)$ by defining

$$\Psi : (0, \infty) \times S^{k-1} \times U \to NM \tag{2.8}$$

by

$$\Psi(\mu, a, y) = \mu \sum_{j=1}^{k} a_j\xi_j(y), \tag{2.9}$$

where the vector $\Psi(\mu, a, y)$ is normal to M at the point $y \in U$.

Then $(E \circ \Psi)(\mu, a, y)$ is the point in $\mathbf{R}^{n+k}$ reached by traversing a distance μ along the geodesic in $\mathbf{R}^{n+k}$ beginning at y and having initial direction

$$\sum_{j=1}^{k} a_j\xi_j(y). \tag{2.10}$$

That is,

$$(E \circ \Psi)(\mu, a, y) = y + \mu \sum_{j=1}^{k} a_j\xi_j(y). \tag{2.11}$$

In this local parametrization, the point $(x, t\xi)$ is equal to $\Psi(t, \epsilon_1, x)$. Evaluating E_* at $(x, t\xi)$ is equivalent to evaluating $(E \circ \Psi)_*$ at the point (t, ϵ_1, x). We now want to express $(E \circ \Psi)_*$ in terms of a basis consisting of $\partial/\partial\mu$ for $(0, \infty)$, $\{\epsilon_j\}$, $2 \leq j \leq k$, for $T_{\epsilon_1}S^{k-1}$, and an orthonormal basis of T_xM consisting of eigenvectors X of A_ξ with corresponding eigenvalues denoted by λ.

We first evaluate $(E \circ \Psi)_*(\partial/\partial\mu)$ at the point (t, ϵ_1, x). We have

$$(E \circ \Psi)_*(\partial/\partial\mu) = \overrightarrow{\beta}(\mu)|_{\mu=t}, \text{ where } \beta(\mu) = x + \mu\, \xi_1(x), \tag{2.12}$$

where $\overrightarrow{\beta}(\mu)$ is the velocity vector (tangent vector) of the curve $\beta(\mu)$.

Thus, we get

$$(E \circ \Psi)_*(\partial/\partial\mu) = \xi_1(x) = \xi. \tag{2.13}$$

Next, the tangent space $T_{\epsilon_1}S^{k-1}$ has an orthonormal basis $\{\epsilon_2, \ldots, \epsilon_k\}$. We want to compute $(E \circ \Psi)_*\epsilon_j$ for $2 \leq j \leq k$. In S^{k-1}, the curve

$$\gamma(s) = \cos s\, \epsilon_1 + \sin s\, \epsilon_j \tag{2.14}$$

has initial point ϵ_1 and initial velocity vector ϵ_j. Thus by equation (2.11), we see that $(E \circ \Psi)_*\epsilon_j$ is the initial velocity vector to the curve

$$\beta(s) = x + t\,(\cos s\, \xi_1(x) + \sin s\, \xi_j(x)). \tag{2.15}$$

Differentiating with respect to s and substituting $s = 0$, we get

$$(E \circ \Psi)_*\epsilon_j = t\xi_j(x). \tag{2.16}$$

Equations (2.13) and (2.16) show that if

$$V = c_1 \left(\frac{\partial}{\partial\mu}\right) + \sum_{j=2}^{k} c_j\epsilon_j, \tag{2.17}$$

then $(E \circ \Psi)_*V = 0$ only if $V = 0$.

Next we compute $(E \circ \Psi)_*X$ for $X \in T_xM$. If $\delta(s)$ is a curve in U with initial point x and initial velocity vector X, then $(E \circ \Psi)_*X$ is the initial velocity vector to the curve

$$\zeta(s) = (E \circ \Psi)\delta(s) = \delta(s) + t\, \xi_1(\delta(s)). \tag{2.18}$$

Differentiating with respect to s and using $\delta(0) = x$ and $\overrightarrow{\delta}(0) = X$, we get

$$\overrightarrow{\zeta}(0) = X + t\, D_X\xi_1. \tag{2.19}$$

We know that $D_X\xi_1 = -A_{\xi_1}X + \nabla^\perp_X\xi_1$, and we have constructed ξ_1 so that $\xi_1(x) = \xi$ and $\nabla^\perp_X\xi_1 = 0$. Hence, we have

$$(E \circ \Psi)_* X = X - tA_\xi X = (I - tA_\xi)X, \tag{2.20}$$

where we are identifying X with its Euclidean parallel translate at the point $p = E(x, t\xi)$.

From equations (2.13), (2.16), and (2.20), we see that for V as in equation (2.17) and a nonzero $X \in T_xM$, we have $(E \circ \Psi)_*(X + V) = 0$ if and only if $V = 0$ and $A_\xi X = (1/t)X$, i.e., $1/t$ is an eigenvalue of A_ξ with eigenvector X. Furthermore, if $\lambda = 1/t$ is an eigenvalue of A_ξ, then the nullity of E_* at $(x, t\xi)$ is equal to the dimension of the eigenspace T_λ, i.e., the multiplicity m of λ. This completes the proof of the theorem. □

2.3 Tubes and Parallel Hypersurfaces

As above, let $f : M^n \to \tilde{M}^{n+k}$ be an immersion into a real space form, and let BM denote the bundle of unit normal vectors to $f(M)$ in $\tilde{M}$. If the codimension k is greater than one, then we define the *tube of radius* $t > 0$ *over* M by the map $f_t : BM \to \tilde{M}$,

$$f_t(x, \xi) = E(x, t\xi). \tag{2.21}$$

If $(x, t\xi)$ is not a critical point of E, then f_t is an immersion in a neighborhood of (x, ξ) in BM. It follows from Theorem 2.1 that given any point $x \in M$, there is a neighborhood U of x in M such that for all $t > 0$ sufficiently small, the restriction of f_t to the unit normal bundle BU over U is an immersion onto an $(n + k - 1)$-dimensional manifold, which is geometrically a tube of radius t over U.

In the case where M is a hypersurface, i.e., the codimension $k = 1$, then BM is a double covering of M. In that case, for local calculations, we can assume that M is orientable with a local field of unit normal vectors ξ. Then we consider the *parallel hypersurface* $f_t : M \to \tilde{M}$ given by

$$f_t(x) = E(x, t\xi), \tag{2.22}$$

for $t \in \mathbf{R}$, rather than defining f_t on the double covering BM. Note that t can take any real value in this case. For a negative value of t, the parallel hypersurface lies locally on the side of M in the direction of the unit normal field $-\xi$, instead of on the side of M in the direction of ξ. For $t = 0$, we have $f_0 = f$, the original hypersurface.

In this section, we will compute the principal curvatures of the tube f_t in terms of the principal curvatures of the original submanifold M. We will treat the case of codimension $k > 1$ here. The case of codimension $k = 1$ is similar and is actually easier, and we omit it here. The formulas in Theorem 2.2 below work for the case of codimension 1 also, except that there are no ϵ_j in that case. We will handle the case $\tilde{M} = \mathbf{R}^{n+k}$ here. The calculations for the other space forms are similar and are left to the reader.

As in the preceding section, in the following local calculation we consider $M^n \subset \mathbf{R}^{n+k}$ as an embedded submanifold and do not mention the embedding f explicitly. We also consider the tangent space T_xM to be a subspace of $T_x\mathbf{R}^{n+k}$. Let (x, ξ) be a point in BM such that f_t is an immersion at (x, ξ), i.e., $(x, t\xi)$ is not a critical point of E. Let $\xi_1, \ldots, \xi_k$ be an orthonormal frame of normal vectors to M at x with $\xi_1 = \xi$. Let U be a normal coordinate neighborhood of x in M. We extend $\xi_1, \ldots, \xi_k$ to orthonormal normal vector fields on U by parallel translation with respect to the normal connection $\nabla^\perp$ along geodesics in U through x. Thus, we have the same setup as for the calculations in the proof of Theorem 2.1.

As in the proof of Theorem 2.1, let $\{\epsilon_1, \ldots, \epsilon_k\}$ be the standard orthonormal basis of $\mathbf{R}^k$. Let S^{k-1} be the unit sphere in $\mathbf{R}^k$ given by

$$S^{k-1} = \{a = \sum_{j=1}^{k} a_j \epsilon_j \mid a_1^2 + \cdots + a_k^2 = 1\}. \tag{2.23}$$

We parametrize the unit normal bundle BM locally in a neighborhood of the point (x, ξ) by defining

$$\Psi : S^{k-1} \times U \to BM \tag{2.24}$$

by

$$\Psi(a, y) = \sum_{j=1}^{k} a_j \xi_j(y), \tag{2.25}$$

where the vector $\Psi(a, y)$ is a unit normal vector to M at the point $y \in U$.

In this local parametrization, the point (x, ξ) in BM is equal to $\Psi(\epsilon_1, x)$. Evaluating $(f_t)_*$ at (x, ξ) is equivalent to evaluating $(f_t \circ \Psi)_*$ at the point (ϵ_1, x). We now want to express $(f_t \circ \Psi)_*$ at (ϵ_1, x) in terms of a basis consisting of $\{\epsilon_j\}$, $2 \leq j \leq k$, for $T_{\epsilon_1}S^{k-1}$, and an orthonormal basis of T_xM consisting of eigenvectors X of A_ξ with corresponding eigenvalues denoted by λ.

The calculations of $(f_t \circ \Psi)_*$ are exactly the same as the calculations of $(E \circ \Psi)_*$ in the proof of Theorem 2.1, except that there is no $\partial/\partial\mu$ term. Specifically, as in equation (2.16), we get

$$(f_t \circ \Psi)_* \epsilon_j = t\xi_j(x). \tag{2.26}$$

Then for $X \in T_xM$, we get as in equation (2.20),

$$(f_t \circ \Psi)_* X = X - tA_\xi X = (I - tA_\xi)X, \tag{2.27}$$

where we are identifying X with its Euclidean parallel translate at the point $p = f_t(x, \xi)$.

Since f_t is an immersion at (x, ξ), there is a neighborhood W of the point (x, ξ) in the unit normal bundle BU such that the restriction of f_t to W is an embedded hypersurface in $\mathbf{R}^{n+k}$. To find the shape operator of $f_t W$, we need to find a local field of unit normals to $f_t W$, and then compute its covariant derivative.

If (u, η) is an arbitrary point of W, then the Euclidean parallel translate of η is a unit normal to the hypersurface $f_t W$ at the point $f_t(u, \eta)$. So we now let η denote a field of unit normals to the hypersurface $f_t W$ on the neighborhood W. We denote the corresponding shape operator of the oriented hypersurface $f_t W$ by A_t.

We use the same local parametrization of BM given above. We can identify the tangent space $T_{(x,\xi)}BM$ with $T_{\epsilon_1}S^{k-1} \times T_xM$ via the parametrization Ψ, and we can consider the shape operator A_t to be defined on $T_{\epsilon_1}S^{k-1} \times T_xM$. In particular, A_t is defined by,

$$(f_t \circ \Psi)_*(A_t Z) = -D_{(f_t \circ \Psi)_* Z}\,\eta, \tag{2.28}$$

for $Z \in T_{\epsilon_1}S^{k-1} \times T_xM$. Note that there is no term involving the normal connection $\nabla^\perp$, since the codimension of $f_t W$ is one.

We first compute $A_t\epsilon_j$ for $2 \leq j \leq k$. As in equation (2.15), we have that $(f_t \circ \Psi)_*\epsilon_j$ is the initial velocity vector to the curve

$$\beta(s) = x + t\,(\cos s\ \xi_1(x) + \sin s\ \xi_j(x)). \tag{2.29}$$

Hence, $D_{(f_t \circ \Psi)_*\epsilon_j}\eta$ is the initial velocity vector $\overrightarrow{\eta}\,(0)$ to the curve

$$\eta(\beta(s)) = \cos s\ \xi_1(x) + \sin s\ \xi_j(x). \tag{2.30}$$

Therefore, we have

$$(f_t \circ \Psi)_*(A_t\epsilon_j) = -\overrightarrow{\eta}\,(0) = -\xi_j(x). \tag{2.31}$$

Since we have $(f_t \circ \Psi)_*\epsilon_j = t\xi_j(x)$ by equation (2.26), we get

$$A_t\epsilon_j = -\frac{1}{t}\epsilon_j. \tag{2.32}$$

Thus, ϵ_j is a principal vector of A_t with corresponding principal curvature $-1/t$, where t is the radius of the tube.

Next we find A_tX for a vector $X \in T_xM$. Let $\delta(s)$ be a curve in M with initial point $\delta(0) = x$ and initial velocity vector $\overrightarrow{\delta}(0) = X$. Then $(f_t \circ \Psi)_*X$ is the initial velocity vector to the curve

$$\zeta(s) = \delta(s) + t\,\xi_1(\delta(s)). \tag{2.33}$$

Along this curve $\zeta(s)$, the unit normal field η to the tube is given by

$$\eta(\delta(s)) = \xi_1(\delta(s)). \tag{2.34}$$

Then $D_{(f_t \circ \Psi)_*X}\eta$ is the initial velocity vector to this curve $\xi_1(\delta(s))$, which is just $D_X\xi_1$, where again we are identifying parallel vectors in $\mathbf{R}^{n+k}$. Then using the fact that $\nabla^\perp_X \xi_1 = 0$, we get from equation (2.4)

$$D_{(f_t \circ \Psi)_*X}\eta = D_X\xi_1 = -A_\xi X, \tag{2.35}$$

since $\xi_1(x) = \xi$. Thus we have from equation (2.28) that $(f_t \circ \Psi)_*(A_tX) = A_\xi X$. Then it follows from equation (2.27) for $(f_t \circ \Psi)_*X$ that

$$A_tX = (I - tA_\xi)^{-1}A_\xi X. \tag{2.36}$$

In the case of a principal vector X such that $A_\xi X = \lambda X$, this reduces to

$$A_tX = \frac{\lambda}{1 - t\lambda}X. \tag{2.37}$$

Therefore, X is a principal vector of A_t with corresponding principal curvature $\lambda/(1 - t\lambda)$.

Principal curvatures of a tube

In summary, we have the following theorem for the shape operators of a tube over a submanifold of Euclidean space $\mathbf{R}^{n+k}$. Similar computations to those above yield the results for submanifolds of S^{n+k} and H^{n+k}, which are also stated in the theorem. In the case $k = 1$, the theorem gives the formula for the shape operator of a parallel hypersurface f_tM. In that case, there are no terms $A_t\epsilon_j$.

Theorem 2.2. *Let M^n be a submanifold of a real space form $\tilde{M}^{n+k}$ and ξ a unit normal vector to M at x such that $f_t : BM \to \tilde{M}^{n+k}$ is an immersion at the point $(x, \xi) \in BM$. Let $\{X_1, \ldots, X_n\}$ be a basis of T_xM consisting of principal vectors of A_ξ with $A_\xi X_i = \lambda_i X_i$ for $1 \leq i \leq n$. In terms of the local parametrization of BM given in this section, the shape operator A_t of the tube f_t of radius t over M at the point (x, ξ) is given in terms of its principal vectors as follows:*

For submanifolds of $\mathbf{R}^{n+k}$,

(1) *For* $2 \leq j \leq k$, $A_t \epsilon_j = -\frac{1}{t}\, \epsilon_j$,
(2) *For* $1 \leq i \leq n$, $A_t X_i = \frac{\lambda_i}{1-t\lambda_i}\, X_i$.

For submanifolds of S^{n+k},

(1) *For* $2 \leq j \leq k$, $A_t \epsilon_j = -\cot t\, \epsilon_j$,
(2) *For* $1 \leq i \leq n$, $A_t X_i = \cot(\theta_i - t)\, X_i$, *if* $\lambda_i = \cot \theta_i$, $0 < \theta_i < \pi$.

For submanifolds of H^{n+k},

(1) *For* $2 \leq j \leq k$, $A_t \epsilon_j = -\coth t\, \epsilon_j$,
(2) *For* $1 \leq i \leq n$,
   (a) $A_t X_i = \coth(\theta_i - t)\, X_i$, *if* $|\lambda_i| > 1$, *and* $\lambda_i = \coth \theta_i$,
   (b) $A_t X_i = \pm X_i$, *if* $\lambda_i = \pm 1$,
   (c) $A_t X_i = \tanh(\theta_i - t)\, X_i$, *if* $|\lambda_i| < 1$, *and* $\lambda_i = \tanh \theta_i$.

As a consequence of Theorems 2.1 and 2.2, we obtain the following useful result. In the case where M has codimension $k > 1$, the points of M are focal points of the tube $f_t M$ corresponding to the principal curvature $\mu = -1/t$ of A_t in the case $\tilde{M}^{n+k} = \mathbf{R}^{n+k}$, $\mu = -\cot t$ in the case $\tilde{M}^{n+k} = S^{n+k}$, and $\mu = -\coth t$ in the case $\tilde{M}^{n+k} = H^{n+k}$.

Theorem 2.3. *Let* M^n *be a submanifold of a real space form* $\tilde{M}^{n+k}$ *and* t *a real number such that* $f_t M$ *is a hypersurface.*

(a) *If* M *is a hypersurface, then the focal set of the parallel hypersurface* $f_t M$ *is the focal set of* M.
(b) *If* M *has codimension greater than one, then the focal set of the tube* $f_t M$ *consists of the union of the focal set of* M *with* M *itself.*

2.4 Focal Submanifolds

In this section, we find a natural manifold structure for the sheet of the focal set of a hypersurface of a real space form corresponding to a principal curvature of constant multiplicity. By considering tubes and using Theorems 2.2 and 2.3, this also enables us to give a manifold structure to a sheet of the focal set of a submanifold of codimension greater than one. These results were originally obtained in the paper of Cecil and Ryan [88], and they were suggested by the work of Nomizu [403], who obtained similar results for the sheets of the focal set of an isoparametric hypersurface. See also the related work of Reckziegel [457–459].

Let $f : M^n \to \tilde{M}^{n+1}$ be an immersed hypersurface in a real space form $\tilde{M}$. For the following local considerations, we assume that $f(M)$ is orientable with a global field of unit normals ξ and corresponding shape operator $A = A_\xi$. If the principal curvature functions on M are ordered as

$$\lambda_1 \geq \lambda_2 \geq \cdots \geq \lambda_n, \tag{2.38}$$

then each λ_i is a continuous function (see Ryan [468, p. 371]). Furthermore, if a continuous principal curvature function λ has constant multiplicity m on M, then λ is a smooth function, and its m-dimensional distribution T_λ of principal vectors is also smooth on M (see, for example, Nomizu [402], Reckziegel [457, 458], or Singley [486]). We will show in this section that T_λ is also integrable, and so it is an m-dimensional foliation on M called the *principal foliation* corresponding to the principal curvature λ. Using this fact, we will then show that if λ is constant along each leaf of T_λ, then the sheet of the focal set of M corresponding to λ is a smooth $(n-m)$-dimensional submanifold of $\tilde{M}$.

Remark 2.4 (An example of principal curvature functions that are not smooth). If a continuous principal curvature function does not have constant multiplicity, then it is not necessarily a smooth function. Consider, for example, the behavior of the principal curvature functions of the monkey saddle in $\mathbf{R}^3$ given as the graph of the function

$$z = \frac{x^3 - 3xy^2}{3}. \tag{2.39}$$

This surface has two distinct principal curvatures at each point except at the umbilic point at the origin. In terms of polar coordinates (r, θ) on $\mathbf{R}^2$, the principal curvatures are given by the formula,

$$(1 + r^4)^{3/2}\,\lambda = -r^5 \cos 3\theta \pm 2r\left(1 + r^4 + \frac{r^8}{4}\cos^2 3\theta\right)^{1/2}. \tag{2.40}$$

As r approaches zero, the two principal curvature functions are asymptotically equal to $\pm 2r$, so these functions are continuous, but not smooth at the origin. (See [96] or [95, pp. 134–135] for more detail.)

If a principal curvature function λ has constant multiplicity m on M, then we can define a smooth *focal map* f_λ from an open subset $U \subset M$ (defined below) onto the sheet of the focal set of M determined by λ. Using Theorem 2.1 for the location of the focal points, we define the map f_λ by the formulas,

$$\begin{aligned} f_\lambda(x) &= f(x) + \frac{1}{\lambda}\,\xi(x), \\ f_\lambda(x) &= \cos\theta\, f(x) + \sin\theta\,\xi(x), \text{ where } \cot\theta = \lambda, \\ f_\lambda(x) &= \cosh\theta\, f(x) + \sinh\theta\,\xi(x), \text{ where } \coth\theta = \lambda, \end{aligned} \tag{2.41}$$

for $\tilde{M}$ equal to $\mathbf{R}^{n+1}$, S^{n+1}, and H^{n+1}, respectively.

In the case of $\mathbf{R}^{n+1}$, the domain U of f_λ is the set of points in M where $\lambda \neq 0$. In hyperbolic space, the domain U of f_λ is the set of points where $|\lambda| > 1$. In the case of S^{n+1}, at each point $x \in M$ the principal curvature λ gives rise to two antipodal focal points in S^{n+1} determined by substituting $\theta = \cot^{-1}\lambda$ and $\theta = \cot^{-1}\lambda + \pi$ into equation (2.41). Thus, λ gives rise to two antipodal focal maps into S^{n+1}.

For a point x in the domain U of f_λ, the hypersphere $K_\lambda(x)$ in $\tilde{M}$ through x and centered at the focal point $f_\lambda(x)$ is called the *curvature sphere* determined by λ at x. This curvature sphere is tangent to $f(M)$ at the point $f(x)$. These curvature spheres play a crucial role in the study of Dupin hypersurfaces in the context of Lie sphere geometry.

Remark 2.5 (On the definition of a curvature sphere). Note that the definition of a curvature sphere does not require that λ have constant multiplicity or be a smooth function. It can be defined pointwise. If x is in the domain U of the map f_λ defined in equation (2.41), then the curvature sphere at x corresponding to the principal curvature λ is the hypersphere $K_\lambda(x)$ in $\tilde{M}$ through x and centered at the focal point $f_\lambda(x)$.

Conformal transformations of the ambient space

The condition that a principal curvature function λ has constant multiplicity on M is important in the study of Dupin hypersurfaces. This consideration is preserved by conformal transformations of the ambient space, as the following considerations show.

Let $(\tilde{M}, g)$ and $(\tilde{M}', g')$ be two Riemannian manifolds, and suppose that $\psi : \tilde{M} \to \tilde{M}'$ is a conformal diffeomorphism such that

$$g'(\psi_* X, \psi_* Y) = e^{2h(x)} g(X, Y), \tag{2.42}$$

for all X, Y tangent to $\tilde{M}$ at x, where h is a smooth function on $\tilde{M}$. Let M be a submanifold of $\tilde{M}$, and let ξ be a local field of unit normals to M in a neighborhood of x. Then $\xi' = \psi_*(e^{-h}\xi)$ is a field of unit normals to $\psi(M)$ near $\psi(x)$ and the corresponding shape operators are related by the equation,

$$B_{\xi'} = e^{-h}(A_\xi - g(\operatorname{grad} h, \xi) I). \tag{2.43}$$

A direct calculation then yields the following relationship between the principal curvatures of M in $\tilde{M}$ and those of $\psi(M)$ in $\tilde{M}'$.

Theorem 2.6. *Let $\psi : (\tilde{M}, g) \to (\tilde{M}', g')$ be a conformal diffeomorphism of Riemannian manifolds with $g'(\psi_* X, \psi_* Y) = e^{2h(x)} g(X, Y)$ for all X, Y tangent to $\tilde{M}$ at x. Let M be an oriented hypersurface in $\tilde{M}$, and let λ be a smooth principal curvature function of constant multiplicity m on M. Then*

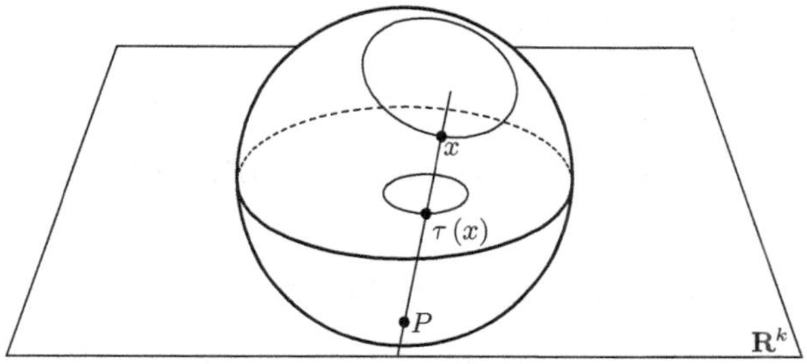

Fig. 2.1 Stereographic projection

$$\mu = e^{-h}(\lambda - g(\text{grad}\, h, \xi))$$

is a smooth principal curvature function of multiplicity m on $\psi(M)$, and the respective principal distributions of λ and μ coincide on M.

Remark 2.7 (Stereographic projection and inversions in spheres). We want to apply Theorem 2.6 to the case of hypersurfaces in real space forms by considering the conformal transformation given by stereographic projection from S^k or H^k into $\mathbf{R}^k$, for any positive integer k. In the spherical case, let P be an arbitrary point of the unit sphere $S^k \subset \mathbf{R}^{k+1}$, and let

$$\mathbf{R}^k = \{x \in \mathbf{R}^{k+1} \mid \langle x, P\rangle = 0\}, \tag{2.44}$$

where $\langle\ ,\ \rangle$ is the Euclidean inner product on $\mathbf{R}^{k+1}$. Then stereographic projection with pole P is the map $\tau : S^k - \{P\} \to \mathbf{R}^k$ defined geometrically as follows. For $x \in S^k - \{P\}$, the ray from x through P intersects $\mathbf{R}^k$ in exactly one point which is $\tau(x)$ (see Figure 2.1). Analytically, this is given by

$$\tau(x) = P + \frac{1}{1 - \langle x, P\rangle}\,(x - P). \tag{2.45}$$

In terms of our conformal geometric considerations, this can be written as

$$\tau(x) = P + e^{h(x)}(x - P), \tag{2.46}$$

where $e^{-h(x)} = 1 - \langle x, P\rangle$. It is easily shown that τ is a conformal diffeomorphism with $\langle \tau_* X, \tau_* Y\rangle = e^{2h(x)}\langle X, Y\rangle$, for all X, Y tangent to S^k at x.

Recall from equation (2.2) that our model of k-dimensional hyperbolic space is given by

$$H^k = \{x \in \mathbf{R}^{k+1}_1 \mid \langle x, x\rangle = -1,\ x_{k+1} \geq 1\},$$

where $\langle\ ,\ \rangle$ is the Lorentz metric,

$$\langle x, y\rangle = x_1 y_1 + \cdots + x_k y_k - x_{k+1} y_{k+1},$$

on $\mathbf{R}_1^{k+1}$. Let P be a point in $\mathbf{R}_1^{k+1}$ such that $-P \in H^k$. Let D^k be the k-dimensional disk

$$D^k = \{x \in \mathbf{R}_1^{k+1} \mid \langle x, P\rangle = 0,\ \langle x, x\rangle < 1\}, \tag{2.47}$$

on which the metric $\langle\ ,\ \rangle$ restricts to a Euclidean metric, which we denote by g. Then we define stereographic projection $\tau : H^k \to D^k$ with pole P as follows. For $x \in H^k$, the ray from P through x intersects D^k in exactly one point which is $\tau(x)$. Analytically, this is given by

$$\tau(x) = P + \frac{1}{1 + \langle x, P\rangle}(x - P). \tag{2.48}$$

In terms of our conformal geometry, this can be written as,

$$\tau(x) = P + e^{h(x)}(x - P), \tag{2.49}$$

where $e^{-h(x)} = 1 + \langle x, P\rangle$. One can easily show that τ is a conformal diffeomorphism with $g(\tau_* X, \tau_* Y) = e^{2h(x)}\langle X, Y\rangle$, for all X, Y tangent to H^k at the point x.

Another important type of conformal transformation is inversion,

$$\sigma : \mathbf{R}^{n+1} - \{p\} \to \mathbf{R}^{n+1} - \{p\}, \tag{2.50}$$

in a sphere centered at $p \in \mathbf{R}^{n+1}$ with radius $r > 0$. The map σ takes a point $q \in \mathbf{R}^{n+1} - \{p\}$ to the point $\sigma(q)$ on the ray from p through q such that $|q - p|\,|\sigma(q) - p| = r^2$.

We now return to the case of a hypersurface $f : M^n \to \tilde{M}^{n+1}$ in a real space form and consider the question of when the image of a focal map f_λ is a submanifold of the ambient space $\tilde{M}$, where λ is a principal curvature of constant multiplicity m on M. Here we will make the calculations only for $\tilde{M} = \mathbf{R}^{n+1}$. The proofs for the other space forms are similar.

Theorem 2.8. *Let $f : M^n \to \tilde{M}^{n+1}$ be an oriented hypersurface of a real space form. Suppose that λ is a smooth principal curvature function of constant multiplicity $m \geq 1$ in a neighborhood of a point x in the domain of f_λ. Then the rank of the focal map f_λ at x equals $n - m + 1$ if there exists $X \in T_\lambda(x)$ such that $X\lambda \neq 0$, and it equals $n - m$ otherwise.*

Proof. Here we consider the case $f : M^n \to \mathbf{R}^{n+1}$. Let ξ be the field of unit normals on M. On a neighborhood W of x on which λ is nonzero and has constant multiplicity m, we have from equation (2.41) that

$$f_\lambda(y) = f(y) + \rho(y)\xi(y),$$

for $y \in W$, where $\rho = 1/\lambda$. For $X \in T_xM$, we compute the differential of f_λ applied to X,

$$(f_\lambda)_*(X) = X + (X\rho)\,\xi + \rho D_X\xi, \tag{2.51}$$

where again we are identifying vectors that are Euclidean parallel. If $X \in T_\mu(x)$ for a principal curvature $\mu \neq \lambda$, then since $D_X\xi = -\mu X$, equation (2.51) yields

$$(f_\lambda)_*(X) = (1 - \frac{\mu}{\lambda})X + (X\rho)\,\xi, \text{ for } X \in T_\mu, \tag{2.52}$$

and thus $(f_\lambda)_*$ is injective on T_μ. This is true for all principal curvatures μ not equal to λ, and so $(f_\lambda)_*$ is injective on $T_\lambda^\perp(x)$, which is the direct sum of the principal spaces corresponding to the other principal curvatures. On the other hand, if $X \in T_\lambda(x)$, then equation (2.51) yields

$$(f_\lambda)_*(X) = (X\rho)\xi = \frac{-X\lambda}{\lambda^2}\,\xi, \text{ for } X \in T_\lambda. \tag{2.53}$$

Thus, if $X\lambda \neq 0$ for some $X \in T_\lambda(x)$, then the range of $(f_\lambda)_*$ is the $(n - m + 1)$-dimensional space spanned by $T_\lambda^\perp(x)$ and ξ, while if $X\lambda = 0$ for all $X \in T_\lambda(x)$, then the range of $(f_\lambda)_*$ is the $(n - m)$-dimensional space $(f_\lambda)_*(T_\lambda^\perp(x))$. □

This proof shows that at a point x where the focal map f_λ has rank equal to $n - m + 1$, a vector parallel to the normal vector $\xi(x)$ is tangent to the image of f_λ at the point $f_\lambda(x)$. Thus, it generalizes the classical result that the normal to a surface M in $\mathbf{R}^3$ is tangent to the evolute surface (focal set) when f_λ has rank two (see, for example, Goetz [175]). In the classical case, if a principal curvature λ has constant multiplicity one, and $X\lambda \neq 0$ on M for a corresponding nonzero principal vector field X, then the sheet of the focal submanifold $f_\lambda(M)$ is also an immersed surface. More generally, if λ has constant multiplicity one, then $f_\lambda(M)$ is a surface with singularities at the images under f_λ of points where $X\lambda = 0$. For example, the evolute of an ellipse in a plane has singularities at the images of the four vertices.

Another consequence of the proof of Theorem 2.8 is the following corollary. Recall that for x in the domain U of f_λ, the curvature sphere $K_\lambda(x)$ in $\tilde{M}$ is the hypersphere through x centered at the focal point $f_\lambda(x)$. Thus, the curvature sphere map K_λ is constant along a leaf of T_λ in U if and only if the focal map f_λ is constant along that leaf.

Corollary 2.9. *Let $f : M^n \to \tilde{M}^{n+1}$ be an oriented hypersurface of a real space form. Suppose that λ is a smooth principal curvature function of constant multiplicity $m \geq 1$ on M, and let U be the domain of the focal map f_λ. Then the following conditions are equivalent on U:*

(1) *λ is constant along each leaf of its principal foliation T_λ.*
(2) *The focal map f_λ is constant along each leaf of T_λ in U.*
(3) *The curvature sphere map K_λ is constant along each leaf of T_λ in U.*

Returning to the general situation of a hypersurface $f : M^n \to \tilde{M}^{n+1}$, it follows from the "constant rank theorem" (see, for example, Conlon [120, p. 39]) that the sheet $f_\lambda(U)$ of the focal set will be a submanifold of $\tilde{M}$ locally if f_λ has constant rank on U. From Theorem 2.8, we see that this is contingent on the value of $X\lambda$ for principal vectors X corresponding to the principal curvature λ. The following theorem shows that in the case where λ has constant multiplicity $m > 1$ on M, the derivative $X\lambda$ is always zero for every principal vector X corresponding to λ at every point of M, and thus f_λ has constant rank $n - m$ on U. However, this is not the case if λ has constant multiplicity $m = 1$ on M, and so we will handle the cases $m > 1$ and $m = 1$ separately, beginning with the case $m > 1$.

Integrability of the principal distribution when m > 1

Theorem 2.10. *Let $f : M^n \to \tilde{M}^{n+1}$ be an oriented hypersurface of a real space form. Suppose that λ is a smooth principal curvature function of constant multiplicity $m > 1$ on M. Then the principal distribution T_λ is integrable, and $X\lambda = 0$ for every $X \in T_\lambda$ at every point of M.*

Proof. We use the Codazzi equation, which for an oriented hypersurface in a real space form takes the form $(\nabla_X A)Y = (\nabla_Y A)X$ (see [283, Vol. II, p. 26]), that is,

$$\nabla_X(AY) - A(\nabla_X Y) = \nabla_Y(AX) - A(\nabla_Y X), \tag{2.54}$$

for vector fields X and Y tangent to M. If take X and Y to be linearly independent (local) vector fields in the principal distribution T_λ, then the Codazzi equation (2.54) becomes

$$(X\lambda)Y + \lambda\nabla_X Y - A(\nabla_X Y) = (Y\lambda)X + \lambda\nabla_Y X - A(\nabla_Y X). \tag{2.55}$$

Since the Levi-Civita connection has zero torsion, the Lie bracket $[X, Y] = \nabla_X Y - \nabla_Y X$, and equation (2.55) reduces to

$$(X\lambda)Y - (Y\lambda)X = (A - \lambda I)\,[X, Y]. \tag{2.56}$$

Since the left side of this equation is in T_λ, while the right side is $T_\lambda^\perp$, both sides are equal to zero. Thus, T_λ is integrable by the Frobenius Theorem (see, for example, [283, Vol I., p. 10]), since $[X, Y]$ is in T_λ. Furthermore, $X\lambda$ and $Y\lambda$ are both zero on M, since X and Y are linearly independent. □

Thus, in the case where λ has constant multiplicity $m > 1$ on M, the distribution T_λ is a foliation on M, which we call the *principal foliation* corresponding to λ. We next prove that the leaves of a principal foliation are m-dimensional totally umbilic submanifolds of $\tilde{M}$, where a submanifold V of a space form $\tilde{M}$ is said to be *totally umbilic* if for each $x \in V$, there is a real-valued linear function ω on $T_x^\perp V$ such that the shape operator B_η of V satisfies $B_\eta = \omega(\eta)I$ for every $\eta \in T_x^\perp V$.

In all three space forms $\tilde{M}^{n+1}$, a totally umbilic m-dimensional submanifold always lies in a totally geodesic $(m+1)$-dimensional submanifold of $\tilde{M}^{n+1}$. Thus it suffices to describe the totally umbilic hypersurfaces in each of the three space forms. In $\mathbf{R}^{m+1}$, a connected totally umbilic hypersurface is an open subset of an m-plane or an m-dimensional metric sphere. In S^{m+1}, a connected totally umbilic hypersurface is an open subset of a great or small hypersphere in S^{m+1}. Finally, in hyperbolic space H^{m+1}, a connected totally umbilic hypersurface is an open subset of a totally geodesic hyperplane, an equidistant hypersurface from a hyperplane, a horosphere, or a metric sphere (see, for example, [283, Vol. II, pp. 30–32] or Spivak [495, Vol. 4, pp. 110–114]).

Theorem 2.11. *Let $f : M^n \to \tilde{M}^{n+1}$ be an oriented hypersurface of a real space form. Suppose that λ is a smooth principal curvature function of constant multiplicity $m > 1$ on M. Then the leaves of the principal foliation T_λ are m-dimensional totally umbilic submanifolds of $\tilde{M}$.*

Proof. Let V be a leaf of the principal foliation T_λ. The normal space $T_x^\perp V$ to V in $\tilde{M}$ at a point $x \in V$ can be decomposed as

$$T_x^\perp V = T_x^\perp M \oplus T_\lambda^\perp(x),$$

where $T_\lambda^\perp(x)$ is the orthogonal complement to $T_\lambda(x)$ in T_xM. For a unit vector $\eta \in T_x^\perp V$, let B_η denote the shape operator of V corresponding to η. If η is the normal vector ξ to M at x with associated shape operator A, then $B_\eta X = AX = \lambda X$, for $X \in T_\lambda(x)$, and thus $B_\eta = \lambda I$.

Next let $\eta \in T_\lambda^\perp(x)$ be a unit length principal vector of A with corresponding principal curvature μ, so that $A\eta = \mu\eta$ for $\mu \neq \lambda$. Extend η to a vector field $Y \in T_\lambda^\perp$ on a neighborhood W of x. Then there exists a unique vector field $Z \in T_\lambda^\perp$ such that $\langle Z, Y\rangle = 0$ and

$$AY = \mu Y + Z, \tag{2.57}$$

for some smooth function μ on W. This is possible since $T_\lambda^\perp$ is invariant under A, even though the eigenvalues of A need not be smooth.

We now find the shape operator B_η. Let X be a vector field in T_λ on the neighborhood W. Since the vector field $Z = 0$ at x, one can easily show that $\nabla_X Z \in T_\lambda^\perp$ at x. Using equation (2.57), we see that the Codazzi equation (2.54) becomes

$$(X\mu)Y - (Y\lambda)X + \nabla_X Z = (A - \mu I)\nabla_X Y - (A - \lambda I)\nabla_Y X. \tag{2.58}$$

If we consider the T_λ-component of both sides of this equation, we see that the T_λ-component of $\nabla_X Y$ at x is

$$\frac{-(Y\lambda)X}{\lambda - \mu}. \tag{2.59}$$

If $\tilde{\nabla}$ is the Levi-Civita connection on $\tilde{M}^{n+1}$, we have the basic equation,

$$\tilde{\nabla}_X Y = \nabla_X Y + \langle AX, Y\rangle \xi.$$

Since $\langle AX, Y\rangle = 0$, the T_λ-component of $\tilde{\nabla}_X Y$ at x is equal to the T_λ-component of $\nabla_X Y$ at x, which is given by equation (2.59). Since $\eta = Y$ at x, the vector $-B_\eta X$ is by definition equal to the T_λ-component of $\tilde{\nabla}_X Y$, and so we have

$$B_\eta X = \frac{(\eta\lambda)X}{\lambda - \mu}. \tag{2.60}$$

This completes the proof of the theorem. □

A manifold structure for the focal set

As we saw in Theorem 2.1, the domain U of the focal map f_λ is the set where $\lambda \neq 0$ in the case $\tilde{M} = \mathbf{R}^{n+1}$, and the set where $|\lambda| > 1$ for $\tilde{M} = H^{n+1}$. At all such points, the leaf of the principal foliation T_λ through the point is an open subset of an m-dimensional metric sphere in $\tilde{M}$.

By Theorems 2.8 and 2.10, we know that in this case of multiplicity $m > 1$, the focal map f_λ is constant along each leaf of T_λ, and so it factors through a map of the space U/T_λ of leaves of T_λ, where U is the domain of f_λ. This enables us to place a manifold structure on the sheet of the focal set $f_\lambda(U)$ as follows.

Theorem 2.12. *Let $f : M^n \to \tilde{M}^{n+1}$ be an oriented hypersurface of a real space form. Suppose that λ is a smooth principal curvature function of constant multiplicity $m > 1$ on M^n. Then the focal map $f_\lambda : U \to \tilde{M}^{n+1}$ factors through an immersion of the (possibly non-Hausdorff) $(n-m)$-dimensional manifold U/T_λ into $\tilde{M}^{n+1}$. If M^n is complete with respect to the induced metric, then the manifold U/T_λ is Hausdorff.*

Proof. Since the leaves of T_λ are totally umbilic submanifolds of $\tilde{M}^{n+1}$, the foliation T_λ is regular as defined by Palais [425, p. 13], that is, every point has a coordinate chart distinguished by the foliation such that each leaf intersects the chart in at most one m-dimensional slice. This implies that the space of leaves U/T_λ is an $(n-m)$-dimensional manifold in the sense of Palais, which may not be Hausdorff. By Theorems 2.8 and 2.10, the focal map f_λ factors through a map $g_\lambda : U/T_\lambda \to \tilde{M}^{n+1}$, and the map g_λ is an immersion, since the rank of g_λ equals the rank f_λ, which is $n-m$ at each point. Finally, the regularity of the foliation T_λ implies that each leaf is a closed subset of M (see Palais [425, p. 18]). Thus, if M is complete, then each leaf is also complete (see, for example, [283, Vol. I, p. 179]). Therefore, each leaf of T_λ that intersects the domain U of f_λ is an m-dimensional metric sphere in $\tilde{M}^{n+1}$ and is thus compact. This implies that the leaf space U/T_λ is Hausdorff [425, p. 16]. □

Remark 2.13 (An example in which $f_\lambda(M)$ is not a Hausdorff manifold). The following example ([88, p. 34] or [95, p. 143]) shows that the leaf space of a regular foliation is not necessarily Hausdorff and that the image of a focal map with constant rank is not necessarily a Hausdorff manifold. Let $\phi(t)$ be the real-valued function on $\mathbf{R}$ defined by $\phi(t) = e^{-1/t}$ if $t > 0$ and $\phi(t) = 0$ if $t \leq 0$. Let K be a tube of constant radius one in $\mathbf{R}^3$ over the curve,

$$\gamma(t) = (t, 0, \phi(t)), \ t \in (-1, 1).$$

Then the curve γ itself is the sheet of the focal set of K corresponding to the principal curvature $\lambda = 1$ with appropriate choice of unit normal field.

Let N be the intersection of K with the closed upper half-space given by $z \geq 0$, with the points satisfying $z = 0, \ x \geq 0$ removed. Let M be the union of N with its mirror image in the plane $z = 0$. Then $\lambda = 1$ is still a constant principal curvature on all of M. However, the leaf space M/T_λ is not Hausdorff, since the two open semi-circular leaves L_1 and L_2 in the plane $x = 0$ cannot be separated by disjoint neighborhoods in the quotient topology. The corresponding sheet of the focal set $f_\lambda(M)$ consists of the union of the curve γ with its mirror image in the plane $z = 0$, and it is not a Hausdorff 1-dimensional manifold in a neighborhood of the origin. Nevertheless, the rank of the focal map f_λ is one on all of M.

In this example, λ has constant multiplicity $m = 1$. One can produce similar examples where $m > 1$ by imitating the construction above in $\mathbf{R}^n$ for $n > 3$.

The case of a principal curvature of multiplicity m = 1

We now consider the case where a principal curvature λ has constant multiplicity $m = 1$ on an oriented hypersurface $f : M^n \to \tilde{M}^{n+1}$. This case differs greatly from the case of multiplicity greater than one, since λ is not necessarily constant along the leaves of its principal foliation T_λ, i.e., along its lines of curvature. In fact, by Theorem 2.8, the rank of the focal map f_λ is n at points x where $X\lambda \neq 0$ for a nonzero vector $X \in T_\lambda(x)$, and it is $n-1$ at points x where $X\lambda = 0$ for all $X \in T_\lambda(x)$. Thus, in general, the sheet of the focal set $f_\lambda(U)$, where U is the domain of f_λ, is a hypersurface with singularities at points where $X\lambda = 0$ for all $X \in T_\lambda(x)$.

We are interested in the case where the sheet $f_\lambda(U)$ of the focal set is a submanifold of dimension $n - 1$ in $\tilde{M}^{n+1}$. As noted above, this is not always the case, as it was in the case of higher multiplicity, and so the main result is formulated in terms of conditions that are equivalent to the condition that $f_\lambda(U)$ is a submanifold of dimension $n - 1$.

In this case of multiplicity one, it is significant that for hypersurfaces in hyperbolic space H^{n+1}, the domain U of f_λ does not include those points $x \in M$ where $|\lambda(x)| \leq 1$. In fact, conditions (1–3) of Theorem 2.14 below are equivalent on U, but not on all of M. Specifically, conditions (1) and (2) are equivalent on M and they imply (3). However, one can construct a surface M in H^3 such that focal set $f_\lambda(U)$ is a curve (and so condition (3) holds), and yet not all of the lines of curvature corresponding to λ are plane curves of constant curvature. This is done by beginning with a surface $N \subset H^3$ on which all three conditions are satisfied, and then modifying N on the set where $|\lambda| < 1$ (which is disjoint from U) so as to destroy condition (2), but introduce no new focal points and thereby preserve condition (3).

Theorem 2.14. *Let $f : M^n \to \tilde{M}^{n+1}$ be an oriented hypersurface of a real space form. Suppose that λ is a smooth principal curvature function of constant multiplicity $m = 1$ on M^n. Then the following conditions are equivalent on $\tilde{M}^{n+1}$ if $\tilde{M}^{n+1} = \mathbf{R}^{n+1}$ or S^{n+1}, and on the domain U of f_λ if $\tilde{M}^{n+1} = H^{n+1}$.*

(1) *λ is constant along each leaf of its principal foliation T_λ.*
(2) *The leaves of T_λ are plane curves of constant curvature.*
(3) *The rank of the focal map f_λ is identically equal to $n-1$ on its domain U, and f_λ factors through an immersion of the $(n-1)$-dimensional space of leaves U/T_λ into $\tilde{M}^{n+1}$.*

We first give the proof in the Euclidean case and then handle the other cases via stereographic projection.

Proof (Euclidean case). (1) $\Leftrightarrow$ (3) This follows immediately from Theorem 2.8 concerning the rank of the focal map f_λ and from the connectedness of the leaves of the foliation T_λ.

(2) $\Rightarrow$ (1) This follows easily from the Frenet equations for plane curves.

(1) $\Rightarrow$ (2) Since T_λ is a 1-dimensional foliation on M, in a neighborhood of any point of M, we can find a local coordinates (t, v) given by the coordinate chart $\phi : (-\varepsilon, \varepsilon) \times V \to M$, where V is an open subset in $\mathbf{R}^{n-1}$, such that the leaves of T_λ that intersect the image $W \subset M$ of ϕ are precisely the images under ϕ of the curves $v =$ constant in $(-\varepsilon, \varepsilon) \times V$.

We first consider the case where λ is nonzero on W. By an appropriate choice of the unit normal field ξ, we may arrange that $\lambda > 0$ on W. Thus, λ is a positive constant on each leaf of T_λ that passes through W. By condition (1) and Theorem 2.8, the focal map f_λ on W is constant on each leaf of T_λ, and so the functions $g_\lambda = f_\lambda \circ \phi$ and $\rho = (1/\lambda) \circ \phi$ are functions of the coordinate $v \in V$ alone, and g_λ is an immersion on V, since it has rank $n-1$.

The point $q = f(\phi(t, v))$ lies on the hypersphere in $\mathbf{R}^{n+1}$ determined by v given by the equation

$$|z - g_\lambda(v)| = \rho(v), \ z \in \mathbf{R}^{n+1}. \tag{2.61}$$

Since the normal line to $f(M)$ at q is the same as the normal line to the sphere given in equation (2.61) at q (i.e., $f(M)$ is the envelope of the $(n-1)$-parameter family of spheres parametrized by v), one can show that for any Y tangent to V at the point v, the point q also lies on the hyperplane in $\mathbf{R}^{n+1}$ given by the equation:

$$\langle z - g_\lambda(v), (g_\lambda)_*(Y) \rangle = -\rho(v) Y(\rho), \ z \in \mathbf{R}^{n+1}. \tag{2.62}$$

Thus, for any given value of v, the leaf of T_λ in W determined by v lies on the circle obtained by intersecting the hypersphere in equation (2.61) with the 2-plane determined by equation (2.62), as Y ranges over the $(n-1)$-dimensional tangent space $T_v V$. Hence, each leaf of T_λ on which λ is nonzero lies locally on a circle, and by connectedness, it is an arc of a circle.

Finally, suppose that γ is a leaf of T_λ on which λ is identically zero. By Theorem 2.6, there exists an inversion σ of $\mathbf{R}^{n+1}$ in a hypersphere such that $(\sigma \circ f)(\gamma)$ is a leaf of the principal foliation T_μ on which the associated principal principal curvature μ, as in Theorem 2.6, is a nonzero constant. By the argument above, $(\sigma \circ f)(\gamma)$ lies on a circle, and so $f(\gamma)$ itself lies on a circle or a straight line. This completes the proof of the theorem in the Euclidean case. □

We now discuss the proof of Theorem 2.14 in the non-Euclidean cases. As in the Euclidean case, (1) $\Leftrightarrow$ (3) follows immediately from Theorem 2.8.

To prove (1) $\Leftrightarrow$ (2), we use stereographic projection τ, as defined in Remark 2.7. Let $f : M^n \to \tilde{M}^{n+1}$ be an oriented hypersurface with field of unit normals ξ in a real space form $\tilde{M}^{n+1} = S^{n+1}$ or H^{n+1}, and let τ be the appropriate stereographic projection for $\tilde{M}^{n+1}$. If λ is a principal curvature of $f(M)$ of multiplicity one, then $\mu = e^{-h}(\lambda - g(\text{grad } h, \xi))$ is a principal curvature of multiplicity one of the hypersurface $(\tau \circ f)(M)$ in $\mathbf{R}^{n+1}$ (or $D^{n+1} \subset \mathbf{R}^{n+1}$ in the hyperbolic case) by Theorem 2.6. By a direct calculation, one can show that the leaves of T_λ are plane curves of constant curvature in $\tilde{M}^{n+1}$ if and only if the leaves of T_μ are plane curves of constant curvature in $\mathbf{R}^{n+1}$ (or D^{n+1}). Thus, the equivalence of conditions (1) and (2) follows from the equivalence of (1) and (2) in the Euclidean case and the following important lemma.

Lemma 2.15. *Let $f : M^n \to S^{n+1}$ (respectively, H^{n+1}) be an oriented hypersurface with field of unit normals ξ. Suppose that λ is a smooth principal curvature function of constant multiplicity $m = 1$ on M^n, and let X denote a field of unit principal vectors of λ on M. Let*

$$\mu = e^{-h}(\lambda - \langle \text{grad } h, \xi \rangle)$$

be the corresponding principal curvature of multiplicity $m = 1$ of the hypersurface $(\tau \circ f) : M \to \mathbf{R}^{n+1}$ (respectively, D^{n+1}), where τ is stereographic projection. Then $X\lambda = 0$ at $x \in M$ if and only if $X\mu = 0$ at x.

Proof. We will do the proof for a hypersurface in S^{n+1}, and the proof in H^{n+1} is quite similar. This is a local calculation, so we will consider M as an embedded hypersurface in S^{n+1} and suppress the mention of the embedding f. We use stereographic projection $\tau : S^{n+1} - \{P\} \to \mathbf{R}^{n+1}$ with pole P as given in equation (2.46), that is

$$\tau(x) = P + e^{h(x)}(x - P),$$

where $e^{-h(x)} = 1 - \langle x, P\rangle$. Then, a direct calculation yields,

$$\text{grad}\, h = e^h(P - \langle x, P\rangle\, x).$$

Using the fact that $\langle x, \xi\rangle = 0$ for $x \in M$ and ξ the local field of unit normals to M, we get

$$\mu = e^{-h}(\lambda - \langle e^h P, \xi\rangle) = e^{-h}\lambda - \langle P, \xi\rangle,$$

and

$$X\mu = -e^{-h}(Xh)\lambda + e^{-h}(X\lambda) - \langle P, D_X\xi\rangle, \tag{2.63}$$

where D is the Euclidean covariant differentiation on $\mathbf{R}^{n+2}$. Since $\langle X, \xi\rangle = 0$, it follows that $D_X\xi = \tilde{\nabla}_X\xi$, where $\tilde{\nabla}$ is the Levi-Civita connection on S^{n+1}. Then, we know that $\tilde{\nabla}_X\xi = -AX = -\lambda X$, so $D_X\xi = -\lambda X$. This and the fact that

$$Xh = \langle \text{grad}\, h, X\rangle = e^h\langle P, X\rangle,$$

enable us to rewrite the expression for $X\mu$ in equation (2.63) as,

$$X\mu = -\langle P, X\rangle\lambda + e^{-h}(X\lambda) + \langle P, X\rangle\lambda = e^{-h}(X\lambda).$$

From this it is clear that $X\mu = 0$ if and only if $X\lambda = 0$.

For a hypersurface M in H^{n+1}, we have from equation (2.49) that stereographic projection $\tau : H^{n+1} \to D^{n+1}$ with pole P is given by

$$\tau(x) = P + e^{h(x)}(x - P),$$

where $e^{-h(x)} = 1 + \langle x, P\rangle$. From this we can compute that

$$\text{grad}\, h = -e^h(P + \langle x, P\rangle\, x),$$

where $\langle\ ,\ \rangle$ is the Lorentz metric, and the rest of the proof follows in a way similar to the spherical case. □

Remark 2.16. In the context of Lie sphere geometry (see, for example, Pinkall [446] or the book [77, p. 67]), it is easy to prove that the property that a curvature sphere map is constant along each leaf of its principal foliation is invariant under Möbius (conformal) transformations, and more generally, under Lie sphere transformations. From this, Lemma 2.15 follows easily.

In the case where the hypersurface M is complete with respect to the metric induced from $\tilde{M}$, Theorem 2.14 enables us to give a manifold structure to the sheet $f_\lambda(U)$ of the focal set, where U is the domain of f_λ, similar to that obtained in Theorem 2.12 for the case of higher multiplicity.

Theorem 2.17. *Let $f : M^n \to \tilde{M}^{n+1}$ be an oriented hypersurface of a real space form which is complete with respect to the induced metric. Suppose that λ is a smooth principal curvature function of constant multiplicity $m = 1$ on M^n such that the equivalent conditions* (1)–(3) *of Theorem 2.14 are satisfied on the domain U of the focal map f_λ. Then f_λ factors through an immersion of the $(n-1)$-dimensional manifold U/T_λ into $\tilde{M}^{n+1}$.*

Proof. The proof is almost identical to the proof of Theorem 2.12 for the case of higher multiplicity. As in that case, the completeness of M implies that each leaf of T_λ is complete with respect to the induced metric. This implies that each leaf of T_λ in M is a covering space of the metric circle in $\tilde{M}^{n+1}$ on which it lies (see, for example, [283, Vol. I, p. 176]). However, since the circle is not simply connected, this does not guarantee that the leaf itself is compact, as in the multiplicity $m > 1$ case. Even so, using the fact that each leaf of T_λ is a covering of the circle on which it lies, one can produce a direct argument (which we omit here) that the leaf space U/T_λ is Hausdorff. □

We close this section with three results that have proven to be valuable in the study of isoparametric and Dupin hypersurfaces. Theorems 2.11 and 2.14 show that if a principal curvature λ of a hypersurface $f : M^n \to \tilde{M}^{n+1}$ has constant multiplicity $m \geq 1$ on M and is constant along the leaves of its principal foliation, then the leaves of T_λ are totally umbilic submanifolds in $\tilde{M}^{n+1}$. The next result shows that the case where λ assumes a critical value along a certain leaf of T_λ has even more geometric significance.

Theorem 2.18. *Let $f : M^n \to \tilde{M}^{n+1}$ be an oriented hypersurface of a real space form. Suppose that λ is a smooth principal curvature function of constant multiplicity $m \geq 1$ on M^n which is constant along each leaf of its principal foliation T_λ. Then λ assumes a critical value along a leaf γ of T_λ if and only if γ is totally geodesic in M.*

Proof. Let x be a point on the leaf γ. Then the normal space to γ in M at x is $T_\lambda^\perp(x)$. We know that $T_\lambda^\perp(x)$ is the direct sum of the principal spaces $T_\mu(x)$, where μ ranges over the principal curvatures of M at x that are not equal to λ. Let $\eta \in T_\mu(x)$, for $\mu \neq \lambda$. By the same calculation used to obtain equation (2.60), we get that the shape operator B_η of γ in M at x has the form

$$B_\eta X = \frac{(\eta\lambda)X}{\lambda - \mu},$$

for $X \in T_\lambda(x)$. The leaf γ is totally geodesic in M if and only if $B_\eta = 0$ for each $\eta \in T_\mu(x)$ for each $\mu \neq \lambda$ for all $x \in \gamma$. This occurs precisely when $\eta\lambda = 0$ for all $\eta \in T_\lambda^\perp(x)$ for all $x \in \gamma$. Since λ is assumed constant along γ, this happens precisely when λ assumes a critical value along γ. □

Since the principal curvatures are constant on an isoparametric hypersurface, the following corollary follows immediately from Theorems 2.11, 2.14, and 2.18, as was first shown by Nomizu [403].

Corollary 2.19. *Let $f : M^n \to \tilde{M}^{n+1}$ be an isoparametric hypersurface of a real space form. Then for each principal curvature λ, the leaves of the principal foliation T_λ are totally umbilic in $\tilde{M}^{n+1}$ and totally geodesic in M^n.*

The following result is similar to Theorem 2.18, and it is useful in the study of Dupin hypersurfaces. Recall from Corollary 2.9 that if a principal curvature λ has constant multiplicity $m \geq 1$, then λ is constant along each leaf of T_λ in the domain U of f_λ if and only if f_λ itself is constant along each leaf of T_λ in U. In that case, the curvature sphere map K_λ is also constant along each leaf of T_λ in U. As in Theorem 2.18, the case where λ has a critical value along a leaf of T_λ has special geometric significance.

Theorem 2.20. *Let $f : M^n \to \tilde{M}^{n+1}$ be an oriented hypersurface of a real space form. Suppose that λ is a smooth principal curvature function of constant multiplicity $m \geq 1$ on M^n which is constant along each leaf of its principal foliation T_λ. Then λ assumes a critical value along a leaf γ of T_λ in the domain U of f_λ if and only if γ is totally geodesic in the curvature sphere K_λ determined by γ.*

Proof. A leaf γ of T_λ in U is a submanifold of the curvature sphere K_λ, and its normal space in K_λ is $T_\lambda^\perp$, since K_λ is tangent to M along γ. The rest of the proof is then exactly the same as the proof of Theorem 2.18. □

2.5 Curvature Surfaces and Dupin Hypersurfaces

Let $f : M^n \to \tilde{M}^{n+1}$ be an oriented hypersurface of a real space form. A connected submanifold S of M^n is called a *curvature surface* if at each $x \in S$, the tangent space T_xS is equal to some principal space $T_\lambda(x)$. In that case, the corresponding principal curvature $\lambda : S \to \mathbf{R}$ is a smooth function on S.

For example, if dim T_λ is constant on an open subset U of M^n, then each leaf of the principal foliation T_λ is a curvature surface on U. Curvature surfaces are plentiful, since the results of Reckziegel [458] and Singley [486] imply that there is an open dense (possibly not connected) subset Ω of M^n on which the multiplicities of the principal curvatures are locally constant. On Ω, each leaf of each principal foliation is a curvature surface.

Remark 2.21 (Curvature surfaces of submanifolds of codimension $k > 1$). Reckziegel [458] generalized the notion of a curvature surface to the case of an immersed submanifold $f : M^n \to \tilde{M}^{n+k}$ of a space form $\tilde{M}^{n+k}$ with codimension $k > 1$. In that case, Reckziegel defines a curvature surface to be a connected submanifold $S \subset M^n$ for which there is a parallel (with respect to the normal connection) section $\eta : S \to B^{n+k-1}$ of the unit normal bundle B^{n+k-1} such that for each $x \in S$, the tangent space T_xS is equal to some eigenspace of $A_{\eta(x)}$. In that case, the corresponding principal curvature $\lambda : S \to \mathbf{R}$ is a smooth function on S.

It is also possible to have a curvature surface S that is not a leaf of a principal foliation, because the multiplicity of the corresponding principal curvature is not constant on a neighborhood of S, as the following example due to Pinkall [447] shows.

Example 2.22 (A curvature surface that is not a leaf of a principal foliation). Let T^2 be a torus of revolution in $\mathbf{R}^3$, and embed $\mathbf{R}^3$ into $\mathbf{R}^4 = \mathbf{R}^3 \times \mathbf{R}$. Let η be a field of unit normals to T^2 in $\mathbf{R}^3$. Let M^3 be a tube of sufficiently small radius $\varepsilon > 0$ around T^2 in $\mathbf{R}^4$, so that M^3 is a compact smooth embedded hypersurface in $\mathbf{R}^4$. The normal space to T^2 in $\mathbf{R}^4$ at a point $x \in T^2$ is spanned by $\eta(x)$ and $e_4 = (0,0,0,1)$. The shape operator A_η of T^2 has two distinct principal curvatures at each point of T^2, while the shape operator A_{e_4} of T^2 is identically zero. Thus the shape operator A_ζ for the normal

$$\zeta = \cos\theta\ \eta(x) + \sin\theta\ e_4,$$

at a point $x \in T^2$, is given by

$$A_\zeta = \cos\theta\ A_{\eta(x)}.$$

From the formulas for the principal curvatures of a tube in Theorem 2.2, we see that at all points of M^3 where $x_4 \neq \pm\varepsilon$, there are three distinct principal curvatures of multiplicity one, which are constant along their corresponding lines of curvature (curvature surfaces of dimension one). However, on the two tori, $T^2 \times \{\pm\varepsilon\}$, the principal curvature $\kappa = 0$ has multiplicity two. These two tori are curvature surfaces for this principal curvature, since the principal space corresponding to κ is tangent to each torus at every point.

Theorem 2.10 has the following generalization to curvature surfaces of submanifolds of arbitrary codimension. The proof is the same as that of Theorem 2.10 with obvious minor modifications.

Theorem 2.23. *Suppose that S is a curvature surface of dimension $m > 1$ of a submanifold $f : M^n \to \tilde{M}^{n+k}$ for $k \geq 1$. Then the corresponding principal curvature is constant along S.*

An oriented hypersurface $f : M^n \to \tilde{M}^{n+1}$ is called a *Dupin hypersurface* if:

(a) along each curvature surface, the corresponding principal curvature is constant.

Furthermore, a Dupin hypersurface M is called *proper Dupin* if, in addition to Condition (a), the following condition is satisfied:

(b) the number g of distinct principal curvatures is constant on M.

By Theorem 2.23, Condition (a) is automatically satisfied along a curvature surface of dimension $m > 1$, and thus the key case is when the dimension of the curvature surface equals one.

Condition (b) is equivalent to requiring that each continuous principal curvature function has constant multiplicity on M^n. The torus T^2 in Example 2.22 above is a proper Dupin hypersurface of $\mathbf{R}^3$, but the tube M^3 over T^2 in $\mathbf{R}^4$ is Dupin, but not proper Dupin, since the number of distinct principal curvatures is not constant on M^3.

Remark 2.24 (On the terms: Dupin and proper Dupin). In some early papers on the subject (see, for example, Thorbergsson [533], and Grove–Halperin [184]) and in the book [95, p. 166], a hypersurface which satisfies Conditions (a) and (b) was called "Dupin" instead of "proper Dupin." Pinkall introduced the term "proper Dupin" in his paper [446], and that has become the standard terminology in the subject. In the book [95, p. 189], hypersurfaces such as the tube M^3 over T^2 in $\mathbf{R}^4$ were called "semi-Dupin."

Pinkall's local construction of proper Dupin hypersurfaces

The following local construction due to Pinkall [446] shows that proper Dupin hypersurfaces are very plentiful.

Theorem 2.25. *Given positive integers $m_1, \ldots, m_g$ with*

$$m_1 + \cdots + m_g = n - 1,$$

there exists a proper Dupin hypersurface in $\mathbf{R}^n$ with g distinct principal curvatures having respective multiplicities $m_1, \ldots, m_g$.

Proof. The proof is by an inductive local construction which will be clear once the first few steps are done. The proof uses the fact that the proper Dupin property is preserved by inversion of $\mathbf{R}^n$ in a hypersphere $S \subset \mathbf{R}^n$ (see Remark 2.7). This follows from Theorems 2.6 and 2.10, and an argument similar to the proof of Lemma 2.15 for stereographic projection. This construction does not, in general, result in a compact proper Dupin hypersurface.

Let $M_1 \subset \mathbf{R}^{m_1+1}$ be a sphere of radius one centered at the origin. Construct a cylinder

$$M_1 \times \mathbf{R}^{m_2} \subset \mathbf{R}^{m_1+1} \times \mathbf{R}^{m_2} = \mathbf{R}^{m_1+m_2+1}$$

over the submanifold $M_1 \subset \mathbf{R}^{m_1+1} \subset \mathbf{R}^{m_1+m_2+1}$. This cylinder has two distinct principal curvatures at each point, $\lambda_1 = 1$ of multiplicity m_1, and $\lambda_2 = 0$ of multiplicity m_2. The next step is to invert the cylinder $M_1 \times \mathbf{R}^{m_2}$ in a hypersphere $S_1 \subset \mathbf{R}^{m_1+m_2+1}$ chosen so that the image of the cylinder under the inversion has an open subset M_2 on which neither of its two principal curvatures equals zero at any point. Then M_2 is a proper Dupin hypersurface in $\mathbf{R}^{m_1+m_2+1}$ having two distinct nonzero principal curvatures with multiplicities m_1 and m_2.

Next construct a cylinder

$$M_2 \times \mathbf{R}^{m_3} \subset \mathbf{R}^{m_1+m_2+1} \times \mathbf{R}^{m_3} = \mathbf{R}^{m_1+m_2+m_3+1}$$

over the submanifold $M_2 \subset \mathbf{R}^{m_1+m_2+1}$. This cylinder has three distinct principal curvatures at each point with respective multiplicities m_1, m_2, m_3, where m_3 is the multiplicity of the principal curvature that is identically zero. As above, invert the cylinder $M_2 \times \mathbf{R}^{m_3}$ in a hypersphere $S_2 \subset \mathbf{R}^{m_1+m_2+m_3+1}$ chosen so that the image of the cylinder under the inversion has an open subset M_3 on which none of its three principal curvatures equals zero at any point. One continues the process by constructing the cylinder

$$M_3 \times \mathbf{R}^{m_4} \subset \mathbf{R}^{m_1+m_2+m_3+1} \times \mathbf{R}^{m_4} = \mathbf{R}^{m_1+m_2+m_3+m_4+1}$$

and so on, until one finally obtains the desired proper Dupin hypersurface $M_g \subset \mathbf{R}^{m_1+\cdots+m_g+1}$ with g distinct principal curvatures having respective multiplicities $m_1, \ldots, m_g$. □

As noted above, the proper Dupin hypersurfaces constructed in Theorem 2.25 are not compact, in general, and compact proper Dupin hypersurfaces are much more rare.

An important class of compact proper Dupin hypersurfaces consists of the isoparametric hypersurfaces in spheres S^n and those hypersurfaces in $\mathbf{R}^n$ obtained from isoparametric hypersurfaces in S^n via stereographic projection. For example, the well-known cyclides of Dupin in $\mathbf{R}^3$ are obtained from a standard product torus $S^1(r) \times S^1(s) \subset S^3$, $r^2 + s^2 = 1$, in this way. These examples will be discussed in more detail in later chapters.

In fact, Thorbergsson [533] proved that the number g of distinct principal curvatures of a compact proper Dupin hypersurface M embedded in S^n (or $\mathbf{R}^n$) can only be $1, 2, 3, 4$, or 6, the same restriction as for an isoparametric hypersurface in S^n. There are also restrictions on the multiplicities of the principal curvatures due to Stolz [502] and Grove and Halperin [184] (see Sections 3.7 and 5.8 for more detail).

We will see in Chapter 4 that both the Dupin and proper Dupin conditions are invariant under Lie sphere transformations. Because of this, Lie sphere geometry has proven to be a useful setting for the study of Dupin hypersurfaces, and we will use Lie sphere geometry extensively in Chapter 5 on Dupin hypersurfaces.

Remark 2.26 (Dupin submanifolds of higher codimension). In the case of an immersed submanifold $f : M^n \to \tilde{M}^{n+k}$ of a space form $\tilde{M}^{n+k}$ with codimension $k > 1$, Pinkall defined $f(M^n)$ to be Dupin if along each curvature surface (in the sense of Remark 2.21), the corresponding principal curvature is constant. In that case, $f(M^n)$ is called proper Dupin if the number of distinct principal curvatures is constant on the unit normal bundle B^{n+k-1}. One can show that Pinkall's definition is equivalent to the definition of a Dupin submanifold given in Section 4.4 in the context of Lie sphere geometry (see Remark 4.10 on page 217).

2.6 Height Functions and Tight Submanifolds

In this section, we give a brief review of the aspects of the theory of tight submanifolds that will be needed later in the book. For more complete coverage of the topic, the reader is referred to Chapter 1 of the book [95] or the survey articles of Kuiper [302, 303], or Banchoff and Kühnel [24]. Our treatment here is based on the book [95, pp. 6–33].

We begin with a review of the critical point theory needed in the theory of tight and taut submanifolds (see Milnor [359, pp. 4–6] for more detail). Let $f : M_1 \to M_2$ be a smooth function between manifolds M_1 and M_2. A point $x \in M_1$ is called a *critical point* of f if the derivative map

$$f_* : T_x M_1 \to T_{f(x)} M_2$$

at x is not surjective. If $y \in M_2$ is the image of a critical point x under f, then y is called a *critical value* of f. All other points in the image of f are called *regular values* of f. Note that if M_1 and M_2 have the same dimension, then x is a critical point of f if and only if f_* is singular at x.

Suppose $\phi : M \to \mathbf{R}$ is a smooth function on a manifold M; then $x \in M$ is a critical point of ϕ if and only if $\phi_* = 0$ at x. If $(x_1, \ldots, x_n)$ are local coordinates on M in a neighborhood of x, then x is a critical point of ϕ if and only if

$$\frac{\partial \phi}{\partial x_1}(x) = \cdots = \frac{\partial \phi}{\partial x_n}(x) = 0. \tag{2.64}$$

If x is a critical point of ϕ, then the behavior of ϕ near x is determined by the *Hessian* H_x of ϕ at x, which is given in local coordinates by the symmetric matrix

$$H_x = \left[\frac{\partial^2 \phi}{\partial x_i \partial x_j} \right]. \tag{2.65}$$

A critical point x of ϕ is said to be *degenerate* if the rank of the Hessian H_x is less than $n = \dim M$. If rank $H_x = n$, then x is called a *nondegenerate* critical point. The *index* of a nondegenerate critical point x is the number of negative eigenvalues of the symmetric matrix H_x. The behavior of ϕ in a neighborhood of a nondegenerate critical point is determined by the index according to the following lemma (see, for example, Milnor [359, p. 6]).

Lemma 2.27 (Lemma of Morse). *Let p be a nondegenerate critical point of index k of a function $\phi : M \to \mathbf{R}$. Then there is a local coordinate system $(x_1, \ldots, x_n)$ in a neighborhood U with origin at p such that the identity*

$$\phi = \phi(p) - x_1^2 - \cdots - x_k^2 + x_{k+1}^2 + \cdots + x_n^2 \tag{2.66}$$

holds throughout U.

From the lemma we see that a critical point of index n is a local maximum of ϕ, and a critical point of index 0 is a local minimum of ϕ. All other nondegenerate critical points are various types of saddle points.

A real-valued function ϕ is called a *Morse function* or *nondegenerate function* if all of its critical points are nondegenerate. From the lemma, we see that if M is compact, then a Morse function ϕ on M can only have a finite number of critical points, since the critical points are isolated.

Let $\phi : M \to \mathbf{R}$ be a Morse function such that the sublevel set,

$$M_r(\phi) = \{x \in M \mid \phi(x) \leq r\}, \tag{2.67}$$

is compact for all $r \in \mathbf{R}$. Of course, this is always true if M itself is compact. Let $\mu_k(\phi, r)$ be the number of critical points of ϕ of index k in $M_r(\phi)$. For compact M, let $\mu_k(\phi)$ be the number of critical points of ϕ of index k in M, and let $\mu(\phi)$ be the total number of critical points of ϕ on M. For a field $\mathbf{F}$, let

$$\beta_k(\phi, r, \mathbf{F}) = \dim_{\mathbf{F}} H_k(M_r(\phi), \mathbf{F}), \tag{2.68}$$

where $H_k(M_r(\phi), \mathbf{F})$ is the k-th homology group of $M_r(\phi)$ over the field $\mathbf{F}$. That is, $\beta_k(\phi, r, \mathbf{F})$ is the k-th $\mathbf{F}$-Betti number of $M_r(\phi)$. Further, let

$$\beta_k(M, \mathbf{F}) = \dim_{\mathbf{F}} H_k(M, \mathbf{F}) \tag{2.69}$$

be the k-th $\mathbf{F}$-Betti number of M. The *Morse inequalities* (see, for example, Morse–Cairns [379, p. 270]) state that

$$\mu_k(\phi, r) \geq \beta_k(\phi, r, \mathbf{F}), \tag{2.70}$$

for all $\mathbf{F}, r, k$. For a compact M, the *Morse number* $\gamma(M)$ of M is defined by

$$\gamma(M) = \min\{\mu(\phi) \mid \phi \text{ is a Morse function on } M\}. \tag{2.71}$$

The Morse inequalities imply that

$$\gamma(M) \geq \beta(M, \mathbf{F}) = \sum_{k=0}^{n} \beta_k(M, \mathbf{F}), \tag{2.72}$$

for any field $\mathbf{F}$. If there exists a field $\mathbf{F}$ such that $\mu(\phi) = \beta(M, \mathbf{F})$, then ϕ is called a *perfect* Morse function. In that case, ϕ has the minimum number of critical points possible in view of the Morse inequalities.

Kuiper [301] noted the following reformulation of the condition that the Morse inequalities are actually equalities, and he used it very effectively in his papers on tight and taut immersions.

Theorem 2.28. *Let ϕ be a Morse function on a compact manifold M. For a given field $\mathbf{F}$, the equality $\mu_k(\phi, r) = \beta_k(\phi, r, \mathbf{F})$ holds for all k, r if and only if the map on homology*

$$H_*(M_r(\phi), \mathbf{F}) \to H_*(M, \mathbf{F})$$

induced by the inclusion $M_r(\phi) \subset M$ is injective for all r.

This theorem follows immediately from Theorem 29.2 of Morse–Cairns [379, p. 260], and we will omit it here, although we will make a few comments about some of the key ideas in the proof.

Suppose that p is a nondegenerate critical point of index k of a Morse function ϕ on M and $\phi(p) = r$. For the sake of simplicity, assume that p is the only critical point at the critical level r. A fundamental result in critical point theory (see, for example, Milnor [359, pp. 12–24] or Morse–Cairns [379, pp. 184–202]) states that $M_r(\phi)$ has the homotopy type of $M_r^-(\phi)$ with a k-cell attached, where $M_r^-(\phi)$ consists of all points in M for which $\phi < r$. Morse and Cairns [379, pp. 258–261] characterize the effect of attaching this k-cell as follows. Let

$$\triangle\beta_i(r) = \beta_i(M_r(\phi)) - \beta_i(M_r^-(\phi)).$$

Then the $\triangle\beta_i(r)$ are 0 for all i, except that $\triangle\beta_k(r) = 1$ if the critical point is of "linking type," and $\triangle\beta_{k-1}(r) = -1$ if the critical point p if of "non-linking type." From this, it is clear that the two conditions in Theorem 2.28 are equivalent, and they hold precisely when every critical point of ϕ is of linking type.

Let $f : M^n \to \mathbf{R}^m$ be a smooth immersion, and let S^{m-1} denote the unit sphere in $\mathbf{R}^m$. For $p \in S^{m-1}$, the *linear height function* $l_p : \mathbf{R}^m \to \mathbf{R}$ is defined by the formula

$$l_p(q) = \langle p, q \rangle, \tag{2.73}$$

where $\langle\ ,\ \rangle$ is the usual Euclidean inner product on $\mathbf{R}^m$. This induces a smooth function l_p defined on M by $l_p(x) = l_p(f(x))$.

Critical points of height functions

The critical point behavior of linear height functions is related to the shape operator of $f(M^n)$ according to the following well-known theorem.

Theorem 2.29. *Let $f : M \to \mathbf{R}^m$ be a smooth immersion of an n-dimensional manifold M into $\mathbf{R}^m$, and let $p \in S^{m-1}$.*

(a) *A point $x \in M$ is a critical point of l_p if and only if p is orthogonal to T_xM.*
(b) *Suppose l_p has a critical point at x. Then for $X, Y \in T_xM$, the Hessian H_x of l_p at x satisfies $H_x(X, Y) = \langle A_pX, Y\rangle$.*

Proof. (a) Let $x \in M$, and let U be a neighborhood of x on which f is an embedding. We omit the mention of f in the following local calculation. Let $X \in T_xM$, and let $\gamma(t)$ be a curve in U with initial point $\gamma(0) = x$ and initial tangent vector $\vec{\gamma}(0) = X$. By definition we have

$$Xl_p(x) = \frac{d}{dt} l_p(\gamma(t))|_{t=0} = \langle p, \vec{\gamma}(t) \rangle|_{t=0} = \langle p, X \rangle. \tag{2.74}$$

Thus, $Xl_p(x) = 0$ if and only if $\langle p, X \rangle = 0$, and so x is a critical point of l_p if and only if p is orthogonal to T_xM.

(b) To compute the Hessian, let X and Y be tangent to M at a critical point x of l_p, and extend Y to a vector field tangent to M on the neighborhood U of x. It is easy to show that the Hessian of l_p at the critical point x is given by $H_x(X, Y) = X(Yl_p)$ at the point x. Using part (a), we compute

$$H_x(X, Y) = X(Yl_p) = X\langle Y, p \rangle = \langle D_X Y, p \rangle. \tag{2.75}$$

Let ξ be a field of unit normals on U with $\xi(x) = p$. Then $\langle Y, \xi \rangle = 0$ on U, and thus

$$\begin{aligned} 0 = D_X\langle Y, \xi \rangle &= \langle D_X Y, \xi \rangle + \langle Y, D_X \xi \rangle \\ &= \langle D_X Y, \xi \rangle + \langle Y, -A_\xi X \rangle = \langle D_X Y, p \rangle + \langle Y, -A_p X \rangle. \end{aligned} \tag{2.76}$$

From equations (2.75) and (2.76), we get $H_x(X, Y) = \langle A_p X, Y \rangle$. □

As an immediate consequence, we get the following corollary, which is an "Index Theorem" for height functions.

Corollary 2.30. *Let $f : M \to \mathbf{R}^m$ be a smooth immersion of an n-dimensional manifold M into $\mathbf{R}^m$, and suppose that p is a unit vector orthogonal to T_xM.*

(a) *The function l_p has a degenerate critical point at x if and only if the shape operator A_p is singular.*
(b) *If l_p has a nondegenerate critical point at x, then the index of l_p at x is equal to the number of negative eigenvalues of A_p.*

We next consider the *Gauss map* $\nu : BM \to S^{m-1}$, where BM is the unit normal bundle of M, defined by $\nu(x, \xi) = \xi$. The following well-known theorem is obtained by a direct calculation using coordinates on the unit normal bundle BM similar to those used in the proof of Theorem 2.1 on page 11, and we omit the proof here.

Theorem 2.31. *The nullity of the Gauss map ν at a point $\xi \in BM$ is equal to the nullity of the shape operator A_ξ. In particular, ξ is a critical point of ν if and only if A_ξ is singular.*

From Theorem 2.31 and Corollary 2.30, we immediately obtain the following theorem.

Theorem 2.32. *For $p \in S^{m-1}$, the height function l_p is a Morse function on M if and only if p is a regular value of the Gauss map ν.*

Since BM and S^{m-1} are manifolds of the same dimension, Sard's Theorem (see, for example, Milnor [360, p. 10]) implies the following corollary.

Corollary 2.33. (a) *For almost all $p \in S^{m-1}$, the function l_p is a Morse function.*
(b) *Suppose l_p has a nondegenerate critical point of index k at $x \in M$. Then there is a Morse function l_q having a critical point $y \in M$ of index k (q and y can be chosen as close to p and x, respectively, as desired).*

Proof. (a) This follows from Theorem 2.32 and Sard's Theorem.

(b) By Theorem 2.29 (a), we know that $p = \nu(\xi)$, where ξ is a unit normal vector to M at x. Since l_p has a nondegenerate critical point of index k at x, the derivative map ν_* is nonsingular at (x, ξ), and A_ξ has k negative eigenvalues and $n-k$ positive eigenvalues. Thus, there is a neighborhood V of (x, ξ) in BM such that ν_* is nonsingular on V, and the restriction of ν to V is a diffeomorphism of V onto a neighborhood U of p in S^{m-1}. Let $q \in U$ be a regular value of ν. Then $q = \nu(y, \eta)$ for some (y, η) in V, and l_q is a Morse function having a critical point at $y \in M$. Furthermore, since ν_* is nonsingular on V, the number of negative eigenvalues of A_η equals the number of negative eigenvalues of A_ξ, so the index of l_q at y is also k. By Sard's Theorem, the points q and y can be chosen to be as close to p and x, respectively, as desired. □

Tight immersions

Suppose now that M is compact. An immersion $f : M \to \mathbf{R}^m$ is said to be a *tight immersion* if there exists a field $\mathbf{F}$ such that every nondegenerate linear height function l_p has $\beta(M, \mathbf{F})$ critical points on M, i.e., every nondegenerate height function is a perfect Morse function. By Theorem 2.28 above, we see that f is tight if and only if for every nondegenerate linear height function l_p, the map on homology

$$H_*(M_r(l_p), \mathbf{F}) \to H_*(M, \mathbf{F}) \tag{2.77}$$

induced by the inclusion $M_r(l_p) \subset M$ is injective for all r. Note that

$$M_r(l_p) = \{x \in M \mid \langle p, f(x) \rangle \leq r\}. \tag{2.78}$$

Thus, $M_r(l_p)$ is the inverse image under f of the half-space in $\mathbf{R}^m$ determined by the inequality $l_p(q) \leq r$. In this formulation, one requires the map on homology in equation (2.77) to be injective for all half-spaces determined by nondegenerate height functions.

Remark 2.34 (Immersions of minimal total absolute curvature). For smooth immersions of manifolds into Euclidean space, tightness is closely related to the property that the immersion has minimal total absolute curvature in the sense of Chern and Lashof [103, 104] (see [95, pp. 9–17] for more detail).

A celebrated result in the theory of tight immersions is the Chern–Lashof Theorem [103, 104] which states that a tight immersion of a sphere S^n is a convex hypersurface $M^n \subset \mathbf{R}^{n+1} \subset \mathbf{R}^m$. This was generalized to tight topological immersions by Kuiper [302], and so we will state the theorem in that generality. (See also [95, p. 86] for a proof.)

Recall that a map f of a topological space X into $\mathbf{R}^m$ is said to be *substantial* if the image $f(X)$ is not contained in any hyperplane in $\mathbf{R}^m$.

Theorem 2.35. *(Chern–Lashof Theorem) Let $f : S^n \to \mathbf{R}^m$ be a substantial topological immersion such that almost all linear height functions have exactly two critical points. Then $m = n+1$, and f embeds S^n as a convex hypersurface in $\mathbf{R}^{n+1}$.*

Remark 2.36. Kuiper [301] first used the term "convex immersions" for tight immersions, because of the Chern–Lashof Theorem. Banchoff [19] was the first to use the term "tight" for such immersions, in conjunction with his introduction of the two-piece property.

An important advance in the theory due to Kuiper [303] was to remove the restriction mentioned above that the half-space be determined by a nondegenerate linear height function, so that one can use all half-spaces. Kuiper accomplished this by using Čech homology and its "continuity property," as we will now describe.

Kuiper's formulation of tightness then generalizes to continuous maps on compact topological spaces, and so we will define it in that context. For the sake of definiteness, we will use the field $\mathbf{F} = \mathbf{Z}_2$, which has been satisfactory in almost all known applications of the theory of tight immersions thus far.

A map f of a compact topological space X into $\mathbf{R}^m$ is called a *tight map* if for every closed half-space h in $\mathbf{R}^m$, the induced homomorphism

$$H_*(f^{-1}h) \to H_*(X) \tag{2.79}$$

in Čech homology with $\mathbf{Z}_2$ coefficients is injective. A subset of $\mathbf{R}^m$ is called a *tight set* if the inclusion map $f : X \to \mathbf{R}^m$ given by $f(x) = x$ is a tight map.

Remark 2.37 (On the use of Čech homology). Kuiper used Čech homology instead of singular homology because of its continuity property. In particular, this property is used in Kuiper's proof (see Theorem 2.41) that for a tight immersion of smooth manifold, one can use all half-spaces instead of only those that are determined by nondegenerate height functions. This fact simplifies many arguments in the theory of tight immersions and maps. Of course, for triangulable spaces (and thus for smooth manifolds), Čech homology agrees with singular homology.

Remark 2.38 (Tightness is a projective property). In the definition of a tight map above, one does not use the Euclidean metric on $\mathbf{R}^m$, but only the underlying affine space A^m. We can consider A^m as the complement of a hyperplane in real projective space $\mathbf{RP}^m$. If $f : X \to A^m$ is a tight map, and $\sigma : \mathbf{RP}^m \to \mathbf{RP}^m$ is a projective transformation such that the image $\sigma(f(X))$ lies in A^m, then $\sigma f : X \to A^m$ is also a tight map. This follows immediately from the definition, since for every half-space h in A^m, the set $(\sigma f)^{-1}h = f^{-1}(\sigma^{-1}h) = f^{-1}h'$, for the appropriate half-space h'.

Remark 2.39 (Orthogonal projections of tight maps). Suppose that $f : X \to \mathbf{R}^m$ is a tight map, and $\phi : \mathbf{R}^m \to \mathbf{R}^k$ is orthogonal projection onto a Euclidean subspace of $\mathbf{R}^m$. Then $\phi \circ f : X \to \mathbf{R}^k$ is also tight. To see this, let h be the closed half-space in $\mathbf{R}^k$ given by the inequality $l_p \leq r$, for $p \in \mathbf{R}^k$ and $r \in \mathbf{R}$. Then the same inequality in $\mathbf{R}^m$ gives a half-space h' in $\mathbf{R}^m$ such that $\phi^{-1}h = h'$. Thus, $(\phi \circ f)^{-1}h = f^{-1}h'$, and the tightness of $\phi \circ f$ follows from the tightness of f. Similarly, if $f : X \to \mathbf{R}^m$ is tight and $i : \mathbf{R}^m \to \mathbf{R}^{m+j}$ is inclusion of $\mathbf{R}^m$ into a higher dimensional Euclidean space, then $i \circ f : X \to \mathbf{R}^{m+j}$ is also tight. In that case, if h is a half-space in $\mathbf{R}^{m+j}$ and $h' = h \cap \mathbf{R}^m$, then $(i \circ f)^{-1}h = f^{-1}h'$.

We will show in Theorem 2.41 that if an immersion $f : M \to \mathbf{R}^m$ is a tight immersion in the sense that every nondegenerate height function is a perfect Morse function on M, then f is a tight map as defined above. The important point here is to show that the injectivity condition on homology in equation (2.77) holds for half-spaces determined by degenerate height functions, as well as those determined by nondegenerate height functions.

The main ingredients of the proof are the continuity property of Čech homology and the following lemma due to Kuiper [303] (see also [95, pp. 24–26]).

Lemma 2.40. *Let $f : M \to \mathbf{R}^m$ be an immersion of a compact manifold. Suppose U is an open subset of M containing $M_r(l_p)$ for some $p \in S^{m-1}$ and real number r. Then there exists a nondegenerate height function l_q and a real number s such that*

$$M_r(l_p) \subset M_s^-(l_q) \subset M_s(l_q) \subset U. \tag{2.80}$$

Proof. Since $M_r(l_p)$ is compact and U is open, one can easily show that there exists $\varepsilon > 0$ such that $M_{r+\varepsilon}(l_p) \subset U$. Let K be the maximum absolute value that any linear height function assumes on M.

We will use the spherical metric $d(p,z) = \cos^{-1}\langle p, z\rangle$ on S^{m-1}. If $d(p,z) = \alpha$, then there is a unit vector p' is orthogonal to p such that

$$z = \cos\alpha\, p + \sin\alpha\, p'. \tag{2.81}$$

Then

$$p = \sec\alpha\, z - \tan\alpha\, p'. \tag{2.82}$$

For any $x \in M$, we have

$$|l_p(x) - l_z(x)| = |\langle p - z, f(x)\rangle| = |\langle(\sec\alpha - 1)z - \tan\alpha\, p', f(x)\rangle|$$
$$\leq |\sec\alpha - 1|\, |l_z(x)| + |\tan\alpha|\, |l_{p'}(x)| \leq (|\sec\alpha - 1| + |\tan\alpha|)K. \tag{2.83}$$

Choose the positive number α sufficiently small so that

$$|\sec\alpha - 1| < \varepsilon/4K \text{ and } |\tan\alpha| < \varepsilon/4K. \tag{2.84}$$

Then

$$|l_p(x) - l_z(x)| < \varepsilon/2 \tag{2.85}$$

for any $x \in M$, and therefore

$$M_r(l_p) \subset M^-_{r+\varepsilon/2}(l_z) \subset M_{r+\varepsilon/2}(l_z) \subset M_{r+\varepsilon}(l_p) \subset U. \tag{2.86}$$

Let W be the open disk in S^{m-1} centered at p of radius α. If $q \in W$, then we can write

$$q = \cos\theta\, p + \sin\theta\, p', \tag{2.87}$$

for some unit vector p' orthogonal to p, where $0 \leq \theta < \alpha$. We can replace z by q and α by θ in equation (2.83), and get

$$|l_p(x) - l_q(x)| \leq (|\sec\theta - 1| + |\tan\theta|)K. \tag{2.88}$$

Since $0 \leq \theta < \alpha$, we have

$$|\sec\theta - 1| < |\sec\alpha - 1| < \varepsilon/4K \text{ and } |\tan\theta| < |\tan\alpha| < \varepsilon/4K, \tag{2.89}$$

and we still get $|l_p(x) - l_q(x)| < \varepsilon/2$ for all $x \in M$. Thus we have

$$M_r(l_p) \subset M^-_{r+\varepsilon/2}(l_q) \subset M_{r+\varepsilon/2}(l_q) \subset M_{r+\varepsilon}(l_p) \subset U. \tag{2.90}$$

This holds for any point q in the open neighborhood W of p in S^{m-1}. Since the set of regular values of the Gauss map is dense in S^{m-1}, there exists a point q in W such that l_q is nondegenerate, and equation (2.80) holds for that q and $s = r + \varepsilon/2$ by equation (2.90). □

With this lemma, we can prove the following important result due to Kuiper [303]. The proof given here is similar to the proof of Theorem 5.4 of [95, pp. 25–26].

Theorem 2.41. *Let $f : M \to \mathbf{R}^m$ be an immersion of a compact, connected manifold. Suppose that every nondegenerate linear height function l_p has $\beta(M, \mathbf{Z}_2)$ critical points on M. Then for every closed half-space h in $\mathbf{R}^m$, the induced homomorphism*

$$H_*(f^{-1}h) \to H_*(M) \tag{2.91}$$

in Čech homology with $\mathbf{Z}_2$ coefficients is injective.

Proof. For a given half-space h, we have $f^{-1}h = M_r(l_p)$ for some $p \in S^{m-1}$, $r \in \mathbf{R}$. If l_p is nondegenerate, then the map in equation (2.91) is injective by Theorem 2.28, since l_p has $\beta(M, \mathbf{Z}_2)$ critical points on M.

Suppose now that $f^{-1}h = M_r(l_p)$ for a degenerate height function l_p, and some $r \in \mathbf{R}$. We need to show that the map in equation (2.91) is injective in that case also. Here we use the continuity property of Čech homology. We will produce a nested sequence of half-spaces $h_i, i = 1, 2, 3, \ldots$ satisfying

$$f^{-1}(h_i) \supset f^{-1}(h_{i+1}) \supset \cdots \supset \bigcap_{j=1}^{\infty} f^{-1}(h_j) = M_r(l_p), \ i = 1, 2, 3, \ldots \tag{2.92}$$

such that the homomorphism in $\mathbf{Z}_2$-homology

$$H_*(f^{-1}(h_i)) \to H_*M \text{ is injective}, \ i = 1, 2, 3, \ldots \tag{2.93}$$

If equations (2.92) and (2.93) are satisfied, then the map

$$H_*(f^{-1}(h_i)) \to H_*(f^{-1}(h_j)) \text{ is injective for all } i > j. \tag{2.94}$$

The continuity property of Čech homology (see Eilenberg–Steenrod [145, p. 261]) says that

$$H_*(M_r(l_p)) = \overset{\leftarrow}{\lim_{i \to \infty}} H_*(f^{-1}(h_i)). \tag{2.95}$$

Equations (2.94) and Theorem 3.4 of Eilenberg–Steenrod [145, p. 216] on inverse limits imply that the map

$$H_*(M_r(l_p)) \to H_*(f^{-1}(h_i)) \tag{2.96}$$

is injective for each i. Thus, the map

$$H_*(M_r(l_p)) \to H_*(M) \tag{2.97}$$

is injective, as needed.

It remains to construct the sequence $\{h_i\}$. This is done by an inductive procedure using Lemma 2.40 to find at each step a nondegenerate height function l_q and a real number s such that

$$M_r(l_p) \subset M_s^-(l_q) \subset M_s(l_q) \subset M_{r+(1/i)}^-(l_p). \tag{2.98}$$

At each step, the set U in Lemma 2.40 should be taken to be the previous $M_s^-(l_q)$, except for $i = 1$ when $U = M_{r+1}^-(l_p)$. We take h_i to be the half-space $l_q \leq s$ constructed at the i-th step. Note that equation (2.93) is satisfied since each l_q is nondegenerate, and the half-spaces h_i are nested as in equation (2.92). Finally, since

$$f^{-1}(h_{i+1}) \subset M_{r+(1/i)}(l_p), \tag{2.99}$$

we get

$$\bigcap_{j=1}^{\infty} f^{-1}(h_j) = M_r(l_p), \tag{2.100}$$

and the theorem is proven. □

The two-piece property

Another important idea in the theory of tight immersions is the two-piece property due to Banchoff [21]. A continuous map $f : X \to \mathbf{R}^m$ of a compact, connected topological space is said to have the *two-piece property* (TPP) if $f^{-1}h$ is connected for every closed half-space h in $\mathbf{R}^m$. A compact, connected space $X \subset \mathbf{R}^m$ embedded in $\mathbf{R}^m$ is said to have the TPP if the inclusion map $f : X \to \mathbf{R}^m$ has the TPP. In that case, the TPP means that every hyperplane in $\mathbf{R}^m$ cuts X into at most two pieces, whence the name "two-piece property." The following result is immediate.

Theorem 2.42. *Let $f : X \to \mathbf{R}^m$ be continuous map of a compact, connected space X into $\mathbf{R}^m$. If f is tight, then f has the TPP.*

Proof. If the map $f : X \to \mathbf{R}^m$ is tight, then f has the TPP, since tightness implies that $\beta_0(f^{-1}h, \mathbf{Z}_2)$ is less than or equal to one, and $\beta_0(f^{-1}h, \mathbf{Z}_2)$ is equal to the number of connected components of $f^{-1}h$. □

More generally, a map f of a compact connected topological space X into $\mathbf{R}^m$ is said to be *k-tight* if for every closed half-space h in $\mathbf{R}^m$ and for every integer $0 \leq i \leq k$, the induced homomorphism $H_i(f^{-1}h) \to H_i(X)$ in Čech homology with $\mathbf{Z}_2$ coefficients is injective. Thus, 0-tightness is just the two-piece property. If $f : M \to \mathbf{R}^m$ is a smooth immersion of a compact, connected manifold, then f is k-tight if and only if every nondegenerate height function l_p has exactly $\beta_i(M, \mathbf{Z}_2)$ critical points of index i for every integer i such that $0 \leq i \leq k$.

In the setting of smooth immersions of compact manifolds into $\mathbf{R}^m$, we have the following theorem due to Banchoff.

Theorem 2.43. *Let $f : M \to \mathbf{R}^m$ be an immersion of a smooth compact, connected manifold. Then f has the TPP if and only if every nondegenerate linear height function l_p has exactly one minimum and one maximum on M.*

The basic idea here is that if a hyperplane determined by a height function l_p cuts $f(M)$ into more than two pieces, then l_p must have either more than one maximum or more than one minimum, since l_p has a maximum or a minimum on each piece. Conversely, if l_p has more than one maximum or more than one minimum, then there exists a hyperplane determined by l_p that cuts $f(M)$ into more than two pieces (see [95, pp. 29–31] for a complete proof).

Banchoff [21] also noted the following corollary in the case where M is a 2-dimensional surface.

Corollary 2.44. *A TPP immersion $f : M^2 \to \mathbf{R}^m$ of a smooth compact, connected manifold 2-dimensional surface M^2 is tight.*

Proof. Let l_p be a nondegenerate linear height function on M^2. Let $\mu_k(l_p)$ be the number of critical points of l_p of index k. Since f has the TPP, we know that $\mu_0(l_p) = \beta_0(M^2, \mathbf{Z}_2) = 1$ and $\mu_2(l_p) = \beta_2(M^2, \mathbf{Z}_2) = 1$, i.e., l_p has one minimum and one maximum on M^2. Then the Morse relation involving the Euler characteristic $\chi(M^2)$ (see, for example, Milnor [359, p. 29]),

$$\sum_{k=0}^{2}(-1)^k\mu_k(l_p) = \sum_{k=0}^{2}(-1)^k\beta_k(M^2, \mathbf{Z}_2) = \chi(M^2), \tag{2.101}$$

implies that $\mu_1(l_p) = \beta_1(M^2, \mathbf{Z}_2)$ as well, and thus f is a tight immersion. □

Bound on the codimension of a substantial TPP immersion

Another important result in the theory of tight immersions concerns the upper bound on the codimension of a substantial smooth TPP immersion. Kuiper [300] proved part (a) of Theorem 2.46 below, and we will give the proof as in [95, pp. 33–34]. The proof of part (b) of Theorem 2.46 is much more difficult. It is due to Kuiper [301] for $n = 2$, and to Little and Pohl [333] for higher dimensions, and we refer the reader to [95, p. 105] for a complete proof.

The standard embeddings of projective spaces mentioned in part (b) are described in detail in Section 2.9 (see also Tai [505] or [95, pp. 87–98]). The standard embedding of $\mathbf{RP}^2$ into $S^4 \subset \mathbf{R}^5$ is the well-known Veronese surface, which we will now describe.

Remark 2.45 (Veronese surface). Let S^2 be the unit sphere in $\mathbf{R}^3$ given by the equation

$$x^2 + y^2 + z^2 = 1. \tag{2.102}$$

Define a map from S^2 into $\mathbf{R}^6$ by

$$(x, y, z) \mapsto (x^2, y^2, z^2, \sqrt{2}yz, \sqrt{2}zx, \sqrt{2}xy). \tag{2.103}$$

One can easily check that this map takes the same value on antipodal points of S^2, and so it induces a map $\phi : \mathbf{RP}^2 \to \mathbf{R}^6$. An elementary calculation then proves that ϕ is a smooth embedding of $\mathbf{RP}^2$. Furthermore, if $(u_1, \ldots, u_6)$ are the standard coordinates on $\mathbf{R}^6$, then the image of ϕ lies in the Euclidean hyperplane $\mathbf{R}^5 \subset \mathbf{R}^6$ given by the equation:

$$u_1 + u_2 + u_3 = 1, \tag{2.104}$$

since $x^2+y^2+z^2 = 1$. One can easily show further that ϕ is a substantial embedding into $\mathbf{R}^5$, and that the image of ϕ is contained in the unit sphere $S^5 \subset \mathbf{R}^6$ given by the equation,

$$u_1^2 + \cdots + u_6^2 = 1. \tag{2.105}$$

Thus, ϕ is a substantial embedding of $\mathbf{RP}^2$ into the 4-sphere $S^4 = S^5 \cap \mathbf{R}^5$ (see Section 2.9 for more detail).

To see that ϕ has the TPP, note that a hyperplane in $\mathbf{R}^5$ given by an equation

$$a_1u_1 + \cdots + a_6u_6 = c, \tag{2.106}$$

for $c \in \mathbf{R}$, cuts $\mathbf{RP}^2$ in a conic. Such a conic does not separate $\mathbf{RP}^2$ into more than two pieces, and so ϕ has the TPP. Since $\mathbf{RP}^2$ has dimension two, ϕ is also tight by Corollary 2.44. Finally, since ϕ is tight and spherical, it is a taut embedding of $\mathbf{RP}^2$ into $S^4 \subset \mathbf{R}^5$ by Theorem 2.69, which will be proven in the next section.

From the Veronese embedding ϕ, we can obtain a tight substantial embedding of $\mathbf{RP}^2$ into a 4-dimensional Euclidean space $\mathbf{R}^4$ in two different ways. First let $\tau : S^4 - \{P\} \to \mathbf{R}^4$ be stereographic projection with pole P not in the image of ϕ (see Remark 2.7). Then by Theorem 2.70 (see page 61), $\tau \circ \phi$ is a taut (and hence tight by Theorem 2.55 on page 55) embedding of $\mathbf{RP}^2$ into $\mathbf{R}^4$. Secondly, we can compose ϕ with orthogonal projection of $\mathbf{R}^6$ onto the 4-space $\mathbf{R}^4$ spanned by the vectors $\{(e_1 - e_2)/\sqrt{2}, e_4, e_5, e_6\}$, where $\{e_1, \ldots, e_6\}$ is the standard basis of $\mathbf{R}^6$. This gives a parametrization

$$(x, y, z) \mapsto \left(\frac{x^2 - y^2}{\sqrt{2}}, \sqrt{2}yz, \sqrt{2}zx, \sqrt{2}xy\right), \tag{2.107}$$

which induces an embedding $f : \mathbf{RP}^2 \to \mathbf{R}^4$. Since tightness is preserved by orthogonal projections (see Remark 2.39), f is a tight embedding of $\mathbf{RP}^2$ into $\mathbf{R}^4$.

TPP immersions with maximal codimension

In the following theorem, the term "up to a projective transformation of $\mathbf{R}^m$" means in the sense defined in Remark 2.38 on page 42.

Theorem 2.46. *Let $f : M^n \to \mathbf{R}^m$ be a substantial smooth immersion of a compact, connected n-dimensional manifold.*

(a) *If f has the TPP, then $m \leq n(n+3)/2$.*
(b) *If f has the TPP and $m = n(n+3)/2$ for $n \geq 2$, then f is a standard embedding $f : \mathbf{RP}^n \to \mathbf{R}^m$ of a projective space, up to a projective transformation of $\mathbf{R}^m$.*

Proof. (a) Let l_p be a nondegenerate linear height function on M with an absolute maximum at a point $x \in M$. After a translation, we can assume that $f(x)$ is the origin of our coordinate system on $\mathbf{R}^m$, so that $l_p(x) = 0$. Since l_p has a maximum at $f(x)$, we know by Theorem 2.29 that the vector p is normal to $f(M)$ at $f(x)$, and the Hessian $H(X,Y) = \langle A_p X, Y\rangle$ of l_p at x is negative-definite. Let $T_x^\perp M$ denote the normal space to $f(M)$ at $f(x)$, and let V be the vector space of symmetric bilinear forms on T_xM. Define a linear map $\phi : T_x^\perp M \to V$ by $\phi(q) = A_q$, i.e.,

$$\phi(q)(X,Y) = \langle A_q X, Y\rangle, \quad X, Y \in T_xM. \tag{2.108}$$

The dimension of $T_x^\perp M$ is $m - n$, and the dimension of V is $n(n+1)/2$. Thus, if $m - n > n(n+1)/2$, i.e., $m > n(n+3)/2$, then the kernel of ϕ contains a nonzero vector.

We now complete the proof by showing that if f has the TPP, then the kernel of ϕ contains only the zero vector, and thus $m \leq n(n+3)/2$. Suppose there exists a vector $q \neq 0$ in $T_x^\perp M$ with $A_q = 0$. Let $z(t) = p + tq$. Then $z(t) \in T_x^\perp M$ for all t, and

$$A_{z(t)} = A_p + tA_q = A_p,$$

for all t. Thus, $l_{z(t)}$ has a nondegenerate maximum at x for all t. Note that $l_{z(t)}(x) = 0$ for all t, since $f(x)$ is at the origin of the coordinate system.

On the other hand, since f is substantial, there exists a point $y \in M$ such that $l_q(y) \neq 0$. Then we have

$$l_{z(t)}(y) = l_p(y) + t l_q(y),$$

and thus $l_{z(t)}(y) > 0$ for a suitable choice of t. For that value of t, the function $l_{z(t)}$ does not assume its absolute maximum at x. Thus, f does not have the TPP, since if h is the half-space determined by the inequality $l_{z(t)}(u) \geq 0$, for $u \in \mathbf{R}^m$, then $f^{-1}h$ has at least two components, the single point $\{x\}$ and a component containing y.
(b) For a proof of part (b), see Kuiper [301] for $n = 2$, and Little and Pohl [333] for higher dimensions (see also [95, pp. 98–105] for a complete proof). □

Remark 2.47 (The case of dimension $n = 1$). For $n = 1$, part (a) of Theorem 2.46 states that if $f : S^1 \to \mathbf{R}^m$ is a substantially immersed closed curve with the TPP, then $m \leq 2$, and hence the curve is a plane curve. In fact, much more can be said. For $n = 1$, the TPP is equivalent to requiring that the closed curve have total absolute curvature equal to 2π, the minimum value possible. Fenchel [155] proved that a closed curve $f : S^1 \to \mathbf{R}^3$ with total absolute curvature equal to 2π is an embedded convex plane curve, and Borsuk [48] obtained the same conclusion for curves in $\mathbf{R}^m$ with $m > 3$. (See also Chern [102] for a proof of Fenchel's Theorem.) If the curve f is knotted, then the total absolute curvature is greater than 4π (see Fary [153] and Milnor [357, 358]). For a related result regarding the total curvature of a knotted torus, see the paper of Kuiper and Meeks [306].

Kuiper [300] also proved the following generalization of part (a) of Theorem 2.46 which is useful in determining the possible codimensions of tight immersions of projective planes (see Theorem 2.95 on page 81). For this theorem, we need the following notation. Let $(\beta_0, \ldots, \beta_n)$ be an $(n+1)$-tuple of nonnegative integers. Let $c(\beta_0, \ldots, \beta_n)$ be the maximal dimension of a linear family of symmetric bilinear forms in n variables which contains a positive definite form and such that no form in the family has index k if $\beta_k = 0$. Note that $c(\beta_0, \ldots, \beta_n) \leq n(n+1)/2$, the dimension of the space of all symmetric bilinear forms in n variables.

In our applications, of course, the $(n+1)$-tuple $(\beta_0, \ldots, \beta_n)$ will be the $\mathbf{Z}_2$-Betti numbers of a compact manifold M. In the case where M is $\mathbf{FP}^2$ for $\mathbf{F} = \mathbf{R}, \mathbf{C}, \mathbf{H}$, or $\mathbf{O}$ (Cayley numbers), the number $c(\beta_0, \ldots, \beta_n)$ can be computed, and so the following theorem can be used to give bounds on the codimension of a tight immersion of these projective planes into Euclidean spaces (see Theorem 2.95 on page 81).

Theorem 2.48. *Let $f : M^n \to \mathbf{R}^m$ be a substantial tight immersion of a compact, connected n-dimensional manifold, and let β_k denote the k-th $\mathbf{Z}_2$-Betti number of M. Then*

$$m - n \leq c(\beta_0, \ldots, \beta_n) \leq n(n+1)/2.$$

Proof. We will use the notation of the proof of Theorem 2.46. Let V be the vector space of all symmetric bilinear forms on T_xM. Consider the linear map $\phi : T_x^{\perp}M \to V$ defined by $\phi(q) = A_q$, that is,

$$\phi(q)(X, Y) = \langle A_q X, Y \rangle, \quad X, Y \in T_xM. \tag{2.109}$$

Since f is tight, it has the TPP, and so the proof of Theorem 2.46 shows that ϕ is injective on $T_x^{\perp}M$, which has dimension $m - n$. Thus, we have

$$m - n = \dim\,(\text{Image}\,\phi). \tag{2.110}$$

The image of ϕ is a vector space that contains a positive definite bilinear form. Furthermore, if $\beta_k = 0$, then no bilinear form in Image ϕ can have index k, for if

$\phi(q)$ has index k, then l_q has a nondegenerate critical point of index k at x. Then by Corollary 2.33, there exists a nondegenerate linear height function l_z having a critical point y of index k, contradicting tightness, since $\beta_k = 0$.

Thus, the space Image ϕ contains a positive definite form and no form of index k if $\beta_k = 0$. Then by definition, the dimension of Image ϕ is less than or equal to $c(\beta_0, \ldots, \beta_n)$, which is less than or equal to $n(n+1)/2$, the dimension of the space V of all symmetric bilinear forms in n variables. So we have

$$m - n = \dim\,(\text{Image}\ \phi) \leq c(\beta_0, \ldots, \beta_n) \leq n(n+1)/2,$$

as needed. □

The product of two tight immersions is tight

We close this section with a proof of the fact that a product of two tight immersions is tight. This was first noted by Kuiper [300], and we follow the presentation given in [95, pp. 43–46].

Let $f : M \to \mathbf{R}^m$ and $g : M' \to \mathbf{R}^{m'}$ be immersions of compact, connected manifolds, each having dimension greater than or equal to 1. The product immersion,

$$f \times g : M \times M' \to \mathbf{R}^m \times \mathbf{R}^{m'} = \mathbf{R}^{m+m'}, \tag{2.111}$$

is defined by

$$(f \times g)(x, y) = (f(x), g(y)). \tag{2.112}$$

Let p be a unit vector in $\mathbf{R}^{m+m'}$. We can decompose p in a unique way as,

$$p = \cos\theta\ q + \sin\theta\ q', \tag{2.113}$$

for $q \in \mathbf{R}^m$, $q' \in \mathbf{R}^{m'}$, and $0 \leq \theta \leq \pi/2$.

Lemma 2.49. *Let* $p = \cos\theta\ q + \sin\theta\ q'$ *for* $0 \leq \theta \leq \pi/2$.

(a) l_p *is nondegenerate on* $M \times M'$ *if and only if* $0 < \theta < \pi/2$, *and* l_q, $l_{q'}$ *are nondegenerate on* M, M', *respectively.*
(b) *If* l_p *is nondegenerate, then the number* $\mu(l_p)$ *of critical points of* l_p *on* $M \times M'$ *is given by* $\mu(l_p) = \mu(l_q)\mu(l_{q'})$

Proof. Let $(x, y) \in M \times M'$, and let $X \in T_xM$, $Y \in T_yM'$. Then a straightforward calculation yields

$$(l_p)_*(X, Y) = \cos\theta\ (l_q)_*X + \sin\theta\ (l_{q'})_*Y. \tag{2.114}$$

If $\theta = 0$, the set of critical points of l_p is the set of all points of the form (x, y), where x is a critical point of l_q, and y is any point in M'. Similarly, if $\theta = \pi/2$, the set of critical points of l_p is the set of all points of the form (x, y), where x any point in M, and y is a critical point of $l_{q'}$. In either case, the critical points of l_p are not isolated, so they are degenerate critical points, and l_p is not a Morse function.

If $0 < \theta < \pi/2$, then we see from equation (2.114) that l_p has a critical point at (x, y) if and only if l_q has a critical point at x and $l_{q'}$ has a critical point at y. We now compute the Hessian of l_p at such a critical point (x, y). Let $(u_1, \ldots, u_n)$ and $(v_1, \ldots, v_{n'})$ be local coordinates in neighborhoods of x in M, and y in M', respectively. Clearly, with respect to the local coordinates $(u_1, \ldots, u_n, v_1, \ldots, v_{n'})$ in a neighborhood of (x, y) in $M \times M'$, the Hessian of l_p at the critical point (x, y) has the form

$$H_{(x,y)}(l_p) = \begin{bmatrix} \cos\theta\, H_x(l_q) & 0 \\ 0 & \sin\theta\, H_y(l_{q'}) \end{bmatrix}. \tag{2.115}$$

Thus, $H_{(x,y)}(l_p)$ is nonsingular if and only if $H_x(l_q)$ and $H_y(l_{q'})$ are nonsingular. Hence, l_p is a nondegenerate function if and only if l_q and $l_{q'}$ are nondegenerate functions. Furthermore, from equation (2.115), we see that the index of l_p at a nondegenerate critical point (x, y) is equal to the sum of the indices of l_q at x and $l_{q'}$ at y. Thus, for $0 \leq k \leq n + n'$, we have

$$\mu_k(l_p) = \sum_{i+j=k} \mu_i(l_q)\mu_j(l_{q'}). \tag{2.116}$$

From this, we compute

$$\begin{aligned} \mu(l_p) &= \sum_{k=0}^{n+n'} \mu_k(l_p) \\ &= \sum_{k=0}^{n+n'} \sum_{i+j=k} \mu_i(l_q)\mu_j(l_{q'}) \\ &= \mu(l_q)\, \mu(l_{q'}). \end{aligned}$$

□

Theorem 2.50. *Suppose $f : M \to \mathbf{R}^m$ and $g : M' \to \mathbf{R}^{m'}$ are tight immersions of compact manifolds. Then $f \times g$ is a tight immersion of $M \times M'$ into $\mathbf{R}^{m+m'}$.*

Proof. In the notation of Lemma 2.49, we have for any nondegenerate height function l_p on $M \times M'$,

$$\mu(l_p) = \mu(l_q)\mu(l_{q'}),$$

for appropriate nondegenerate functions l_q and $l_{q'}$. Since f and g are tight, we know that

$$\mu(l_q) = \beta(M, \mathbf{Z}_2) \text{ and } \mu(l_{q'}) = \beta(M', \mathbf{Z}_2),$$

for every such pair of nondegenerate functions l_q and $l_{q'}$. Thus,

$$\mu(l_p) = \beta(M, \mathbf{Z}_2)\, \beta(M', \mathbf{Z}_2) = \beta(M \times M', \mathbf{Z}_2),$$

where the last equality is due to the Künneth formula (see, for example, Greenberg [183, p. 198]). Hence, $f \times g$ is tight. □

2.7 Distance Functions and Taut Submanifolds

In this section, we give a brief review of the theory of distance functions and taut submanifolds in Euclidean space. Various aspects of this theory will be treated in more detail later in this book. For more complete coverage of the topic, the reader is referred to Chapter 2 of the book [95] or the survey article [76]. Our treatment here is based on [95, pp. 113–127].

Let $f : M \to \mathbf{R}^m$ be a smooth immersion of an n-dimensional manifold M into Euclidean space $\mathbf{R}^m$. For $p \in \mathbf{R}^m$, the *Euclidean distance function* L_p is defined on $\mathbf{R}^m$ by

$$L_p(q) = |p - q|^2. \tag{2.117}$$

The restriction of L_p to M gives a real-valued function $L_p : M \to \mathbf{R}$ defined by $L_p(x) = |p - f(x)|^2$. As with linear height functions, Sard's Theorem implies that for almost all $p \in \mathbf{R}^m$, the function L_p is a Morse function on M.

We recall the normal exponential map $E : NM \to \mathbf{R}^m$ defined in Section 2.2 by

$$E(x, \zeta) = f(x) + \zeta, \tag{2.118}$$

where ζ is a normal vector to $f(M)$ at $f(x)$. As in Section 2.2, the focal set of M is the set of critical values of the map E. Hence, by Sard's Theorem, the focal set of M has measure zero in $\mathbf{R}^m$. As noted in Theorem 2.1, if $p = E(x, t\xi)$, where $|\xi| = 1$, then p is a focal point of (M, x) of multiplicity $\nu > 0$ if and only if $1/t$ is a principal curvature of multiplicity ν of the shape operator A_ξ.

Index Theorem for distance functions

The critical point behavior of the L_p functions is described by the following well-known *Index Theorem* (see Milnor [359, pp. 32–38] for a proof).

Theorem 2.51. *Let $f : M \to \mathbf{R}^m$ be a smooth immersion of an n-dimensional manifold M into Euclidean space $\mathbf{R}^m$, and let $p \in \mathbf{R}^m$.*

(a) *A point $x \in M$ is a critical point of L_p if and only if $p = E(x, \zeta)$ for some $\zeta \in T_x^\perp M$.*
(b) *L_p has a degenerate critical point at x if and only if p is a focal point of (M, x).*
(c) *If L_p has a nondegenerate critical point at x, then the index of L_p at x is equal to the number of focal points of (M, x) (counting multiplicities) on the segment from $f(x)$ to p.*

The following corollary, due to Nomizu and Rodriguez [405, p. 199], follows from Theorem 2.51 in the same way that Corollary 2.33 follows from Theorem 2.29 for height functions. In the proof of Corollary 2.52, one uses the normal exponential map E instead of the Gauss map ν. In particular, part (a) of the corollary follows from part (b) of Theorem 2.51, since the focal set of M has measure zero in $\mathbf{R}^m$.

Corollary 2.52. *Let $f : M \to \mathbf{R}^m$ be a smooth immersion of an n-dimensional manifold M into Euclidean space $\mathbf{R}^m$.*

(a) *For almost all $p \in \mathbf{R}^m$, L_p is a Morse function on M.*
(b) *Suppose L_p has a nondegenerate critical point of index k at $x \in M$. Then there is a Morse function L_q having a critical point $y \in M$ of index k (q and y may be chosen as close to p and x, respectively, as desired).*

Taut immersions

We can now define taut immersions in a similar way to how we defined tight immersions in the previous section. Suppose first that M is a compact n-dimensional manifold. An immersion $f : M \to \mathbf{R}^m$ is said to be a *taut immersion* if there exists a field $\mathbf{F}$ such that every nondegenerate Euclidean distance function L_p has $\beta(M, \mathbf{F})$ critical points on M, i.e., every nondegenerate distance function is a perfect Morse function. By Theorem 2.28 in the previous section, we see that f is taut if and only if for every nondegenerate Euclidean distance function L_p, the map on homology

$$H_*(M_r(L_p), \mathbf{F}) \to H_*(M, \mathbf{F}) \tag{2.119}$$

induced by the inclusion $M_r(L_p) \subset M$ is injective for all r. Note that

$$M_r(L_p) = \{x \in M \mid |p - f(x)|^2 \leq r\}. \tag{2.120}$$

Thus, $M_r(L_p)$ is the inverse image under f of the closed ball in $\mathbf{R}^m$ with center p and radius $\sqrt{r}$. In this formulation of tautness, one requires the map on homology in equation (2.119) to be injective for all closed balls determined by nondegenerate distance functions, that is, the map

$$H_*(f^{-1}B, \mathbf{F}) \to H_*(M, \mathbf{F}) \tag{2.121}$$

is injective for every closed ball B centered at a point $p \in \mathbf{R}^m$ such that L_p is a nondegenerate function.

Remark 2.53 ($\mathbf{F}$-taut implies $\mathbf{Z}_2$-taut). Grove and Halperin [185]), and independently Terng and Thorbergsson [531], extended the notion of tautness to submanifolds of complete Riemannian manifolds. Their definition agrees with the definition of tautness above for submanifolds of Euclidean space. Recently Wiesendorf [554] showed that if a compact submanifold of a complete Riemannian manifold is taut with respect to some field $\mathbf{F}$, then it is also $\mathbf{Z}_2$-taut. Thus, we will use $\mathbf{Z}_2$-tautness at all times.

As with tight immersions, if we use $\mathbf{Z}_2$–Čech homology and its continuity property, we can prove results similar to Lemma 2.40 and Theorem 2.41 which imply that we can use all closed balls B in $\mathbf{R}^m$ in equation (2.121) and not just those determined by nondegenerate distance functions.

Next one shows that if the injectivity condition in equation (2.121) holds for all closed balls B, then it also holds for all closed half-spaces h and for all complements of open balls in $\mathbf{R}^m$. For half-spaces, this comes from approximating $f^{-1}h$ by $f^{-1}B$, for an appropriate large closed ball B, in a manner similar to the proof of Lemma 2.40. For complements of open balls, one uses the fact that $L_p(x) \geq r$ if and only if $-L_p(x) \leq -r$, and L_p is a perfect Morse function on M if and only if $-L_p$ is a perfect Morse function on M. Using these ideas and techniques similar to those in the proofs of Lemma 2.40 and Theorem 2.41, one can prove the following theorem, and we omit the proof here.

Theorem 2.54. *Let $f : M \to \mathbf{R}^m$ be an immersion of a compact, connected manifold. Suppose that every nondegenerate Euclidean distance function L_p has $\beta(M, \mathbf{Z}_2)$ critical points on M. Then for every closed ball, complement of an open ball, and closed half-space Ω in $\mathbf{R}^m$, the induced homomorphism,*

$$H_*(f^{-1}\Omega) \to H_*(M) \tag{2.122}$$

in Čech homology with $\mathbf{Z}_2$ coefficients is injective.

If S^{m-1} is the metric hypersphere in $\mathbf{R}^m$ with center p and radius r, then S^{m-1} is a taut subset of $\mathbf{R}^m$. To see this, note that if q is any point in $\mathbf{R}^m$ other than p, then L_q is a nondegenerate function having exactly two critical points on S^{m-1} at the two points where the line determined by p and q intersects the sphere S^{m-1}. We will show later (see Theorems 2.73 and 2.74 on page 63) that every taut immersion of an $(m-1)$-sphere into $\mathbf{R}^m$ is a metric hypersphere. This result was proven for $m = 2$ and $m = 3$ by Banchoff [20], and for higher dimensions by Carter and West [61], and independently by Nomizu and Rodriguez [405] using a different proof.

As in the case of tightness, one can generalize the notion of tautness to continuous maps of compact spaces as follows. A map f of a compact topological space X into

$\mathbf{R}^m$ is called a *taut map* if for every closed ball, complement of an open ball, and closed half-space Ω in $\mathbf{R}^m$, the induced homomorphism

$$H_*(f^{-1}\Omega) \to H_*(X) \tag{2.123}$$

in Čech homology with $\mathbf{Z}_2$ coefficients is injective. A subset of $\mathbf{R}^m$ is called a *taut set* if the inclusion map $f : X \to \mathbf{R}^m$ is a taut map.

An obvious consequence of this definition and the definition of a tight map is the following theorem, since tightness only requires that the map on homology in equation (2.123) be injective when Ω is a closed half-space.

Theorem 2.55. *Let $f : X \to \mathbf{R}^m$ be continuous map of a compact, connected space X into $\mathbf{R}^m$. If f is taut, then f is tight.*

Of course, Theorem 2.54 shows that if an immersion $f : M \to \mathbf{R}^m$ of a compact, connected manifold M is taut in the sense that every nondegenerate L_p-function is a perfect Morse function, then f is a taut map as defined above.

The spherical two-piece property

As is the case with tightness, there is a two-piece property associated to tautness due to Banchoff [20]. A continuous map $f : X \to \mathbf{R}^m$ of a compact, connected topological space is said to have the *spherical two-piece property* (STPP) if $f^{-1}\Omega$ is connected for every closed ball, complement of an open ball, and closed half-space Ω in $\mathbf{R}^m$. A compact, connected space $X \subset \mathbf{R}^m$ embedded in $\mathbf{R}^m$ is said to have the STPP if the inclusion map $f : X \to \mathbf{R}^m$ has the STPP. In that case, the STPP means that every hypersphere and hyperplane in $\mathbf{R}^m$ cuts X into at most two pieces, whence the name "spherical two-piece property."

The following results can be proven in a way to very similar Theorem 2.42, Theorem 2.43, and Corollary 2.44 for the two-piece property in the last section, and we omit the proofs here. These are due to Banchoff [20].

Theorem 2.56. *Let $f : X \to \mathbf{R}^m$ be continuous map of a compact, connected space X into $\mathbf{R}^m$. If f is taut, then f has the STPP.*

Theorem 2.57. *Let $f : M \to \mathbf{R}^m$ be an immersion of a smooth compact, connected manifold. Then f has the STPP if and only if every nondegenerate distance function L_p has exactly one minimum and one maximum on M.*

Corollary 2.58. *An STPP immersion $f : M^2 \to \mathbf{R}^m$ of a smooth compact, connected manifold 2-dimensional surface M^2 is taut.*

Carter and West [61] introduced the term "taut immersion" in a paper published in 1972. They also noted that tautness can be defined for proper immersions of non-compact manifolds as follows. Recall that a map $f : X \to \mathbf{R}^m$ of a topological space X is called *proper* if $f^{-1}K$ is compact for every compact subset K of $\mathbf{R}^m$.

If $f : M \to \mathbf{R}^m$ is a proper immersion of a smooth manifold, then $f^{-1}B$ is compact for every closed ball B in $\mathbf{R}^m$, and so the Morse inequalities (2.68) for a nondegenerate distance function L_p can be applied.

We say that such a proper immersion of a non-compact manifold is *taut* if for every closed ball B in $\mathbf{R}^m$, the map $H_*(f^{-1}B) \to H_*M$ (in $\mathbf{Z}_2$–Čech homology) is injective. This definition agrees with the definition of a taut immersion of a compact manifold by Theorem 2.54, since in that case if the map $H_*(f^{-1}B) \to H_*M$ is injective for all closed balls B, then the map $H_*(f^{-1}\Omega) \to H_*M$ is also injective for all closed half-spaces and all complements of open balls Ω.

From Theorem 2.28 on page 38, we see that a proper immersion of a non-compact manifold is taut if and only if for every nondegenerate L_p, the equation

$$\mu_k(L_p, r) = \beta_k(L_p, r, \mathbf{Z}_2) \tag{2.124}$$

holds for all $r \in \mathbf{R}$, $k \in \mathbf{Z}$.

More generally, a proper immersion $f : M \to \mathbf{R}^m$ of a smooth connected manifold is said to be *k-taut* if for every closed ball B in $\mathbf{R}^m$ and for every integer $i \leq k$, the induced homomorphism $H_i(f^{-1}B) \to H_i(M)$ in Čech homology with $\mathbf{Z}_2$ coefficients is injective. If M is compact and connected, then 0-tautness is equivalent to the STPP, since in that case if the map $H_0(f^{-1}B) \to H_*M$ is injective for all closed balls B, then the map $H_0(f^{-1}\Omega) \to H_*M$ is also injective for all closed half-spaces and all complements of open balls Ω, and so f has the STPP.

Next we have the following important consequence of the 0-tautness for smooth immersions due to Banchoff [20], and Carter and West [61].

Theorem 2.59. *Let $f : M \to \mathbf{R}^m$ be a proper immersion of a smooth connected manifold. If f is* 0*-taut, then f is an embedding.*

Proof. Suppose $f(x_1) = f(x_2) = p$ for two distinct points x_1, x_2 in M. Since f is an immersion, there exist neighborhoods U_1 and U_2 of x_1 and x_2, respectively, on which f is an embedding. Let B be the closed ball of radius 0 centered at p. Then $f^{-1}B$ has at least two connected components, $\{x_1\} \subset U_1$ and $\{x_2\} \subset U_2$, contradicting the assumption that f is 0-taut. □

Constructions preserving tautness

In the following three remarks, we discuss three important constructions which preserve tautness: cylinders over taut submanifolds, tubes over taut submanifolds, and hypersurfaces of revolution with a taut submanifold as the profile submanifold.

Remark 2.60 (Cylinders over taut submanifolds). An example of a taut embedding of a non-compact manifold is a circular cylinder in $\mathbf{R}^3$, or more generally a *spherical cylinder* defined by the product embedding,

$$f \times g : S^k \times \mathbf{R}^{n-k} \to \mathbf{R}^{k+1} \times \mathbf{R}^{n-k} = \mathbf{R}^{n+1}, \tag{2.125}$$

where f embeds S^k as a metric sphere, and g is the identity map on $\mathbf{R}^{n-k}$. Every nondegenerate L_p function has two critical points on $S^k \times \mathbf{R}^{n-k}$, one of index 0 and one of index k. More generally, suppose that $f : M \to \mathbf{R}^k$ is a taut immersion of a compact, connected manifold M, and g is the identity map on $\mathbf{R}^{m-k}$ then

$$f \times g : M \times \mathbf{R}^{m-k} \to \mathbf{R}^k \times \mathbf{R}^{m-k} = \mathbf{R}^m, \tag{2.126}$$

is taut. To see this, note that if $p = (p_1, p_2)$ and $x = (x_1, x_2)$ are points in $\mathbf{R}^k \times \mathbf{R}^{m-k}$, then

$$L_p(x) = L_{p_1}(x_1) + L_{p_2}(x_2). \tag{2.127}$$

Thus, L_p has a critical point at x if and only if L_{p_1} has a critical point at x_1 and $p_2 = x_2$. Such a critical point is nondegenerate if and only if the critical point of L_{p_1} at x_1 is nondegenerate, and in that case, these two critical points have the same index. Since $M \times \mathbf{R}^{m-k}$ and M have the same Betti numbers, f is taut if and only if $f \times g$ is taut.

Remark 2.61 (Parallel hypersurfaces and tubes over taut submanifolds). Carter and West [61] and Pinkall [447, p. 83] pointed out that constructing parallel hypersurfaces or tubes over taut submanifolds preserves tautness.

First suppose that $f : M \to \mathbf{R}^{n+1}$ is an embedded compact, connected oriented hypersurface with global field of unit normals ξ. Suppose that t is a real number such that the parallel map $f_t : M \to \mathbf{R}^{n+1}$ given by

$$f_t(x) = f(x) + t\xi(x) \tag{2.128}$$

is an embedding. Then f_t is a parallel hypersurface of the original embedding $f_0 = f$. By Theorem 2.3 on page 18, the parallel hypersurfaces f_t and f have the same focal set. Suppose that $p \in \mathbf{R}^{n+1}$ is not a focal point of these hypersurfaces. Let L_p denote the restriction of the distance function determined by p to the original embedding f, and let $\tilde{L}_p$ denote its restriction to the parallel hypersurface f_t. By Theorem 2.51, L_p has a critical point at $x \in M$ if and only if $\tilde{L}_p$ has a critical point at x, since the normal line to $f(M)$ at $f(x)$ is the same as the normal line to $f_t(M)$ at $f_t(x)$. So the functions L_p and $\tilde{L}_p$ have the same number of critical points on M. Thus, f_t is taut if and only if f is taut.

Next consider the case where $f : M \to \mathbf{R}^{n+k}$ is a tautly embedded compact, connected submanifold of codimension $k > 1$ in $\mathbf{R}^{n+k}$. Again consider $t > 0$ sufficiently small so that the tube $f_t : BM \to \mathbf{R}^{n+k}$ is an embedded hypersurface, where BM is the unit normal bundle of $f(M)$ in $\mathbf{R}^{n+k}$.

By Theorem 2.3, the focal set of the tube f_t is the union of the focal set of $f(M)$ with $f(M)$ itself. Let $p \in \mathbf{R}^{n+k}$ be a point that is not a focal point of the tube f_t. Let L_p denote the restriction of the distance function determined by p to the original embedding f, and let $\tilde{L}_p$ denote its restriction to the tube f_t. Each critical point $x \in M$ of L_p corresponds to two critical points of $\tilde{L}_p$ on the tube f_t at points where the line from p to $f(x)$ intersects the tube. These critical points are

$$z_1 = f(x) + t\xi, \text{ and } z_2 = f(x) - t\xi, \tag{2.129}$$

where $\xi = (p - f(x))/|p - f(x)|$. Thus, the number of critical points of $\tilde{L}_p$ is equal to twice the number of critical points of L_p. (Note that if $p \in f(M)$, then $\tilde{L}_p$ is a degenerate function, whereas L_p may be nondegenerate. This will not affect tautness since $f(M)$ has measure zero in $\mathbf{R}^{n+k}$.)

Therefore, f is taut if and only if f_t is taut, since the sum of the $\mathbf{Z}_2$-Betti numbers of the unit normal bundle BM (the domain of the tube f_t) is equal to twice the sum of the $\mathbf{Z}_2$-Betti numbers of M. This last fact follows from the Gysin sequence of the unit normal bundle BM of M (see, for example, [418, Lemma 4.7, p. 264]), as was pointed out by Pinkall [447, p. 83]). Thus we have the following theorem due to Pinkall.

Theorem 2.62. *Let $f : M \to \mathbf{R}^n$ be a compact, connected embedded submanifold of $\mathbf{R}^n$ of codimension greater than one, and let $t > 0$ be sufficiently small so that the tube $f_t : BM \to \mathbf{R}^n$ is a compact, connected embedded hypersurface in $\mathbf{R}^n$. Then $f(M)$ is taut with respect to $\mathbf{Z}_2$ coefficients if and only if the tube $f_t(M)$ is taut with respect to $\mathbf{Z}_2$ coefficients.*

Remark 2.63 (Taut hypersurfaces of revolution). Suppose M is a taut compact, connected hypersurface embedded in $\mathbf{R}^{k+1}$ which is disjoint from a hyperplane $\mathbf{R}^k \subset \mathbf{R}^{k+1}$ through the origin. Let e_{k+1} be a unit normal to the hyperplane $\mathbf{R}^k$ in $\mathbf{R}^{k+1}$. Embed $\mathbf{R}^{k+1}$ in $\mathbf{R}^{n+1}$, and let $\mathbf{R}^{n-k+1}$ be the orthogonal complement of $\mathbf{R}^k$ in $\mathbf{R}^{n+1}$. Let $SO(n-k+1)$ denote the group of isometries in $SO(n+1)$ that keep $\mathbf{R}^k$ pointwise fixed. If we consider $\mathbf{R}^{n+1}$ as $\mathbf{R}^k \times \mathbf{R}^{n-k+1}$, then each point of $M \subset \mathbf{R}^{k+1}$ has the form (x, y), where $y = ce_{k+1}$ for some $c > 0$. Let

$$W = \{(x, Ay) \mid (x, y) \in M,\ A \in SO(n-k+1)\} \tag{2.130}$$

be the hypersurface in $\mathbf{R}^{n+1}$ obtained by rotating M about the axis $\mathbf{R}^k$. Then W is diffeomorphic to $M \times S^{n-k}$, and the sum of the $\mathbf{Z}_2$-Betti numbers of W satisfies $\beta(W) = 2\beta(M)$.

We now show that W is taut in $\mathbf{R}^{n+1}$. First, if $p \in \mathbf{R}^k \subset \mathbf{R}^{n+1}$, then L_p has an absolute minimum on M at some point $z \in M$. Hence, L_p has critical points at all the points of W in the orbit of z under the action of $SO(n-k+1)$. Since these critical points are not isolated, they are degenerate critical points. Thus, every point $p \in \mathbf{R}^k$ is a focal point of W. Next consider any $(k+1)$-plane of the form

$$V = \mathbf{R}^k \oplus \text{Span}\,\{Ae_{k+1}\}, \tag{2.131}$$

for a fixed $A \in SO(n-k+1)$. Then $W \cap V$ consists of two disjoint congruent copies of M. If $z \in W \cap V$, then the normal line to W through z lies in V. Now suppose that L_p, for $p \in \mathbf{R}^{n+1}$, is a nondegenerate function on W. Then p does not lie in the axis $\mathbf{R}^k$, so p lies in the space V spanned by $\mathbf{R}^k$ and p itself. All of the critical points of L_p on W lie in $W \cap V$. Since M is taut, L_p has exactly $\beta(M)$ critical points on each of the two copies of M in $W \cap V$. Thus, L_p has $\beta(W) = 2\beta(M)$ critical points on W. This is true for all Morse functions of the form L_p on W, and so W is tautly embedded in $\mathbf{R}^{n+1}$.

Basic results on taut embeddings

We now follow the development of the theory in Carter and West [61]. The first theorem is essentially Banchoff's [20] observation that for an STPP embedding, every local support sphere is a global support sphere.

Theorem 2.64. (a) *Let $f : M \to \mathbf{R}^m$ be a 0-taut embedding of a connected manifold M. Suppose p is the first focal point of (M,x) on a normal ray to $f(M)$ at $f(x)$. If q is any point except p on the closed segment from $f(x)$ to p, then L_q has a strict absolute minimum on M at x. Further, the function L_p itself has an absolute minimum at x.*

(b) *Let $f : M \to \mathbf{R}^m$ be an STPP embedding of a compact, connected n-dimensional manifold. Suppose that p is a focal point of (M,x) such that the sum of the multiplicities of the focal points of (M,x) on the closed segment from $f(x)$ to p is n. If q is any point beyond p on the normal ray from $f(x)$ through p, then L_q has a strict absolute maximum at p. Further, the function L_p itself has an absolute maximum at x.*

Proof. (a) For any point $q \neq p$ on the closed segment from $f(x)$ to p, the function L_q has a strict local minimum at x by the Index Theorem (Theorem 2.51). Since the intersection of $f(M)$ with the closed ball centered at q through $f(x)$ is connected by 0-tautness, this intersection consists of the point $f(x)$ alone. Therefore, $f(M)$ lies outside the corresponding open ball centered at q with radius equal to the length of the segment from q to $f(x)$. Thus $f(M)$ lies outside the union of these open balls as q varies from $f(x)$ to p, and so $f(M)$ lies outside the open ball centered at p through $f(x)$. Therefore, L_p has an absolute minimum at x.
(b) This is proven in a way similar to (a) using maxima rather than minima. □

This theorem has the following three useful corollaries. Here l_ξ denotes the linear height function in the direction ξ.

Corollary 2.65. *Let $f : M \to \mathbf{R}^m$ be a 0-taut embedding of a connected manifold M. Suppose there are no focal points of (M,x) on the normal ray to $f(M)$ in the direction ξ at $f(x)$. Then $f(M)$ lies in the closed half-space determined by the inequality $l_\xi \leq l_\xi(x)$.*

Proof. For all q on the normal ray in question, part (a) of Theorem 2.64 above implies that the set $f(M)$ is disjoint from the open sphere centered at q of radius $|q - f(x)|$. Hence, $f(M)$ is disjoint from the union of such open balls, which is the open half-space determined by the inequality $l_\xi(u) > l_\xi(x)$, for $u \in \mathbf{R}^m$. □

From Theorem 2.64 and Corollary 2.65, we see that the existence of a normal vector ξ such that $A_\xi = \lambda I$ has strong implications for an STPP embedding, as the next two corollaries show.

Corollary 2.66. *Let $f : M \to \mathbf{R}^m$ be a 0-taut embedding of a connected manifold M. If $A_\xi = 0$ for some unit normal ξ to $f(M)$ at a point $f(x)$, then $f(M)$ lies in the hyperplane in $\mathbf{R}^m$ determined by the condition $l_\xi = l_\xi(x)$.*

Proof. Since $A_\xi = 0$, there do not exist any focal points on the normal line determined by ξ. If we apply Corollary 2.65 to each of the normal rays determined by ξ, we get that $f(M)$ lies in the intersection of the two closed half spaces determined by the hyperplane with equation $l_\xi = l_\xi(x)$, and so $f(M)$ lies in that hyperplane. □

Corollary 2.67. *Let $f : M \to \mathbf{R}^m$ be an STPP embedding of a compact, connected manifold M. If $A_\xi = \lambda I, \lambda \neq 0$, for some unit normal ξ to $f(M)$ at a point $f(x)$, then $f(M)$ lies in the hypersphere in $\mathbf{R}^m$ centered at the focal point $p = f(x) + (1/\lambda)\xi$ with radius $1/|\lambda|$.*

Proof. Let q be a point on the open segment from $f(x)$ to p. By Theorem 2.64 (a), the set $f(M)$ does not intersect the open ball centered at q of radius $|q - f(x)|$. Hence $f(M)$ is disjoint from the union of such open balls, i.e., the open ball centered at p of radius $1/|\lambda|$. Similarly, by part (b) of Theorem 2.64, the set $f(M)$ is disjoint from the complement of the closed ball centered at p with radius $1/|\lambda|$. Thus $f(M)$ lies in the hypersphere centered at p with radius $1/|\lambda|$. □

This has the following immediate corollary in the case where M is a hypersurface. This was first proven by Banchoff [20] in the case of where M is a 2-dimensional surface.

Corollary 2.68. *Let $f : M^n \to \mathbf{R}^{n+1}$ be a codimension one STPP embedding of a compact, connected manifold M. If $f(M)$ has one umbilic point, then f embeds M as a metric sphere in $\mathbf{R}^{n+1}$.*

Proof. By Corollaries 2.66 and 2.67, if $f(M)$ has one umbilic point, then $f(M)$ is contained in a hyperplane in $\mathbf{R}^{n+1}$ or in a metric hypersphere $S^n \subset \mathbf{R}^{n+1}$. The image of the compact n-dimensional manifold M cannot be contained in a hyperplane, and so $f(M)$ is a compact, connected n-dimensional submanifold of S^n. Thus, $f(M)$ is S^n itself. □

The relationship between tight and taut maps

We now wish to explore the relationship between tightness and tautness further. The first result is that a tight, spherical map is taut. Here $f : X \to \mathbf{R}^m$ is *spherical* if the image of f lies in a metric hypersphere in $\mathbf{R}^m$.

Theorem 2.69. *Let $f : M \to S^m \subset \mathbf{R}^{m+1}$ be a tight spherical map of a compact topological space X. Then f is a taut map into $\mathbf{R}^{m+1}$.*

Proof. Let Ω be a closed ball or the complement of an open ball in $\mathbf{R}^{m+1}$. Then $\Omega \cap S^m = h \cap S^m$ for some closed half-space in $\mathbf{R}^{m+1}$. Since $f(X)$ is contained in S^m, we have

$$f^{-1}\Omega = f^{-1}(\Omega \cap S^m) = f^{-1}(h \cap S^m) = f^{-1}h.$$

Since f is tight, the map $H_*(f^{-1}h) \to H_*(X)$ is injective, and so the map $H_*(f^{-1}\Omega) \to H_*(X)$ is injective, and f is taut. □

Let $\tau : S^m - \{P\} \to \mathbf{R}^m$ be stereographic projection with pole $P \in S^m$ as in equation (2.45). Via the map τ, the space $S^m - \{P\}$ is conformally equivalent to $\mathbf{R}^m$, or we may consider S^m as $\mathbf{R}^m \cup \{\infty\}$, the one-point compactification of $\mathbf{R}^m$. A conformal transformation of $\mathbf{R}^m \cup \{\infty\}$ takes the collection of all hyperspheres and hyperplanes onto itself. Hence, tautness and the STPP are preserved by such a conformal transformation. We formulate this conformal invariance of tautness and the STPP specifically in the following theorem. In this way, we see that tautness is equivalent to the combination of tight and spherical via stereographic projection.

Theorem 2.70. *Let X be a compact topological space.*

(a) *If $f : X \to \mathbf{R}^m$ is a taut (respectively STPP) map, and φ is a conformal transformation of $\mathbf{R}^m \cup \{\infty\}$ such that $\varphi(f(X)) \subset \mathbf{R}^m$, then $\varphi \circ f$ is a taut (respectively STPP) map of X into $\mathbf{R}^m$.*
(b) *If $f : X \to S^m \subset \mathbf{R}^{m+1}$ is taut (respectively STPP), and $\tau : S^m - \{P\} \to \mathbf{R}^m$ is stereographic projection with pole P not in $f(X)$, then $\tau \circ f$ is a taut (respectively STPP) map of X into $\mathbf{R}^m$.*
(c) *If $f : X \to \mathbf{R}^m$ is taut (respectively STPP) and $\tau^{-1} : \mathbf{R}^m \to S^m \subset \mathbf{R}^{m+1}$ is inverse stereographic projection with respect to any pole P, then $\tau^{-1} \circ f$ is a taut (respectively STPP) map of X into S^m.*

By similar considerations, we see another method to obtain taut embeddings of non-compact manifolds, as in Carter and West [61] (see also [95, pp. 120–121]).

Theorem 2.71. *Suppose that $f : M \to \mathbf{R}^m$ is a taut embedding of a compact, connected manifold M, and φ is a conformal transformation of $\mathbf{R}^m \cup \{\infty\}$ such that $\varphi(f(x)) = \infty$ for some $x \in M$. Then $\varphi \circ f$ is a taut embedding of $M - \{x\}$ into $\mathbf{R}^m$.*

Proof. If B is any closed ball in $\mathbf{R}^m$, then $\varphi^{-1}B$ is a closed ball, the complement of an open ball, or a closed half-space in $\mathbf{R}^m$. Since f is taut, the map $H_*(f^{-1}(\varphi^{-1}B)) \to$

$H_*(M)$ is injective. Since this map factors through the homomorphism $H_*(M - \{x\}) \to H_*(M)$, the map

$$H_*(f^{-1}(\varphi^{-1}B)) \to H_*(M - \{x\})$$

is injective also, as needed. □

Recall that a map f of a topological space X into $\mathbf{R}^m$ is said to be substantial if the image $f(X)$ is not contained in any hyperplane in $\mathbf{R}^m$. From the theorem above, we immediately get a way to obtain more examples of taut submanifolds in $\mathbf{R}^m$ by taking the image under stereographic projection of taut submanifolds in S^m.

Corollary 2.72. *Let M be a compact manifold. Then there exists a substantial, non-spherical taut (respectively STPP) embedding $f : M \to \mathbf{R}^m$ if and only if there exists a substantial taut (respectively STPP) spherical embedding*

$$\tilde{f} : M \to \mathbf{R}^{m+1}.$$

Proof. Let $f : M \to \mathbf{R}^m$ be a substantial, non-spherical taut embedding. Let τ^{-1} be the inverse of stereographic projection with respect to any pole $P \in S^m$. Then $\tilde{f} = \tau^{-1} \circ f$ is a taut embedding of M into $S^m \subset \mathbf{R}^{m+1}$. Furthermore, $\tau^{-1} \circ f$ is substantial in $\mathbf{R}^{m+1}$, since if the image of $\tau^{-1} \circ f$ lies in a hyperplane π in $\mathbf{R}^{m+1}$, then it lies in the hypersphere $\Sigma^{m-1} = \pi \cap S^m$. This implies that the image of f lies in the hyperplane or hypersphere $\tau(\Sigma^{m-1})$ in $\mathbf{R}^m$, contradicting the assumption that f is substantial and non-spherical in $\mathbf{R}^m$.

Conversely, suppose that $\tilde{f} : M \to S^m \subset \mathbf{R}^{m+1}$ is a substantial taut spherical embedding. Since $\tilde{f}$ is substantial in $\mathbf{R}^{m+1}$, the image of $\tilde{f}$ does not lie in a hypersphere in $S^m \subset \mathbf{R}^{m+1}$. Let P be any point in S^m that is not in the image of $\tilde{f}$, and let $\tau : S^m - \{P\} \to \mathbf{R}^m$ be stereographic projection with pole P. Then $f = \tau \circ \tilde{f}$ is a taut embedding of M into $\mathbf{R}^m$, and it is substantial and non-spherical in $\mathbf{R}^m$, since the image of $\tilde{f}$ does not lie in a hypersphere in S^m. □

This corollary is useful, because many important examples of taut submanifolds lie in a sphere S^m in $\mathbf{R}^{m+1}$. In particular, all isoparametric (constant principal curvatures) hypersurfaces and their focal submanifolds in S^m are taut [93], as we will see in Section 3.6 (see Corollary 3.56 on page 139).

Using Theorem 2.70, we can thus obtain many new taut submanifolds in $\mathbf{R}^m$ via stereographic projection. In particular, the cyclides of Dupin in $\mathbf{R}^m$ are obtained from a standard product of two spheres (which is an isoparametric hypersurface in S^m, see Section 3.8.2 on page 148),

$$S^p(r) \times S^{m-1-p}(s) \subset S^m, \quad r^2 + s^2 = 1, \tag{2.132}$$

via stereographic projection, and thus they are taut in $\mathbf{R}^m$.

Taut embeddings of spheres

Banchoff [20], and Carter and West [61] pointed out that Theorem 2.70, when combined with known results for tight immersions, yields theorems for taut immersions. In particular, in conjunction with the Chern–Lashof Theorem (Theorem 2.35), which states that if $f : S^n \to \mathbf{R}^m$ is a tight immersion, then f embeds S^n as a convex hypersurface in a Euclidean space $\mathbf{R}^{n+1} \subset \mathbf{R}^m$, one gets the following theorem.

Theorem 2.73. *Let $f : S^n \to \mathbf{R}^m$ be a substantial taut immersion. Then $m = n + 1$, and f embeds S^n as a metric hypersphere.*

Proof. Since a taut immersion is tight, the Chern–Lashof Theorem implies that $m = n + 1$ and f embeds S^n as a convex hypersurface in $\mathbf{R}^{n+1}$. If $f(S^n)$ were not a metric hypersphere, then by part (c) of Theorem 2.70, the map $\tau^{-1} \circ f$ would be a taut substantial embedding of S^n into $\mathbf{R}^{n+2}$, contradicting the Chern–Lashof Theorem. □

We next give a different proof of Theorem 2.73 due to Nomizu and Rodriguez [405]. This is an important type of proof using the properties of distance functions and the Index Theorem (Theorem 2.51), as opposed to the proof above which is based on the theory of tight immersions. Another key element in this proof is the characterization of spheres as compact, totally umbilical submanifolds in Euclidean space. A similar approach can be used to characterize totally umbilic submanifolds of hyperbolic space (see Cecil–Ryan [90]) and to characterize totally geodesic embeddings of $\mathbf{CP}^n$ and complex quadrics Q^n in complex projective space $\mathbf{CP}^m$ in terms of the critical point behavior of distance functions (see Cecil [71]).

Since the proof of the following theorem relies on the characterization of metric spheres as totally umbilic submanifolds, it only works for S^n with $n \geq 2$. For $n = 1$, one can use the approach of Theorem 2.73 above, or else use Banchoff's [20] elementary direct proof using the spherical two-piece property. The following theorem is due to Nomizu and Rodriguez [405]. Here we are following the proof in [95, p. 126].

Theorem 2.74. *Let $M^n, n \geq 2$, be a connected, complete Riemannian manifold isometrically immersed in $\mathbf{R}^m$. If every nondegenerate distance function L_p has index 0 or n at each of its critical points, then M^n is embedded as a totally geodesic n-plane or a metric n-sphere $S^n \subset \mathbf{R}^{n+1} \subset \mathbf{R}^m$.*

Proof. As noted above, the proof is accomplished by showing that the immersion $f : M^n \to \mathbf{R}^m$ is totally umbilic, that is, for every normal vector ξ to $f(M^n)$ at every point $f(x)$, the shape operator A_ξ is a multiple of the identity endomorphism on T_xM^n.

Let ξ be a unit normal to $f(M^n)$ at a point $f(x)$. If $A_\xi = 0$, then A_ξ is a multiple of the identity as needed. If not, then we may assume that A_ξ has a positive eigenvalue by considering $A_{-\xi} = -A_\xi$, if necessary. Let λ be the largest positive eigenvalue of A_ξ. Let t be a real number such that $1/\lambda < t < 1/\mu$, where μ is the next largest positive eigenvalue of A_ξ (if λ is the only positive eigenvalue, just consider $1/\lambda < t$). Then for $p = f(x) + t\xi$, the Index Theorem (Theorem 2.51) implies that the distance function L_p has a nondegenerate critical point of index k at x, where k is the multiplicity of the eigenvalue λ. While L_p may not be a nondegenerate function, Corollary 2.52 implies that there exists a nondegenerate distance function L_q having a critical point y of index k, where q and y can be chosen to be as close to p and x, respectively, as desired. By the hypothesis of the theorem, since k is greater than 0, we get $k = n$, and so $A_\xi = \lambda I$. Since this is true for any unit normal ξ at any point $x \in M^n$, we have that f is totally umbilical. The result then follows from a theorem of E. Cartan [57] which states that a complete Riemannian n-manifold isometrically and totally umbilically immersed in $\mathbf{R}^m$ is embedded as a totally geodesic n-plane or a metric n-sphere $S^n \subset \mathbf{R}^{n+1} \subset \mathbf{R}^m$. (See also B.Y. Chen [98] or M. Spivak [495, Vol. 4, p. 110] for a proof of Cartan's theorem.) □

Remark 2.75 (Another proof of Theorem 2.73). As a consequence of Theorem 2.74, we get another proof of Theorem 2.73 that a taut immersion $f : S^n \to \mathbf{R}^m$ is an embedding of S^n as a metric hypersphere in $\mathbf{R}^{n+1} \subset \mathbf{R}^m$. Specifically, if $f : S^n \to \mathbf{R}^m$ is taut, then every nondegenerate distance function L_p has exactly one maximum and one minimum. Thus all of the critical points of L_p have index 0 or n, and so f embeds S^n as a metric hypersphere in $\mathbf{R}^{n+1} \subset \mathbf{R}^m$ by Theorem 2.74.

Taut embeddings of maximal codimension

Finally, as with Theorem 2.46 for TPP immersions, there is a bound on the codimension of a substantial STPP embedding of a compact, connected n-dimensional manifold into $\mathbf{R}^m$. This follows fairly directly from Theorem 2.46, Corollary 2.72, and the fact that the STPP implies the TPP. This result is due to Banchoff [20] for $n = 2$ and to Carter and West [61] for $n \geq 3$ (see also [95, pp. 124–125]). The standard embeddings of $\mathbf{RP}^n$ into $\mathbf{R}^m$, $m = n(n + 3)/2$, are described in detail in Section 2.9. The term "projectively equivalent" means up to a projective transformation in the sense defined in Remark 2.38.

Theorem 2.76. *Let $f : M^n \to \mathbf{R}^m$, $n \geq 2$, be a substantial smooth immersion of a compact, connected n-dimensional manifold.*

(a) *If f has the STPP, then $m \leq n(n + 3)/2$.*
(b) *If f has the STPP and $m = n(n + 3)/2$, then f is projectively equivalent to a standard embedding of $\mathbf{RP}^n$ into $\mathbf{R}^m$, and the image $f(M)$ lies in a metric sphere $S^{m-1} \subset \mathbf{R}^m$.*

Proof. Since the STPP implies the TPP, part (a) follows immediately from part (a) of Theorem 2.46.

To prove part (b), suppose that $m = n(n+3)/2$ for $n \geq 2$. If the image $f(M)$ does not lie in a metric sphere in $\mathbf{R}^m$, then by Corollary 2.72 (for the STPP), there exists a substantial spherical STPP embedding $\tilde{f} : M \to \mathbf{R}^{m+1}$. This contradicts part (a) of the theorem. Furthermore, f is a TPP embedding of an n-dimensional manifold into $\mathbf{R}^m$ with $m = n(n+3)/2$ for $n \geq 2$. Thus, by part (b) of Theorem 2.46, f is a standard embedding $f : \mathbf{RP}^n \to \mathbf{R}^m$ of a projective space, up to a projective transformation of $\mathbf{R}^m$. □

We also have the following similar result for taut embeddings of non-compact manifolds into $\mathbf{R}^{m-1}$ due to Carter and West [61] (see also [95, pp. 124–125]).

Theorem 2.77. *Let $g : M^n \to \mathbf{R}^m$, $n \geq 2$, be a substantial smooth proper immersion of a non-compact, connected n-dimensional manifold.*

(a) *If g is taut, then $m \leq \frac{n(n+3)}{2} - 1$.*
(b) *If g is taut and $m = \frac{n(n+3)}{2} - 1$, then $g = \tau \circ f$, where $f : \mathbf{RP}^n \to \mathbf{R}^{m+1}$ is projectively equivalent to a standard embedding and the image of f lies in a metric sphere $S^m \subset \mathbf{R}^{m+1}$, and $\tau : S^m - \{P\} \to \mathbf{R}^m$ is stereographic projection with pole $P \in f(M)$.*

Remark 2.78 (Tight and taut immersions into hyperbolic space). In hyperbolic space H^m there are three types of totally umbilic hypersurfaces: spheres, horospheres, and equidistant hypersurfaces (those at a fixed oriented distance from a totally geodesic hyperplane, including hyperplanes themselves). These have constant sectional curvature which is positive, zero, or negative, for spheres, horospheres, and equidistant hypersurfaces, respectively. Thus, there are three natural types of distance functions L_p, L_h, and L_π, which measure the distance from a given point p, horosphere h, or hyperplane π, respectively. Just as in Euclidean space (see Theorem 2.74), the totally umbilic hypersurfaces of H^m can be characterized in terms of the critical point behavior of these distance functions as follows (see Cecil–Ryan [90]).

Theorem 2.79. *Let $M^n, n \geq 2$, be a connected, complete Riemannian manifold isometrically immersed in H^m. Every Morse function of the form L_p or L_π has index 0 or n at all of its critical points if and only if M is embedded as a sphere, horosphere, or equidistant hypersurface in a totally geodesic $H^{n+1} \subset H^m$.*

An immersion $f : M \to H^m$ is called *taut, horo-tight*, or *tight*, respectively, if every nondegenerate function L_p, L_h, or L_π, has the minimum number of critical points required by the Morse inequalities. See Cecil and Ryan [90, 91], [95, pp. 233–236], and Izumiya et al. [227, 228], for more on these conditions.

2.8 The Relationship between Taut and Dupin

In this section, we examine the relationship between taut and Dupin submanifolds. We begin with a theorem of Thorbergsson [533] which states that the proper Dupin condition implies tautness for complete embedded hypersurfaces in real space forms. After stating the theorem, we will make some comments regarding

Thorbergsson's approach to proving this result, and we refer the reader to [533] for a complete proof.

Theorem 2.80. *Let $M^n \subset \tilde{M}^{n+1}$ be a complete, connected proper Dupin hypersurface embedded in a real space form $\tilde{M}^{n+1}$. Then M is taut.*

We now discuss Thorbergsson's method of proof. Let $p \in \tilde{M}^{n+1}$ and let L_p be the distance function $L_p(x) = d(p,x)^2$, where $d(p,x)$ is the distance from p to x in $\tilde{M}^{n+1}$. By Sard's Theorem, the restriction of L_p to M is a Morse function for almost all $p \in \tilde{M}^{n+1}$.

To prove that M is tautly embedded, one must show that every nondegenerate (Morse) function of the form L_p has the minimum number of critical points required by the Morse inequalities on M. Equivalently, one can show that every critical point of every nondegenerate distance function is of linking type (see, Morse–Cairns [379, p. 258] and the comments after Theorem 2.28 on page 38). That is the method used by Thorbergsson, as we now discuss.

Specifically, since M is a complete embedded hypersurface in a (simply connected) real space form, M is orientable, and we take ξ to be a field of unit normal vectors on M. Since M is proper Dupin, it has g distinct principal curvatures at each point, and thus we have g smooth principal curvature functions:

$$\lambda_1 > \lambda_2 > \cdots > \lambda_g, \tag{2.133}$$

with respective constant multiplicities $m_1, \ldots, m_g$ on M.

Let $E : NM \to \tilde{M}^{n+1}$ be the normal exponential map of M as defined in Section 2.2. By Theorem 2.1, a point $p = E(x, t\xi(x))$ is a focal point of (M, x) of multiplicity $m > 0$ if and only if there is a principal curvature λ of A_ξ of multiplicity m such that:

$$\begin{aligned} \lambda &= 1/t, \text{ if } \tilde{M}^{n+1} = \mathbf{R}^{n+1}, \\ \lambda &= \cot t, \text{ if } \tilde{M}^{n+1} = S^{n+1}, \\ \lambda &= \coth t, \text{ if } \tilde{M}^{n+1} = H^{n+1}. \end{aligned} \tag{2.134}$$

Thus, as noted earlier, if a principal curvature function λ has constant multiplicity m on M, then we can define a smooth focal map f_λ from an open subset $U \subset M$ (defined below) onto the sheet of the focal set of M determined by λ. Using equation (2.134) for the location of the focal points, we define the map f_λ by the formulas:

$$\begin{aligned} f_\lambda(x) &= f(x) + \frac{1}{\lambda}\, \xi(x), \\ f_\lambda(x) &= \cos\theta\, f(x) + \sin\theta\, \xi(x), \text{ where } \cot\theta = \lambda, \\ f_\lambda(x) &= \cosh\theta\, f(x) + \sinh\theta\, \xi(x), \text{ where } \coth\theta = \lambda, \end{aligned} \tag{2.135}$$

for $\tilde{M}$ equal to $\mathbf{R}^{n+1}$, S^{n+1}, and H^{n+1}, respectively.

In the case of $\mathbf{R}^{n+1}$, the domain U of the focal map f_λ is the set of points in M where $\lambda \neq 0$. In hyperbolic space, the domain U of f_λ is the set of points where $|\lambda| > 1$. In the case of S^{n+1}, at each point $x \in M$ the principal curvature λ gives rise to two antipodal focal points in S^{n+1} determined by substituting $\theta = \cot^{-1}\lambda$ and $\theta = \cot^{-1}\lambda + \pi$ into equation (2.135). Thus, λ gives rise to two antipodal focal maps into S^{n+1}.

Define $\rho_i(x) = 1/\lambda_i(x)$ if $\tilde{M}^{n+1} = \mathbf{R}^{n+1}$ and $\lambda_i(x) \neq 0$; $\rho_i(x) = \cot^{-1}\lambda_i(x)$ if $\tilde{M}^{n+1} = S^{n+1}$; and $\rho_i(x) = \coth^{-1}\lambda_i(x)$ if $\tilde{M}^{n+1} = H^{n+1}$ and $|\lambda_i(x)| > 1$. Then the focal point $f_i(x)$ corresponding to the principal curvature $\lambda_i(x)$ is $f_i(x) = E(x, \rho_i(x)\xi(x))$.

For $x \in M$, let $S_i(x)$ denote the leaf of the principal foliation T_i determined by λ_i through the point x. If x is in the domain U of the focal map f_i, then by Theorems 2.11 and 2.14, the Dupin condition implies that the leaf $S_i(x)$ is a compact m_i-dimensional metric sphere contained in a totally geodesic $(m_i + 1)$-dimensional submanifold of $\tilde{M}^{n+1}$ (which does not necessarily contain the focal point $f_i(x)$). The m_i-sphere $S_i(x)$ is also contained in the metric hypersphere (the curvature sphere) in $\tilde{M}^{n+1}$ with center $f_i(x)$ and radius $|\rho_i(x)|$, and $S_i(x)$ is either a great or small sphere in this curvature sphere.

Using these facts about the principal foliations T_i by m_i-spheres on the domain U of f_i, Thorbergsson gave an inductive procedure using iterated sphere bundles to construct concrete $\mathbf{Z}_2$-cycles in M to show that every critical point of every nondegenerate distance function L_p is of linking type, and thus M is taut. (See Thorbergsson's paper [533] for the detailed construction.)

As noted earlier in Theorem 2.62, using the Gysin sequence of the unit normal bundle of the submanifold M, Pinkall [447] proved the following result concerning submanifolds of codimension greater than one. We restate the theorem here for the sake of completeness.

Theorem 2.81. *Let $f : M \to \mathbf{R}^n$ be a compact, connected embedded submanifold of $\mathbf{R}^n$ of codimension greater than one, and let $t > 0$ be sufficiently small so that the tube $f_t : BM \to \mathbf{R}^n$ is a compact, connected embedded hypersurface in $\mathbf{R}^n$. Then $f(M)$ is taut with respect to $\mathbf{Z}_2$ coefficients if and only if the tube $f_t(M)$ is taut with respect to $\mathbf{Z}_2$ coefficients.*

We can use this to generalize Theorem 2.80 to submanifolds of higher codimension as follows. Recall from Remark 2.21 that if $f : M \to \mathbf{R}^n$ is an immersed submanifold of $\mathbf{R}^n$ with codimension greater than one, then a connected submanifold $S \subset M$ is called a curvature surface of $f(M)$ if there exists a parallel (with respect to the normal connection) section of the unit normal bundle $\eta : S \to B^{n-1}$ such that for each $x \in S$, the tangent space T_xS is equal to some eigenspace of $A_{\eta(x)}$. As in Remark 2.26, the submanifold $f(M)$ is called Dupin if along each curvature surface, the corresponding principal curvature is constant. In that case, $f(M)$ is called proper Dupin if the number of distinct principal curvatures is constant on the unit normal bundle B^{n-1}. We can now prove the following result due to Pinkall [447].

Theorem 2.82. *Let M be a compact, connected proper Dupin submanifold of codimension greater than one embedded in $\mathbf{R}^n$. Then M is taut with respect to $\mathbf{Z}_2$ coefficients.*

Proof. Let $f : M \to \mathbf{R}^n$ be the embedding of M as a compact, connected proper Dupin submanifold, and let $t > 0$ be sufficiently small so that the tube $f_t : BM \to \mathbf{R}^n$ is a compact, connected embedded hypersurface in $\mathbf{R}^n$. Since $f(M)$ is proper Dupin, the multiplicities of its principal curvatures are constant on its unit normal bundle B^{n-1}. Then by using Theorem 2.2 (page 17) regarding the shape operators of a tube, one easily shows that $f_t(M)$ is a proper Dupin hypersurface embedded in $\mathbf{R}^n$. By Theorem 2.80, $f_t(M)$ is a taut hypersurface, and thus by Theorem 2.81, $f(M)$ is also taut. □

More generally, tautness has been established for *Dupin submanifolds with constant multiplicities* of higher codimension by Terng [527] and [529, p. 467]. These are Dupin submanifolds $M \subset \mathbf{R}^n$ of codimension greater than one such that the multiplicities of the principal curvatures of any parallel normal field $\xi(t)$ along any piecewise smooth curve on M are constant.

Taut implies Dupin

In the opposite direction of Theorem 2.80, Pinkall [447] and Miyaoka [364] (for hypersurfaces) independently proved the following theorem, which is also valid for submanifolds of S^n. We give Pinkall's proof below, following the presentation given in [95, pp. 194–196].

Theorem 2.83. *Every taut submanifold $M \subset \mathbf{R}^n$ is Dupin (but not necessarily proper Dupin).*

Remark 2.84. Although a taut submanifold is always Dupin, it need not be proper Dupin, as we see from Example 2.22 (page 33). In that example, the tube M^3 of sufficiently small radius ϵ over a torus of revolution $T^2 \subset \mathbf{R}^3 \subset \mathbf{R}^4$ is taut (see Remark 2.61), but it is not proper Dupin, since there are only two distinct principal curvatures on the set $T^2 \times \{\pm\epsilon\}$, but three distinct principal curvatures elsewhere on M.

To begin the proof of Theorem 2.83, let $M \subset \mathbf{R}^n$ be a connected taut submanifold of arbitrary codimension. To prove that M is Dupin, we must show that along any curvature surface the corresponding principal curvature is constant. As shown in Theorem 2.23, this is always true if the dimension of the curvature surface is greater than one. Thus, the proof consists in showing that along any 1-dimensional curvature surface (line of curvature), the corresponding principal curvature is constant.

Let γ be a line of curvature in M. By definition γ is a connected 1-dimensional submanifold of M for which there is a parallel (with respect to the normal connection $\nabla^{\perp}$) unit normal field ξ defined along γ such that for each $x \in \gamma$, the tangent space $T_x\gamma$ is a principal space of the shape operator A_ξ. Assuming that $\lambda(x) \neq 0$ for some $x \in \gamma$, the curvature sphere determined by λ at x is the hypersphere in $\mathbf{R}^n$ with center at the focal point

$$f_\lambda(x) = x + \frac{1}{\lambda(x)}\xi(x), \tag{2.136}$$

and radius $1/|\lambda(x)|$.

The following lemma is a generalization of the classical result that if the curvature of a plane curve has nonvanishing derivative on a parameter interval, then the corresponding one-parameter family of osculating circles is nested one within another (see, for example, Stoker [501, p. 31]).

Lemma 2.85. *Let $\gamma(s)$ be a unit speed parametrization of a line of curvature of a submanifold $M \subset \mathbf{R}^n$ with corresponding principal curvature function λ. Suppose that λ and its derivative λ' are both nonzero along γ. Then along γ, the family of curvature spheres determined by λ is nested.*

Proof. By appropriate choice of sign of the parallel unit normal field ξ and the direction of the unit speed parametrization, we can assume that $\lambda < 0$ and $\lambda' > 0$ on γ, where the prime denotes differentiation with respect to s. Let $\xi(s)$ denote the normal vector field $\xi(\gamma(s))$. Let s_1 and s_2 be any two parameter values with $s_1 < s_2$, and let p_1 and p_2 be the λ-focal points of $x_1 = \gamma(s_1)$ and $x_2 = \gamma(s_2)$, as in equation (2.136). Let $\alpha(s)$ be the evolute curve (focal curve)

$$\alpha(s) = \gamma(s) + \frac{1}{\lambda(s)}\xi(s). \tag{2.137}$$

Using the fact that $\nabla^{\perp}\xi = 0$, we can compute that the velocity vector $\vec{\xi}(s)$ to the curve $\xi(s)$ is given by

$$\vec{\xi}(s) = -A_\xi(\vec{\gamma}(s)) = -\lambda(s)\vec{\gamma}(s). \tag{2.138}$$

Using this, we calculate that the velocity vector to the curve $\alpha(s)$ is

$$\vec{\alpha}(s) = \left(\frac{1}{\lambda(s)}\right)' \xi(s). \tag{2.139}$$

Thus, the arc-length of the evolute curve from p_1 to p_2 is

$$\frac{1}{\lambda(s_1)} - \frac{1}{\lambda(s_2)} = \sigma - \rho, \tag{2.140}$$

where ρ is the Euclidean distance $d(x_1, p_1)$ and $\sigma = d(x_2, p_2)$. The left side of equation (2.140) equals $\sigma - \rho$, since $\lambda(s_1)$ and $\lambda(s_2)$ are both negative. We know that α is not a straight line segment, since ξ is not constant along γ. Thus, $d(p_1, p_2) < \sigma - \rho$, and by the triangle inequality, the closed ball $B_\rho(p_1)$ with center p_1 and radius ρ is contained in the interior of the closed ball $B_\sigma(p_2)$. □

Proof (of Theorem 2.83). As noted earlier, to prove that M is Dupin we must show that if γ is any line of curvature on M, then the corresponding principal curvature λ is constant along γ. If λ is identically zero on γ, then λ is constant along γ as needed. Otherwise, there exists a unit speed parametrization $\gamma(s)$ on a real parameter interval (a, b) with $\lambda(s) < 0$ and $\lambda'(s) > 0$ for all $s \in (a, b)$, as in Lemma 2.85. For each s in the interval (a, b), let B_s be the closed ball of radius $1/|\lambda(s)|$ centered at the λ-focal point $\alpha(s)$ given in equation (2.137). Let

$$\beta(s) = \dim H_*(M \cap B_s, \mathbf{Z}_2). \tag{2.141}$$

By the tautness of M, the number $\beta(s)$ is a finite integer for each s in (a, b). We will obtain a contradiction by proving that the function $\beta(s)$ is strictly increasing on the parameter interval (a, b), which is clearly impossible for an integer-valued function.

To see this, let s_1 and s_2 be any two parameter values in (a, b) with $s_1 < s_2$, and let B_1 and B_2 be the corresponding closed balls centered at the λ-focal points $\alpha(s_1)$ and $\alpha(s_2)$, respectively. We will prove that the homomorphism,

$$j : H_*(M \cap B_1) \to H_*(M \cap B_2), \tag{2.142}$$

induced by the inclusion $B_1 \subset B_2$ is injective, but not surjective, and thus $\beta(s_1) < \beta(s_2)$.

The injectivity of the map j follows immediately from the tautness of M, since the injective map

$$H_*(M \cap B_1) \to H_*(M), \tag{2.143}$$

factors through the sequence

$$H_*(M \cap B_1) \xrightarrow{j} H_*(M \cap B_2) \to H_*(M). \tag{2.144}$$

To show that j is not surjective, consider any parameter value s_0 with $s_1 < s_0 < s_2$. Let $p_0 = \alpha(s_0)$ denote the λ-focal point of $\gamma(s_0)$, and let B_0 be the closed ball centered at p_0 of radius $1/|\lambda(s_0)|$. Let q be a point on the normal ray from $\gamma(s_0)$ to p_0 such that q is beyond p_0 and before the next focal point (if any exist) of $(M, \gamma(s_0))$ on the normal ray. The point q can be chosen arbitrarily close to p_0. By Corollary 2.52, there exists a point $p \in \mathbf{R}^n$ arbitrarily near to q (and hence to p_0 also) such that L_p is a Morse function having a nondegenerate critical point x arbitrarily near to $\gamma(s_0)$ and no other critical points at the same level. Let $r = d(p, x)$. By Lemma 2.85, we have

$$B_1 \subset \text{int}(B_0), \quad B_0 \subset \text{int}(B_2), \tag{2.145}$$

where $\mathrm{int}(B_0)$ denotes the interior of B_0. Since p and x can be chosen arbitrarily close to p_0 and $\gamma(s_0)$, respectively, there exists a $\delta > 0$ such that

$$B_1 \subset \mathrm{int}(B_{r-\delta}(p)), \quad B_{r+\delta}(p) \subset \mathrm{int}(B_2). \tag{2.146}$$

Let k be the index of L_p at the critical point x. Since M is taut, the k-th Betti number increases by one as the critical point x is passed, and thus the homomorphism

$$H_k(M \cap B_{r-\delta}(p)) \to H_k(M \cap B_{r+\delta}(p)) \tag{2.147}$$

is injective but not surjective. The map j factors through the sequence of homomorphisms induced by inclusions

$$H_k(M \cap B_1) \to H_k(M \cap B_{r-\delta}(p)) \to H_k(M \cap B_{r+\delta}(p)) \to H_k(M \cap B_2).$$

By tautness, all of the maps in this sequence are injective, but the middle one is not surjective, as shown in equation (2.147). Thus, the map j is not surjective, and so $\beta(s_1) < \beta(s_2)$. This is true for all $s_1 < s_2$ in the interval (a, b), which is impossible for the integer-valued function β. This completes the proof of Theorem 2.83. □

Ozawa's Theorem

In the case where M is compact, we can use a theorem of Ozawa [421] to obtain a result which is slightly stronger than Theorem 2.83, as was noted in [76]. Note that the definition of a Dupin hypersurface in Section 2.5 does not require that given a principal space T_λ at a point $x \in M$, there exists a curvature surface S through x whose tangent space at x is T_λ. However, using the following result of Ozawa [421], we can show that tautness does imply that this property holds on M (see Corollary 2.88 below). We first state Ozawa's result and then use it to derive this corollary. Ozawa proved his result using Morse–Bott critical point theory (see [49]) and a careful analysis of the critical submanifolds, and we refer the reader to Ozawa's paper for a complete proof.

Theorem 2.86. *Let M be a taut compact, connected submanifold of $\mathbf{R}^n$, and let L_p be a Euclidean distance function on M. Let $x \in M$ be a critical point of L_p and let S be the connected component of the critical set of L_p which contains x. Then S is*

(a) *a smooth compact manifold of dimension equal to the nullity of the Hessian of L_p at the critical point x,*
(b) *nondegenerate as a critical manifold,*
(c) *taut in $\mathbf{R}^n$.*

Part (a) of the theorem implies that for each $p \in \mathbf{R}^n$, the critical set of L_p is a union of smooth, compact submanifolds of $\mathbf{R}^n$. Note that the critical set of L_p is the pre-image of p under the normal exponential map of the submanifold M. Thus, part (a) of the theorem implies that for each $p \in \mathbf{R}^n$, the pre-image of p under the normal exponential map is a union of submanifolds.

Remark 2.87 (Taut embeddings into complete Riemannian manifolds). Using different approaches, Grove and Halperin [185]), and independently, Terng and Thorbergsson [531], extended the notion of tautness to properly embedded submanifolds of complete Riemannian manifolds. Specifically, a submanifold M of a complete Riemannian manifold N is said to be taut if there exists a field $\mathbf{F}$ such that each energy functional:

$$E_p(\gamma) = \int_0^1 |\gamma'(t)|^2 dt, \tag{2.148}$$

on the space $\mathcal{P}(N, M \times p)$ of H^1-paths $\gamma : [0, 1] \to N$ from M to a fixed point $p \in N$ is a perfect Morse function with respect to $\mathbf{F}$, if p is not a focal point of M. (Here a path is H^1 if it is absolutely continuous and the length of its derivative is square integrable.)

This definition can be shown to agree with the usual definition of tautness for submanifolds of Euclidean space. Terng and Thorbergsson [531] showed that many of the important properties of taut embeddings into Euclidean space have natural analogues in this more general setting.

In a recent paper, Wiesendorf [554] proved that a compact, connected submanifold M embedded in a complete Riemannian manifold N is taut if and only if for each point p in N, the pre-image of p under the normal exponential map of M is a union of submanifolds, as in Ozawa's Theorem above.

Wiesendorf also proved that if M is taut with respect to any field $\mathbf{F}$, then M is also taut with respect to $\mathbf{Z}_2$. In addition, Wiesendorf proved several results concerning singular Riemannian foliations, all of whose leaves are taut (see also Lytchak [338, 339], Lytchak and Thorbergsson [340, 341]).

In the context of taut submanifolds of complete Riemannian manifolds, Taylor [524] gave a classification of immersions of S^{n-1} into a complete Riemannian manifold N^n which have odd order in homotopy and are taut. (See also Hebda [191, 192], Kahn [232], and Ruberman [467] for related results.)

Using Ozawa's theorem, we can prove the following corollary (as in [76, p. 154]).

Corollary 2.88. *Let M be a taut compact, connected submanifold of $\mathbf{R}^n$. Then*

(a) *M is a Dupin submanifold.*
(b) *Given a principal space T_λ of a shape operator A_ξ at a point $x \in M$, there exists a curvature surface S through x whose tangent space at x is equal to T_λ, and λ is constant along S.*

Proof. Note that part (b) implies part (a), so we will prove part (b). Let $f : M \to \mathbf{R}^n$ be a taut embedding. Let ξ be a unit normal vector at an arbitrary point $x \in M$, and let λ be a principal curvature of A_ξ. We first consider the case where $\lambda \neq 0$. Let

$p = f(x) + (1/\lambda)\xi$ be the focal point of (M, x) determined by the principal curvature λ of A_ξ. Then the distance function L_p has a degenerate critical point at x and the nullity of the Hessian of L_p at x is equal to the multiplicity m of λ as an eigenvalue of A_ξ (see [359, p. 36]). By Ozawa's theorem, the connected component S of the critical set of L_p containing x is a smooth submanifold (a critical submanifold) of dimension m. We will now show that S is the desired curvature surface and that the corresponding principal curvature is constant along S.

The function L_p has a constant value, which is $1/\lambda^2$, on the critical submanifold S. Thus, for every point $y \in S$, the vector $p - f(y)$ is normal to $f(M)$ at $f(y)$, and it has length $1/|\lambda|$. So we can extend the normal vector ξ to a unit normal vector field to $f(M)$ along S, which we also denote by ξ, by setting $\xi(y) = \lambda(p - f(y))$. Note that p is a focal point of (M, y) for every point $y \in S$, and Ozawa's theorem implies that the number λ is an eigenvalue of $A_{\xi(y)}$ of multiplicity $m = \dim S$ for every point $y \in S$. Thus, the principal curvature λ is constant along S. We next show that T_yS equals the principal space $T_\lambda(y)$ at each point $y \in S$, and that the normal field ξ is parallel along S with respect to the normal connection. Consider the focal map,

$$f_\lambda(y) = f(y) + \frac{1}{\lambda}\xi(y),$$

for $y \in S$. Then $f_\lambda(y) = p$ for all $y \in S$. Let X be any tangent vector to S at any point $y \in S$. Then $(f_\lambda)_*X = 0$, since f_λ is constant on S. On the other hand,

$$(f_\lambda)_*X = f_*X + \frac{1}{\lambda}\xi_*X,$$

and $\xi_*X = D_X\xi = f_*(-A_\xi X) + \nabla_X^\perp\xi$. Therefore,

$$(f_\lambda)_*X = f_*(X - \frac{1}{\lambda}A_\xi X) + \frac{1}{\lambda}\nabla_X^\perp\xi.$$

Since $(f_\lambda)_*X = 0$, we see that $A_\xi X = \lambda X$ and $\nabla_X^\perp\xi = 0$. Thus, ξ is parallel along S and $T_yS \subset T_\lambda(y)$. Since T_yS and $T_\lambda(y)$ have the same dimension, they are equal. So S is the curvature surface through y corresponding to the principal curvature λ, which is constant along S.

Now suppose that $\lambda = 0$ is an eigenvalue of A_ξ at x. Let $\sigma : \mathbf{R}^{n+1} - \{q\} \to \mathbf{R}^{n+1} - \{q\}$ be an inversion, as in equation (2.50), centered at a point $q \in \mathbf{R}^n$ chosen so that $q \notin f(M)$, and so that the principal curvature μ of the embedding $\sigma f : M \to \mathbf{R}^n$ corresponding to λ by Theorem 2.6 is not zero. Since σf is taut by Theorem 2.70 and $\mu \neq 0$, the argument above shows that there exists a curvature surface V of σf through x whose tangent space at x is equal to T_μ, and μ is constant along V. Applying the inversion σ again, we get a curvature surface $S = \sigma(V)$ corresponding to the principal curvature λ of $f = \sigma^2 f$, and λ is constant along S, as needed in part (b) of the theorem. This completes the proof. □

Remark 2.89 (On the relationship between taut and "semi-Dupin"). In the book [95, p. 189], a Dupin (but not necessarily proper Dupin) hypersurface which satisfies Condition (b) in Corollary 2.88 was called "semi-Dupin." Corollary 2.88 gives an affirmative answer to one direction of Conjecture 6.19 in [95, p. 189], that is, taut implies semi-Dupin for a compact, connected submanifold of $\mathbf{R}^n$. Perhaps the converse can be proved using the approach of Wiesendorf [554] .

2.9 Standard Embeddings of Projective Spaces

In this section, we consider the standard embeddings of projective spaces into Euclidean space. These are important in the theory of tight and taut submanifolds, as well as in the theory of isoparametric hypersurfaces (see Subsection 3.8.3, page 151), and we will present some of the associated results here also. This section is based on the paper of Tai [505] (see also Section 9 of Chapter 1 of [95, pp. 87–98]).

As noted in Theorem 2.46 on page 48, Kuiper [300] showed that if $f : M^n \to \mathbf{R}^m$ is a substantial TPP immersion, then $m \leq n(n+3)/2$. In a much deeper result, he also showed that a substantial TPP immersion $f : M^2 \to \mathbf{R}^5$ (so having maximal codimension) is a Veronese surface (see Remark 2.45), which is a standard embedding $f : \mathbf{RP}^2 \to \mathbf{R}^5$, up to a projective transformation (as defined in Remark 2.38). Kuiper's result was then generalized by Little and Pohl [333], who showed that a TPP immersion $f : M^n \to \mathbf{R}^m$, $m = n(n+3)/2$, is a standard embedding of $\mathbf{RP}^n$, up to projective transformation. (See also [95, pp. 98–108] for a proof of the result of Little and Pohl).

Kuiper and Pohl [307] also generalized some of these results to the topological category by proving that if $f : \mathbf{RP}^2 \to \mathbf{R}^m$, $m \geq 5$, is a substantial TPP topological embedding, then $m = 5$, and f is either a smooth standard embedding (up to projective transformation) or the TPP polyhedral embedding of Banchoff [22] (see also Example 5.21 of [95, p. 37]).

We now begin our presentation of the standard embeddings, following the approach and using the notation of Tai [505]. Let $\mathbf{F}$ be one of the division algebras, $\mathbf{R}$, $\mathbf{C}$ or $\mathbf{H}$ (quaternions). For $q \in \mathbf{H}$, we can write,

$$q = r_0 + r_1 i + r_2 j + r_3 k, \tag{2.149}$$

where r_0, r_1, r_2, r_3 are real numbers, and the *conjugate* of q is defined by

$$\bar{q} = r_0 - r_1 i - r_2 j - r_3 k. \tag{2.150}$$

The norm of q is given by $|q| = (q\bar{q})^{1/2}$. If $q \in \mathbf{C}$, then $\bar{q}$ is the usual complex conjugate, and if $q \in \mathbf{R}$, then $\bar{q} = q$. We let $d = 1, 2, 4$, respectively, for the algebras $\mathbf{R}$, $\mathbf{C}$, $\mathbf{H}$. If A is a matrix with coefficients in $\mathbf{F}$, we define $A^* = \bar{A}^T$, where A^T denotes the transpose of A. Then it is easy to check that the following two equations hold, whenever the indicated operations make sense,

$$(AB)^* = B^* A^*, \tag{2.151}$$

$$\Re\,(\text{trace}\,(AB)) = \Re\,(\text{trace}\,(BA)). \tag{2.152}$$

Here $\Re$ denotes the real part.

Let $M(n+1, \mathbf{F})$ denote the space of all $(n+1) \times (n+1)$ matrices over $\mathbf{F}$. Let

$$H(n+1, \mathbf{F}) = \{A \in M(n+1, \mathbf{F}) \mid A^* = A\} \tag{2.153}$$

be the space of *Hermitian* matrices over $\mathbf{F}$. If A is Hermitian, then the off-diagonal entries in A are in $\mathbf{F}$, while the diagonal entries are in $\mathbf{R}$. Thus, $H(n+1, \mathbf{F})$ is a real vector space with dimension given by

$$\dim H(n+1, \mathbf{F}) = \frac{n(n+1)d}{2} + n + 1. \tag{2.154}$$

Let

$$U(n+1, \mathbf{F}) = \{A \in M(n+1, \mathbf{F}) \mid AA^* = I\}. \tag{2.155}$$

Then for $\mathbf{F} = \mathbf{R}, \mathbf{C}, \mathbf{H}$, respectively, $U(n+1, \mathbf{F})$ is equal to $O(n+1)$, $U(n+1)$, $Sp(n+1)$, respectively.

The space $\mathbf{F}^{n+1}$ is a Euclidean space of real dimension $(n+1)d$. The usual Euclidean inner product on $\mathbf{F}^{n+1} = \mathbf{R}^{(n+1)d}$ is given by

$$\langle x, y \rangle = \Re\,(x^* y), \tag{2.156}$$

where x and y in $\mathbf{F}^{n+1}$ are considered as column vectors, such as,

$$x = \begin{bmatrix} x_0 \\ x_1 \\ \cdot \\ \cdot \\ \cdot \\ x_n \end{bmatrix}, \tag{2.157}$$

and thus $x^* = (\bar{x}_0, \ldots, \bar{x}_n)$, a row vector. Then

$$\langle x, Ay \rangle = \langle A^* x, y \rangle, \tag{2.158}$$

for all $A \in M(n+1, \mathbf{F})$.

The space $M(n+1, \mathbf{F})$ can be considered as a Euclidean space of real dimension $(n+1)^2 d$, and the usual Euclidean inner product is given by

$$\langle A, B \rangle = \Re\,(\text{trace}\,(AB^*)), \tag{2.159}$$

for $A, B \in M(n+1, \mathbf{F})$. On the subspace $H(n+1, \mathbf{F})$, this simplifies to

$$\langle A, B \rangle = \Re\,(\text{trace}\,(AB)). \tag{2.160}$$

Let $S^{(n+1)d-1}$ be the unit sphere in $\mathbf{F}^{n+1}$, and let $\mathbf{FP}^n$ be the quotient space of $S^{(n+1)d-1}$ under the equivalence relation,

$$(x_0, \ldots, x_n) \simeq (x_0\lambda, \ldots, x_n\lambda), \quad \lambda \in \mathbf{F}, \quad |\lambda| = 1. \tag{2.161}$$

Consider the map from $S^{(n+1)d-1}$ into $H(n+1, \mathbf{F})$ given by

$$x \mapsto xx^* = \begin{bmatrix} |x_0|^2 & x_0\bar{x}_1 & \cdots & x_0\bar{x}_n \\ x_1\bar{x}_0 & |x_1|^2 & \cdots & x_1\bar{x}_n \\ \cdots & \cdots & \cdots & \cdots \\ x_n\bar{x}_0 & x_n\bar{x}_1 & \cdots & |x_n|^2 \end{bmatrix} \tag{2.162}$$

for x a column vector as in equation (2.157) with $|x| = 1$. Note that if $y = x\lambda$ for $\lambda \in \mathbf{F}$ with $|\lambda| = 1$, then $xx^* = yy^*$. Furthermore, if $xx^* = yy^*$, then multiplication of this equation by x on the right gives

$$x = yy^*x = y\lambda, \tag{2.163}$$

where $\lambda = y^*x$ is in $\mathbf{F}$ and $|\lambda| = 1$. Thus, the map in equation (2.162) induces a well-defined, injective map $\phi : \mathbf{FP}^n \to H(n+1, \mathbf{F})$.

The image of ϕ consists precisely of those matrices in $M(n+1, \mathbf{F})$ satisfying the equation,

$$A = A^* = A^2, \quad \text{rank}\, A = 1. \tag{2.164}$$

In fact, $\phi(x)$ is just the matrix representation of orthogonal projection of $\mathbf{F}^{n+1}$ onto the $\mathbf{F}$-line spanned by the vector x. One can verify that ϕ is a smooth immersion on $\mathbf{FP}^n$ by a direct calculation, or else deduce this fact as a consequence of the equivariance given in Theorem 2.91 below. Thus, ϕ is a smooth embedding of $\mathbf{FP}^n$ into $H(n+1, \mathbf{F})$. We can, and often will, consider $(x_0, \ldots, x_n)$ to be homogeneous coordinates on $\mathbf{FP}^n$.

This embedding $\phi : \mathbf{FP}^n \to H(n+1, \mathbf{F})$ is often called the *standard embedding* of $\mathbf{FP}^n$ into the Euclidean space $H(n+1, \mathbf{F})$. In the case $\mathbf{F} = \mathbf{R}$, the formula in equation (2.162) agrees with the formula $x \mapsto xx^T$ for the Veronese embedding, but this is not true for $\mathbf{F} = \mathbf{C}$ or $\mathbf{H}$, since x^* does not equal x^T in those cases.

The condition $|x| = 1$ is equivalent to the condition trace $\phi(x) = 1$. Hence, the image of ϕ lies in the hyperplane in $H(n+1,\mathbf{F})$ given by the linear equation trace $A = 1$. We now show that the image of ϕ does not lie in any lower dimensional plane, and hence ϕ is a substantial map into the space

$$\mathbf{R}^N = \{A \in H(n+1,\mathbf{F}) \mid \text{trace}\, A = 1\}, \tag{2.165}$$

where

$$N = \frac{n(n+1)d}{2} + n. \tag{2.166}$$

For the remainder of this section, N will always have the value given in equation (2.166).

Theorem 2.90. *The standard embedding $\phi : \mathbf{FP}^n \to \mathbf{R}^N$ is substantial in $\mathbf{R}^N$, and its image lies in a metric sphere in $\mathbf{R}^N$.*

Proof. Let p be an arbitrary point in the unit sphere $S^{(n+1)d-1}$, and let X be a unit tangent vector to $S^{(n+1)d-1}$ at p. Consider the curve,

$$\alpha(t) = \cos t\, p + \sin t\, X. \tag{2.167}$$

Then

$$\phi_*(X) = \frac{d}{dt}[\alpha(t)\, \alpha^*(t)]|_{t=0} = pX^* + Xp^*. \tag{2.168}$$

Let $\{e_0, \ldots, e_n\}$ be the standard basis of $\mathbf{F}^{n+1}$ as a vector space over the field $\mathbf{F}$. If we take $p = e_i$ and $X = e_j u$, for $j \neq i$ and u a unit length element of $\mathbf{F}$, equation (2.168) implies that $\phi_*(X)$ is a matrix which is zero except for u in the (j,i) position and $\bar{u}$ in the (i,j) position. This shows that all off-diagonal elements of $H(n+1,\mathbf{F})$ occur as tangent vectors to ϕ.

If we take $p = e_0, X = e_j$ and evaluate at $t = \pi/4$, we get

$$\phi_*(X) = e_j e_j^* - e_0 e_0^*, \tag{2.169}$$

showing that all real diagonal matrices with trace zero also occur. Thus, ϕ embeds $\mathbf{FP}^n$ substantially into the Euclidean space $\mathbf{R}^N$ given in equation (2.165).

Finally, note that

$$\langle xx^*, xx^* \rangle = \text{trace}\, [(xx^*)^2] = \text{trace}\, [xx^*] = 1, \tag{2.170}$$

so that the image of ϕ lies in the intersection of $\mathbf{R}^N$ with the unit sphere in $M(n+1,\mathbf{F})$, which is a metric sphere in $\mathbf{R}^N$. □

We next show that the embedding $\phi : \mathbf{FP}^n \to H(n+1,\mathbf{F})$ is equivariant with respect to the linear action of $U(n+1,\mathbf{F})$ on $M(n+1,\mathbf{F})$ defined by

$$U(A) = UAU^*, \tag{2.171}$$

for $U \in U(n+1,\mathbf{F})$ and $A \in M(n+1,\mathbf{F})$. An elementary calculation shows that this group action preserves the inner product on $M(n+1,\mathbf{F})$. Further, we have

$$\phi(Ux) = (Ux)(Ux)^* = Uxx^*U^* = U(\phi(x)), \tag{2.172}$$

for $x \in \mathbf{FP}^n$ and $U \in U(n+1,\mathbf{F})$. Thus we have the following theorem.

Theorem 2.91. *The embedding $\phi : \mathbf{FP}^n \to H(n+1,\mathbf{F})$ is equivariant with respect to and invariant under the action of $U(n+1,\mathbf{F})$, i.e.,*

$$\phi(Ux) = U(\phi(x)) \in \phi(\mathbf{FP}^n), \tag{2.173}$$

for all $x \in \mathbf{FP}^n$, and $U \in U(n+1,\mathbf{F})$.

The standard embeddings are taut

As noted in Sections 2.6 and 2.7, the standard embeddings of projective spaces play a special role in the theory of tight and taut immersions of manifolds into Euclidean spaces. We now prove that these standard embeddings are taut, substantial embeddings of $\mathbf{FP}^n$ into $\mathbf{R}^N$.

Theorem 2.92. *The embedding $\phi : \mathbf{FP}^n \to H(n+1,\mathbf{F})$ is taut. Hence, the embedding $\phi : \mathbf{FP}^n \to \mathbf{R}^N$ is taut and substantial.*

Proof. We will prove that the embedding $\phi : \mathbf{FP}^n \to \mathbf{R}^N \subset H(n+1,\mathbf{F})$ is tight, and since ϕ is spherical by Theorem 2.90, ϕ is also taut by Theorem 2.69. We already know that the embedding $\phi : \mathbf{FP}^n \to \mathbf{R}^N$ is substantial by Theorem 2.90.

To establish the tightness if ϕ, we will prove that every nondegenerate linear height function l_A, for $A \in H(n+1,\mathbf{F})$, has the minimum number of critical points required by the Morse inequalities. Thus $\phi : \mathbf{FP}^n \to H(n+1,\mathbf{F})$ is tight. Since $\mathbf{R}^N$ is a Euclidean subspace of $H(n+1,\mathbf{F})$, every height function in $\mathbf{R}^N$ corresponds to a height function in $H(n+1,\mathbf{F})$, and so ϕ is also tight as an embedding into $\mathbf{R}^N$.

Let $A \in H(n+1,\mathbf{F})$, and let x be a point in the sphere $S^{(n+1)d-1}$. Then x is also a homogeneous coordinate vector of the point in $\mathbf{FP}^n$ corresponding to the $\mathbf{F}$-line in $\mathbf{F}^{n+1}$ determined by x. We compute the value of the linear height function l_A at x as,

$$l_A(x) = \langle A, \phi(x)\rangle = \langle A, xx^*\rangle = \Re \text{ trace } (Axx^*) = \Re \text{ trace } (x^*Ax) = \langle x, Ax\rangle.$$

Thus, if X is a tangent vector to the sphere at x, we have

$$Xl_A = \langle X, Ax\rangle + \langle x, AX\rangle = 2\langle X, Ax\rangle. \tag{2.174}$$

Therefore, l_A has a critical point at x if and only if $\langle X, Ax\rangle = 0$ for all X tangent to the sphere at x. This means that the vector Ax is normal to the sphere $S^{(n+1)d-1}$ at x, and so Ax is a real multiple of the vector x. Thus, the critical points of the height function l_A on the sphere correspond to real eigenvectors of the matrix A. The usual inductive process (maximizing $\langle x, Ax\rangle$) can be used to produce $n+1$ real eigenvalues (not necessarily distinct) of A, each with a d-dimensional eigenspace. In fact, if x and $\lambda \in \mathbf{R}$ are such that $Ax = \lambda x$, and u is a unit length element in $\mathbf{F}$, then $A(xu) = \lambda(xu)$, and thus xu is also an eigenvector of A corresponding to the real eigenvalue λ. However, for any given $x \in S^{(n+1)d-1}$, all the points xu in $S^{(n+1)d-1}$ determine the same point of the projective space $\mathbf{FP}^n$. Thus, l_A has precisely $n+1$ critical points on $\mathbf{FP}^n$ provided that the $n+1$ real eigenvalues of A are distinct.

We now compute the Hessian of l_A at a point x such that $Ax = \lambda x$ for $\lambda \in \mathbf{R}$. Let X and Y be tangent to the sphere at x. To get $H(X, Y)$, we differentiate equation (2.174) in the direction Y. We use the decomposition of the Euclidean covariant derivative:

$$D_Y X = \nabla_Y X - \langle X, Y\rangle x, \tag{2.175}$$

where $\nabla_Y X$ is the component tangent to the sphere $S^{(n+1)d-1}$, and the normal component is $-\langle X, Y\rangle x$. We extend X to a vector field tangent to the sphere in a neighborhood of x and then differentiate the expression $2\langle X, Ax\rangle$ in the direction Y to get

$$\begin{aligned} H(X,Y) &= Y(2\langle X, Ax\rangle) = 2(\langle D_Y X, Ax\rangle + \langle X, AY\rangle) \\ &= 2\left(\langle \nabla_Y X, Ax\rangle - \langle X, Y\rangle\langle x, Ax\rangle + \langle X, AY\rangle\right) \\ &= 2\left(-\langle X, Y\rangle\lambda + \langle AX, Y\rangle\right) = 2\left\langle (A-\lambda I)X, Y\right\rangle, \end{aligned} \tag{2.176}$$

since the term $\langle \nabla_Y X, Ax\rangle$ equals zero, because $\nabla_Y X$ is tangent to the sphere, while $Ax = \lambda x$ is normal to the sphere at x. Equation (2.176) shows that the Hessian is nondegenerate if and only if all of the eigenvalues of A with eigenspaces orthogonal to the $\mathbf{F}$-line determined by x are distinct from λ. In particular, l_A is a Morse function on $\mathbf{FP}^n$ if and only if all $n+1$ eigenvalues are distinct. In that case, a consideration of the Hessian shows that l_A has one critical point of index k for each of the following values,

$$k = 0, d, 2d, \ldots, nd. \tag{2.177}$$

Thus, every Morse function of the form l_A has $n+1$ critical points with indices given in equation (2.177). This shows that the embedding $\phi : \mathbf{FP}^n \to H(n+1, \mathbf{F})$ is tight. In the case of $\mathbf{F} = \mathbf{R}$, this follows from the well-known fact that the $\mathbf{Z}_2$-Betti numbers of $\mathbf{RP}^n$ are as follows:

$$\beta_i(\mathbf{RP}^n, \mathbf{Z}_2) = 1, \ \text{for } 0 \le i \le n. \tag{2.178}$$

In the cases of $\mathbf{F} = \mathbf{C}$ or $\mathbf{F} = \mathbf{H}$, the construction of a Morse function on $\mathbf{FP}^n$ having exactly one critical point of index k for each k in equation (2.177) and no other critical points determines the Betti numbers of these spaces as follows,

$$\beta_i(\mathbf{FP}^n, \mathbf{Z}_2) = 1, \text{ for } i = 0, d, 2d, \ldots, nd, \text{ and } 0 \text{ otherwise.} \tag{2.179}$$

This follows from the *lacunary principle* in Morse theory (see, for example, Morse–Cairns [379, p. 272] or Milnor [359, p. 31]), which states if a Morse function $f : M \to \mathbf{R}$ on a compact manifold has no critical points of index $i-1$ and no critical points of index $i+1$, then for any field $\mathbf{K}$, the $\mathbf{K}$-Betti numbers of M satisfy $\beta_{i-1} = \beta_{i+1} = 0$, and $\beta_i = \mu_i$, where μ_i is the number of critical points of f of index i on M. □

Tight embeddings of projective spaces

Kuiper [300] presented a variation of the standard embeddings of projective spaces due to H. Hopf [201] which gives tight substantial embeddings of $\mathbf{FP}^n$ into lower dimensional Euclidean spaces produced by composing ϕ with orthogonal projections onto certain subspaces of $H(n+1, \mathbf{F})$.

Theorem 2.93. *There exists a tight substantial embedding of* $\mathbf{FP}^n$ *into* $\mathbf{R}^m$ *for*

$$(2n-1)d + 1 \leq m \leq N, \text{ where } N = \frac{n(n+1)d}{2} + n.$$

Proof. The embeddings are obtained by projecting the standard embedding onto an appropriate subspace $\mathbf{R}^m$ of $H(n+1, \mathbf{F})$. Define the following quadratic functions in the homogeneous coordinates $(x_0, \ldots, x_n)$ of $\mathbf{FP}^n$,

$$z_k = \sum_{\substack{i+j=k \\ i \leq j}} x_i \bar{x}_j, \; k = 0, \ldots, 2n-1. \tag{2.180}$$

The values of z_k are real for $k = 0$ and are in $\mathbf{F}$ for $k > 0$. These functions are easily shown to be linearly independent, and so the mapping $\psi : \mathbf{FP}^n \to \mathbf{R}^K$, where $K = (2n-1)d + 1$, given by

$$\psi(x) = (z_0, \ldots, z_{2n-1}) \tag{2.181}$$

is a substantial map of $\mathbf{FP}^n$ into $\mathbf{R}^K$. Furthermore, the values of all the homogeneous coordinates $(x_0, \ldots, x_n)$ can be recovered by knowing $(z_0, \ldots, z_{2n-1})$, so the mapping ψ is injective on $\mathbf{FP}^n$. Finally, one can compute that ψ is an immersion, and thus ψ is a substantial embedding of $\mathbf{FP}^n$ into $\mathbf{R}^K$.

The embedding ψ is related to the standard embedding ϕ as follows. For each k, $0 \leq k \leq 2n-1$, let M_k be the matrix having a 1 in the (i,j) position for $i+j=k$, $i \leq j$, and zero elsewhere. The M_k are mutually orthogonal, and so we can write,

$$\psi(x) = \sum_{k=0}^{2n-1} z_k M_k / |M_k|. \tag{2.182}$$

Note that $\psi = \sigma \circ \phi$, where ϕ is the standard embedding and σ is the orthogonal projection of $H(n+1, \mathbf{F})$ onto the Euclidean subspace $\mathbf{R}^K$ determined by real multiples of M_0 and $\mathbf{F}$-multiples of the other M_k. By Remark 2.39 on page 42, the map $\psi = \sigma \circ \phi$ is tight, since it is an orthogonal projection of a tight map.

To obtain a tight substantial embedding of $\mathbf{FP}^n$ into $\mathbf{R}^m$ for $K < m < N$, one needs to adjoin appropriate coordinates of the embedding ϕ which are linearly independent from the coordinates of the embedding ψ, i.e., project $\phi(\mathbf{FP}^n)$ into a subspace $\mathbf{R}^m$ of $H(n+1, \mathbf{F})$ that contains $\mathbf{R}^K$. Such an embedding is tight and substantial for the same reasons as those given for ψ. □

Remark 2.94 (Taut embeddings of Grassmann manifolds). The standard embeddings of projective spaces can be generalized to produce taut embeddings of Grassmann manifolds over $\mathbf{F} = \mathbf{R}$, $\mathbf{C}$, or $\mathbf{H}$ into $\mathbf{R}^m$ (see, for example, Kuiper [303, p. 113]).

For projective planes, one can get even sharper results. From Theorem 2.93 with $n = 2$, we get the existence of substantial tight embeddings of $\mathbf{FP}^2$ into $\mathbf{R}^m$ for

$$3d + 1 \leq m \leq 3d + 2. \tag{2.183}$$

In fact, we can also obtain taut embeddings of $\mathbf{FP}^2$ into $\mathbf{R}^m$ for these values of m as follows. First of all, the standard embedding ϕ of $\mathbf{FP}^2$ into $\mathbf{R}^{3d+2}$ is taut and spherical, as was shown in Theorems 2.90 and 2.92. By composing ϕ with stereographic projection with respect to a pole not in the image of ϕ, we obtain a taut, non-spherical embedding of $\mathbf{FP}^2$ into $\mathbf{R}^{3d+1}$ by Corollary 2.72.

Using methods similar to those employed in the proof of Theorem 2.92, Tai [505] showed that the analogous embedding of $\mathbf{OP}^2$ (Cayley projective plane) into $\mathbf{R}^{26}$ is tight and spherical, and thus taut. Again by Corollary 2.72, we can obtain a substantial non-spherical taut embedding of $\mathbf{OP}^2$ into $\mathbf{R}^{25}$ via stereographic projection.

Kuiper [302, pp. 215–217] proved that these are the only dimensions possible for tight substantial embeddings of these projective planes as follows.

Theorem 2.95. *There exist tight substantial embeddings of the projective planes $\mathbf{FP}^2$ into $\mathbf{R}^m$ for precisely the following dimensions.*

(a) $\mathbf{RP}^2$ *into* $\mathbf{R}^4$ *or* $\mathbf{R}^5$,
(b) $\mathbf{CP}^2$ *into* $\mathbf{R}^7$ *or* $\mathbf{R}^8$,
(c) $\mathbf{HP}^2$ *into* $\mathbf{R}^{13}$ *or* $\mathbf{R}^{14}$,
(d) $\mathbf{OP}^2$ *into* $\mathbf{R}^{25}$ *or* $\mathbf{R}^{26}$.

Proof. The existence of tight embeddings into the spaces listed in the theorem has been noted above. There do not exist embeddings of $\mathbf{FP}^2$ into lower dimensional Euclidean spaces because the normal Stiefel–Whitney class $\bar{w}_d(\mathbf{FP}^2) \neq 0$, where $d = 1, 2, 4, 8$ for $\mathbf{F} = \mathbf{R}, \mathbf{C}, \mathbf{H}, \mathbf{O}$, respectively. (See, for example, Husemöller [212, p. 263] and Borel–Hirzebruch [47, p. 533] for the case $\mathbf{F} = \mathbf{O}$.)

The upper bound in the case $\mathbf{F} = \mathbf{R}$ is given in Theorem 2.46 (due to Kuiper) on page 48 regarding tight immersions of maximal codimension. For the other division algebras $\mathbf{C}, \mathbf{H}, \mathbf{O}$, we need to use Theorem 2.48 (also due to Kuiper) on page 49 to obtain the upper bound as follows.

The $\mathbf{Z}_2$-Betti numbers of $\mathbf{FP}^2$ are known to be as follows,

$$\beta_i(\mathbf{FP}^2, \mathbf{Z}_2) = 1 \text{ for } i = 0, d, 2d, \quad \beta_i(\mathbf{FP}^2, \mathbf{Z}_2) = 0 \text{ for } i \neq 0, d, 2d. \tag{2.184}$$

By Theorem 2.48, we know that the substantial codimension of a tight smooth immersion is less than or equal to $c(\beta_0, \dots, \beta_{2d})$, which is the maximal dimension of a linear family of symmetric bilinear forms in $2d$ variables which contains a positive definite form and such that no form of the family has index k if $\beta_k = 0$. Thus, we will complete the proof if we show that for the β_i given in equation (2.184), we have $c(\beta_0, \dots, \beta_{2d}) = 4, 6, 10$, for $d = 2, 4, 8$, respectively.

The result that we need is contained in Hurwitz [211] (see also Kuiper [302, pp. 232–234]). There it is shown that the desired linear family of symmetric bilinear forms with maximal dimension can be represented by the set of symmetric matrices of the form

$$\begin{bmatrix} \lambda I & B \\ B^T & \mu I \end{bmatrix}, \tag{2.185}$$

where B is the 2×2, 4×4, or 8×8 matrix in the upper left corner of the matrix in equation (2.186) below, depending on whether $\mathbf{F} = \mathbf{C}, \mathbf{H}, \mathbf{O}$, respectively. From this, we see that $c(\beta_0, \dots, \beta_{2d})$ has the desired values.

$$\begin{bmatrix} x_1 & -x_2 & -x_3 & -x_4 & -x_5 & -x_6 & -x_7 & -x_8 \\ x_2 & x_1 & -x_4 & x_3 & -x_6 & x_5 & -x_8 & x_7 \\ x_3 & x_4 & x_1 & -x_2 & -x_7 & x_8 & x_5 & -x_6 \\ x_4 & -x_3 & x_2 & x_1 & x_8 & x_7 & -x_6 & -x_5 \\ x_5 & x_6 & x_7 & -x_8 & x_1 & -x_2 & -x_3 & x_4 \\ x_6 & -x_5 & -x_8 & -x_7 & x_2 & x_1 & x_4 & x_3 \\ x_7 & x_8 & -x_5 & x_6 & x_3 & -x_4 & x_1 & -x_2 \\ x_8 & -x_7 & x_6 & x_5 & -x_4 & -x_3 & x_2 & x_1 \end{bmatrix} \tag{2.186}$$

□

Remark 2.96 (Manifolds which are like projective planes). Recall that the Morse number of a compact manifold M is the minimum number of critical points that any Morse function has on M. A compact, connected manifold with Morse number 3 was called a *manifold which is like a projective plane* by Eells and Kuiper [144], who gave many examples of such manifolds M^{2k}, all necessarily of dimensions $2k = 2, 4, 8$, or 16. They are obtained from $\mathbf{R}^{2k}$ under compactification by a k-sphere. Of course, the projective planes $\mathbf{FP}^2$ for $\mathbf{F} = \mathbf{R}, \mathbf{C}, \mathbf{H}, \mathbf{O}$ are examples. Kuiper [303, p. 132] showed that if $f : M^{2k} \to \mathbf{R}^m$ is a tight substantial topological embedding of a manifold which is like a projective plane, then $m \leq 3k + 2$. Moreover, if $f : M^{2k} \to \mathbf{R}^{3k+2}$ is a tight smooth substantial embedding of a manifold like a projective plane, then M^{2k} is embedded as an algebraic submanifold. For $k = 1, 2$, respectively, Kuiper showed that M^{2k} is $\mathbf{RP}^2$, $\mathbf{CP}^2$, respectively, and f is a standard embedding up to a real projective transformation of $\mathbf{R}^{3k+2}$. The hypothesis of smoothness is necessary in these results, as the piecewise linear embeddings of $\mathbf{RP}^2$ into $\mathbf{R}^5$ due to Banchoff [19], and of $\mathbf{CP}^2$ into $\mathbf{R}^8$ due to Kühnel and Banchoff [299] show.

In Theorem 2.46 on page 48, we showed that the substantial codimension of a tight immersion of an n-manifold always satisfies the inequality $1 \leq k \leq n(n+1)/2$. In the following theorem, we show that every value k in this interval can be realized.

Theorem 2.97. *For every integer k satisfying* $1 \leq k \leq n(n+1)/2$, *there exists a tight substantial embedding of an n-dimensional manifold M into* $\mathbf{R}^{n+k}$.

Proof. For $k = 1$, we have the embedding of S^n as a metric hypersphere in $\mathbf{R}^{n+1}$. For $k = 2$, take the standard product embedding of $S^{n-1} \times S^1$ into $\mathbf{R}^{n+2}$, which is tight by Theorem 2.50 on page 51 concerning a product of tight immersions. More generally, for $2 \leq k \leq n$, we can take the standard product of S^{n-k+1} with $k - 1$ copies of S^1,

$$S^{n-k+1} \times S^1 \times \cdots \times S^1 \subset \mathbf{R}^{n-k+2} \times \mathbf{R}^2 \times \cdots \times \mathbf{R}^2 = \mathbf{R}^{n+k}. \tag{2.187}$$

Finally, for codimensions $n+1 \leq k \leq n(n+1)/2$, we can use the tight embeddings of $\mathbf{RP}^n$ given in Theorem 2.93. □

Chapter 3
Isoparametric Hypersurfaces

This chapter is devoted to the basic theory of isoparametric hypersurfaces in real space forms developed primarily by Cartan [53–56] and Münzner [381, 382]. In Section 3.1, we describe the aspects of the theory that are common to all three space forms, and then prove the classification of isoparametric hypersurfaces M^n in Euclidean space $\mathbf{R}^{n+1}$ and in hyperbolic space H^{n+1} using Cartan's formula involving the principal curvatures of M^n.

The rest of the chapter is devoted to the theory of isoparametric hypersurfaces in the sphere S^{n+1}. In Sections 3.2–3.6, we present Münzner's theory, including the proof that an isoparametric hypersurface in $S^{n+1} \subset \mathbf{R}^{n+2}$ with g distinct principal curvatures is contained in a level set of a homogeneous polynomial of degree g on $\mathbf{R}^{n+2}$ satisfying certain differential equations on the length of its gradient and its Laplacian, known as the Cartan–Münzner differential equations.

As a result of this construction, each isoparametric hypersurface M^n in S^{n+1} has two focal submanifolds of codimension greater than one. These codimensions are determined by the multiplicities of the principal curvatures of M^n. From this one can show that M^n separates S^{n+1} into two ball bundles over these two focal submanifolds. Münzner then used cohomology theory to show that this topological situation implies that the number g of distinct principal curvatures of M^n can only be $1, 2, 3, 4$, or 6. At about the same time as Münzner's work, Takagi and Takahashi [511] classified homogeneous isoparametric hypersurfaces and found examples having g distinct principal curvatures for each of the values $g = 1, 2, 3, 4$, or 6.

Thorbergsson [533] then applied Münzner's theory to show that the number of distinct principal curvatures of a compact proper Dupin hypersurface M^n embedded in S^{n+1} can only be $1, 2, 3, 4$, or 6, since M^n also separates S^{n+1} into two ball bundles over two focal submanifolds of M^n. Several authors used this same topological information to find a complete list of possibilities for the multiplicities of the principal curvatures of a compact proper Dupin hypersurface in S^{n+1}. This is discussed in Section 3.7.

T.E. Cecil, P.J. Ryan, *Geometry of Hypersurfaces*, Springer Monographs in Mathematics, DOI 10.1007/978-1-4939-3246-7_3

In Section 3.8, we describe many important examples of isoparametric hypersurfaces in S^{n+1} from various points of view, both algebraic and geometric. Then in Section 3.9, we study the important construction of isoparametric hypersurfaces with $g = 4$ principal curvatures based on representations of Clifford algebras due to Ferus, Karcher, and Münzner [160]. At the end of that section, we discuss progress that has been made on the classification of isoparametric hypersurfaces with four principal curvatures (see Subsection 3.9.1, page 180).

Isoparametric hypersurfaces in spheres have also occurred in considerations of several concepts in Riemannian geometry, such as the spectrum of the Laplacian, constant scalar curvature, and Willmore submanifolds. These applications and others are discussed in Section 3.10.

The reader is also referred to the survey articles by Ferus [159], Thorbergsson [538], and Cecil [78] on isoparametric hypersurfaces and related topics.

3.1 Isoparametric Hypersurfaces in Real Space Forms

As noted in Chapter 1, the original definition of an isoparametric family of hypersurfaces in a real space form $\tilde{M}^{n+1}$ was formulated in terms of the level sets of an isoparametric function, as we now describe. Let $F : \tilde{M}^{n+1} \to \mathbf{R}$ be a nonconstant smooth function. The classical Beltrami differential parameters of F are defined by

$$\Delta_1 F = |\text{grad } F|^2, \quad \Delta_2 F = \Delta F \text{ (Laplacian of } F). \tag{3.1}$$

The function F is said to be *isoparametric* if there exist smooth functions ϕ_1 and ϕ_2 from $\mathbf{R}$ to $\mathbf{R}$ such that

$$\Delta_1 F = \phi_1(F), \quad \Delta_2 F = \phi_2(F). \tag{3.2}$$

That is, both of the Beltrami differential parameters are constant on each level set of F. This is the origin of the term *isoparametric*. The collection of level sets of an isoparametric function is called an *isoparametric family* of $\tilde{M}^{n+1}$. (See Thorbergsson [538, pp. 965–967], Q.-M. Wang [548, 549], Ge and Tang [169, 170], and Ge, Tang, and Yan [171] for more discussion of isoparametric functions.)

We now show that if M is a level set of an isoparametric function F on which grad F is nonzero, then M has constant principal curvatures. To do this, we need to express the shape operator and principal curvatures of M in terms of the function F and its derivatives.

Theorem 3.1. *Let* $F : \tilde{M}^{n+1} \to \mathbf{R}$ *be a smooth function defined on a real space form* $\tilde{M}^{n+1}$. *Suppose that* grad F *does not vanish on the level set* $M = F^{-1}(0)$. *Then the shape operator* A *of the hypersurface* M *satisfies the equation*

$$\langle AX, Y\rangle = -\frac{H_F(X,Y)}{|\mathrm{grad}\ F|}, \tag{3.3}$$

where X and Y are tangent vectors to M and H_F is the Hessian of the function F.

Proof. The vector field $\xi = \mathrm{grad}\ F/|\mathrm{grad}\ F|$ is a field of unit normals to M, and its corresponding shape operator A is determined by the equation

$$\langle AX, Y\rangle = -\langle\tilde{\nabla}_X\xi, Y\rangle,$$

where $\tilde{\nabla}$ is the Levi-Civita connection on $\tilde{M}^{n+1}$. If we take $\rho = |\mathrm{grad}\ F|$, we have $\mathrm{grad}\ F = \rho\xi$, and we compute,

$$\langle\tilde{\nabla}_X(\rho\xi), Y\rangle = (X\rho)\langle\xi, Y\rangle + \rho\langle\tilde{\nabla}_X\xi, Y\rangle = -\rho\langle AX, Y\rangle. \tag{3.4}$$

On the other hand, we have

$$\begin{aligned}\langle\tilde{\nabla}_X\mathrm{grad}\ F, Y\rangle &= \tilde{\nabla}_X\langle\mathrm{grad}\ F, Y\rangle - \langle\mathrm{grad}\ F, \tilde{\nabla}_X Y\rangle \\ &= X(YF) - (\tilde{\nabla}_X Y)(F) = H_F(X,Y).\end{aligned} \tag{3.5}$$

Equating the right sides of these two equations and dividing by ρ, we get equation (3.3). □

Remark 3.2 (On the definition of the Hessian). In the proof above, the Hessian H_F is the symmetric tensor field of type $(0,2)$ defined by the equation

$$H_F(X,Y) = X(YF) - (\tilde{\nabla}_X Y)(F). \tag{3.6}$$

At a critical point of F, this definition reduces to the definition given in equation (2.65) on page 36. In the calculation above, F is constant on M, so that $X(YF) = 0$, but

$$\begin{aligned}(\tilde{\nabla}_X Y)(F) &= (\nabla_X Y)F + \langle AX, Y\rangle\xi F \\ &= \langle AX, Y\rangle\langle\mathrm{grad}\ F, \xi\rangle = \langle AX, Y\rangle\rho,\end{aligned} \tag{3.7}$$

need not be zero.

We next express the mean curvature of the level hypersurface M in terms of the function F and its derivatives.

Theorem 3.3. *With the notation of the preceding theorem, the mean curvature h of the level hypersurface M is given by*

$$h = \frac{1}{n\rho^2}(\langle\mathrm{grad}\ F,\ \mathrm{grad}\ \rho\rangle - \rho\Delta F), \tag{3.8}$$

where $\rho = |\mathrm{grad}\ F|$.

Proof. At a given point $x \in M$, let $\{e_1, \ldots, e_n\}$ be an orthonormal basis for the tangent space T_xM. Using equation (3.4), we compute

$$\begin{aligned} nh &= -\frac{1}{\rho}\sum_{i=1}^{n}\langle \tilde{\nabla}_{e_i}\text{grad } F, e_i\rangle \\ &= -\frac{1}{\rho}(\Delta F - \langle \tilde{\nabla}_{\xi}\text{grad } F, \xi\rangle) \\ &= -\frac{1}{\rho}(\Delta F - \frac{1}{\rho^2}\langle \tilde{\nabla}_{\text{grad } F}\text{ grad } F, \text{grad } F\rangle) \\ &= -\frac{1}{\rho}(\Delta F - \frac{1}{2\rho^2}\tilde{\nabla}_{\text{grad } F}|\text{grad } F|^2) \\ &= -\frac{1}{\rho}(\Delta F - \frac{2\rho}{2\rho^2}\tilde{\nabla}_{\text{grad } F}\rho) \\ &= -\frac{1}{\rho}(\Delta F - \frac{1}{\rho}\langle \text{grad } F,\ \text{grad } \rho\rangle) \\ &= \frac{1}{\rho^2}(\langle \text{grad } F,\ \text{grad } \rho\rangle - \rho\Delta F). \end{aligned}$$

□

The following theorem is important in showing that the level sets of an isoparametric function have constant principal curvatures.

Theorem 3.4. *If $F : \tilde{M}^{n+1} \to \mathbf{R}$ is an isoparametric function on a real space form, then each level hypersurface of F has constant mean curvature.*

Proof. Since F is an isoparametric function, there exist smooth real-valued functions T and S such that

$$|\text{grad } F|^2 = T \circ F, \quad \Delta F = S \circ F. \tag{3.9}$$

Using Theorem 3.3 to compute the mean curvature of the level hypersurface $M = F^{-1}(0)$, we get

$$\begin{aligned} nh &= \frac{1}{\rho^2}(\langle \text{grad } F,\ \text{grad } \rho\rangle - \rho\Delta F) \\ &= \frac{1}{\rho^2}(\langle \text{grad } F, \frac{T' \circ F}{2\rho}\text{ grad } F\rangle - \rho\, S \circ F) \\ &= \frac{T' \circ F}{2\rho} - \frac{S \circ F}{\rho} = \left(\frac{1}{2\sqrt{T}}(T' - 2S)\right) \circ F. \end{aligned}$$

This shows that h is constant on the level hypersurface M, and a similar proof shows that h is constant on any level hypersurface of F. □

The vector field $\xi = \text{grad } F/|\text{grad } F|$ is defined on the open subset of $\tilde{M}^{n+1}$ on which grad F is nonzero. We now show that the integral curves of ξ are geodesics in $\tilde{M}^{n+1}$.

Theorem 3.5. *Let $F : \tilde{M}^{n+1} \to \mathbf{R}$ be a function for which $|\text{grad } F|$ is a function of F. Then on the subset of $\tilde{M}^{n+1}$ where $|\text{grad } F|$ is nonzero, the integral curves of the vector field $\xi = \text{grad } F/|\text{grad } F|$ are geodesics in $\tilde{M}^{n+1}$.*

Proof. We need to show that $\tilde{\nabla}_\xi \xi = 0$ on the subset of $\tilde{M}^{n+1}$ where $|\text{grad } F|$ is nonzero. Since ξ has constant length, we know that $\langle \tilde{\nabla}_\xi \xi, \xi \rangle = 0$. Let X be a vector field in a neighborhood U of a point where $|\text{grad } F|$ is nonzero such that X is orthogonal to ξ at each point of U. Thus, X is tangent to a level surface of F at each point of U. We must show that $\langle \tilde{\nabla}_\xi \xi, X \rangle = 0$. We have

$$\langle \tilde{\nabla}_\xi \xi, X \rangle = -\langle \xi, \tilde{\nabla}_\xi X \rangle = -\langle \xi, \tilde{\nabla}_X \xi + [X, \xi] \rangle,$$

where $[\ ,\ ]$ is the Lie bracket. Since ξ has constant length, we know that $\langle \xi, \tilde{\nabla}_X \xi \rangle = 0$, and so

$$\langle \tilde{\nabla}_\xi \xi, X \rangle = -\langle \xi, [X, \xi] \rangle.$$

To complete the proof, we show that $[X, \xi]F = 0$ at all points in U, and so $[X, \xi]$ is orthogonal to grad F and thus to ξ. We have

$$[X, \xi]F = X(\xi F) - \xi(XF).$$

Since X is orthogonal to grad F at all points of U, we know that XF is identically zero, and so $\xi(XF) = 0$. On the other hand, since $\xi F = |\text{grad } F|$, and $|\text{grad } F|$ is a function of F, we have $X(\xi F) = 0$. □

Theorem 3.5 shows that a family of level hypersurfaces of an isoparametric function is a family of parallel hypersurfaces in $\tilde{M}^{n+1}$ modulo reparametrization to take into account the possibility that $|\text{grad } F|$ is not identically equal to one. We now show that each of these level hypersurfaces has constant principal curvatures. This follows from Theorem 3.4 and the next theorem.

Theorem 3.6. *Let $f_t : M \to \tilde{M}^{n+1}$, $-\varepsilon < t < \varepsilon$, be a family of parallel hypersurfaces in a real space form. Then $f_0 M$ has constant principal curvatures if and only if each $f_t M$ has constant mean curvature.*

Proof. We do the proof for the case $\tilde{M}^{n+1} = \mathbf{R}^{n+1}$ following Nomizu [403, p. 192] (see Cecil–Ryan [95, p. 272] for the case $\tilde{M}^{n+1} = S^{n+1}$ and a similar proof can be constructed for $\tilde{M}^{n+1} = H^{n+1}$). Let λ_i, $1 \leq i \leq n$, be the principal curvature functions of $f_0 M$, where we are not assuming that the λ_i are necessarily distinct. By Theorem 2.2 on page 17, the principal curvature functions of $f_t M$ are $\lambda_i/(1 - t\lambda_i)$ for $1 \leq i \leq n$. Since the mean curvature of $f_t M$ is constant on M, we have that

$$\sum_{i=1}^{n} \frac{\lambda_i(x)}{1 - t\lambda_i(x)} = \phi(t) \tag{3.10}$$

is a function of t alone, even though the functions $\lambda_i(x)$ are assumed to depend on $x \in M$. Evaluating $\phi(t), d\phi/dt, d^2\phi/dt^2, \ldots, d^n\phi/dt^n$ at $t = 0$, we get that

$$s_k(x) = \sum_{i=1}^{n} \lambda_i^k(x) = c_k, \quad 1 \leq k \leq n, \tag{3.11}$$

where the c_k are constants. By Newton's identities (see Van der Waerden [541, p. 81]), the coefficients of the characteristic polynomial of the shape operator A of f_0M are polynomials in the $s_k(x)$, which we have just shown to be constant on M. Therefore, the principal curvatures, which are the roots of this characteristic polynomial, are also constant on M. □

As a consequence of Theorems 3.4–3.6, we have the following corollary.

Corollary 3.7. *If $F : \tilde{M}^{n+1} \to \mathbf{R}$ is an isoparametric function on a real space form, then each level hypersurface of F has constant principal curvatures.*

Conversely, let $f_t : M \to \tilde{M}^{n+1}$, $-\varepsilon < t < \varepsilon$, be a family of parallel hypersurfaces such that f_0 has constant principal curvatures. Then each f_tM has constant principal curvatures by Theorem 2.2 on page 17, and thus each f_tM has constant mean curvature. Then the function F defined by $F(x) = t$, if $x \in f_tM$, is a smooth function defined on an open subset of $\tilde{M}^{n+1}$ such that grad $F = \xi$ is a unit length vector field with the property that $\tilde{\nabla}_\xi \xi = 0$. Furthermore, since the function $\rho = |\text{grad } F|$ is constant, we see from Theorem 3.3 that the constancy of the mean curvature h on each level hypersurface f_tM implies that the Laplacian of F is also constant on each f_tM, and therefore F is an isoparametric function.

Thus, the analytic definition of an isoparametric family of hypersurfaces in terms of level sets of an isoparametric function on a real space form $\tilde{M}^{n+1}$ is equivalent to the geometric definition as a family of parallel hypersurfaces to a hypersurface with constant principal curvatures. Note that this is not true if $\tilde{M}^{n+1}$ is only assumed to be a Riemannian manifold, as can be seen by examples in complex projective space due to Q.-M. Wang [547] (see also Section 8.7 and Thorbergsson [538]).

For hypersurfaces in real space forms we now adopt the following official definition of an isoparametric hypersurface.

Definition 3.8. A connected hypersurface M^n immersed in a real space form $\tilde{M}^{n+1}$ is said to be isoparametric if it has constant principal curvatures.

Note that the definition does not include the condition that M is complete. It follows from the work of Cartan [53–56] and Münzner [381, 382] that any connected isoparametric hypersurface embedded in a real space form is contained in a unique complete isoparametric hypersurface, as we shall see.

We now continue the study of isoparametric hypersurfaces from the geometric point of view. Let $f : M^n \to \tilde{M}^{n+1}$ be a connected, oriented isoparametric hypersurface in a real space form with field of unit normals ξ, and g distinct constant principal curvatures,

$$\lambda_1 > \cdots > \lambda_g, \tag{3.12}$$

with respective multiplicities $m_1, \ldots, m_g$.

Since the principal curvatures are constant, we see from Theorem 2.1 on page 11 that the focal points along the normal geodesic to M at a point $x \in M$ occur for the same values of t independent of the point x. If t is not one of those values, then the map $f_t : M^n \to \tilde{M}^{n+1}$ defined in equation (2.22) on page 14 is an immersion, and $f_t(M)$ is a parallel hypersurface to $f(M)$. By Theorem 2.2 on page 17, $f_t(M)$ also has constant principal curvatures, and so it is an isoparametric hypersurface, too. On the other hand, if t is a value such that f_t is not an immersion, then $f_t(M)$ is a focal submanifold of dimension $n - m$, where m is the multiplicity of the principal curvature corresponding to the parameter value t as in Theorem 2.1. It is important that isoparametric hypersurfaces always come as a family of parallel hypersurfaces together with their focal submanifolds.

An isoparametric hypersurface M^n in $\mathbf{R}^{n+1}$ is an open subset of a hyperplane, a hypersphere or a spherical cylinder $S^k \times \mathbf{R}^{n-k}$, as we will show in Theorem 3.12 on page 96. This was first shown for $n = 2$ by Somigliana [492] (see also B. Segre [479] and Levi-Civita [315]), and for arbitrary n by B. Segre [480]. In the late 1930s, shortly after the publication of the papers of Levi-Civita and Segre, Élie Cartan [52–55] began a study of isoparametric hypersurfaces in arbitrary real space forms which produced a classification of isoparametric hypersurfaces in hyperbolic space, and made great progress in the study of isoparametric hypersurfaces in spheres. We now describe Cartan's work as it is applicable to hypersurfaces in $\mathbf{R}^{n+1}$ and H^{n+1}, and prove the classification of isoparametric hypersurfaces in those space forms using Cartan's theory.

Cartan's formula

A crucial element in Cartan's work is the following equation, known as *Cartan's formula* or *Cartan's identity*, involving the distinct principal curvatures $\lambda_1, \ldots, \lambda_g$ and their multiplicities $m_1, \ldots, m_g$ for an isoparametric hypersurface $f : M^n \to \tilde{M}^{n+1}(c)$ in a space form of constant sectional curvature c. If the number g of distinct principal curvatures is greater than one, Cartan showed that for each i, $1 \le i \le g$, the following equation holds,

$$\sum_{j \neq i} m_j \frac{c + \lambda_i \lambda_j}{\lambda_i - \lambda_j} = 0. \tag{3.13}$$

We now give a proof of Cartan's formula that is valid for all real space forms (see also Cartan [55] or Ferus [159, p. 10]). In this proof, $\langle\ ,\ \rangle$ denotes the usual Euclidean inner product in the cases $\mathbf{R}^{n+1}$ and S^{n+1}, and it denotes the Lorentz metric given in equation (2.1) on page 10 in the case H^{n+1}.

We first make a few preliminary observations. If X and Y are vector fields on M, then $\nabla_X A$ is the tensor field of type $(1,1)$ defined by

$$(\nabla_X A)Y = \nabla_X(AY) - A(\nabla_X Y). \tag{3.14}$$

Since A is symmetric, we can show that $\nabla_X A$ is symmetric as follows. First note that for vector fields X, Y and Z on M, we have

$$X\langle AY, Z\rangle = \langle \nabla_X(AY), Z\rangle + \langle AY, \nabla_X Z\rangle, \tag{3.15}$$

and thus,

$$\langle \nabla_X(AY), Z\rangle = X\langle AY, Z\rangle - \langle AY, \nabla_X Z\rangle. \tag{3.16}$$

Using equations (3.14) and (3.16) and the symmetry of A, we compute

$$\begin{aligned} \langle(\nabla_X A)Y, Z\rangle &= \langle \nabla_X(AY), Z\rangle - \langle A(\nabla_X Y), Z\rangle \\ &= X\langle AY, Z\rangle - \langle AY, \nabla_X Z\rangle - \langle \nabla_X Y, AZ\rangle \\ &= X\langle Y, AZ\rangle - \langle \nabla_X Y, AZ\rangle - \langle Y, A(\nabla_X Z)\rangle \\ &= \langle Y, \nabla_X(AZ) - A(\nabla_X Z)\rangle = \langle Y, (\nabla_X A)Z\rangle, \end{aligned} \tag{3.17}$$

and so $\nabla_X A$ is symmetric.

A crucial element in our calculations is the Codazzi equation, which for an oriented hypersurface in a real space form takes the following form (see Kobayashi–Nomizu [283, Vol. II, p. 26]),

$$(\nabla_X A)Y = (\nabla_Y A)X. \tag{3.18}$$

As our final preliminary calculation, suppose that λ and μ are distinct constant principal curvatures with corresponding principal foliations T_λ and T_μ. If $X \in T_\lambda$ and $Y \in T_\mu$, then one easily verifies that

$$\langle(\nabla_Z A)X, Y\rangle = (\lambda - \mu)\langle \nabla_Z X, Y\rangle \tag{3.19}$$

for all vectors Z tangent to M.

Lemma 3.9. *Let M be an isoparametric hypersurface in a space form $\tilde{M}^{n+1}(c)$ of constant sectional curvature c. For all principal curvatures λ, μ, we have*

(1) $\nabla_X Y \in T_\lambda$ *for all* X, Y *in* T_λ,
(2) $\nabla_X Y \perp T_\lambda$ *if* $X \in T_\lambda$, $Y \in T_\mu$, $\lambda \neq \mu$.

Proof. Let X and Y be in T_λ, and take any $\mu \neq \lambda$ and choose any $Z \in T_\mu$. By the Codazzi equation and equation (3.19),

$$\begin{aligned} 0 &= \langle \big((\nabla_X A)Z - (\nabla_Z A)X\big), Y\rangle \\ &= (\mu - \lambda)\langle \nabla_X Z, Y\rangle - \langle \big(\nabla_Z(\lambda X) - A\nabla_Z X\big), Y\rangle \\ &= (\mu - \lambda)\langle \nabla_X Z, Y\rangle - (Z\lambda)\langle X, Y\rangle - \lambda\langle \nabla_Z X, Y\rangle + \lambda\langle \nabla_Z X, Y\rangle \\ &= (\lambda - \mu)\langle \nabla_X Y, Z\rangle - (Z\lambda)\langle X, Y\rangle. \end{aligned} \tag{3.20}$$

Since $Z\lambda = 0$ we have $\langle \nabla_X Y, Z\rangle = 0$ and thus, $\nabla_X Y \in T_\lambda$. Note that the second assertion follows from the first since $\langle \nabla_X Z, Y\rangle = -\langle \nabla_X Y, Z\rangle$. □

In the following version of Cartan's formula, we denote the principal curvatures by μ_i for $1 \leq i \leq n$, that is, we allow some of the principal curvatures to equal each other. This is for ease of notation in the calculations required in the proof. The following version of Cartan's formula is clearly equivalent to the version given above in equation (3.13), which is written in terms of the distinct principal curvatures.

Lemma 3.10 (Cartan's formula). *Let M be an isoparametric hypersurface in a space form $\tilde{M}^{n+1}(c)$. Let X be a unit principal vector at a point p and λ the associated principal curvature. For any principal orthonormal basis $\{e_i\}_{i=1}^n$ satisfying $Ae_i = \mu_i e_i$, we have*

$$\sum_{\mu_i \neq \lambda} \frac{c + \lambda\mu_i}{\lambda - \mu_i} = 0. \tag{3.21}$$

Proof. We first assume that the number g of distinct principal curvatures is greater than two. After we complete that proof, we will give the modifications necessary in the case $g = 2$.

We first give an outline of the main steps of the proof in the case $g > 2$ and then follow with the details. Let Y be a second unit principal vector at p with corresponding principal curvature $\mu \neq \lambda$. Extend X and Y to be principal vector fields near p. The main steps of the proof are as follows:

1. Using the Codazzi equation we show that

$$\langle(\nabla_{[X,Y]}A)X, Y\rangle = (\lambda - \mu)\ \langle \nabla_X Y, \nabla_Y X\rangle. \tag{3.22}$$

2. Using the Gauss equation, show that

$$\langle R(X, Y)Y, X\rangle = c + \lambda\mu. \tag{3.23}$$

3. Using the definition of the curvature tensor, show that

$$\langle R(X,Y)Y,X\rangle = \langle \nabla_X Y, \nabla_Y X\rangle + \frac{1}{\lambda-\mu}\langle (\nabla_{[X,Y]}A)X, Y\rangle. \tag{3.24}$$

4. Using the Codazzi equation, show that for a unit principal vector Z corresponding to a principal curvature ν not equal to λ or μ, we have

$$(\lambda-\nu)(\mu-\nu)\langle \nabla_X Y, Z\rangle\langle \nabla_Y X, Z\rangle = \langle (\nabla_Z A)X, Y\rangle^2 = \langle (\nabla_Z A)Y, X\rangle^2. \tag{3.25}$$

5. Express $\langle \nabla_X Y, \nabla_Y X\rangle$ in terms of the orthonormal principal basis as follows:

$$\langle \nabla_X Y, \nabla_Y X\rangle = \sum_{\mu_i \neq \lambda,\mu} \langle \nabla_X Y, e_i\rangle\langle \nabla_Y X, e_i\rangle. \tag{3.26}$$

6. Use the results of Steps 1, 2, and 3 to show that

$$2\sum_{\mu_i \neq \lambda,\mu} \frac{\langle (\nabla_{e_i}A)Y, X\rangle^2}{(\lambda-\mu_i)(\mu-\mu_i)} = c + \lambda\mu. \tag{3.27}$$

To complete the proof of the lemma, note that for any j with $\mu_j \neq \lambda$, we have by setting $Y = e_j$ in equation (3.27) and dividing by $\lambda - \mu_j$,

$$\frac{c+\lambda\mu_j}{\lambda-\mu_j} = 2\sum_{\mu_i \neq \lambda,\mu_j} \frac{\langle (\nabla_{e_i}A)e_j, X\rangle^2}{(\lambda-\mu_j)(\lambda-\mu_i)(\mu_j-\mu_i)}. \tag{3.28}$$

Summing this over all j for which $\mu_j \neq \lambda$, we have

$$\sum_{\mu_j \neq \lambda} \frac{c+\lambda\mu_j}{\lambda-\mu_j} = 2\sum_{\mu_j \neq \lambda}\ \sum_{\mu_i \neq \lambda,\mu_j} \frac{\langle (\nabla_{e_i}A)e_j, X\rangle^2}{(\lambda-\mu_j)(\lambda-\mu_i)(\mu_j-\mu_i)}. \tag{3.29}$$

Since the summand on the right side of equation (3.29) is skew-symmetric in $\{i,j\}$, the value of the sum is 0, and so the sum on the left equals 0.

We now give the details of the proofs of the steps listed above.

1. Using the Codazzi equation and the fact that $\nabla_X A$ is symmetric, we compute

$$\begin{aligned}\langle (\nabla_{[X,Y]}A)X, Y\rangle &= \langle (\nabla_X A)[X,Y], Y\rangle = \langle [X,Y], (\nabla_X A)Y\rangle \\ &= \langle [X,Y], (\nabla_Y A)X\rangle.\end{aligned} \tag{3.30}$$

But now,

$$\langle \nabla_X Y, (\nabla_Y A)X\rangle = \langle (\lambda I - A)\nabla_Y X, \nabla_X Y\rangle,$$

while

$$\langle \nabla_Y X, (\nabla_Y A)X \rangle = \langle \nabla_Y X, (\nabla_X A)Y \rangle = \langle (\mu I - A)\nabla_Y X, \nabla_X Y \rangle.$$

Thus

$$\langle [X, Y], (\nabla_Y A)X \rangle = (\lambda - \mu)\langle \nabla_X Y, \nabla_Y X \rangle. \tag{3.31}$$

On substituting in (3.30) we obtain (3.22) as desired.

2. This is immediate from the Gauss equation (see, for example, [283, Vol. II, p. 23]),

$$R(X, Y)Z = c\,(\langle Y, Z\rangle X - \langle X, Z\rangle Y) - (\langle AX, Z\rangle AY - \langle AY, Z\rangle AX). \tag{3.32}$$

3. First note that $\langle \nabla_Y Y, X \rangle = 0$ by Lemma 3.9, and so

$$\langle \nabla_X \nabla_Y Y, X \rangle = -\langle \nabla_Y Y, \nabla_X X \rangle,$$

which vanishes, again by Lemma 3.9. Similarly, $\langle \nabla_X Y, X \rangle = 0$, so that

$$\langle \nabla_Y \nabla_X Y, X \rangle = -\langle \nabla_X Y, \nabla_Y X \rangle.$$

Finally,

$$\langle (\nabla_{[X,Y]} A)X, Y \rangle = (\lambda - \mu)\langle \nabla_{[X,Y]} X, Y \rangle.$$

Using these three equations, we compute $\langle R(X, Y)Y, X \rangle$ to be

$$\langle \nabla_X \nabla_Y Y - \nabla_Y \nabla_X Y - \nabla_{[X,Y]}) Y, X \rangle = \langle \nabla_X Y, \nabla_Y X \rangle + \frac{1}{\lambda - \mu}\langle (\nabla_{[X,Y]} A)X, Y \rangle,$$

which gives equation (3.24).

4. From the Codazzi equation and the symmetry of $\nabla_X A$, we have

$$\langle (\nabla_Z A)X, Y \rangle = \langle (\nabla_X A)Z, Y \rangle = \langle Z, (\nabla_X A)Y \rangle = (\mu - \nu)\langle Z, \nabla_X Y \rangle.$$

The same calculation with X and Y interchanged gives

$$\langle (\nabla_Z A)X, Y \rangle = (\lambda - \nu)\langle Z, \nabla_Y X \rangle.$$

Multiplying these two equations together gives the first equation in (3.25). The last equation in (3.25) follows from the symmetry of $\nabla_Z A$.

5. To establish equation (3.26), we need only observe that the terms omitted from the full summation of the terms $\langle \nabla_X Y, e_i \rangle \langle \nabla_Y X, e_i \rangle$ (i.e., those i for which $\mu_i = \lambda$ or $\mu_i = \mu$) actually vanish and make no contribution to the sum. This is easily checked using Lemma 3.9.

6. Combine equations (3.22), (3.23), and (3.24) to get

$$2\langle \nabla_X Y, \nabla_Y X \rangle = c + \lambda\mu.$$

Using this and the result of equation (3.25) in equation (3.26), we get equation (3.27). This completes the proof of Lemma 3.10 in the case where the number g of distinct principal curvatures is greater than two.

In the case $g = 2$, we can take the two distinct principal curvatures to be λ and μ, as in the proof above. In that case, Cartan's formula reduces to the single equation $c + \lambda\mu = 0$. By steps 1–3 in the proof above, we get

$$c + \lambda\mu = \langle R(X,Y)Y, X \rangle = 2\langle \nabla_X Y, \nabla_Y X \rangle. \tag{3.33}$$

The right side of this equation vanishes by part 2 of Lemma 3.9, and thus we have $c + \lambda\mu = 0$. This completes the proof of Lemma 3.10. □

For $\tilde{M}^{n+1}(c) = S^{n+1}$, we will also give a different proof of Cartan's formula due to Münzner [381] in Corollary 3.24 on page 106. In the spherical case, Cartan's formula is equivalent to the minimality of the focal submanifolds of M^n in S^{n+1}. Münzner proved that the focal submanifolds are minimal (see Corollary 3.23 on page 106). This was also established independently by Nomizu [403, 404] by a different proof than that of Münzner.

Remark 3.11 (Generalizations of Cartan's formula). Berndt found a generalization of Cartan's formula for real hypersurfaces with constant principal curvatures in complex space forms [27] (see Theorem 8.6), and in quaternionic space forms [28]. Nomizu [403] proved a version of Cartan's formula for isoparametric hypersurfaces in Lorentzian forms, and Ooguri [413] found one for equiaffine isoparametric hypersurfaces in affine differential geometry.

Later Abe and Hasegawa [1] extended these results to a more general setting. Recently, Koike [286] found a Cartan type formula for isoparametric hypersurfaces in symmetric spaces. (See also Koike [287].)

Isoparametric hypersurfaces in Euclidean space

Using his formula, Cartan was able to classify isoparametric hypersurfaces in $\mathbf{R}^{n+1}$ and H^{n+1}, and we will do that now. We first consider an isoparametric hypersurface in Euclidean space. This is a local theorem so we consider the hypersurface to be embedded in $\mathbf{R}^{n+1}$.

Theorem 3.12. *Let $M^n \subset \mathbf{R}^{n+1}$ be a connected isoparametric hypersurface. Then M^n is an open subset of a flat hyperplane, a metric hypersphere, or a spherical cylinder $S^k(r) \times \mathbf{R}^{n-k}$.*

Proof. If the number g of distinct principal curvatures of M^n is one, then M^n is totally umbilic, and it is well known that M^n is an open subset of a hyperplane or hypersphere in $\mathbf{R}^{n+1}$ (see, for example, [495, Vol. 4, p. 110]).

If $g \geq 2$, then by taking an appropriate choice of unit normal field ξ, one can assume that at least one of the principal curvatures is positive. If λ_i is the smallest positive principal curvature, then each term $\lambda_i\lambda_j/(\lambda_i - \lambda_j)$ in Cartan's formula (3.13) is non-positive, and thus it equals zero. Hence, there can be at most two distinct principal curvatures, and if there are two, then one of them equals zero.

In the case $g = 2$, suppose that the principal curvatures are $\lambda_1 > 0$ with multiplicity $m_1 = k$, and $\lambda_2 = 0$ with multiplicity $m_2 = n - k$. Then for $t = 1/\lambda_1$, the focal submanifold $V = f_t(M^n)$ has dimension $n - k$, and it is totally geodesic in $\mathbf{R}^{n+1}$, since the same calculations given in the proof of Theorem 2.2 (page 17) show that for every unit normal η to V at every point p of V, the shape operator A_η has one distinct principal curvature given by

$$\frac{\lambda_2}{(1 - t\lambda_2)} = 0. \tag{3.34}$$

Thus, this focal submanifold V is contained in a totally geodesic submanifold $\mathbf{R}^{n-k} \subset \mathbf{R}^{n+1}$, and M^n is an open subset of a tube of radius $1/\lambda_1$ over $\mathbf{R}^{n-k}$. Such a tube is a spherical cylinder $S^k(r) \times \mathbf{R}^{n-k}$, where $S^k(r)$ is a k-dimensional sphere of radius $r = 1/\lambda_1$ in a totally geodesic $\mathbf{R}^{k+1} \subset \mathbf{R}^{n+1}$ orthogonal to $\mathbf{R}^{n-k}$. □

Remark 3.13. A formal proof of the fact that for every unit normal η at every point $p \in V$, the shape operator A_η has one principal curvature equal to zero as in equation (3.34) can be constructed in the same way as the proofs of Theorem 3.21 and Corollary 3.22 in the spherical case in Section 3.3 on page 105. Specifically, one first gets that $A_\eta = 0$ on an open subset of the m_1-sphere $S_p^\perp V$ of unit normals to V at p. Then, as in Corollary 3.22, since the characteristic polynomial of A_η is analytic as a function of η, we get that $A_\eta = 0$ for all $\eta \in S_p^\perp V$.

Isoparametric hypersurfaces in hyperbolic space

Next we consider an isoparametric hypersurface $M^n \subset H^{n+1}$ in hyperbolic space ($c = -1$). Again this is a local result so we consider the hypersurface to be embedded.

Theorem 3.14. *Let $M^n \subset \mathbf{H}^{n+1}$ be a connected isoparametric hypersurface. Then M^n is an open subset of a totally geodesic hyperplane, an equidistant hypersurface, a horosphere, a metric hypersphere, or a tube over a totally geodesic submanifold of codimension greater than one in H^{n+1}.*

Proof. Let g be the number of distinct principal curvatures of M^n. If $g = 1$, then M^n is totally umbilic, and it is an open subset of a totally geodesic hyperplane, an equidistant hypersurface, a horosphere or a metric hypersphere in H^{n+1} (see, for example, [495, Vol. 4, p. 114]).

If $g \geq 2$, then by an appropriate choice of the unit normal field ξ, we can arrange that at least one of the principal curvatures is positive. Then there exists a positive principal curvature λ_i such that no principal curvature lies between λ_i and $1/\lambda_i$. (In fact, λ_i is either the largest principal curvature between 0 and 1, or the smallest principal curvature greater than or equal to one.) For this λ_i, each term $(-1 + \lambda_i\lambda_j)/(\lambda_i - \lambda_j)$ in Cartan's formula (3.13) with $c = -1$ is negative unless $\lambda_j = 1/\lambda_i$. Therefore, there can be at most two distinct principal curvatures, and if there are two, they are reciprocals of each other.

In that case, suppose that the two principal curvatures are $\lambda_1 = \coth\theta$ with multiplicity $m_1 = k$, and $\lambda_2 = 1/\lambda_1 = \tanh\theta$ with multiplicity $m_2 = n - k$. If we take $t = \theta$, then $V = f_t(M^n)$ is a focal submanifold of dimension $n - k$, and it is totally geodesic in H^{n+1}, since the same calculations that prove Theorem 2.2 on page 17 in the hyperbolic case show that for every unit normal η to V at every point of V, the shape operator A_η has one distinct principal curvature given by

$$\tanh(\theta - t) = \tanh(t - t) = 0. \tag{3.35}$$

Thus, this focal submanifold V is contained in a totally geodesic submanifold $H^{n-k} \subset H^{n+1}$, and $f(M^n)$ is an open subset of a tube of radius $t = \theta$ over H^{n-k}. Such a tube is standard Riemannian product $S^k(c_1) \times H^{n-k}(c_2)$ in hyperbolic space H^{n+1}, where $c_1 = 1/\sinh^2\theta$ and $c_2 = -1/\cosh^2\theta$ are the constant sectional curvatures of the sphere $S^k(c_1)$ and the hyperbolic space $H^{n-k}(c_2)$, respectively (see Ryan [469]). □

Remark 3.15. As in the Euclidean case (see Remark 3.13), a formal proof of the fact that for every unit normal η at every point $p \in V$, the shape operator A_η has one principal curvature equal to zero as in equation (3.35) can be constructed in the same way as the proofs of Theorem 3.21 and Corollary 3.22 in the spherical case in Section 3.3 on page 105.

Isoparametric hypersurfaces in the sphere

In the sphere S^{n+1}, however, Cartan's formula does not lead to the conclusion that $g \leq 2$, and in fact Cartan produced examples with $g = 1, 2, 3$ or 4 distinct principal curvatures. Moreover, he classified isoparametric hypersurfaces $M^n \subset S^{n+1}$ with $g \leq 3$ as follows.

In the case $g = 1$, the hypersurface M^n is totally umbilic, and it is well known that M^n is an open subset of a great or small hypersphere in S^{n+1} (see Subsection 3.8.1 on page 144 and [495, Vol. 4, p. 112]).

If $g = 2$, then M^n is an open subset of a standard product of two spheres,

$$S^p(r) \times S^q(s) \subset S^{n+1}(1) \subset \mathbf{R}^{p+1} \times \mathbf{R}^{q+1} = \mathbf{R}^{n+2}, \quad r^2 + s^2 = 1, \tag{3.36}$$

where $n = p + q$, and $r > 0$, $s > 0$. The proof of this result is similar to the proofs of Theorems 3.12 and 3.14 above, i.e., one shows that a focal submanifold is totally geodesic (see Theorem 3.29 on page 111).

The case of $g = 3$ distinct principal curvatures is much more difficult, and it is a highlight of Cartan's work. Ultimately, Cartan [54] showed that in the case $g = 3$, all the principal curvatures have the same multiplicity $m = 1, 2, 4$ or 8, and M^n is an open subset of a tube of constant radius over a standard embedding of a projective plane $\mathbf{FP}^2$ into S^{3m+1}, where $\mathbf{F}$ is the division algebra $\mathbf{R}$, $\mathbf{C}$, $\mathbf{H}$ (quaternions), $\mathbf{O}$ (Cayley numbers), for $m = 1, 2, 4, 8$, respectively. (See Section 2.9 on page 74 and Subsection 3.8.3 on page 151 for more detail on the standard embeddings.)

Thus, up to congruence, there is only one such family of isoparametric hypersurfaces for each value of m. For each of these hypersurfaces, the focal set of M^n consists of two antipodal standard embeddings of $\mathbf{FP}^2$, and M^n is a tube of constant radius over each focal submanifold.

More generally, Cartan showed that any isoparametric family with g distinct principal curvatures of the same multiplicity can be defined by an equation of the form $F = \cos gt$ (restricted to S^{n+1}), where F is a harmonic homogeneous polynomial of degree g on $\mathbf{R}^{n+1}$ satisfying

$$|\operatorname{grad} F|^2 = g^2 r^{2g-2}, \tag{3.37}$$

where $r = |x|$ for $x \in \mathbf{R}^{n+2}$, and grad F is the gradient of F in $\mathbf{R}^{n+2}$. This was a forerunner of Münzner's general result that every isoparametric hypersurface is algebraic, and its defining polynomial satisfies certain differential equations (see Theorem 3.32 on page 115), which generalize those that Cartan found in this special case where all the principal curvatures have the same multiplicity.

In the case $g = 4$, Cartan [56] produced isoparametric hypersurfaces with four principal curvatures of multiplicity one in S^5, and four principal curvatures of multiplicity two in S^9. Cartan claimed without writing the proof that these are the only examples of isoparametric hypersurfaces with four principal curvatures of the same multiplicity m. Cartan's claim was shown to be true in the case $m = 1$ by Takagi [510], and in the case $m = 2$ by Ozeki and Takeuchi [423]. The fact that no other values of m are possible was shown by Grove and Halperin [184].

Cartan noted that all of his examples are homogeneous, each being an orbit of a point under an appropriate closed subgroup of $SO(n + 2)$. Based on his results and the properties of his examples, Cartan asked the following three questions [54], all of which were answered in the 1970s, as we will describe below.

Cartan's questions

1. For each positive integer g, does there exist an isoparametric family with g distinct principal curvatures of the same multiplicity?

2. Does there exist an isoparametric family of hypersurfaces with more than three distinct principal curvatures such that the principal curvatures do not all have the same multiplicity?

3. Does every isoparametric family of hypersurfaces admit a transitive group of isometries?

In the early 1970s, Nomizu [401, 402] wrote two papers describing the highlights of Cartan's work. He also generalized Cartan's example with four principal curvatures of multiplicity one to produce examples with four principal curvatures having multiplicities $m_1 = m_3 = m$, and $m_2 = m_4 = 1$, for any positive integer m. This answered Cartan's Question 2 in the affirmative.

Nomizu also proved that every focal submanifold of every isoparametric hypersurface is a minimal submanifold of S^{n+1}. This also follows from Münzner's work (see Corollary 3.23 on page 106), and Münzner's proof is different than that of Nomizu.

In 1972, Takagi and Takahashi [511] gave a complete classification of all homogeneous isoparametric hypersurfaces in S^{n+1}, based on the work of Hsiang and Lawson [195]. Takagi and Takahashi showed that each homogeneous isoparametric hypersurface in S^{n+1} is a principal orbit of the isotropy representation of a Riemannian symmetric space of rank 2, and they gave a complete list of examples [511, p. 480]. This list contains examples with $g = 6$ principal curvatures as well as those with $g = 1, 2, 3, 4$ principal curvatures. In some cases with $g = 4$, the principal curvatures do not all have the same multiplicity, so this also provided an affirmative answer to Cartan's Question 2.

At about the same time as the papers of Nomizu and Takagi-Takahashi, Münzner published two preprints that greatly extended Cartan's work and have served as the basis for much of the research in the field since that time. The preprints were eventually published as papers [381, 382] in 1980–1981. Of course, one of Münzner's primary results is that the number g of distinct principal curvatures of an isoparametric hypersurface in a sphere equals $1, 2, 3, 4$ or 6 (see Theorem 3.49 on page 136), and thus the answer to Cartan's Question 1 is negative.

Finally, the answer to Cartan's Question 3 is also negative, as was first shown by the construction of inhomogeneous isoparametric hypersurfaces with $g = 4$ principal curvatures by Ozeki and Takeuchi [422] in 1975. Their construction was then generalized in 1981 to yield even more inhomogeneous examples by Ferus, Karcher, and Münzner [160] (see Section 3.9 on page 162).

Isoparametric submanifolds of higher codimension

Remark 3.16 (Isoparametric submanifolds of codimension greater than one). There is also an extensive theory of isoparametric submanifolds of codimension greater than one in the sphere, due primarily to Carter and West [66–68, 553], Terng [525–529], and Hsiang, Palais and Terng [203]. (See also Harle [189] and Strübing [503].) Terng [525] formulated the definition as follows: a connected, complete submanifold V in a real space form $\tilde{M}^{n+1}$ is said to be *isoparametric* if it has flat normal bundle and if for any parallel section of the unit normal bundle $\eta : V \to B^n$, the principal curvatures of A_η are constant. Note that Terng's definition does include the assumption of completeness.

After considerable development of the theory, Thorbergsson [537] showed that a compact, irreducible isoparametric submanifold M substantially embedded in S^{n+1} with codimension greater than one is homogeneous. Thus, M is a principal orbit of an isotropy representation of a symmetric space (also called s-representations), as in the codimension one case. Orbits of isotropy representations of symmetric spaces are also known as *generalized flag manifolds* or *R-spaces*. (See Bott-Samelson [49] and Takeuchi-Kobayashi [513]). This is an important theory, but it is not the focus of this book. See the books by Palais and Terng [426], and Berndt, Console, and Olmos [33] for a thorough treatment of this subject. (See also Subsection 3.8.6 for more detail and further generalizations.)

Remark 3.17 (Isoparametric hypersurfaces in semi-Riemannian spaces). Nomizu [403] began the study of isoparametric hypersurfaces in semi-Riemannian space forms by proving a generalization of Cartan's formula for spacelike hypersurfaces in a Lorentzian space form $\tilde{M}^n_1(c)$ of constant sectional curvature c. As a consequence of this formula, Nomizu showed that a spacelike isoparametric hypersurface in $\tilde{M}^n_1(c)$ can have at most two distinct principal curvatures if $c \geq 0$. Later Li and Xie [325] proved that this conclusion also holds for spacelike isoparametric hypersurfaces in $\tilde{M}^n_1(c)$ for $c < 0$. Magid [351] studied isoparametric hypersurfaces in Lorentz space whose shape operator is not diagonalizable, and Hahn [187] did an extensive study of isoparametric hypersurfaces in semi-Riemannian space forms of arbitrary signatures.

Xiao [557] studied isoparametric hypersurfaces in the anti-de Sitter space H^{n+1}_1, and Li and Wang [316] studied isoparametric surfaces in the de Sitter space S^3_1 (see also [324] for further developments in this area).

Working in a different direction, Niebergall and Ryan [395–398] extended the notions of isoparametric and Dupin to Blaschke hypersurfaces in affine differential geometry, where the eigenvalues of the affine shape operator are considered as the principal curvatures. (See also the related papers of Cecil [75] and Cecil-Magid-Vrancken [87].) Then Koike [284, 285] extended the definition of Dupin to the setting of equiaffine hypersurfaces in equiaffine spaces and proved several theorems in that context that are analogous to well-known results about Dupin hypersurfaces in Euclidean space.

3.2 Parallel Hypersurfaces in the Sphere

We now begin a thorough treatment of Münzner's theory, which is contained in Sections 3.2–3.6. For the following local calculations, we consider a connected, oriented hypersurface $M^n \subset S^{n+1} \subset \mathbf{R}^{n+2}$ with field of unit normals ξ. Assume that M has g distinct principal curvatures at each point, which we label as in Section 2.3 by

$$\lambda_i = \cot \theta_i, \ 0 < \theta_i < \pi, \ 1 \leq i \leq g, \tag{3.38}$$

where the θ_i form an increasing sequence, and λ_i has constant multiplicity m_i on M. We denote the corresponding principal distribution by

$$T_i(x) = \{X \in T_xM \mid AX = \lambda_i X\}, \tag{3.39}$$

where A is the shape operator determined by the field of unit normals ξ. By Theorem 2.10 on page 24 in the case of multiplicity $m_i > 1$, and by the theory of ordinary equations in the case $m_i = 1$, each T_i is a foliation of M with leaves of dimension m_i.

We consider the parallel hypersurface $f_t : M \to S^{n+1}$ defined as in Section 2.3 by

$$f_t(x) = \cos t \, x + \sin t \, \xi(x), \tag{3.40}$$

that is, $f_t(x)$ is the point in S^{n+1} at an oriented distance t along the normal geodesic in S^{n+1} to M through the point x. Note that f_0 is the original embedding f (whose mention we are suppressing at present).

In the following calculations, we show that f_t is an immersion at x if and only if $\cot t$ is not a principal curvature of M at x. In that case, we then find the principal curvatures of f_t in terms of the principal curvatures of the original embedding f.

Let $X \in T_xM$. Then differentiating equation (3.40) in the direction X, we get

$$(f_t)_*X = \cos t \, X + \sin t \, D_X\xi = \cos t \, X - \sin t \, AX = (\cos t \, I - \sin t \, A)X,$$

where on the right side we are identifying X with its Euclidean parallel translate at $f_t(x)$. If $X \in T_i(x)$, this yields

$$(f_t)_*X = (\cos t - \sin t \cot \theta_i) \, X = \frac{\sin(\theta_i - t)}{\sin \theta_i} X. \tag{3.41}$$

Since T_xM is the direct sum of the principal spaces $T_i(x)$, we see that $(f_t)_*$ is injective on T_xM, unless $t = \theta_i \pmod{\pi}$ for some i, that is, $f_t(x)$ is a focal point of (M, x).

If the principal curvatures $\lambda_i = \cot \theta_i$ are all constant on M, then we see that a parallel hypersurface f_tM is an immersed hypersurface if $t \neq \theta_i \pmod{\pi}$ for any i. In that case, we want to find the principal curvatures of f_tM. These are given in

Theorem 2.2 on page 17, where we did the computations in the ambient space $\mathbf{R}^{n+1}$. We include these calculations here for the sake of completeness, and because they are important in the development of the theory.

Theorem 3.18. *Let $M^n \subset S^{n+1}$ be an oriented hypersurface with a principal curvature $\lambda = \cot\theta$ of multiplicity m on M, and suppose that f_t is an immersion in a neighborhood of a point $x \in M$. Then the parallel hypersurface f_tM has a principal curvature $\tilde{\lambda} = \cot(\theta - t)$ at $f_t(x)$ having the same multiplicity m and (up to parallel translation in $\mathbf{R}^{n+2}$) the same principal space $T_\lambda(x)$ as λ at x.*

Proof. Let W be a neighborhood of x in M on which f_t is an immersion. Let ξ be the field of unit normals on M^n, and denote ξ at a point y by ξ_y. Then for $y \in W$, one can easily show that the vector

$$\tilde{\xi}_y = -\sin t\, y + \cos t\, \xi_y, \tag{3.42}$$

when translated to $\tilde{y} = f_t(y)$, is a unit normal to the hypersurface f_tW at the point $\tilde{y}$. We want to find the shape operator A_t determined by this field of unit normals $\tilde{\xi}$ on f_tW. Let $X \in T_\lambda(x)$. Since $\langle X, \xi\rangle = 0$, we have

$$D_X\xi = \tilde{\nabla}_X\xi - \langle X, \xi\rangle\, x = \tilde{\nabla}_X\xi = -AX = -\lambda X = -\cot\theta\, X, \tag{3.43}$$

where D is the Euclidean covariant derivative on $\mathbf{R}^{n+2}$, and $\tilde{\nabla}$ is the induced Levi-Civita connection on S^{n+1}. By definition, the shape operator A_t is given by

$$(f_t)_*(A_tX) = -\tilde{\nabla}_{(f_t)_*X}\tilde{\xi} = -D_{(f_t)_*X}\tilde{\xi}, \tag{3.44}$$

since $\langle (f_t)_*X, \tilde{\xi}\rangle = 0$. To compute this, let x_u be a curve in M with initial point $x_0 = x$ and initial tangent vector $\overrightarrow{x_0} = X$. Then we have using equation (3.43),

$$\begin{aligned} D_{(f_t)_*X}\tilde{\xi} &= \frac{d}{du}(\tilde{\xi}_{x_u})|_{u=0} = -\sin t\, X - \cos t \cot\theta\, X \\ &= \frac{-\cos(\theta - t)}{\sin\theta} X, \end{aligned} \tag{3.45}$$

where we again identify X on the right with its Euclidean parallel translate at $f_t(x)$. If we compare this with equation (3.41) and use equation (3.44), we see that

$$A_tX = \cot(\theta - t)X, \tag{3.46}$$

for $X \in T_\lambda(x)$, and so $\tilde{\lambda} = \cot(\theta - t)$ is a principal curvature of A_t with the same principal space $T_\lambda(x)$ and same multiplicity m as λ at x. □

As a consequence of Theorem 3.18, we get the following corollary regarding the principal curvatures of a family of parallel isoparametric hypersurfaces.

Corollary 3.19. *Let $M^n \subset S^{n+1}$ be a connected isoparametric hypersurface having g distinct principal curvatures $\lambda_i = \cot\theta_i$, $1 \leq i \leq g$, with respective multiplicities m_i. If t is any real number not congruent to any θ_i (mod π), then the map f_t immerses M as an isoparametric hypersurface with principal curvatures $\tilde{\lambda}_i = \cot(\theta_i - t)$, $1 \leq i \leq g$, with the same multiplicities m_i. Furthermore, for each i, the principal foliation corresponding to $\tilde{\lambda}_i$ is the same as the principal foliation T_i corresponding to λ_i on M.*

Remark 3.20. It follows from Münzner's theory that if M is an isoparametric hypersurface embedded in S^{n+1}, then each parallel isoparametric hypersurface f_tM is also embedded in S^{n+1} and not just immersed. This is because M and its parallel hypersurfaces are level sets of the restriction to S^{n+1} of a certain polynomial function on $\mathbf{R}^{n+2}$, as will be discussed later in the chapter in Theorem 3.32 on page 115.

3.3 Focal Submanifolds

The geometry of the focal submanifolds of an isoparametric hypersurface is a crucial element in this theory. In this section, we obtain some important basic results about isoparametric hypersurfaces and their focal submanifolds due to Münzner [381] (see also Chapter 3 of [95]).

As in the previous section, let $M^n \subset S^{n+1}$ be a connected, oriented isoparametric hypersurface with field of unit normals ξ having g distinct constant principal curvatures,

$$\lambda_i = \cot\theta_i,\ 0 < \theta_i < \pi,\ 1 \leq i \leq g, \tag{3.47}$$

where the θ_i form an increasing sequence, and denote the multiplicity of λ_i by m_i.

By Theorems 2.11, 2.12, and 2.14 of Section 2.4, the leaves of the principal foliation T_i are open subsets of m_i-dimensional metric spheres in S^{n+1}, and the space of leaves M/T_i is an $(n - m_i)$-dimensional manifold with the quotient topology.

If $t = \theta_i$, then the map f_t has constant rank $n - m_i$ on M, and f_t factors through an immersion $\psi_i : M/T_i \to S^{n+1}$, i.e., $f_t = \psi_i \circ \pi$, where π is the projection from M onto M/T_i. Thus, f_t is a focal map, as in Theorems 2.12 and 2.14 of Section 2.4, and we denote image of ψ_i by V_i.

We now want to find the principal curvatures of this focal submanifold V_i. This is similar to the calculation for parallel hypersurfaces in the preceding section, although we must make some adjustments because V_i has codimension greater than one.

Let $x \in M$. Then we have the orthogonal decomposition of the tangent space T_xM as

$$T_xM = T_i(x) \oplus T_i^{\perp}(x), \tag{3.48}$$

where $T_i^{\perp}(x)$ is the direct sum of the spaces $T_j(x)$, for $j \neq i$. By equation (3.41), $(f_t)_* = 0$ on $T_i(x)$, and $(f_t)_*$ is injective on $T_i^{\perp}(x)$.

We consider a map $h : M \to S^{n+1}$ given by

$$h(x) = -\sin t\, x + \cos t\, \xi(x). \tag{3.49}$$

This is basically the same function that we considered in equation (3.42) for the field $\tilde{\xi}$ of unit normals to f_tM in the case where f_t is an immersion.

In the case $t = \theta_i$, we see that $\langle f_t(x), h(x)\rangle = 0$, and so the vector $h(x)$ is tangent to the sphere S^{n+1} at the point $p = f_t(x)$. Furthermore, $\langle h(x), X\rangle = 0$ for all $X \in T_i^{\perp}(x)$, and so by equation (3.41), $h(x)$ is normal to the focal submanifold V_i at the point p.

We can use the map h to find the shape operator of a normal vector to V_i as follows. Let p be an arbitrary point in V_i. Then the set $C = f_t^{-1}(p)$ is an open subset of an m_i-sphere in S^{n+1}. For each $x \in C$, the vector $h(x)$ is a unit normal to the focal submanifold V_i at p. Thus, the restriction of h to C is a map from C into the m_i-sphere $S_p^{\perp}V_i$ of unit normal vectors to V_i at p. At a point $x \in C$, the tangent space T_xC equals $T_i(x)$. Since $t = \theta_i$, we have for $x \in C$ and a nonzero vector $X \in T_i(x)$ that

$$\begin{aligned} h_*(X) &= -\sin tX + \cos t(-AX) = -\sin\theta_i X + \cos\theta_i(-\cot\theta_i)X \\ &= \frac{-1}{\sin\theta_i}X \neq 0. \end{aligned} \tag{3.50}$$

Thus, h_* has full rank m_i on C, and so h is a local diffeomorphism of open subsets of m_i-spheres. This enables us to prove the following important result due to Münzner [381].

Theorem 3.21. *Let $M \subset S^{n+1}$ be a connected isoparametric hypersurface, and let $V_i = f_tM$ for $t = \theta_i$ be a focal submanifold of M. Let η be a unit normal vector to V_i at a point $p \in V_i$, and suppose that $\eta = h(x)$ for some $x \in f_t^{-1}(p)$. Then the shape operator A_η of V_i is given in terms of its principal vectors by*

$$A_\eta X = \cot(\theta_j - \theta_i)X, \text{ for } X \in T_j(x),\ j \neq i. \tag{3.51}$$

(As before we are identifying $T_j(x)$ with its Euclidean parallel translate at p.)

Proof. Let $\eta = h(x)$ for some $x \in C = f_t^{-1}(p)$ for $t = \theta_i$. The same calculation used in proving Theorem 3.18 is valid here, and it leads to equation (3.46), which we write as

$$A_\eta X = \cot(\theta_j - \theta_i)X, \text{ for } X \in T_j(x),\ j \neq i.$$

□

Corollary 3.22. *Let $M \subset S^{n+1}$ be a connected isoparametric hypersurface, and let $V_i = f_t M$, for $t = \theta_i$, be a focal submanifold of M. Then for every unit normal vector η at every point $p \in V_i$, the shape operator A_η has principal curvatures* $\cot(\theta_j - \theta_i)$ *with multiplicities m_j, for $j \neq i$, $1 \leq j \leq g$.*

Proof. By Theorem 3.21, the corollary holds on the open subset $h(C)$ of the m_i-sphere $S_p^\perp V_i$ of unit normal vectors to V_i at p. Consider the characteristic polynomial $P_u(\eta) = \det(A_\eta - uI)$ as a function of η on the normal space $T_p^\perp V_i$. Since A_η is linear in η, we have for each fixed $u \in \mathbf{R}$ that the function $P_u(\eta)$ is a polynomial of degree $n - m_i$ on the vector space $T_p^\perp V_i$. Thus, the restriction of $P_u(\eta)$ to the sphere $S_p^\perp V_i$ is an analytic function of η. Then since $P_u(\eta)$ is constant on the open subset $h(C)$ of $S_p^\perp V_i$, it is constant on all of $S_p^\perp V_i$. □

Minimality of the focal submanifolds

Münzner also obtained the following consequence of Corollary 3.22. This result was obtained independently with a different proof by Nomizu [403].

Corollary 3.23. *Let $M \subset S^{n+1}$ be a connected isoparametric hypersurface. Then each focal submanifold V_i of M is a minimal submanifold in S^{n+1}.*

Proof. Let η be a unit normal vector to a focal submanifold V_i of M. Then $-\eta$ is also a unit normal vector to V_i. By Corollary 3.22, the shape operators A_η and $A_{-\eta}$ have the same eigenvalues with the same multiplicities. So

$$\operatorname{trace} A_{-\eta} = \operatorname{trace} A_\eta .$$

On the other hand, $\operatorname{trace} A_{-\eta} = -\operatorname{trace} A_\eta$, since $A_{-\eta} = -A_\eta$. Thus, we have $\operatorname{trace} A_\eta = -\operatorname{trace} A_\eta$, and so $\operatorname{trace} A_\eta = 0$. Since this is true for all unit normal vectors η, we conclude that V_i is a minimal submanifold in S^{n+1}. □

As a consequence of Theorem 3.21, we can give a proof of Cartan's formula for isoparametric hypersurfaces in S^n that is different than the proof given for Lemma 3.10 on page 93.

Corollary 3.24 (Cartan's formula). *Let $M \subset S^{n+1}$ be a connected isoparametric hypersurface with g principal curvatures*

$$\lambda_i = \cot\theta_i, \quad 0 < \theta_1 < \cdots < \theta_g < \pi,$$

with respective multiplicities m_i. Then for each i, $1 \leq i \leq g$, the following formula holds

$$\sum_{j \neq i} m_j \frac{1 + \lambda_i \lambda_j}{\lambda_i - \lambda_j} = 0. \tag{3.52}$$

Proof. We will show that for each i and for any unit normal η to the focal submanifold V_i, the left side of equation (3.52) equals trace A_η, which equals zero by Corollary 3.23. Represent the principal curvatures of M as $\lambda_i = \cot\theta_i$, $0 < \theta_1 < \cdots < \theta_g < \pi$, with respective multiplicities m_i. Let V_i be the focal submanifold corresponding to λ_i. By Corollary 3.22, we have

$$0 = \text{trace}\, A_\eta = \sum_{j \neq i} m_j \cot(\theta_j - \theta_i) = \sum_{j \neq i} m_j \frac{1 + \cot\theta_i \cot\theta_j}{\cot\theta_i - \cot\theta_j} \tag{3.53}$$

$$= \sum_{j \neq i} m_j \frac{1 + \lambda_i \lambda_j}{\lambda_i - \lambda_j}.$$

□

Remark 3.25 (Cartan's formula in $\mathbf{R}^{n+1}$ *and* H^{n+1}*).* The same type of proof for Cartan's formula is possible for isoparametric hypersurfaces in $\mathbf{R}^{n+1}$. In that case the sectional curvature $c = 0$ and Cartan's formula becomes

$$\sum_{j \neq i} m_j \frac{\lambda_i \lambda_j}{\lambda_i - \lambda_j} = 0, \tag{3.54}$$

for each $i, 1 \leq i \leq g$. If the principal curvature $\lambda_i = 0$, then both sides of the equation are zero, and the formula holds. If $\lambda_i \neq 0$, let V_i be the focal submanifold $f_t M$, for $t = 1/\lambda_i$. Then calculations similar to those in Theorem 2.2 on page 17 and an argument similar to that in Corollary 3.22 above show that for each unit normal η to V_i, the shape operator A_η has distinct principal curvatures

$$\frac{\lambda_j}{1 - t\lambda_j} = \frac{\lambda_j}{1 - (\lambda_j/\lambda_i)} = \frac{\lambda_i \lambda_j}{\lambda_i - \lambda_j}, \tag{3.55}$$

with multiplicities m_j, for $j \neq i, 1 \leq j \leq g$. Thus the expression on the left side of equation (3.54) is the trace of the shape operator A_η. Therefore, Cartan's formula is equivalent to the minimality of the focal submanifolds V_i, $1 \leq i \leq g$. One can prove that these focal submanifolds are minimal by the argument given in Corollary 3.23, and thus Cartan's formula can be proven by this method for isoparametric hypersurfaces in $\mathbf{R}^{n+1}$.

For an isoparametric hypersurface M in H^{n+1}, the situation is more complicated, since by Theorem 2.1 on page 11, if $|\lambda_i| \leq 1$ for a principal curvature λ_i of M, then there is no focal submanifold in H^{n+1} corresponding to λ_i. Thus, one cannot consider the minimality of the focal submanifolds corresponding to such principal curvatures. If $|\lambda_i| > 1$, then there does exist a focal submanifold V_i corresponding to λ_i, and one can show that Cartan's formula for that λ_i is equivalent to the minimality of V_i. Of course, we already have a different proof of Cartan's formula which is valid for all real space forms (see Lemma 3.10, page 93).

Formula for the principal curvatures

We now return to our consideration of isoparametric hypersurfaces in S^{n+1}. The following remarkable result of Münzner shows that the principal curvatures of such an isoparametric hypersurface have a very specific form.

Theorem 3.26. *Let $M \subset S^{n+1}$ be a connected isoparametric hypersurface with g principal curvatures $\lambda_i = \cot\theta_i$, $0 < \theta_1 < \cdots < \theta_g < \pi$, with respective multiplicities m_i. Then*

$$\theta_i = \theta_1 + (i-1)\frac{\pi}{g}, \quad 1 \leq i \leq g, \tag{3.56}$$

and the multiplicities satisfy $m_i = m_{i+2}$ (subscripts mod g). For any point $x \in M$, there are $2g$ focal points of (M,x) along the normal geodesic to M through x, and they are evenly distributed at intervals of length π/g.

Proof. If $g = 1$, then the theorem is trivially true, so we now consider $g = 2$. Let V_1 be the focal submanifold determined by the map f_t for $t = \theta_1$. By Corollary 3.22, the principal curvature $\cot(\theta_2 - \theta_1)$ of the shape operator A_η is the same for every choice of unit normal η at every point $p \in V_1$. Since $A_{-\eta} = -A_\eta$, this says that

$$\cot(\theta_2 - \theta_1) = -\cot(\theta_2 - \theta_1).$$

Thus, $\cot(\theta_2 - \theta_1) = 0$, so $\theta_2 - \theta_1 = \pi/2$ as desired. In the case $g = 2$, there is no restriction on the multiplicities.

Next we consider the case $g \geq 3$. For a fixed value of i, $1 \leq i \leq g$, let V_i be the focal submanifold determined by the map f_t for $t = \theta_i$. By Corollary 3.22, the set

$$\{\cot(\theta_j - \theta_i) \mid j \neq i\}$$

of principal curvatures of the shape operator A_η is the same for every choice of unit normal η at every point $p \in V_i$. Since $A_{-\eta} = -A_\eta$, this says that the two sets

$$\{\cot(\theta_j - \theta_i) \mid j \neq i\} \text{ and } \{-\cot(\theta_j - \theta_i) \mid j \neq i\}$$

are the same. In the case $2 \leq i \leq g-1$, the largest principal curvature of A_η is $\cot(\theta_{i+1} - \theta_i)$ with multiplicity m_{i+1}, while the largest principal curvature of $A_{-\eta}$ is $\cot(\theta_i - \theta_{i-1})$ with multiplicity m_{i-1}. Since these two largest principal curvatures and their respective multiplicities are equal, we conclude that

$$\theta_{i+1} - \theta_i = \theta_i - \theta_{i-1}, \quad m_{i+1} = m_{i-1}, \quad 2 \leq i \leq g-1. \tag{3.57}$$

If $i = 1$, the largest principal curvature of A_η is $\cot(\theta_2 - \theta_1)$ with multiplicity m_2, and the largest principal curvature of $A_{-\eta}$ is

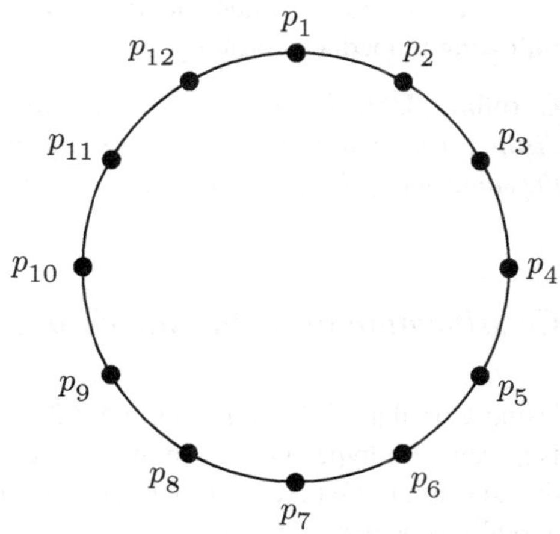

Fig. 3.1 Focal points on a normal geodesic, $g = 6$

$$\cot(\theta_1 - \theta_g) = \cot(\theta_1 - (\theta_g - \pi)),$$

with multiplicity m_g, and we have

$$\theta_2 - \theta_1 = \theta_1 - (\theta_g - \pi), \quad m_2 = m_g. \tag{3.58}$$

If we let $\theta_2 - \theta_1 = \delta$, then equation (3.57) implies that $\theta_g - \theta_1 = (g-1)\delta$, while equation (3.58) implies that $\theta_g - \theta_1 = \pi - \delta$. Combining these two equations, we get that $g\delta = \pi$, and thus $\delta = \pi/g$. From this we get the formula in equation (3.56) for θ_i. The formula for the multiplicities in the theorem follows from equations (3.57) and (3.58).

If x is any point of M, then each principal curvature $\cot\theta_i$ of M gives rise to a pair of antipodal focal points along the normal geodesic to M through x. Thus, there are $2g$ focal points of (M, x) along this normal geodesic, and they are evenly distributed at intervals of length π/g by equation (3.56). □

Figure 3.1 illustrates the case $g = 6$. In the figure, the two antipodal focal points labeled p_i and p_{i+6} are determined by the same principal curvature λ_i for $1 \leq i \leq 6$.

Remark 3.27 (Isoparametric submanifolds and their Coxeter groups). It follows from Theorem 3.26 that the set of focal points along a normal circle to $M \subset S^{n+1}$ is invariant under the dihedral group D_g of order $2g$ that acts on the normal circle and is generated by reflections in the focal points. This is a fundamental idea that generalizes to isoparametric submanifolds of higher codimension in the sphere. Specifically, for an isoparametric submanifold M^n of codimension $k > 1$ in S^{n+k}, Carter and West [66] (in the case $k = 2$) and Terng [525] for arbitrary $k > 1$ found a Coxeter group (finite group generated by reflections) that acts in a way similar to this dihedral group in the codimension one case. This Coxeter group is important in the overall development of the theory in the case of higher codimension.

Since the multiplicities satisfy $m_i = m_{i+2}$ (subscripts mod g), we have the following immediate corollary.

Corollary 3.28. *Let $M \subset S^{n+1}$ be a connected isoparametric hypersurface with g distinct principal curvatures. If g is odd, then all of the principal curvatures have the same multiplicity. If g is even, then there are at most two distinct multiplicities.*

Classification of isoparametric hypersurfaces with g = 2

Using Corollary 3.22 and Theorem 3.26, we can now derive the classification of isoparametric hypersurfaces in the sphere with $g = 2$ distinct principal curvatures due to Cartan. We first want to describe the basic example in the form of a tube over a totally geodesic submanifold of S^{n+1}.

We consider $\mathbf{R}^{n+2} = \mathbf{R}^{p+1} \times \mathbf{R}^{q+1}$, where $p + q = n$. A standard product of a p-sphere and a q-sphere has the form,

$$S^p(r) \times S^q(s) = \{(x, y) \in \mathbf{R}^{p+1} \times \mathbf{R}^{q+1} \mid |x|^2 = r^2, |y|^2 = s^2, \quad r^2 + s^2 = 1\}, \tag{3.59}$$

where $r > 0$ and $s > 0$. Since $r^2 + s^2 = 1$, we have

$$S^p(r) \times S^q(s) \subset S^{n+1}(1) \subset \mathbf{R}^{p+1} \times \mathbf{R}^{q+1} = \mathbf{R}^{n+2}. \tag{3.60}$$

We want to show that such a standard product of two spheres is a tube over a totally geodesic p-sphere,

$$V = \{(u, v) \mid v = 0\} = S^p \times \{0\}, \tag{3.61}$$

where S^p is the unit sphere in $\mathbf{R}^{p+1}$.

We now construct the tube of radius t, $0 < t < \pi/2$, over V, as in Section 2.3. Let $(u, 0)$ be a point in V. Then every unit normal to V in S^{n+1} at $(u, 0)$ has the form $(0, v)$, where $v \in S^q$, the unit sphere in $\mathbf{R}^{q+1}$. Thus, the unit normal bundle BV is given by

$$BV = \{((u, 0), (0, v)) \mid u \in S^p, \ v \in S^q\}. \tag{3.62}$$

Then the map $f_t : BV \to S^{n+1}$ to the tube of (spherical) radius t over V is given by

$$f_t((u, 0), (0, v)) = \cos t\,(u, 0) + \sin t\,(0, v) = (\cos t\, u, \sin t\, v) = (x, y), \tag{3.63}$$

where $x = \cos t\, u$ and $y = \sin t\, v$. We see that the point (x, y) lies in the standard product of spheres $S^p(r) \times S^q(s)$ given in equation (3.59) for the values $r = \cos t$ and $s = \sin t$.

Conversely, if (x, y) in any point in $S^p(r) \times S^q(s)$, then

$$(x, y) = f_t((u, 0), (0, v)),$$

where $u = x/r$, $v = y/s$, and $t = \cos^{-1} r$. Thus, the standard product of spheres $S^p(r) \times S^q(s)$ is precisely the tube of radius t over V in S^{n+1}.

We can find the principal curvatures of a tube M_t of radius t over $V = S^p \times \{0\}$ from the formula in Theorem 2.2 on page 17 for the shape operators of a tube. Since V is totally geodesic, we see from Theorem 2.2 that M_t has two constant principal curvatures,

$$\lambda_1 = \cot\left(\frac{\pi}{2} - t\right) = \tan t, \quad \lambda_2 = -\cot t,$$

with respective multiplicities p and q.

We now prove the classification of isoparametric hypersurfaces with $g = 2$ principal curvatures due to Cartan.

Theorem 3.29. *Let $M \subset S^{n+1}$ be a connected isoparametric hypersurface with $g = 2$ distinct principal curvatures. Then M is an open subset of a standard product of two spheres.*

Proof. Let $\lambda_1 = \cot\theta_1$ and $\lambda_2 = \cot\theta_2$, $0 < \theta_1 < \theta_2 < \pi$, be the two principal curvatures of M with respective multiplicities $m_1 = q$ and $m_2 = p$ with $p + q = n$. Let $V_1 = f_{\theta_1}M$ be the focal submanifold of M corresponding to the principal curvature λ_1. Then by Corollary 3.22, V_1 is a p-dimensional submanifold of S^{n+1} such that for every unit normal vector η to V_1, the shape operator A_η has one distinct principal curvature $\cot(\theta_2 - \theta_1)$. By Theorem 3.26, we have $\theta_2 - \theta_1 = \pi/2$, and so the one principal curvature is zero. Therefore, V_1 is a p-dimensional totally geodesic submanifold of S^{n+1}, which is thus a p-dimensional great sphere $S^p \subset \mathbf{R}^{p+1} \subset \mathbf{R}^{n+2}$. The connected hypersurface M lies on a tube of radius θ_1 over V_1, and by what we have shown above, such a tube is a standard product of two spheres. □

Other characterizations of standard products of spheres are given by Alías, Brasil and Perdomo [10], by Brasil, Gervasio and Palmas [50], and by Adachi and Maeda [5]. For a survey of results concerning flat tori in S^3, see Kitigawa [280].

3.4 Isoparametric Functions

As noted at the beginning of this chapter, the original definition of an isoparametric family of hypersurfaces in a real space form $\tilde{M}^{n+1}$ was formulated in terms of the level sets of an isoparametric function, where a smooth function $F : \tilde{M}^{n+1} \to \mathbf{R}$ is called *isoparametric* if both of the classical Beltrami differential parameters,

$$\Delta_1 F = |\text{grad } F|^2, \quad \Delta_2 F = \Delta F \text{ (Laplacian } F), \tag{3.64}$$

are functions of F itself, and are therefore constant on the level sets of F in $\tilde{M}^{n+1}$.

In the case of an isoparametric hypersurface M in the sphere S^{n+1} in $\mathbf{R}^{n+2}$, Münzner showed that the corresponding isoparametric function $V : S^{n+1} \to \mathbf{R}$ is the restriction to S^{n+1} of a homogeneous polynomial $F : \mathbf{R}^{n+2} \to \mathbf{R}$ satisfying certain differential equations. Thus, it is useful to be able to relate the Beltrami differential parameters of a homogeneous function F on $\mathbf{R}^{n+2}$ to those of its restriction V to S^{n+1}.

The type of polynomial that arises in Münzner's theory is homogeneous of degree g, where g is the number of distinct principal curvatures of the corresponding isoparametric hypersurface $M \subset S^{n+1}$. Recall that a function $F : \mathbf{R}^{n+2} \to \mathbf{R}$ is *homogeneous of degree g* if $F(tx) = t^g F(x)$, for all $t \in \mathbf{R}$ and $x \in \mathbf{R}^{n+2}$. By Euler's Theorem, we know that for any $x \in \mathbf{R}^{n+2}$,

$$\langle \mathrm{grad}^E F, x \rangle = gF(x). \tag{3.65}$$

Here the superscript E is used to denote the Euclidean gradient of F, and the gradient of the restriction V of F to S^{n+1} will be denoted by $\mathrm{grad}^S F$. Similarly, we will denote the respective Laplacians by $\Delta^E F$ and $\Delta^S F$. The following theorem relates the various differential operators for a homogeneous function F of degree g.

Theorem 3.30. *Let $F : \mathbf{R}^{n+2} \to \mathbf{R}$ be a homogeneous function of degree g. Then*

(a) $|\mathrm{grad}^S F|^2 = |\mathrm{grad}^E F|^2 - g^2 F^2$,
(b) $\Delta^S F = \Delta^E F - g(g-1)F - g(n+1)F$.

Proof. (a) At a point $x \in S^{n+1}$, the vector $\mathrm{grad}^S F$ is obtained from $\mathrm{grad}^E F$ by subtracting off the component of $\mathrm{grad}^E F$ normal to S^{n+1} at x. Thus, we have

$$\mathrm{grad}^S F = \mathrm{grad}^E F - \langle \mathrm{grad}^E F, x \rangle x. \tag{3.66}$$

Then, using equation (3.65), we compute

$$\begin{aligned} |\mathrm{grad}^S F|^2 &= |\mathrm{grad}^E F|^2 - 2gF(x)\langle \mathrm{grad}^E F, x\rangle + g^2 F^2(x)|x|^2 \\ &= |\mathrm{grad}^E F|^2 - 2g^2F^2(x) + g^2F^2(x) = |\mathrm{grad}^E F|^2 - g^2 F^2(x). \end{aligned} \tag{3.67}$$

(b) Let D and $\tilde{\nabla}$ denote the Levi-Civita connections on $\mathbf{R}^{n+2}$ and S^{n+1}, respectively. The Laplacian $\Delta^S F$ is the trace of the operator on $T_x S^{n+1}$ given by

$$X \mapsto \tilde{\nabla}_X \, \mathrm{grad}^S F. \tag{3.68}$$

For $X \in T_x S^{n+1}$, we know that $\tilde{\nabla}_X \, \mathrm{grad}^S F$ is the component of $D_X \, \mathrm{grad}^S F$ that is tangent to S^{n+1}, and thus

$$\tilde{\nabla}_X \, \mathrm{grad}^S F = D_X \, \mathrm{grad}^S F - \langle D_X \, \mathrm{grad}^S F, x \rangle \, x. \tag{3.69}$$

We take the covariant derivative of the formula for $\text{grad}^S F$ given in equation (3.66) and use equation (3.65) to compute

$$D_X \,\text{grad}^S F = D_X \,\text{grad}^E F - gFX - g(XF)x. \tag{3.70}$$

Next we take the component of equation (3.70) tangent to S^{n+1} to get

$$\tilde{\nabla}_X \,\text{grad}^S F = D_X \,\text{grad}^E F - \langle D_X \,\text{grad}^E F, x\rangle\, x - gFX. \tag{3.71}$$

From this we can compute the Laplacian as the trace of the operator given in equation (3.68). Let $\{e_1, \ldots, e_{n+1}\}$ be an orthonormal basis for $T_x S^{n+1}$. Then $\{e_1, \ldots, e_{n+1}, x\}$ is an orthonormal basis for $T_x \mathbf{R}^{n+2}$. We work with the terms on the right side of equation (3.71) to compute the Laplacian. First,

$$\sum_{i=1}^{n+1} \langle D_{e_i} \,\text{grad}^E F, e_i\rangle = \Delta^E F - \langle D_x \,\text{grad}^E F, x\rangle. \tag{3.72}$$

Recall that

$$\langle D_x \,\text{grad}^E F, x\rangle = D_x \langle \text{grad}^E F, x\rangle - \langle \text{grad}^E F, x\rangle. \tag{3.73}$$

Using equation (3.65), we have

$$D_x \langle \text{grad}^E F, x\rangle = D_x(gF) = g\langle \text{grad}^E F, x\rangle. \tag{3.74}$$

Substituting this into equation (3.73) and using equation (3.65), we get

$$\langle D_x \,\text{grad}^E F, x\rangle = (g-1)\langle \text{grad}^E F, x\rangle = (g-1)gF(x). \tag{3.75}$$

Next we compute that

$$\sum_{i=1}^{n+1} \langle D_{e_i} \,\text{grad}^E F, x\rangle \langle x, e_i\rangle = 0, \tag{3.76}$$

and the trace of the map $X \mapsto -gFX$ on $T_x S^{n+1}$ is clearly $-g(n+1)F$. Thus, using equations (3.71)–(3.76), we have computed

$$\Delta^S F = \Delta^E F - g(g-1)F - g(n+1)F, \tag{3.77}$$

as needed to prove part (b) of the theorem. □

Examples of isoparametric functions

We now consider some examples of homogeneous polynomials with the property that $|\mathrm{grad}^S F|^2$ and $\Delta^S F$ are both functions of F itself. Thus, the restriction of F to S^{n+1} is an isoparametric function on S^{n+1}.

Example 3.31 (Examples of isoparametric functions on S^{n+1}).

(a) Let $F : \mathbf{R}^{n+2} \to \mathbf{R}$ be the linear height function,

$$F(z) = \langle z, p\rangle, \quad p \in \mathbf{R}^{n+2}, \ |p| = 1. \tag{3.78}$$

Then F is a homogeneous polynomial of degree $g = 1$, and

$$|\mathrm{grad}^E F|^2 = 1, \quad \Delta^E F = 0. \tag{3.79}$$

Consequently, by Theorem 3.30, we have

$$|\mathrm{grad}^S F|^2 = 1 - F^2, \quad \Delta^S F = -(n+1)F, \tag{3.80}$$

so that the restriction of F to S^{n+1} is an isoparametric function on S^{n+1}. The level sets

$$M_t = \{z \in S^{n+1} \mid F(z) = t\}, \quad -1 < t < 1, \tag{3.81}$$

form an isoparametric family of n-spheres in hyperplanes orthogonal to p. The two focal submanifolds of this family are the 1-point sets $\{p\}$ and $\{-p\}$.

(b) Decompose $\mathbf{R}^{n+2}$ as $\mathbf{R}^{n+2} = \mathbf{R}^{p+1} \times \mathbf{R}^{q+1}$, where p and q are positive integers such that $p + q = n$. For any point $z = (x, y)$ in $\mathbf{R}^{p+1} \times \mathbf{R}^{q+1}$, define

$$F(z) = |x|^2 - |y|^2. \tag{3.82}$$

Then F is a homogeneous polynomial of degree $g = 2$, and for $r^2 = |x|^2 + |y|^2$, we have

$$\mathrm{grad}^E F = 2(x, -y), \quad |\mathrm{grad}^E F|^2 = 4r^2, \quad \Delta^E F = 2(p - q). \tag{3.83}$$

Thus by Theorem 3.30, we have

$$|\mathrm{grad}^S F|^2 = 4(1 - F^2), \quad \Delta^S F = 2(p - q) - 2(n+2)F. \tag{3.84}$$

Therefore, the restriction of F to S^{n+1} is an isoparametric function. The level sets form an isoparametric family of hypersurfaces as follows. Let

$$M_t = \{z \in S^{n+1} \mid F(z) = \cos 2t\}, \quad 0 \leq t \leq \pi/2. \tag{3.85}$$

For $(x, y) \in M_t$, we have

$$|x|^2 - |y|^2 = \cos 2t = \cos^2 t - \sin^2 t,$$

and

$$|x|^2 + |y|^2 = 1 = \cos^2 t + \sin^2 t,$$

and so we get $|x|^2 = \cos^2 t$ and $|y|^2 = \sin^2 t$. Thus, for $0 < t < \pi/2$, the level set is the Cartesian product of a p-sphere of radius $\cos t$ and a q-sphere of radius $\sin t$. The two focal submanifolds of the family of isoparametric hypersurfaces are

$$M_0 = \{(x, y) \mid y = 0\} = S^p \times \{0\}, \quad M_{\pi/2} = \{(x, y) \mid x = 0\} = \{0\} \times S^q.$$

Note that one can obtain the same family of level sets from the polynomial $G(x, y) = |x|^2$, since $F = 2G - 1$ on S^{n+1}.

3.5 Cartan–Münzner Polynomials

In this section, we describe Münzner's [381] work concerning the algebraic nature of isoparametric hypersurfaces in spheres. Münzner's primary result in this regard is the following theorem (see also Ferus [159] and Cecil–Ryan [95, pp. 255–267]).

Theorem 3.32. *Let $M \subset S^{n+1} \subset \mathbf{R}^{n+2}$ be a connected isoparametric hypersurface with g principal curvatures $\lambda_i = \cot \theta_i$, $0 < \theta_i < \pi$, with respective multiplicities m_i. Then M is an open subset of a level set of the restriction to S^{n+1} of a homogeneous polynomial F on $\mathbf{R}^{n+2}$ of degree g satisfying the differential equations,*

$$|\mathrm{grad}^E F|^2 = g^2 r^{2g-2}, \tag{3.86}$$

$$\Delta^E F = c r^{g-2}, \tag{3.87}$$

where $r = |x|$, and $c = g^2(m_2 - m_1)/2$.

Remark 3.33. Recall from Corollary 3.28 that there are at most two distinct multiplicities m_1, m_2, and the multiplicities satisfy $m_{i+2} = m_i$ (subscripts mod g).

Münzner called F the *Cartan polynomial* of M, and now F is usually referred to as the *Cartan–Münzner polynomial* of M. Equations (3.86)–(3.87) are called the *Cartan–Münzner differential equations*. By Theorem 3.30 the restriction V of F to S^{n+1} satisfies the differential equations,

$$|\mathrm{grad}^S V|^2 = g^2(1 - V^2), \tag{3.88}$$

$$\Delta^S V = c - g(n+g)V, \tag{3.89}$$

where $c = g^2(m_2 - m_1)/2$. Thus V is an isoparametric function in the sense of Cartan, since both $|\text{grad}^S V|^2$ and $\Delta^S V$ are functions of V itself.

We now describe Münzner's construction of this polynomial F in detail. Let $M \subset S^{n+1}$ be a connected, oriented isoparametric hypersurface with g distinct principal curvatures $\lambda_i = \cot\theta_i$, $0 < \theta_1 < \cdots < \theta_g < \pi$, with respective multiplicities m_i. The normal bundle NM of M in S^{n+1} is trivial and is therefore diffeomorphic to $M \times \mathbf{R}$. Thus we can consider the normal exponential map $E : M \times \mathbf{R} \to S^{n+1}$ defined by

$$E(x,t) = f_t(x) = \cos t\, x + \sin t\, \xi(x), \tag{3.90}$$

where ξ is the field of unit normals to M in S^{n+1}.

By Theorem 2.1 on page 11, we know that the differential of E has rank $n+1$ at $(x,t) \in M \times \mathbf{R}$ unless $\cot t$ is a principal curvature of M at x. Thus, for any noncritical point (x,t) of E, there is an open neighborhood U of (x,t) in $M \times \mathbf{R}$ on which E restricts to a diffeomorphism onto an open subset $\tilde{U} = E(U)$ in S^{n+1}. We define a function $\tau : \tilde{U} \to \mathbf{R}$ by

$$\tau(p) = \theta_1 - \pi_2(E^{-1}(p)), \tag{3.91}$$

where π_2 is projection onto the second coordinate. That is, if $p = E(x,t)$, then

$$\tau(p) = \theta_1 - t. \tag{3.92}$$

Then we define a function $V : \tilde{U} \to \mathbf{R}$ by,

$$V(p) = \cos(g\tau(p)). \tag{3.93}$$

Clearly, τ and V are constant on each parallel hypersurface $M_t = f_t(M)$ in $\tilde{U}$.

The number $\tau(p)$ is the oriented distance from p to the first focal point along the normal geodesic to the parallel hypersurface M_t through p. Thus, if we begin the construction with a parallel hypersurface M_t near M rather than with M itself, we get the same functions τ and V. If we begin the construction with the opposite field of unit normals $-\xi$ instead of ξ, then we obtain the function $-V$ instead of V.

We next extend V to a homogeneous function of degree g on the cone in $\mathbf{R}^{n+2}$ over $\tilde{U}$ by the formula

$$F(rp) = r^g \cos(g(\tau(p)), \; p \in \tilde{U}, \; r > 0. \tag{3.94}$$

The first step in the proof of Theorem 3.32 is to show that the function F in equation (3.94) satisfies the Cartan–Münzner differential equations (3.86)–(3.87). One then completes the proof of Theorem 3.32 by showing that F is the restriction to the cone over $\tilde{U}$ of a homogeneous polynomial of degree g. These two steps involve lengthy calculations based on the formula for the principal curvatures of an isoparametric hypersurface given in Theorem 3.26.

Remark 3.34 (Consequences of Theorem 3.32). Before giving the proof of Theorem 3.32, we will make a few remarks concerning some important consequences of the theorem. From equation (3.88), we see that the range of the restriction V of F to S^{n+1} is contained in the closed interval $[-1, 1]$, since the left side of the equation is nonnegative. We can see that the range of V is all of the interval $[-1, 1]$ as follows. Since V is not constant on S^{n+1}, it has distinct maximum and minimum values on S^{n+1}. By equation (3.88) these maximum and minimum values are 1 and -1, respectively, since $\text{grad}^S V$ is nonzero at any point where V is not equal to ± 1. For any s in the open interval $(-1, 1)$, the level set $V^{-1}(s)$ is a compact hypersurface, since $\text{grad}^S V$ is never zero on $V^{-1}(s)$. Münzner also proves (see Theorem 3.44 in Section 3.6) that each level set of V is connected, and therefore, the original connected isoparametric hypersurface M is contained in a unique compact, connected isoparametric hypersurface.

For $s = \pm 1$, $\text{grad}^S V$ is identically equal to zero on $V^{-1}(s)$, and the sets $M_+ = V^{-1}(1)$ and $M_- = V^{-1}(-1)$ are submanifolds of codimension greater than one in S^{n+1}. We will show in Section 3.6 that M_+ and M_- are connected, and that they are the focal submanifolds of any isoparametric hypersurface $V^{-1}(s)$, $-1 < s < 1$, in the family of isoparametric hypersurfaces. Thus, there are only two focal submanifolds regardless of the number g of distinct principal curvatures. By Theorem 3.26, there are $2g$ focal points evenly distributed along each normal geodesic to the family $\{V^{-1}(s)\}$ of isoparametric hypersurfaces. We will see in Section 3.6 that these focal points lie alternately on the two focal submanifolds M_+ and M_-.

We now consider the isoparametric hypersurface $V^{-1}(0)$. From equation (3.93), we see that the function τ equals $\pi/2g$ on $V^{-1}(0)$. The function τ is the distance from a point x in $V^{-1}(0)$ to the first focal point along the normal geodesic through x. By Theorem 3.26, this means that the largest principal curvature of $V^{-1}(0)$ is $\cot(\pi/2g)$, and the principal curvatures of $V^{-1}(0)$ are given by $\cot\theta_i$, where

$$\theta_i = \frac{\pi}{2g} + \frac{(i-1)}{g}\pi, \quad 1 \le i \le g, \tag{3.95}$$

with multiplicities satisfying $m_{i+2} = m_i$ (subscripts mod g).

Remark 3.35 (Transnormal systems). A parallel family of isoparametric hypersurfaces in S^{n+1} together with its focal submanifolds forms what Bolton [45] defines to be a *transnormal system* in that any geodesic in S^{n+1} meets the submanifolds in the family orthogonally at either none or all of its points. With just this geometric hypothesis, Bolton recovered many facts about an isoparametric family, in particular, that there are only two submanifolds of codimension greater than one in a transnormal system that contains at least one hypersurface. See the recent papers of Miyaoka [375, 376] for more on transnormal systems.

Proof of Münzner's Theorem

In the remainder of this section, we will give a proof of Münzner's Theorem 3.32. Our proof follows the proof in the book [95, pp. 256–267] closely. That proof is based on Münzner's original proof [381, pp. 62–65] (see also Ferus [159, pp. 18–20]).

Let $p = \cos t\ x + \sin t\ \xi(x)$ be a point in the open set $\tilde{U}$ defined above. As in equation (3.42), the vector,

$$\tilde{\xi}(p) = -\sin t\, x + \cos t\, \xi(x),$$

is a unit normal vector to the parallel hypersurface $M_t = f_t(M)$ at p. Moreover,

$$\tilde{\xi}(p) = (E_*)_{(x,t)}\left(\frac{\partial}{\partial t}\right), \tag{3.96}$$

and it is easy to check that

$$\tilde{\xi}(p) = -\mathrm{grad}^S \tau. \tag{3.97}$$

If z is point in the cone over $\tilde{U}$, we see from equation (3.94) that

$$F(z) = |z|^g \cos g(\tau(z/|z|)).$$

Thus, if we define $\sigma : \mathbf{R}^{n+2} - \{0\} \to S^{n+1}$ by $\sigma(z) = z/|z| = z/r$, then we can re-write the equation above in terms of functions as

$$F = r^g \cos g(\tau \circ \sigma). \tag{3.98}$$

One can easily compute that $\mathrm{grad}^E r = \sigma$. Further, if $Z \in T_z\mathbf{R}^{n+2}$, then

$$\langle Z, \mathrm{grad}^E(\tau \circ \sigma)\rangle = \tau_*\sigma_* Z.$$

Then we compute

$$\sigma_*(Z) = \frac{1}{r}(Z - \langle Z, \sigma\rangle\sigma), \tag{3.99}$$

and $\tau_*(\sigma_* Z) = \langle \mathrm{grad}^S \tau, \sigma_* Z\rangle$. Using equations (3.97) and (3.99) and the fact that $\langle \mathrm{grad}^S \tau, \sigma\rangle = 0$, we get

$$\tau_*(\sigma_* Z) = -\frac{1}{r}\langle \tilde{\xi} \circ \sigma, Z\rangle.$$

Thus, we get

$$\text{grad}^E(\tau \circ \sigma) = -\frac{1}{r}\,\tilde{\xi} \circ \sigma. \tag{3.100}$$

Hence, we have

$$\text{grad}^E F = gr^{g-1}(\cos g(\tau \circ \sigma)\sigma + \sin g(\tau \circ \sigma)\tilde{\xi} \circ \sigma). \tag{3.101}$$

Since $\sigma(z)$ and $\tilde{\xi}(\sigma(z))$ are orthonormal at $\sigma(z)$, we get

$$|\text{grad}^E F|^2 = g^2 r^{2g-2}, \tag{3.102}$$

and so equation (3.86) is satisfied.

Next we compute the Laplacian $\Delta^E F$. This is a lengthy computation using the formula for the principal curvatures of M obtained in Theorem 3.26. From equation (3.101), we have $\text{grad}^E F = gr^{g-1} W$, where W is the vector field

$$W = \cos g(\tau \circ \sigma)\sigma + \sin g(\tau \circ \sigma)\tilde{\xi} \circ \sigma.$$

In the rest of the proof, all gradients are with respect to Euclidean space $\mathbf{R}^{n+2}$, and we often omit the superscript E in the notation for gradients. By definition,

$$\Delta^E F = \text{div grad}^E F = \langle \text{grad}\,(gr^{g-1}), W\rangle + gr^{g-1}\text{div}\, W. \tag{3.103}$$

We compute the first term on the right side of the equation above to be

$$\begin{aligned}\langle \text{grad}\,(gr^{g-1}), W\rangle &= g(g-1)r^{g-2}\langle \text{grad}\, r, W\rangle = g(g-1)r^{g-2}\langle \sigma, W\rangle \\ &= g(g-1)r^{g-2}\cos g(\tau \circ \sigma).\end{aligned} \tag{3.104}$$

The last term on the right side of equation (3.103) is

$$\begin{aligned}\text{div}\, W = {}& \langle \text{grad}\ \cos g(\tau \circ \sigma), \sigma\rangle + \langle \text{grad}\ \sin g(\tau \circ \sigma), \tilde{\xi} \circ \sigma\rangle \\ &+ \cos g(\tau \circ \sigma)\ \text{div}\, \sigma + \sin g(\tau \circ \sigma)\ \text{div}\,(\tilde{\xi} \circ \sigma).\end{aligned} \tag{3.105}$$

We will handle the terms on the right side of equation (3.105) one at a time. Using equation (3.100), we compute the first term as follows

$$\begin{aligned}\langle \text{grad}\ \cos g(\tau \circ \sigma), \sigma\rangle &= -g \sin g(\tau \circ \sigma)\langle \text{grad}\ \tau \circ \sigma, \sigma\rangle \\ &= -g \sin g(\tau \circ \sigma)\langle (-1/r)\tilde{\xi} \circ \sigma, \sigma\rangle = 0.\end{aligned} \tag{3.106}$$

Similarly, the second term on the right side of equation (3.105) is

$$\begin{aligned}\langle \text{grad}\ \sin g(\tau \circ \sigma), \tilde{\xi} \circ \sigma\rangle &= g \cos g(\tau \circ \sigma)\langle (-1/r)\tilde{\xi} \circ \sigma, \tilde{\xi} \circ \sigma\rangle \\ &= (-g/r) \cos g(\tau \circ \sigma).\end{aligned} \tag{3.107}$$

For the third term on the right side of equation (3.105), we compute

$$\mathrm{div}\,\sigma = -\frac{1}{r^2}\langle \mathrm{grad}\, r, z\rangle + \frac{1}{r}\,\mathrm{div}\, z = -\frac{1}{r} + \frac{n+2}{r} = \frac{n+1}{r}. \tag{3.108}$$

For the final term in equation (3.105), we need the following lemma.

Lemma 3.36. (a) $\mathrm{div}\,(\tilde{\xi}\circ\sigma) = \frac{1}{r}(\mathrm{div}\,\tilde{\xi})\circ\sigma$,
(b) $\mathrm{div}\,\tilde{\xi} = -\sum_{i=1}^{g} m_i \cot(\tau + \frac{(i-1)}{g}\pi)$,
(c) $\mathrm{div}\,\tilde{\xi} = -n\cot g\tau - \frac{(m_1-m_2)g}{2\sin g\tau}$.

Proof. (a) Let Z be any tangent vector to $\mathbf{R}^{n+2}$ at an arbitrary point z in the cone over $\tilde{U}$, and let z_u be a curve in $\mathbf{R}^{n+2}$ with initial position z and initial tangent vector Z. Then the Euclidean covariant derivative $D_Z(\tilde{\xi}\circ\sigma)$ equals the initial tangent vector $\overrightarrow{\alpha_0}$ to the curve $\alpha_u = (\tilde{\xi}\circ\sigma)(z_u)$. Note that $\overrightarrow{\alpha_0} = D_Y\tilde{\xi}$, where $Y = \sigma_* Z$. From equation (3.99), we have

$$\sigma_* Z = (Z - \langle Z, \sigma\rangle\sigma)/r.$$

Thus, for $Z = z/r$, we have $\sigma_* Z = 0$, while for Z parallel to a tangent vector to S^{n+1} at $\sigma(z) = z/r$, we have $\sigma_* Z = Z/r$. Hence, we have

$$D_Z(\tilde{\xi}\circ\sigma) = 0,\ \text{for } Z = \frac{z}{r},$$

$$D_Z(\tilde{\xi}\circ\sigma) = \frac{1}{r}(D_Z\tilde{\xi})\circ\sigma,$$

for Z parallel to a tangent vector to S^{n+1} at $\sigma(z)$. Using this we can compute

$$\mathrm{div}\,(\tilde{\xi}\circ\sigma) = \frac{1}{r}\sum_{i=1}^{n+1}\langle (D_{e_i}\tilde{\xi})\circ\sigma, e_i\rangle e_i,$$

where $e_1,\ldots,e_{n+1}$ are orthonormal and tangent to S^{n+1} at $\sigma(z)$. Thus, we have

$$\mathrm{div}\,(\tilde{\xi}\circ\sigma) = \frac{1}{r}(\mathrm{div}\,\tilde{\xi})\circ\sigma,$$

and part (a) of the lemma is proven.

(b) In proving part (b) of the lemma, we make use of formula (3.56) in Theorem 3.26 for the principal curvatures of an isoparametric hypersurface. Let $\tilde{\nabla}$ be the Levi-Civita connection on S^{n+1}, and let $p = E(x,t)$ be a point in $\tilde{U}$. Let $\{e_1,\ldots,e_n\}$ be an orthonormal basis for the tangent space T_pM_t, where $M_t = f_tM$. Then

$$\operatorname{div} \tilde{\xi} = \sum_{i=1}^{n} \langle \tilde{\nabla}_{e_i} \tilde{\xi}, e_i \rangle + \langle \tilde{\nabla}_{\tilde{\xi}} \tilde{\xi}, \tilde{\xi} \rangle.$$

The last term in the equation above is zero, since $\tilde{\xi}$ has constant length. If we choose the e_i to be principal vectors for the shape operator A_t of M_t, then we have

$$\operatorname{div} \tilde{\xi} = -\sum_{i=1}^{n} \langle A_t e_i, e_i \rangle = -\operatorname{trace} A_t.$$

Using Corollary 3.19 and Theorem 3.26, we get

$$\begin{aligned} \operatorname{div} \tilde{\xi} &= -\sum_{i=1}^{g} m_i \cot(\theta_i - t) = -\sum_{i=1}^{g} m_i \cot((\theta_i - \theta_1) - (t - \theta_1)) \\ &= -\sum_{i=1}^{g} m_i \cot(\tau + \frac{(i-1)}{g}\pi). \end{aligned} \tag{3.109}$$

(c) We know that either all of the multiplicities m_i are equal or else there exist two distinct multiplicities m_1 and m_2. In the case where all of the multiplicities are equal, the formula above for $\operatorname{div} \tilde{\xi}$ becomes

$$\begin{aligned} -\sum_{i=1}^{g} m_i \cot(\tau + \frac{(i-1)}{g}\pi) &= -\frac{n}{g} \sum_{i=1}^{g} \cot(\tau + \frac{(i-1)}{g}\pi) \\ &= -\frac{n}{g}\, g \cot g\tau = -n \cot g\tau, \end{aligned}$$

which proves (c) in this case.

In the case where there are two distinct multiplicities, we know that g is even and that $m_1 = m_3 = \cdots = m_{g-1}$, while $m_2 = m_4 = \cdots = m_g$. For simplicity, we set $g = 2l$. Then we compute the last line in equation (3.109) to be

$$\sum_{i=1}^{g} m_i \cot(\tau + \frac{(i-1)}{g}\pi) \tag{3.110}$$

$$\begin{aligned} &= m_1 \sum_{j=1}^{l} \cot(\tau + \frac{(j-1)}{l}\pi) + m_2 \sum_{j=1}^{l} \cot(\tau + \frac{\pi}{2l} + \frac{(j-1)}{l}\pi) \\ &= m_1\, l \cot(l\tau) + m_2\, l \cot(l\tau + \frac{\pi}{2}) = m_1\, l \cot(l\tau) - m_2\, l \tan(l\tau). \end{aligned}$$

Then we can find ω so that

$$\cos^2 \omega = m_1/(m_1 + m_2), \quad \sin^2 \omega = m_2/(m_1 + m_2).$$

The total number n of principal curvatures of M is $(m_1 + m_2)g/2$. Since $l = g/2$, we have $n = (m_1 + m_2)l$, and the last expression in equation (3.110) can be written as

$$\begin{aligned}
m_1\, l \cot l\tau - m_2\, l \tan l\tau &= n\,(\cos^2\omega \cot l\tau - \sin^2\omega \tan l\tau) \\
&= n\,\frac{\cos^2\omega(1+\cos g\tau) - \sin^2\omega(1-\cos g\tau)}{\sin g\tau} \\
&= n\,\frac{(\cos 2\omega + \cos g\tau)}{\sin g\tau} = n\cot g\tau + \frac{g(m_1 - m_2)}{2\sin g\tau},
\end{aligned}$$

and the lemma is proven. □

Remark 3.37. Since $\mathrm{grad}^S \tau = -\tilde{\xi}$, part (c) of the lemma implies that

$$\Delta^S \tau = n\cot g\tau + \frac{(m_1 - m_2)g}{2\sin g\tau}.$$

We can now complete the calculation of $\Delta^E F$. From equations (3.103)–(3.108) and Lemma 3.36, we get

$$\begin{aligned}
\Delta^E F &= g(g-1)r^{g-2}\cos g(\tau\circ\sigma) + gr^{g-1}\left(-\frac{g}{r}\cos g(\tau\circ\sigma)\right) \\
&\quad + gr^{g-1}\cos g(\tau\circ\sigma)\left(\frac{n+1}{r}\right) \\
&\quad + gr^{g-1}\sin g(\tau\circ\sigma)\left(\frac{1}{r}\right)\left(-n\cot g(\tau\circ\sigma) - \frac{(m_1 - m_2)g}{2\sin g(\tau\circ\sigma)}\right) \\
&= gr^{g-2}\cos g(\tau\circ\sigma)(g-1-g+n+1-n) + gr^{g-2}\frac{(-(m_1-m_2))g}{2} \\
&= g^2 r^{g-2}\frac{(m_2 - m_1)}{2}.
\end{aligned}$$

Proof that F is a homogeneous polynomial of degree g

This completes the proof that the function F defined on the cone over the open set $\tilde{U} \subset S^{n+1}$ satisfies the Cartan–Münzner differential equations (3.86)–(3.87). In order to complete the proof of Münzner's Theorem 3.32, we must show that F is the restriction to the cone over $\tilde{U}$ of a homogeneous polynomial of degree g on $\mathbf{R}^{n+2}$. We first need the following elementary lemma.

Lemma 3.38. *$\Delta r^k = k(k+n)r^{k-2}$ on $\mathbf{R}^{n+2}$ for any positive integer k.*

Proof. We first compute that $\mathrm{grad}\, r^k = kr^{k-1}\,\mathrm{grad}\, r = kr^{k-1}\sigma$, where $\sigma(z) = z/r$, as defined above. Then

$$\begin{aligned}\Delta r^k &= \text{div grad } r^k = k\langle \text{grad } r^{k-1}, \sigma\rangle + kr^{k-1}\text{div } \sigma \\ &= k(k-1)r^{k-2}\langle \sigma, \sigma\rangle + kr^{k-1}(n+1)/r \\ &= k(k+n)r^{k-2}.\end{aligned}$$

□

Consider the function $G = F - ar^g$ on the cone over $\tilde{U}$, where

$$a = \frac{g(m_2 - m_1)}{2(g+n)}.$$

From equation (3.87) for the Laplacian of F and by Lemma 3.38, we see that G is a harmonic function. Thus, all the partial derivatives of G of all orders are also harmonic. Next we show that the Laplacian of order g of the function $|\text{grad } G|^2$ is zero.

Lemma 3.39. $\Delta^g |\text{grad } G|^2 = 0$, *where* $\Delta^1 = \Delta$, *and* $\Delta^k = \Delta \circ \Delta^{k-1}$.

Proof. Since $G = F - ar^g$, we compute grad G = grad $F - agr^{g-1}\sigma$, as in the preceding lemma. Then

$$|\text{grad } G|^2 = |\text{grad } F|^2 + a^2g^2r^{2g-2} - 2agr^{g-1}\langle \text{grad } F, \sigma\rangle.$$

Using equation (3.98) for F, equation (3.101) for grad F, and equation (3.102) for $|\text{grad } F|^2$, the equation above yields

$$\begin{aligned}|\text{grad } G|^2 &= g^2r^{2g-2}(1+a^2) - (2agr^{g-1})(gr^{g-1}\cos g(\tau\circ\sigma)) \\ &= g^2r^{2g-2}(1+a^2) - 2ag^2r^{g-2}F.\end{aligned}$$

Thus, we have since $F = G + ar^g$,

$$\begin{aligned}|\text{grad } G|^2 + 2ag^2r^{g-2}G &= g^2r^{2g-2}(1+a^2) - 2ag^2r^{g-2}ar^g \\ &= g^2r^{2g-2}(1-a^2).\end{aligned} \tag{3.111}$$

Now using Lemma 3.38, we can compute

$$\begin{aligned}\Delta(r^kG) &= \text{div } (r^k \text{ grad } G) + \text{div } (G \text{ grad } r^k) \\ &= 2\langle \text{grad } r^k, \text{grad } G\rangle + r^k\Delta G + G(\Delta r^k) \\ &= 2kr^{k-1}\langle \sigma, \text{grad } F - agr^{g-1}\sigma\rangle + k(k+n)r^{k-2}G \\ &= 2kgr^{g+k-2}\cos g(\tau\circ\sigma) - 2kagr^{k+g-2} + k(k+n)r^{k-2}G \\ &= 2kgr^{k-2}(F - ar^g) + k(k+n)r^{k-2}G \\ &= 2kgr^{k-2}G + k(k+n)r^{k-2}G \\ &= kr^{k-2}G(2g + (k+n)).\end{aligned}$$

Consider the expression in equation (3.111). If the multiplicities m_1 and m_2 are not equal, then g is even, and the calculation above shows that $g/2$ applications of Δ will reduce the term $2ag^2r^{g-2}G$ on the left side of equation (3.111) to zero. The term $g^2r^{2g-2}(1-a^2)$ on the right side of equation (3.111) will be reduced to zero by g applications of Δ, and so the lemma follows in that case.

On the other hand, if all the multiplicities are equal, then $a = 0$, and the term $2ag^2r^{g-2}G$ on the left side of equation (3.111) vanishes. As in the previous case, g applications of Δ reduce the right side of the equation to zero, and so the lemma follows in this case also. □

Next we compute the following formula for $\Delta^g|\text{grad } G|^2$.

Lemma 3.40. *For any harmonic function G,*

$$\Delta^g|\text{grad } G|^2 = 2^g \sum \left(\frac{\partial^{g+1} G}{\partial x_{i_1} \partial x_{i_2} \cdots \partial x_{i_{g+1}}} \right)^2,$$

where the sum takes place over all $(g+1)$*-tuples* $(i_1, \ldots, i_{g+1})$ *with each* i_j *in* $\{1, \ldots, n+2\}$. *(The* i_j*'s are not necessarily distinct from one another.)*

Proof. We have

$$|\text{grad } G|^2 = \sum_{i=1}^{n+2} \left(\frac{\partial G}{\partial x_i} \right)^2.$$

For any function $f : \mathbf{R}^{n+2} \to \mathbf{R}$, we have the formula

$$\Delta f^2 = 2\,|\text{grad} f|^2 + 2f\Delta f.$$

Using this and the fact that $\Delta G = 0$, we get

$$\Delta|\text{grad } G|^2 = \sum_{i=1}^{n+2} 2 \left| \text{grad } \frac{\partial G}{\partial x_i} \right|^2 = \sum_{i=1}^{n+2} \sum_{j=1}^{n+2} \left(\frac{\partial^2 G}{\partial x_i \partial x_j} \right)^2.$$

For the rest of the proof, the sums are all from 1 to $n+2$. Using the fact that all of the partial derivatives of G are harmonic, we repeat the step above to get

$$\begin{aligned} \Delta^2|\text{grad } G|^2 &= 2^2 \sum \sum \left| \text{grad } \frac{\partial^2 G}{\partial x_i \partial x_j} \right|^2 \\ &= 2^2 \sum \sum \sum \left(\frac{\partial^3 G}{\partial x_i \partial x_j \partial x_k} \right)^2. \end{aligned}$$

Continuing this process, we eventually obtain

$$\Delta^g |\text{grad } G|^2 = 2^g \sum \left(\frac{\partial^{g+1} G}{\partial x_{i_1} \partial x_{i_2} \cdots \partial x_{i_{g+1}}} \right)^2,$$

with the sum over all $(g+1)$-tuples $(i_1, \ldots, i_{g+1})$ with each $i_j \in \{1, \ldots, n+2\}$. □

Combining the results of Lemmas 3.39 and 3.40, we get that all partial derivatives of G of order $g+1$ are zero. Further, we can see that $\Delta^{g-1}|\text{grad } G|^2$ is a nonzero constant as follows. If the quantity $a = g(m_2 - m_1)/2(g+n)$ equals zero, then we see from equation (3.111) that

$$|\text{grad } G|^2 = g^2 r^{2g-2},$$

and

$$\Delta^{g-1}|\text{grad } G|^2 = (2g)^{g-1}(g-1)!(n+2(g-1))(n+2(g-2))\cdots(n+2), \qquad (3.112)$$

which is a nonzero constant. On the other hand, if $a \neq 0$, then g is even, say $g = 2l$, and the second term on the left side of equation (3.111) is handled by noting that $\Delta^l(r^{g-2}G) = 0$, so that $\Delta^{g-1}|\text{grad } G|^2$ is equal to $1 - a^2$ times the term on the right side of equation (3.112), and so it is still a nonzero constant, since $|a| < 1$.

Thus G is a polynomial of degree g. Since G is a homogeneous function on the cone over $\tilde{U}$, it is a homogeneous polynomial of degree g. This completes the proof of Münzner's Theorem 3.32.

A converse result to Münzner's Theorem

Münzner also proved a result in the converse direction by showing that any homogeneous polynomial on $\mathbf{R}^{n+2}$ that satisfies the differential equations (3.86)–(3.87) is related to the Cartan–Münzner polynomial of an isoparametric hypersurface in a very specific way given in the following theorem.

Theorem 3.41. *Let $\tilde{F} : \mathbf{R}^{n+2} \to \mathbf{R}$ be a homogeneous polynomial of degree $\tilde{g}$ which satisfies the Cartan–Münzner differential equations (3.86)–(3.87) with parameters $\tilde{g}$ and $\tilde{c}$, such that the restriction $\tilde{V}$ of $\tilde{F}$ to S^{n+1} is not constant. Then zero is a regular value of $\tilde{V}$ and $\tilde{V}^{-1}(0)$ is an oriented isoparametric hypersurface with normal field* $\text{grad}^S \tilde{V}$. *Let F be the Cartan–Münzner polynomial of a connected component of $\tilde{V}^{-1}(0)$. Then either $\tilde{F} = F$ or $\tilde{F} = \pm(2F^2 - r^{2g})$, in which case $c = 0$, $\tilde{g} = 2g$ and $\tilde{c} = \mp \tilde{g}n$.*

Remark 3.42. There exist other polynomials which when restricted to S^{n+1} have a family of isoparametric hypersurfaces and their focal submanifolds as level sets. For example, the polynomial G in Example 3.31 (b) defined on $\mathbf{R}^{n+2} = \mathbf{R}^{p+1} \times \mathbf{R}^{q+1}$

by $G(x, y) = |x|^2$ has the same family of level sets in S^{n+1} as the Cartan–Münzner polynomial $F(x, y) = |x|^2 - |y|^2$, since $F = 2G - 1$ on S^{n+1}. However, G does not satisfy the Cartan–Münzner differential equations.

Proof (of Theorem 3.41). Suppose that $\tilde{F}$ is a homogeneous polynomial of degree $\tilde{g}$ that satisfies the Cartan–Münzner differential equations,

$$|\mathrm{grad}^E \tilde{F}|^2 = \tilde{g}^2 r^{2\tilde{g}-2}, \tag{3.113}$$

$$\Delta^E \tilde{F} = \tilde{c} r^{\tilde{g}-2}, \tag{3.114}$$

for some constant $\tilde{c}$, such that the restriction $\tilde{V}$ of $\tilde{F}$ to S^{n+1} is not constant. By Theorem 3.30, $\tilde{V}$ satisfies the differential equations

$$|\mathrm{grad}^S \tilde{V}|^2 = \tilde{g}^2(1 - \tilde{V}^2), \tag{3.115}$$

$$\Delta^S \tilde{V} = \tilde{c} - \tilde{g}(n + \tilde{g})\tilde{V}. \tag{3.116}$$

Since $\tilde{V}$ is not constant on S^{n+1}, it has a maximum value of 1 and a minimum value of -1 by equation (3.115), as noted in Remark 3.34.

Let M_0 be a connected component of the level hypersurface $\tilde{V}^{-1}(0)$. Let $x_0 \in M_0$, and let U_0 be a neighborhood of x_0 in S^{n+1} on which $|\tilde{V}| \neq 1$. By Corollary 3.7, M_0 has constant principal curvatures, and so we can construct the Cartan–Münzner polynomial F of M_0, as in equations (3.92)–(3.94). In particular, we can define the function τ as in equation (3.92) on U_0. By Theorem 3.26, we have $\tau(x_0) = \theta_1$, where $0 < \theta_1 < \pi/g$, where g is the number of distinct principal curvatures of M_0. By equation (3.97), we have for each $x \in M_0 \cap U_0$,

$$(\mathrm{grad}\ \tau)_p = -\tilde{\xi}_p, \tag{3.117}$$

where $\tilde{\xi}_p$ is a unit normal at the point $p = f_t(x)$ to the parallel hypersurface $f_t(M_0)$, where $t = \theta_1 - \tau(p)$. That is,

$$\tilde{\xi}_p = -\sin\,(\theta_1 - \tau(p))\, x + \cos\,(\theta_1 - \tau(p))\, \xi_x, \tag{3.118}$$

where we may assume for convenience that the unit normal ξ_x to M_0 at x is given by

$$\xi_x = \frac{\mathrm{grad}\ \tilde{V}(x)}{|\mathrm{grad}\ \tilde{V}(x)|} = \frac{\mathrm{grad}\ \tilde{V}(x)}{\tilde{g}}, \tag{3.119}$$

since $|\mathrm{grad}\ \tilde{V}(x)| = \tilde{g}$ for $x \in M_0$ by equation (3.115).

By the procedure in the proof of Theorem 3.32, the function τ leads to the Cartan–Münzner polynomial F of M_0, whose restriction to S^{n+1} will be denoted by V. We now compare the two functions V and $\tilde{V}$. Note that on U_0, we have $\tilde{\tau} = (\cos^{-1}\ \tilde{V})/\tilde{g}$. Then we compute

$$\text{grad } \tilde{\tau} = \frac{1}{\tilde{g}} \left(\frac{-1}{(1-\tilde{V}^2)^{1/2}} \right) \text{grad } \tilde{V}. \tag{3.120}$$

Since $|\text{grad } \tilde{V}| = \tilde{g}(1-\tilde{V}^2)^{1/2}$ by equation (3.115), the right side of equation (3.120) is a unit vector field. Furthermore, it is normal to the appropriate level hypersurface of $\tilde{V}$ at each point $z \in U_0$. By Theorem 3.5 on page 89, each level hypersurface of $\tilde{V}$ is a parallel hypersurface of M_0, and thus its unit normal field coincides with the normal field $\tilde{\xi}$ defined above. Hence, if $p = f_t(x)$ for $x \in M_0$ as above, we have

$$\tilde{\xi}_p = \frac{1}{\tilde{g}} \frac{\text{grad } \tilde{V}(p)}{(1-\tilde{V}^2(p))^{1/2}}. \tag{3.121}$$

By equations (3.117), (3.120), and (3.121), the functions τ and $\tilde{\tau}$ have the same gradient, and therefore they differ by a constant on U_0, i.e., $\tilde{\tau} = \tau + a$, where $|a|$ is less than the maximum of $\{\pi/g, \pi/\tilde{g}\}$.

Since grad τ = grad $\tilde{\tau} = -\tilde{\xi}$, it follows from Remark 3.37 that

$$\Delta\tilde{\tau} = \Delta\tau = \frac{1}{\sin\, g\tau} \left(n \cos g\tau - \frac{c}{g} \right), \tag{3.122}$$

where $c = (m_2 - m_1)g^2/2$, and m_1, m_2 have the usual meaning as multiplicities of the principal curvatures of M_0. On the other hand, using equations (3.115) and (3.120), we compute that

$$\begin{aligned} \Delta\tilde{\tau} &= \text{div grad } \tilde{\tau} \qquad (3.123) \\ &= \frac{-1}{\tilde{g}} (\langle \text{grad } (1-\tilde{V}^2)^{1/2}, \text{grad } \tilde{V} \rangle + (1-\tilde{V}^2)^{-1/2} \Delta\tilde{V}) \\ &= \frac{-1}{\tilde{g}} ((-1/2)((1-\tilde{V}^2)^{-3/2}(-2\tilde{V})|\text{grad } \tilde{V}|^2 + (1-\tilde{V}^2)^{-1/2} \Delta\tilde{V}) \\ &= -(1-\tilde{V}^2)^{-1/2} \tilde{g}\tilde{V} + (1-\tilde{V}^2)^{-1/2}(n+\tilde{g})\tilde{V} - \frac{\tilde{c}}{\tilde{g}}(1-\tilde{V}^2)^{-1/2} \\ &= (1-\tilde{V}^2)^{-1/2}(n\tilde{V} - \tilde{c}/\tilde{g}) \\ &= \frac{1}{\sin \tilde{g}\tilde{\tau}}(n \cos \tilde{g}\tilde{\tau} - \tilde{c}/\tilde{g}). \end{aligned}$$

Here $\tilde{g}$ is the degree of $\tilde{F}$, and $\tilde{c}$ is the constant given in equation (3.114). We know that $\tilde{\tau} = \tau + a$ for some constant a on U_0. Thus, by equating the two expressions for $\Delta\tilde{\tau}$ in equations (3.122) and (3.123), we get

$$\frac{n \cos\, g\tau - c/g}{\sin\, g\tau} = \frac{n \cos\, \tilde{g}(\tau + a) - \tilde{c}/\tilde{g}}{\sin\, \tilde{g}(\tau + a)}. \tag{3.124}$$

Since a whole interval of values of τ are covered along a suitable normal geodesic in U_0, the equation,

$$(n\cos\, g\tau - c/g)\sin\, \tilde{g}(\tau + a) = (n\cos\, \tilde{g}(\tau + a) - \tilde{c}/\tilde{g})\sin\, g\tau, \tag{3.125}$$

must hold as an identity in τ, since each side is an analytic expression in τ.

We now derive some consequences of equation (3.125). First note that the right side is zero at integral multiples of π/g, while since $n = (m_1 + m_2)g/2$, we get

$$n\cos\, g\tau - \frac{c}{g} = \pm\frac{(m_1 + m_2)}{2}g - \frac{(m_2 - m_1)}{2}g \neq 0, \tag{3.126}$$

when τ is an integral multiple of π/g. Thus $\sin\, g(\tau + a)$ vanishes at such points. Therefore, for any integer k, we have

$$\tilde{g}\left(\frac{k\pi}{g} + a\right) = l\pi, \tag{3.127}$$

for some integer l. If we set $k = 0$, we see that a is an integral multiple of $\pi/\tilde{g}$. If we set $k = 1$, we get that π/g is an integral multiple of $\pi/\tilde{g}$. Thus, we have $\tilde{g} = \alpha g$ for some positive integer α. The restriction on $|a|$ reduces to $|a| < \pi/g$, since $g \leq \tilde{g}$.

We now examine the possibilities for the positive integer α. If $\alpha = 1$, then we have $a = 0$, so that $\tau = \tilde{\tau}$, and thus $V = \tilde{V}$, $F = \tilde{F}$ and $c = \tilde{c}$. Suppose now that $\alpha > 1$. Substituting $a = l\pi/\tilde{g}$ for an integer l into equation (3.125) yields

$$(n\cos\, g\tau - c/g)\sin\,(\alpha g\tau + l\pi) = (n\cos\,(\alpha g\tau + l\pi) - \tilde{c}/\tilde{g})(\sin\, g\tau).$$

Thus, we have either

$$\frac{\sin\, \alpha g\tau}{\sin\, g\tau} = \frac{n\cos\, \alpha g\tau - \tilde{c}/\tilde{g}}{n\cos\, g\tau - c/g}, \tag{3.128}$$

or

$$\frac{\sin\, \alpha g\tau}{\sin\, g\tau} = \frac{n\cos\, \alpha g\tau + \tilde{c}/\tilde{g}}{n\cos\, g\tau - c/g}, \tag{3.129}$$

depending on whether l is even or odd. Set $\tau = \pi/\alpha g$. If (3.128) holds, then we have

$$-n - \tilde{c}/\tilde{g} = 0, \text{ i.e., } \tilde{c} = -\alpha g n, \text{ for } l \text{ even}. \tag{3.130}$$

Similarly, if equation (3.129) holds, we get

$$\tilde{c} = \alpha g n, \text{ for } l \text{ odd}. \tag{3.131}$$

On the other hand, taking $\tau = 2\pi/\alpha g$ in equation (3.128) yields

$$(\sin\, 2\pi/\alpha)(n - \tilde{c}/\tilde{g}) = 0, \text{ for } l \text{ even}, \tag{3.132}$$

and equation (3.129) gives

$$(\sin\, 2\pi/\alpha)(n + \tilde{c}/\tilde{g}) = 0, \text{ for } l \text{ odd}. \tag{3.133}$$

We see that equations (3.132) and (3.133) contradict equations (3.130) and (3.131), unless $\sin\, 2\pi/\alpha = 0$, i.e., $\alpha = 2$. Thus, $\tilde{g} = 2g$. Then the equation

$$|a| = |l\pi/2g| < \pi/g \tag{3.134}$$

implies that $l = 0, 1$ or -1.

First consider $l = 0$. Then we have from equations (3.128) and (3.130) that

$$\frac{\sin\, 2g\tau}{\sin\, g\tau} = \frac{n(\cos\, 2g\tau + 1)}{n\cos\, g\tau - c/g}, \tag{3.135}$$

which, using half-angle formulas, gives

$$2\cos\, g\tau = \frac{2n\cos^2 g\tau}{n\cos\, g\tau - c/g}. \tag{3.136}$$

Thus, we get $c = 0$, $\tilde{c} = -2ng = -n\tilde{g}$, $\tau = \tilde{\tau}$ and

$$\tilde{V} = \cos\, \tilde{g}\tilde{\tau} = \cos\, 2g\tau = 2\cos^2 g\tau - 1 = 2V^2 - 1. \tag{3.137}$$

There is at most one homogeneous polynomial that extends $\tilde{V}$, namely, the polynomial $2F^2 - r^{2g}$, so we have $\tilde{F} = 2F^2 - r^{2g}$ in that case.

Next we consider the possibility that $l = \pm 1$. In that case, equations (3.129) and (3.131) imply that

$$\frac{\sin\, 2g\tau}{\sin\, g\tau} = \frac{n(\cos 2g\tau + 1)}{n\cos\, g\tau - c/g}, \tag{3.138}$$

so that $c = 0$, $\tilde{c} = 2ng = n\tilde{g}$, and $\tilde{\tau} = \tau \pm \pi/2g$. Then, we get

$$\tilde{V} = \cos\, \tilde{g}\tilde{\tau} = \cos\, (2g\tau \pm \pi) = -\cos\, 2g\tau = -(2V^2 - 1). \tag{3.139}$$

In that case, we have $\tilde{F} = -(2F^2 - r^{2g})$, as stated in the theorem. In both of the last two cases, the conditions $c = 0$, $\tilde{g} = 2g$ and $\tilde{c} = \pm\tilde{g}n$ are satisfied. □

Corollary 3.43. *Let $F : \mathbf{R}^{n+2} \to \mathbf{R}$ be a homogeneous polynomial of degree g which satisfies the Cartan–Münzner differential equations (3.86)–(3.87) with parameters g and c. Suppose that the restriction V of F to S^{n+1} is not constant. Then the following statements are equivalent.*

(a) *F is the Cartan–Münzner polynomial of a connected, oriented isoparametric hypersurface.*
(b) $c \neq \pm gn$.
(c) *F is the Cartan–Münzner polynomial of each component of* $V^{-1}(0)$.

Proof. Suppose that F is the Cartan–Münzner polynomial of a connected, oriented isoparametric hypersurface. We know that $n = (m_1 + m_2)g/2$, and so

$$\pm\, gn = \pm(m_1 + m_2)g^2/2. \tag{3.140}$$

Since F is a Cartan–Münzner polynomial, we have by Theorem 3.32 that

$$c = (m_2 - m_1)g^2/2. \tag{3.141}$$

Thus, $c = \pm gn$ is not possible, since m_1 and m_2 are nonzero. So (a) implies (b). Next suppose that (b) is true. By Theorem 3.41, F is the Cartan–Münzner polynomial of each connected component of $V^{-1}(0)$, so (c) is true. Finally, it is obvious that (c) implies (a). □

3.6 Global Structure Theorems

In this section, we discuss several results concerning the global structure of an isoparametric family of hypersurfaces in S^{n+1} that stem from Münzner's construction of the Cartan–Münzner polynomials as discussed in the previous section.

Let $F : \mathbf{R}^{n+2} \to \mathbf{R}$ be the Cartan–Münzner polynomial of degree g constructed from a connected isoparametric hypersurface in $S^{n+1} \subset \mathbf{R}^{n+2}$ with g distinct principal curvatures as in Theorem 3.32, and let V denote the restriction of F to S^{n+1}. As noted in Remark 3.34 on page 117, the range of the function V is the closed interval $[-1, 1]$, and each level set $M_t = V^{-1}(t), -1 < t < 1$, is an isoparametric hypersurface in S^{n+1}.

For the sake of concreteness, let $M = M_0 = V^{-1}(0)$ be the isoparametric hypersurface discussed in the previous section. Denote the two (possibly equal) multiplicities of M by $m_+ = m_1$ and $m_- = m_{-1}$. We next want to prove that M and indeed all of the hypersurfaces $M_t, -1 < t < 1$, as well as the two focal submanifolds, are connected.

For $x \in M$, we have $V(x) = 0$, and so the function τ in equation (3.92) satisfies $\tau(x) = \pi/2g$. Thus, the largest principal curvature of M is $\cot\theta_1 = \cot(\pi/2g)$ (see also Remark 3.34). The focal map $f_+ : M \to S^{n+1}$ onto the focal submanifold $f_+(M)$ is given by

$$f_+(x) = E\left(x, \frac{\pi}{2g}\right) = \cos\left(\frac{\pi}{2g}\right)x + \sin\left(\frac{\pi}{2g}\right)\xi(x), \tag{3.142}$$

where ξ is the field of unit normals to M, and E is the normal exponential map. The focal map f_+ factors through an immersion of the space of leaves M/T_1 into S^{n+1}, as in Theorems 2.12 and 2.14 (see page 26), and so $f_+(M)$ is a smooth submanifold of S^{n+1} of dimension $n - m_+$.

By Theorem 3.26 on page 108, we have $\lambda_g = \cot\theta_g$, where

$$\theta_g = \pi - (\pi/2g) = -\pi/2g \ (\mathrm{mod}\ \pi),$$

and so the focal map $f_- : M \to S^{n+1}$ onto the focal submanifold $f_-(M)$ is given by

$$f_-(x) = E\left(x, \frac{-\pi}{2g}\right) = \cos\left(\frac{-\pi}{2g}\right)x + \sin\left(\frac{-\pi}{2g}\right)\xi(x). \tag{3.143}$$

The focal map f_- factors through an immersion of the space of leaves M/T_g into S^{n+1}, and so $f_-(M)$ is a smooth submanifold of S^{n+1} of dimension $n - m_-$.

Connectedness of the level sets of F

Theorem 3.44. *Let $F : \mathbf{R}^{n+2} \to \mathbf{R}$ be a Cartan–Münzner polynomial of degree g and V its restriction to S^{n+1}. Then each isoparametric hypersurface*

$$M_t = V^{-1}(t), \ -1 < t < 1,$$

is connected. Moreover, $M_+ = V^{-1}(1)$ and $M_- = V^{-1}(-1)$ are the focal submanifolds $f_+(M)$ and $f_-(M)$, respectively, and they are also connected.

Proof. We begin by defining a map $d : M \times \mathbf{R} \to S^{n+1}$ by

$$d(x, \tau) = E\left(x, \frac{\pi}{2g} - \tau\right) = \cos\left(\frac{\pi}{2g} - \tau\right)x + \sin\left(\frac{\pi}{2g} - \tau\right)\xi(x). \tag{3.144}$$

From the construction of the Cartan–Münzner polynomials, we have on an open neighborhood of $M \times \{\frac{\pi}{2g}\}$,

$$V(d(x, \tau)) = \cos(g\tau), \tag{3.145}$$

since the largest principal curvature $\theta_1 = \pi/2g$ on M. By analyticity, this relation (3.145) holds on all of $M \times \mathbf{R}$. By equation (3.145), we see that $V = 1$ on the focal submanifold $f_+(M) = d(M \times \{0\})$, so that $f_+(M) \subset M_+$. Similarly, $V = -1$ on the focal submanifold $f_-(M) = d(M \times \{\frac{\pi}{g}\})$, so that $f_-(M) \subset M_-$. We will show that, in fact, $f_+(M) = M_+$ and $f_-(M) = M_-$.

The restriction of d to the set $\{x\} \times (0, \frac{\pi}{g})$, for $x \in M$, is a geodesic in S^{n+1} which intersects each hypersurface $M_t, -1 < t < 1$, in exactly one point by equation (3.145). This geodesic intersects each hypersurface orthogonally, and at the point of intersection x of this geodesic with M, the tangent vector to the geodesic is $-\xi(x)$. Hence $V^{-1}(-1, 1)$ is the disjoint union of these geodesics as x varies over M. The map d is a local diffeomorphism on $M\times(0, \pi/g)$, since its image contains no focal points. Since $V^{-1}(-1, 1)$ is a disjoint union of these normal geodesic segments through M with tangent vectors $-\xi(x)$ at $x \in M$, the map d is a diffeomorphism of $M \times (0, \pi/g)$ onto $V^{-1}(-1, 1)$. Note that the open set $V^{-1}(-1, 1)$ is dense in S^{n+1}, since if $V = \pm 1$ on an open set in S^{n+1}, then $V = \pm 1$ on all of S^{n+1} by the analyticity of V. The image $d(M \times [0, \pi/g])$ is compact and it contains the dense open subset $V^{-1}(-1, 1)$, so it equals all of S^{n+1}.

Since $M_+ = V^{-1}(1)$ is a subset of $d(M\times[0, \pi/g])$, equation (3.145) implies that

$$M_+ = d(M \times \{0\}) = E(M \times \{\frac{\pi}{2g}\}) = f_+(M).$$

In a similar way, we get $f_-(M) = M_-$. By Theorems 2.12 and 2.14 (see page 26), the focal submanifolds $f_+(M)$ and $f_-(M)$ have dimensions $n - m_+$ and $n - m_-$, respectively. Thus, both focal submanifolds have codimension greater than one in S^{n+1}. Therefore,

$$S^{n+1} - (M_+ \cup M_-) = V^{-1}(-1, 1)$$

is connected. Since $V^{-1}(-1, 1)$ is diffeomorphic to $M \times (0, \pi/g)$, we conclude that M is connected. Furthermore, for $-1 \leq t \leq 1$, the submanifold M_t equals $d(M\times\{s\})$ for an appropriate choice of $s \in [0, \pi/g]$, and so each M_t is also connected, including the two focal submanifolds. □

As we noted earlier, a consequence of Theorem 3.44 is that any connected isoparametric hypersurface M lies in a unique compact, connected isoparametric hypersurface of the form $V^{-1}(t), -1 < t < 1$, where V is the restriction to S^{n+1} of the Cartan–Münzner polynomial of M. For the rest of this section, we will assume that each isoparametric hypersurface and each focal submanifold is compact and connected.

Münzner's Structure Theorem

We now prove Münzner's important structure theorem for an isoparametric family of hypersurfaces which states that a compact, connected isoparametric hypersurface M divides the sphere S^{n+1} into two ball bundles over the two focal submanifolds M_+ and M_-, which lie on different sides of M in S^{n+1}. The precise wording of the theorem is as follows.

Theorem 3.45. *Let $k = \pm 1$, and let Z be a normal vector to the focal submanifold M_k in S^{n+1}. Let* $\exp : NM_k \to S^{n+1}$ *denote the normal exponential map for M_k. Then*

(a) $V(\exp Z) = k\cos(g|Z|)$.
(b) *Let* $B_k = \{q \in S^{n+1} \mid kV(q) \geq 0\}$, *and let* $(B^\perp M_k, S^\perp M_k)$ *be the bounded unit ball bundle in* NM_k. *Then*

$$\psi_k : (B^\perp M_k, S^\perp M_k) \to (B_k, M),$$

where $M = V^{-1}(0)$ *and* $\psi_k(Z) = \exp(\frac{\pi}{2g}Z)$, *is a diffeomorphism of manifolds with boundary.*

Proof. Let p be a point in the focal submanifold M_k and let Z be a unit normal vector to M_k at p, i.e., $Z \in S_p^\perp M_k$. Then the focal map $f_k : M \to M_k$ maps an m_k-dimensional sphere (which is a leaf of a principal foliation on M) to the point p. By considering the map $h : f_k^{-1}(p) \to S_p^\perp M_k$ used in equation (3.49) on page 105, we see that there exists a point $x \in f_k^{-1}(p)$ such that Z is the tangent vector to the geodesic starting at x with initial tangent vector $\xi(x)$. Thus, $\exp(\tau Z)$ and $d(x,\tau)$ traverse the same geodesic in S^{n+1} with unit speed as τ varies over the real numbers **R**. From equation (3.145), we see that

$$V(\exp \tau Z) = \cos(g(\tau + \tau')), \tag{3.146}$$

for some $\tau' \in [0, 2\pi/g)$. For $\tau = 0$, the left side of the equation above equals k, and so $\tau' = 0$ if $k = 1$, and $\tau' = \pi/g$ if $k = -1$. This proves part (a) of the theorem.

To prove (b), note that part (a) implies that the family of all normal geodesics to M_k is the same as the family of all normal geodesics to M. Each point in the complement $B_k - M_k$ lies on precisely one level hypersurface of V, and therefore it lies on exactly one normal geodesic to M_k. Thus, the map ψ_k is bijective, and so it is a diffeomorphism, since it is clearly a local diffeomorphism. □

As a consequence of the fact that the set of normal geodesics to each focal submanifold M_k, $k = \pm 1$, is the same as the set of normal geodesics to each of the parallel isoparametric hypersurfaces, we immediately obtain the following corollary.

Corollary 3.46. *Let* M_k, $k = \pm 1$, *be a focal submanifold of an isoparametric hypersurface* M. *Then the focal set of* M_k *is the same as the focal set of* M, *i.e., it is* $M_k \cup M_{-k}$.

Münzner's restriction on the number of principal curvatures

Münzner's major result is that the number g of distinct principal curvatures of an isoparametric hypersurface M in S^{n+1} is $1, 2, 3, 4$, or 6. This is a lengthy and delicate computation involving the cohomology rings of the hypersurface M and its two focal submanifolds M_+ and M_-. The structure of these rings is determined by the basic topological fact that a compact, connected isoparametric hypersurface $M \subset S^{n+1}$ divides S^{n+1} into two ball bundles over the two focal submanifolds as in Theorem 3.45 (b).

Theorem 3.47 below does not assume that M is isoparametric, but only that it divides the sphere into two ball bundles. This is important, since the theorem can be applied to more general settings, in particular, the case of a compact, connected proper Dupin hypersurface embedded in S^{n+1}, as we will see in the next section. Using methods of algebraic topology, Münzner [382] proved the theorem below, and we refer the reader to Münzner's paper for the proof.

Theorem 3.47. *Let M be a compact, connected hypersurface in S^{n+1} which divides S^{n+1} into two ball bundles over submanifolds M_+ and M_-. Then $\alpha = (1/2)\dim H^*(M, R)$ can only assume the values $1, 2, 3, 4$, and 6. (The ring R of coefficients is $\mathbf{Z}$ if both M_+ and M_- are orientable, and $\mathbf{Z}_2$ otherwise.)*

Münzner then proved Theorem 3.48 below regarding the cohomology of an isoparametric hypersurface and its focal submanifolds. Since all of the parallel hypersurfaces $M_t = V^{-1}(t)$ are diffeomorphic, it is sufficient to consider the case $M = V^{-1}(0)$. In that case, M has two focal submanifolds $M_1 = V^{-1}(1)$ of dimension $n - m_1$ and $M_{-1} = V^{-1}(-1)$ of dimension $n - m_{-1}$, where m_1 and m_{-1} are the two (possibly equal) multiplicities of the principal curvatures of M. Then by Theorem 3.45, the sets,

$$B_1 = \{q \in S^{n+1} \mid V(q) \geq 0\}, \quad B_{-1} = \{q \in S^{n+1} \mid V(q) \leq 0\},$$

are $(m_k + 1)$-ball bundles over the focal submanifolds M_k, for $k = 1, -1$, respectively.

The dimension n of M is equal to $g(m_1 + m_{-1})/2$, and so $g = 2n/\mu$, where $\mu = m_1 + m_{-1}$. Thus, an isoparametric hypersurface M satisfies the hypothesis of the following theorem of Münzner [382] (the presentation of Münzner's proof here follows [95, pp. 289–292] closely).

Theorem 3.48. *Let M be a compact, connected hypersurface in S^{n+1} such that:*

(a) *S^{n+1} is divided into two manifolds (B_1, M) and (B_{-1}, M) with boundary along M.*
(b) *For $k = \pm 1$, the manifold B_k has the structure of a differentiable ball bundle over a compact, connected manifold M_k of dimension $n - m_k$.*

Let the ring R of coefficients be $\mathbf{Z}$ if both M_1 and M_{-1} are orientable, and $\mathbf{Z}_2$ otherwise. Let $\mu = m_1 + m_{-1}$. Then $\alpha = 2n/\mu$ is an integer, and for $k = \pm 1$,

$$H^q(M_k) = \begin{cases} R \text{ for } q \equiv 0 \pmod{\mu},\ 0 \leq q < n, \\ R \text{ for } q \equiv m_{-k} \pmod{\mu},\ 0 \leq q < n, \\ 0 \text{ otherwise.} \end{cases}$$

Further,

$$H^q(M) = \begin{cases} R \textit{ for } q = 0, n, \\ H^q(M_1) \oplus H^q(M_{-1}), \textit{ for } 1 \leq q \leq n-1. \end{cases}$$

Proof. By part (b) of the hypothesis, the submanifold M_k is a deformation retract of B_k and also of the open ball bundle $B_k - M$ for $k = \pm 1$. Thus, for all q, we have

$$H^q(B_k) = H^q(M_k) = H^q(B_k - M), \; k = \pm 1. \tag{3.147}$$

Suppose now that $1 \leq q \leq n-1$. Then $H^q(S^{n+1}) = H^{q+1}(S^{n+1}) = 0$. Hence, from the Mayer-Vietoris sequence for the exact triad (S^{n+1}, B_1, B_{-1}), we get using equation (3.147) that

$$H^q(M) \simeq H^q(B_1) \oplus H^q(B_{-1}) = H^q(M_1) \oplus H^q(M_{-1}). \tag{3.148}$$

Since $S^{n+1} - B_k = B_{-k} - M$, the Alexander duality theorem and equation (3.147) imply that for $1 \leq q \leq n-1$,

$$H^q(M_k) \simeq H^q(B_k) \simeq H_{n-q}(B_{-k} - M) = H_{n-q}(M_{-k}), \; k = \pm 1. \tag{3.149}$$

Our convention on the coefficient ring R allows us to use Poincaré duality on the $(n - m_{-k})$-dimensional manifold M_{-k}. This and equation (3.149) give

$$H^q(M_k) \simeq H_{n-q}(M_{-k}) \simeq H^{q-m_{-k}}(M_{-k}), \; k = \pm 1. \tag{3.150}$$

Two repetitions of the formula above lead to the equation

$$H^q(M_k) \simeq H^{q-m_1-m_{-1}}(M_k), \text{ for } 1 + m_{-k} \leq q \leq n-1. \tag{3.151}$$

Since M_k is connected, we have $H^0(M_k) = R$. Since M_{-k} has dimension $n - m_{-k}$, we have $H_q(M_{-k}) = 0$ for $q > n - m_{-k}$. Further, by our convention on R, we have $H_{n-m_{-k}}(M_{-k}) \simeq R$. From this and equation (3.149), we get

$$H^{m_{-k}}(M_k) \simeq H_{n-m_{-k}}(M_{-k}) \simeq R, \quad H^q(M_k) = 0, \text{ for } 0 < q < m_{-k}.$$

Using this and equation (3.150) or (3.151), we get

$$H^q(M_k) = 0 \text{ for } m_{-k} < q < \mu, \; H^\mu(M_k) \simeq R.$$

Repeated use of formula (3.151) then gives the desired cohomology for M_k. Then we get the cohomology of M immediately from equation (3.148) and the fact that M is connected.

Finally, we show that α is an integer. From the formulas for the cohomology of M_k, we know that

$$\dim M_k \equiv 0 \pmod{\mu} \text{ or } \dim M_k \equiv m_{-k} \pmod{\mu}.$$

We consider two cases. First suppose that

$$\dim M_k \equiv 0 \pmod{\mu} \text{ for } k = 1 \text{ and } k = -1.$$

Then $n - m_1 = j\mu$ and $n - m_{-1} = \ell\mu$, for some integers j and ℓ. Hence,

$$\mu = 2n - (n - m_1) - (n - m_{-1}) = 2n - j\mu - \ell\mu = 2n - (j + \ell)\mu.$$

Thus, $\alpha = 2n/\mu = j + \ell + 1$. Note that since $|j\mu - \ell\mu| = |m_{-1} - m_1| < \mu$, we have $j = \ell$. Hence, α is odd and $m_1 = m_{-1}$. In the other case, we can assume without loss of generality that $\dim M_1 \equiv m_{-1} \pmod{\mu}$. Then, we have

$$n = \dim M_1 + m_1 \equiv 0 \pmod{\mu},$$

and we see that $\alpha = 2n/\mu$ is, in fact, an even integer. □

For a compact, connected isoparametric hypersurface $M \subset S^{n+1}$ with g distinct principal curvatures, we have

$$\dim M = n = g(m_1 + m_{-1})/2 = g\mu/2.$$

Thus, $\alpha = 2n/\mu = g$. By Theorem 3.48, we see that α is also equal to $\dim_R H^*(M, R)/2$. Hence by Münzner's Theorem 3.47, $g = \alpha$ can only assume the values $1, 2, 3, 4$ or 6, and we have Münzner's major theorem.

Theorem 3.49. *Let $M \subset S^{n+1}$ be a connected isoparametric hypersurface with g distinct principal curvatures. Then g is $1, 2, 3, 4$ or 6.*

Note that we do not have to assume that M is compact in the theorem, because any connected isoparametric hypersurface is contained in a unique compact, connected isoparametric hypersurface to which the arguments above can be applied. We also note that there do exist isoparametric hypersurfaces for each of the values of g in the theorem, as mentioned in Section 3.1 and Subsection 3.8.5 (to follow).

Remark 3.50 (Crystallographic groups). A consequence of Münzner's Theorem 3.49 is that the dihedral group D_g associated with M (see Remark 3.27) is crystallographic (see L.C. Grove and C.T. Benson [186, pp. 21–22]). A direct proof of this fact could possibly give a simpler proof of Theorem 3.49 (see also K. Grove and S. Halperin [184, pp. 437–438]).

Remark 3.51 (Multiplicities of the principal curvatures). Cartan [52–55] classified isoparametric hypersurfaces with $g \leq 3$ principal curvatures, as discussed in

Section 3.1. In the cases $g = 4$ and $g = 6$, many results concerning the possible multiplicities of the principal curvatures have been obtained from the topological situation given in Theorem 3.48, i.e., that a compact, connected isoparametric hypersurface M in S^{n+1} divides S^{n+1} into two ball bundles over its two focal submanifolds. In the case of four principal curvatures, several mathematicians, including Münzner [381, 382], Abresch [2], Grove and Halperin [184], Tang [514, 515] and Fang [149, 150], found restrictions on the multiplicities (m_1, m_2). This series of results culminated with the paper of Stolz [502], who proved the following theorem.

Theorem 3.52. *The multiplicities (m_1, m_2) of the principal curvatures of an isoparametric hypersurface with $g = 4$ principal curvatures are the same as those in the known examples due to Ferus, Karcher, and Münzner [160] or the two homogeneous examples with $(m_1, m_2) = (2, 2)$ or $(4, 5)$ that are not of FKM-type.*

In the case of $g = 6$ principal curvatures, Münzner [382] showed that all of the principal curvatures have the same multiplicity m, and Abresch [2] showed that m equals 1 or 2. Thus we have:

Theorem 3.53. *For an isoparametric hypersurface with $g = 6$ principal curvatures, all the principal curvatures have the same multiplicity m, and m equals 1 or 2.*

Isoparametric hypersurfaces are taut

We next give a proof that compact, connected isoparametric hypersurfaces are taut. For the sake of completeness, we briefly recall some basic results from Sections 2.6 and 2.7 related to the theory of taut embeddings. Compact, connected isoparametric submanifolds of higher codimension (see Section 3.8.6) were also shown to be taut by Terng [525].

Let M be a compact, connected submanifold of arbitrary codimension embedded in an m-dimensional sphere S^m. For $p \in S^m$ and $x \in M$, the spherical distance function L_p is defined by

$$L_p(x) = d(p,x)^2 = (\cos^{-1}\langle p, x\rangle)^2. \tag{3.152}$$

There is an Index Theorem for spherical distance functions similar to the well-known Index Theorem (Theorem 2.51, page 53) for Euclidean distance functions. The Index Theorem for spherical distance functions states that L_p has a critical point at $x \in M$ if and only if p lies on the normal geodesic to M at x. This critical point is nondegenerate if and only if p is not a focal point of (M, x). The index of L_p at a nondegenerate critical point x is equal to the number of focal points (counting multiplicities) of (M, x) on the shortest geodesic segment from p to x.

A function $\phi : M \to \mathbf{R}$ on a compact, connected manifold M is said to be a nondegenerate function, or a Morse function, if every critical point of ϕ is nondegenerate. If any of the critical points of ϕ is degenerate, then we will say that ϕ is a degenerate function on M.

A compact, connected submanifold $M \subset S^m$ is taut if there exists a field $\mathbf{F}$ such that every nondegenerate function of the form $L_p, p \in S^m$, has $\beta(M, \mathbf{F})$ critical points on M, where $\beta(M, \mathbf{F})$ is the sum of the $\mathbf{F}$-Betti numbers of M. By the Morse inequalities (see, for example, Morse-Cairns [379, p. 270]), the number of critical points of any nondegenerate function on M is at least $\beta(M, \mathbf{F})$ for any field $\mathbf{F}$. In the theory of taut submanifolds, the field $\mathbf{F} = \mathbf{Z}_2$ has proved to be optimal, and we will used it exclusively here. (See Remark 2.53, page 54.)

Note that a spherical distance function L_p has the same critical points on M as the linear height function,

$$l_p(x) = \langle p, x \rangle, \quad p \in S^m. \tag{3.153}$$

A compact, connected submanifold $M \subset \mathbf{R}^{m+1}$ is tight if there exists a field $\mathbf{F}$ such that every nondegenerate height function $l_p, p \in S^m$, has $\beta(M, \mathbf{F})$ critical points on M (see Section 2.6, page 36). For a compact, connected submanifold $M \subset S^m \subset \mathbf{R}^{m+1}$, tautness is equivalent to tightness, that is, the combination of tight and spherical is equivalent to tautness (see Theorems 2.69 and 2.70, and Corollary 2.72). At times it is more convenient to use linear height functions than spherical distance functions, and we will do this when it is appropriate.

As a consequence of the Index Theorem, we immediately obtain the following result.

Theorem 3.54. *Let $M \subset S^{n+1}$ be a compact, connected hypersurface. If a spherical distance function L_p is nondegenerate on M, then the number of critical points of L_p on M is equal to the number of pre-images of p under the normal exponential map E in the set $M \times (-\pi, \pi]$.*

If $M \subset S^{n+1}$ is a compact, connected isoparametric hypersurface, then Corollary 3.46 implies that a spherical distance function L_p is nondegenerate on M if and only if it is nondegenerate on each focal submanifold of M. The following theorem gives the number of critical points of any nondegenerate spherical distance function on M and its focal submanifolds.

Theorem 3.55. *Let $M \subset S^{n+1}$ be a compact, connected isoparametric hypersurface with g distinct principal curvatures, and let M_+ and M_- be the two focal submanifolds of M. Then*

(a) *Every nondegenerate spherical distance function L_p has $2g$ critical points on M.*
(b) *Every nondegenerate L_p has g critical points on M_+ and g critical points on M_-.*

Proof. (a) Let ξ be a field of unit normals to M, and let $\lambda_1 = \cot\theta_1$, where $0 < \theta_1 < \pi$, be the largest principal curvature of the shape operator A determined by ξ. As in the proof of Theorem 3.44, we define a map $d : M \times \mathbf{R} \to S^{n+1}$ by

$$d(x, \tau) = E(x, \theta_1 - \tau) = \cos(\theta_1 - \tau)x + \sin(\theta_1 - \tau)\xi(x),$$

where E is the normal exponential map on M. The restriction V of the Cartan–Münzner polynomial of M to S^{n+1} satisfies the equation

$$V(d(x, \tau)) = \cos g\tau. \tag{3.154}$$

As in Theorem 3.44, the restriction of the map d to $M \times (0, \pi/g)$ is a diffeomorphism onto $V^{-1}(-1, 1) = S^{n+1} - (M_+ \cup M_-)$. The same is true if $(0, \pi/g)$ is replaced by any interval of the form $(j\pi/g, (j+1)\pi/g)$, for any integer j. Thus, as τ varies over the interval $(-\pi, \pi)$, each point p in $S^{n+1} - (M_+ \cup M_-)$ is covered $2g$ times by d, and so L_p has $2g$ critical points on M.

(b) As we noted above, L_p is nondegenerate on M_+ or M_- if and only if p is in the set $S^{n+1} - (M_+ \cup M_-)$. In that case, p lies on precisely one of the isoparametric hypersurfaces $M_t, -1 < t < 1$. There is exactly one geodesic γ through p that is normal to each isoparametric hypersurface in the family. This geodesic γ is also the only geodesic through p which intersects M_+ or M_- orthogonally. The geodesic γ can be parametrized as $\gamma(\tau) = d(p, \tau)$, for $\tau \in \mathbf{R}$. As τ varies over the interval $(-\pi, \pi]$, equation (3.154) implies that the geodesic γ alternately intersects M_+ and M_- at values $\tau = j\pi/g$, where j is an integer. Thus, γ meets M_+ and M_- exactly g times, and so L_p has g critical points on each of the focal submanifolds. □

As a consequence, we have the following result first obtained in [93].

Corollary 3.56. *Let $M \subset S^{n+1}$ be a compact, connected isoparametric hypersurface. Then M and its two focal submanifolds M_+ and M_- are taut submanifolds of S^{n+1}.*

Proof. As noted after the proof of Theorem 3.48, the quantity $\alpha = 2n/\mu = g$. From Theorem 3.48, we see that the sum of the Betti numbers $\beta(M, \mathbf{Z}_2)$ of M is $2\alpha = 2g$, whereas $\beta(M_+, \mathbf{Z}_2) = \beta(M_-, \mathbf{Z}_2) = g$. Since every nondegenerate L_p has $2g$ critical points on M and g critical points on M_+ and M_-, the corollary is proved. □

Totally focal embeddings

We close this section with a discussion of the notion of totally focal embeddings introduced by Carter and West [62–64, 68]. An embedding $f : M \to S^m$ of a compact, connected manifold is said to be *totally focal* if every spherical distance function L_p is either nondegenerate or has only degenerate critical points. (Carter and West [62, 63] also considered totally focal embeddings into Euclidean space $\mathbf{R}^m$ which are defined in a similar way.) Cecil and Ryan [93, p. 102] proved the following corollary of Theorem 3.55.

Corollary 3.57. *Let $M \subset S^{n+1}$ be a compact, connected isoparametric hypersurface. Then M is totally focal in S^{n+1}.*

Proof. Using the notation of Theorem 3.55, suppose that L_p is a degenerate function on M, i.e., suppose that p is a focal point of M. Suppose first that $p \in M_+$ so that $V(p) = 1$. If p is also equal to $d(y, \tau)$ for some $y \neq x$ in M, then by equation (3.154), τ is an even integral multiple of π/g. If p were in M_-, then τ would have to be an odd multiple of π/g. In either case, p is a focal point of (M, y). Thus, all of the critical points of L_p on M are degenerate. Therefore, if L_p is a degenerate function on M, then all of the critical points of L_p are degenerate, and so M is totally focal. □

Remark 3.58. Note that the focal submanifolds M_+ and M_- are not totally focal. If $p \in M_+$, then p is a focal point of M_+, but L_p has a nondegenerate absolute minimum at p itself. A similar proof shows that M_- is not totally focal.

Remark 3.59 (On the converse to Corollary 3.57). Carter and West [64] proved the converse of Corollary 3.57, i.e., a compact, connected totally focal hypersurface $M \subset S^{n+1}$ is isoparametric.

In a later paper [68], Carter and West presented a proof that a compact, connected totally focal submanifold of codimension greater than one in S^{n+1} is an isoparametric submanifold as defined by Terng [525] (see Subsection 3.8.6). However, Terng and Thorbergsson [531, p. 197] noted a gap in the proof of Theorem 5.1 of that paper [68, p. 619].

For simplicity, we describe the gap in the situation of a proper Dupin hypersurface $M \subset S^{n+1}$. The gap involves the assertion that for any leaf C of any principal foliation T_λ through an arbitrary point $x \in M$, the normal geodesic to M at x intersects M again at the point $y \in M$ antipodal to x in the leaf C.

This implies that C is totally geodesic in the corresponding curvature sphere at x. By Theorem 2.20 on page 32, this means that the corresponding principal curvature function λ has a critical value along the leaf C. Since x is an arbitrary point in M, this means that every point of M is a critical point of λ, and thus λ is constant on M. This holds for any principal curvature λ, and so M is isoparametric.

The gap in the proof is the assertion that the normal geodesic to M at any point $x \in M$ always intersects M again at the point $y \in M$ antipodal to x in the leaf C. This is not necessarily true if M is proper Dupin but not isoparametric. There is a similar gap in the proof given in [95, Theorem 9.25, p. 231].

Attempting to eliminate the gap in the case of codimension greater than one remains as an open problem.

3.7 Applications to Dupin Hypersurfaces

In this section, we discuss a result due to Thorbergsson [533] that the restriction $g = 1, 2, 3, 4$, or 6 on the number of distinct principal curvatures of an isoparametric hypersurface in S^{n+1} also holds for a compact, connected proper Dupin hypersurface

M embedded in S^{n+1}. This leads to restrictions on the multiplicities of the principal curvatures of a compact, connected proper Dupin hypersurface due to Stolz [502] and Grove and Halperin [184].

Let $M^n \subset S^{n+1} \subset \mathbf{R}^{n+2}$ be a compact, connected proper Dupin hypersurface embedded in S^{n+1}. Then M is orientable (see, for example, [471]), and let ξ denote a global field of unit normals on M. Assume that M has g distinct smooth principal curvature functions, which we label as in Section 3.2 on page 102,

$$\lambda_i = \cot \theta_i, \ 0 < \theta_i < \pi, \ 1 \leq i \leq g, \tag{3.155}$$

where the θ_i form an increasing sequence. Then λ_i has constant multiplicity on M which we denote by m_i.

Let $f_+ : M \to S^{n+1}$ denote the focal map of M onto the first focal submanifold M_+ reached by going a distance θ_1 from M in the direction of ξ, that is,

$$f_+(x) = E(x, \theta_1(x)) = \cos(\theta_1(x))x + \sin(\theta_1(x))\xi(x), \tag{3.156}$$

where E is the normal exponential map on M. The focal map f_+ factors through an immersion of the space of leaves M/T_1 into S^{n+1}, as in Theorems 2.12 and 2.14 (see page 26), and so M_+ is a smooth immersed submanifold of S^{n+1} of dimension $n - m_1$. Similarly, we have the focal map $f_- : M \to S^{n+1}$ onto the first focal submanifold M_- in the direction $-\xi$. Then M_- is a smooth immersed submanifold of S^{n+1} of dimension $n - m_{-1}$, where $m_{-1} = m_g$.

We now discuss Thorbergsson's argument that M divides S^{n+1} into a union of two ball bundles over these two focal submanifolds M_+ and M_-. This leads to the restriction on the number of distinct principal curvatures and to many restrictions on the possible multiplicities of the principal curvatures, as noted in Remark 3.51.

Thorbergsson's proof relies on the following theorem which he proved in [533]. His theorem is valid in all the real space forms, Euclidean space $\mathbf{R}^{n+1}$, the sphere S^{n+1}, and hyperbolic space H^{n+1}. Here tautness means with respect to $\mathbf{Z}_2$-homology as usual.

Theorem 3.60. *Let $M^n \subset \tilde{M}^{n+1}$ be a complete, connected proper Dupin hypersurface embedded in a real space form $\tilde{M}^{n+1}$. Then M is taut.*

This theorem was also stated earlier as Theorem 2.80 on page 66, where we discussed Thorbergsson's method of proof. As noted there, Thorbergsson's proof is different than the proof of tautness in the isoparametric case given in Corollary 3.56. Thorbergsson used the principal foliations to construct concrete $\mathbf{Z}_2$-cycles in M, which enabled him to show that every critical point of every nondegenerate distance function is of linking type (see, Morse-Cairns [379, p. 258] and the discussion after Theorem 2.28), and thus M is taut. See Thorbergsson's paper [533] for a complete proof.

Remark 3.61 (Focal submanifolds of Dupin hypersurfaces may not be taut). In contrast to the case of an isoparametric hypersurface (see Corollary 3.56), the focal submanifolds of a compact, connected proper Dupin hypersurface embedded in S^{n+1} or $\mathbf{R}^{n+1}$ are not necessarily taut. For example, for a compact ring cyclide of Dupin M in $\mathbf{R}^3$ that is not a torus of revolution (see Figure 5.3, page 277), one focal submanifold is an ellipse, which is not taut in $\mathbf{R}^3$, although it is tight in $\mathbf{R}^3$. More generally, using techniques from Lie sphere geometry, Buyske [51] showed that if a hypersurface M in $\mathbf{R}^{n+1}$ is Lie equivalent to an isoparametric hypersurface in S^{n+1}, then each compact focal submanifold of M is tight in $\mathbf{R}^{n+1}$, although not necessarily taut.

We now use Theorem 3.60 to derive an important theorem due to Thorbergsson [533]. Here M_+ and M_- are the first focal submanifolds on either side of M in S^{n+1}, as defined in the paragraph containing equation (3.156).

Theorem 3.62. *Let $M \subset S^{n+1}$ be a compact, connected proper Dupin hypersurface. Then M divides S^{n+1} into a union of two ball bundles over the two focal submanifolds M_+ and M_-.*

Proof. We first show that M_+ and M_- lie in different components of $S^{n+1} - M$, the complement of M in S^{n+1}. Let $p = f_+(x)$ for some point $x \in M$. Since M is tautly embedded, the spherical distance function L_p has an absolute minimum at x by Theorem 2.64 on page 59. Therefore, the normal geodesic segment $[x, p]$ does not intersect M except at x, and so p lies in the component W_+ of $S^{n+1} - M$ to which the normal field ξ points. Thus, M_+ is contained in W_+, and similarly, M_- is contained in the other component W_-.

Since M is proper Dupin and compact, each leaf of each principal foliation is an embedded metric sphere of the appropriate dimension by Theorems 2.11 and 2.14 (see page 25). Thus, the inverse image $f_+^{-1}(p)$ of any focal point p in M_+ consists of a discrete union of m_1-dimensional spheres which are leaves of the principal foliation T_1. Each such leaf lies on the n-sphere in S^{n+1} with center p and radius r, where r is the minimum value of L_p on M. Since M is taut, it has the spherical two-piece property (STPP) of Banchoff [20] (see Section 2.7), and so this discrete collection of m_1-dimensional spheres consists of only one leaf L of T_1. Thus, the immersion of the space of leaves M/T_1 onto M_+ given in Theorems 2.12 and 2.14 is injective, and M_+ is an embedded $(n - m_1)$-dimensional submanifold of S^{n+1}.

For each point $x \in M$, the geodesic segment $[x, f_+(x)]$ lies in the closure $\overline{W_+}$ of W_+. Furthermore, if q is any point of $\overline{W_+} - M_+$, then the distance function L_q has a nondegenerate minimum on M which is unique by tautness. Hence, q lies on exactly one of the segments $[x, f_+(x)]$ from a point $x \in M$ to the focal point $f_+(x)$ on M_+. Thus $\overline{W_+}$ is the union of the segments $[x, f_+(x)]$ as x ranges over M, and the map $\pi : \overline{W_+} \to M_+$ which takes the segment $[x, f_+(x)]$ to the point $f_+(x)$ is a ball bundle projection. The same proof shows that $\overline{W_-}$ is a ball bundle over the embedded $(n - m_{-1})$-dimensional focal submanifold M_-. □

Thus, the hypotheses of Theorem 3.48 on page 134 are satisfied by M and the focal submanifolds M_+ and M_-, and therefore the $\mathbf{Z}_2$-cohomology of M, M_+ and M_- is determined by that theorem. In particular, the number $\alpha = (1/2) \dim H^*(M, \mathbf{Z}_2)$ can only assume the values $1, 2, 3, 4$, or 6.

Restriction on g for compact proper Dupin hypersurfaces

By counting the number of cycles that he constructed in his proof of Theorem 3.60, Thorbergsson showed that dim $H^*(M, \mathbf{Z}_2) = 2g$, where g is the number of distinct principal curvatures of M, and thus $\alpha = g$. Hence, Thorbergsson obtained the following theorem (see also Grove-Halperin [184, pp. 437–438] for another proof of Theorem 3.63).

Theorem 3.63. *Let $M \subset S^{n+1}$ be a compact, connected proper Dupin hypersurface. Then the number g of distinct principal curvatures of M is $1, 2, 3, 4$ or 6.*

Thorbergsson [538, p. 980] pointed out that his construction of the cycles in [533] can be used to show that the multiplicities of the principal curvatures of a compact, connected proper Dupin hypersurface satisfy the relation $m_k = m_{k+2}$ (subscripts mod g), as in Theorem 3.26 for isoparametric hypersurfaces. The proof is different than in the isoparametric case. In the Dupin case, it is accomplished by calculating the homology of M by using the cycles constructed by Thorbergsson and comparing that with the calculation of the homology based on the fact that M divides S^{n+1} into two ball bundles, as in Theorem 3.62.

In fact, the following theorem regarding the multiplicities of a compact, connected proper Dupin hypersurface embedded in S^{n+1} has been proven.

Theorem 3.64. *For each value of $g = 1, 2, 3, 4$ or 6, the possible multiplicities of the principal curvatures of a compact, connected proper Dupin hypersurface embedded in S^{n+1} are the same as the possible multiplicities for an isoparametric hypersurface with the same number of principal curvatures (see Remark 3.51, page 136).*

This was shown in the case $g = 4$ by Stolz [502] and for $g = 6$ by Grove and Halperin [184]. Both of these papers involve sophisticated topological arguments. Note that for $g = 1$ and $g = 2$, there are no restrictions on the multiplicities. In the case $g = 3$, Miyaoka [363] proved that a compact, connected proper Dupin hypersurface M is Lie equivalent to an isoparametric hypersurface, and thus it has the same multiplicities as an isoparametric hypersurface. That is, all the multiplicities are equal to a certain integer m, and m is $1, 2, 4$ or 8.

Proper Dupin hypersurfaces are algebraic

Another important result is that like isoparametric hypersurfaces, proper Dupin hypersurfaces are algebraic. This was formulated by Cecil, Chi, and Jensen [84] as follows.

Theorem 3.65. *Every connected proper Dupin hypersurface $f : M \to \mathbf{R}^n$ embedded in $\mathbf{R}^n$ is contained in a connected component of an irreducible algebraic subset of $\mathbf{R}^n$ of dimension $n - 1$.*

The main idea of the proof of this theorem is due to Pinkall, who sent a letter [444] to T. Cecil in 1984 that contained a sketch of the proof. However, Pinkall did not publish a proof, and a full proof based on Pinkall's sketch was not published until 2008 by Cecil, Chi, and Jensen [84]. The proof makes use of the various principal foliations whose leaves are open subsets of spheres to construct an analytic algebraic parametrization of a neighborhood of $f(x)$ for each point $x \in M$. From this, one can get the final conclusion stated above by using methods of real algebraic geometry.

In contrast to the situation for isoparametric hypersurfaces, however, a connected proper Dupin hypersurface in $\mathbf{R}^n$ or S^n does not necessarily lie in a compact connected proper Dupin hypersurface, as the tube M^3 over a torus $T^2 \subset \mathbf{R}^3 \subset \mathbf{R}^4$ in Example 2.22 on page 33 illustrates. The tube M^3 is a compact algebraic Dupin hypersurface that contains the open subset $U \subset M^3$ on which there are three distinct principal curvatures. Each connected component of U is a proper Dupin hypersurface contained in the compact algebraic hypersurface M^3, but M^3 itself is only Dupin and not proper Dupin.

The algebraicity, and hence analyticity, of proper Dupin hypersurfaces was useful in clarifying certain fine points in the 2007 paper [82] of Cecil, Chi, and Jensen on proper Dupin hypersurfaces with four principal curvatures.

In 1984, Kuiper [305] asked whether all taut submanifolds are algebraic. Of course, taut implies Dupin but not necessarily proper Dupin, so Theorem 3.65 does not completely answer Kuiper's question. In their paper, Cecil, Chi, and Jensen [84] proved that a compact taut submanifold M^n is algebraic if $n \leq 4$. This was known in dimensions $n \leq 2$ by Banchoff's [20] classification of taut compact curves (must be a metric circle) and surfaces (see Theorem 5.49, page 326). The main result of a recent preprint of Chi [110] is that the answer to Kuiper's question is affirmative.

3.8 Examples of Isoparametric Hypersurfaces

In this section, we discuss some examples of isoparametric hypersurfaces in S^{n+1} based on the number g of distinct principal curvatures.

3.8.1 *The case $g = 1$*

If a connected isoparametric hypersurface $M^n \subset S^{n+1}$ has $g = 1$ principal curvature, then M^n is a totally umbilic hypersurface in S^{n+1}. As such, M^n is an open subset of a great or small hypersphere in S^{n+1}. Each great or small hypersphere is the set where S^{n+1} intersects a hyperplane in $\mathbf{R}^{n+2}$. The spheres lying in hyperplanes perpendicular to a given diameter of S^{n+1} make up an isoparametric family whose focal set consists of the two end-points (poles) of the diameter. We now describe these hypersurfaces in terms of the general theory of isoparametric hypersurfaces that we have developed.

Let p be a unit vector in $\mathbf{R}^{n+2}$ which we will use to determine a diameter of S^{n+1}. Let F be the linear height function,

$$F(z) = \langle z, p \rangle, \tag{3.157}$$

for $z \in \mathbf{R}^{n+2}$. Then

$$\text{grad}^E F = p, \quad |\text{grad}^E F|^2 = 1, \quad \Delta^E F = 0,$$

so that F satisfies the Cartan–Münzner differential equations (3.86)–(3.87) with $g = 1$ and $c = 0$. It follows from Theorem 3.30 on page 112 that the restriction V of F to S^{n+1} satisfies

$$|\text{grad}^S V|^2 = 1 - V^2, \quad \Delta^S V = -(n+1)V, \tag{3.158}$$

so that V is an isoparametric function on S^{n+1}. In fact, it is useful to note that at any $z \in S^{n+1}$, we have

$$\text{grad}^S V = p - \langle z, p \rangle z. \tag{3.159}$$

We consider the level sets

$$M_s = \{z \in S^{n+1} \mid \langle z, p \rangle = \cos s\}, \quad 0 \leq s \leq \pi.$$

Except for the two focal submanifolds, $M_0 = \{p\}$ and $M_\pi = \{-p\}$, each level set is an n-sphere with radius $\sin s$ lying in the hyperplane situated $1 - \cos s$ units below the north pole p. From the point of view of the intrinsic geometry of S^{n+1}, the set M_s is the geodesic hypersphere $S(p, s)$ centered at the point p having radius s. Note that M_s may also be regarded as the sphere $S(-p, \pi - s)$. Finally, the collection of all M_s forms a system of parallel hypersurfaces with

$$M_{s_0 - t} = f_t M_{s_0}, \quad s_0 - \pi < t < s_0,$$

provided that $\xi = \text{grad}^S V / |\text{grad}^S V|$ is used as the field of unit normals to M_s in the definition of f_t.

We now compute the shape operator A of M_{s_0}. We first compute for $z \in M_{s_0}$, $X \in T_z M_{s_0}$,

$$\tilde{\nabla}_X(\text{grad}^S V) = -(\langle X, p \rangle z + \langle z, p \rangle X).$$

Then, using Theorem 3.1 on page 86, we have for $X, Y \in T_z M_{s_0}$ and $\rho = |\text{grad}^S V|$,

$$
\begin{aligned}
\langle AX, Y\rangle &= \frac{-1}{\rho} H_F(X,Y) \\
&= \frac{-1}{\sqrt{1-F^2}} \langle \tilde{\nabla}_X \operatorname{grad}^S V, Y\rangle \\
&= \frac{1}{\sqrt{1-\cos^2 s_0}} (\langle X,p\rangle\langle z,Y\rangle + \langle z,p\rangle\langle X,Y\rangle) \\
&= \frac{\cos s_0}{\sin s_0} \langle X, Y\rangle.
\end{aligned}
$$

Thus M_s is totally umbilic with principal curvature $\cot s_0$. The sectional curvature is therefore $1+\cot^2 s_0 = \csc^2 s_0$, as is appropriate for a sphere of radius $\sin s_0$. All this agrees with our general calculations of principal curvatures of parallel hypersurfaces in Theorem 2.2 on page 17, which would assign the principal curvature $\cot(s_0 - t)$ to the parallel hypersurface $f_t(M_{s_0}) = M_{s_0-t}$.

Orbits of a group action

The family $\{M_s\}$ may also be realized as the set of orbits of a group action as follows. Pick an arbitrary point $p \in S^{n+1}$ and consider the subgroup G of $SO(n+2)$ of transformations that leave p fixed. For $z \in S^{n+1}$, $z \neq \pm p$, the isotropy subgroup,

$$G_z = \{g \in G \mid gz = z\},$$

is a naturally embedded copy of $SO(n)$, so that the orbit $M = Gz$ has codimension 1 in S^{n+1}.

Take an orthonormal basis $\{e_1, \ldots, e_{n+2}\}$ with $e_1 = p$, and e_2 chosen so that z lies in the plane spanned by $\{e_1, e_2\}$. Thus, $z = z_1 e_1 + z_2 e_2$ and $z_2 \neq 0$, since $z \neq \pm p$. With respect to this orthonormal basis, let A_0 be an arbitrary skew-symmetric $(n+1)\times(n+1)$ matrix, and let A be the $(n+2)\times(n+2)$ matrix,

$$A = \begin{bmatrix} 0 & 0 \\ 0 & A_0 \end{bmatrix}. \tag{3.160}$$

Then $g_t = \exp tA$ is a curve in G with $g_0 = I$ and initial tangent vector $\vec{g}_0 = A$. The tangent space to the orbit M of z is the set of vectors of the form,

$$\frac{d}{dt}(g_t z)|_{t=0} = \frac{d}{dt}\exp(tA)z = Az,$$

for A as in equation (3.160).

Let $\xi = \lambda p + \mu z + w$, where $\langle w, p\rangle = \langle w, z\rangle = 0$. Then

$$\langle Az, \xi\rangle = 0 \Leftrightarrow \langle z, A\xi\rangle = 0 \Leftrightarrow \langle z, 0 + \mu Az + Aw\rangle = 0 \Leftrightarrow \langle z, Aw\rangle = 0,$$

since $Ap = 0$, and $\langle z, Az\rangle = 0$, because A is skew-symmetric.

If ξ is normal to M at z, then we have $\langle Az, \xi \rangle = \langle z, Aw \rangle = 0$ for all skew-symmetric A of the form described above. In particular, the function A such that $A(e_j) = 0$, for $j \neq 2, k$, $A(e_2) = e_k$, $A(e_k) = -e_2$ is skew-symmetric and

$$Aw = A(\sum_{i=3}^{n+2} w_i e_i) = -w_k e_2.$$

Thus,

$$0 = \langle z, Aw \rangle = \langle z_1 e_1 + z_2 e_2, Aw \rangle = \langle z_1 e_1 + z_2 e_2, -w_k e_2 \rangle = -z_2 w_k,$$

and so $w_k = 0$ for all k, since $z_2 \neq 0$. We conclude that

$$\xi = \lambda p + \mu z.$$

We also have $\langle \xi, z \rangle = 0$, since ξ is tangent to the sphere at z. Thus, we have

$$\lambda \langle z, p \rangle + \mu = 0,$$

so that $\mu = -\lambda \langle z, p \rangle$. Let $\langle z, p \rangle = \cos s_0$, $0 < s_0 < \pi$. Then

$$\xi = \lambda p - \lambda \langle z, p \rangle z = \lambda (p - \langle z, p \rangle z),$$

and

$$|\xi|^2 = \lambda^2 (1 + \langle z, p \rangle^2 - 2 \langle z, p \rangle^2) = \lambda^2 (1 - \langle z, p \rangle^2) = \lambda^2 \sin^2 s_0.$$

Thus, a unit normal vector ξ_0 to M at z has the form,

$$\xi_0 = \pm \frac{1}{\sin s_0} (p - \cos s_0 \, z), \tag{3.161}$$

and we take the positive sign in order to agree with the level set approach.

Recall that at a point $z_0 \in M$, we have

$$T_{z_0} M = \{ A z_0 \mid A \in \mathfrak{g} \},$$

where $\mathfrak{g}$ is the set of matrices of the form (3.160). For any $A \in \mathfrak{g}$, we have $\exp(tA) \in G$, and the curve

$$z_t = \exp(tA) z_0$$

has initial tangent vector $\overrightarrow{z}_0 = A z_0$. The vector ξ_0 in equation (3.161) with $z = z_0$ (where we take the positive sign) is

$$\xi_0 = \frac{1}{\sin s_0}(p - \cos s_0 \, z_0), \tag{3.162}$$

where $\langle z_0, p\rangle = \cos s_0$. This is a unit normal to M at the point z_0.

For any $g \in G$, at the point $z = gz_0$ we have $T_z M = gT_{z_0}M$ and $\xi_z = g\xi_0$. Then,

$$D_{Az_0}\xi = \frac{d}{dt}(\exp tA)\xi_0|_{t=0} = A\xi_0 = A\left(\frac{p - \cos s_0 z_0}{\sin s_0}\right) = -\cot s_0(Az_0),$$

which shows that M is umbilic at z_0 with principal curvature $\cot s_0$.

Now let $z = gz_0$ be any point in M. Then

$$gAz_0 = \frac{d}{dt}g(\exp tA)z_0|_{t=0},$$

so that

$$D_{gAz_0}\xi = \frac{d}{dt}(g\exp(tA)\xi_0) = gA\xi_0 = g(-\cot s_0)Az_0 = -\cot s_0(gAz_0),$$

so that M is totally umbilic with constant principal curvature $\cot s_0$.

3.8.2 The case $g = 2$

We know from Theorem 3.29 that an isoparametric hypersurface $M \subset S^{n+1}$ with two distinct principal curvatures is a standard product of two spheres in complementary orthogonal Euclidean spaces. In this section, we describe these hypersurfaces from various points of view.

We consider $\mathbf{R}^{n+2} = \mathbf{R}^{p+1} \times \mathbf{R}^{q+1}$, where $p + q = n$. For $z = (x, y)$ in $\mathbf{R}^{p+1} \times \mathbf{R}^{q+1}$, set

$$F(z) = |x|^2 - |y|^2. \tag{3.163}$$

Then we compute

$$\mathrm{grad}^E F = 2(x, -y), \quad |\mathrm{grad}^E F|^2 = 4r^2, \quad \Delta^E F = 2(p - q), \tag{3.164}$$

so that F satisfies the Cartan–Münzner differential equations (3.86)–(3.87) with $g = 2$ and $c = 2(p - q)$. Note that since $c = g^2(m_2 - m_1)/2 = 2(m_2 - m_1)$, we have $m_2 - m_1 = p - q$. Since we also have $m_1 + m_2 = p + q = n$, we conclude that

$$m_1 = q, \quad m_2 = p. \tag{3.165}$$

The restriction V of F to S^{n+1} satisfies

$$\operatorname{grad}^S V = 2((1-V)x, -(1+V)y), \quad |\operatorname{grad}^S V|^2 = 4(1-V^2), \qquad (3.166)$$
$$\Delta^S V = 2(p-q) - 2(n+2)V,$$

so that V is an isoparametric function. Let

$$M_s = \{z \in S^{n+1} \mid F(z) = \cos 2s\}, \quad 0 \le s \le \frac{\pi}{2}.$$

The two focal submanifolds are

$$M_0 = \{(x,y) \mid y = 0\} = S^p \times \{0\}, \quad M_{\frac{\pi}{2}} = \{(x,y) \mid x = 0\} = \{0\} \times S^q.$$

Note that $\dim M_0 = p = n - q = n - m_1$, and $\dim M_{\pi/2} = q = n - p = n - m_2$.

Products of spheres

Except for these two focal submanifolds, each level set is the Cartesian product of a p-sphere of radius $\cos s$ with a q-sphere of radius $\sin s$. To see this, note that if $z = (x,y)$ is in M_s, then

$$|x|^2 - |y|^2 = \cos 2s = \cos^2 s - \sin^2 s. \qquad (3.167)$$

On the other hand, since $(x,y) \in S^{n+1}$, we have

$$|x|^2 + |y|^2 = 1 = \cos^2 s + \sin^2 s. \qquad (3.168)$$

Adding these two equations, we get $|x|^2 = \cos^2 s$, while subtracting equation (3.167) from equation (3.168) yields $|y|^2 = \sin^2 s$.

Finally, the M_s form a system of parallel hypersurfaces with

$$M_{s_0 - t} = f_t(M_{s_0}), \quad s_0 - \frac{\pi}{2} < t < s_0,$$

provided that $\xi = \operatorname{grad}^S V / |\operatorname{grad}^S V|$ is used as unit normal to M_{s_0}.

As in the previous example, we compute the shape operator A at $z = (x,y)$ by using

$$\tilde{\nabla}_X(\operatorname{grad}^S V) = 2(1 - \cos 2s)X, \quad X \in \mathbf{R}^{p+1} \cap T_z S^{n+1},$$

$$\tilde{\nabla}_Y(\operatorname{grad}^S V) = -2(1 + \cos 2s)Y, \quad Y \in \mathbf{R}^{q+1} \cap T_z S^{n+1}.$$

Note that

$$T_z M = \{(X,Y) \mid \langle X, x\rangle = 0, \ \langle Y, y\rangle = 0\}$$

naturally decomposes into principal subspaces corresponding to principal curvatures,

$$\frac{-2(1-\cos 2s)}{2\sqrt{1-\cos^2 2s}} = \frac{-2\sin^2 s}{2\sin s\cos s} = -\tan s = \cot(s+\frac{\pi}{2})$$

and

$$\frac{2(1+\cos 2s)}{2\sqrt{1-\cos^2 2s}} = \cot s,$$

with respective multiplicities p and q.

Homogeneity in the case $g = 2$

The family $\{M_s\}$ may also be realized as the set of orbits of a group action. Pick $z_0 = (x_0, y_0)$ in S^{n+1} with neither x_0 nor y_0 equal to zero. Then

$$G = SO(p+1) \times SO(q+1)$$

is naturally embedded in $SO(n+2)$. The Lie algebra $\mathfrak{g}$ of G decomposes into the direct sum $\mathfrak{o}(p+1) \oplus \mathfrak{o}(q+1)$, and an element of $\mathfrak{g}$ may be written in the form

$$\begin{bmatrix} A & 0 \\ 0 & B \end{bmatrix}, \tag{3.169}$$

where A is a skew-symmetric $(p+1)\times(p+1)$ matrix and B is a skew-symmetric $(q+1)\times(q+1)$ matrix. The isotropy subgroup at $z_0 = (x_0, y_0)$ is $SO(p)\times SO(q)$, so that each orbit has codimension two in $\mathbf{R}^{n+2}$, and hence codimension one in S^{n+1}.

Again considering the exponential map of $\mathfrak{o}(n+2)$, we see that

$$T_{z_0}M = \{(Ax_0, By_0) \mid A \in \mathfrak{o}(p+1),\ B \in \mathfrak{o}(q+1)\},$$

so that $\xi = (u, v)$ is a unit normal to M at z_0 if and only if $\langle \xi, z_0\rangle = 0$, $|\xi| = 1$, and

$$\langle Ax_0, u\rangle = \langle By_0, v\rangle = 0,$$

for all A, B. In other words,

$$\xi = \pm(\tan s\, x_0, -\cot s\, y_0),$$

where $s \in (0, \pi/2)$ is chosen so that $|x_0| = \cos s$, $|y_0| = \sin s$. We take the positive sign to agree with the level set approach.

As in the $g = 1$ case, we compute for $Z = (Ax_0, By_0)$,

$$D_Z\xi = \frac{d}{dt}\begin{bmatrix} \exp tA & 0 \\ 0 & \exp tB \end{bmatrix}\begin{bmatrix} \tan s\, x_0 \\ -\cot s\, y_0 \end{bmatrix} = -\begin{bmatrix} -\tan s\, Ax_0 \\ \cot s\, By_0 \end{bmatrix},$$

so that the principal vectors and principal curvatures at z_0 are what we expected. Then the same situation holds at every point of M by homogeneity.

Note that the orbit under the G action of a point $(x, 0)$ with $|x| = 1$ is the focal submanifold M_0, while the orbit of a point of the form $(0, y)$ with $|y| = 1$ is the focal submanifold $M_{\pi/2}$.

Since the focal submanifolds are totally geodesic, we can also find the principal curvatures of M_s from the formula in Theorem 2.2 on page 17 for the shape operators of a tube. Consider M_s as a tube of radius s over the totally geodesic focal submanifold $M_0 = S^p \times \{0\}$. From Theorem 2.2, we see that M_s has two constant principal curvatures,

$$\lambda_1 = \cot\left(\frac{\pi}{2} - s\right) = \tan s, \quad \lambda_2 = -\cot s, \tag{3.170}$$

with respective multiplicities p and q. If we choose the opposite field of unit normals, then these principal curvatures agree with those obtained by the level set approach above.

3.8.3 The case $g = 3$

Isoparametric hypersurfaces with $g = 3$ principal curvatures are not as easily described as the hyperspheres and the products of spheres. We recall from Corollary 3.28 that $g = 3$ requires that the all the principal curvatures have the same multiplicity m so that the dimension n of M is a multiple of 3. Thus, for the first time, we have a restriction on the dimension in which isoparametric hypersurfaces with a given g can occur. As we have noted earlier, Cartan [55] proved that m is one of the numbers $1, 2, 4, 8$, and therefore n is 3, 6, 12, or 24.

In the examples presented so far, each isoparametric family can be viewed in the following three ways.

(1) as the level sets of an isoparametric function,
(2) as the set of tubes over its focal submanifolds,
(3) as the orbits of a certain group action.

We will present the case $g = 3$ in the same three ways. Note that description (3) is only possible if the hypersurfaces in the family are homogeneous, and this is not necessarily true in the case of $g = 4$ principal curvatures.

Cartan [54] showed that an isoparametric family of hypersurfaces with $g = 3$ principal curvatures of multiplicity m is determined as the family of level sets of the Cartan–Münzner polynomial $F(x, y, X, Y, Z)$ on $\mathbf{R}^{3m+2}$ given by

$$x^3 - 3xy^2 + \frac{3}{2}x(X\overline{X} + Y\overline{Y} - 2Z\overline{Z}) + \frac{3\sqrt{3}}{2}y(X\overline{X} - Y\overline{Y}) + \frac{3\sqrt{3}}{2}(XYZ + \overline{ZYX}). \qquad (3.171)$$

In this formula, x and y are real parameters, while X, Y, Z are coordinates in the division algebra $\mathbf{F} = \mathbf{R}, \mathbf{C}, \mathbf{H}$ (quaternions), $\mathbf{O}$ (Cayley numbers, i.e., octonians), for $m = 1, 2, 4, 8$, respectively. Note that the sum $XYZ + \overline{ZYX}$ is twice the real part of the product XYZ. In the case of the Cayley numbers, multiplication is not associative, but the real part of XYZ is the same whether one interprets the product as $(XY)Z$ or $X(YZ)$.

The isoparametric hypersurfaces in the family are the level sets M_t in S^{3m+1} determined by the equation $F = \cos 3t$, $0 < t < \pi/3$, where F is the polynomial in equation (3.171). The focal submanifolds are obtained by taking $t = 0$ and $t = \pi/3$. These focal submanifolds are a pair of antipodal standard embeddings of the projective plane $\mathbf{FP}^2$, for the appropriate division algebra $\mathbf{F}$ (see Section 2.9, [303, 505] or [95, pp. 87–90]). In the case of $\mathbf{F} = \mathbf{R}$, these focal submanifolds are standard Veronese surfaces in $S^4 \subset \mathbf{R}^5$. For the cases $\mathbf{F} = \mathbf{R}, \mathbf{C}, \mathbf{H}$, Cartan gave a specific parametrization of the focal submanifold M_0 where the polynomial $F = 1$ as follows:

$$X = \sqrt{3}v\overline{w}, \quad Y = \sqrt{3}w\overline{u}, \quad Z = \sqrt{3}u\overline{v},$$
$$x = \frac{\sqrt{3}}{2}(|u|^2 - |v|^2), \quad y = |w|^2 - \frac{|u|^2 + |v|^2}{2},$$

where u, v, w are in $\mathbf{F}$, and $|u|^2 + |v|^2 + |w|^2 = 1$. This map is invariant under the equivalence relation

$$(u, v, w) \sim (u\lambda, v\lambda, w\lambda), \quad \lambda \in \mathbf{F}, \ |\lambda| = 1.$$

Thus, it is well-defined on $\mathbf{FP}^2$, and it is easily shown to be injective on $\mathbf{FP}^2$. Therefore, it is an embedding of $\mathbf{FP}^2$ into S^{3m+1}. This parametrization differs slightly from that given in Section 2.9 for the standard embeddings of projective spaces.

Principal curvatures of the focal submanifolds

We now determine the principal curvatures of the focal submanifolds and the isoparametric hypersurfaces using Münzner's theory. For simplicity, we consider the case $\mathbf{F} = \mathbf{R}$. We first consider the focal submanifold M_0. By Corollary 3.22 and Theorem 3.26, for any unit normal ξ at any point $p \in M_0$, the shape operator A_ξ has the form,

$$A_\xi = \begin{bmatrix} \cot(2\pi/3) & 0 \\ 0 & \cot(\pi/3) \end{bmatrix} = \begin{bmatrix} -1/\sqrt{3} & 0 \\ 0 & 1/\sqrt{3} \end{bmatrix}. \tag{3.172}$$

For $0 < t < \pi/3$, the hypersurface M_t is a tube f_t of radius t over the Veronese surface M_0 in S^4, as can be seen by the definition of the Cartan–Münzner polynomial on page 116. By Theorem 2.2 on page 17 concerning the shape operator of a tube, the shape operator of M_t at the point $f_t(p, \xi)$ has the matrix form

$$A_t = \begin{bmatrix} \cot(\frac{2\pi}{3} - t) & 0 & 0 \\ 0 & \cot(\frac{\pi}{3} - t) & 0 \\ 0 & 0 & -\cot t \end{bmatrix}, \tag{3.173}$$

with respect to the appropriate orthonormal basis. For $t = \pi/6$, and thus $F = 0$, we get Cartan's minimal isoparametric hypersurface $M_{\pi/6}$, which has shape operator,

$$A_{\pi/6} = \begin{bmatrix} \cot\frac{\pi}{2} & 0 & 0 \\ 0 & \cot\frac{\pi}{6} & 0 \\ 0 & 0 & -\cot\frac{\pi}{6} \end{bmatrix} = \begin{bmatrix} 0 & 0 & 0 \\ 0 & \sqrt{3} & 0 \\ 0 & 0 & -\sqrt{3} \end{bmatrix}. \tag{3.174}$$

Note that $M_{\pi/6}$ is the unique minimal hypersurface in the isoparametric family, as can be seen by taking the trace of A_t given in equation (3.173). In 2006, Adachi and Maeda gave a characterization of a minimal Cartan hypersurface and certain products of spheres in terms of extrinsic properties of their geodesics (see [5] for more details).

Due to the nonassociativity of the Cayley numbers, there is no corresponding parametrization of the standard embedding of the Cayley projective plane. However, it can be described (see, for example, [162, 303]) as the submanifold

$$V = \{A \in M_{3\times 3}(\mathbf{O}) \mid \overline{A}^T = A = A^2, \text{ trace } A = 1\},$$

where $M_{3\times 3}(\mathbf{O})$ is the space of 3×3 matrices of Cayley numbers. This submanifold V lies in a sphere S^{25} in a 26-dimensional real subspace of $M_{3\times 3}(\mathbf{O})$.

Cartan's results imply that up to congruence, there is only one isoparametric family of hypersurfaces with $g = 3$ principal curvatures for each value of m. This classification is closely related to various characterizations of these standard embeddings of $\mathbf{FP}^2$. (See Ewert [148], Little [332], and Knarr-Kramer [282].)

See the papers of Knarr and Kramer [282], and Console and Olmos [121], for alternative proofs of Cartan's classification of isoparametric hypersurfaces with $g = 3$ principal curvatures. In a related paper, Sanchez [473] studied Cartan's isoparametric hypersurfaces from an algebraic point of view. (See also the paper of Giunta and Sanchez [174].)

Homogeneity of Cartan's hypersurfaces

Cartan showed that in each case $m = 1, 2, 4$ or 8, the isoparametric hypersurfaces and the two focal submanifolds are homogeneous, that is, they are orbits of points in S^{n+1} under the action of a closed subgroup of $SO(n+2)$. Here we give a presentation of this fact for the case $\mathbf{F} = \mathbf{R}$ as in [95, pp. 297–299]. An analogous construction can be made for the other algebras.

We consider $\mathbf{R}^9$ as the space $M_{3\times 3}(\mathbf{R})$ of 3×3 real matrices with standard inner product

$$\langle A, B\rangle = \text{trace}\, AB^T.$$

We consider the 5-dimensional subspace $\mathbf{R}^5$ of symmetric matrices with trace zero, and let S^4 be the unit sphere in $\mathbf{R}^5$. That is,

$$S^4 = \{A \in M_{3\times 3}(\mathbf{R}) \mid A = A^T,\ \text{trace}\, A = 0,\ |A| = 1\}.$$

The group $SO(3)$ acts on S^4 by conjugation. This action is isometric and thus preserves S^4. For every $A \in S^4$, there exists a matrix $U \in SO(3)$ such that UAU^T is diagonal. In fact, a direct calculation shows that every orbit of this action contains a representative of the form B_t, where B_t is a diagonal matrix whose diagonal entries are

$$\sqrt{\frac{2}{3}}\,\{\cos(t - \frac{\pi}{3}),\ \cos(t + \frac{\pi}{3}),\ \cos(t + \pi)\}.$$

If all the eigenvalues of B_t are distinct, then the orbit of B_t is 3-dimensional. For example, consider

$$B_{\pi/6} = \text{diagonal}\,\{1/\sqrt{2}, 0, -1/\sqrt{2}\}.$$

The isotropy subgroup of $B_{\pi/6}$ under this group action is the set of matrices in $SO(3)$ that commute with $B_{\pi/6}$. One can easily compute that this group consists of diagonal matrices in $SO(3)$ with entries ± 1 along the diagonal. This group is isomorphic to the Klein 4-group $\mathbf{Z}_2 \times \mathbf{Z}_2$, and thus the orbit $M_{\pi/6}$ is isomorphic to $SO(3)/\mathbf{Z}_2 \times \mathbf{Z}_2$. The hypersurface $M_{\pi/6}$ is the unique minimal hypersurface in the isoparametric family.

The two focal submanifolds are lower-dimensional orbits, and they occur when B_t has a repeated eigenvalue. For example, when $t = 0$, we have

$$B_0 = \text{diagonal}\,\{1/\sqrt{6}, 1/\sqrt{6}, -2/\sqrt{6}\}.$$

The isotropy subgroup for B_0 is the subgroup $S(O(2) \times O(1))$ consisting of all matrices of the form

$$\begin{bmatrix} A & 0 \\ 0 & \pm 1 \end{bmatrix}, \quad A \in O(2),$$

having determinant one. Thus, M_0 is diffeomorphic to $SO(3)/S(O(2) \times O(1))$ which is the real projective plane $\mathbf{RP}^2$. In fact, M_0 is a standard Veronese surface, as noted above. The other focal submanifold $M_{\pi/3}$ is the orbit of

$$B_{\pi/3} = \text{diagonal } \{2/\sqrt{6}, -1/\sqrt{6}, -1/\sqrt{6}\}.$$

This is also a standard Veronese surface antipodal to M_0.

Remark 3.66 (Irreducible proper Dupin hypersurfaces with $g = 3$). Working in the context of Lie sphere geometry, Pinkall [442, 445] showed that a connected irreducible proper Dupin hypersurface $M^3 \subset S^4$ with $g = 3$ principal curvatures is Lie equivalent to an isoparametric hypersurface in S^4. Later without using Cartan's classification of isoparametric hypersurfaces with $g = 3$ in S^4, Cecil and Chern [80] (see also [77, pp. 182–186]) showed directly that a connected irreducible proper Dupin hypersurface $M^3 \subset S^4$ with $g = 3$ principal curvatures is Lie equivalent to an open subset of a tube over a spherical Veronese surface $V^2 \subset S^4$.

Cecil and Jensen [85] later proved that if M^{n-1} is a connected irreducible proper Dupin hypersurface in S^n with three distinct principal curvatures of multiplicities m_1, m_2, m_3, then $m_1 = m_2 = m_3$, and M^{n-1} is Lie equivalent to an isoparametric hypersurface in S^n. Thus, by Cartan's classification of isoparametric hypersurfaces with $g = 3$, the common multiplicity m of the principal curvatures satisfies $m = 1, 2, 4$, or 8.

Remark 3.67 (Tight polyhedral models of isoparametric families). Banchoff and Kühnel [25] constructed tight polyhedral models of isoparametric families in spheres with $g \leq 3$ principal curvatures. In the case $g = 3$, their models are obtained as deleted joins of minimal (number of vertices) triangulated projective planes with their Alexander duals. This constructions works for the real and complex projective planes, but the cases of the quaternionic and Cayley projective planes are unresolved, since it is not clear if such minimal triangulations of these projective planes exist (see also [23]). Banchoff and Kühnel also introduced a definition of PL-tautness and showed that their models are PL-taut.

3.8.4 A homogeneous example with $g = 4$

In this section, we describe an important class of homogeneous isoparametric hypersurfaces with $g = 4$ distinct principal curvatures. This example is due in full generality to Nomizu [403, 404], and it was given by Cartan [56] in the case of dimension M equal to 4. This is an important example, because it is the simplest case of the isoparametric hypersurfaces constructed by Ferus, Karcher and Münzner which will be discussed in detail in Section 3.9. Our treatment here follows Nomizu [403, 404] (see also [95, pp. 299–303]).

We consider the $(m+1)$-dimensional complex vector space $\mathbf{C}^{m+1}$ as a real vector space $\mathbf{C}^{m+1} = \mathbf{R}^{m+1} \oplus i\mathbf{R}^{m+1}$. The real inner product on $\mathbf{C}^{m+1}$ is given by

$$\langle z, w \rangle = \langle x, u \rangle + \langle y, v \rangle,$$

for $z = x + iy$, $w = u + iv$ for $x, y, u, v \in \mathbf{R}^{m+1}$. The unit sphere in $\mathbf{C}^{m+1}$ is

$$S^{2m+1} = \{z \in \mathbf{C}^{m+1} \mid |z| = 1\}.$$

In the following construction, we assume that $m \geq 2$. In the case $m = 1$, this construction reduces to a product of two circles in S^3. Consider the homogeneous polynomial F of degree 4 on $\mathbf{C}^{m+1}$ given by

$$F(z) = |\sum_{k=0}^{m} z_k^2|^2 = (|x|^2 - |y|^2)^2 + 4\langle x, y \rangle^2, \text{ for } z = x + iy. \tag{3.175}$$

A direct computation shows that

$$|\text{grad}^E F|^2 = 16r^2 F, \quad \Delta^E F = 16r^2. \tag{3.176}$$

By Theorem 3.30, the restriction V of F to S^{2m+1} satisfies

$$|\text{grad}^S V|^2 = 16V(1 - V), \quad \Delta^S V = 16 - V(16 + 8m), \tag{3.177}$$

and so V is an isoparametric function on S^{2m+1}.

Remark 3.68. Note that F does not satisfy the Cartan–Münzner differential equations (3.86)–(3.87). However, as noted by Takagi [510], the polynomial $\tilde{F} = r^4 - 2F$ has the same level sets as F on S^{2m+1}, since the restriction $\tilde{V}$ of $\tilde{F}$ to S^{2m+1} satisfies $\tilde{V} = 1 - 2V$. The function $\tilde{F}$ satisfies the equations,

$$|\text{grad}^E \tilde{F}|^2 = 16r^6, \quad \Delta^E \tilde{F} = 8(m - 2)r^2, \tag{3.178}$$

and so it does satisfy the Cartan–Münzner differential equations with $g = 4$ and multiplicities $m_2 = m - 1$ and $m_1 = 1$, since $c = g^2(m_2 - m_1)/2 = 8(m - 2)$.

Focal submanifolds

We now continue our discussion of this example using the functions F and V. From equation (3.177), we see that the focal submanifolds occur when $V = 0$ or 1. From equation (3.175), we that $V = 1$ is equivalent to the condition,

$$|\sum_{k=0}^{m} z_k^2| = 1.$$

This is easily seen to be equivalent to the condition that z lies in the set

$$M_0 = \{e^{i\theta}x \mid x \in S^m\},$$

where S^m is the unit sphere in the first factor $\mathbf{R}^{m+1}$. For $x \in S^m$, we have

$$T_xM_0 = T_xS^m \oplus \text{Span}\,\{ix\}.$$

Thus, the normal space to M_0 at x is

$$T_x^{\perp}M_0 = \{iy \mid y \in S^m,\ \langle x, y\rangle = 0\}.$$

The normal geodesic to M_0 through x in the direction iy can be parametrized as

$$\cos t\, x + \sin t\, iy. \tag{3.179}$$

At the point $e^{i\theta}x$ in the focal submanifold M_0, one can easily show that

$$T^{\perp}M_0 = \{e^{i\theta}y \mid y \in S^m,\ \langle x, y\rangle = 0\}.$$

Thus, the normal geodesic to M_0 through the point $e^{i\theta}x$ in the direction $e^{i\theta}iy$ can be parametrized as

$$\cos t\, e^{i\theta}x + \sin t\, e^{i\theta}iy = e^{i\theta}(\cos t\, x + \sin t\, iy). \tag{3.180}$$

Let $V_{m+1,2}$ be the Stiefel manifold of orthonormal pairs of vectors (x, y) in $\mathbf{R}^{m+1}$. By equations (3.179) and (3.180), we see that the tube M_t of radius t over the focal submanifold M_0 is given by

$$M_t = \{e^{i\theta}(\cos t\, x + \sin t\, iy) \mid (x, y) \in V_{m+1,2}\}. \tag{3.181}$$

In fact, the map $f_t : S^1 \times V_{m+1,2} \to S^{2m+1}$ given by

$$f_t(e^{i\theta}, (x, y)) = e^{i\theta}(\cos t\, x + \sin t\, iy), \tag{3.182}$$

is an immersion that is a double covering of the tube M_t, since

$$f_t(e^{i\theta}, (x, y)) = f_t(e^{i(\theta+\pi)}, (-x, -y)).$$

Substituting equation (3.182) into the defining formula (3.175) for F shows that for $z \in M_t$, the restriction V of F to S^{2m+1} satisfies

$$V(z) = (\cos^2 t - \sin^2 t)^2 = \cos^2 2t.$$

Therefore, the other focal submanifold determined by the equation $V = 0$ occurs when $t = \pi/4$. By equations (3.181), we get that the focal submanifold $M_{\pi/4}$ consists of points of the form $e^{i\theta}(x+iy)/\sqrt{2}$ for $(x, y) \in V_{m+1,2}$. On the other hand, the equation $V = 0$ implies that $|x| = |y|$ and $\langle x, y\rangle = 0$, and we conclude that

$$M_{\pi/4} = \{(x + iy)/\sqrt{2} \mid (x, y) \in V_{m+1,2}\}.$$

This is an embedded image of the Stiefel manifold $V_{m+1,2}$ which has dimension $2m - 1$.

Remark 3.69. The fact that one of the focal submanifolds is a Stiefel manifold is an important feature of the general construction of Ferus, Karcher, and Münzner [160] (see Section 3.9), in which case one of the focal submanifolds is always a so-called Clifford–Stiefel manifold determined by a corresponding Clifford algebra.

Since adjacent focal points along a normal geodesic are at a distance $\pi/4$ apart, we know from Münzner's Theorem 3.26 on page 108 that the tube M_t is an isoparametric hypersurface with four principal curvatures,

$$\cot t,\ \cot(t + \frac{\pi}{4}),\ \cot(t + \frac{\pi}{2}),\ \cot(t + \frac{3\pi}{4}). \tag{3.183}$$

The focal submanifolds M_0 and $M_{\pi/4}$ have respective dimensions $m+1$ and $2m - 1$. Thus, the principal curvatures in equation (3.183) have respective multiplicities $(m - 1), 1, (m - 1), 1$. This agrees with the information that we obtained from the Cartan-Münzner polynomial in Remark 3.68.

From equation (3.182), we see that M_t admits a transitive group of isometries isomorphic to $SO(2) \times SO(m + 1)$, and hence each M_t is an orbit hypersurface. This is the fifth example on the list of Takagi and Takahashi [511] of homogeneous isoparametric hypersurfaces. Later Takagi [510] showed that if an isoparametric hypersurface M in S^{2m+1} has four principal curvatures with multiplicities $(m - 1)$, $1, (m-1), 1$, then M is congruent to a hypersurface M_t of this example. In particular, M is homogeneous.

Regarding the other focal submanifold M_0, the map $f_0 : S^1 \times S^m \to M_0$ given by $f_0(e^{i\theta}, x) = e^{i\theta}x$ is a double covering of M_0, since

$$f_0(e^{i\theta}, x) = f_0(e^{i(\theta+\pi)}, -x).$$

Hence, M_0 can be considered as a quotient manifold with identifications given by the map f_0. The two spheres $\{1\} \times S^m$ and $\{-1\} \times S^m$ are attached via the antipodal map of S^m. Thus, M_0 is orientable if and only if the antipodal map on S^m preserves orientation on S^m. Therefore, M_0 is orientable if m is odd and non-orientable if m is even.

This example illustrates Münzner's discussion of the orientability of the focal submanifolds in Theorem C [381, p. 59]. Münzner shows that in the case $g = 4$ there are two possibilities. In the first situation, one of the focal submanifolds

M_+, M_- is orientable and the other is not, in which case, if M_+ is orientable, then the multiplicity $m_+ = 1$. In the second situation, both focal submanifolds are orientable, in which case $m_+ + m_-$ is odd or else $m_+ = m_-$ is even.

In our example, $m_+ = 1$, $m_- = m - 1$ and $m_+ + m_- = m$. Thus, if m is even, Münzner's theorem implies that the focal submanifold $M_- = M_0$ corresponding to the principal curvature of multiplicity $m - 1$ is non-orientable, as we have shown above. In the case where m is odd, then both focal submanifolds in our example are orientable, which is consistent with Münzner's theorem, although not a consequence of it.

Remark 3.70 (Real hypersurfaces in complex projective space). Note that in our example, each M_t, including the two focal submanifolds, is invariant under the S^1-action on S^{2m+1} given by multiplication by $e^{i\theta}$. Hence, under the projection $\pi : S^{2m+1} \to \mathbf{CP}^m$ of the Hopf fibration, each submanifold M_t projects to a real submanifold of $\mathbf{CP}^m$ of dimension one less than the dimension of M_t. Under the projection π, the image of M_0 is a naturally embedded totally geodesic $\mathbf{RP}^m$ in $\mathbf{CP}^m$, whereas the image of $M_{\pi/4}$ is the complex quadric hypersurface Q^{m-1} (complex dimension) given by the equation

$$z_0^2 + \cdots + z_m^2 = 0.$$

The image under π of each hypersurface $M_t, 0 < t < \pi/4$, is a tube of constant radius in $\mathbf{CP}^m$ over $\mathbf{RP}^m$ and also over Q^{m-1}. It is a real hypersurface with three distinct constant principal curvatures in $\mathbf{CP}^m$. Real hypersurfaces with constant principal curvatures in $\mathbf{CP}^m$ will be studied in more detail in Chapters 6–8. (see also Takagi [507–509], Cecil–Ryan [94] and Niebergall-Ryan [399]).

3.8.5 Homogeneous examples and the case $g = 6$

All of the examples that we have given so far are homogeneous. In 1972, R. Takagi and T. Takahashi [511] published a complete classification of homogeneous isoparametric hypersurfaces in spheres based on the work of Hsiang and Lawson [202]. Takagi and Takahashi showed that each homogeneous isoparametric hypersurface M in S^{n+1} is a principal orbit of the isotropy representation of a Riemannian symmetric space of rank 2, and they gave a complete list of examples [511, p. 480]. (See also H. Takagi [506] for necessary and sufficient conditions for an isoparametric hypersurface to be homogeneous in terms of derivatives of the second fundamental form.)

Takagi and Takahashi found homogeneous examples with $g = 1, 2, 3, 4$, or 6 principal curvatures, i.e., all values of g possible by Münzner's Theorem 3.49. The examples with $g = 1, 2, 3$ have been discussed earlier in this section. For $g = 4$, there exist homogeneous examples with multiplicities (m_1, m_2) as follows (in this list m can be any positive integer): $(1, m)$ (as in Section 3.8.4), $(2, 2m - 1)$,

$(4, 4m-1)$, $(2, 2)$, $(4, 5)$, $(9, 6)$. As we will see in Section 3.9, there are inhomogeneous isometric hypersurfaces with multiplicities different from those on this list due to Ozeki and Takeuchi [422, 423], and Ferus, Karcher, and Münzner [160].

In the case of $g = 6$ principal curvatures, Münzner [382] showed that all of the principal curvatures have the same multiplicity m, and Abresch [2] showed that m equals 1 or 2. Takagi and Takahashi found homogeneous isoparametric families in both cases $m = 1$ and $m = 2$, and they showed that up to congruence, there is only one homogeneous isoparametric family in each case.

Peng and Hou [429] gave explicit forms for the Cartan–Münzner polynomials of degree six for the homogeneous isoparametric hypersurfaces with $g = 6$, while Grove and Halperin [184], and Fang [151] proved several results concerning the topology of isoparametric and compact proper Dupin hypersurfaces with six principal curvatures.

R. Miyaoka [367] gave a geometric description of the case $g = 6$, $m = 1$. She proved that a homogeneous isoparametric $M^6 \subset S^7$ with six principal curvatures can be obtained as the inverse image under the Hopf fibration $h : S^7 \to S^4$ of an isoparametric hypersurface $W^3 \subset S^4$ with three principal curvatures, as discussed in Subsection 3.8.3. As noted earlier, the two focal submanifolds of W^3 in S^4 are a pair of antipodal Veronese surfaces. Miyaoka showed that the two focal submanifolds of M^6 in S^7 are not congruent, even though they are lifts under h^{-1} of the two congruent antipodal Veronese surfaces in S^4. Thus, these two focal submanifolds of M^6 in S^7 are two non-congruent minimal homogeneous embeddings of $\mathbf{RP}^2 \times S^3$ in S^7.

In a later paper, Miyaoka [371] gave a geometric description of the homogeneous case $g = 6$, $m = 2$, by considering adjoint orbits of the exceptional compact Lie group G_2 on S^{13}, where G_2 acts on its Lie algebra $\mathfrak{g} \simeq \mathbf{R}^{14}$ as an isometry with respect to the bi-invariant metric. Miyaoka characterizes the orbits as fibered spaces over S^6 with fibers given by Cartan hypersurfaces with three principal curvatures of multiplicity two, i.e., the case $\mathbf{F} = \mathbf{C}$ in Subsection 3.8.3. This connects the case $(g, m) = (6, 2)$ with the case $(g, m) = (3, 2)$. The fibrations on the two singular orbits M_+ and M_- are diffeomorphic to the twistor fibrations on S^6 and on the quaternionic Kähler manifold $G_2/SO(4)$, respectively. From the viewpoint of symplectic geometry, Miyaoka shows that there exists a 2-parameter family of Lagrangian submanifolds on every orbit.

Dorfmeister and Neher [139] proved that every isoparametric hypersurface $M^6 \subset S^7$ with $g = 6$ principal curvatures of multiplicity $m = 1$ is homogeneous. The proof of Dorfmeister and Neher is very algebraic in nature. See Miyaoka [370] and Siffert [484] for alternate approaches.

The classification of isoparametric hypersurfaces with six principal curvatures of multiplicity $m = 2$ has been a long-standing problem in the field, and it was listed together with the classification of isoparametric hypersurfaces with $g = 4$ as Problem 34 on Yau's [562] list of important open problems in geometry in 1990. Recently, R. Miyaoka [373] (see also the errata [374]) published a proof that every isoparametric hypersurface $M^{12} \subset S^{13}$ with $g = 6$ principal curvatures of multiplicity $m = 2$ is homogeneous. The errata, which have been accepted for publication, pertain to an error in the original proof that was pointed out by Abresch and Siffert (see also [484]).

3.8.6 Isoparametric submanifolds of higher codimension

As noted in Remark 3.16 on page 101, there is also an extensive theory of isoparametric to submanifolds of codimension greater than one in the sphere. This theory was developed by several authors, including Carter and West [66–68, 553], Terng [525–529], Hsiang, Palais and Terng [203], Strübing [503], and Harle [189].

By definition, a connected, complete submanifold V in a real space form $\tilde{M}^n$ is said to be *isoparametric* if it has flat normal bundle and if for any parallel section of the unit normal bundle $\eta : V \to B^{n-1}$, the principal curvatures of the shape operator A_η are constant.

Several authors, in particular Terng [525], who showed that compact, connected isoparametric submanifolds of higher codimension are taut, made important contributions to this theory. This culminated with the work of Thorbergsson [537], who used the theory of Tits buildings to show that all irreducible isoparametric submanifolds which are substantially embedded in S^n with codimension greater than one are homogeneous. Thus, they are principal orbits of isotropy representations of symmetric spaces (also known as standard embeddings of R-spaces or generalized flag manifolds), as was the case for homogeneous isoparametric hypersurfaces in the sphere. Subsequent to Thorbergsson's paper, Olmos [409], and Heintze and Liu [195] published alternate proofs of Thorbergsson's result.

The theory of R-spaces was developed extensively in the papers of Bott and Samelson [49], and Takeuchi and Kobayashi [513]. (See also the papers of Heintze, Olmos and Thorbergsson [196], Thorbergsson [538], and the books by Palais and Terng [426], Kramer [297], and Berndt, Console and Olmos [33] for thorough treatments of these topics.) The R-spaces were shown to be taut submanifolds by Bott and Samelson [49] (see also Takeuchi and Kobayashi [513]).

An isoparametric submanifold of codimension greater than one in the sphere is always Dupin, but it may not be proper Dupin (see Terng [529, pp. 464–469] for more detail). Pinkall [446, p. 439] proved that every extrinsically symmetric submanifold of a real space form is Dupin. Takeuchi [512] then determined which of these are proper Dupin.

In a further generalization, Heintze, Olmos, and Thorbergsson [196] defined a submanifold $\phi : V \to \mathbf{R}^n$ (or S^n) to have *constant principal curvatures* if for any smooth curve γ on V and any parallel normal vector field $\xi(t)$ along γ, the shape operator $A_{\xi(t)}$ has constant eigenvalues along γ. If the normal bundle $N(M)$ is flat, then having constant principal curvatures is equivalent to being isoparametric. They then showed that a submanifold with constant principal curvatures is either isoparametric or a focal submanifold of an isoparametric submanifold.

The theory of isoparametric submanifolds of codimension greater than one was generalized to submanifolds of hyperbolic space by Wu [555] and Zhao [563]. Verhóczki [543] then developed a theory of isoparametric submanifolds for Riemannian manifolds which do not have constant curvature.

Terng and Thorbergsson [531] (and independently Grove and Halperin [185]) gave a definition of tautness for submanifolds of arbitrary complete Riemannian manifolds, and they discussed the notions of isoparametric, equifocal and Dupin submanifolds in that setting.

West [553] and Mullen [380] formulated a theory of isoparametric systems on symmetric spaces, and Terng and Thorbergsson [530] studied compact isoparametric submanifolds of symmetric spaces using the related notion of equifocal submanifolds.

Christ [119] then generalized Thorbergsson's result for submanifolds of S^n by showing that a complete connected irreducible equifocal submanifold of codimension greater than one in a simply connected compact symmetric space is homogeneous. In a related paper, Tang [515] studied the possible multiplicities of the focal points of equifocal hypersurfaces in symmetric spaces (see also Fang [152]). A promising recent generalization of the theory of isoparametric submanifolds is the theory of singular Riemannian foliations admitting sections (see Alexandrino [9], Töben [540], Lytchak [338, 339], Lytchak and Thorbergsson [340, 341], Thorbergsson [539], and Wiesendorf [554]).

Terng [529] considered isoparametric submanifolds in infinite-dimensional Hilbert spaces and generalized many results from the finite-dimensional case to that setting. Pinkall and Thorbergsson [450] then gave more examples of such submanifolds, and Heintze and Liu [195] generalized the finite-dimensional homogeneity result of Thorbergsson [534] to the infinite-dimensional case. Mare [352] obtained descriptions of the cohomology ring of an isoparametric hypersurface in Hilbert space in terms of multiplicities and characteristic classes of the curvature distributions. (See also the recent paper of Koike [288] on infinite-dimensional isoparametric submanifolds.)

The survey paper of Thorbergsson [538] has a good account of all the topics mentioned in this section.

3.9 Isoparametric Hypersurfaces of FKM-type

In a paper published in 1981, Ferus, Karcher, and Münzner [160] constructed an infinite class of isoparametric hypersurfaces with $g = 4$ principal curvatures that includes all known examples with $g = 4$ except for two homogeneous examples with multiplicities $(2, 2)$ and $(4, 5)$. This construction is based on representations of Clifford algebras, and the classification of such representations is an important element in the construction. The FKM construction is a generalization of an earlier construction of Ozeki and Takeuchi [422, 423], who also used representations of certain Clifford algebras.

In this section, we describe the construction of Ferus, Karcher, and Münzner following their original paper [160] closely. (See also the notes of Ferus [159], and Section 4.7 of the book [77], which is also based on the original paper [160].) At the end of this section, we will also discuss the known classification results for isoparametric hypersurfaces with four principal curvatures.

First we recall some facts about Clifford algebras and their representations. For each integer $m \geq 0$, the *Clifford algebra* C_m is the associative algebra over $\mathbf{R}$ that is generated by a unity 1 and the elements $e_1, \ldots, e_m$ subject only to the relations

$$e_i^2 = -1, \quad e_i e_j = -e_j e_i, \quad i \neq j, \quad 1 \leq i,j \leq m. \tag{3.184}$$

One can show that the set

$$\{1, e_{i_1} \cdots e_{i_r} \mid i_1 < \cdots < i_r, \quad 1 \leq r \leq m\}, \tag{3.185}$$

forms a basis for the underlying vector space C_m, and thus $\dim C_m = 2^m$.

The Clifford algebra C_0 is isomorphic to $\mathbf{R}$, and C_1 is isomorphic to the complex numbers $\mathbf{C}$ with e_1 equal to the complex number i. Atiyah, Bott, and Shapiro [16] explicitly determined all the Clifford algebras and their representations as formulated in the table in equation (3.188) below.

We first need some terminology in order to understand the table. Let $R(q)$ denote the algebra of $q \times q$ matrices with entries from the algebra R. The multiplication in $R(q)$ is matrix multiplication defined using the operations of addition and multiplication in the algebra R. The direct sum $R_1 \oplus R_2$ is the Cartesian product $R_1 \times R_2$ with all algebra operations defined coordinatewise.

A representation of a Clifford algebra on $\mathbf{R}^q$ corresponds to a set $E_1, \ldots, E_m$ of skew-symmetric $q \times q$ matrices satisfying

$$E_i^2 = -I, \quad E_i E_j = -E_j E_i, \quad i \neq j, \quad 1 \leq i,j \leq m. \tag{3.186}$$

Note that the skew-symmetry and equation (3.186) imply that the E_i is also orthogonal, since for all $v, w \in \mathbf{R}^q$,

$$\langle E_i v, E_i w \rangle = \langle v, E_i^T E_i w \rangle = \langle v, (-E_i) E_i w \rangle = \langle v, Iw \rangle = \langle v, w \rangle. \tag{3.187}$$

Atiyah, Bott, and Shapiro determined all of the Clifford algebras according to the table below. Moreover, they showed that the Clifford algebra C_{m-1} has an irreducible representation of degree q if and only if $q = \delta(m)$ as in the table.

m	C_{m-1}	$\delta(m)$
1	$\mathbf{R}$	1
2	$\mathbf{C}$	2
3	$\mathbf{H}$	4
4	$\mathbf{H} \oplus \mathbf{H}$	4
5	$\mathbf{H}(2)$	8
6	$\mathbf{C}(4)$	8
7	$\mathbf{R}(8)$	8
8	$\mathbf{R}(8) \oplus \mathbf{R}(8)$	8
$k+8$	$C_{k-1}(16)$	$16\delta(k)$

(3.188)

Clifford algebras and the degree of an irreducible representation

One can obtain reducible representations of C_{m-1} on $\mathbf{R}^q$ for $q = k\delta(m)$, $k > 1$, by taking a direct sum of k irreducible representations of C_{m-1} on $\mathbf{R}^{\delta(m)}$.

Clifford systems of symmetric operators

Ferus, Karcher, and Münzner used Clifford systems of symmetric operators that are closely related to representations of Clifford algebras. Let $H(n, \mathbf{R})$ be the space of symmetric $n \times n$ matrices with real entries on which we have the standard inner product,

$$\langle A, B \rangle = \text{trace}\,(AB)/n. \tag{3.189}$$

For positive integers l and m, the $(m + 1)$-tuple $(P_0, \ldots, P_m)$ with $P_i \in H(2l, \mathbf{R})$ is called a *(symmetric) Clifford system* on $\mathbf{R}^{2l}$ if the P_i satisfy

$$P_i^2 = I, \quad P_iP_j = -P_jP_i, \quad i \neq j, \quad 0 \leq i, j \leq m. \tag{3.190}$$

Note that the transformations P_i in a Clifford system are orthogonal since for all $x, y \in \mathbf{R}^{2l}$,

$$\langle P_ix, P_iy \rangle = \langle x, P_i^2 y \rangle = \langle x, Iy \rangle = \langle x, y \rangle. \tag{3.191}$$

If $(P_0, \ldots, P_m)$ is a Clifford system on $\mathbf{R}^{2l}$ and $(Q_0, \ldots, Q_m)$ is a Clifford system on $\mathbf{R}^{2n}$, then we can define a Clifford system $(P_0 \oplus Q_0, \ldots, P_m \oplus Q_m)$ on $\mathbf{R}^{2l} \oplus \mathbf{R}^{2n} = \mathbf{R}^{2(l+n)}$ by defining $(P_i \oplus Q_i)(x, y) = (P_ix, Q_iy)$. This system is called the *direct sum* of $(P_0, \ldots, P_m)$ and $(Q_0, \ldots, Q_m)$.

A Clifford system $(P_0, \ldots, P_m)$ on $\mathbf{R}^{2l}$ is called *irreducible* if it is not possible to write $\mathbf{R}^{2l}$ as a direct sum of two positive-dimensional subspaces that are invariant under all of the P_i.

There is an explicit correspondence between Clifford systems on $\mathbf{R}^{2l}$ and representations of Clifford algebras on $\mathbf{R}^l$ which we now describe. First suppose that $E_1, \ldots, E_m$ are skew-symmetric $l \times l$ real matrices that satisfy equation (3.186) and thereby determine a representation of the Clifford algebra C_{m-1} on $\mathbf{R}^l$. We write $\mathbf{R}^{2l} = \mathbf{R}^l \oplus \mathbf{R}^l$ and define symmetric transformations $P_0, \ldots, P_m$ by

$$\begin{aligned} P_0(u, v) &= (u, -v), \quad P_1(u, v) = (v, u), \\ P_{1+i}(u, v) &= (E_iv, -E_iu), \quad 1 \leq i \leq m - 1. \end{aligned} \tag{3.192}$$

We can show that $(P_0, \ldots, P_m)$ is a Clifford system as follows. First the transformations P_0 and P_1 are clearly symmetric. To check that P_{1+i} is symmetric for $1 \leq i \leq m - 1$, note that

$$\langle P_{1+i}(u,v),(x,y)\rangle = \langle (E_i v, -E_i u),(x,y)\rangle = \langle E_i v, x\rangle - \langle E_i u, y\rangle =$$
$$\langle -v, E_i x\rangle + \langle u, E_i y\rangle = \langle (u,v),(E_i y, -E_i x)\rangle = \langle (u,v), P_{1+i}(x,y)\rangle.$$

To show that the P_i satisfy equation (3.190), note that $P_0^2 = P_1^2 = I$ is clearly true. Then we compute for $1 \leq i \leq m-1$,

$$P_{1+i}^2(u,v) = P_{1+i}(E_i v, -E_i u) = (-E_i^2 u, -E_i^2 v) = (u,v), \tag{3.193}$$

so that $P_{1+i}^2 = I$. In a similar way, one can show that $P_i P_j = -P_j P_i$ for $i \neq j$, $0 \leq i,j \leq m$, and so $(P_0, \ldots, P_m)$ is a Clifford system.

Conversely, suppose that $(P_0, \ldots, P_m)$ is a Clifford system on $\mathbf{R}^{2l}$. Since $P_i^2 = I$, the eigenvalues of P_i can only be ± 1 for $0 \leq i \leq m$. We denote the eigenspaces by

$$\begin{aligned} E_+(P_i) &= \{x \in \mathbf{R}^{2l} \mid P_i x = x\}, \\ E_-(P_i) &= \{x \in \mathbf{R}^{2l} \mid P_i x = -x\}. \end{aligned} \tag{3.194}$$

The equation $P_i P_j + P_j P_i = 0$, for $i \neq j$, implies that P_j interchanges the eigenspaces $E_+(P_i)$ and $E_-(P_i)$ for each $i \neq j$. Thus, $E_+(P_i)$ and $E_-(P_i)$ both have dimension l for each i. This shows that each P_i has trace zero. In particular, for $1 \leq j \leq m$, the transformation P_j interchanges the spaces $E_+(P_0)$ and $E_-(P_0)$. Thus, the space $E_+(P_0)$ is invariant under the transformations $P_1 P_{1+i}$, $1 \leq i \leq m-1$, since P_{1+i} maps $E_+(P_0)$ to $E_-(P_0)$, and then P_1 maps $E_-(P_0)$ to $E_+(P_0)$.

We identify $\mathbf{R}^l$ with $E_+(P_0)$ and define the transformation $E_i : \mathbf{R}^l \to \mathbf{R}^l$ $1 \leq i \leq m-1$, to be the restriction to $E_+(P_0)$ of the transformation $P_1 P_{1+i}$. Then a direct calculation using equation (3.190) shows that $E_1, \ldots, E_{m-1}$ are skew-symmetric and that they determine a representation of the Clifford algebra C_{m-1} on $\mathbf{R}^l$ (see [77, pp. 100–101] for more detail).

Using this correspondence between Clifford systems and representations of Clifford algebras, Ferus, Karcher, and Münzner deduce several important facts about Clifford systems from known results about representations of Clifford algebras. In particular, a Clifford system is irreducible if and only if the corresponding Clifford algebra representation is irreducible, and thus there exists an irreducible Clifford system $(P_0, \ldots, P_m)$ on $\mathbf{R}^{2l}$ if and only if $l = \delta(m)$ as in equation (3.188).

A next logical question is when are two Clifford systems equivalent in some sense. Ferus, Karcher, and Münzner define two Clifford systems $(P_0, \ldots, P_m)$ and $(Q_0, \ldots, Q_m)$ on $\mathbf{R}^{2l}$ to be *algebraically equivalent* if there exists an orthogonal transformation $A \in O(2l)$ such that $Q_i = A P_i A^T$, for $0 \leq i \leq m$. Two Clifford systems are said to be *geometrically equivalent* if there exists

$$B \in O(\text{ Span } \{P_0, \ldots, P_m\} \subset H(2l, \mathbf{R}))$$

such that $(Q_0, \ldots, Q_m)$ and $(BP_0, \ldots, BP_m)$ are algebraically equivalent.

Ferus, Karcher, and Münzner show that for $m \not\equiv 0 \pmod 4$, there exists exactly one algebraic equivalence class of irreducible Clifford systems. Thus, in this case, there can be only one geometric equivalence class also. Hence, for each positive integer k there exists exactly one algebraic (or geometric) equivalence class of Clifford systems $(P_0, \ldots, P_m)$ on $\mathbf{R}^{2l}$ with $l = k\delta(m)$.

For $m \equiv 0 \pmod 4$, there exist exactly two algebraic classes of irreducible Clifford systems. These can be distinguished from each other by the choice of sign in the equation

$$\text{trace } (P_0 \cdots P_m) = \pm \text{ trace } I = \pm\, 2\delta(m). \tag{3.195}$$

In this case, there is also only one geometric equivalence class of irreducible Clifford systems. This can be seen by replacing P_0 by $-P_0$. The absolute trace,

$$|\text{ trace } (P_0 \cdots P_m)|, \tag{3.196}$$

is obviously an invariant under geometric equivalence. If one constructs all possible direct sums using both of the algebraic equivalence classes of irreducible Clifford systems with altogether k summands, then this invariant takes on $[k/2] + 1$ different values, where $[k/2]$ is the greatest integer less than or equal to $k/2$. Thus, for $m \equiv 0$ (mod 4), there are exactly $[k/2]+1$ distinct geometric equivalence classes of Clifford systems on $\mathbf{R}^{2l}$ with $l = k\delta(m)$.

The Clifford sphere

An important object in the theory of Ferus, Karcher, and Münzner is the *Clifford sphere* determined by a Clifford system $(P_0, \ldots, P_m)$ on $\mathbf{R}^{2l}$. This is defined to be the unit sphere in the space Span $\{P_0, \ldots, P_m\} \subset H(2l, \mathbf{R})$. Ferus, Karcher, and Münzner [160, p. 484] show that the Clifford sphere, denoted $\Sigma(P_0, \ldots, P_m)$, has the following important properties. The proof here follows the original proof of Ferus, Karcher, and Münzner (see also [77, pp. 102–105]).

Theorem 3.71 (Properties of the Clifford sphere). *The Clifford sphere* $\Sigma(P_0, \ldots, P_m)$ *has the following properties.*

(a) *For each* $P \in \Sigma(P_0, \ldots, P_m)$*, we have* $P^2 = I$*. Conversely, if* Σ *is the unit sphere in a linear subspace* W *spanned by* Σ *in* $H(2l, \mathbf{R})$ *such that* $P^2 = I$ *for all* $P \in \Sigma$*, then every orthonormal basis of* W *is a Clifford system on* $\mathbf{R}^{2l}$.

(b) *Two Clifford systems are geometrically equivalent if and only if their Clifford spheres are conjugate to one another under an orthogonal transformation of* $\mathbf{R}^{2l}$.

(c) *The function,*

$$H(x) = \sum_{i=0}^{m} \langle P_i x, x\rangle^2, \tag{3.197}$$

depends only on $\Sigma(P_0,\ldots,P_m)$ *and not on the choice of orthonormal basis* $(P_0,\ldots,P_m)$. *For* $P \in \Sigma(P_0,\ldots,P_m)$, *we have*

$$H(Px) = H(x), \tag{3.198}$$

for all x *in* $\mathbf{R}^{2l}$.

(d) *For an orthonormal set* $\{Q_1,\ldots,Q_r\}$ *in* $\Sigma(P_0,\ldots,P_m)$, *since* $Q_iQ_j = -Q_jQ_i$, *for* $i \neq j$, *we have*

$$\begin{aligned} &Q_1\cdots Q_r \text{ is symmetric if } r \equiv 0,1 \mod 4, \\ &Q_1\cdots Q_r \text{ is skew-symmetric if } r \equiv 2,3 \mod 4. \end{aligned} \tag{3.199}$$

Furthermore, the product $Q_1\cdots Q_r$ *is uniquely determined by a choice of orientation of Span* $\{Q_1,\ldots.Q_r\}$.

(e) *For* $P,Q \in$ *Span* $\{P_0,\ldots,P_m\}$ *and* $x \in \mathbf{R}^{2l}$, *we have*

$$\langle Px, Qx\rangle = \langle P,Q\rangle\langle x,x\rangle. \tag{3.200}$$

Proof. (a) Let $P = \sum_{i=0}^m a_iP_i$ with $\sum_{i=0}^m a_i^2 = 1$. Then

$$\begin{aligned} P^2 &= \left(\sum_{i=0}^m a_iP_i\right)\left(\sum_{j=0}^m a_jP_j\right) = \sum_{i=0}^m\sum_{j=0}^m a_ia_jP_iP_j \\ &= \sum_{i=0}^m a_i^2P_i^2 + \sum_{i=0}^m\sum_{j\neq i} a_ia_jP_iP_j \\ &= \sum_{i=0}^m a_i^2I + \sum_{i=0}^m\sum_{j>i} a_ia_j(P_iP_j + P_jP_i) = \sum_{i=0}^m a_i^2I = I. \end{aligned} \tag{3.201}$$

Conversely, let $\{Q_0,\ldots,Q_m\}$ be an orthonormal basis for W. By hypothesis, $Q_i^2 = I$ for all i. We must show that $Q_iQ_j + Q_jQ_i = 0$ for all $i \neq j$. Let $Q = (1/\sqrt{2})(Q_i + Q_j)$, for $i \neq j$. Then Q has length 1, so $Q^2 = I$. On the other hand,

$$\begin{aligned} Q^2 &= (Q_i^2 + (Q_iQ_j + Q_jQ_i) + Q_j^2)/2 \\ &= (I + (Q_iQ_j + Q_jQ_i) + I)/2 = I + \frac{1}{2}(Q_iQ_j + Q_jQ_i), \end{aligned} \tag{3.202}$$

and so $(Q_iQ_j + Q_jQ_i) = 0$.

(b) Two Clifford systems $(P_0, \ldots, P_m)$ and $(Q_0, \ldots, Q_m)$ on $\mathbf{R}^{2l}$ are geometrically equivalent if there exists an orthogonal transformation

$$B \in O(\text{ Span } \{P_0, \ldots, P_m\} \subset H(2l, \mathbf{R}))$$

such that

$$Q_i = A(BP_i)A^T, \quad \text{for } A \in O(2l). \tag{3.203}$$

Let $R_i = BP_i$. Then the Clifford spheres $\Sigma(P_0, \ldots, P_m)$ and $\Sigma(R_0, \ldots, R_m)$ are equal because B is orthogonal, and $\Sigma(Q_0, \ldots, Q_m)$ and $\Sigma(R_0, \ldots, R_m)$ are clearly conjugate by equation (3.203).

Conversely, if the Clifford spheres $\Sigma(P_0, \ldots, P_m)$ and $\Sigma(Q_0, \ldots, Q_m)$ are conjugate, then there exists $A \in O(2l)$ such that $\{Q_0, \ldots, Q_m\}$ is an orthonormal frame in the Clifford sphere

$$\Sigma(AP_0A^T, \ldots, AP_mA^T) = \Sigma(Q_0, \ldots, Q_m).$$

So there exists an orthogonal transformation $B \in O(\text{ Span } \{Q_0, \ldots, Q_m\})$ such that $BQ_i = AP_iA^T$, for $0 \leq i \leq m$, and so the Clifford systems $(P_0, \ldots, P_m)$ and $(Q_0, \ldots, Q_m)$ are geometrically equivalent.

(c) Suppose that $\{Q_0, \ldots, Q_m\}$ is another orthonormal basis for

$$\text{Span } \{P_0, \ldots, P_m\}.$$

Then

$$Q_i = \sum_{j=0}^{m} b_i^j P_j, \quad [b_i^j] \in O(m+1).$$

Then we have

$$\sum_{i=0}^{m} \langle Q_i x, x\rangle^2 = \sum_{i=0}^{m} \left(\langle \sum_{j=0}^{m} b_i^j P_j x, x\rangle \right)^2 = \sum_{i=0}^{m} \left(\sum_{j=0}^{m} b_i^j \langle P_j x, x\rangle \right)^2. \tag{3.204}$$

Fix $x \in \mathbf{R}^{2l}$, and let $a_j = \langle P_j x, x\rangle$. Then the sum on the right side of equation (3.204) becomes

$$\sum_{i=0}^{m} (\sum_{j=0}^{m} b_i^j a_j)^2 = \sum_{i=0}^{m} a_i^2 = \sum_{i=0}^{m} \langle P_i x, x\rangle^2,$$

since $[b_i^j] \in O(m+1)$. So $H(x) = \sum_{i=0}^{m} \langle P_i x, x\rangle^2$ does not depend on the choice of orthonormal basis.

To show that $H(Px) = H(x)$ for $P \in \Sigma(P_0, \ldots, P_m)$, choose an orthonormal basis $\{Q_0, \ldots, Q_m\}$ for Span $\{P_0, \ldots, P_m\}$ with $Q_0 = P$. Then

$$\begin{aligned} H(Px) &= \sum_{i=0}^{m} \langle Q_i(Px), Px \rangle^2 = \sum_{i=0}^{m} \langle Q_i Q_0 x, Q_0 x \rangle^2 \qquad (3.205) \\ &= \langle Q_0^2 x, Q_0 x \rangle^2 + \sum_{i=1}^{m} \langle -Q_0 Q_i x, Q_0 x \rangle^2 \\ &= \langle x, Q_0 x \rangle^2 + \sum_{i=1}^{m} \langle Q_0 Q_i x, Q_0 x \rangle^2 \\ &= \langle x, Q_0 x \rangle^2 + \sum_{i=1}^{m} \langle Q_i x, x \rangle^2 = \sum_{i=0}^{m} \langle Q_i x, x \rangle^2 = H(x), \end{aligned}$$

where we used the fact that Q_0 is orthogonal in going from the second to last line to the last line.

(d) Let $\{Q_1, \ldots, Q_r\}$ be an orthonormal set in $\Sigma(P_0, \ldots, P_m)$. Since the Q_i are symmetric, we have

$$\langle Q_1 \cdots Q_r x, y \rangle = \langle x, Q_r \cdots Q_1 y \rangle. \qquad (3.206)$$

We use the equation $Q_i Q_j = -Q_j Q_i$ for $i \neq j$, to change $Q_r \cdots Q_1$ into $Q_1 \cdots Q_r$. The number of switches required is

$$(r-1) + (r-2) + \cdots + 1 = (r-1)r/2,$$

and this is even for $r \equiv 0, 1 \bmod 4$, and odd for $r \equiv 2, 3 \bmod 4$. Thus $Q_1 \cdots Q_r$ is symmetric for $r \equiv 0, 1 \bmod 4$, and skew-symmetric for $r \equiv 2, 3 \bmod 4$.

To see that $Q_1 \cdots Q_r$ is determined by an orientation of Span $\{Q_1, \ldots, Q_r\}$, note that $SO(r)$ is generated by rotations in two-dimensional coordinate planes. Since any two of the Q_i can be brought next to each other through interchanges using $Q_i Q_j = -Q_j Q_i$, it suffices to do the proof for $r = 2$, and this can be easily done by a direct calculation.

(e) First, it suffices to show the equation (3.200) for P_i and P_j in the orthonormal basis $\{P_0, \ldots, P_m\}$, since if

$$P = \sum_{i=0}^{m} a_i P_i, \quad Q = \sum_{j=0}^{m} b_j P_j$$

we have using equation (3.200) for P_i and P_j,

$$\begin{aligned}\langle Px, Qx\rangle &= \langle\sum_{i=0}^{m} a_i P_i x, \sum_{j=0}^{m} b_j P_j x\rangle = \sum_{i=0}^{m}\sum_{j=0}^{m} a_i b_j \langle P_i x, P_j x\rangle \\ &= \sum_{i=0}^{m}\sum_{j=0}^{m} a_i b_j \langle P_i, P_j\rangle \langle x, x\rangle = \langle P, Q\rangle \langle x, x\rangle,\end{aligned} \tag{3.207}$$

since

$$\langle P, Q\rangle = \sum_{i=0}^{m}\sum_{j=0}^{m} a_i b_j \langle P_i, P_j\rangle.$$

Next we show that equation (3.200) holds for P_i and P_j. First, if $i = j$, since P_i is orthogonal, we have

$$\langle P_i x, P_i x\rangle = \langle x, x\rangle = 1\langle x, x\rangle = \langle P_i, P_i\rangle \langle x, x\rangle.$$

Now suppose that $i \neq j$. Then $\langle P_i, P_j\rangle = 0$, so we must show that $\langle P_i x, P_j x\rangle = 0$ for all $x \in \mathbf{R}^{2l}$. Then $\langle P_i x, P_j x\rangle = \langle x, P_i P_j x\rangle$ and $\langle P_i x, P_j x\rangle = \langle P_j P_i x, x\rangle$, so

$$2\langle P_i x, P_j x\rangle = \langle x, (P_i P_j + P_j P_i)x\rangle = \langle x, 0\rangle = 0,$$

as needed. □

The examples of Ferus, Karcher and Münzner

We now give the examples of Ferus, Karcher, and Münzner in the following theorem. We will describe the examples in more detail after the proof of the theorem. Note that by part (c) of Theorem 3.71, the function F in the theorem below depends only on the Clifford sphere $\Sigma(P_0, \ldots, P_m)$ of the Clifford system $(P_0, \ldots, P_m)$. (See [77, pp.106–107] for more detail in the calculations in the proof.)

Theorem 3.72. *Let $(P_0, \ldots, P_m)$ be a Clifford system on $\mathbf{R}^{2l}$. Let $m_1 = m$ be a positive integer, $m_2 = l - m - 1$, and $F : \mathbf{R}^{2l} \to \mathbf{R}$ be defined by*

$$F(x) = \langle x, x\rangle^2 - 2\sum_{i=0}^{m} \langle P_i x, x\rangle^2. \tag{3.208}$$

Then F satisfies the Cartan–Münzner differential equations (3.86)–(3.87),

$$|\text{grad } F|^2 = g^2 r^{2g-2},$$

$$\Delta F = c r^{g-2},$$

where $g = 4$ and $c = g^2(m_2 - m_1)/2$. If $m_2 > 0$, then the level sets of F on S^{2l-1} form a family of isoparametric hypersurfaces with $g = 4$ principal curvatures with multiplicities (m_1, m_2).

Proof. By differentiating equation (3.208), we calculate that

$$\text{grad } F = 4\langle x, x\rangle x - 8\sum_{i=0}^{m}\langle P_i x, x\rangle P_i x. \tag{3.209}$$

Thus,

$$\begin{aligned} |\text{ grad } F|^2 &= 16\,\langle x, x\rangle^3 - 64\,\langle x, x\rangle \sum_{i=0}^{m}\langle P_i x, x\rangle^2 \\ &\quad + 64\left\langle \sum_{i=0}^{m}\langle P_i x, x\rangle\, P_i x, \sum_{j=0}^{m}\langle P_j x, x\rangle\, P_j x\right\rangle. \end{aligned} \tag{3.210}$$

Then using equation (3.200) with $P = P_i, Q = P_j$, so that $\langle P_i, P_j\rangle = \delta_{ij}$, we have

$$\begin{aligned} \left\langle \sum_{i=0}^{m}\langle P_i x, x\rangle P_i x, \sum_{j=0}^{m}\langle P_j x, x\rangle P_j x \right\rangle &= \sum_{i=0}^{m}\sum_{j=0}^{m}\langle P_i x, x\rangle\langle P_j x, x\rangle\,\langle P_i x, P_j x\rangle \\ &= \sum_{i=0}^{m}\sum_{j=0}^{m}\langle P_i x, x\rangle\langle P_j x, x\rangle\,\langle P_i, P_j\rangle\,\langle x, x\rangle \\ &= \sum_{i=0}^{m}\langle P_i x, x\rangle^2\langle x, x\rangle. \end{aligned} \tag{3.211}$$

Substituting equation (3.211) into equation (3.210), we get

$$\begin{aligned} |\text{ grad } F|^2 &= 16\,\langle x, x\rangle^3 - 64\,\langle x, x\rangle \sum_{i=0}^{m}\langle P_i x, x\rangle^2 + 64\,\langle x, x\rangle \sum_{i=0}^{m}\langle P_i x, x\rangle^2 \\ &= 16\,\langle x, x\rangle^3 = 16 r^6 = g^2 r^{2g-2}, \end{aligned} \tag{3.212}$$

for $g = 4$, and thus we have the first of the Cartan–Münzner differential equations.

To show that the second Cartan–Münzner differential equation (3.87) is satisfied, we use the identity,

$$\Delta h^2 = 2|\text{ grad } h|^2 + 2h\Delta h, \tag{3.213}$$

which holds for any smooth function $h : \mathbf{R}^{2l} \to \mathbf{R}$. We have

$$\Delta F = \Delta \langle x, x\rangle^2 - 2\sum_{i=0}^{m} \Delta \langle P_i x, x\rangle^2. \tag{3.214}$$

We can use the identity in equation (3.213) on each term on the right side of equation (3.214). First, we take $h = \langle x, x\rangle$. Then grad $h = 2x$, and $\Delta h = 4l$, so

$$\begin{aligned} \Delta \langle x, x\rangle^2 &= 2|\operatorname{grad} h|^2 + 2h\Delta h \\ &= 8\langle x, x\rangle + 2\langle x, x\rangle 4l \\ &= 8(l+1)\langle x, x\rangle. \end{aligned} \tag{3.215}$$

Next we consider the term

$$\sum_{i=0}^{m} \Delta \langle P_i x, x\rangle^2,$$

in equation (3.214). Let $h_i = \langle P_i x, x\rangle$. Then by equation (3.213),

$$\Delta h_i^2 = 2|\operatorname{grad} h_i|^2 + 2h_i \Delta h_i. \tag{3.216}$$

We compute grad $h_i = 2P_i x$, so

$$|\operatorname{grad} h_i|^2 = 4\langle P_i x, P_i x\rangle = 4\langle x, x\rangle. \tag{3.217}$$

Next we calculate

$$\Delta h_i = \operatorname{trace} P_i = 0. \tag{3.218}$$

Thus the terms $h_i \Delta h_i$ in equation (3.216) are all zero. So from equations (3.216)–(3.218), we have

$$\sum_{i=0}^{m} \Delta \langle P_i x, x\rangle^2 = \sum_{i=0}^{m} \Delta h_i^2 = \sum_{i=0}^{m} 8\langle x, x\rangle = 8(m+1)\langle x, x\rangle. \tag{3.219}$$

Combining equations (3.214), (3.215), and (3.219), we get

$$\begin{aligned} \Delta F &= 8(l+1)\langle x, x\rangle - 16(m+1)\langle x, x\rangle = 8((l-m-1)-m)\langle x, x\rangle \\ &= 8(m_2 - m_1)\langle x, x\rangle = g^2 \left(\frac{m_2 - m_1}{2}\right) r^{g-2} = cr^{g-2}, \end{aligned} \tag{3.220}$$

so the second Cartan–Münzner differential equation is satisfied.

We have $c = 8(m_2-m_1)$, $n = 2(m_1+m_2)$ and $g = 4$. Thus, $c = \pm gn$ implies that $m_1 = 0$ or $m_2 = 0$. Hence, if m_1 and m_2 are both positive, Corollary 3.43 (page 129) implies that level sets of F on S^{n+1} form a family of isoparametric hypersurfaces having $g = 4$ principal curvatures with multiplicities (m_1, m_2), and F is the Cartan–Münzner polynomial of this isoparametric family. □

The isoparametric hypersurfaces obtained by this construction of Ferus, Karcher, and Münzner are usually referred to as isoparametric hypersurfaces of *FKM-type*. Since these examples are a generalization of the work of Ozeki and Takeuchi [422]–[423], they are sometimes referred to as isoparametric hypersurfaces of *OT-FKM-type*.

As we know from Münzner's general theory (see Remark 3.34 on page 117), the restriction V of the polynomial F in Theorem 3.72 to S^{2l-1} takes values in the closed interval $[-1, 1]$, and the two focal submanifolds are $M_+ = V^{-1}(1)$ and $M_- = V^{-1}(-1)$. We will primarily focus our attention on M_+, which turns out to be a so-called Clifford–Stiefel manifold as described below.

We first state a theorem concerning the focal submanifold M_-. We refer the reader to the paper of Ferus, Karcher, and Münzner [160, pp. 485–487] for a proof.

Theorem 3.73. *With the notation as in Theorem 3.72, let V be the restriction of F to S^{2l-1} and let $\Sigma = \Sigma(P_0, \dots, P_m)$. For $M_- = V^{-1}(-1)$, we have*

$$M_- = \{x \in S^{2l-1} \mid \text{there exists } P \in \Sigma \text{ with } x \in E_+(P)\}. \tag{3.221}$$

In the case $m_2 < 0$, then $V = -1$, and thus $M_- = S^{2l-1}$; this is only possible for $m \in \{1, 2, 4, 8\}$.

In the case $m_2 \geq 0$, then M_- is diffeomorphic to the total space of an $(l-1)$-sphere bundle,

$$\Gamma = \{(x, P) \mid x \in S^{2l-1},\ P \in \Sigma,\ x \in E_+(P)\} \xrightarrow{\pi} \Sigma,\ (x, P) \mapsto P.$$

The diffeomorphism from Γ onto M_- is furnished by $(x, P) \mapsto x$. In particular, if V is not constant, then M_- is a – trivially connected – submanifold of codimension $m_2 + 1$ in the sphere S^{2l-1}.

In the case $m_2 = 0$, then M_- is a hypersurface; this is only possible for $m \in \{1, 3, 7\}$. In the case $m_2 > 0$, then M_- is the focal submanifold corresponding to the principal curvatures of the isoparametric hypersurfaces having multiplicity m_2. The isoparametric hypersurfaces are m_2-sphere bundles over the connected sphere bundle space M_-.

Suppose $(P_0, \dots, P_m)$ can be extended to a Clifford system $(P_0, \dots, P_{m+1})$. Then $\pi : \Gamma \to \Sigma$ is trivial and M_- is diffeomorphic to $S^{l-1} \times S^m$. For $m \equiv 0 \pmod 4$, the geometrically inequivalent Clifford systems (see page 166) lead to inequivalent sphere bundles $\Gamma \to \Sigma$.

Clifford–Stiefel manifolds

We now consider the case $m_2 > 0$, and study the other focal submanifold M_+ of codimension $m+1$, where $m = m_1$. From the defining equation (3.208) for F, we see that M_+ is the set

$$M_+ = \{x \in S^{2l-1} \mid \langle P_i x, x \rangle = 0, \quad 0 \leq i \leq m\}. \tag{3.222}$$

As we saw in equation (3.192), the Clifford system $(P_0, \ldots, P_m)$ on $\mathbf{R}^{2l}$ is related to a representation of the Clifford algebra C_{m-1} on $\mathbf{R}^l$ determined by skew-symmetric transformations $E_1, \ldots, E_{m-1}$ on $\mathbf{R}^l$ given by the following equations for $(u, v) \in \mathbf{R}^l \times \mathbf{R}^l = \mathbf{R}^{2l}$,

$$\begin{aligned} P_0(u,v) &= (u,-v), \quad P_1(u,v) = (v,u), \\ P_{1+i}(u,v) &= (E_i v, -E_i u), \quad 1 \leq i \leq m-1. \end{aligned} \tag{3.223}$$

Thus, we have

$$\begin{aligned} \langle P_0(u,v),(u,v)\rangle &= |u|^2 - |v|^2, \quad \langle P_1(u,v),(u,v)\rangle = 2\langle u,v\rangle, \\ \langle P_{1+i}(u,v),(u,v)\rangle &= -2\langle E_i u, v\rangle, \quad 1 \leq i \leq m-1. \end{aligned} \tag{3.224}$$

The equations

$$|u|^2 - |v|^2 = 0, \quad |u|^2 + |v|^2 = 1$$

imply that

$$|u|^2 = |v|^2 = 1/2, \tag{3.225}$$

and we see that

$$M_+ = \{(u,v) \in S^{2l-1} \mid |u| = |v| = \frac{1}{\sqrt{2}}, \langle u,v\rangle = 0, \langle E_i u, v\rangle = 0, 1 \leq i \leq m-1\}. \tag{3.226}$$

Pinkall and Thorbergsson [448] called M_+ a *Clifford–Stiefel manifold* $V_2(C_{m-1})$ of Clifford orthogonal 2-frames of length $1/\sqrt{2}$ in $\mathbf{R}^l$, where vectors u and v in $\mathbf{R}^l$ are said to be *Clifford orthogonal* if

$$\langle u,v\rangle = \langle E_1 u, v\rangle = \cdots = \langle E_{m-1} u, v\rangle = 0. \tag{3.227}$$

Many of the results of Ferus, Karcher, and Münzner concerning these examples involve a careful analysis of the shape operators of the focal submanifolds, especially the focal submanifold M_+ of codimension $m+1$ in S^{2l-1}. We now find an explicit description of the principal curvatures and their corresponding principal spaces for M_+.

From equation (3.222), we see that M_+ is determined by the $m+1$ conditions, $\langle P_i x, x\rangle = 0,\ 0 \le i \le m$. If $X \in T_x M_+$, then $X\langle P_i x, x\rangle = 0$. On the other hand, since P_i is linear and symmetric, we have

$$X\langle P_i x, x\rangle = \langle X(P_i x), x\rangle + \langle P_i x, X\rangle = \langle P_i X, x\rangle + \langle P_i x, X\rangle = 2\langle P_i x, X\rangle.$$

Thus $\langle P_i x, X\rangle = 0$ for all tangent vectors X to M_+ at x, and we see that $P_i(x)$ is normal to M_+ for $0 \le i \le m$. Furthermore, the set $\{P_0 x, \ldots, P_m x\}$ is an orthonormal basis for the normal space $T_x^\perp M_+$ to M_+ in the sphere S^{2l-1}, since

$$\langle P_i x, P_i x\rangle = \langle x, P_i^2 x\rangle = \langle x, x\rangle = 1, \tag{3.228}$$

and

$$\langle P_i x, P_j x\rangle = \langle x, P_i P_j x\rangle = -\langle x, P_j P_i x\rangle = -\langle P_j x, P_i x\rangle, \tag{3.229}$$

so $\langle P_i x, P_j x\rangle = 0$ if $i \neq j$.

This shows that the normal bundle of M_+ is trivial with $\{P_0 x, \ldots, P_m x\}$ a global orthonormal frame as x varies over M_+. Hence, the isoparametric hypersurfaces are trivial sphere bundles over M_+. It also implies that

$$T_x^\perp M_+(x) = \{Qx \mid Q \in \text{ Span } \{P_0, \ldots, P_m\}\}, \tag{3.230}$$

and the space of unit normals to M_+ at x is

$$B(x) = \{Px \mid P \in \Sigma(P_0, \ldots, P_m)\}. \tag{3.231}$$

By Corollary 3.22 (page 106) and Theorem 3.26 (page 108), we know that the principal curvatures of a focal submanifold of an isoparametric hypersurface with four principal curvatures are $-1, 0, 1$, and the principal curvatures -1 and 1 have the same multiplicity.

We now want to explicitly find the principal spaces for these principal curvatures on the focal submanifold M_+. Let $\xi = Px$ be a unit normal to M_+ at a point x, where $P \in \Sigma(P_0, \ldots, P_m)$. We can extend ξ to a normal field on M_+ by setting $\xi(y) = Py$, for $y \in M_+$. Then for $X \in T_x M_+$, we know that $A_\xi X$ is the negative of the tangential component of $\tilde{\nabla}_X \xi$, where $\tilde{\nabla}$ is the Levi-Civita connection on S^{2l-1}. Since $\xi(y) = Py$, we have

$$\tilde{\nabla}_X \xi = \tilde{\nabla}_X Py = P(X), \tag{3.232}$$

since P is a linear transformation on $\mathbf{R}^{2l}$, and so

$$A_\xi X = -(\text{ tangential component } P(X)). \tag{3.233}$$

We can now compute the principal curvatures of A_ξ as follows.

Theorem 3.74. *Let x be a point on the focal submanifold M_+, and let $\xi = Px$ be a unit normal vector to M_+ at x, where $P \in \Sigma(P_0, \ldots, P_m)$. Let E_+ and E_- be the l-dimensional eigenspaces of P for the eigenvalues $+1$ and -1, respectively, so that $\mathbf{R}^{2l} = E_+ \oplus E_-$. Then the shape operator A_ξ has principal curvatures $0, 1, -1$ with corresponding principal spaces $T_0(\xi), T_1(\xi), T_{-1}(\xi)$ as follows:*

$$
\begin{aligned}
T_0(\xi) &= \{QPx \mid Q \in \Sigma, \langle Q, P\rangle = 0\}, \\
T_1(\xi) &= \{X \in E_- \mid \langle X, Qx\rangle = 0, \forall Q \in \Sigma\} = E_- \cap T_x M_+, \\
T_{-1}(\xi) &= \{X \in E_+ \mid \langle X, Qx\rangle = 0, \forall Q \in \Sigma\} = E_+ \cap T_x M_+,
\end{aligned}
\tag{3.234}
$$

where $\Sigma = \Sigma(P_0, \ldots, P_m)$. Furthermore,

$$
\dim T_0(\xi) = m, \quad \dim T_1(\xi) = \dim T_{-1}(\xi) = l - m - 1. \tag{3.235}
$$

Proof. Let $X = QPx$ for $Q \in \Sigma$, and $\langle Q, P\rangle = 0$. We first want to show that X is tangent to M_+ at x. To do this, we need to show that X is orthogonal to every vector in the space $T_x^\perp M_+(x)$ given in equation (3.230). First, we have

$$
\begin{aligned}
\langle X, Px\rangle &= \langle QPx, Px\rangle = -\langle PQx, Px\rangle = -\langle Qx, x\rangle = 0, \\
\langle X, Qx\rangle &= \langle QPx, Qx\rangle = \langle Px, x\rangle = 0.
\end{aligned}
\tag{3.236}
$$

Next, suppose that $R \in \Sigma$ such that $\langle R, P\rangle = \langle R, Q\rangle = 0$. Then

$$
\langle X, Rx\rangle = \langle QPx, Rx\rangle = \langle RQPx, x\rangle = -\langle x, RQPx\rangle = -\langle Rx, QPx\rangle = -\langle X, Rx\rangle,
$$

so $\langle X, Rx\rangle = 0$, where we have used the fact that RQP is skew-symmetric by part (d) of Theorem 3.71. Thus $X = QPx$ is tangent to M_+ at x.

We now compute $A_\xi X$, for $\xi = Px$. By equation (3.232), we have

$$
\tilde{\nabla}_X \xi = P(X) = P(QPx) = -(P^2 Qx) = -Qx,
$$

which is normal to M_+ at x. Thus the tangential component of $\tilde{\nabla}_X \xi$ is zero, and so $A_\xi X = 0$. Therefore, the m-dimensional space

$$
\{QPx \mid Q \in \Sigma,\ \langle Q, P\rangle = 0\} \subset T_0(\xi). \tag{3.237}
$$

Later we will see that these two sets are actually equal to each other.

Next for $X \in E_- \cap T_x M_+$, we have $\tilde{\nabla}_X \xi = P(X) = -X$, so that $A_\xi X = X$, and $X \in T_1(\xi)$. Thus

$$
E_- \cap T_x M_+ \subset T_1(\xi). \tag{3.238}
$$

Since

$$E_- \cap T_x M_+ = \{X \in E_- \mid \langle X, Qx \rangle = 0, \ \forall Q \in \Sigma\},$$

this space has dimension $l - (m + 1) = l - m - 1$, and we will show later that $E_- \cap T_x M_+$ is actually equal to $T_1(\xi)$.

Finally, let $X \in E_+ \cap T_x M_+$. Then as above, we can show that $A_\xi X = -X$, and so $X \in T_{-1}(\xi)$, and we have

$$E_+ \cap T_x M_+ \subset T_{-1}(\xi), \tag{3.239}$$

for the $(l-m-1)$-dimensional space $E_+ \cap T_x M_+$. Since the sum of the dimensions of the three mutually orthogonal spaces on the left sides of equations (3.237)–(3.239) is equal to $m+2(l-m-1) = 2l-2 = \dim M_+$, the inclusions in equations (3.237)–(3.239) are all equalities, and the theorem is proved. □

From the formulas for the shape operator of a tube in Theorem 2.2 (page 17), we immediately obtain the following corollary of Theorem 3.74.

Corollary 3.75. *Let M_t be a tube of radius t over the focal submanifold M_+, where $0 < t < \pi$ and $t \notin \{\frac{\pi}{4}, \frac{\pi}{2}, \frac{3\pi}{4}\}$. Then M_t is an isoparametric hypersurface with four distinct principal curvatures,*

$$\cot(-t), \quad \cot\left(\frac{\pi}{4} - t\right), \quad \cot\left(\frac{\pi}{2} - t\right), \quad \cot\left(\frac{3\pi}{4} - t\right),$$

having respective multiplicities $m, l - m - 1, m, l - m - 1$.

Multiplicities of the principal curvatures of FKM-hypersurfaces

As we see in Corollary 3.75, the multiplicities of the FKM-hypersurfaces are $m_1 = m$ and $m_2 = l - m - 1$. Here m can be any positive integer, and l must be such that the Clifford algebra C_{m-1} has a representation on $\mathbf{R}^l$, and therefore $l = k\delta(m)$, where $\delta(m)$ is the unique positive integer such that C_{m-1} has an irreducible representation on $\mathbf{R}^l$ as in equation (3.188). Thus, the multiplicities of an isoparametric hypersurface of FKM-type are

$$m_1 = m, \quad m_2 = k\delta(m) - m - 1, \quad k > 0, \tag{3.240}$$

where k is sufficiently large as to make $m_2 > 0$. In the table below of possible multiplicities of the principal curvatures of an isoparametric hypersurface of FKM-type, the cases where $m_2 \leq 0$ are denoted by a dash.

$\delta(m)$	1	2	4	4	8	8	8	8	16	32	··
$\underline{k}$											
1	–	–	–	–	$(5,2)$	$(6,1)$	–	–	$(9,6)$	$(10,21)$	··
2	–	$(2,1)$	$(3,4)$	$\underline{(4,3)}$	$(5,10)$	$(6,9)$	$(7,8)$	$\underline{(8,7)}$	$(9,22)$	$(10,53)$	··
3	$(1,1)$	$(2,3)$	$(3,8)$	$\underline{(4,7)}$	$(5,18)$	$(6,17)$	$(7,16)$	$\underline{(8,15)}$	$(9,38)$	$(10,85)$	··
4	$(1,2)$	$(2,5)$	$(3,12)$	$\underline{\underline{(4,11)}}$	$(5,26)$	$(6,25)$	$(7,24)$	$\underline{\underline{(8,23)}}$	$(9,54)$	·	··
5	$(1,3)$	$(2,7)$	$(3,16)$	$\underline{\underline{(4,15)}}$	$(5,34)$	$(6,33)$	$(7,32)$	$\underline{\underline{(8,31)}}$	·	·	··
·	·	·	·	·	·	·	·	·	·	·	··
·	·	·	·	·	·	·	·	·	·	·	··
·	·	·	·	·	·	·	·	·	·	·	··

Table FKM: Multiplicities of principal curvatures of FKM-hypersurfaces

From parts (b) and (c) of Theorem 3.71 and from formula (3.208) for the Cartan–Münzner polynomial F, we see that geometrically equivalent Clifford systems determine congruent families of isoparametric hypersurfaces. In the table above, the underlined multiplicities,

$$\underline{(m_1, m_2)}, \quad \underline{\underline{(m_1, m_2)}},$$

denote the two, respectively, three geometrically inequivalent Clifford systems for the multiplicities (m_1, m_2). Ferus, Karcher, and Münzner show that these geometrically inequivalent Clifford systems with $m \equiv 0 \pmod 4$ and $l = k\delta(m)$ actually lead to incongruent families of isoparametric hypersurfaces, of which there are $[k/2] + 1$.

Through a study of the second fundamental forms of the focal submanifolds, Ferus, Karcher, and Münzner show that the families for multiplicities $(2,1), (6,1), (5,2)$ and one of the $(4,3)$-families are congruent to those with multiplicities $(1,2), (1,6), (2,5)$, and $(3,4)$, respectively, and these are the only coincidences under congruence among the FKM-hypersurfaces.

Ferus, Karcher, and Münzner [160, p. 490] (see also Ferus [159, pp. 30–31]) point out the following interesting consequence of the incongruence in the case $m \equiv 0 \pmod 4$ of two families with the same multiplicities. Note first that the FKM-hypersurfaces all have at least three nonzero principal curvatures, so they are rigid in the sphere by the classical rigidity theorem.

Consider two incongruent families of FKM-hypersurfaces in the case $m \equiv 0 \pmod 4$ with the same multiplicities, and choose one hypersurface from each family at the same distance from the corresponding focal submanifolds M_-. Then these two hypersurfaces have the same shape operator and thus from the Gauss equation the same curvature tensor pointwise (as defined below). Nevertheless, these two hypersurfaces are not intrinsically isometric, since any such isometry would extend to an isometry of the whole sphere taking one family to the other, and this does not exist.

Here we define two Riemannian manifolds M and M' to have the same curvature tensor at points $x \in M$ and $x' \in M'$ if there exists a linear isometry $\phi : T_xM \to T_{x'}M'$ such that for the respective curvature tensors R and R', we have

$$\phi(R(X,Y)Z) = R'(\phi X, \phi Y)\phi Z. \tag{3.241}$$

Thus, Ferus, Karcher, and Münzner [160, p. 490] obtained the following theorem as a consequence of the incongruence of various families with the same multiplicities.

Theorem 3.76. *For $m \equiv 0 \ (mod\ 4)$ and any positive integer k, there exist $[k/2]+1$ non-isometric compact Riemannian manifolds with the same curvature tensor (at any two points of any two of them). The dimension of these manifolds is $2k\delta(m)-2$.*

Inhomogeneity of many FKM-hypersurfaces

Regarding the question of homogeneity, Ozeki and Takeuchi [422, 423] were the first to produce examples of inhomogeneous isoparametric hypersurfaces. They used representations of Clifford algebras to produce the FKM-series with multiplicities $(3,4k)$ and $(7,8k)$. Most of these multiplicities are not on the list of Takagi and Takahashi [511] of multiplicities of homogeneous isoparametric hypersurfaces (see Section 3.8.5, page 159). Thus, those examples whose multiplicities are not on the list of Takagi and Takahashi are inhomogeneous.

Ferus, Karcher, and Münzner [160, p. 491] gave a geometric argument which we will now describe to prove that many of these FKM-hypersurfaces are not homogeneous. Let $(P_0,\ldots,P_m)$ be a Clifford system on $\mathbf{R}^{2l}$ with $m_1 = m \geq 3$ and $m_2 = l-m-1$, and let $\Sigma = \Sigma(P_0,\ldots,P_m)$ be the associated Clifford sphere.

Let N_+ be the set of all points x in the focal submanifold M_+ such that there exists orthonormal $Q_0, Q_1, Q_2, Q_3 \in \Sigma$ such that $Q_0Q_1Q_2Q_3x = x$. Ferus, Karcher and Münzner showed that N_+ can also be described as the set of $x \in M_+$ such that there exist orthonormal vectors $\eta_0, \eta_1, \eta_2, \eta_3$ in the normal space $T_x^\perp M_+$ such that

$$\dim \left(\bigcap_{i=0}^{3} \ker A_{\eta_i}\right) \geq 3. \tag{3.242}$$

Ferus, Karcher and Münzner [160, p. 491] (see also Ferus [159, pp. 32–33]) then proved the following theorem which provides a geometric proof of inhomogeneity in many cases.

Theorem 3.77. *Suppose that $9 \leq 3m_1 < m_2 + 9$, and for $m_1 = 4$ suppose also that $P_0 \cdots P_4 \neq \pm I$. Then $\emptyset \neq N_+ \neq M_+$. Thus, the focal submanifold M_+ and the whole isoparametric family are not homogeneously embedded.*

Ferus, Karcher, and Münzner also handle the pairs of multiplicities not covered by Theorem 3.77. As noted by Ferus [159, p. 33], these fall into two categories:

(a) $m_1 \leq 2$, $(4, 4k-1)$ and $P_0 \cdots P_4 = \pm I$, $(5,2)$, $(6,1)$, $(9,6)$
(b) $(4,3)$ and $P_0 \cdots P_4 \neq \pm I$, $(6,9)$, $(7,8)$, $\underline{(8,7)}$, $\underline{(8,15)}$, $(10,21)$.

Ferus, Karcher, and Münzner [160, pp. 490–502] then proved that FKM-hypersurfaces with multiplicities as in Case (a) are homogeneous, while those with multiplicities as in Case (b) are inhomogeneous. They also settled the question of the homogeneity or inhomogeneity of the focal submanifolds in all but a few cases.

Later Q.-M. Wang [550] proved many results about the topology of FKM-hypersurfaces, including the fact that hypersurfaces in two different isoparametric families can be diffeomorphic but not congruent to each other. Wu [556] showed that for each n, there are only finitely many diffeomorphism classes of compact isoparametric hypersurfaces in S^{n+1} with four principal curvatures.

3.9.1 Classification results for $g = 4$

We now turn our attention to classification results for isoparametric hypersurfaces with four principal curvatures. All known examples of isoparametric hypersurfaces with four principal curvatures are of FKM-type with the exception of two homogeneous families, having multiplicities $(2,2)$ and $(4,5)$. Many mathematicians, including Münzner [381, 382], Abresch [2], Grove and Halperin [184], Tang [514, 515], and Fang [149, 150], found restrictions on the multiplicities (m_1, m_2) of the principal curvatures of an isoparametric hypersurface with four principal curvatures. This series of results culminated with the paper of Stolz [502], who proved that the multiplicities (m_1, m_2) must be the same as those in the known examples of FKM-type or the two homogeneous exceptions.

These papers are topological in nature, based on Theorem 3.45 of Münzner (see page 132), which states that an isoparametric hypersurface separates the sphere S^{n+1} into two ball bundles over the two focal submanifolds. In particular, the proof of Stolz is homotopy theoretic, and the main tools used are the Hopf invariant and the EHP-sequence. It is worth noting that the theorem of Stolz is actually valid for the more general case of a compact, connected proper Dupin hypersurface with four principal curvatures embedded in S^{n+1}. Such a result is possible because Thorbergsson [533] had shown earlier that a compact, connected proper Dupin hypersurface $M \subset S^{n+1}$ also separates S^{n+1} into two ball bundles over the first focal submanifolds on either side of M (see Theorem 3.62 on page 142).

In 1976, Takagi [510] showed that if one of the multiplicities $m_1 = 1$, then the isoparametric family is congruent to the example described in Subsection 3.8.4. Therefore, such an isoparametric hypersurface is homogeneous and of FKM-type. At approximately the same time, Ozeki and Takeuchi [422, 423] showed that if one of the multiplicities $m_1 = 2$, then M is of FKM-type unless $(m_1, m_2) = (2,2)$, in which case M is the known homogeneous example of Cartan.

Cecil, Chi, and Jensen [81] then showed in a paper published in 2007 that if the multiplicities (m_1, m_2) satisfy $m_2 \geq 2m_1 - 1$, then M is of FKM-type. Taken together with the results of Takagi and Ozeki-Takeuchi mentioned above, the theorem of Cecil, Chi, and Jensen classifies isoparametric hypersurfaces with four principal curvatures for all possible pairs of multiplicities except for four cases, the homogeneous pair $(4, 5)$, and the FKM pairs $(3, 4)$, $(6, 9)$ and $(7, 8)$.

Following this, Chi published two papers [107, 109] (see also [106, 108]) which resolved the classification problem for all pairs of multiplicities except the $(7, 8)$ case. Specifically, in [107] Chi used more commutative algebra than had been employed in the paper of Cecil, Chi, and Jensen [81] to simplify the last part of the proof of the main result in [81]. Furthermore, Chi showed that in the case of multiplicities $(3, 4)$, the isoparametric hypersurface is of FKM-type, thereby resolving that case.

Next in [109], Chi resolved the $(4, 5)$ and $(6, 9)$ cases. In the $(4, 5)$ case, Chi proved that the isoparametric hypersurface is homogeneous, and so it is congruent to an isoparametric hypersurface in the unique $(4, 5)$-family of parallel isoparametric hypersurfaces on the list of Takagi and Takahashi [511] of homogeneous isoparametric hypersurfaces.

In the $(6, 9)$ case, Chi showed that an isoparametric hypersurface is of FKM-type. It can be either homogeneous or inhomogeneous. If it is homogeneous, then it is congruent to an isoparametric hypersurface in the unique $(9, 6)$-family of parallel isoparametric hypersurfaces on the list of Takagi and Takahashi (the $(9, 6)$ pair on Table FKM). If it is inhomogeneous, then it belongs to the inhomogeneous $(6, 9)$-family constructed by Ferus, Karcher and Münzner (the $(6, 9)$ pair on Table FKM). Thus, only the case of multiplicities $(7, 8)$ remains to be classified.

We now give a brief description of the method of proof of Cecil, Chi, and Jensen [81]. In Sections 8–9 of the paper, Cecil, Chi, and Jensen use Cartan's method of moving frames to find necessary and sufficient conditions for the codimension $m_1 + 1$ focal submanifold M_+ of an isoparametric hypersurface M with four principal curvatures and multiplicities (m_1, m_2) to be a Clifford–Stiefel manifold $V_2(C_{m_1-1})$. These conditions are equations (8.1)–(8.4) of [81]. (Later Chi [105] gave a different proof of the fact that equations (8.1)–(8.4) of [81] are necessary and sufficient to show that M_+ is a Clifford–Stiefel manifold.)

These conditions involve the shape operators A_η of M_+, where η is a unit normal vector to M_+ at a point $x \in M_+$. By Corollary 3.22 (page 106) and Theorem 3.26 (page 108) of Münzner, one knows that every A_η at every point $x \in M_+$ has the same eigenvalues $-1, 0, 1$, with respective multiplicities m_2, m_1, m_2. If η is a unit normal vector to M_+ at $x \in M_+$, then the point η is also in M_+ by Münzner's results, since it lies at a distance $\pi/2$ along the normal geodesic to M_+ beginning at the point x in the direction η.

The shape operators corresponding to an orthonormal basis of normal vectors to M_+ at the point x naturally determine a family of $m_1 + 1$ homogeneous polynomials. Similarly, the shape operators corresponding to an orthonormal basis of normal vectors to M_+ at the point $\eta \in M_+$ determine a family of $m_1 + 1$ homogeneous polynomials.

In Section 10 of [81], Cecil, Chi, and Jensen show that these two families of homogeneous polynomials have the same zero set in projective space by use of a formulation of the Cartan–Münzner polynomial due to Ozeki and Takeuchi [422]. This fact is then shown in Sections 11–13 of [81] to imply that the necessary and sufficient conditions for M_+ to be a Clifford–Stiefel manifold are satisfied if $m_2 \geq 2m_1 - 1$. This completes the proof that M is of FKM-type, since M is a tube of constant radius over the Clifford–Stiefel manifold M_+. The proof in Sections 11–13 of [81] involves techniques from algebraic geometry, and Chi [107] later gave a simpler version of this part of the proof. (See also the expository paper of Miyaoka [369] concerning this proof.)

Later Immervoll [218] used the method of isoparametric triple systems developed by Dorfmeister and Neher [135–140] to give a different proof of the theorem of Cecil, Chi and Jensen [81]. (See also the papers of Immervoll [214–217] concerning various aspects of the theory of isoparametric hypersurfaces with four principal curvatures, including triple systems and smooth generalized quadrangles.)

3.10 Applications to Riemannian geometry

In this section, we discuss several contexts in Riemannian geometry where isoparametric hypersurfaces have been studied.

Remark 3.78 (A geometric characterization of isoparametric hypersurfaces). Kimura and Maeda [279] proved that a connected hypersurface M in a real space form $\tilde{M}$ is isoparametric if and only if for each point $p \in M$ there exists an orthonormal basis $X_1, \ldots, X_r$ of the orthogonal complement of the kernel of the shape operator A of M ($r = \text{rank } A$) such that the geodesics in M through p in the direction X_i. $1 \leq i \leq r$, lie on circles of nonzero curvature in $\tilde{M}$. (Here the authors use the term "circle" in the sense of Riemannian geometry.) See also the paper of Maeda and Tanabe [349] and the survey article of Kimura [273] for related results.

Remark 3.79 (Spectrum of the Laplacian). Solomon [489–491] found results concerning the spectrum of the Laplacian of isoparametric hypersurfaces in S^n with three or four principal curvatures. A conjecture of Yau [561] states that the first eigenvalue of the Laplacian of every compact minimal hypersurface M^n in S^{n+1} is equal to n, the dimension of M^n.

Progress on verifying this conjecture for minimal isoparametric hypersurfaces was made by Muto-Ohnita-Urakawa [387], Kotani [294], Muto [386], and Solomon [489–491], ultimately leading to the verification of Yau's conjecture for all minimal isoparametric hypersurfaces in the sphere by Tang and Yan [520].

Tang and Yan also made progress in determining which focal submanifolds V of isoparametric hypersurfaces with $g = 4$ principal curvatures have the property that the first eigenvalue of the Laplacian is equal to the dimension of V. In a subsequent paper, Tang, Xie, and Yan [517] made further progress on that question, and they also proved results concerning the focal submanifolds in the case $g = 6$. (See also Tang and Yan [521] for more characterizations of the focal submanifolds.)

In a related paper, Tang and Yan [519] studied the critical sets of various eigenfunctions of the Laplacian on an isoparametric hypersurface of FKM-type. In an application related to the Schoen-Yau-Gromov-Lawson surgery theory on metrics of positive scalar curvature, Tang, Xie, and Yan [516] constructed a double manifold associated with a minimal isoparametric hypersurface such that this double manifold carries a metric of positive scalar curvature and an isoparametric foliation as well. See also the paper of Henry and Petean [197] on isoparametric hypersurfaces and metrics of constant scalar curvature.

Remark 3.80 (Chern conjecture for isoparametric hypersurfaces). The Chern conjecture for isoparametric hypersurfaces states that every closed minimal hypersurface immersed into the sphere with constant scalar curvature is isoparametric. See the papers of Scherfner and Weiss [474], and of Scherfner, Weiss and Yau [475], for a survey of progress on this conjecture and its generalizations. See also the paper of Ge and Tang [168].

Remark 3.81 (Applications to Willmore submanifolds). An isometric immersion x of a compact, connected n-dimensional manifold M^n into the unit sphere $S^{n+p} \subset \mathbf{R}^{n+p+1}$ is called *Willmore* if it is an extremal submanifold of the Willmore functional:

$$W(x) = \int_{M^n} (S - nh^2)^{n/2} dv,$$

where S is the norm square of the second fundamental form and h is the mean curvature. In the papers of Tang and Yan [518] and Qian, Tang, and Yan [456], the authors prove that both focal submanifolds of every isoparametric hypersurface with four distinct principal curvatures are Willmore. They also completely determine which focal submanifolds are Einstein for all known isoparametric hypersurfaces with $g = 4$ principal curvatures.

Remark 3.82 (Anisotropic isoparametric hypersurfaces in Euclidean spaces). Ge and Ma [167] gave a generalization of the classification of isoparametric hypersurfaces M^n in Euclidean space $\mathbf{R}^{n+1}$ (Theorem 3.12) to the setting of anisotropic isoparametric hypersurfaces in $\mathbf{R}^{n+1}$ as follows.

Let $F : S^n \to \mathbf{R}^+$ be a smooth positive function defined on the unit sphere S^n in $\mathbf{R}^{n+1}$ satisfying the convexity condition that the matrix $A_F = (D^2F + FI)_u$ is positive-definite for all $u \in S^n$, where D^2F denotes the Hessian of F on S^n, and I denotes the identity on T_uS^n. Let $x : M \to \mathbf{R}^{n+1}$ be an immersed oriented hypersurface, and $\nu : M \to S^n$ be its Gauss map. Then the *anisotropic surface energy of* x is a parametric elliptic functional $\mathcal{F}$ defined by:

$$\mathcal{F}(x) = \int_M F(\nu) dA.$$

If $F \equiv 1$, then $\mathcal{F}(x)$ is just the area of x.

A fundamental result is Wulff's Theorem which states that among all closed hypersurfaces enclosing the same volume, there exists a unique absolute minimizer W_F of $\mathcal{F}$ called the *Wulff shape* of F. Under the convexity condition of F, W_F is a smooth convex hypersurface. In the case $F \equiv 1$, W_F is just the unit sphere S^n. Frequently, the Wulff shape plays the same role as the unit sphere S^n does for the area functional. In a way analogous to the classical situation, one can define the anisotropic shape operator, the anisotropic principal curvatures, and the anisotropic mean curvature. An anisotropic isoparametric hypersurface is one which has constant anisotropic principal curvatures. Ge and Ma prove that a complete hypersurface in Euclidean space $\mathbf{R}^{n+1}$ has constant anisotropic principal curvatures if and only if up to translations and homotheties, it is one of the following:

1. $\mathbf{R}^n \subset \mathbf{R}^{n+1}$,
2. $W_F \subset \mathbf{R}^{n+1}$,
3. $\phi_t : W_F^k \times \mathbf{R}^{n-k} \to \mathbf{R}^{n+1}$, for some $0 < k < n, t \neq 0$.

Here the immersion ϕ_t in the third case is a generalization of a spherical cylinder $S^k(r) \times \mathbf{R}^{n-k} \subset \mathbf{R}^{n+1}$ for each value of t.

Remark 3.83 (Moment maps and isoparametric hypersurfaces). For a survey of the results and open problems on the relationship between isoparametric hypersurfaces in spheres and moment maps, see the papers of Miyaoka [372], Fujii [164], and Fujii and Tamaru [165].

Remark 3.84 (Other applications to Riemannian geometry). In other Riemannian applications, Eschenburg and Schroeder [147] studied the behavior of the Tits metric on isoparametric hypersurfaces. Other similar topics such as homogeneous spaces, Tits buildings, and their relationship to the theory of isoparametric hypersurfaces are discussed in the book by Kramer [297]. (See also the papers of Kramer-Van Maldeghem [298], and Kramer [295, 296].)

Ferapontov [156, 157] studied the relationship between isoparametric and Dupin hypersurfaces and Hamiltonian systems of hydrodynamic type, listing several open research problems in that context. (See also the paper of Miyaoka [368].) Ma and Ohnita [345] published a paper on the Hamiltonian stability of the Gauss images of homogeneous isoparametric hypersurfaces, and J. Kaneko wrote two papers concerning the wave equation and Dupin [235], respectively, isoparametric [236] hypersurfaces.

Shklover [481] studied the relationship between isoparametric hypersurfaces and the Schiffer problem in Riemannian geometry, and Kim and Takahashi [264] found various characterizations of isoparametric hypersurfaces in terms of metric connections. The relationship between the geometry of Lagrangian submanifolds and isoparametric hypersurfaces together with related open problems is discussed in the papers of Ma and Ohnita [344], and of Ohnita [407].

Chapter 4
Submanifolds in Lie Sphere Geometry

This chapter is an outline of the method for studying submanifolds of Euclidean space $\mathbf{R}^n$ or the sphere S^n in the context of Lie sphere geometry. For Dupin hypersurfaces this has proven to be a valuable approach, since Dupin hypersurfaces occur naturally as envelopes of families of spheres, which can be handled well in Lie sphere geometry. Since the Dupin property is invariant under Lie sphere transformations, this is also a natural setting for classification theorems.

In Section 4.5, we give a Lie geometric criterion for a Legendre submanifold to be Lie equivalent to the Legendre lift of an isoparametric hypersurface in S^n, and we develop the important invariants known as Lie curvatures of a Legendre submanifold. Finally, in Section 4.6, we formulate the notion of tautness in the setting of Lie sphere geometry and prove that it is invariant under Lie sphere transformations.

For the early development of Lie sphere geometry, see the paper of Lie [326] and the books of Lie and Scheffers [327], Klein [281], Blaschke [42] and Bol [44]. For a historical treatment of the subject, see the papers of Hawkins [190] and Rowe [466]. For a modern treatment of Möbius geometry, see the book of Hertrich-Jeromin [198]. The material in this chapter is covered in more detail in Chapters 2–4 of the book [77], and the figures in this chapter are also taken from that book.

4.1 Möbius Geometry of Unoriented Spheres

We begin with the "Möbius geometry" of unoriented hyperspheres in Euclidean space $\mathbf{R}^n$ or in the unit sphere S^n in $\mathbf{R}^{n+1}$. We always assume that $n \geq 2$.

We can go back and forth between these two ambient spaces $\mathbf{R}^n$ and S^n via stereographic projection, which we recall here. Let $\mathbf{R}^{n+1}$ have coordinates $x = (x_1, \ldots, x_{n+1})$, and denote the usual inner product in $\mathbf{R}^{n+1}$ by $x \cdot y$, where

T.E. Cecil, P.J. Ryan, *Geometry of Hypersurfaces*, Springer Monographs in Mathematics, DOI 10.1007/978-1-4939-3246-7_4

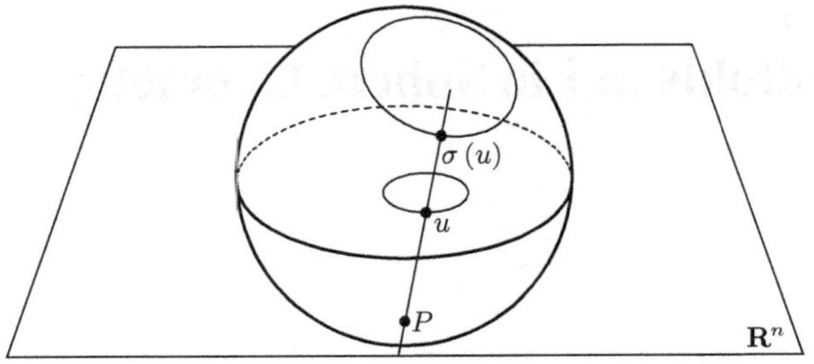

Fig. 4.1 Inverse stereographic projection

$$x \cdot y = x_1 y_1 + \cdots + x_{n+1} y_{n+1}. \tag{4.1}$$

In this chapter, we will use the notation $x \cdot y$ instead of $\langle x, y \rangle$ (as used in the preceding chapters) to denote the Euclidean inner product, because we want to use $\langle x, y \rangle$ for the Lie scalar product, which we will introduce later in this chapter.

The unit sphere S^n is the set of points $x \in \mathbf{R}^{n+1}$ such that $x \cdot x = 1$. We identify $\mathbf{R}^n$ with the hyperplane given by the equation $x_1 = 0$ in $\mathbf{R}^{n+1}$. Let $P = (-1, 0, \ldots, 0)$ be the south pole of S^n.

As in Remark 2.7 on page 21, we define stereographic projection with pole P to be the map $\tau : S^n - \{P\} \to \mathbf{R}^n$ given by the formula,

$$\tau(x_1, \ldots, x_{n+1}) = \left(0, \frac{x_2}{x_1 + 1}, \ldots, \frac{x_{n+1}}{x_1 + 1}\right). \tag{4.2}$$

To describe inverse stereographic projection $\sigma : \mathbf{R}^n \to S^n - \{P\}$ (see Figure 4.1), we write a point $u \in \mathbf{R}^n$ as $u = (u_2, \ldots, u_{n+1})$, that is, we omit the first coordinate 0. Then inverse stereographic projection is given by the formula:

$$\sigma(u) = \left(\frac{1 - u \cdot u}{1 + u \cdot u}, \frac{2u}{1 + u \cdot u}\right). \tag{4.3}$$

Later in this section we will show that stereographic projection τ maps a hypersphere S in S^n that does not contain the point P to a hypersphere $\tau(S)$ in $\mathbf{R}^n$. If S does contain P, then τ maps $S - \{P\}$ to a hyperplane in $\mathbf{R}^n$. Obviously, the inverse map σ has similar properties.

Remark 4.1. Sometimes the map σ is referred to as "stereographic projection," as in the book *Lie Sphere Geometry* [77]. However, in this book, we will call the map τ "stereographic projection," and the map σ "inverse stereographic projection."

To construct the space of unoriented hyperspheres in S^n, we need to consider the Lorentz space $\mathbf{R}^{n+2}_1$ of dimension $n + 2$ endowed with the Lorentz metric (bilinear form) of signature $(1, n+1)$ defined for $x = (x_1, \ldots, x_{n+2})$ and $y = y_1, \ldots, y_{n+2})$ by

$$(x, y) = -x_1 y_1 + x_2 y_2 + \cdots + x_{n+2} y_{n+2}. \tag{4.4}$$

This metric is also referred to as the *Lorentz scalar product*.

We borrow the terminology of relativity theory and say that vector x in $\mathbf{R}^{n+2}_1$ is *spacelike*, *timelike*, or *lightlike*, respectively, depending on whether (x,x) is positive, negative, or zero. We will use this terminology even when we are using a metric of different signature.

In the Lorentz space $\mathbf{R}^{n+2}_1$, the set of all lightlike vectors forms a cone of revolution, called the *light cone* or *isotropy cone*. Lightlike vectors are often called *isotropic* in the literature. Timelike vectors are "inside the cone" and spacelike vectors are "outside the cone."

We identify $\mathbf{R}^{n+1}$ with the spacelike subspace of $\mathbf{R}^{n+2}_1$ determined by the equation $x_1 = 0$, and we consider S^n to be the unit sphere in this space $\mathbf{R}^{n+1}$. We next embed this space $\mathbf{R}^{n+1}$ as an affine subspace of projective space $\mathbf{RP}^{n+1}$ as follows. Define projective space $\mathbf{RP}^{n+1}$ to be the space of lines through the origin in $\mathbf{R}^{n+2}$. Equivalently, $\mathbf{RP}^{n+1}$ is the set of equivalence classes $[x]$ for the equivalence relation $\simeq$ on $\mathbf{R}^{n+2} - \{0\}$ defined by $x \simeq y$ if and only if $y = tx$ for some nonzero real number t.

We embed the space $\mathbf{R}^{n+1}$ determined by the equation $x_1 = 0$ in $\mathbf{R}^{n+2}_1$ as an affine hyperplane in $\mathbf{RP}^{n+1}$ by the map $\phi : \mathbf{R}^{n+1} \to \mathbf{RP}^{n+1}$,

$$\phi(x_2, \ldots, x_{n+2}) = [(1, x_2, \ldots, x_{n+2})]. \tag{4.5}$$

If $x \in \mathbf{R}^{n+2}_1$ is a spacelike, timelike, or lightlike vector, then the corresponding point $[x]$ in $\mathbf{RP}^{n+1}$ will be referred to as spacelike, timelike, or lightlike point, respectively.

Let S^n be the unit sphere in $\mathbf{R}^{n+1}$. The image Σ of S^n under the embedding ϕ consists of all points $[(1,y)]$ for $y \in S^n$. If we compute the Lorentz scalar product on such a point $(1,y)$, we get

$$((1,y),(1,y)) = -1 \cdot 1 + y \cdot y = -1 + 1 = 0.$$

Conversely, if the Lorentz scalar product of $(1,y)$ with itself is zero, then y is in S^n. Thus the image $\Sigma = \phi(S^n)$ consists precisely of the projective classes of lightlike vectors in $\mathbf{R}^{n+2}_1$.

We identify $\mathbf{R}^n$ with the subspace of $\mathbf{R}^{n+1}$ determined by the equation $x_2 = 0$. We next consider the composition of the map ϕ above with inverse stereographic projection σ, that is, $\phi\sigma : \mathbf{R}^n \to \mathbf{RP}^{n+1}$ given by

$$\phi\sigma(u) = \left[\left(1, \frac{1 - u \cdot u}{1 + u \cdot u}, \frac{2u}{1 + u \cdot u}\right)\right] = \left[\left(\frac{1 + u \cdot u}{2}, \frac{1 - u \cdot u}{2}, u\right)\right]. \tag{4.6}$$

Let $(z_1, \ldots, z_{n+2})$ be homogeneous coordinates on $\mathbf{RP}^{n+1}$. Then $\phi\sigma(\mathbf{R}^n)$ is just the set of points in $\mathbf{RP}^{n+1}$ lying on the n-sphere Σ given by the equation $(z,z) = 0$, with the exception of the *improper point* $[(1,-1,0,\ldots,0)]$, that is, the image under ϕ of the south pole $P \in S^n$. We will refer to the points in Σ other than $[(1,-1,0,\ldots,0)]$ as *proper points*, and will call Σ the *Möbius sphere* or *Möbius space*.

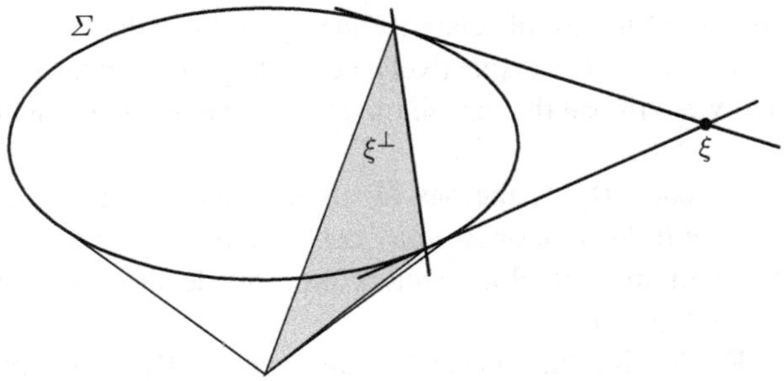

Fig. 4.2 Intersection of Σ with $\xi^{\perp}$

Spheres in Möbius geometry

The basic construction in the Möbius geometry of unoriented spheres is a correspondence between the set of all hyperspheres and hyperplanes in $\mathbf{R}^n$ and the manifold of all spacelike points in projective space $\mathbf{RP}^{n+1}$, and we now give a brief description of this correspondence.

Let ξ be a spacelike vector in $\mathbf{R}^{n+2}_1$. The polar hyperplane $\xi^{\perp}$ of $[\xi]$ in $\mathbf{RP}^{n+1}$ intersects the sphere Σ in an $(n-1)$-sphere S^{n-1} (see Figure 4.2).

This sphere S^{n-1} is the image under $\phi\sigma$ of a hypersphere in $\mathbf{R}^n$, unless it contains the improper point, in which case it is the image under $\phi\sigma$ of a hyperplane in $\mathbf{R}^n$. Thus we have a bijective correspondence between the set of all hyperspheres and hyperplanes in $\mathbf{R}^n$ and the manifold of all spacelike points $\mathbf{RP}^{n+1}$. We next derive the analytic formulas for this correspondence.

The hypersphere in $\mathbf{R}^n$ with center p and radius $r > 0$ has equation

$$(u-p)\cdot(u-p) = r^2. \tag{4.7}$$

A straightforward calculation shows that this is equivalent to the following equation in homogeneous coordinates in $\mathbf{RP}^{n+1}$,

$$(\xi, \phi\sigma(u)) = 0, \tag{4.8}$$

where ξ is the spacelike vector,

$$\xi = \left(\frac{1+p\cdot p - r^2}{2}, \frac{1-p\cdot p + r^2}{2}, p\right), \tag{4.9}$$

and $\phi\sigma(u)$ is given by equation (4.6). Thus the point $u \in \mathbf{R}^n$ lies on the sphere given by equation (4.7) if and only if $\phi\sigma(u)$ lies on the polar hyperplane to $[\xi]$. Since $(\xi,\xi) = r^2 > 0$, the point $[\xi]$ is spacelike. Note also that $\xi_1 + \xi_2 = 1$. The homogeneous coordinates of $[\xi]$ are only determined up to a nonzero scalar multiple, but we can conclude that $\xi_1 + \xi_2 \neq 0$ for any homogeneous coordinates of $[\xi]$.

Conversely, if $[z]$ is a spacelike point in $\mathbf{RP}^{n+1}$ with $z_1 + z_2 \neq 0$, then $[z]$ corresponds to a hypersphere in $\mathbf{R}^n$ as follows. Let $\xi = z/(z_1 + z_2)$ so that $[\xi] = [z]$ is a spacelike point with $\xi_1 + \xi_2 = 1$. Then $(\xi, \xi) = r^2 > 0$ for some $r > 0$, and there exists a unique $p \in \mathbf{R}^n$ such that ξ can be written in the form of equation (4.9). This $p \in \mathbf{R}^n$ and $r > 0$ determine the sphere in $\mathbf{R}^n$ corresponding to $[\xi]$ via equation (4.8).

Next consider the hyperplane in $\mathbf{R}^n$ given by the equation

$$u \cdot N = h, \quad |N| = 1. \tag{4.10}$$

A direct calculation shows that (4.10) is equivalent to the equation

$$(\eta, \phi\sigma(u)) = 0, \text{where } \eta = (h, -h, N). \tag{4.11}$$

Note that $\eta_1 + \eta_2 = 0$, and this is true for any nonzero scalar multiple of η. This condition $\eta_1 + \eta_2 = 0$ is equivalent to the equation

$$(\eta, (1, -1, 0, \ldots, 0)) = 0,$$

and thus the improper point $[(1, -1, 0, \ldots, 0)]$ lies on the hypersphere of Σ obtained by intersecting Σ with the polar hyperplane of η.

Conversely, if $[z]$ is a spacelike point in $\mathbf{RP}^{n+1}$ with $z_1 + z_2 = 0$, then $(z, z) = v \cdot v$, where $v = (z_3, \ldots, z_{n+2})$ is a nonzero vector in $\mathbf{R}^n$. If we take $\eta = z/|v|$, then η has the form $(h, -h, N)$ for some real number h and some unit vector $N \in \mathbf{R}^n$, and the polar hyperplane of $[\eta]$ intersects Σ in an $(n-1)$-sphere corresponding to the hyperplane in $\mathbf{R}^n$ given by equation (4.10).

Thus we have a correspondence between each spacelike point in $\mathbf{RP}^{n+1}$ and a unique hypersphere or hyperplane in $\mathbf{R}^n$. The set of all spacelike points in $\mathbf{RP}^{n+1}$ can be realized as an $(n+1)$-dimensional manifold in the following natural way. Let W^{n+1} be the set of vectors in $\mathbf{R}^{n+2}_1$ satisfying $(\zeta, \zeta) = 1$. This is a hyperboloid of revolution of one sheet in $\mathbf{R}^{n+2}_1$. If $[\xi]$ is a spacelike point in $\mathbf{RP}^{n+1}$, then there are precisely two vectors $\zeta = \pm\xi/\sqrt{(\xi, \xi)}$ in W^{n+1} with $[\zeta] = [\xi]$. Thus the set of all spacelike points in $\mathbf{RP}^{n+1}$ is diffeomorphic to the quotient manifold $W^{n+1}/\simeq$, where $\simeq$ is projective equivalence.

Note that this correspondence also demonstrates that inverse stereographic projection σ maps a hypersphere or hyperplane in $\mathbf{R}^n$ to a hypersphere in the sphere Σ corresponding to the intersection of Σ with the polar hyperplane of the appropriate spacelike point $[\xi]$ or $[\eta]$. Conversely, any hypersphere in Σ is obtained by intersecting Σ with the polar hyperplane of some spacelike point $[\xi]$ or $[\eta]$ in $\mathbf{RP}^{n+1}$, and stereographic projection τ maps this hypersphere in Σ to a hypersphere or hyperplane in $\mathbf{R}^n$ determined by equation (4.7) or (4.10), as the case may be.

The space of hyperspheres in the sphere S^n

Similarly, we can construct a bijective correspondence between the space of all hyperspheres in the unit sphere $S^n \subset \mathbf{R}^{n+1}$ and the manifold of all spacelike points in $\mathbf{RP}^{n+1}$ as follows. The hypersphere S in S^n with center $p \in S^n$ and (spherical) radius $\rho, 0 < \rho < \pi$, is given by the equation

$$p \cdot y = \cos \rho, \quad 0 < \rho < \pi, \tag{4.12}$$

for $y \in S^n$. If we take $[z] = \phi(y) = [(1, y)]$, then

$$p \cdot y = \frac{-(z, (0, p))}{(z, e_1)},$$

where $e_1 = (1, 0, \ldots, 0)$. Thus equation (4.12) is equivalent to the equation

$$(z, (\cos \rho, p)) = 0, \tag{4.13}$$

in homogeneous coordinates in $\mathbf{RP}^{n+1}$. Therefore, y lies on the hypersphere S given by equation (4.12) if and only if $[z] = \phi(y)$ lies on the polar hyperplane in $\mathbf{RP}^{n+1}$ of the spacelike point

$$[\xi] = [(\cos \rho, p)]. \tag{4.14}$$

Remark 4.2 (The space of hyperspheres in hyperbolic space H^n). One can also construct the space of unoriented hyperspheres in hyperbolic space H^n with constant sectional curvature -1. To do this, we let $\mathbf{R}^{n+1}_1$ denote the Lorentz subspace of $\mathbf{R}^{n+2}_1$ spanned by the orthonormal basis $\{e_1, e_3, \ldots, e_{n+2}\}$. Then H^n is the hypersurface

$$\{y \in \mathbf{R}^{n+1}_1 \mid (y, y) = -1, \ y_1 \geq 1\},$$

on which the restriction of the Lorentz metric (,) is a positive definite metric of constant sectional curvature -1 (see Kobayashi–Nomizu [283, Vol. II, p. 268–271] for more detail). The distance between two points p and q in H^n is given by

$$d(p, q) = \cosh^{-1}(-(p, q)).$$

Thus the equation for the unoriented sphere in H^n with center p and radius ρ is

$$(p, y) = -\cosh \rho. \tag{4.15}$$

As with S^n, we first embed $\mathbf{R}^{n+1}_1$ into $\mathbf{RP}^{n+1}$ as an affine space by the map

$$\psi(y) = [y + e_2].$$

Let $p \in H^n$ and let $z = y + e_2$ for $y \in H^n$. Then we have

$$(p, y) = (z, p)/(z, e_2).$$

Thus, the condition (4.15) for y to lie on sphere S with center p and radius ρ is equivalent to the condition that $[z] = [y+e_2]$ lies on the polar hyperplane in $\mathbf{RP}^{n+1}$ to

$$[\xi] = [p + \cosh \rho \, e_2], \tag{4.16}$$

and we can associate the sphere S with the point $[\xi]$.

Orthogonal spheres

Möbius geometry in $\mathbf{R}^n$ or S^n is often identified with the conformal geometry of these spaces via the following considerations. Let S_1 and S_2 denote hyperspheres in $\mathbf{R}^n$ with centers p_1 and p_2 and radii r_1 and r_2, respectively. These two spheres intersect orthogonally (see Figure 4.3) if and only if

$$|p_1 - p_2|^2 = r_1^2 + r_2^2. \tag{4.17}$$

Suppose that S_1 and S_2 correspond to the spacelike points $[\xi_1]$ and $[\xi_2]$ via equation (4.9). Then a straightforward calculation shows that equation (4.17) is equivalent to the condition

$$(\xi_1, \xi_2) = 0, \tag{4.18}$$

in homogeneous coordinates in $\mathbf{RP}^{n+1}$.

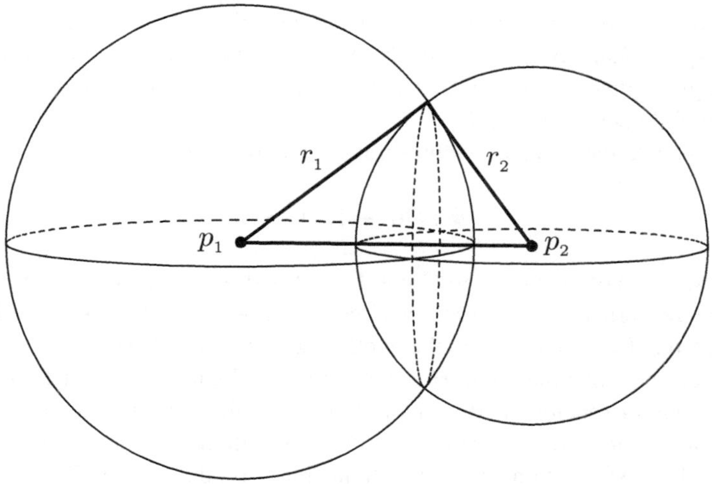

Fig. 4.3 Orthogonal spheres

Similarly, a hyperplane π in $\mathbf{R}^n$ intersects a hypersphere S in $\mathbf{R}^n$ orthogonally if and only if the center p of S lies in the hyperplane π. If π is given by equation (4.10) above, then this condition is $p \cdot N = 0$. One can easily verify that this equation is equivalent to the condition $(\xi, \eta) = 0$ in homogeneous coordinates in $\mathbf{RP}^{n+1}$, where ξ and η correspond to S and π via equations (4.8) or (4.11), respectively. Finally, two hyperplanes π_1 and π_2 in $\mathbf{R}^n$ are orthogonal if and only if their unit normals N_1 and N_2 are orthogonal. A direct calculation shows that this is equivalent to the equation $(\eta_1, \eta_2) = 0$ in homogeneous coordinates for the spacelike points $[\eta_1]$ and $[\eta_2]$ corresponding to π_1 and π_2 via equation (4.11). Thus, in all cases of hyperspheres or hyperplanes in $\mathbf{R}^n$, orthogonal intersection corresponds to a polar relationship in $\mathbf{RP}^{n+1}$ given by equations (4.8) or (4.11).

Möbius transformations

We conclude this section with a discussion of Möbius transformations. Recall that a linear transformation $A \in GL(n+2)$ induces a projective transformation $P(A)$ on $\mathbf{RP}^{n+1}$ defined by $P(A)[x] = [Ax]$. The map P is a homomorphism of $GL(n+2)$ onto the group $PGL(n+1)$ of projective transformations of $\mathbf{RP}^{n+1}$, and its kernel is the group of nonzero multiples of the identity transformation $I \in GL(n+2)$.

A *Möbius transformation* is a projective transformation α of $\mathbf{RP}^{n+1}$ that preserves the condition $(\eta, \eta) = 0$ for $[\eta] \in \mathbf{RP}^{n+1}$, that is, $\alpha = P(A)$, where $A \in GL(n+2)$ maps lightlike vectors in $\mathbf{R}^{n+2}_1$ to lightlike vectors. It can be shown (see, for example, [77, pp. 26–27]) that such a linear transformation A is a nonzero scalar multiple of a linear transformation $B \in O(n+1, 1)$, the orthogonal group for the Lorentz inner product space $\mathbf{R}^{n+2}_1$. Thus, $\alpha = P(A) = P(B)$.

The Möbius transformation $\alpha = P(B)$ induced by an orthogonal transformation $B \in O(n+1, 1)$ maps spacelike points to spacelike points in $\mathbf{RP}^{n+1}$, and it preserves the polarity condition $(\xi, \eta) = 0$ for any two points $[\xi]$ and $[\eta]$ in $\mathbf{RP}^{n+1}$. Therefore by the correspondence given in equations (4.8) and (4.11) above, α maps the set of hyperspheres and hyperplanes in $\mathbf{R}^n$ to itself, and it preserves orthogonality and hence angles between hyperspheres and hyperplanes. A similar statement holds for the set of all hyperspheres in S^n.

Let H denote the group of Möbius transformations and let

$$\psi : O(n+1, 1) \to H \tag{4.19}$$

be the restriction of the map P to $O(n+1, 1)$. The discussion above shows that ψ is onto, and the kernel of ψ is $\{\pm I\}$, the intersection of $O(n+1, 1)$ with the kernel of P. Therefore, H is isomorphic to the quotient group $O(n+1, 1)/\{\pm I\}$.

One can show that the group H is generated by Möbius transformations induced by inversions in spheres in $\mathbf{R}^n$. This follows from the fact that the corresponding orthogonal groups are generated by reflections in hyperplanes. In fact, every orthogonal transformation on an indefinite inner product space $\mathbf{R}^n_k$ is a product of at most n reflections, a result due to Cartan and Dieudonné. (See Cartan [58, pp. 10–12], Chapter 3 of E. Artin's book [15], or [77, pp. 30–34]).

Since a Möbius transformation $\alpha = P(B)$ for $B \in O(n+1,1)$ maps lightlike points to lightlike points in $\mathbf{RP}^{n+1}$ in a bijective way, it induces a diffeomorphism of the n-sphere Σ which is conformal by the considerations given above. It is well known that the group of conformal diffeomorphisms of the n-sphere is precisely the Möbius group.

4.2 Lie Geometry of Oriented Spheres

We now turn to Lie's construction of the space of oriented spheres which is a natural setting for the study of Dupin hypersurfaces. As noted in the previous section, each unoriented hypersphere or hyperplane in $\mathbf{R}^n$ corresponds to a spacelike point $[\xi]$ in $\mathbf{RP}^{n+1}$ via the polarity relationships in equations (4.8) and (4.11). If $[\xi]$ is a spacelike point in $\mathbf{RP}^{n+1}$, then there are precisely two unit length spacelike vectors $\pm\xi/\sqrt{(\xi,\xi)}$ that determine the same spacelike point $[\xi]$ in $\mathbf{RP}^{n+1}$. Thus, as noted earlier, the set of spacelike points in $\mathbf{RP}^{n+1}$ is diffeomorphic to the quotient manifold $W^{n+1}/\simeq$, where W^{n+1} is the set of all unit spacelike vectors in $\mathbf{R}^{n+2}_1$ and $\simeq$ is projective equivalence.

We can associate the two points $\pm\xi/\sqrt{(\xi,\xi)}$ to the two orientations of the hypersphere or hyperplane corresponding to $[\xi]$ by the following construction. We first embed $\mathbf{R}^{n+2}_1$ as an affine space in projective space $\mathbf{RP}^{n+2}$ by the embedding $z \mapsto [(z,1)]$, i.e., we introduce one more coordinate x_{n+3} to give $\mathbf{R}^{n+3}$ and then let $\mathbf{RP}^{n+2}$ be the space of lines through the origin in $\mathbf{R}^{n+3}$. If $\zeta \in W^{n+1}$ is a unit spacelike vector in $\mathbf{R}^{n+2}_1$, then

$$-\zeta_1^2 + \zeta_2^2 + \cdots + \zeta_{n+2}^2 = 1,$$

so the point $[(\zeta,1)]$ in $\mathbf{RP}^{n+2}$ lies on the quadric Q^{n+1} in $\mathbf{RP}^{n+2}$ given in homogeneous coordinates by the equation

$$\langle x,x\rangle = -x_1^2 + x_2^2 + \cdots + x_{n+2}^2 - x_{n+3}^2 = 0, \tag{4.20}$$

which defines the indefinite scalar product $\langle\ ,\ \rangle$ of signature $(n+1,2)$ on the space $\mathbf{R}^{n+3}$, which we now denote as $\mathbf{R}^{n+3}_2$ to indicate the signature of the indefinite scalar product $\langle\ ,\ \rangle$. This scalar product is called the *Lie metric* or *Lie scalar product*, and the quadric Q^{n+1} is called the *Lie quadric*.

We now give the details of how the set of points on the Lie quadric corresponds to the set of all oriented hyperspheres, oriented hyperplanes and point spheres in $\mathbf{R}^n$, or equivalently, to the set of all oriented hyperspheres and point spheres in S^n.

First consider a point $[x] = [(x_1,\ldots,x_{n+3})]$ on Q^{n+1} with last coordinate $x_{n+3} \neq 0$. Then we can divide x by x_{n+3} and represent $[x]$ by a vector of the form

$(\zeta, 1)$ with $\zeta \in W^{n+1}$. Thus, ζ represents an unoriented hypersphere or unoriented hyperplane in $\mathbf{R}^n$ via the Möbius geometric correspondence.

Suppose first that $\zeta_1 + \zeta_2$ is nonzero. Then $[\zeta]$ represents a hypersphere in Möbius geometry via equation (4.8). Specifically, we can divide ζ by $\zeta_1 + \zeta_2$ and get a vector ξ that is projectively equivalent to ζ that satisfies $\xi_1 + \xi_2 = 1$. Then, as in Möbius geometry, $(\xi, \xi) = r^2$ for some $r > 0$, and we can take $p = (\xi_3, \dots, \xi_{n+2})$ in $\mathbf{R}^n$ so that ξ has the form

$$\xi = \left(\frac{1 + p \cdot p - r^2}{2}, \frac{1 - p \cdot p + r^2}{2}, p \right). \tag{4.21}$$

Since $(\xi, \xi) = r^2$ and $(\zeta, \zeta) = 1$, we see that $\zeta = \pm \xi / r$. So the two unit vectors $\pm\zeta$ in $[\xi] \in \mathbf{RP}^{n+1}$ give rise to two points

$$[(\pm\zeta, 1)] = [(\pm\xi/r, 1)] = [(\xi, \pm r)]$$

in the Lie quadric. We associate these two points to the two orientations of the unoriented hypersphere S in $\mathbf{R}^n$ corresponding to $[\xi] = [\zeta]$ as follows. For $p \in \mathbf{R}^n$ and $r > 0$, and ξ given by equation (4.21), the point $[(\xi, r)]$ in Q^{n+1} corresponds to the oriented hypersphere in $\mathbf{R}^n$ with center p, radius r, and orientation given by the inner field of unit normals. The point $[(\xi, -r)]$ corresponds to the same sphere in $\mathbf{R}^n$ with the opposite orientation.

Next we handle the case where $(\zeta, \zeta) = 1$, but $\zeta_1 + \zeta_2 = 0$. In this case, $[(\zeta, 1)]$ corresponds to an oriented hyperplane in $\mathbf{R}^n$ as follows. Since $\zeta_1 + \zeta_2 = 0$, the vector ζ can be written in the form $\zeta = (h, -h, N)$, with $|N| = 1$ since $(\zeta, \zeta) = 1$. Then the two projective points on Q^{n+1} induced by ζ and $-\zeta$ are

$$[(h, -h, N, \pm 1)]. \tag{4.22}$$

These represent the two orientations of the hyperplane in $\mathbf{R}^n$ with equation $u \cdot N = h$. We adopt the convention that $[(h, -h, N, 1)]$ corresponds to the orientation given by the field of unit normals N, while $[(h, -h, N, -1)] = [(-h, h, -N, 1)]$ corresponds to the opposite orientation.

Finally, we consider the case of $[x] = [(x_1, \dots, x_{n+3})]$ in Q^{n+1} with $x_{n+3} = 0$. Then if we take $z = (x_1, \dots, x_{n+2})$, we have

$$0 = \langle x, x \rangle = -x_1^2 + x_2^2 + \dots + x_{n+2}^2 = (z, z),$$

and so $[z] \in \mathbf{RP}^{n+1}$ represents a point in the Möbius sphere Σ, or equivalently a point in $\mathbf{R}^n \cup \{\infty\}$, where ∞ corresponds to the improper point $[(1, -1, 0, \dots, 0)] \in \Sigma$. Thus, $[x]$ represents a *point sphere* or sphere with radius zero in $\mathbf{R}^n \cup \{\infty\}$. Point spheres do not have an orientation assigned to them.

Lie coordinates of oriented spheres

In summary, we have the following bijective correspondence between the set of all oriented hyperspheres, oriented hyperplanes and point spheres in $\mathbf{R}^n \cup \{\infty\}$ and the set of points on the Lie quadric Q^{n+1}.

Euclidean	**Lie**
points : $u \in \mathbf{R}^n$	$\left[\left(\frac{1+u\cdot u}{2}, \frac{1-u\cdot u}{2}, u, 0\right)\right]$
∞	$[(1, -1, 0, 0)]$
spheres: center p, signed radius r	$\left[\left(\frac{1+p\cdot p-r^2}{2}, \frac{1-p\cdot p+r^2}{2}, p, r\right)\right]$
planes: $u \cdot N = h$, unit normal N	$[(h, -h, N, 1)]$

(4.23)

We will use the term *Lie sphere* to denote any oriented hypersphere, oriented hyperplane, or point sphere in $\mathbf{R}^n \cup \{\infty\}$, and we will refer to the coordinates on the right side of the table above as the *Lie coordinates* of the corresponding Lie sphere.

We can begin with a point $[x] = [(x_1, \ldots, x_{n+3})]$ in Q^{n+1} and find the corresponding Euclidean object as follows. If $x_1 + x_2 \neq 0$, then we can divide x by $x_1 + x_2$ to obtain a point $y = (y_1, \ldots, y_{n+3})$ with $y_1 + y_2 = 1$. Then if $y_{n+3} \neq 0$, we can take $r = y_{n+3}$, and $p = (y_3, \ldots, y_{n+2})$, and see that y is in the correct form for the Lie coordinates of the oriented hypersphere with center $p \in \mathbf{R}^n$ and signed radius r. If $y_{n+3} = 0$, then y is in the correct form for the point $u = (y_3, \ldots, y_{n+2})$ in $\mathbf{R}^n$.

Next if $x_1 + x_2 = 0$ and $x_{n+3} \neq 0$, then we can divide x by x_{n+3} to get a vector $y = (h, -h, N, 1)$, which clearly represents an oriented hyperplane in $\mathbf{R}^n$. Finally, if $x_1 + x_2 = 0$ and $x_{n+3} = 0$, then the equation $\langle x, x\rangle = 0$ forces x to have the form $(h, -h, 0, \ldots, 0) \simeq (1, -1, 0, \ldots, 0)$, and so $[x]$ is the improper point corresponding to the point ∞.

Oriented spheres in S^n and H^n

If we wish to consider oriented hyperspheres and point spheres in the unit sphere S^n in $\mathbf{R}^{n+1}$, then the table above can be simplified. First, we have shown that in Möbius geometry, the unoriented hypersphere S in S^n with center $p \in S^n$ and spherical radius ρ, $0 < \rho < \pi$, corresponds to the point $[\xi] = [(\cos\rho, p)]$ in $\mathbf{RP}^{n+1}$. To correspond the two orientations of this sphere to points on the Lie quadric, we first note that

$$(\xi, \xi) = -\cos^2\rho + 1 = \sin^2\rho.$$

Since $\sin\rho > 0$ for $0 < \rho < \pi$, we can divide ξ by $\sin\rho$ and consider the two vectors $\zeta = \pm\xi/\sin\rho$ that satisfy $(\zeta,\zeta) = 1$. We then map these two points into the Lie quadric to get the points

$$[(\zeta, 1)] = [(\xi, \pm\sin\rho)] = [(\cos\rho, p, \pm\sin\rho)].$$

in Q^{n+1}. We can incorporate the sign of the last coordinate into the radius and thereby arrange that the oriented sphere S with signed radius $\rho \neq 0$, where $-\pi < \rho < \pi$, and center p corresponds to the point

$$[x] = [(\cos\rho, p, \sin\rho)]. \tag{4.24}$$

in Q^{n+1}. This formula still makes sense if the radius $\rho = 0$, in which case it yields the point sphere $[(1, p, 0)]$.

We adopt the convention that the positive radius ρ in (4.24) corresponds to the orientation of the sphere given by the field of unit normals which are tangent vectors to geodesics from $-p$ to p, and a negative radius corresponds to the opposite orientation. Each oriented sphere can be considered in two ways, with center p and signed radius ρ, $-\pi < \rho < \pi$, or with center $-p$ and the appropriate signed radius $\rho \pm \pi$.

For a given point $[x]$ in the quadric Q^{n+1}, we can determine the corresponding oriented hypersphere or point sphere in S^n as follows. Multiplying by -1, if necessary, we can arrange that the first coordinate x_1 of x is nonnegative. If x_1 is positive, then it follows from equation (4.24) that the center p and signed radius ρ, $-\pi/2 < \rho < \pi/2$, are given by

$$\tan\rho = x_{n+3}/x_1, \quad p = (x_2, \ldots, x_{n+2})/(x_1^2 + x_{n+3}^2)^{1/2}. \tag{4.25}$$

If $x_1 = 0$, then x_{n+3} is nonzero, and we can divide by x_{n+3} to obtain a point with coordinates $(0, p, 1)$. This corresponds to the oriented hypersphere in S^n with center p and signed radius $\pi/2$, which is a great sphere in S^n.

We can also find a representation for oriented hyperspheres in hyperbolic space H^n. We know from equation (4.16) in Möbius geometry that the unoriented hypersphere S in H^n with center $p \in H^n$ and hyperbolic radius ρ corresponds to the point $[p + \cosh\rho\, e_2]$ in $\mathbf{RP}^{n+1}$. Following exactly the same procedure as in the spherical case, we find that the oriented hypersphere in H^n with center p and signed radius ρ corresponds to a point $[x] \in Q^{n+1}$ given by

$$[x] = [p + \cosh\rho\, e_2 + \sinh\rho\, e_{n+3}]. \tag{4.26}$$

Oriented contact of spheres

As we saw in the previous section, the angle between two spheres is the fundamental geometric quantity in Möbius geometry, and it is the quantity that is preserved by Möbius transformations. In Lie's geometry of oriented spheres, the corresponding fundamental notion is that of oriented contact of spheres. By definition, two oriented spheres S_1 and S_2 in $\mathbf{R}^n$ are in *oriented contact* if they are tangent to each other and they have the same orientation at the point of contact. (See Figures 4.4 and 4.5 for the two possibilities.)

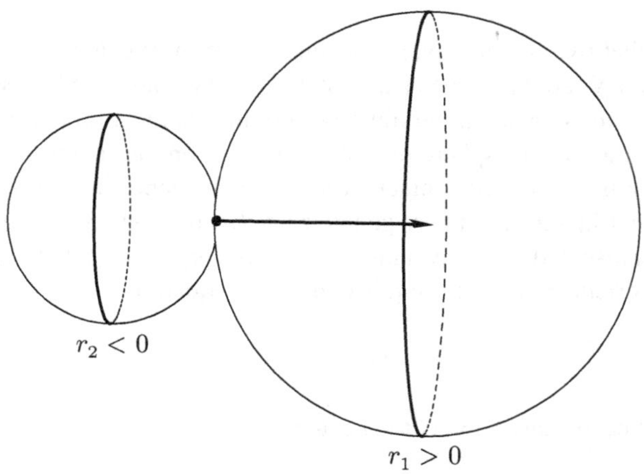

Fig. 4.4 Oriented contact of spheres, first case

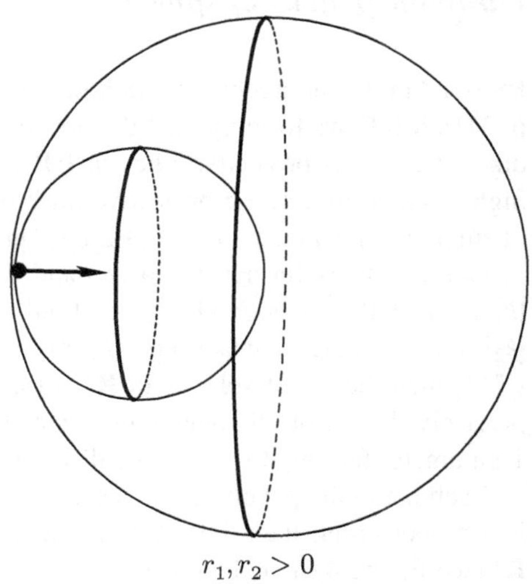

Fig. 4.5 Oriented contact of spheres, second case

If p_1 and p_2 are the respective centers of S_1 and S_2, and r_1 and r_2 are their respective signed radii, then the analytic condition for oriented contact is

$$|p_1 - p_2| = |r_1 - r_2|. \tag{4.27}$$

Similarly, we say that an oriented hypersphere sphere S with center p and signed radius r and an oriented hyperplane π with unit normal N and equation $u \cdot N = h$ are in oriented contact if π is tangent to S and their orientations agree at the point of contact. This condition is given by the equation

$$p \cdot N = r + h. \tag{4.28}$$

Next we say that two oriented planes π_1 and π_2 are in oriented contact if their unit normals N_1 and N_2 are the same. These planes can be considered to be two oriented spheres in oriented contact at the improper point. Finally, a proper point u in $\mathbf{R}^n$ is in oriented contact with a sphere or a plane if it lies on the sphere or plane, and the improper point is in oriented contact with each plane, since it lies on each plane.

An important fact in Lie sphere geometry is that if S_1 and S_2 are two Lie spheres which are represented as in equation (4.23) by $[k_1]$ and $[k_2]$, then the analytic condition for oriented contact is equivalent to the equation

$$\langle k_1, k_2 \rangle = 0. \tag{4.29}$$

This can be checked easily by a direct calculation.

Parabolic pencils of spheres

By standard linear algebra in indefinite inner product spaces (see, for example, [77, p. 21]), it follows from the fact that the signature of $\mathbf{R}^{n+3}_2$ is $(n+1, 2)$ that the Lie quadric contains projective lines in $\mathbf{RP}^{n+2}$, but no linear subspaces of $\mathbf{RP}^{n+2}$ of higher dimension. These projective lines on Q^{n+1} play a crucial role in the theory of submanifolds in the context of Lie sphere geometry.

One can show further that if $[k_1]$ and $[k_2]$ are two points of Q^{n+1}, then the line $[k_1, k_2]$ in $\mathbf{RP}^{n+2}$ lies on Q^{n+1} if and only if the spheres corresponding to $[k_1]$ and $[k_2]$ are in oriented contact, i.e., $\langle k_1, k_2 \rangle = 0$. Moreover, if the line $[k_1, k_2]$ lies on Q^{n+1}, then the set of spheres in $\mathbf{R}^n$ corresponding to points on the line $[k_1, k_2]$ is precisely the set of all spheres in oriented contact with both $[k_1]$ and $[k_2]$. Such a 1-parameter family of spheres is called a *parabolic pencil* of spheres in $\mathbf{R}^n \cup \{\infty\}$.

Each parabolic pencil contains exactly one point sphere, and if that point sphere is a proper point, then the parabolic pencil contains exactly one hyperplane π in $\mathbf{R}^n$ (see Figure 4.6), and the pencil consists of all spheres in oriented contact with a certain oriented plane π at p. Thus, we can associate the parabolic pencil with the point (p, N) in the unit tangent bundle to $\mathbf{R}^n \cup \{\infty\}$, where N is the unit normal to the oriented plane π.

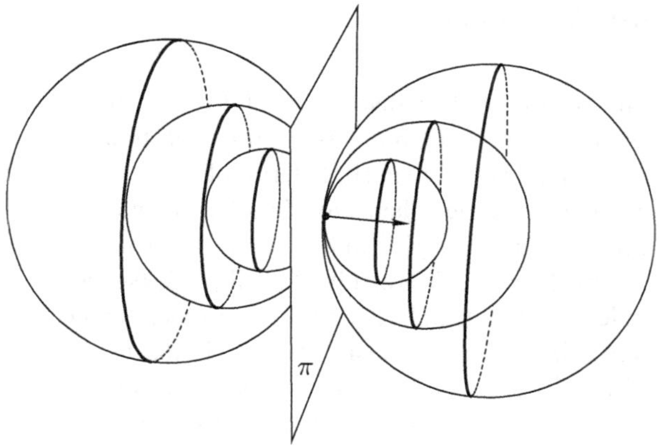

Fig. 4.6 Parabolic pencil of spheres

If the point sphere in the pencil is the improper point, then the parabolic pencil is a family of parallel hyperplanes in oriented contact at the improper point. If N is the common unit normal to all of these planes, then we can associate the pencil with the point (∞, N) in the unit tangent bundle to $\mathbf{R}^n \cup \{\infty\}$.

Similarly, we can establish a correspondence between parabolic pencils and elements of the unit tangent bundle T_1S^n that is expressed in terms of the spherical metric on S^n. If ℓ is a line on the quadric, then ℓ intersects both $e_1^\perp$ and $e_{n+3}^\perp$ at exactly one point, where $e_1 = (1, 0, \ldots, 0)$ and $e_{n+3} = (0, \ldots, 0, 1)$. So the parabolic pencil corresponding to ℓ contains exactly one point sphere (orthogonal to e_{n+3}) and one great sphere (orthogonal to e_1), given respectively by the points,

$$[k_1] = [(1, p, 0)], \quad [k_2] = [(0, \xi, 1)]. \tag{4.30}$$

Since ℓ lies on the quadric we know that $\langle k_1, k_2 \rangle = 0$, and this condition is equivalent to the condition $p \cdot \xi = 0$, i.e., ξ is tangent to S^n at p. Thus, the parabolic pencil of spheres corresponding to the line ℓ can be associated with the point (p, ξ) in T_1S^n. More specifically, the line ℓ can be parametrized as

$$[K_t] = [\cos t\, k_1 + \sin t\, k_2] = [(\cos t, \cos t\, p + \sin t\, \xi, \sin t)].$$

From equation (4.24) above, we see that $[K_t]$ corresponds to the oriented sphere in S^n with center

$$p_t = \cos t\, p + \sin t\, \xi, \tag{4.31}$$

and signed radius t. The pencil consists of all oriented spheres in S^n in oriented contact with the great sphere corresponding to $[k_2]$ at the point (p, ξ) in T_1S^n. Their centers p_t lie along the geodesic in S^n with initial point p and initial velocity vector ξ. Detailed proofs of all these facts are given in [77, pp. 21–23].

Lie sphere transformations

We conclude this section with a discussion of Lie sphere transformations. By definition, a *Lie sphere transformation* is a projective transformation of $\mathbf{RP}^{n+2}$ which maps the Lie quadric Q^{n+1} to itself. In terms of the geometry of $\mathbf{R}^n$ or S^n, a Lie sphere transformation maps Lie spheres to Lie spheres, and since it is a projective transformation, it maps lines on Q^{n+1} to lines on Q^{n+1}. Thus, it preserves oriented contact of spheres in $\mathbf{R}^n$ or S^n. Conversely, Pinkall [443] (see also [77, pp. 28–30]) proved the so-called "Fundamental Theorem of Lie sphere geometry," which states that any line preserving diffeomorphism of Q^{n+1} is the restriction to Q^{n+1} of a projective transformation, that is, a transformation of the space of oriented spheres which preserves oriented contact is a Lie sphere transformation.

By the same type of reasoning given for Möbius transformations, one can show that the group G of Lie sphere transformations is isomorphic to the group $O(n+1,2)/\{\pm I\}$, where $O(n+1,2)$ is the group of orthogonal transformations of $\mathbf{R}^{n+3}_2$. As with the Möbius group, it follows from the theorem of Cartan and Dieudonné (see [77, pp. 30–34]) that the Lie sphere group G is generated by Lie inversions, that is, projective transformations that are induced by reflections in $O(n+1,2)$.

The Möbius group H can be considered to be a subgroup of G in the following manner. Each Möbius transformation on the space of unoriented spheres, naturally induces two Lie sphere transformations on the space Q^{n+1} of oriented spheres as follows. If A is in $O(n+1,1)$, then we can extend A to a transformation B in $O(n+1,2)$ by setting $B = A$ on $\mathbf{R}^{n+2}_1$ and $B(e_{n+3}) = e_{n+3}$. In terms the standard orthonormal basis in $\mathbf{R}^{n+3}_2$, the transformation B has the matrix representation,

$$B = \begin{bmatrix} A & 0 \\ 0 & 1 \end{bmatrix}. \tag{4.32}$$

Although A and $-A$ induce the same Möbius transformation in H, the Lie transformation $P(B)$ is not the same as the Lie transformation $P(C)$ induced by the matrix

$$C = \begin{bmatrix} -A & 0 \\ 0 & 1 \end{bmatrix} \simeq \begin{bmatrix} A & 0 \\ 0 & -1 \end{bmatrix},$$

where $\simeq$ denotes equivalence as projective transformations. Note that $P(B) = \Gamma P(C)$, where Γ is the Lie transformation represented in matrix form by

$$\Gamma = \begin{bmatrix} I & 0 \\ 0 & -1 \end{bmatrix} \simeq \begin{bmatrix} -I & 0 \\ 0 & 1 \end{bmatrix}.$$

From equation (4.23), we see that Γ has the effect of changing the orientation of every oriented sphere or plane. The transformation Γ is called the *change of orientation transformation* or "Richtungswechsel" in German. Hence, the two Lie sphere transformations induced by the Möbius transformation $P(A)$ differ by this change of orientation factor.

Thus, the group of Lie sphere transformations induced from Möbius transformations is isomorphic to $O(n+1,1)$. This group consists of those Lie transformations that map $[e_{n+3}]$ to itself, and it is a double covering of the Möbius group H. Since these transformations are induced from orthogonal transformations of $\mathbf{R}^{n+3}_2$, they also map $e^{\perp}_{n+3}$ to itself, and thereby map point spheres to point spheres. When working in the context of Lie sphere geometry, we will refer to these transformations as "Möbius transformations."

Laguerre transformations

A Lie sphere transformation that maps the improper point to itself is a *Laguerre transformation*. Since oriented contact must be preserved, Laguerre transformations can also be characterized as those Lie sphere transformations that take planes to planes. Like Möbius geometry, Laguerre geometry can be studied on its own, independent of Lie sphere geometry (see, for example, Blaschke [42]). One can show (see, for example, [77, p. 47]) that the group G of Lie sphere transformations is generated by the union of the groups of Möbius and Laguerre.

An important Laguerre transformation in the study of submanifolds is *Euclidean parallel transformation* P_t that adds t to the signed radius of every oriented sphere in $\mathbf{R}^n$ while keeping the center fixed. In terms of the standard basis of $\mathbf{R}^{n+3}_2$, the transformation P_t has the matrix representation,

$$P_t = \begin{bmatrix} 1-(t^2/2) & -t^2/2 & 0\ldots0 & -t \\ t^2/2 & 1+(t^2/2) & 0\ldots0 & t \\ 0 & 0 & I & 0 \\ t & t & 0\ldots0 & 1 \end{bmatrix}. \tag{4.33}$$

One can check that if the column vector consisting of the Lie coordinates (see equation (4.23)) of the oriented sphere with center $p \in \mathbf{R}^n$ and signed radius r is multiplied on the left by this matrix P_t, the result is the column vector consisting of the Lie coordinates of the oriented hypersphere with center p and signed radius $r+t$.

There is also a parallel transformation that adds t to the signed radius of every oriented sphere in S^n or H^n while keeping the center fixed. In the case of S^n, using the fact that $[x] = [(\cos\rho, p, \sin\rho)]$ represents the oriented hypersphere in S^n with center $p \in S^n$ and signed radius ρ, one can check that *spherical parallel transformation* P_t is given by the following transformation in $O(n+1,2)$,

$$\begin{aligned} P_t e_1 &= \cos t\, e_1 + \sin t\, e_{n+3}, \\ P_t e_{n+3} &= -\sin t\, e_1 + \cos t\, e_{n+3}, \\ P_t e_i &= e_i, \quad 2 \leq i \leq n+2. \end{aligned} \tag{4.34}$$

In hyperbolic space, the sphere with center $p \in H^n$ and signed radius ρ corresponds to the point $[p + \cosh\rho\, e_2 + \sinh\rho\, e_{n+3}]$ in Q^{n+1}, and so *hyperbolic parallel transformation* is accomplished by the transformation,

$$\begin{aligned} P_t e_i &= e_i, \quad i = 1, 3, \ldots, n+2. \\ P_t e_2 &= \cosh t\, e_2 + \sinh t\, e_{n+3}, \\ P_t e_{n+3} &= \sinh t\, e_2 + \cosh t\, e_{n+3}. \end{aligned} \tag{4.35}$$

The following theorem of Cecil and Chern [79] (see also [77, p. 49]) demonstrates the important role played by parallel transformations.

Theorem 4.3. *Any Lie sphere transformation α can be written as*

$$\alpha = \phi P_t \psi,$$

where ϕ and ψ are Möbius transformations and P_t is some Euclidean, spherical or hyperbolic parallel transformation.

4.3 Contact Structure and Legendre Submanifolds

The goal of this section is to define a contact structure on the unit tangent bundle T_1S^n and on the $(2n-1)$-dimensional manifold Λ^{2n-1} of projective lines on the Lie quadric Q^{n+1}, and to describe its associated Legendre submanifolds. This will enable us to study submanifolds of $\mathbf{R}^n$ or S^n within the context of Lie sphere geometry in a natural way. This theory was first developed extensively in a modern setting by Pinkall [447] (see also Cecil–Chern [79] or [77, pp. 51–60]).

We consider T_1S^n to be the $(2n-1)$-dimensional submanifold of

$$S^n \times S^n \subset \mathbf{R}^{n+1} \times \mathbf{R}^{n+1}$$

given by

$$T_1S^n = \{(x, \xi) \mid |x| = 1,\ |\xi| = 1,\ x \cdot \xi = 0\}. \tag{4.36}$$

As shown in the previous section, the points on a line ℓ lying on Q^{n+1} correspond to the spheres in a parabolic pencil of spheres in S^n. In particular, as in equation (4.30), ℓ contains one point $[k_1] = [(1, x, 0)]$ corresponding to a point sphere in S^n, and one point $[k_2] = [(0, \xi, 1)]$ corresponding to a great sphere in S^n, where the coordinates are with respect to the standard orthonormal basis $\{e_1, \ldots, e_{n+3}\}$ of $\mathbf{R}^{n+3}_2$. Thus we get a bijective correspondence between the points (x, ξ) of T_1S^n and the space Λ^{2n-1} of lines on Q^{n+1} given by the map:

$$(x, \xi) \mapsto [Y_1(x, \xi), Y_{n+3}(x, \xi)], \tag{4.37}$$

where

$$Y_1(x,\xi) = (1,x,0), \quad Y_{n+3}(x,\xi) = (0,\xi,1). \tag{4.38}$$

We use this correspondence to place a natural differentiable structure on Λ^{2n-1} in such a way as to make the map in equation (4.37) a diffeomorphism.

We now show how to define a contact structure on the manifold T_1S^n. By the diffeomorphism in equation (4.37), this also determines a contact structure on Λ^{2n-1}. Recall that a $(2n-1)$-dimensional manifold V^{2n-1} is said to be a *contact manifold* if it carries a globally defined 1-form ω such that

$$\omega \wedge (d\omega)^{n-1} \neq 0 \tag{4.39}$$

at all points of V^{2n-1}. Such a form ω is called a *contact form*. A contact form ω determines a codimension one distribution (the *contact distribution*) D on V^{2n-1} defined by

$$D_p = \{Y \in T_pV^{2n-1} \mid \omega(Y) = 0\}, \tag{4.40}$$

for $p \in V^{2n-1}$. This distribution is as far from being integrable as possible, in that there exist integral submanifolds of D of dimension $n-1$ but none of higher dimension (see, for example, [77, p. 57]). The distribution D determines the corresponding contact form ω up to multiplication by a nonvanishing smooth function.

A tangent vector to T_1S^n at a point (x,ξ) can be written in the form (X,Z) where

$$X \cdot x = 0, \quad Z \cdot \xi = 0. \tag{4.41}$$

Differentiation of the condition $x \cdot \xi = 0$ implies that (X,Z) also satisfies

$$X \cdot \xi + Z \cdot x = 0. \tag{4.42}$$

We now show that the form ω defined by

$$\omega(X,Z) = X \cdot \xi, \tag{4.43}$$

is a contact form on T_1S^n. At a point (x,ξ), the distribution D is the $(2n-2)$-dimensional space of vectors (X,Z) satisfying $X \cdot \xi = 0$, as well as the equations (4.41) and (4.42). The equation $X \cdot \xi = 0$ together with equation (4.42) implies that

$$Z \cdot x = 0, \tag{4.44}$$

for vectors (X,Z) in D.

Note that if we take $Y_1(x,\xi) = (1,x,0)$, and $Y_{n+3}(x,\xi) = (0,\xi,1)$ as in equation (4.38), then

$$dY_1(X,Z) = (0,X,0), \quad dY_{n+3}(X,Z) = (0,Z,0). \tag{4.45}$$

Thus,

$$\langle dY_1(X,Z), Y_{n+3}(x,\xi)\rangle = X\cdot\xi = \omega(X,Z). \tag{4.46}$$

To prove that the form ω defined by equation (4.43) is a contact form and to study submanifolds in the context of Lie sphere geometry, we use the method of moving frames, as in Cecil–Chern [79] or the book [77]. (See also the paper of Jensen [229] and the forthcoming book of Jensen, Musso and Nicolodi [230].)

Moving frames in Lie sphere geometry

Since we want to define frames on the manifold Λ^{2n-1}, it is better to use frames for which some of the vectors are lightlike, rather than orthonormal frames. For the sake of brevity, we use the following ranges of indices in this section:

$$1 \le a,b,c \le n+3, \quad 3 \le i,j,k \le n+1. \tag{4.47}$$

A *Lie frame* is an ordered set of vectors $\{Y_1,\ldots,Y_{n+3}\}$ in $\mathbf{R}^{n+3}_2$ satisfying the relations

$$\langle Y_a, Y_b\rangle = g_{ab}, \tag{4.48}$$

for

$$[g_{ab}] = \begin{bmatrix} J & 0 & 0 \\ 0 & I_{n-1} & 0 \\ 0 & 0 & J \end{bmatrix}, \tag{4.49}$$

where I_{n-1} is the $(n-1)\times(n-1)$ identity matrix and

$$J = \begin{bmatrix} 0 & 1 \\ 1 & 0 \end{bmatrix}. \tag{4.50}$$

If $(y_1,\ldots,y_{n+3})$ are homogeneous coordinates on $\mathbf{RP}^{n+2}$ with respect to a Lie frame, then the Lie metric has the form

$$\langle y,y\rangle = 2(y_1y_2 + y_{n+2}y_{n+3}) + y_3^2 + \cdots + y_{n+1}^2. \tag{4.51}$$

The space of all Lie frames can be identified with the group $O(n+1,2)$ of which the Lie sphere group G, being isomorphic to $O(n+1,2)/\{\pm I\}$, is a quotient group. In this space, we use the *Maurer–Cartan forms* ω_a^b defined by the equation

$$dY_a = \sum \omega_a^b Y_b, \tag{4.52}$$

and we adopt the convention that the sum is always over the repeated index. Differentiating equation (4.48), we get

$$\omega_{ab} + \omega_{ba} = 0, \tag{4.53}$$

where

$$\omega_{ab} = \sum g_{bc}\omega_a^c. \tag{4.54}$$

Equation (4.53) says that the following matrix is skew-symmetric,

$$[\omega_{ab}] = \begin{bmatrix} \omega_1^2 & \omega_1^1 & \omega_1^i & \omega_1^{n+3} & \omega_1^{n+2} \\ \omega_2^2 & \omega_2^1 & \omega_2^i & \omega_2^{n+3} & \omega_2^{n+2} \\ \omega_j^2 & \omega_j^1 & \omega_j^i & \omega_j^{n+3} & \omega_j^{n+2} \\ \omega_{n+2}^2 & \omega_{n+2}^1 & \omega_{n+2}^i & \omega_{n+2}^{n+3} & \omega_{n+2}^{n+2} \\ \omega_{n+3}^2 & \omega_{n+3}^1 & \omega_{n+3}^i & \omega_{n+3}^{n+3} & \omega_{n+3}^{n+2} \end{bmatrix}. \tag{4.55}$$

Taking the exterior derivative of equation (4.52) yields the *Maurer–Cartan equations*,

$$d\omega_a^b = \sum \omega_a^c \wedge \omega_c^b. \tag{4.56}$$

To show that the form defined by equation (4.43) is a contact form on T_1S^n we want to choose a local frame $\{Y_1,\ldots,Y_{n+3}\}$ on T_1S^n with Y_1 and Y_{n+3} given by equation (4.38). When we transfer this frame to Λ^{2n-1}, it will have the property that for each point $\lambda \in \Lambda^{2n-1}$, the line $[Y_1, Y_{n+3}]$ of the frame at λ is the line on the quadric Q^{n+1} corresponding to λ.

On a sufficiently small open subset U in T_1S^n, we can find smooth mappings,

$$v_i : U \to \mathbf{R}^{n+1}, \quad 3 \le i \le n+1,$$

such that at each point $(x,\xi) \in U$, the vectors $v_3(x,\xi),\ldots,v_{n+1}(x,\xi)$ are unit vectors orthogonal to each other and to x and ξ. By equations (4.41) and (4.42), we see that the vectors

$$\{(v_i,0),(0,v_i),(\xi,-x)\}, \quad 3 \le i \le n+1, \tag{4.57}$$

form a basis to the tangent space to T_1S^n at (x, ξ). We now define a Lie frame on U as follows:

$$\begin{aligned} Y_1(x,\xi) &= (1,x,0), \\ Y_2(x,\xi) &= (-1/2,x/2,0), \\ Y_i(x,\xi) &= (0,v_i(x,\xi),0), \quad 3 \le i \le n+1, \\ Y_{n+2}(x,\xi) &= (0,\xi/2,-1/2) \\ Y_{n+3}(x,\xi) &= (0,\xi,1). \end{aligned} \tag{4.58}$$

Note that Y_1 and Y_{n+3} are defined on all of T_1S^n. We compute the derivatives dY_1 and dY_{n+3} and find

$$\begin{aligned} dY_1(v_i,0) &= (0,v_i,0) = Y_i, \\ dY_1(0,v_i) &= (0,0,0), \\ dY_1(\xi,-x) &= (0,\xi,0) = Y_{n+2} + (1/2)Y_{n+3}, \end{aligned} \tag{4.59}$$

and

$$\begin{aligned} dY_{n+3}(v_i,0) &= (0,0,0), \\ dY_{n+3}(0,v_i) &= (0,v_i,0) = Y_i, \\ dY_{n+3}(\xi,-x) &= (0,-x,0) = (-1/2)Y_1 - Y_2. \end{aligned} \tag{4.60}$$

Comparing these equations with the equation (4.52), we see that the 1-forms,

$$\{\omega_1^i, \omega_{n+3}^i, \omega_1^{n+2}\}, \quad 3 \le i \le n+1, \tag{4.61}$$

form the dual basis to the basis given in (4.57) for the tangent space to T_1S^n at (x, ξ). Furthermore,

$$\omega_1^{n+2}(X,Z) = \langle dY_1(X,Z), Y_{n+3}(x,\xi)\rangle = X \cdot \xi = \omega(X,Z), \tag{4.62}$$

so ω_1^{n+2} is the form ω in equation (4.43).

To prove that ω_1^{n+2} satisfies the condition (4.39) for a contact form, we use the Maurer–Cartan equations and the skew-symmetry of the matrix in equation (4.55) to show by a straightforward calculation that

$$\begin{aligned} \omega_1^{n+2} \wedge (d\omega_1^{n+2})^{n-1} &= \omega_1^{n+2} \wedge \left(\sum \omega_1^i \wedge \omega_i^{n+2}\right)^{n-1} \\ &= (-1)^{n-1}(n-1)! \quad \omega_1^{n+2} \wedge \omega_1^3 \wedge \omega_{n+3}^3 \wedge \cdots \wedge \omega_1^{n+1} \wedge \omega_{n+3}^{n+1} \neq 0. \end{aligned} \tag{4.63}$$

Here the last form is nonzero because the set (4.61) is a basis for the cotangent space to T_1S^n at (x, ξ). We can use the diffeomorphism given in (4.37) to transfer this contact form ω_1^{n+2} to the manifold Λ^{2n-1} of lines on the Lie quadric.

Finally, suppose that

$$Z_1 = \alpha Y_1 + \beta Y_{n+3}, \quad Z_{n+3} = \gamma Y_1 + \delta Y_{n+3}, \tag{4.64}$$

for smooth functions $\alpha, \beta, \gamma, \delta$ with $\alpha\delta - \beta\gamma \neq 0$ on T_1S^n, so that the line $[Z_1, Z_{n+3}]$ equals the line $[Y_1, Y_{n+3}]$ at all points of T_1S^n. Let θ_1^{n+2} be the 1-form defined by $\theta_1^{n+2} = \langle dZ_1, Z_{n+3}\rangle$. Then using equation (4.48), we can compute

$$\begin{aligned} \theta_1^{n+2} &= \langle dZ_1, Z_{n+3}\rangle = \langle d(\alpha Y_1 + \beta Y_{n+3}), \gamma Y_1 + \delta Y_{n+3}\rangle \\ &= \alpha\delta\langle dY_1, Y_{n+3}\rangle + \beta\gamma\langle dY_{n+3}, Y_1\rangle = (\alpha\delta - \beta\gamma)\langle dY_1, Y_{n+3}\rangle \\ &= (\alpha\delta - \beta\gamma)\omega_1^{n+2}. \end{aligned} \tag{4.65}$$

Thus, θ_1^{n+2} is also a contact form on T_1S^n.

Legendre submanifolds

Returning briefly to the general theory, let V^{2n-1} be a contact manifold with contact form ω and corresponding contact distribution D, as in equation (4.40). An immersion $\phi : W^k \to V^{2n-1}$ of a smooth k-dimensional manifold W^k into V^{2n-1} is called an *integral submanifold* of the distribution D if $\phi^*\omega = 0$ on W^k, i.e., for each tangent vector Y at each point $w \in W$, the vector $d\phi(Y)$ is in the distribution D at the point $\phi(w)$. (See Blair [41, p. 36].) It is well known (see, for example, [77, p. 57]) that the contact distribution D has integral submanifolds of dimension $n-1$, but none of higher dimension. These integral submanifolds of maximal dimension are called *Legendre submanifolds* of the contact structure.

In our specific case, we now formulate conditions for a smooth map $\mu : M^{n-1} \to T_1S^n$ to be a Legendre submanifold. We consider T_1S^n as a submanifold of $S^n \times S^n$ as in equation (4.36), and so we can write $\mu = (f, \xi)$, where f and ξ are both smooth maps from M^{n-1} to S^n. We have the following theorem (see [77, p. 58]) giving necessary and sufficient conditions for μ to be a Legendre submanifold.

Theorem 4.4. *A smooth map $\mu = (f, \xi)$ from an $(n-1)$-dimensional manifold M^{n-1} into T_1S^n is a Legendre submanifold if and only if the following three conditions are satisfied.*

(1) *Scalar product conditions: $f \cdot f = 1, \quad \xi \cdot \xi = 1, \quad f \cdot \xi = 0$.*
(2) *Immersion condition: there is no nonzero tangent vector X at any point $x \in M^{n-1}$ such that $df(X)$ and $d\xi(X)$ are both equal to zero.*
(3) *Contact condition: $df \cdot \xi = 0$.*

Note that by equation (4.36), the scalar product conditions are precisely the conditions necessary for the image of the map $\mu = (f, \xi)$ to be contained in $T_1 S^n$. Next, since $d\mu(X) = (df(X), d\xi(X))$, Condition (2) is necessary and sufficient for μ to be an immersion. Finally, from equation (4.43), we see that $\omega(d\mu(X)) = df(X) \cdot \xi(x)$, for each $X \in T_x M^{n-1}$. Hence Condition (3) is equivalent to the requirement that $\mu^*\omega = 0$ on M^{n-1}.

We now want to translate these conditions into the projective setting, and find necessary and sufficient conditions for a smooth map $\lambda : M^{n-1} \to \Lambda^{2n-1}$ to be a Legendre submanifold. We again make use of the diffeomorphism defined in equation (4.37) between $T_1 S^n$ and Λ^{2n-1}.

For each $x \in M^{n-1}$, we know that $\lambda(x)$ is a line on the quadric Q^{n+1}. This line contains exactly one point $[Y_1(x)] = [(1, f(x), 0)]$ corresponding to a point sphere in S^n, and one point $[Y_{n+3}(x)] = [(0, \xi(x), 1)]$ corresponding to a great sphere in S^n. These two formulas define maps f and ξ from M^{n-1} to S^n which depend on the choice of orthonormal basis $\{e_1, \ldots, e_{n+2}\}$ for the orthogonal complement of e_{n+3}.

The map $[Y_1]$ from M^{n-1} to Q^{n+1} is called the *Möbius projection* or *point sphere map* of λ, and the map $[Y_{n+3}]$ from M^{n-1} to Q^{n+1} is called the *great sphere map*. The maps f and ξ are called the *spherical projection* of λ, and the *spherical field of unit normals* of λ, respectively.

In this way, λ determines a map $\mu = (f, \xi)$ from M^{n-1} to $T_1 S^n$, and because of the diffeomorphism (4.37), λ is a Legendre submanifold if and only if μ satisfies the conditions of Theorem 4.4.

It is often useful to have conditions for when λ determines a Legendre submanifold that do not depend on the special parametrization of λ in terms of the point sphere and great sphere maps, $[Y_1]$ and $[Y_{n+3}]$. In fact, in many applications of Lie sphere geometry to submanifolds of S^n or $\mathbf{R}^n$, it is better to consider $\lambda = [Z_1, Z_{n+3}]$, where Z_1 and Z_{n+3} are not the point sphere and great sphere maps.

Legendre submanifolds in Lie sphere geometry

Pinkall [447] gave the following projective formulation of the conditions needed for a Legendre submanifold. In his paper, Pinkall referred to a Legendre submanifold as a "Lie geometric hypersurface." The proof that the three conditions of the theorem below are equivalent to the three conditions of Theorem 4.4 can be found in [77, pp. 59–60].

Theorem 4.5. *Let $\lambda : M^{n-1} \to \Lambda^{2n-1}$ be a smooth map with $\lambda = [Z_1, Z_{n+3}]$, where Z_1 and Z_{n+3} are smooth maps from M^{n-1} into $\mathbf{R}^{n+3}_2$. Then λ determines a Legendre submanifold if and only if Z_1 and Z_{n+3} satisfy the following conditions.*

(1) *Scalar product conditions: for each $x \in M^{n-1}$, the vectors $Z_1(x)$ and $Z_{n+3}(x)$ are linearly independent and*

$$\langle Z_1, Z_1 \rangle = 0, \quad \langle Z_{n+3}, Z_{n+3} \rangle = 0, \quad \langle Z_1, Z_{n+3} \rangle = 0.$$

(2) *Immersion condition: there is no nonzero tangent vector X at any point $x \in M^{n-1}$ such that $dZ_1(X)$ and $dZ_{n+3}(X)$ are both in*

$$\text{Span}\,\{Z_1(x), Z_{n+3}(x)\}.$$

(3) *Contact condition:* $\langle dZ_1, Z_{n+3}\rangle = 0$.

These conditions are invariant under a reparametrization $\lambda = [W_1, W_{n+3}]$, where $W_1 = \alpha Z_1 + \beta Z_{n+3}$ and $W_{n+3} = \gamma Z_1 + \delta Z_{n+3}$, for smooth functions $\alpha, \beta, \gamma, \delta$ on M^{n-1} with $\alpha\delta - \beta\gamma \neq 0$.

The Legendre lift of a submanifold of a real space form

Every oriented hypersurface in a real space form S^n, $\mathbf{R}^n$ or H^n naturally induces a Legendre submanifold of Λ^{2n-1}, as does every submanifold of codimension $m > 1$ in these spaces. Conversely, a Legendre submanifold naturally induces a smooth map into S^n which may have singularities. We now study the details of these maps.

Let $f : M^{n-1} \to S^n$ be an immersed oriented hypersurface with field of unit normals $\xi : M^{n-1} \to S^n$. The induced Legendre submanifold is given by the map $\lambda : M^{n-1} \to \Lambda^{2n-1}$ defined by $\lambda(x) = [Y_1(x), Y_{n+3}(x)]$, where

$$Y_1(x) = (1, f(x), 0), \quad Y_{n+3}(x) = (0, \xi(x), 1). \tag{4.66}$$

The map λ is called the *Legendre lift* of the immersion f with field of unit normals ξ.

To show that λ is a Legendre submanifold, we check the conditions of Theorem 4.5. Condition (1) is satisfied since both f and ξ are maps into S^n, and $\xi(x)$ is tangent to S^n at $f(x)$ for each x in M^{n-1}. Since f is an immersion, $dY_1(X) = (0, df(X), 0)$ is not in Span $\{Y_1(x), Y_{n+3}(x)\}$, for any nonzero vector $X \in T_xM^{n-1}$, and so Condition (2) is satisfied. Finally, Condition (3) is satisfied since

$$\langle dY_1(X), Y_{n+3}(x)\rangle = df(X) \cdot \xi(x) = 0,$$

because ξ is a field of unit normals to f.

In the case of a submanifold $\phi : V \to S^n$ of codimension $m + 1$ greater than one, the domain of the Legendre lift is be the unit normal bundle B^{n-1} of the submanifold $\phi(V)$. We consider B^{n-1} to be the submanifold of $V \times S^n$ given by

$$B^{n-1} = \{(x, \xi) | \phi(x) \cdot \xi = 0,\ d\phi(X) \cdot \xi = 0,\ \text{for all } X \in T_xV\}.$$

The *Legendre lift* $\phi(V)$ (or the *Legendre submanifold induced by* ϕ) is the map $\lambda : B^{n-1} \to \Lambda^{2n-1}$ defined by

$$\lambda(x, \xi) = [Y_1(x, \xi), Y_{n+3}(x, \xi)], \tag{4.67}$$

where

$$Y_1(x,\xi) = (1, \phi(x), 0), \quad Y_{n+3}(x,\xi) = (0, \xi, 1). \tag{4.68}$$

Geometrically, $\lambda(x,\xi)$ is the line on the quadric Q^{n+1} corresponding to the parabolic pencil of spheres in S^n in oriented contact at the contact element $(\phi(x), \xi) \in T_1 S^n$. In [77, pp. 61–62], we show that λ satisfies the conditions of Theorem 4.5,

Similarly, suppose that $F : M^{n-1} \to \mathbf{R}^n$ is an oriented hypersurface with field of unit normals $\eta : M^{n-1} \to \mathbf{R}^n$, where we identify $\mathbf{R}^n$ with the subspace of $\mathbf{R}^{n+3}_2$ spanned by $\{e_3, \dots, e_{n+2}\}$. The Legendre lift of (F, η) is the map $\lambda : M^{n-1} \to \Lambda^{2n-1}$ defined by $\lambda = [Y_1, Y_{n+3}]$, where

$$Y_1 = (1 + F \cdot F, 1 - F \cdot F, 2F, 0)/2, \quad Y_{n+3} = (F \cdot \eta, -(F \cdot \eta), \eta, 1). \tag{4.69}$$

By equation (4.23), $[Y_1(x)]$ corresponds to the point sphere and $[Y_{n+3}(x)]$ corresponds to the hyperplane in the parabolic pencil determined by the line $\lambda(x)$ for each $x \in M^{n-1}$. One can easily verify that Conditions (1)–(3) of Theorem 4.5 are satisfied in a manner similar to the spherical case. In the case of a submanifold $\psi : V \to \mathbf{R}^n$ of codimension greater than one, the Legendre lift of ψ is the map λ from the unit normal bundle B^{n-1} to Λ^{2n-1} defined by $\lambda(x,\eta) = [Y_1(x,\eta), Y_{n+3}(x,\eta)]$, where

$$\begin{aligned} Y_1(x,\eta) &= (1 + \psi(x) \cdot \psi(x), 1 - \psi(x) \cdot \psi(x), 2\psi(x), 0)/2, \\ Y_{n+3}(x,\eta) &= (\psi(x) \cdot \eta, -(\psi(x) \cdot \eta), \eta, 1). \end{aligned} \tag{4.70}$$

The verification that the pair $\{Y_1, Y_{n+3}\}$ satisfies conditions (1)–(3) of Theorem 4.5 is similar to that for submanifolds of S^n of codimension greater than one.

Finally, as in Section 4.1, we consider H^n to be the submanifold of the Lorentz space $\mathbf{R}^{n+1}_1$ spanned by $\{e_1, e_3, \dots, e_{n+2}\}$ defined by:

$$H^n = \{y \in \mathbf{R}^{n+1}_1 | (y,y) = -1, \quad y_1 \geq 1\},$$

where $(\ ,\ )$ is the Lorentz metric on $\mathbf{R}^{n+1}_1$ obtained by restricting the Lie metric. Let $h : M^{n-1} \to H^n$ be an oriented hypersurface with field of unit normals $\zeta : M^{n-1} \to \mathbf{R}^{n+1}_1$. The Legendre lift of (h, ζ) is given by the map $\lambda = [Y_1, Y_{n+3}]$, where

$$Y_1(x) = h(x) + e_2, \quad Y_{n+3}(x) = \zeta(x) + e_{n+3}. \tag{4.71}$$

Note that $(h,h) = -1$, so $\langle Y_1, Y_1 \rangle = 0$, while $(\zeta, \zeta) = 1$, so $\langle Y_{n+3}, Y_{n+3} \rangle = 0$. One can easily check that the conditions (1)–(3) are satisfied. Finally, if $\gamma : V \to H^n$ is an immersed submanifold of codimension greater than one, then the Legendre submanifold $\lambda : B^{n-1} \to \Lambda^{2n-1}$ is again defined on the unit normal bundle B^{n-1} in the obvious way.

Conversely, suppose that $\lambda : M^{n-1} \to \Lambda^{2n-1}$ is an arbitrary Legendre submanifold. We have seen above that we can parametrize λ as $\lambda = [Y_1, Y_{n+3}]$, where

$$Y_1 = (1, f, 0), \quad Y_{n+3} = (0, \xi, 1). \tag{4.72}$$

for the spherical projection f and spherical field of unit normals ξ. Both f and ξ are smooth maps, but neither need be an immersion or even have constant rank (see Example 4.6 below). The Legendre lift of an oriented hypersurface in S^n is the special case where the spherical projection f is an immersion, i.e., f has constant rank $n-1$ on M^{n-1}. In the case of the Legendre lift of a submanifold $\phi : V^k \to S^n$, the spherical projection $f : B^{n-1} \to S^n$ defined by $f(x, \xi) = \phi(x)$ has constant rank k.

If the range of the point sphere map $[Y_1]$ does not contain the improper point $[(1, -1, 0, \ldots, 0)]$, then λ also determines a *Euclidean projection* $F : M^{n-1} \to \mathbf{R}^n$, and a *Euclidean field of unit normals*, $\eta : M^{n-1} \to \mathbf{R}^n$. These are defined by the equation $\lambda = [Z_1, Z_{n+3}]$, where

$$Z_1 = (1 + F \cdot F, 1 - F \cdot F, 2F, 0)/2, \quad Z_{n+3} = (F \cdot \eta, -(F \cdot \eta), \eta, 1). \tag{4.73}$$

Here $[Z_1(x)]$ corresponds to the unique point sphere in the parabolic pencil determined by $\lambda(x)$, and $[Z_{n+3}(x)]$ corresponds to the unique plane in this pencil. As in the spherical case, the smooth maps F and η need not have constant rank.

Finally, if the range of the Euclidean projection F lies inside some disk Ω in $\mathbf{R}^n$, then one can define a hyperbolic projection and hyperbolic field of unit normals by placing a hyperbolic metric on Ω.

There are, however, many Dupin submanifolds whose spherical (or Euclidean) projection is not an immersion and does not have constant rank. Examples of this type can be obtained by applying a parallel transformation P_t to a Dupin submanifold λ whose spherical or Euclidean projection is an immersion, where P_t is chosen in such a way that the spherical or Euclidean projection of $P_t\lambda$ contains a focal point of the original hypersurface. In particular, consider the following example from [77, pp. 63–64].

Example 4.6. A Euclidean projection F that is not an immersion.
An example where the Euclidean (or spherical) projection does not have constant rank is illustrated by the cyclide of Dupin in Figure 4.7. Here the corresponding Legendre submanifold is a map $\lambda : T^2 \to \Lambda^5$, where T^2 is a 2-dimensional torus. The Euclidean projection $F : T^2 \to \mathbf{R}^3$ maps the circle S^1 containing the points A, B, C and D to the point P. However, the map λ into the space of lines on the quadric (corresponding to contact elements) is an immersion. The four arrows in Figure 4.7 represent the contact elements corresponding under the map λ to the four points indicated on the circle S^1.

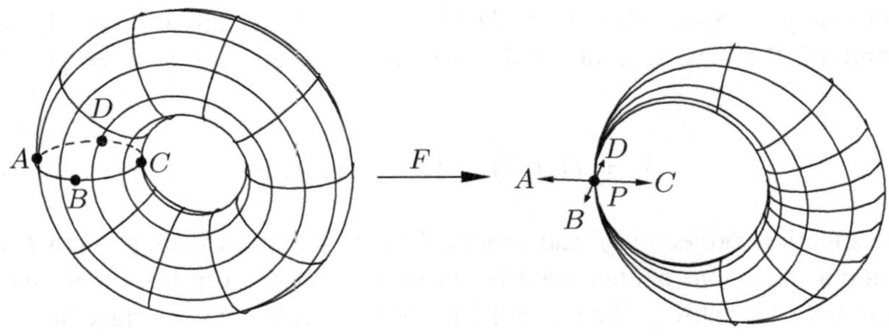

Fig. 4.7 A Euclidean projection F with a singularity

4.4 Curvature Spheres and Dupin Submanifolds

In this section, we discuss the notions of curvature spheres and Dupin hypersurfaces in the context of Lie sphere geometry, and we prove that the Dupin property is invariant under Lie sphere transformations.

We begin with the case of an oriented hypersurface $f : M^{n-1} \to S^n$ with field of unit normals $\xi : M^{n-1} \to S^n$. As we showed in Section 2.2, a point

$$f_t(x) = \cos t\, f(x) + \sin t\, \xi(x) \tag{4.74}$$

is a focal point of (M^{n-1}, x) of multiplicity $m > 0$ if and only if $\cot t$ is a principal curvature of multiplicity m at x. Note that each principal curvature $\kappa = \cot t = \cot(t + \pi)$ produces two distinct antipodal focal points on the normal geodesic to $f(M^{n-1})$ at f with parameter values t and $t + \pi$. The oriented hypersphere centered at a focal point p and in oriented contact with $f(M^{n-1})$ at $f(x)$ is called a *curvature sphere* of f at x. The two antipodal focal points determined by κ are the two centers of the corresponding curvature sphere. Thus, the correspondence between principal curvatures and curvature spheres is bijective. The multiplicity of the curvature sphere is by definition equal to the multiplicity of the corresponding principal curvature.

Curvature spheres in Lie sphere geometry

We now formulate the notion of curvature sphere in the context of Lie sphere geometry. As in equation (4.66), the Legendre lift $\lambda : M^{n-1} \to \Lambda^{2n-1}$ of the oriented hypersurface (f, ξ) is given by $\lambda = [Y_1, Y_{n+3}]$, where

$$Y_1 = (1, f, 0), \quad Y_{n+3} = (0, \xi, 1). \tag{4.75}$$

For each $x \in M^{n-1}$, the points on the line $\lambda(x)$ can be parametrized as

$$[K_t(x)] = [\cos t \, Y_1(x) + \sin t \, Y_{n+3}(x)] = [(\cos t, f_t(x), \sin t)], \tag{4.76}$$

where f_t is given in equation (4.74) above. By equation (4.24), the point $[K_t(x)]$ in Q^{n+1} corresponds to the oriented sphere in S^n with center $f_t(x)$ and signed radius t. This sphere is in oriented contact with the oriented hypersurface $f(M^{n-1})$ at $f(x)$. Given a tangent vector $X \in T_xM^{n-1}$, we have

$$dK_t(X) = (0, df_t(X), 0). \tag{4.77}$$

Thus, $dK_t(X) = (0, 0, 0)$ for a nonzero vector $X \in T_xM^{n-1}$ if and only if $df_t(X) = 0$, i.e., $p = f_t(x)$ is a focal point of f at x corresponding to the principal curvature $\cot t$. The vector X is a principal vector corresponding to the principal curvature $\cot t$, and it is also called a principal vector corresponding to the curvature sphere $[K_t]$.

This characterization of curvature spheres depends on the parametrization of $\lambda = [Y_1, Y_{n+3}]$ given by the point sphere and great sphere maps $[Y_1]$ and $[Y_{n+3}]$, and it has only been defined in the case where the spherical projection f is an immersion. We now give a projective formulation of the definition of a curvature sphere that is independent of the parametrization of λ and is valid for an arbitrary Legendre submanifold.

Let $\lambda : M^{n-1} \to \Lambda^{2n-1}$ be a Legendre submanifold parametrized by the pair $\{Z_1, Z_{n+3}\}$, as in Theorem 4.5. Let $x \in M^{n-1}$ and $r, s \in \mathbf{R}$ with at least one of r and s not equal to zero. The sphere,

$$[K] = [rZ_1(x) + sZ_{n+3}(x)],$$

is called a *curvature sphere* of λ at x if there exists a nonzero vector X in T_xM^{n-1} such that

$$r \, dZ_1(X) + s \, dZ_{n+3}(X) \in \text{Span}\,\{Z_1(x), Z_{n+3}(x)\}. \tag{4.78}$$

The vector X is called a *principal vector* corresponding to the curvature sphere $[K]$. This definition is invariant under a change of parametrization of the form considered in Theorem 4.5 on page 208. Furthermore, if we take the special parametrization $Z_1 = Y_1$, $Z_{n+3} = Y_{n+3}$ given in equation (4.75), then condition (4.78) holds if and only if $r \, dY_1(X) + s \, dY_{n+3}(X)$ actually equals $(0, 0, 0)$.

From equation (4.78), it is clear that the set of principal vectors corresponding to a given curvature sphere $[K]$ at x is a subspace of T_xM^{n-1}. This set is called the *principal space* corresponding to the curvature sphere $[K]$. Its dimension is the *multiplicity* of $[K]$.

Lie equivalent Legendre submanifolds

We next show that a Lie sphere transformation maps curvature spheres to curvature spheres. We first need to discuss the notion of Lie equivalent Legendre submanifolds. Let $\lambda : M^{n-1} \to \Lambda^{2n-1}$ be a Legendre submanifold parametrized by $\lambda = [Z_1, Z_{n+3}]$. Suppose $\beta = P(B)$ is the Lie sphere transformation induced by an orthogonal transformation B in the group $O(n+1, 2)$. Since B is orthogonal, the maps, $W_1 = BZ_1$, $W_{n+3} = BZ_{n+3}$, satisfy the Conditions (1)–(3) of Theorem 4.5, and thus $\gamma = [W_1, W_{n+3}]$ is a Legendre submanifold which we denote by $\beta\lambda : M^{n-1} \to \Lambda^{2n-1}$. We say that the Legendre submanifolds λ and $\beta\lambda$ are *Lie equivalent*. In terms of submanifolds of real space forms, we say that two immersed submanifolds of $\mathbf{R}^n$, S^n, or H^n are *Lie equivalent* if their Legendre lifts are Lie equivalent.

Theorem 4.7. *Let $\lambda : M^{n-1} \to \Lambda^{2n-1}$ be a Legendre submanifold and β a Lie sphere transformation. The point $[K]$ on the line $\lambda(x)$ is a curvature sphere of λ at x if and only if the point $\beta[K]$ is a curvature sphere of the Legendre submanifold $\beta\lambda$ at x. Furthermore, the principal spaces corresponding to $[K]$ and $\beta[K]$ are identical.*

Proof. Let $\lambda = [Z_1, Z_{n+3}]$ and $\beta\lambda = [W_1, W_{n+3}]$ as above. For a tangent vector $X \in T_xM^{n-1}$ and real numbers r and s, at least one of which is not zero, we have

$$\begin{aligned} r\,dW_1(X) + s\,dW_{n+3}(X) &= r\,d(BZ_1)(X) + s\,d(BZ_{n+3})(X) \\ &= B(r\,dZ_1(X) + s\,dZ_{n+3}(X)), \end{aligned} \tag{4.79}$$

since B is a constant linear transformation. Thus, we see that

$$r\,dW_1(X) + s\,dW_{n+3}(X) \in \text{Span}\,\{W_1(x), W_{n+3}(x)\}$$

if and only if

$$r\,dZ_1(X) + s\,dZ_{n+3}(X) \in \text{Span}\,\{Z_1(x), Z_{n+3}(x)\}.$$

□

We next consider the case when the Lie sphere transformation β is a spherical parallel transformation P_t given in equation (4.34), that is,

$$\begin{aligned} P_te_1 &= \cos t\, e_1 + \sin t\, e_{n+3}, \\ P_te_{n+3} &= -\sin t\, e_1 + \cos t\, e_{n+3}, \\ P_te_i &= e_i, \quad 2 \leq i \leq n+2. \end{aligned} \tag{4.80}$$

Recall that P_t has the effect of adding t to the signed radius of each oriented sphere in S^n while keeping the center fixed.

If $\lambda : M^{n-1} \to \Lambda^{2n-1}$ is a Legendre submanifold parametrized by the point sphere map $Y_1 = (1, f, 0)$ and the great sphere map $Y_{n+3} = (0, \xi, 1)$, then $P_t\lambda = [W_1, W_{n+3}]$, where

$$W_1 = P_t Y_1 = (\cos t, f, \sin t), \quad W_{n+3} = P_t Y_{n+3} = (-\sin t, \xi, \cos t). \tag{4.81}$$

Note that W_1 and W_{n+3} are not the point sphere and great sphere maps for $P_t\lambda$. Solving for the point sphere map Z_1 and the great sphere map Z_{n+3} of $P_t\lambda$, we find

$$\begin{aligned} Z_1 &= \cos t\, W_1 - \sin t\, W_{n+3} = (1, \cos t\, f - \sin t\, \xi, 0), \\ Z_{n+3} &= \sin t\, W_1 + \cos t\, W_{n+3} = (0, \sin t\, f + \cos t\, \xi, 1). \end{aligned} \tag{4.82}$$

From this, we see that $P_t\lambda$ has spherical projection and spherical unit normal field given, respectively, by

$$\begin{aligned} f_{-t} &= \cos t\, f - \sin t\, \xi = \cos(-t) f + \sin(-t)\xi, \\ \xi_{-t} &= \sin t\, f + \cos t\, \xi = -\sin(-t) f + \cos(-t)\xi. \end{aligned} \tag{4.83}$$

The minus sign occurs because P_t takes a sphere with center $f_{-t}(x)$ and radius $-t$ to the point sphere $f_{-t}(x)$. We call $P_t\lambda$ a *parallel submanifold* of λ. Formula (4.83) shows the close correspondence between these parallel submanifolds and the parallel hypersurfaces f_t to f, in the case where f is an immersed hypersurface.

In the case where the spherical projection f is an immersion at a point $x \in M^{n-1}$, we know that the number of values of t in the interval $[0, \pi)$ for which f_t is not an immersion is at most $n-1$, the maximum number of distinct principal curvatures of f at x. Pinkall [446, p. 428] proved that this statement is also true for an arbitrary Legendre submanifold, even if the spherical projection f is not an immersion at x by proving the following theorem (see also [77, pp. 68–72] for a proof).

Theorem 4.8. *Let $\lambda : M^{n-1} \to \Lambda^{2n-1}$ be a Legendre submanifold with spherical projection f and spherical unit normal field ξ. Then for each $x \in M^{n-1}$, the parallel map,*

$$f_t = \cos t\, f + \sin t\, \xi,$$

fails to be an immersion at x for at most $n-1$ values of $t \in [0, \pi)$.

As a consequence of Pinkall's theorem, one can pass to a parallel submanifold to obtain the following important corollary. Note that parts (a)–(c) of the corollary are pointwise statements, while (d)–(e) hold on an open set U if they can be shown to hold in a neighborhood of each point of U.

Now let x be an arbitrary point of M^{n-1}. If the spherical projection f of λ is an immersion at x, then it is an immersion on a neighborhood of x, and the corollary holds on this neighborhood by known results concerning hypersurfaces in S^n given in Chapter 2, and by the correspondence between the curvature spheres of λ and the principal curvatures of f. If the spherical projection f is not an immersion at x, then by Theorem 4.8, there exists parallel transformation P_{-t} such that the spherical projection f_t of the Legendre submanifold $P_{-t}\lambda$ is an immersion at x, and hence on a neighborhood of x. So the corollary holds for $P_{-t}\lambda$ on this neighborhood of x, and by Theorem 4.7, the corollary also holds for λ on this neighborhood x.

Corollary 4.9. *Let $\lambda : M^{n-1} \to \Lambda^{2n-1}$ be a Legendre submanifold. Then:*

(a) *at each point $x \in M^{n-1}$, there are at most $n-1$ distinct curvature spheres $K_1, \ldots, K_g$,*
(b) *the principal vectors corresponding to a curvature sphere K_i form a subspace T_i of the tangent space T_xM^{n-1},*
(c) *the tangent space $T_xM^{n-1} = T_1 \oplus \cdots \oplus T_g$,*
(d) *if the dimension of a given T_i is constant on an open subset U of M^{n-1}, then the principal distribution T_i is integrable on U,*
(e) *if $\dim T_i = m > 1$ on an open subset U of M^{n-1}, then the curvature sphere map K_i is constant along the leaves of the principal foliation T_i.*

We can also generalize the notion of a curvature surface defined in Section 2.5 (page 32) for hypersurfaces in real space forms to Legendre submanifolds. Specifically, let $\lambda : M^{n-1} \to \Lambda^{2n-1}$ be a Legendre submanifold. A connected submanifold S of M^{n-1} is called a *curvature surface* if at each $x \in S$, the tangent space T_xS is equal to some principal space T_i. For example, if $\dim T_i$ is constant on an open subset U of M^{n-1}, then each leaf of the principal foliation T_i is a curvature surface on U. It is also possible to have a curvature surface S which is not a leaf of a principal foliation as in Example 2.22 on page 33.

Dupin submanifolds in Lie sphere geometry

Next we generalize the definition of a Dupin hypersurface in a real space form to the setting of Legendre submanifolds in Lie sphere geometry. We say that a Legendre submanifold $\lambda : M^{n-1} \to \Lambda^{2n-1}$ is a *Dupin submanifold* if:

(a) along each curvature surface, the corresponding curvature sphere map is constant.

The Dupin submanifold λ is called *proper Dupin* if, in addition to Condition (a), the following condition is satisfied:

(b) the number g of distinct curvature spheres is constant on M.

In the case of the Legendre lift $\lambda : M^{n-1} \to \Lambda^{2n-1}$ of an immersed Dupin hypersurface $f : M^{n-1} \to S^n$, the submanifold λ is a Dupin submanifold, since a curvature sphere map of λ is constant along a curvature surface if and only if the corresponding principal curvature map of f is constant along that curvature surface. Similarly, λ is proper Dupin if and only if f is proper Dupin, since the number of distinct curvatures spheres of λ at a point $x \in M^{n-1}$ equals the number of distinct principal curvatures of f at x. Particularly important examples of proper Dupin submanifolds are the Legendre lifts of isoparametric hypersurfaces in S^n.

Remark 4.10 (Relationship to the Euclidean definition of Dupin). Reckziegel [458] gives a definition of principal curvatures and curvature surfaces in the case of an immersed submanifold $\phi : V \to S^n$ of codimension $\nu + 1 > 1$. In that case, Reckziegel defines a curvature surface to be a connected submanifold $S \subset V$ for which there is a parallel section of the unit normal bundle $\eta : S \to B^{n-1}$ such that for each $x \in S$, the tangent space T_xS is equal to some eigenspace of $A_{\eta(x)}$. The corresponding principal curvature function $\kappa : S \to \mathbf{R}$ is then a smooth function on S. As noted in Remark 2.26 on page 35, Pinkall [447] calls a submanifold $\phi(V)$ of codimension greater than one Dupin if along each curvature surface (in the sense of Reckziegel), the corresponding principal curvature is constant. A Dupin submanifold $\phi(V)$ is proper Dupin if the number of distinct principal curvatures is constant on the unit normal bundle B^{n-1}. One can show that Pinkall's definition is equivalent to requiring that the Legendre lift $\lambda : B^{n-1} \to \Lambda^{2n-1}$ of the submanifold $\phi(V)$ is a proper Dupin submanifold in the sense of Lie sphere geometry, as defined above.

Lie invariance of the Dupin condition

By Theorem 4.7 both the Dupin and proper Dupin conditions are invariant under Lie sphere transformations (see Theorem 4.11 below), and many important classification results for Dupin submanifolds have been obtained in the setting of Lie sphere geometry, as we will see in Chapter 5.

Theorem 4.11. *Let $\lambda : M^{n-1} \to \Lambda^{2n-1}$ be a Legendre submanifold and β a Lie sphere transformation.*

(a) *If λ is Dupin, then $\beta\lambda$ is Dupin.*
(b) *If λ is proper Dupin, then $\beta\lambda$ is proper Dupin.*

Proof. By Theorem 4.7, a point $[K]$ on the line $\lambda(x)$ is a curvature sphere of λ at $x \in M$ if and only if the point $\beta[K]$ is a curvature sphere of $\beta\lambda$ at x, and the principal spaces corresponding $[K]$ and $\beta[K]$ are identical. Since these principal spaces are the same, if S is a curvature surface of λ corresponding to a curvature sphere map $[K]$, then S is also a curvature surface of $\beta\lambda$ corresponding to a curvature sphere map $\beta[K]$, and clearly $[K]$ is constant along S if and only if $\beta[K]$ is constant along S. This proves part (a) of the theorem. Part (b) also follows immediately from Theorem 4.7, since for each $x \in M$, the number g of distinct curvature spheres of λ at x equals the number of distinct curvatures spheres of $\beta\lambda$ at x. So if this number g is constant on M for λ, then it is constant on M for $\beta\lambda$. □

4.5 Lie Curvatures and Isoparametric Hypersurfaces

In this section,we introduce certain natural Lie invariants, known as Lie curvatures, due to R. Miyaoka [365], that have been important in the study of Dupin and isoparametric hypersurfaces in the context of Lie sphere geometry. We also find a criterion (Theorem 4.16) for when a Legendre submanifold is Lie equivalent to the Legendre lift of an isoparametric hypersurface in S^n. This theorem has been used in proving various classification results for Dupin hypersurfaces.

Let $\lambda : M^{n-1} \to \Lambda^{2n-1}$ be an arbitrary Legendre submanifold. As before, we can write $\lambda = [Y_1, Y_{n+3}]$, where

$$Y_1 = (1, f, 0), \quad Y_{n+3} = (0, \xi, 1), \tag{4.84}$$

where f and ξ are the spherical projection and spherical field of unit normals, respectively.

For $x \in M^{n-1}$, the points on the line $\lambda(x)$ can be written in the form,

$$\mu Y_1(x) + Y_{n+3}(x), \tag{4.85}$$

that is, we take μ as an inhomogeneous coordinate along the projective line $\lambda(x)$. Then the point sphere $[Y_1]$ corresponds to $\mu = \infty$. The next two theorems give the relationship between the coordinates of the curvature spheres of λ and the principal curvatures of f, in the case where f has constant rank. In the first theorem, we assume that the spherical projection f is an immersion on M^{n-1}. By Theorem 4.8, we know that this can always be achieved locally by passing to a parallel submanifold.

Theorem 4.12. *Let $\lambda : M^{n-1} \to \Lambda^{2n-1}$ be a Legendre submanifold whose spherical projection $f : M^{n-1} \to S^n$ is an immersion. Let Y_1 and Y_{n+3} be the point sphere and great sphere maps of λ as in equation (4.84). Then the curvature spheres of λ at a point $x \in M^{n-1}$ are*

$$[K_i] = [\kappa_i Y_1 + Y_{n+3}], \quad 1 \leq i \leq g,$$

where $\kappa_1, \ldots, \kappa_g$ are the distinct principal curvatures at x of the oriented hypersurface f with field of unit normals ξ. The multiplicity of the curvature sphere $[K_i]$ equals the multiplicity of the principal curvature κ_i.

Proof. Let X be a nonzero vector in T_xM^{n-1}. Then for any real number μ,

$$d(\mu Y_1 + Y_{n+3})(X) = (0, \mu\, df(X) + d\xi(X), 0).$$

This vector is in Span $\{Y_1(x), Y_{n+3}(x)\}$ if and only if

$$\mu\, df(X) + d\xi(X) = 0,$$

i.e., μ is a principal curvature of f with corresponding principal vector X. □

We next consider the case where the point sphere map Y_1 is a curvature sphere of constant multiplicity m on M^{n-1}. By Corollary 4.9, the corresponding principal distribution is a foliation, and the curvature sphere map $[Y_1]$ is constant along the leaves of this foliation. Thus the map $[Y_1]$ factors through an immersion $[W_1]$ from the space of leaves V of this foliation into Q^{n+1}. We can write $[W_1] = [(1, \phi, 0)]$, where $\phi : V \to S^n$ is an immersed submanifold of codimension $m+1$. The manifold M^{n-1} is locally diffeomorphic to an open subset of the unit normal bundle B^{n-1} of the submanifold ϕ, and λ is essentially the Legendre lift of $\phi(V)$, as defined in Section 4.3. The following theorem relates the curvature spheres of λ to the principal curvatures of ϕ. Recall that the point sphere and great sphere maps for λ are given as in equation (4.68) by

$$Y_1(x, \xi) = (1, \phi(x), 0), \quad Y_{n+3}(x, \xi) = (0, \xi, 1). \tag{4.86}$$

Theorem 4.13. *Let $\lambda : B^{n-1} \to \Lambda^{2n-1}$ be the Legendre lift of an immersed submanifold $\phi(V)$ in S^n of codimension $m+1$. Let Y_1 and Y_{n+3} be the point sphere and great sphere maps of λ as in equation (4.86). Then the curvature spheres of λ at a point $(x, \xi) \in B^{n-1}$ are*

$$[K_i] = [\kappa_i Y_1 + Y_{n+3}], \quad 1 \leq i \leq g,$$

where $\kappa_1, \ldots, \kappa_{g-1}$ are the distinct principal curvatures of the shape operator A_ξ, and $\kappa_g = \infty$. For $1 \leq i \leq g-1$, the multiplicity of the curvature sphere $[K_i]$ equals the multiplicity of the principal curvature κ_i, while the multiplicity of $[K_g]$ is m.

The proof of this theorem is similar to that of Theorem 4.12, but one must introduce local coordinates on the unit normal bundle to get a complete proof (see [77, p. 74]).

Given these two theorems, we define a *principal curvature* of a Legendre submanifold $\lambda : M^{n-1} \to \Lambda^{2n-1}$ at a point $x \in M^{n-1}$ to be a value κ in the set $\mathbf{R} \cup \{\infty\}$ such that $[\kappa Y_1(x) + Y_{n+3}(x)]$ is a curvature sphere of λ at x, where Y_1 and Y_{n+3} are as in equation (4.84).

Lie curvatures and Möbius curvatures

The principal curvatures of a Legendre submanifold are not Lie invariants, and they depend on the special parametrization for λ given in equation (4.84). However, R. Miyaoka [365] pointed out that the cross-ratios of the principal curvatures are Lie invariants. This is due to the fact that a projective transformation preserves the cross-ratio of four points on a projective line.

We now formulate Miyaoka's theorem specifically. Let $\lambda : B^{n-1} \to \Lambda^{2n-1}$ be a Legendre submanifold, and let β be a Lie sphere transformation. The Legendre submanifold $\beta\lambda$ has point sphere and great sphere maps which we denote by

$$Z_1 = (1, h, 0), \quad Z_{n+3} = (0, \zeta, 1),$$

where h and ζ are the spherical projection and spherical field of unit normals of $\beta\lambda$. Let

$$[K_i] = [\kappa_i Y_1 + Y_{n+3}], \quad 1 \leq i \leq g,$$

denote the distinct curvature spheres of λ at a point $x \in M^{n-1}$. By Theorem 4.7, the points $\beta[K_i]$, $1 \leq i \leq g$, are the distinct curvature spheres of $\beta\lambda$ at x. We can write

$$\beta[K_i] = [\gamma_i Z_1 + Z_{n+3}], \quad 1 \leq i \leq g.$$

Then these γ_i are the principal curvatures of $\beta\lambda$ at x.

Next recall that the *cross-ratio* of four distinct numbers a, b, c, d in $\mathbf{R} \cup \{\infty\}$ is given by

$$[a, b; c, d] = \frac{(a-b)(d-c)}{(a-c)(d-b)}. \tag{4.87}$$

We use the usual conventions involving operations with ∞. For example, if $d = \infty$, then the expression $(d-c)/(d-b)$ evaluates to one, and the cross-ratio $[a, b; c, d]$ equals $(a-b)/(a-c)$.

Miyaoka's theorem can now be stated as follows.

Theorem 4.14. *Let $\lambda : M^{n-1} \to \Lambda^{2n-1}$ be a Legendre submanifold and β a Lie sphere transformation. Suppose that $\kappa_1, \ldots, \kappa_g, g \geq 4$, are the distinct principal curvatures of λ at a point $x \in M^{n-1}$, and $\gamma_1, \ldots, \gamma_g$ are the corresponding principal curvatures of $\beta\lambda$ at x. Then for any choice of four numbers h, i, j, k from the set $\{1, \ldots, g\}$, we have*

$$[\kappa_h, \kappa_i; \kappa_j, \kappa_k] = [\gamma_h, \gamma_i; \gamma_j, \gamma_k]. \tag{4.88}$$

Proof. The left side of equation (4.88) is the cross-ratio, in the sense of projective geometry, of the four points $[K_h], [K_i], [K_j], [K_k]$ on the projective line $\lambda(x)$. The right side of equation (4.88) is the cross-ratio of the images of these four points under β. The theorem now follows from the fact that the projective transformation β preserves the cross-ratio of four points on a line. □

The cross-ratios of the principal curvatures of λ are called the *Lie curvatures* of λ. There is also a set of similar invariants for the Möbius group defined as follows. Here we consider a Möbius transformation to be a Lie sphere transformation that takes point spheres to point spheres. Hence the transformation β in Theorem 4.14 is a Möbius transformation if and only if $\beta[Y_1] = [Z_1]$. This leads to the following corollary of Theorem 4.14.

Corollary 4.15. *Let $\lambda : M^{n-1} \rightarrow \Lambda^{2n-1}$ be a Legendre submanifold and β a Möbius transformation. Then for any three distinct principal curvatures $\kappa_h, \kappa_i, \kappa_j$ of λ at a point $x \in M^{n-1}$, none of which equals ∞, we have*

$$\Phi(\kappa_h, \kappa_i, \kappa_j) = (\kappa_h - \kappa_i)/(\kappa_h - \kappa_j) = (\gamma_h - \gamma_i)/(\gamma_h - \gamma_j), \tag{4.89}$$

where γ_h, γ_i, and γ_j are the corresponding principal curvatures of $\beta\lambda$ at the point x.

Proof. Note that we are using equation (4.89) to define the ratio Φ, which is called a *Möbius curvature* of λ. Since β is a Möbius transformation, the point $[Y_1]$, corresponding to $\mu = \infty$, is taken by β to the point Z_1 with coordinate $\gamma = \infty$. Since β preserves cross-ratios, we have

$$[\kappa_h, \kappa_i; \kappa_j, \infty] = [\gamma_h, \gamma_i; \gamma_j, \infty]. \tag{4.90}$$

Since the cross-ratio on the left in equation (4.90) equals the left side of equation (4.89), and the cross-ratio on the right in equation (4.90) equals the right side of equation (4.89), the corollary holds. □

Criterion for Lie equivalence to an isoparametric hypersurface

We close this section with a local Lie geometric characterization of Legendre submanifolds that are Lie equivalent to the Legendre lift of an isoparametric hypersurface in S^n (see Cecil [73]). Here a line in $\mathbf{RP}^{n+2}$ is called *timelike* if it contains only timelike points. This means that an orthonormal basis for the 2-plane in $\mathbf{R}^{n+3}_2$ determined by the timelike line consists of two timelike vectors. An example is the line $[e_1, e_{n+3}]$.

Theorem 4.16. *Let $\lambda : M^{n-1} \rightarrow \Lambda^{2n-1}$ be a Legendre submanifold with g distinct curvature spheres $[K_1], \ldots, [K_g]$ at each point. Then λ is Lie equivalent to the Legendre lift of an isoparametric hypersurface in S^n if and only if there exist g points $[P_1], \ldots, [P_g]$ on a timelike line in $\mathbf{RP}^{n+2}$ such that*

$$\langle K_i, P_i \rangle = 0, \quad 1 \leq i \leq g.$$

Proof. If λ is the Legendre lift of an isoparametric hypersurface in S^n, then all the spheres in a family $[K_i]$ have the same radius ρ_i, where $0 < \rho_i < \pi$. By formula (4.24), this is equivalent to the condition $\langle K_i, P_i \rangle = 0$, where

$$P_i = \sin \rho_i \, e_1 - \cos \rho_i \, e_{n+3}, \quad 1 \leq i \leq g, \tag{4.91}$$

are g points on the timelike line $[e_1, e_{n+3}]$. Since a Lie sphere transformation preserves curvature spheres, timelike lines and the polarity relationship, the same is true for any image of λ under a Lie sphere transformation.

Conversely, suppose that there exist g points $[P_1], \ldots, [P_g]$ on a timelike line ℓ such that $\langle K_i, P_i \rangle = 0$, for $1 \leq i \leq g$. Let β be a Lie sphere transformation that maps ℓ to the line $[e_1, e_{n+3}]$. Then the curvature spheres $\beta[K_i]$ of $\beta\lambda$ are orthogonal to the points $[Q_i] = \beta[P_i]$ on the line $[e_1, e_{n+3}]$. This means that the spheres corresponding to $\beta[K_i]$ have constant radius on M^{n-1}. By applying a parallel transformation P_t, if necessary, we can arrange that none of these curvature spheres has radius zero. Then $P_t\beta\lambda$ is the Legendre lift of an isoparametric hypersurface in S^n. □

Remark 4.17. In the case where λ is Lie equivalent to the Legendre lift of an isoparametric hypersurface in S^n, one can say more about the position of the points $[P_1], \ldots, [P_g]$ on the timelike line ℓ. By Theorem 3.26 (page 108) due to Münzner, the radii ρ_i of the curvature spheres of an isoparametric hypersurface satisfy the equation

$$\rho_i = \rho_1 + (i-1)\frac{\pi}{g}, \quad 1 \leq i \leq g, \tag{4.92}$$

for some $\rho_1 \in (0, \pi/g)$. Hence, after Lie sphere transformation, the $[P_i]$ have the form (4.91) for ρ_i as in equation (4.92).

On an isoparametric hypersurface, the distinct principal curvatures have the form

$$\cot \rho_i, \quad 1 \leq i \leq g, \tag{4.93}$$

for ρ_i as in equation (4.92). From this, we can determine the Lie curvatures of an isoparametric hypersurface, which are obviously constant.

For the sake of definiteness, we make the calculation as follows. First we order the principal curvatures so that

$$\kappa_1 < \cdots < \kappa_g, \tag{4.94}$$

and so the κ_i decrease as the ρ_i increase.

We first consider the case $g = 4$. Then the ordering of the principal curvatures in equation (4.94) leads to a unique Lie curvature Ψ defined by

$$\Psi = [\kappa_1, \kappa_2; \kappa_3, \kappa_4] = (\kappa_1 - \kappa_2)(\kappa_4 - \kappa_3)/(\kappa_1 - \kappa_3)(\kappa_4 - \kappa_2). \tag{4.95}$$

With this ordering of the principal curvatures, the Lie curvature Ψ satisfies the inequality $0 < \Psi < 1$. Using equations (4.93) and (4.95), one can compute that $\Psi = 1/2$ on any isoparametric hypersurface with $g = 4$ principal curvatures, i.e., the four curvature spheres form a *harmonic set* in the sense of projective geometry (see, for example, [472, p. 59]).

Computation of the Lie curvature

There is, however, a simpler way to compute Ψ by considering the focal submanifolds. By Theorem 3.44 (page 131), each isoparametric hypersurface M^{n-1} embedded in S^n has two distinct focal submanifolds, each of codimension greater than one. The hypersurface M^{n-1} is a tube of constant radius over each of these focal submanifolds. Therefore, the Legendre lift of M^{n-1} is obtained from the Legendre lift of either focal submanifold by parallel transformation. Thus, the Legendre lift of M^{n-1} has the same Lie curvature as the Legendre lift of either focal submanifold.

Let $\phi : V \to S^n$ be one of the focal submanifolds of an isoparametric hypersurface M^{n-1} with $g = 4$ principal curvatures. By Theorem 3.21 (page 105) and Theorem 3.26 (page 108), we see that if ξ is any unit normal to $\phi(V)$ at any point, then the shape operator A_ξ has three distinct principal curvatures,

$$\kappa_1 = -1, \quad \kappa_2 = 0, \quad \kappa_3 = 1.$$

By Theorem 4.13, the Legendre lift of ϕ has a fourth principal curvature $\kappa_4 = \infty$. Thus, the Lie curvature of this Legendre lift is

$$\Psi = (-1-0)(\infty - 1)/(-1-1)(\infty - 0) = 1/2, \tag{4.96}$$

as stated above.

We can determine the Lie curvatures of an isoparametric hypersurface M^{n-1} in S^n with $g = 6$ principal curvatures in the same way. Let $\phi(V)$ be one of the focal submanifolds of M^{n-1}. By Münzner's formula (4.92) and Theorem 3.21 (page 105), the Legendre lift of $\phi(V)$ has six constant principal curvatures,

$$\kappa_1 = -\sqrt{3},\ \kappa_2 = -1/\sqrt{3},\ \kappa_3 = 0,\ \kappa_4 = 1/\sqrt{3},\ \kappa_5 = \sqrt{3},\ \kappa_6 = \infty,$$

as in Theorem 4.13. The corresponding six curvature spheres $[K_1], \ldots, [K_6]$ are situated symmetrically on a projective line, as in Figure 4.8.

There are only three geometrically distinct configurations which can obtained by choosing four of the six curvature spheres. These give the cross-ratios:

$$[\kappa_3, \kappa_4; \kappa_5, \kappa_6] = 1/3, \quad [\kappa_2, \kappa_3; \kappa_5, \kappa_6] = 1/4, \quad [\kappa_2, \kappa_3; \kappa_4, \kappa_6] = 1/2.$$

Of course, if a certain cross-ratio has the value r, then one can obtain the values,

$$\{r,\ 1/r,\ 1-r,\ 1/(1-r),\ (r-1)/r,\ r/(r-1)\}, \tag{4.97}$$

by permuting the order of the spheres (see, for example, Samuel [472, p. 58]).

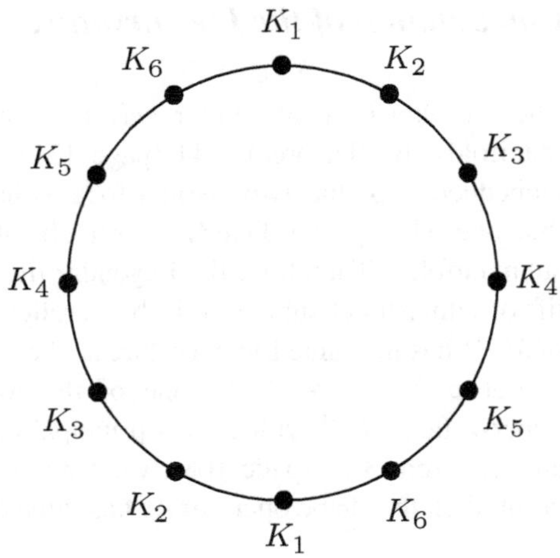

Fig. 4.8 Curvature spheres on a projective line, $g = 6$

4.6 Lie Invariance of Tautness

In this section, we discuss the notion of tautness for Legendre submanifolds in the context of Lie sphere geometry. This was introduced in a paper of Cecil and Chern [79], although the approach taken here is due to Álvarez Paiva [14], who used functions whose level sets form a parabolic pencil of spheres rather than the usual distance functions or height functions to formulate tautness. This approach leads to a natural proof of the invariance of tautness under Lie sphere transformations. In this section, we follow Section 4.6 of the book [77] closely, although we will omit some of the calculations given there. (See also another paper of Álvarez Paiva [13] that extends the notion of tautness to symplectic geometry.)

In the proof of the Lie invariance of tautness, it is more convenient to consider embeddings of compact, connected manifolds into S^n rather than $\mathbf{R}^n$. Theorem 2.70 on page 61 shows that these two theories are equivalent.

As noted in Theorem 2.28 on page 38, Kuiper [301] reformulated tightness and tautness in terms of an injectivity condition on homology which has turned out be very useful. Let f be a nondegenerate function on a manifold V. We consider the sublevel set

$$V_r(f) = \{x \in V \mid f(x) \leq r\}, \quad r \in \mathbf{R}. \tag{4.98}$$

The next theorem, which follows immediately from Theorem 29.2 of Morse–Cairns [379, p. 260] was a key to Kuiper's formulation of these conditions. (This is the same as Theorem 2.28, see page 38 for more discussion).

Theorem 4.18. *Let f be a nondegenerate function on a compact, connected manifold V. For a given field $\mathbf{F}$, the number $\mu(f)$ of critical points of f equals the sum $\beta(V,\mathbf{F})$ of the $\mathbf{F}$-Betti numbers of V if and only if the map on homology,*

$$H_*(V_r(f),\mathbf{F}) \to H_*(V,\mathbf{F}), \tag{4.99}$$

induced by the inclusion $V_r(f) \subset V$ is injective for all $r \in \mathbf{R}$.

Of course, for an embedding $\phi : V \to S^n$ and a height function ℓ_p, the set $V_r(\ell_p)$, is equal to $\phi^{-1}(B)$, where B is the closed ball in S^n obtained by intersecting S^n with the half-space in $\mathbf{R}^{n+1}$ determined by the inequality $\ell_p(q) \leq r$. Kuiper [304] used the continuity property of $\mathbf{Z}_2$-Čech homology to formulate tautness in terms of $\phi^{-1}(B)$, for all closed balls B in S^n, not just those centered at non-focal points of ϕ. Thus, Kuiper proved the following theorem (see also Theorem 2.54 on page 54 for the Euclidean version).

Theorem 4.19. *Let $\phi : V \to S^n$ be an embedding of a compact, connected manifold V into S^n. Then ϕ is taut if and only if for every closed ball B in S^n, the induced homomorphism $H_*(f^{-1}(B)) \to H_*(V)$ in $\mathbf{Z}_2$-Čech homology is injective.*

The key to the approach of Álvarez Paiva [14] is to formulate tautness of Legendre submanifolds in terms of functions whose level sets form a parabolic pencil of unoriented spheres, instead of using linear height functions. This is quite natural in the context of Lie sphere geometry, and it is equivalent to the usual formulation of tautness in the case of the Legendre lift of an embedding $\phi : V \to S^n$.

The specific construction is as follows (see [77, pp. 83–84]). Given a contact element $(p, \xi) \in T_1 S^n$, we want to define a function

$$r_{(p,\xi)} : S^n - \{p\} \to (0, \pi),$$

whose level sets are unoriented spheres in the parabolic pencil of unoriented spheres determined by (p, ξ). (We will often denote $r_{(p,\xi)}$ simply by r when the context is clear.) Every point x in $S^n - \{p\}$ lies on precisely one sphere S_x in the pencil as the spherical radius r of the spheres in the pencil varies from 0 to π. The radius $r_{(p,\xi)}(x)$ of S_x is defined implicitly by the equation

$$\cos r = x \cdot (\cos r\, p + \sin r\, \xi). \tag{4.100}$$

This equation says that x lies in the unoriented sphere S_x in the pencil with center

$$q = \cos r\, p + \sin r\, \xi, \tag{4.101}$$

and spherical radius $r \in (0, \pi)$ (see Figure 4.9).

This defines a smooth function

$$r_{(p,\xi)} : S^n - \{p\} \to (0, \pi). \tag{4.102}$$

Note that the contact element $(p, -\xi)$ determines the same pencil of unoriented spheres and the function $r_{(p,-\xi)} = \pi - r_{(p,\xi)}$. Some sample values of the function $r_{(p,\xi)}$ are

$$r_{(p,\xi)}(\xi) = \pi/4, \quad r_{(p,\xi)}(-p) = \pi/2, \quad r_{(p,\xi)}(-\xi) = 3\pi/4.$$

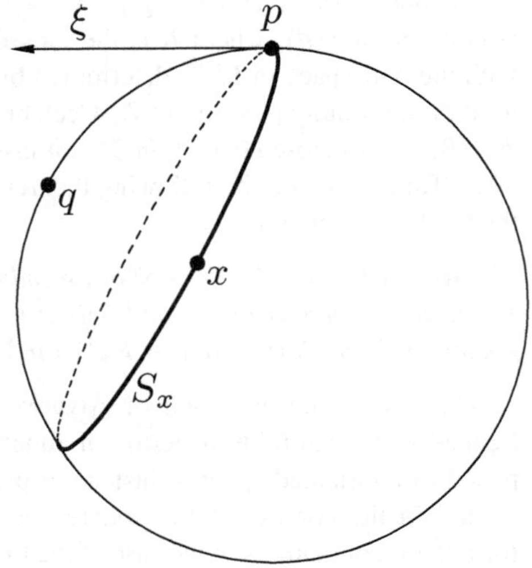

Fig. 4.9 The sphere S_x in the parabolic pencil determined by (p, ξ)

Consider an immersion $\phi : V \to S^n$, where V is a k-dimensional manifold with $k < n$. If $x \in V$, we say that the sphere S_x and $\phi(V)$ are *tangent* at $\phi(x)$ if

$$d\phi(T_x V) \subset T_{\phi(x)} S_x, \tag{4.103}$$

where $d\phi$ is the differential of ϕ.

Critical point behavior

The following lemma describes the critical point behavior of a function of the form $r_{(p,\xi)}$ on an immersed submanifold $\phi : V \to S^n$. This lemma is similar to the Index Theorem L_p functions (Theorem 2.51 on page 53), and it is proven by a direct calculation of the first and second derivatives of r. We will omit the proof here and refer the reader to [77, pp. 84–88] for a complete proof.

Lemma 4.20. *Let $\phi : V \to S^n$ be an immersion of a connected manifold V with $\dim V < n$ into S^n, and let $(p, \xi) \in T_1 S^n$ such that $p \notin \phi(V)$.*

(a) *A point $x_0 \in V$ is a critical point of the function $r_{(p,\xi)}$ if and only if the sphere S_{x_0} containing $\phi(x_0)$ in the parabolic pencil of unoriented spheres determined by (p, ξ) and the submanifold $\phi(V)$ are tangent at $\phi(x_0)$.*
(b) *If $r_{(p,\xi)}$ has a critical point at $x_0 \in V$, then this critical point is degenerate if and only if the sphere S_{x_0} is a curvature sphere of $\phi(V)$ at x_0.*

Next we show that except for (p, ξ) in a set of measure zero in T_1S^n, the function $r_{(p,\xi)}$ is a Morse function on $\phi(V)$. This is accomplished using Sard's Theorem in a manner similar to the proof of Corollary 2.33 on page 40.

In our particular case, from Lemma 4.20 we know that the function $r_{(p,\xi)}$, for $p \notin \phi(V)$, is a Morse function on $\phi(V)$ unless the parabolic pencil of unoriented spheres determined by (p, ξ) contains a curvature sphere of $\phi(V)$. We now show that the set of (p, ξ) in T_1S^n such that the parabolic pencil determined by (p, ξ) contains a curvature sphere of $\phi(V)$ has measure zero in T_1S^n.

Let B^{n-1} denote the unit normal bundle of the submanifold $\phi(V)$ in S^n. Note that in the case where $\phi(V)$ is a hypersurface, B^{n-1} is a two-sheeted covering of V. We first recall the normal exponential map,

$$q : B^{n-1} \times (0, \pi) \to S^n, \tag{4.104}$$

defined as follows. For a point (x, N) in B^{n-1} and $r \in (0, \pi)$, we define

$$q((x, N), r) = \cos r\, x + \sin r\, N. \tag{4.105}$$

Next we define a $(2n-1)$-dimensional manifold W^{2n-1} by

$$W^{2n-1} = \{((x, N), r, \eta) \in B^{n-1} \times (0, \pi) \times S^n \mid \eta \cdot q((x, N), r) = 0\}. \tag{4.106}$$

The manifold W^{2n-1} is a fiber bundle over $B^{n-1} \times (0, \pi)$ with fiber diffeomorphic to S^{n-1}. For each point $((x, N), r) \in B^{n-1} \times (0, \pi)$, the fiber consists of all unit vectors η in $\mathbf{R}^{n+1}$ that are tangent to S^n at the point $q((x, N), r)$.

We define a map,

$$F : W^{2n-1} \to T_1S^n, \tag{4.107}$$

by

$$F((x, N), r, \eta) = (\cos r\, q + \sin r\, \eta,\ \sin r\, q - \cos r\, \eta), \tag{4.108}$$

where $q = q((x, N), r)$ is defined in equation (4.105).

The next lemma shows that if the parabolic pencil of unoriented spheres determined by $(p, \xi) \in T_1S^n$ contains a curvature sphere of $\phi(V)$, then (p, ξ) is a critical value of F. Since the set of critical values of F has measure zero by Sard's Theorem (see, for example, Milnor [359, p. 33]), this will give the desired conclusion. The proof of this lemma is a fairly straightforward calculation of the differential of the map F, and we refer the reader to [77, pp. 89–91] for a detailed proof.

Lemma 4.21. *Let $\phi : V \to S^n$ be an immersion of a connected manifold V with* $\dim V < n$ *into S^n, and let B^{n-1} be the unit normal bundle of $\phi(V)$. Define*

$$F : W^{2n-1} \to T_1S^n,$$

as in equation (4.108). If the parabolic pencil of unoriented spheres determined by (p,ξ) in T_1S^n contains a curvature sphere of $\phi(V)$, then (p,ξ) is a critical value of F. Thus, the set of such (p,ξ) has measure zero in T_1S^n.

Corollary 4.22. *Let $\phi : V \to S^n$ be an immersion of a connected manifold V with $\dim V < n$ into S^n. For almost all $(p,\xi) \in T_1S^n$, the function $r_{(p,\xi)}$ is a Morse function on V.*

Proof. By Lemma 4.20, the function $r_{(p,\xi)}$ is a Morse function on V if and only if $p \notin \phi(V)$ and the parabolic pencil of unoriented spheres determined by (p,ξ) does not contain a curvature sphere of $\phi(V)$. The set of (p,ξ) such that $p \in \phi(V)$ has measure zero, since $\phi(V)$ is a submanifold of codimension at least one in S^n. The set of (p,ξ) such that the parabolic pencil determined by (p,ξ) contains a curvature sphere of $\phi(V)$ has measure zero by Lemma 4.21. Thus, except for (p,ξ) in the set of measure zero obtained by taking the union of these two sets, the function $r_{(p,\xi)}$ is a Morse function on V. □

Tautness in Lie sphere geometry

We will now formulate the definition of tautness for Legendre submanifolds in Lie sphere geometry. Recall the diffeomorphism from T_1S^n to the space Λ^{2n-1} of lines on the Lie quadric Q^{n+1} given by equations (4.37) and (4.38),

$$(p,\xi) \mapsto [(1,p,0),(0,\xi,1)] = \ell \in \Lambda^{2n-1}. \tag{4.109}$$

Under this correspondence, an oriented sphere S in S^n belongs to the parabolic pencil of oriented spheres determined by $(p,\xi) \in T_1S^n$ if and only if the point $[k]$ in Q^{n+1} corresponding to S lies on the line ℓ. Thus, the parabolic pencil of oriented spheres determined by a contact element (p,ξ) contains a curvature sphere S of a Legendre submanifold $\lambda : B^{n-1} \to \Lambda^{2n-1}$ if and only if the corresponding line ℓ contains the point $[k]$ corresponding to S.

A compact, connected Legendre submanifold $\lambda : B^{n-1} \to \Lambda^{2n-1}$ is said to be *Lie-taut* if for almost every line ℓ on the Lie quadric Q^{n+1}, the number of points $x \in B^{n-1}$ such that $\lambda(x)$ intersects ℓ is $\beta(B^{n-1},\mathbf{Z}_2)/2$, i.e., one-half the sum of the $\mathbf{Z}_2$-Betti numbers of B^{n-1}. Here by "almost every," we mean except for a set of measure zero.

Equivalently, this definition says that for almost every contact element (p,ξ) in T_1S^n, the number of points $x \in B^{n-1}$ such that the contact element corresponding to $\lambda(x)$ is in oriented contact with some sphere in the parabolic pencil of oriented spheres determined by (p,ξ) is $\beta(B^{n-1},\mathbf{Z}_2)/2$.

The property of Lie-tautness is clearly invariant under Lie sphere transformations, i.e., if $\lambda : B^{n-1} \to \Lambda^{2n-1}$ is Lie-taut and α is a Lie sphere transformation, then the Legendre submanifold $\alpha\lambda : B^{n-1} \to \Lambda^{2n-1}$ is also Lie-taut. This follows from the fact that the line $\lambda(x)$ intersects a line ℓ if and only if the line $\alpha(\lambda(x))$

intersects the line $\alpha(\ell)$, and α maps the complement of a set of measure zero in Λ^{2n-1} to the complement of a set of measure zero in Λ^{2n-1}.

Remark 4.23 (Comments on the definition of Lie-tautness). The factor of one-half in the definition comes from the fact that Lie sphere geometry deals with oriented contact and not just unoriented tangency, as we will see in the proof of Theorem 4.24 below. Recall that if $\phi : V \to S^n$ is an embedding of a compact, connected manifold V into S^n and B^{n-1} is the unit normal bundle of $\phi(V)$, then the Legendre lift of ϕ is defined to be the Legendre submanifold $\lambda : B^{n-1} \to \Lambda^{2n-1}$ given by

$$\lambda(x,N) = [(1,\phi(x),0),(0,N,1)], \tag{4.110}$$

where N is a unit normal vector to $\phi(V)$ at $\phi(x)$. If V has dimension $n-1$, then B^{n-1} is a two-sheeted covering of V. If V has dimension less than $n-1$, then B^{n-1} is diffeomorphic to a tube W^{n-1} of sufficiently small radius over $\phi(V)$ so that W^{n-1} is an embedded hypersurface in S^n. In either case,

$$\beta(B^{n-1},\mathbf{Z}_2) = 2\beta(V,\mathbf{Z}_2).$$

This is obvious in the case where V has dimension $n-1$, and it was proved by Pinkall [447] in the case where V has dimension less than $n-1$.

Since Lie-tautness is invariant under Lie sphere transformations, the following theorem establishes that tautness is Lie invariant. Recall that a taut immersion $\phi : V \to S^n$ is in fact an embedding (see Theorem 2.59 on page 56). Here we use the proof of Theorem 4.28 of the book [77, pp. 93–95].

Theorem 4.24. *Let $\phi : V \to S^n$ be an embedding of a compact, connected manifold V with $\dim V < n$ into S^n. Then $\phi(V)$ is a taut submanifold in S^n if and only if the Legendre lift $\lambda : B^{n-1} \to \Lambda^{2n-1}$ of ϕ is Lie-taut.*

Proof. Suppose that $\phi(V)$ is a taut submanifold in S^n, and let

$$\lambda : B^{n-1} \to \Lambda^{2n-1}$$

be the Legendre lift of ϕ. Let $(p,\xi) \in T_1S^n$ such that $p \notin \phi(V)$ and such that the parabolic pencil of unoriented spheres determined by (p,ξ) does not contain a curvature sphere of $\phi(V)$. By Lemma 4.21, the set of such (p,ξ) is the complement of a set of measure zero in T_1S^n. For such (p,ξ), the function $r_{(p,\xi)}$ is a Morse function on V, and the sublevel set

$$V_s(r_{(p,\xi)}) = \{x \in V \mid r_{(p,\xi)}(x) \leq s\} = \phi(V) \cap B, \quad 0 < s < \pi, \tag{4.111}$$

is the intersection of $\phi(V)$ with a closed ball $B \subset S^n$. By tautness and Theorem 4.19, the map on $\mathbf{Z}_2$-Čech homology,

$$H_*(V_s(r_{(p,\xi)})) = H_*(\phi^{-1}(B)) \to H_*(V), \tag{4.112}$$

is injective for every $s \in \mathbf{R}$, and so by Theorem 4.18, the function $r_{(p,\xi)}$ has $\beta(V, \mathbf{Z}_2)$ critical points on V.

By Lemma 4.20, a point $x \in V$ is a critical point of $r_{(p,\xi)}$ if and only if the unoriented sphere S_x in the parabolic pencil determined by (p, ξ) containing x is tangent to $\phi(V)$ at $\phi(x)$. At each such point x, exactly one contact element $(x, N) \in B^{n-1}$ is in oriented contact with the oriented sphere $\tilde{S}_x$ through x in the parabolic pencil of oriented spheres determined by (p, ξ). Thus, the number of critical points of $r_{(p,\xi)}$ on V equals the number of points $(x, N) \in B^{n-1}$ such that (x, N) is in oriented contact with an oriented sphere in the parabolic pencil of oriented spheres determined by (p, ξ).

Thus there are

$$\beta(V, \mathbf{Z}_2) = \beta(B^{n-1}, \mathbf{Z}_2)/2$$

points $(x, N) \in B^{n-1}$ such that (x, N) is in oriented contact with an oriented sphere in the parabolic pencil of oriented spheres determined by (p, ξ). This means that there are $\beta(B^{n-1}, \mathbf{Z}_2)/2$ points $(x, N) \in B^{n-1}$ such that the line $\lambda(x, N)$ intersects the line ℓ on Q^{n+1} corresponding to the contact element (p, ξ). Since this true for almost every $(p, \xi) \in T_1S^n$, the Legendre lift λ of ϕ is Lie-taut.

To prove the converse, we use a Čech homology argument similar to that of Kuiper [303] used in the proof of Theorem 2.41 on page 44. Suppose that the Legendre lift $\lambda : B^{n-1} \to \Lambda^{2n-1}$ of ϕ is Lie-taut. Then for all $(p, \xi) \in T_1S^n$ except for a set Z of measure zero, the number of points $(x, N) \in B^{n-1}$ that are in oriented contact with some sphere in the parabolic pencil of oriented spheres determined by (p, ξ) is $\beta(B^{n-1}, \mathbf{Z}_2)/2 = \beta(V, \mathbf{Z}_2)$. This means that the corresponding function $r_{(p,\xi)}$ has $\beta(V, \mathbf{Z}_2)$ critical points on V. By Theorem 4.18, this implies that for a closed ball $B \subset S^n$ such that $\phi^{-1}(B) = V_s(r_{(p,\xi)})$ for $(p, \xi) \notin Z$ and $s \in \mathbf{R}$, the map on homology,

$$H_*(\phi^{-1}(B)) \to H_*(V), \tag{4.113}$$

is injective. On the other hand, if B is a closed ball corresponding to a sublevel set of $r_{(p,\xi)}$ for $(p, \xi) \in Z$, then since Z has measure zero, one can produce a nested sequence,

$$\{B_i\}, \quad i = 1, 2, 3, \ldots,$$

of closed balls (coming from $r_{(p,\xi)}$ for $(p, \xi) \notin Z$) satisfying

$$\phi^{-1}(B_i) \supset \phi^{-1}(B_{i+1}) \supset \cdots \supset \cap_{j=1}^{\infty} \phi^{-1}(B_j) = \phi^{-1}(B), \tag{4.114}$$

for $i = 1, 2, 3, \ldots$, such that the homomorphism in $\mathbf{Z}_2$-homology,

$$H_*(\phi^{-1}(B_i)) \to H_*(V), \text{ is injective for } i = 1, 2, 3, \ldots \tag{4.115}$$

If equations (4.114) and (4.115) are satisfied, then the map

$$H_*(\phi^{-1}(B_i)) \to H_*(\phi^{-1}(B_j)) \text{ is injective for all } i > j. \tag{4.116}$$

The continuity property of Čech homology (see Eilenberg–Steenrod [145, p. 261]) says that

$$H_*(\phi^{-1}(B)) = \overleftarrow{\lim_{i\to\infty}}\, H_*(\phi^{-1}(B_i)).$$

Equation (4.116) and Theorem 3.4 of Eilenberg–Steenrod [145, p. 216] on inverse limits imply that the map

$$H_*(\phi^{-1}(B)) \to H_*(\phi^{-1}(B_i))$$

is injective for each i. Thus, from equation (4.115), we get that the map

$$H_*(\phi^{-1}(B)) \to H_*(V)$$

is also injective. Since this holds for all closed balls B in S^n, the embedding $\phi(V)$ is taut by Theorem 4.19. □

Another formulation of the Lie invariance of tautness is the following corollary, as in [77, p. 95].

Corollary 4.25. *Let $\phi : V \to S^n$ and $\psi : V \to S^n$ be two embeddings of a compact, connected manifold V with $\dim V < n$ into S^n, such that their corresponding Legendre lifts are Lie equivalent. Then ϕ is taut if and only if ψ is taut.*

Proof. Since the Legendre lifts of ϕ and ψ are Lie equivalent, the unit normal bundles of $\phi(V)$ and $\psi(V)$ are diffeomorphic, and we will denote them both by B^{n-1}. Now let $\lambda : B^{n-1} \to \Lambda^{2n-1}$ and $\mu : B^{n-1} \to \Lambda^{2n-1}$ be the Legendre lifts of ϕ and ψ, respectively. By Theorem 4.24, ϕ is taut if and only if λ is Lie-taut, and ψ is taut if and only if μ is Lie-taut. Further, since λ and μ are Lie equivalent, λ is Lie-taut if and only if μ is Lie-taut, so it follows that ϕ is taut if and only if ψ is taut. □

Chapter 5
Dupin Hypersurfaces

In this chapter, we study Dupin hypersurfaces in a real space form $\tilde{M}^n$. As noted earlier, Dupin hypersurfaces can also be studied in the context of Lie sphere geometry, and many classification results have been obtained in that setting. In this chapter, we will use the viewpoint of the metric geometry of $\tilde{M}^n$ as well as that of Lie sphere geometry to obtain results about Dupin hypersurfaces.

As we saw in Lemma 2.15 on page 29, the theory of Dupin hypersurfaces in the three space forms $\mathbf{R}^n$, S^n or H^n, is essentially the same via stereographic projection. We will use whichever ambient space is most convenient for the discussion at hand.

We begin by recalling some definitions and results that were discussed in Chapter 2. Let $f : M \to \tilde{M}^n$ be a connected immersed hypersurface, and let ξ be a locally defined field of unit normals to $f(M)$. A *curvature surface* of M is a smooth submanifold S such that for each point $x \in S$, the tangent space T_xS is equal to a principal space of the shape operator A of M at x.

An oriented hypersurface $f : M^n \to \tilde{M}^{n+1}$ is called a *Dupin hypersurface* if:

(a) along each curvature surface, the corresponding principal curvature is constant.

Furthermore, a Dupin hypersurface M is called *proper Dupin* if, in addition to Condition (a), the following condition is satisfied:

(b) the number g of distinct principal curvatures is constant on M.

We showed in Theorem 2.23 on page 33 that as a result of the Codazzi equation, Condition (a) is automatically satisfied on a curvature surface S of dimension greater than one.

Note that Condition (b) is equivalent to requiring that each continuous principal curvature function has constant multiplicity on M. Furthermore, the number of distinct principal curvatures is locally constant on a dense open subset of any hypersurface in S^n (see Reckziegel [457, 458] or Singley [486]).

T.E. Cecil, P.J. Ryan, *Geometry of Hypersurfaces*, Springer Monographs in Mathematics, DOI 10.1007/978-1-4939-3246-7_5

Next, if a continuous principal curvature function μ has constant multiplicity m on M, then μ is a smooth function on M, and the distribution T_μ of principal vectors corresponding to μ is a smooth distribution on M (see, for example, Nomizu [402], Reckziegel [457, 458], or Singley [486]). Furthermore, by using the Codazzi equation, we showed in Theorem 2.10 on page 24 that T_μ is integrable, and the leaves of the foliation T_μ are curvature surfaces of M.

In Theorems 2.11 and 2.14 (see page 25), we showed that a principal curvature μ of constant multiplicity m is constant along each of its curvature surfaces in M if and only if these curvature surfaces are open subsets of m-dimensional totally umbilic submanifolds of $\tilde{M}^n$ in the case $m > 1$, and they are plane curves of constant curvature in the case $m = 1$. Furthermore, on the open subset $U \subset M$ on which the focal map $f_\mu : U \to \tilde{M}^n$ is defined, f_μ factors through an immersion of the (possibly non-Hausdorff) $(n-1-m)$-dimensional manifold U/T_μ into $\tilde{M}^n$ (see Theorems 2.12 and 2.14). Furthermore, if M is complete with respect to the induced metric, then the leaf space U/T_λ is a Hausdorff manifold.

The *curvature sphere* $K_\mu(x)$ corresponding to the principal curvature μ at a point $x \in U$ is the hypersphere in $\tilde{M}^n$ through $f(x)$ centered at the focal point $f_\mu(x)$. Thus, $K_\mu(x)$ is tangent to $f(M)$ at $f(x)$. By Corollary 2.9 on page 23, the principal curvature μ is constant along each of its curvature surfaces in U if and only if the curvature sphere map K_μ is constant along each of these curvature surfaces.

An important class of proper Dupin hypersurfaces is the set of all isoparametric hypersurfaces in S^n, and those hypersurfaces in $\mathbf{R}^n$ obtained from isoparametric hypersurfaces in S^n via stereographic projection. For example, the well-known cyclides of Dupin in $\mathbf{R}^3$ are obtained from a standard product torus $S^1(r) \times S^1(s) \subset S^3$, $r^2 + s^2 = 1$, in this way. These examples will be discussed in more detail in Section 5.5.

As noted in Chapter 1, there are both local and global aspects to the theory of proper Dupin hypersurfaces with quite different results. For example, Thorbergsson [533] proved that the restriction $g = 1, 2, 3, 4$, or 6 on the number of distinct principal curvatures of an isoparametric hypersurface in S^n also holds for compact proper Dupin hypersurfaces embedded in S^n (see Theorem 3.63 on page 143). However, as we saw in Theorem 2.25 on page 34, there exist noncompact proper Dupin hypersurfaces with any given number g of distinct principal curvatures having any prescribed multiplicities.

In Section 5.1, we discuss Pinkall's standard constructions for producing a proper Dupin hypersurface in $\mathbf{R}^{n+m}$, $m \geq 1$, with $g + 1$ distinct principal curvatures from a Dupin hypersurface M^{n-1} in $\mathbf{R}^n \subset \mathbf{R}^{n+m}$ with g distinct principal curvatures. These constructions involving building a tube, cylinder or a surface of revolution over M^{n-1} in $\mathbf{R}^{n+m}$, and they were used in proving Pinkall's Theorem 2.25.

In Section 5.2, we discuss the notion of a principal coordinate system, and find a local criterion for a hypersurface to have a principal coordinate system. The examples constructed by Pinkall in proving Theorem 2.25 all have such a local principal coordinate system, and several authors have considered principal coordinate systems in their work on Dupin hypersurfaces.

The notion of reducibility due to Pinkall [446, p. 438] is important in the theory of Dupin hypersurfaces, and it is a concept that is best formulated in the context of Lie sphere geometry. A proper Dupin submanifold $\lambda : M^{n-1} \to \Lambda^{2n-1}$ is said to be *reducible* if it is locally Lie equivalent to the Legendre lift of a proper Dupin hypersurface in $\mathbf{R}^n$ obtained by one of Pinkall's constructions. In Section 5.3, we discuss reducible proper Dupin hypersurfaces in detail and develop Lie geometric criteria for reducibility.

In Section 5.4, we study the method of moving frames in Lie sphere geometry. This method has been used to obtain local classifications of proper Dupin hypersurfaces with 2, 3, or 4 distinct principal curvatures. In Section 5.5, we use this method to give a complete local classification of proper Dupin hypersurfaces with $g = 2$ distinct principal curvatures, i.e., the cyclides of Dupin. This is a nineteenth century result for $n = 3$, and it was obtained in dimensions $n > 3$ by Pinkall [446] in 1985. In Sections 5.6 and 5.7, we discuss local classification results for the cases $g = 3$ and $g = 4$, respectively, that have been obtained using the moving frames approach.

As demonstrated by Thorbergsson's theorem (Theorem 3.63 on page 143), compact proper Dupin hypersurfaces in S^n are far less plentiful, and several important classification results have been obtained for them. These results are discussed in detail in Section 5.8 together with the important counterexamples of Pinkall–Thorbergsson [448] and Miyaoka–Ozawa [377] to the conjecture that every compact proper Dupin hypersurface embedded in S^n is Lie equivalent to an isoparametric hypersurface.

As noted in Section 2.8, the notions of Dupin and taut for submanifolds of real space forms are closely related. In Sections 5.9 and 5.10, we discuss important classification results that have been obtained for taut submanifolds in Euclidean space $\mathbf{R}^n$. Many of these have been proven by using classifications of compact proper Dupin hypersurfaces.

5.1 Pinkall's Standard Constructions

In this section, we discuss three standard constructions introduced by Pinkall [446] for producing a Dupin hypersurface in $\mathbf{R}^{n+m}$, $m \geq 1$, with $g + 1$ distinct principal curvatures from a Dupin hypersurface M^{n-1} in $\mathbf{R}^n \subset \mathbf{R}^{n+m}$ with g distinct principal curvatures. These constructions involving building a tube, cylinder, or a surface of revolution over M in $\mathbf{R}^{n+m}$.

These constructions can be generalized to the context of Lie sphere geometry (see [77, pp. 127–141]), and some of the problems involving singularities that occur in the Euclidean setting are handled much better in the setting of Lie sphere geometry. Even so, the constructions are simpler to formulate in the Euclidean setting, and we will do that here.

For all of the constructions, we begin with an immersed hypersurface $f : M^{n-1} \to \mathbf{R}^n$. We then embed $\mathbf{R}^n$ as a linear subspace of $\mathbf{R}^{n+m}, m \geq 1$, and consider the immersion $f : M^{n-1} \to \mathbf{R}^n \subset \mathbf{R}^{n+m}$. We then construct a tube, cylinder or surface of revolution starting with this given f.

Remark 5.1 (Pinkall's cone construction). When Pinkall [446] introduced his constructions, he also listed the following cone construction. Begin with a proper Dupin hypersurface $M^{n-1} \subset S^n \subset \mathbf{R}^{n+1}$. Then construct a cone C^n over M^{n-1} in $\mathbf{R}^{n+1}$ with vertex at the origin. This cone is also a Dupin hypersurface in $\mathbf{R}^{n+1}$. Pinkall pointed out that this cone construction is locally Lie equivalent to the tube construction (see also [77, p. 144]). Thus one does not need both the tube and cone constructions when considering Dupin hypersurfaces in the context of Lie sphere geometry. In this section, we will describe the tube construction in detail, but not the cone construction.

We first set some notation common to all three constructions. Let

$$\{e_1, \ldots, e_{n+m}\} \tag{5.1}$$

be the standard orthonormal basis for $\mathbf{R}^{n+m}$, and let $\mathbf{R}^n$ be the linear subspace of $\mathbf{R}^{n+m}$ spanned by $\{e_1, \ldots, e_n\}$. For all three constructions, we begin with an oriented immersed hypersurface $f : M^{n-1} \to \mathbf{R}^n \subset \mathbf{R}^{n+m}$ with field of unit normal vectors ξ to $f(M)$ in $\mathbf{R}^n$. The vector fields $\xi, e_{n+1}, \ldots, e_{n+m}$ form a basis to the normal space to $f(M)$ in $\mathbf{R}^{n+m}$ at $f(x)$ for each $x \in M$. The unit normal bundle B^{n+m-1} of $f(M)$ in $\mathbf{R}^{n+m}$ is diffeomorphic to $M \times S^m$ as follows. Let

$$S^m = \{(y_0, y_1, \ldots, y_m) \mid y_0^2 + \cdots + y_m^2 = 1\}. \tag{5.2}$$

For $x \in M$ and $y = (y_0, y_1, \ldots, y_m) \in S^m$, let

$$\eta(x, y) = y_0 \xi(x) + y_1 e_{n+1} + \cdots + y_m e_{n+m}. \tag{5.3}$$

The map,

$$(x, y) \mapsto (x, \eta(x, y)), \tag{5.4}$$

is a diffeomorphism from $M \times S^m$ onto B^{n+m-1}. With this notation set, we now formulate Pinkall's three constructions.

Tubes

For $t > 0$, we define a map $f_t : B^{n+m-1} \to \mathbf{R}^{n+m}$ which gives the tube of radius t around the submanifold $f(M)$ in $\mathbf{R}^{n+m}$. Using the diffeomorphism between $M \times S^m$ and B^{n+m-1} in equation (5.4), we can consider f_t as a map from $M \times S^m$ into $\mathbf{R}^{n+m}$ given by

$$f_t(x, y) = f(x) + t\eta(x, y), \tag{5.5}$$

where $\eta(x, y)$ is given in equation (5.3).

In the following local calculation, we consider $M^{n-1} \subset \mathbf{R}^n \subset \mathbf{R}^{n+m}$ as an embedded submanifold, and we do not specifically mention the embedding f. We consider the tangent space T_xM at a point $x \in M$ to be a subspace of $T_x\mathbf{R}^{n+m}$. Then the tangent space to $M \times S^m$ at a point (x, y) is given by

$$T_{(x,y)}(M \times S^m) = \{(X, Y) \mid X \in T_xM,\ Y \in T_yS^m\}. \tag{5.6}$$

We first compute the differential $(f_t)_*$ of f_t in order to determine where f_t is an immersion. We begin by computing $(f_t)_*(0, Y)$ at the point $(x, y) \in M \times S^m$, where $Y = (Y_0, Y_1, \ldots, Y_m)$ is a tangent vector to S^m at $y \in S^m$. Let $\alpha(s) = (\alpha_0(s), \ldots, \alpha_m(s))$ be a curve in S^m with $\alpha(0) = y$ and initial tangent vector $\overrightarrow{\alpha}(0) = Y$. Then $(f_t)_*(0, Y)$ is the initial tangent vector to the curve

$$\beta(s) = f_t(x, \alpha(s)) = x + t(\alpha_0(s)\xi(x) + \alpha_1(s)e_{n+1} + \cdots + \alpha_m(s)e_{n+m}). \tag{5.7}$$

Using the fact that $\overrightarrow{\alpha}(0) = Y$, we compute

$$(f_t)_*(0, Y) = \overrightarrow{\beta}(0) = t\,(Y_0\xi(x) + Y_1e_{n+1} + \cdots + Y_me_{n+m}). \tag{5.8}$$

Next we calculate $(f_t)_*(X, 0)$ for $X \in T_xM$. Let $\delta(s)$ be a curve in M with initial point $\delta(0) = x$ and initial tangent vector $\overrightarrow{\delta}(0) = X$. Then $(f_t)_*(X, 0)$ is the initial tangent vector to the curve

$$\begin{aligned} \varepsilon(s) &= f_t(\delta(s), y) = \delta(s) + t\eta(\delta(s), y) \\ &= \delta(s) + t(y_0\xi(\delta(s)) + y_1e_{n+1} + \cdots + y_me_{n+m}). \end{aligned} \tag{5.9}$$

Differentiating with respect to s and evaluating at $s = 0$, we get

$$\begin{aligned} (f_t)_*(X, 0) &= \overrightarrow{\varepsilon}(0) = X + ty_0D_X\xi = X - ty_0AX \\ &= (I - ty_0A)X, \end{aligned} \tag{5.10}$$

where we are identifying X with its Euclidean parallel translate at the point $f_t(x, y)$, and A is the shape operator of M determined by ξ, i.e., $AX = -D_X\xi$.

From equations (5.8) and (5.10), we see that $(f_t)_*(0, Y)$ is parallel to a normal vector to M at x, and $(f_t)_*(X, 0)$ is parallel to a tangent vector to M at x. Since $(f_t)_*(0, Y) \neq 0$ for a nonzero vector $Y \in T_yS^m$, we see that $(f_t)_*(X, Y) \neq 0$ if $Y \neq 0$.

Next we check when $(f_t)_*(X, 0)$ equals zero. From equation (5.10), we see that for a nonzero vector $X \in T_xM$, the vector $(f_t)_*(X, 0)$ equals zero if and only if

$$(I - ty_0A)X = 0. \tag{5.11}$$

In the case where $y_0 \neq 0$, this holds if and only if $1/ty_0$ is a principal curvature of A at x with corresponding principal vector X. Note that this happens precisely when

$$p = f_t(x, y) = x + t\eta(x, y). \tag{5.12}$$

is a focal point of (M, x) in $\mathbf{R}^{n+m}$, since $A_\eta = y_0 A$ by equation (5.3), and so $1/ty_0$ is a principal curvature of A at x if and only if $1/t$ is an eigenvalue of the shape operator A_η at x. Thus, this calculation agrees with Theorem 2.1 on page 11 regarding the location of focal points.

Shape operator of a tube

We now wish to find the shape operator A_t of the tube f_t at points (x, y) where $f_t(x, y)$ is not a focal point of (M, x) in $\mathbf{R}^{n+m}$. Let W be the open subset of $M \times S^m$ on which $(f_t)_*$ has rank $n + m - 1$. Then $f_t W$ is an immersed hypersurface in $\mathbf{R}^{n+m}$, and we want to find the shape operator and principal curvatures of $f_t W$. For $(x, y) \in W$, the vector field,

$$\eta(x, y) = y_0 \xi(x) + y_1 e_{n+1} + \cdots + y_m e_{n+m}, \tag{5.13}$$

is a field of unit normals on the tube $f_t W$, as can be seen from equations (5.8) and (5.10). The shape operator A_t determined by the unit normal field η on the tube $f_t W$ is defined by the equation,

$$(f_t)_*(A_t(X, Y)) = -D_{(f_t)_*(X,Y)}\eta, \tag{5.14}$$

where η is the field of unit normals given in equation (5.13).

We first compute A_t for vectors of the form $(0, Y)$ in $T_{(x,y)}M \times S^m$, and then for vectors of the form $(X, 0)$. As we saw earlier, $(f_t)_*(0, Y)$ is the initial tangent vector to the curve

$$\beta(s) = f_t(x, \alpha(s)) = x + t(\alpha_0(s)\xi(x) + \alpha_1(s)e_{n+1} + \cdots + \alpha_m(s)e_{n+m}),$$

as in equation (5.7) above. Hence $D_{(f_t)_*(0,Y)}\eta$ is the initial tangent vector to the curve

$$\eta(\beta(s)) = \alpha_0(s)\xi(x) + \alpha_1(s)e_{n+1} + \cdots + \alpha_m(s)e_{n+m},$$

and so,

$$(f_t)_*(A_t(0, Y)) = -\ (Y_0\xi(x) + Y_1 e_{n+1} + \cdots + Y_m e_{n+m}). \tag{5.15}$$

Comparing this with equation (5.8), we see that

$$A_t(0, Y) = -\frac{1}{t}(0, Y). \tag{5.16}$$

Thus, every vector of the form $(0, Y)$ is a principal vector of A_t corresponding to the principal curvature $-1/t$. This new principal curvature arises because $f(M)$ has codimension $m + 1 > 1$ in $\mathbf{R}^{n+m}$. This agrees with Theorem 2.2 (page 17) for the shape operator of a tube.

Next we find $A_t(X, 0)$ for $X \in T_xM$. Let $\delta(s)$ be a curve in M with initial point $\delta(0) = x$ and initial tangent vector $\overrightarrow{\delta}(0) = X$, as above. Then $(f_t)_*(X, 0)$ is the initial tangent vector to the curve,

$$\varepsilon(s) = f_t(\delta(s), y) = \delta(s) + t(y_0\xi(\delta(s)) + y_1e_{n+1} + \cdots + y_me_{n+m}),$$

as in equation (5.9) above. Along this curve $\varepsilon(s)$, the unit normal to the tube f_tW is given by

$$\eta(\varepsilon(s)) = y_0\xi(\delta(s)) + y_1e_{n+1} + \cdots + y_me_{n+m}.$$

Then $D_{(f_t)_*(X,0)}\eta$ is the initial tangent vector to the curve $\eta(\varepsilon(s))$, which is just $D_X(y_0\xi)$, since the other terms in the formula for $\eta(\varepsilon(s))$ are constants. So we have

$$D_{(f_t)_*(X,0)}\eta = y_0D_x\xi = -y_0AX. \tag{5.17}$$

where we are identifying AX with its Euclidean parallel translate at the point $f_t(x, y)$. Comparing this with equation (5.10),

$$(f_t)_*(X, 0) = (I - ty_0A)X,$$

and using equation (5.14), we get

$$A_t(X, 0) = ((I - ty_0A)^{-1}y_0AX, 0). \tag{5.18}$$

In the case of a principal vector X such that $AX = \lambda X$ at a point (x, y) with $y_0 \neq 0$, this reduces to

$$A_t(X, 0) = ((I - ty_0A)^{-1}y_0AX, 0) = \left(\frac{y_0\lambda}{1 - ty_0\lambda}X, 0\right). \tag{5.19}$$

Thus, $(X, 0)$ is a principal vector with corresponding principal curvature

$$\mu = \frac{y_0\lambda}{1 - ty_0\lambda}. \tag{5.20}$$

We now assume that the original hypersurface $f : M^{n-1} \to \mathbf{R}^n$ is a proper Dupin hypersurface with g distinct principal curvatures λ_i, $1 \leq i \leq g$, at each point. We first list the principal curvatures of the tube $f_t W$ and check the Dupin condition at a point $(x, y) \in W$. As we have seen above, this depends on whether or not the coordinate y_0 is zero, that is, whether or not the vector $\eta(x, y)$ in equation (5.3) is orthogonal to $\mathbf{R}^n$ in $\mathbf{R}^{n+m}$.

We first treat the case where $y_0 \neq 0$, that is, when the vector $\eta(x, y)$ in equation (5.3) is not orthogonal to $\mathbf{R}^n$ in $\mathbf{R}^{n+m}$. From equations (5.16) and (5.20), we see that A_t has $g + 1$ distinct principal curvatures at such points,

$$\mu_i = \frac{y_0 \lambda_i}{1 - t y_0 \lambda_i}, \quad 1 \leq i \leq g, \tag{5.21}$$

and $\mu_{g+1} = -1/t$ of multiplicity m. For $1 \leq i \leq g$, the principal space corresponding to μ_i is

$$T_{\mu_i} = \{(X, 0) \mid X \in T_{\lambda_i}\} = T_{\lambda_i} \times \{0\}. \tag{5.22}$$

So the principal curvature μ_i has the same multiplicity as λ_i. The leaves of the principal foliation T_{μ_i} have the form $S \times \{y\}$, where S is a leaf of the principal foliation T_{λ_i} on M. Furthermore, for a vector $X \in T_{\lambda_i}$, the derivative $(X, 0)\mu_i = 0$ if and only if $X\lambda_i = 0$, and this condition holds, because f is Dupin. Thus, μ_i is constant along its curvatures surfaces. In addition, the new principal curvature $\mu_{g+1} = -1/t$ is constant on W, and therefore it is constant along its curvature surfaces, which have the form $\{x\} \times S^m$ for $x \in M$. Thus, the Dupin Condition (a) is satisfied at a point (x, y) with $y_0 \neq 0$.

Next we consider points (x, y) in $W \subset M \times S^m$ where $y_0 = 0$, that is, where the vector $\eta(x, y)$ in equation (5.3) is orthogonal to $\mathbf{R}^n$ in $\mathbf{R}^{n+m}$. From equations (5.10) and (5.19), we see that $(f_t)_*(X, 0) = X$ and $A_t(X, 0) = (0, 0)$ for every $X \in T_x M$. Thus $\kappa = 0$ is a principal curvature of multiplicity $n - 1$, and its curvature surfaces are of the form $M \times \{y\}$, where $y_0 = 0$. The other principal curvature at such points is $\mu = -1/t$ having multiplicity m, and its curvature surfaces have the form $\{x\} \times S^m$, for $x \in M$. So the number of distinct principal curvatures is two at such points, and the Dupin Condition (a) is satisfied at these points also. However, the tube $f_t W$ is not proper Dupin unless the number g of distinct principal curvatures of M is one, since $g + 1 \neq 2$ unless $g = 1$.

Number of distinct principal curvatures

We summarize these results in the following proposition.

Proposition 5.2. *Suppose that $f : M^{n-1} \to \mathbf{R}^n \subset \mathbf{R}^{n+m}$ is a proper Dupin hypersurface in $\mathbf{R}^n$ with g distinct principal curvatures. Then the tube construction*

f_t yields a Dupin hypersurface $f_t : W \to \mathbf{R}^{n+m}$, where W is the open subset of the unit normal bundle B^{n+m-1} on which f_t is an immersion. The number $\gamma(x,\eta)$ of distinct principal curvatures of f_t at a point $(x,\eta) \in B^{n+m-1}$ is as follows:

(a) $\gamma(x,\eta) = 2$, *if η is orthogonal to $\mathbf{R}^n$ in $\mathbf{R}^{n+m}$.*
(b) $\gamma(x,\eta) = g + 1$, *otherwise.*

Remark 5.3. The tube over a torus $T^2 \subset \mathbf{R}^3 \subset \mathbf{R}^4$ in Example 2.22 on page 33 provides a good example of Proposition 5.2.

The tube construction can be formulated in the context of Lie sphere geometry (see [77, pp. 127–133]). In that context, there is no need to exclude the points where f_t is not an immersion, and the tube construction can be defined on the whole unit normal bundle B^{n+m-1}. The precise formulation in that context is the following proposition from [77, p. 131].

Proposition 5.4. *Suppose that $\lambda : M^{n-1} \to \Lambda^{2n-1}$ is a proper Dupin submanifold with g distinct curvature spheres such that the Euclidean projection f is an immersion of M^{n-1} into $\mathbf{R}^n \subset \mathbf{R}^{n+m}$. Then the tube construction yields a Dupin submanifold μ defined on the unit normal bundle B^{n+m-1} of $f(M^{n-1})$ in $\mathbf{R}^{n+m}$. The number $\gamma(x,\eta)$ of distinct curvature spheres of μ at a point $(x,\eta) \in B^{n+m-1}$ is as follows:*

(a) $\gamma(x,\eta) = 2$, *if η is orthogonal to $\mathbf{R}^n$ in $\mathbf{R}^{n+m}$.*
(b) $\gamma(x,\eta) = g + 1$, *otherwise.*

In this case, the new curvature sphere in Case (b) corresponding to the principal curvature $-1/t$ in equation (5.16) arises because $f(M)$ has codimension $m + 1 > 1$ in $\mathbf{R}^{n+m}$.

Another situation in which the tube construction can be applied is that of an immersed proper Dupin submanifold $\psi : V \to \mathbf{R}^n$, where $\psi(V)$ has codimension $\nu + 1 > 1$ in $\mathbf{R}^n$. Recall from equation (4.70) on page 210 that the Legendre lift of ψ is the Legendre submanifold $\lambda : B^{n-1} \to \Lambda^{2n-1}$ defined by $\lambda(x,\eta) = [Y_1(x,\eta), Y_{n+3}(x,\eta)]$, where

$$Y_1(x,\eta) = (1 + \psi(x)\cdot\psi(x), 1 - \psi(x)\cdot\psi(x), 2\psi(x), 0)/2, \qquad (5.23)$$
$$Y_{n+3}(x,\eta) = (\psi(x)\cdot\eta, -(\psi(x)\cdot\eta), \eta, 1),$$

where B^{n-1} is the unit normal bundle of $\psi(V)$ in $\mathbf{R}^n$. The submanifold $\psi(V)$ is said to be Dupin, respectively, proper Dupin, if its Legendre lift λ is Dupin, respectively proper Dupin, as defined in Section 4.4 (see page 212).

Specifically, the Legendre submanifold $\lambda : B^{n-1} \to \Lambda^{2n-1}$ is a Dupin submanifold if:

(a) along each curvature surface, the corresponding curvature sphere map is constant.

Furthermore, λ is called proper Dupin if, in addition to Condition (a), the following condition is satisfied:

(b) the number g of distinct curvature spheres is constant on B^{n-1}.

As noted Remark 4.10 on page 217, the definition of a proper Dupin submanifold $\psi : V \to \mathbf{R}^n$ of codimension $\nu + 1 > 1$ in $\mathbf{R}^n$ can also be formulated in terms of Euclidean submanifold theory (see Pinkall [447]). That formulation is equivalent to the Lie sphere geometric formulation given above.

The application of the tube construction to the Legendre lift $\lambda : B^{n-1} \to \Lambda^{2n-1}$ of an immersed proper Dupin submanifold $\psi : V \to \mathbf{R}^n$ of codimension $\nu + 1 > 1$, yields a Dupin submanifold according to the following proposition [77, p. 132].

Proposition 5.5. *Suppose that $\lambda : B^{n-1} \to \Lambda^{2n-1}$ is a proper Dupin submanifold with g distinct curvature spheres induced by an immersed submanifold $\phi(V)$ of codimension $\nu + 1$ in $\mathbf{R}^n$, where $\mathbf{R}^n \subset \mathbf{R}^{n+m}$. Then the tube construction yields a Dupin submanifold μ defined on the unit normal bundle B^{n+m-1} to $\phi(V)$ in $\mathbf{R}^{n+m}$. The number $\gamma(x, \eta)$ of distinct curvature spheres of μ at a point $(x, \eta) \in B^{n+m-1}$ is as follows:*

(a) $\gamma(x, \eta) = 2$, *if η is orthogonal to $\mathbf{R}^n$ in $\mathbf{R}^{n+m}$.*
(b) $\gamma(x, \eta) = g$, *otherwise.*

In this situation, the number of distinct curvature spheres does not increase in Case (b), because the point sphere map of λ is already a curvature sphere map, since $\phi(V)$ has codimension $\nu + 1 > 1$ in $\mathbf{R}^n$ (see [77, pp. 131–132] for more detail).

Cylinders

For the cylinder construction, we again begin with an immersion

$$f : M^{n-1} \to \mathbf{R}^n \subset \mathbf{R}^{n+m},$$

with field of unit normals ξ to $f(M)$ in $\mathbf{R}^n$. In the following local calculation, we consider $M^{n-1} \subset \mathbf{R}^n \subset \mathbf{R}^{n+m}$ as an embedded submanifold, and we do not specifically mention the embedding f. We consider the tangent space T_xM at a point $x \in M$ to be a subspace of $T_x\mathbf{R}^{n+m}$. Then the tangent space to $M \times \mathbf{R}^m$ at a point (x, z), where $z = (z_1, \ldots, z_m)$, is given by

$$T_{(x,z)}(M \times \mathbf{R}^m) = \{(X, Z) \mid X \in T_xM,\ Z \in T_z\mathbf{R}^m\}. \tag{5.24}$$

The cylinder built over M in $\mathbf{R}^{n+m}$ is defined by the map $F : M \times \mathbf{R}^m \to \mathbf{R}^{n+m}$ given by the formula

$$F(x, z) = x + z_1 e_{n+1} + \cdots + z_m e_{n+m}. \tag{5.25}$$

The field $\eta(x,z)$ of unit normals to the cylinder is given by

$$\eta(x,z) = \xi(x), \tag{5.26}$$

that is, the normal field is constant along the rulings of the cylinder given by setting x equal to a constant. For (X,Z) tangent to $M \times \mathbf{R}^m$ at a point (x,z), where $Z = (Z_1, \ldots, Z_m)$, we compute

$$F_*(X,Z) = X + Z_1 e_{n+1} + \cdots + Z_m e_{n+m}, \tag{5.27}$$

where we identify vectors that are parallel in $\mathbf{R}^{n+m}$. Thus, F is an immersion on $M \times \mathbf{R}^m$.

We next find the shape operator B of the cylinder built over M in $\mathbf{R}^{n+m}$ at a point (x,z) in terms of the shape operator A of $M \subset \mathbf{R}^n$ at x. First we compute using equation (5.26),

$$F_*(B(X,0)) = -D_{F_*(X,0)}\eta = -D_X\xi = AX, \tag{5.28}$$

where we are again identifying parallel vectors in $\mathbf{R}^{n+m}$. Next we compute

$$F_*(B(0,Z)) = -D_{F_*(0,Z)}\eta = 0, \tag{5.29}$$

since the field of unit normals η is constant along the rulings of the cylinder through x. From equations (5.27)–(5.29), we see that if X is a principal vector of M in $\mathbf{R}^n$ with $AX = \lambda X$, then $(X,0)$ is a principal vector of the cylinder at (x,z) with $B(X,0) = \lambda(X,0)$ for every $z \in \mathbf{R}^m$. Furthermore, $B(0,Z) = 0$ for every $Z \in T_z\mathbf{R}^m$.

From this we see that there are two situations in which the cylinder construction applied to a proper Dupin hypersurface M in $\mathbf{R}^n$ leads to a proper Dupin cylinder in $\mathbf{R}^{n+m}$. The first is the following.

Proposition 5.6. *Suppose that $f : M^{n-1} \to \mathbf{R}^n \subset \mathbf{R}^{n+m}$ is a proper Dupin hypersurface in $\mathbf{R}^n$ with g distinct principal curvatures such that zero is not a principal curvature at any point of M^{n-1}. Then the cylinder construction yields a proper Dupin hypersurface $F : M \times \mathbf{R}^m \to \mathbf{R}^{n+m}$ with $g+1$ distinct principal curvatures at each point.*

Proof. Let λ_i, $1 \leq i \leq g$, denote the distinct principal curvature functions of the hypersurface $f(M)$ in $\mathbf{R}^n$ with corresponding m_i-dimensional principal foliations T_i, where m_i is the multiplicity of λ_i. By equations (5.28)–(5.29), the principal curvature functions of the cylinder $F(M \times \mathbf{R}^m)$ in $\mathbf{R}^{n+m}$ are $\mu_i(x,z) = \lambda_i(x)$, $1 \leq i \leq g$, and $\mu_{g+1} = 0$. Since none of the λ_i ever equal zero on M, these $g+1$ principal curvature functions $\mu_1, \ldots, \mu_g, \mu_{g+1}$ are distinct at each point of $M \times \mathbf{R}^m$, and they have multiplicities $m_1, \ldots, m_g, m_{g+1}$, respectively, where $m_{g+1} = m$. For $1 \leq i \leq g$, the principal curvature function λ_i is constant along the leaves of its principal foliation T_i, since $f(M)$ is proper Dupin. Therefore, μ_i is constant along the leaves of its principal foliation in $M \times \mathbf{R}^m$, which have the form $S \times \{z\}$, where S is a leaf of T_i in M. The principal curvature $\mu_{g+1} = 0$ is constant on $M \times \mathbf{R}^m$, so it is clearly constant along the leaves of its principal foliation, which have the form $\{x\} \times \mathbf{R}^m$ in $M \times \mathbf{R}^m$. Thus the cylinder is a proper Dupin hypersurface in $M \times \mathbf{R}^m$ with $g + 1$ distinct principal curvature functions at each point. □

The other situation in which the cylinder construction leads to a proper Dupin hypersurface in $\mathbf{R}^{n+m}$ is when one of the principal curvature functions λ_k of M is identically zero on M. In that case, we get the following.

Proposition 5.7. *Suppose that* $f : M^{n-1} \to \mathbf{R}^n \subset \mathbf{R}^{n+m}$ *is a proper Dupin hypersurface in* $\mathbf{R}^n$ *with g distinct principal curvature functions* λ_i, $1 \leq i \leq g$, *such that one particular principal curvature function* λ_k *is identically zero on* M^{n-1}. *Then the cylinder construction yields a proper Dupin hypersurface* $F : M \times \mathbf{R}^m \to \mathbf{R}^{n+m}$ *with g distinct principal curvatures at each point.*

Proof. From equations (5.28)–(5.29), we see that at a point $(x, z) \in M \times \mathbf{R}^m$, the cylinder has g distinct principal curvatures, $\mu_i(x, z) = \lambda_i(x)$, for $1 \leq i \leq g$. For $i \neq k$, the principal curvature μ_i has the same multiplicity as λ_i, but the multiplicity of the principal curvature $\mu_k = 0$ is $m_k + m$, where m_k is the multiplicity of the principal curvature $\lambda_k = 0$ on M. As in the previous proposition, for $i \neq k$, the fact that λ_i is constant along the leaves of its principal foliation T_i implies that μ_i is constant along the leaves of its principal foliation in $M \times \mathbf{R}^m$, which have the form $S \times \{z\}$, where S is a leaf of T_i in M. For $i = k$, the principal curvature $\mu_k = 0$ is constant on $M \times \mathbf{R}^m$, so it is clearly constant along the leaves of its principal foliation, which have the form $S \times \mathbf{R}^m$, where S is a leaf of the foliation T_k on M. □

Remark 5.8 (Cylinder construction in Lie sphere geometry). The cylinder construction can be formulated in the context of Lie sphere geometry (see [77, pp. 133–136]). In that context, the cylinder construction can be extended to $M \times S^m$ in the case where none of the principal curvatures of the hypersurface $f : M \to \mathbf{R}^n$ is ever equal to zero on M.

Surfaces of revolution

As in the other two constructions, we begin with an immersion $f : M^{n-1} \to \mathbf{R}^n \subset \mathbf{R}^{n+m}$ with field of unit normals $\xi : M^{n-1} \to S^{n-1} \subset \mathbf{R}^n$. We want to construct the hypersurface of revolution in $\mathbf{R}^{n+m}$ obtained by revolving the profile submanifold $f(M)$ about an axis $\mathbf{R}^{n-1} \subset \mathbf{R}^n$, where $\mathbf{R}^{n-1}$ is determined by the equation $x_n = 0$. We will not insist that $f(M)$ be disjoint from the axis $\mathbf{R}^{n-1}$, although the hypersurface of revolution will have singularities at points where $f(M)$ intersects $\mathbf{R}^{n-1}$. In keeping with the usual terminology, we will refer to a hypersurface of revolution as a "surface of revolution" in this section.

First we decompose the maps f and ξ into components along the axis $\mathbf{R}^{n-1}$ and orthogonal to $\mathbf{R}^{n-1}$, and we write

$$f(x) = \hat{f}(x) + f_n(x)e_n, \quad \hat{f}(x) \in \mathbf{R}^{n-1}, \tag{5.30}$$

$$\xi(x) = \hat{\xi}(x) + \xi_n(x)e_n, \quad \hat{\xi}(x) \in \mathbf{R}^{n-1}. \tag{5.31}$$

For $x \in M$ and $y = (y_0, \ldots, y_m) \in S^m$, we define

$$F(x,y) = \hat{f}(x) + f_n(x)(y_0e_n + y_1e_{n+1} + \cdots + y_me_{n+m}), \tag{5.32}$$

$$\eta(x,y) = \hat{\xi}(x) + \xi_n(x)(y_0e_n + y_1e_{n+1} + \cdots + y_me_{n+m}). \tag{5.33}$$

Note that for $y = (1, 0, \ldots, 0)$, we have

$$F(x, (1, 0, \ldots, 0)) = f(x), \quad \eta(x, (1, 0, \ldots, 0)) = \xi(x),$$

that is, we have the profile submanifold for the surface of revolution.

For a fixed point $x \in M$, the points $F(x,y)$, $y \in S^m$, form an m-dimensional sphere of radius $|f_n(x)|$ obtained by revolving the point $f(x)$ about the axis $\mathbf{R}^{n-1}$ in $\mathbf{R}^{n+m}$, provided that $f_n(x)$ is not zero. If $f_n(x) = 0$, then F maps all of the set $\{x\} \times S^m$ to the point $f(x)$, and thus F has a singularity at all such points.

For a vector $X \in T_xM$ and $Y = (Y_0, \ldots, Y_m) \in T_yS^m$, we compute

$$F_*(X, 0) = \hat{f}_*(X) + (Xf_n)(y_0e_n + y_1e_{n+1} + \cdots + y_me_{n+m}), \tag{5.34}$$

$$F_*(0, Y) = f_n(x)(Y_0e_n + Y_1e_{n+1} + \cdots + Y_me_{n+m}). \tag{5.35}$$

Note that when $y = (1, 0, \ldots, 0)$, we have

$$F_*(X, 0) = f_*(X).$$

From equations (5.34)–(5.35), we see that $F_*(X,0)$ is orthogonal to $F_*(0,Y)$, since Y is orthogonal to y in S^m. These equations imply that F is an immersion at points (x,y) where $f_n(x) \neq 0$, i.e., points where $f(x)$ does not lie on the axis of revolution $\mathbf{R}^{n-1}$. If $f(x)$ does lie on the axis, then $F_*(0,Y) = 0$ for all $Y \in T_yS^m$, and F has a singularity at (x,y) for all $y \in S^m$. In that case, the point $F(x,y)$ is fixed as the profile submanifold is revolved around the axis of revolution.

Shape operator of a surface of revolution

Now we want to find the shape operator of the surface of revolution. First we recall that if $X \in T_xM$, then the shape operator A of the surface $f : M^{n-1} \to \mathbf{R}^n$ applied to X is determined by the equation

$$f_*(AX) = -D_{f_*(X)}\xi = -\xi_*(X) = -(\hat{\xi}_*(X) + (X\xi_n)e_n). \tag{5.36}$$

Since

$$f_*(X) = \hat{f}_*(X) + (Xf_n)e_n, \tag{5.37}$$

we see that $AX = \lambda X$ if and only if

$$\hat{\xi}_*(X) = -\lambda \hat{f}_*(X), \text{ and } X\xi_n = -\lambda(Xf_n). \tag{5.38}$$

From equations (5.34)–(5.35), we see that the vector $\eta(x,y)$ is normal to the surface of revolution F at the point $F(x,y)$. Thus, η is a field of unit normals to the surface of revolution. By definition, the shape operator B determined by the field of unit normal η satisfies the equation,

$$F_*(B(X,0)) = -D_{F_*(X,0)}\eta, \tag{5.39}$$

which we can compute as follows. Take a curve $\delta(s)$ in M with $\delta(0) = x$ and initial tangent vector $\overrightarrow{\delta}(0) = X$. Let $\varepsilon(s) = F(\delta(s), y)$. Then $D_{F_*(X,0)}\eta$ is the initial tangent vector to the curve

$$\eta(\varepsilon(s)) = \hat{\xi}(\delta(s)) + \xi_n(\delta(s))(y_0 e_n + \cdots + y_m e_{n+m}). \tag{5.40}$$

Therefore,

$$F_*(B(X,0)) = -\overrightarrow{\eta}(0) = -(\hat{\xi}_*(X) + (X\xi_n)(y_0 e_n + \cdots + y_m e_{n+m})). \tag{5.41}$$

From equations (5.34) and (5.41), we see that $B(X,0) = \lambda(X,0)$ if and only if

$$\hat{\xi}_*(X) = -\lambda f_*(X), \text{ and } X\xi_n = -\lambda(Xf_n), \tag{5.42}$$

that is, if and only if equation (5.38) holds.

Thus, we have shown that $(X,0)$ is a principal vector of the shape operator B of F with corresponding principal curvature λ if and only if X is a principal vector of the shape operator A of f with the same corresponding principal curvature λ. The principal space of the principal curvature λ of B at (x,y) is $T_\lambda \times \{0\}$, where T_λ is the principal space of the principal curvature λ of A at x. The curvature surface of the principal curvature λ of B through the point (x,y) has the form $S \times \{y\}$, where $S \subset M$ is the curvature surface of the principal curvature λ of A through x. Clearly, the principal curvature function λ of B is constant along the curvature surface $S \times \{y\}$ if and only if the principal curvature function λ of A is constant along S in M.

Next we turn to the new principal curvature created by the surface of revolution construction. For these considerations, we insist that $f_n(x) \neq 0$ at the point $x \in M$, that is, the surface of revolution map F is an immersion at points (x,y) for all $y \in S^m$.

For $Y = (Y_0, \ldots, Y_m) \in T_y S^m$, we have

$$F_*(B(0,Y)) = -D_{F_*(0,Y)}\eta, \tag{5.43}$$

which can be computed as follows. Let $\alpha(s) = (\alpha_0(s), \ldots, \alpha_m(s))$ be a curve in S^m with $\alpha(0) = y$ and initial tangent vector

$$\vec{\alpha}(0) = (\alpha_0'(0), \ldots, \alpha_m'(0)) = (Y_0, \ldots, Y_m) = Y,$$

where $\alpha_i'(0)$ is the derivative of the coordinate function $\alpha_i(s)$ at $s = 0$.

Let $\beta(s) = F(x, \alpha(s))$. Then $D_{F_*(0,Y)}\eta$ is the initial tangent vector to the curve,

$$\eta(\beta(s)) = \hat{\xi}(x) + \xi_n(x)(\alpha_0(s)e_n + \cdots + \alpha_m(s)e_{n+m}), \tag{5.44}$$

and thus

$$D_{F_*(0,Y)}\eta = \xi_n(x)(Y_0 e_n + \cdots + Y_m e_{n+m}). \tag{5.45}$$

Therefore, we have

$$F_*(B(0,Y)) = -\xi_n(x)(Y_0 e_n + \cdots + Y_m e_{n+m}). \tag{5.46}$$

Comparing this with equation (5.35) for $F_*(0,Y)$, we see that at points where $f_n(x) \neq 0$,

$$B(0,Y) = -\frac{\xi_n(x)}{f_n(x)}(0,Y). \tag{5.47}$$

Hence, each vector $(0,Y)$ in $T_x M \times T_y S^m$ is a principal vector of B with corresponding principal curvature $\mu(x,y) = -\xi_n(x)/f_n(x)$, which only depends on x. This principal curvature μ is constant on the set $\{x\} \times S^m$.

Note that if $\mu(x) \neq 0$, the focal point of the surface of revolution corresponding to this principal curvature is

$$\begin{aligned}
F_\mu(x,y) &= F(x,y) + (1/\mu)\eta(x,y) \\
&= \hat{f}(x) + f_n(x)(y_0 e_n + \cdots + y_m e_{n+m}) \qquad (5.48)\\
&\quad - \frac{f_n(x)}{\xi_n(x)}(\hat{\xi}(x) + \xi_n(x)(y_0 e_n + \cdots + y_m e_{n+m})) \qquad (5.49)\\
&= \hat{f}(x) - \frac{f_n(x)}{\xi_n(x)}\hat{\xi}(x).
\end{aligned}$$

This point is the point of intersection of the normal line to the surface of revolution at $F(x,y)$ with the axis of revolution $\mathbf{R}^{n-1}$. Thus, the center of the curvature sphere corresponding to μ at $F(x,y)$ lies on the axis of revolution, and the curvature sphere itself is orthogonal to the axis of revolution. If $\mu(x) = 0$, then there is no focal point corresponding to μ, and the corresponding curvature sphere at $F(x,y)$ is the tangent hyperplane to the surface of revolution at $F(x,y)$. In that case, $\xi_n(x) = 0$, so that $\eta(x,y) = \hat{\xi}(x)$, which is parallel to the axis of revolution $\mathbf{R}^{n-1}$. Thus the tangent plane to the surface of revolution at $F(x,y)$ is orthogonal to $\mathbf{R}^{n-1}$ at such points.

Therefore, there are two possibilities for the number $\gamma(x,y)$ of distinct curvature spheres of F at (x,y). If none of the principal curvatures of $f(M)$ at x is equal to the new principal curvature $\mu = -\xi_n(x)/f_n(x)$, then we have $\gamma(x,y) = g+1$ for all $y \in S^m$, and the curvature surface of F through (x,y) corresponding to the new principal curvature μ has the form $\{x\} \times S^m$. On the other hand, if μ equals one of the principal curvatures λ of $f(M)$ at x, then $\gamma(x,y) = g$ for all $y \in S^m$, and the curvature surface of F through (x,y) has the form $S \times S^m$, where S is the curvature surface of $f(M)$ through x corresponding to the principal curvature λ. We summarize these results in the following proposition.

Proposition 5.9. *Suppose that $f : M^{n-1} \to \mathbf{R}^n$ is a proper Dupin hypersurface with g distinct principal curvatures at each point. The surface of revolution construction $F : M^{n-1} \times S^m \to \mathbf{R}^{n+m}$ yields a Dupin hypersurface defined on all points of $M^{n-1} \times S^m$, except those points where $f(x)$ lies in the axis of revolution $\mathbf{R}^{n-1}$. For (x,y) in the domain of definition of the surface of revolution, the number $\gamma(x,y)$ of distinct principal curvatures of F at (x,y) is as follows:*

(a) $\gamma(x,y) = g+1$, *if none of the principal curvatures of f at x is equal to the new principal curvature $\mu = -\xi_n(x)/f_n(x)$.*
(b) $\gamma(x,y) = g$, *otherwise.*

Remark 5.10 (Surface of revolution construction in Lie sphere geometry). For a formulation of the surface of revolution construction in the context of Lie sphere geometry, see [77, pp. 136–141].

5.2 Principal Coordinate Systems

In his dissertation, Pinkall ([442] and [445]) classified all proper Dupin hypersurfaces in $\mathbf{R}^4$ with $g = 3$ principal curvatures up to Lie equivalence. He showed that the only compact proper Dupin hypersurfaces with $g = 3$ in $\mathbf{R}^4$ are those that are Lie equivalent to an isoparametric hypersurface M^3 in S^4 (see Theorem 5.29). Pinkall also showed that those are the only proper Dupin hypersurfaces in $\mathbf{R}^4$ with $g = 3$ principal curvatures for which the lines of curvature cannot serve as coordinates of a local parametrization.

In this section, we show that if $M^n \subset S^{n+1}$ is an isoparametric hypersurface with $g \geq 3$ principal curvatures, then there does not exist any local principal coordinate system on M^n (see Pinkall [442, p. 42] and Cecil–Ryan [95, pp. 180–184]). We then give necessary and sufficient conditions for a hypersurface in a real space form with a fixed number g of distinct principal curvatures to have a local principal coordinate system in a neighborhood of each of its points.

Let M^n be a hypersurface in a real space form $\tilde{M}^{n+1}$ with g distinct principal curvatures $\lambda_1, \dots, \lambda_g$ at each point having respective multiplicities $m_1, \dots, m_g$. For each i, let T_i be the principal foliation corresponding to λ_i. A *principal coordinate system* is a local coordinate system

$$(x_{11}, \dots, x_{1m_1}, x_{21}, \dots, x_{2m_2}, \dots, x_{g1}, \dots, x_{gm_g}) \tag{5.50}$$

defined on a connected open set $U \subset M^n$ such that

$$T_i = \text{Span}\,\{\partial/\partial x_{i1}, \dots, \partial/\partial x_{im_i}\}, \quad 1 \leq i \leq g. \tag{5.51}$$

That is, for any set of constants $c_{jk}, j \neq i, 1 \leq j \leq g, 1 \leq k \leq m_j$, the equations

$$x_{jk} = c_{jk}, \quad j \neq i, \quad 1 \leq j \leq g, \quad 1 \leq k \leq m_j, \tag{5.52}$$

determine an integral manifold of T_i.

Two foliations T' and T'' on a manifold M^n are said to be *complementary* if the sum of their dimensions equals the dimension of M^n. Kobayashi and Nomizu [283, Vol. I, p. 182] proved that if two foliations T' and T'' are complementary on a manifold M^n, then for each point $y \in M^n$, there exists a local coordinate system $(x_1, \dots, x_n)$ with origin at y such that $(\partial/\partial x_1, \dots, \partial/\partial x_k)$ is a local basis for T' and $(\partial/\partial x_{k+1}, \dots, \partial/\partial x_n)$ is a local basis for T''. Thus, if $M^n \subset \tilde{M}^{n+1}$ is a hypersurface in a real space form with $g = 2$ distinct principal curvatures at each point, then the principal distributions T_1 and T_2 are complementary, and there exists a principal coordinate system in a neighborhood of every point of M^n.

In the following theorem due to Pinkall [442, p. 42] (see also Cecil–Ryan [95, pp. 180–184]), we show that there cannot exist a local principal coordinate system on an isoparametric hypersurface in S^n with $g \geq 3$ distinct principal curvatures.

Theorem 5.11. *Let $M \subset S^{n+1}$ be an isoparametric hypersurface with $g \geq 3$ distinct principal curvatures. Then there cannot exist a local principal coordinate system on M.*

Proof. Suppose there exists such a local coordinate system on an open subset $U \subset M$. Let λ and μ be two distinct principal curvatures of M, and let $X \in T_\lambda$ and $Y \in T_\mu$ be coordinate vector fields in this principal coordinate system. Since they are coordinate vector fields, the Lie bracket

$$[X, Y] = \nabla_X Y - \nabla_Y X = 0. \tag{5.53}$$

The Codazzi equation,

$$(\nabla_X A)Y = (\nabla_Y A)X, \tag{5.54}$$

then yields

$$\nabla_X(\mu Y) - \nabla_Y(\lambda X) = A(\nabla_X Y - \nabla_Y X) = A([X,Y]) = 0. \tag{5.55}$$

Thus, using equation (5.53), we have

$$0 = \mu \nabla_X Y - \lambda \nabla_Y X = (\mu - \lambda)\nabla_X Y, \tag{5.56}$$

so that $\nabla_X Y = 0$, since $\lambda \neq \mu$. On the other hand, since the Euclidean inner product $\langle X, Y\rangle = 0$, we have

$$\langle \nabla_X X, Y\rangle = -\langle X, \nabla_X Y\rangle = 0. \tag{5.57}$$

Thus, $\nabla_X X$ lies in $T_\mu^\perp$ for all $\mu \neq \lambda$, and so

$$\nabla_X X \in T_\lambda. \tag{5.58}$$

Using the Gauss equation (see, for example, [468, p. 366]), we find

$$R(X,Y)Y = (\lambda\mu + 1)(\langle Y, Y\rangle X - \langle X, Y\rangle Y) = (\lambda\mu + 1)\langle Y, Y\rangle X. \tag{5.59}$$

On the other hand,

$$R(X,Y)Y = \nabla_X \nabla_Y Y - \nabla_Y \nabla_X Y - \nabla_{[X,Y]} Y, \tag{5.60}$$

which, since $\nabla_X Y = 0$ and $[X, Y] = 0$, becomes

$$R(X,Y)Y = \nabla_X \nabla_Y Y. \tag{5.61}$$

Thus we get

$$\langle R(X,Y)Y, X\rangle = \langle \nabla_X \nabla_Y Y, X\rangle = X\langle \nabla_Y Y, X\rangle - \langle \nabla_Y Y, \nabla_X X\rangle = 0, \tag{5.62}$$

since equation (5.58) also applies to Y and T_μ. In view of (5.59), we have $\lambda\mu + 1 = 0$. Thus any principal curvature μ distinct from λ satisfies $\mu = -1/\lambda$ and so $g = 2$, contradicting the assumption that $g \geq 3$. □

Remark 5.12 (Proper Dupin hypersurfaces and principal coordinate systems). Principal coordinate systems arose in the work of T. Otsuki [419, 420] and R. Miyaoka [361, 362] on minimal hypersurfaces in the sphere. Otsuki [420, p. 17] gave examples of minimal hypersurfaces in the sphere with three distinct *non-simple* (having constant multiplicity greater than one) principal curvatures. Since all of the multiplicities are greater than one, these hypersurfaces are proper Dupin by Theorem 2.10 on page 24. Otsuki's examples have the property that each orthogonal complement $T_i^\perp$ of a principal foliation is integrable. Theorem 5.13 below shows that this is equivalent to the assumption that each point of the hypersurface M has a principal coordinate neighborhood. Otsuki showed that his examples are not isoparametric and cannot be complete. In fact, Miyaoka [362] showed that if M is a complete hypersurface with constant mean curvature and three non-simple principal curvatures in a real space form $\tilde{M}^{n+1}(c)$ with constant sectional curvature $c \geq 0$, then $c > 0$ and M is isoparametric.

Using a different approach based on the theory of higher-dimensional Laplace invariants due to Kamran and Tenenblat [234], Riveros and Tenenblat [463, 464] gave a local classification of proper Dupin hypersurfaces M^4 in $\mathbf{R}^5$ with four distinct principal curvatures which are parametrized by lines of curvature. In a related result, Riveros, Rodrigues, and Tenenblat [462] proved that a proper Dupin hypersurface $M^n \subset \mathbf{R}^{n+1}$, $n \geq 4$, with n distinct principal curvatures and constant Möbius curvatures cannot be parametrized by lines of curvature. They also showed that up to Möbius transformations, there is a unique proper Dupin hypersurface $M^3 \subset \mathbf{R}^4$ with three principal curvatures and constant Möbius curvature that is parametrized by lines of curvature. This M^3 is a cone in $\mathbf{R}^4$ over a standard flat torus in the unit sphere $S^3 \subset \mathbf{R}^4$. In a recent paper, Riveros [461] gave a characterization of a class of proper Dupin hypersurfaces in $\mathbf{R}^4$ that satisfy an additional condition on their higher-dimensional invariants.

Conditions for the existence of a principal coordinate system

The following theorem taken from Cecil–Ryan [95, pp. 182–184] gives necessary and sufficient conditions for a hypersurface in a real space form with a fixed number g of distinct principal curvatures to have a local principal coordinate system in a neighborhood of each of its points.

Theorem 5.13. *Let M^n be a hypersurface in a real space form $\tilde{M}^{n+1}$ with g distinct principal curvatures $\lambda_1, \ldots, \lambda_g$ at each point. Then each point of M^n has a principal coordinate neighborhood if and only if each $T_i^{\perp}$ is integrable on M^n.*

Proof. Suppose there exists a principal coordinate system on an open set $U \subset M^n$. Then for a fixed i, the equations,

$$x_{ik} = c_{ik}, \quad 1 \leq k \leq m_i, \tag{5.63}$$

determine a manifold whose tangent space is the direct sum of the T_j for $j \neq i$. This direct sum is equal to $T_i^{\perp}$, and so $T_i^{\perp}$ is integrable on U. Since M^n is covered by such neighborhoods, $T_i^{\perp}$ is integrable on M^n.

Conversely, suppose that each $T_i^{\perp}$ is integrable. Then for each i, the two foliations T_i and $T_i^{\perp}$ are complementary, and so by the theorem of Kobayashi and Nomizu [283, Vol. I, p. 182] mentioned above, for each point $p \in M^n$, there exists a system of local coordinates,

$$(x_{i1}, \ldots, x_{im_i}, y_{i1}, \ldots, y_{i(n-m_i)}), \tag{5.64}$$

with origin at p such that the leaves of $T_i^{\perp}$ are given by the equations

$$x_{ik} = c_{ik}, \quad 1 \leq k \leq m_i. \tag{5.65}$$

It can then be shown by induction that

$$(x_{11}, \ldots, x_{1m_1}, x_{21}, \ldots, x_{2m_2}, \ldots, x_{g1}, \ldots, x_{gm_g}) \tag{5.66}$$

is a coordinate system by using the coordinate systems in equation (5.64) and verifying that for each h, $1 \leq h \leq g$,

$$(x_{11}, \ldots, x_{1m_1}, x_{21}, \ldots, x_{2m_2}, \ldots, x_{h1}, \ldots, x_{hm_h}, y_{hj_1}, \ldots, y_{hj_r}),$$

where

$$r = n - \sum_{i=1}^{h} m_i,$$

is a coordinate system for a suitable subset $\{j_1, \ldots, j_r\}$ of $\{1, \ldots, n - m_h\}$. If L is the manifold given by the equations,

$$x_{jk} = c_{jk}, \quad j \neq i, \quad 1 \leq j \leq g, \quad 1 \leq k \leq m_j, \tag{5.67}$$

then at each point $q \in L$, we have

$$T_q L \subset \bigcap_{j \neq i} T_j^{\perp}(q) = T_i(q). \tag{5.68}$$

Since $T_q L$ has the same dimension as $T_i(q)$, we get $T_q L = T_i(q)$, and thus the coordinate system in equation (5.66) is a principal coordinate system with origin at p. □

5.3 Reducible Dupin Hypersurfaces

The notion of reducibility due to Pinkall [446, p. 438] is important in the theory of Dupin hypersurfaces, and it is a concept that is best formulated in the context of Lie sphere geometry. A proper Dupin submanifold,

$$\lambda : M^{n-1} \to \Lambda^{2n-1}, \tag{5.69}$$

is said to be *reducible* if it is locally Lie equivalent to the Legendre lift of a proper Dupin hypersurface in $\mathbf{R}^n$ obtained by one of Pinkall's constructions given in Section 5.1.

More specifically, a Dupin submanifold η that is obtained from a Dupin submanifold λ by one of Pinkall's standard constructions is *reducible to* λ. Further, a Dupin submanifold μ that is Lie equivalent to such a Dupin submanifold η is also said to be *reducible to* λ.

A detailed study of Pinkall's constructions in the context of Lie sphere geometry (see [446] or [77, pp. 127–148]) yields the main results of this section. For the sake of completeness, we want to state these results here. However, we will omit several of the proofs since they can be found in these two references.

The following result characterizes the case when the application of one of the standard constructions to a proper Dupin submanifold with g distinct curvature spheres produces a proper Dupin submanifold with $g+1$ distinct curvature spheres defined on an open subset of $M^{n-1} \times S^m$.

Theorem 5.14. *For $g \geq 1$, a proper Dupin submanifold $\mu : W^{d-1} \to \Lambda^{2d-1}$ with $g+1$ distinct curvature spheres is reducible to a proper Dupin submanifold λ with g distinct curvature spheres if and only if μ has a curvature sphere $[K]$ of multiplicity $m \geq 1$ that lies in a $(d+1-m)$-dimensional linear subspace of $\mathbf{RP}^{d+2}$.*

As Proposition 5.5 on page 242 shows, the application of one of the standard constructions does not always increase the number of curvature spheres, even though the resulting Dupin submanifold is still reducible. Pinkall [446, p. 438] also formulated his local criterion for reducibility to handle the case where the number of distinct curvature spheres of μ is the same as the number of distinct curvature spheres of λ. This criterion does not take into account the multiplicity of the curvature sphere $[K]$, as was done in Theorem 5.14 (see also [77, p. 143]).

Theorem 5.15. *A connected proper Dupin submanifold $\mu : W^{d-1} \to \Lambda^{2d-1}$ is reducible if and only if there exists a curvature sphere $[K]$ of μ that lies in a linear subspace of $\mathbf{RP}^{d+2}$ of codimension at least two.*

Tube over a Veronese surface in $S^4 \subset S^5$

The following example taken from [77, p. 132] is important in the theory of reducibility, and it illustrates some of the problems that can occur in attempting to characterize reducible Dupin submanifolds. The key point here is that the number of distinct curvatures spheres does not necessarily increase when one of the standard constructions is applied to a proper Dupin submanifold (as illustrated by Proposition 5.5 on page 242).

Example 5.16 (Tube over a Veronese surface in $S^4 \subset S^5$). In this example, we consider the case where V^2 is a Veronese surface embedded in $S^4 \subset S^5$, where S^4 is a great sphere in S^5. Of course, $V^2 \subset S^4$ is one of the focal submanifolds of an isoparametric hypersurface in S^4 with three distinct principal curvatures of multiplicity one described in Section 3.8.3 (see page 151). We first recall the details of the Veronese surface. Let S^2 be the unit sphere in $\mathbf{R}^3$ given by the equation

$$u^2 + v^2 + w^2 = 1.$$

Consider the map from S^2 into the unit sphere $S^4 \subset \mathbf{R}^5$ given by

$$(u, v, w) \mapsto \left(\sqrt{3}vw, \sqrt{3}wu, \sqrt{3}uv, \frac{\sqrt{3}}{2}(u^2 - v^2), w^2 - \frac{u^2 + v^2}{2} \right).$$

This map takes the same value on antipodal points of S^2, so it induces a map $\phi : \mathbf{RP}^2 \to S^4$, and one can show that ϕ is an embedding. The surface $V^2 = \phi(\mathbf{RP}^2)$ is a Veronese surface. One can show (see Section 3.8.3) that a tube over V^2 of radius ε, for $0 < \varepsilon < \pi/3$, in the spherical metric of S^4 is an isoparametric hypersurface M^3 with $g = 3$ distinct principal curvatures. This isoparametric hypersurface M^3 is not a reducible Dupin hypersurface, because the Veronese surface is substantial (does not lie in a hyperplane) in $\mathbf{R}^5$, so M^3 is not obtained as a result of the tube construction as described in Section 5.1. In terms of Lie sphere geometry, the point sphere map of the Legendre lift of the submanifold $V^2 \subset S^4$ lies in a linear subspace of codimension one in the projective space $\mathbf{RP}^6$, but not in a linear subspace of codimension two.

Now embed $\mathbf{R}^5$ as a hyperplane through the origin in $\mathbf{R}^6$ and let e_6 be a unit normal vector to $\mathbf{R}^5$ in $\mathbf{R}^6$. The surface V^2 is a subset of the unit sphere $S^5 \subset \mathbf{R}^6$. By Proposition 5.5 on page 242 (see [77, pp. 131–132] for a proof), we see that a tube over V^2 of radius ε in S^5 is not an isoparametric hypersurface, nor is it even a proper Dupin hypersurface, because the number of distinct principal curvatures is not constant on the unit normal bundle B^4 to V^2 in S^5. Specifically, if μ is the Legendre submanifold induced by the submanifold $V^2 \subset S^5$, then μ has two distinct curvature spheres at points in B^4 of the form $(x, \pm e_6)$, and three distinct curvature spheres at all other points of B^4. A tube W^4 over V^2 in S^5 is a reducible Dupin hypersurface, but it is not proper Dupin. At points of W^4 corresponding to the points $(x, \pm e_6)$ in B^4, there are two principal curvatures, both of multiplicity two. At the other points of W^4, there are three distinct principal curvatures, one of multiplicity two, and the others of multiplicity one. Thus, W^4 has an open dense subset U which is a reducible proper Dupin hypersurface with three principal curvatures at each point, but W^4 itself is not proper Dupin.

The following remark taken from [77, pp. 131–132] demonstrates the subtlety of the notion of reducibility of Dupin hypersurfaces even further.

Remark 5.17 (Weak reducibility). In the paper [124], Dajczer, Florit, and Tojeiro studied reducibility in the context of Riemannian geometry. They formulated a concept of weak reducibility for proper Dupin submanifolds that have a flat normal bundle including proper Dupin hypersurfaces. For hypersurfaces, their definition can be formulated as follows. A proper Dupin hypersurface $f : M^{n-1} \to \mathbf{R}^n$ (or S^n) is said to be *weakly reducible* if, for some principal curvature κ_i with corresponding principal space T_i, the orthogonal complement $T_i^{\perp}$ is integrable. Dajczer, Florit, and Tojeiro show that if a proper Dupin hypersurface $f : M^{n-1} \to \mathbf{R}^n$ is Lie equivalent to a proper Dupin hypersurface with $g + 1$ distinct principal curvatures that is obtained via one of the standard constructions from a proper Dupin hypersurface with g distinct principal curvatures, then f is weakly reducible. Thus, reducible implies weakly reducible for such hypersurfaces.

However, one can show that the open set U of the tube W^4 over V^2 in S^5 in Example 5.16 on which there are three principal curvatures at each point is reducible but not weakly reducible, because none of the orthogonal complements of the principal spaces is integrable. Of course, U is not constructed from a proper Dupin submanifold with two curvature spheres, but rather from one with three curvature spheres, so this does not violate the theorem of Dajczer, Florit, and Tojeiro.

Irreducibility

In two papers by Cecil and Jensen [85, 86], the notion of local irreducibility was used in the formulation of the main classification results. Specifically, a proper Dupin submanifold $\lambda : M^{n-1} \to \Lambda^{2n-1}$ is said to be *locally irreducible* if there does not exist any open subset $U \subset M^{n-1}$ such that the restriction of λ to U is reducible. Theoretically, this is a stronger condition than irreducibility of λ itself. However, using the analyticity of proper Dupin submanifolds (see Theorem 3.65 on page 143 and [84]), Cecil, Chi, and Jensen [82] proved the following proposition which shows that the concepts of local irreducibility and irreducibility are equivalent. (See also [77, pp. 145–146] for a complete proof of this result.)

Proposition 5.18. *Let $\lambda : M^{n-1} \to \Lambda^{2n-1}$ be a connected, proper Dupin submanifold. If the restriction of λ to an open subset $U \subset M^{n-1}$ is reducible, then λ is reducible. Thus, a connected proper Dupin submanifold is locally irreducible if and only if it is irreducible.*

Every connected proper Dupin hypersurface in $\mathbf{R}^n$ with $g = 2$ principal curvatures, i.e., a cyclide of Dupin, is reducible, as we shall see in the next section. It is Lie equivalent to a hypersurface obtained by applying one of Pinkall's constructions to an open subset of a metric sphere $S^q \subset \mathbf{R}^{q+1} \subset \mathbf{R}^n$, where $1 \leq q \leq n-2$. Thus, there exist reducible compact proper Dupin hypersurfaces in $\mathbf{R}^n$ or S^n with $g = 2$ principal curvatures, such as a standard product of two spheres in S^n. However, Cecil, Chi, and Jensen [82] proved that this is not possible for $g > 2$ in the following theorem (see also [77, pp. 146–147] for a proof).

Theorem 5.19. *Let W^{d-1} be a compact, connected proper Dupin hypersurface immersed in $\mathbf{R}^d$ with $g > 2$ distinct principal curvatures. Then W^{d-1} is irreducible. That is, the Legendre submanifold induced by the hypersurface W^{d-1} is irreducible.*

Remark 5.20 (Comments on irreducibility). We now make a few comments on irreducibility following the book [77, p. 147]. Since the proper Dupin property is invariant under stereographic projection, Theorem 5.19 implies that a compact, connected isoparametric hypersurface in S^d is irreducible as a Dupin hypersurface if the number g of distinct principal curvatures is greater than two. This was proved earlier by Pinkall in his dissertation [442]. Of course, compactness is not really a restriction for an isoparametric hypersurface, since any connected isoparametric hypersurface is contained in a unique compact, connected isoparametric hypersurface by Münzner's results. The same is not true for proper Dupin hypersurfaces, since a compact Dupin hypersurface containing a connected non-compact proper Dupin hypersurface may not be proper Dupin, as we see with the tube M^3 over a torus $T^2 \subset \mathbf{R}^3 \subset \mathbf{R}^4$ in Example 2.22 on page 33. The tube M^3 contains the open subset U of M^3 on which there are three distinct principal curvatures of multiplicity one. The set U is a proper Dupin hypersurface (with two connected components), but M^3 is only Dupin, but not proper Dupin.

Another geometric consequence of Theorem 5.19 is the following. Münzner showed that an isoparametric hypersurface $M^{n-1} \subset S^n \subset \mathbf{R}^{n+1}$ is a tube of constant radius in S^n over each of its two focal submanifolds. If $g = 2$, then the isoparametric hypersurface M^{n-1} is a standard product of two spheres,

$$S^q(r) \times S^{n-q-1}(s) \subset S^n, \ r^2 + s^2 = 1, \ 1 \leq q \leq n-2,$$

and the two focal submanifolds are both totally geodesic spheres, $S^q(1) \times \{0\}$ and $\{0\} \times S^{n-q-1}(1)$ (see Theorem 3.29 on page 111). The isoparametric hypersurface M^{n-1} is reducible in two ways, since it can be obtained as a tube of constant radius over each of these focal submanifolds, which are not substantial in $\mathbf{R}^{n+1}$. On the other hand, if an isoparametric hypersurface M^{n-1} has $g \geq 3$ distinct principal curvatures, then each of its focal submanifolds is substantial in $\mathbf{R}^{n+1}$. Otherwise, M^{n-1} would be reducible to such a non-substantial focal submanifold by the tube construction, contradicting Theorem 5.19.

5.4 Moving Frames in Lie Sphere Geometry

In this section, we develop the framework for studying Legendre submanifolds in the context of Lie sphere geometry by using the method of moving frames. In particular, we show that the assumption that the Legendre submanifold is proper Dupin leads to a Lie frame with certain special properties. We follow the approach of Cecil and Chern [79, 80], also given in the book [77, pp. 159–165]. (See also the paper of Jensen [229] and the forthcoming book of Jensen, Musso and Nicolodi [230].)

We first recall the basic definitions introduced in Section 4.3. Specifically, we use the following range of indices, and all summations are over the repeated index or indices:

$$1 \leq a, b, c \leq n+3, \quad 3 \leq i, j, k \leq n+1. \tag{5.70}$$

A Lie frame is an ordered set of vectors $\{Y_1, \ldots, Y_{n+3}\}$ in $\mathbf{R}^{n+3}_2$ satisfying the relations,

$$\langle Y_a, Y_b \rangle = g_{ab}, \tag{5.71}$$

for

$$[g_{ab}] = \begin{bmatrix} J & 0 & 0 \\ 0 & I_{n-1} & 0 \\ 0 & 0 & J \end{bmatrix}, \tag{5.72}$$

where I_{n-1} is the $(n-1) \times (n-1)$ identity matrix and

$$J = \begin{bmatrix} 0 & 1 \\ 1 & 0 \end{bmatrix}. \tag{5.73}$$

The Maurer–Cartan forms ω_a^b are defined by the equation

$$dY_a = \sum \omega_a^b Y_b. \tag{5.74}$$

Let $\lambda : M^{n-1} \to \Lambda^{2n-1}$ be an arbitrary Legendre submanifold. Let $\{Y_a\}$ be a smooth Lie frame on an open subset $U \subset M^{n-1}$ such that for each $x \in U$, we have

$$\lambda(x) = [Y_1(x), Y_{n+3}(x)].$$

We will pull back these Maurer–Cartan forms to U using the map λ^* and omit the symbols of such pull-backs for simplicity. Recall that the following matrix of forms is skew-symmetric,

$$[\omega_{ab}] = \begin{bmatrix} \omega_1^2 & \omega_1^1 & \omega_1^i & \omega_1^{n+3} & \omega_1^{n+2} \\ \omega_2^2 & \omega_2^1 & \omega_2^i & \omega_2^{n+3} & \omega_2^{n+2} \\ \omega_j^2 & \omega_j^1 & \omega_j^i & \omega_j^{n+3} & \omega_j^{n+2} \\ \omega_{n+2}^2 & \omega_{n+2}^1 & \omega_{n+2}^i & \omega_{n+2}^{n+3} & \omega_{n+2}^{n+2} \\ \omega_{n+3}^2 & \omega_{n+3}^1 & \omega_{n+3}^i & \omega_{n+3}^{n+3} & \omega_{n+3}^{n+2} \end{bmatrix}, \tag{5.75}$$

and that the forms satisfy the Maurer–Cartan equations,

$$d\omega_a^b = \sum \omega_a^c \wedge \omega_c^b. \tag{5.76}$$

By Theorem 4.8 on page 215, there are at most $n-1$ distinct curvature spheres along each line $\lambda(x)$. Thus, we can choose the Lie frame locally so that neither

Y_1 nor Y_{n+3} is a curvature sphere at any point of U. We now examine what those conditions mean in terms of our Lie frame.

The form $\omega_1^2 = 0$ by the skew-symmetry of the matrix in equation (5.75), and $\omega_1^{n+2} = 0$ by the contact condition (3) of Theorem 4.5 (page 208) for λ. Thus, for any $X \in T_x M^{n-1}$ at any point $x \in U$, we have

$$\begin{aligned} dY_1(X) &= \omega_1^1(X)Y_1 + \sum \omega_1^i(X)Y_i + \omega_1^{n+3}(X)Y_{n+3} \\ &\equiv \sum \omega_1^i(X)Y_i, \quad \operatorname{mod}\{Y_1, Y_{n+3}\}. \end{aligned} \tag{5.77}$$

The assumption that Y_1 is not a curvature sphere means that there does not exist any nonzero tangent vector X at any point $x \in U$ such that $dY_1(X)$ is congruent to zero $\operatorname{mod}\{Y_1, Y_{n+3}\}$. By equation (5.77), this assumption is equivalent to the condition that the forms $\{\omega_1^3, \ldots, \omega_1^{n+1}\}$ are linearly independent, i.e., they satisfy the regularity condition,

$$\omega_1^3 \wedge \cdots \wedge \omega_1^{n+1} \neq 0, \tag{5.78}$$

on U. Similarly, the condition that Y_{n+3} is not a curvature sphere is equivalent to the condition

$$\omega_{n+3}^3 \wedge \cdots \wedge \omega_{n+3}^{n+1} \neq 0. \tag{5.79}$$

Second fundamental form of a Legendre submanifold

We next consider the curvature spheres of λ in the context of Lie frames. The Legendre condition (3) for λ is equivalent to $\omega_1^{n+2} = 0$. Exterior differentiation of this equation using equations (5.75)–(5.76) yields the equation

$$\sum \omega_1^i \wedge \omega_{n+3}^i = 0. \tag{5.80}$$

Hence, by Cartan's Lemma and the regularity condition (5.78), we get that for each i,

$$\omega_{n+3}^i = \sum h_{ij}\omega_1^j, \quad \text{with } h_{ij} = h_{ji}. \tag{5.81}$$

The quadratic differential form

$$II(Y_1) = \sum h_{ij}\omega_1^i \omega_1^j, \tag{5.82}$$

defined up to a nonzero factor and dependent on the choice of Y_1, is called the *second fundamental form* of λ determined by Y_1.

This second fundamental form is related to the usual Euclidean second fundamental form as follows. Suppose that Y_1 and Y_{n+3} are given by

$$Y_1 = (1 + f \cdot f, 1 - f \cdot f, 2f, 0)/2, \quad Y_{n+3} = (f \cdot \xi, -f \cdot \xi, \xi, 1), \tag{5.83}$$

where f is the Euclidean projection of λ, and ξ is the Euclidean field of unit normals. The condition in equation (5.78) is equivalent to assuming that f is an immersion on U. Since f is an immersion, we can choose the Lie frame vectors $Y_3, \ldots, Y_{n+1}$ to satisfy

$$Y_i = dY_1(X_i) = (f \cdot df(X_i), -f \cdot df(X_i), df(X_i), 0), \quad 3 \leq i \leq n+1, \tag{5.84}$$

where $X_3, \ldots, X_{n+1}$ are smooth vector fields on U. Then we have

$$\omega_1^i(X_j) = \langle dY_1(X_j), Y_i \rangle = \langle Y_j, Y_i \rangle = \delta_{ij}. \tag{5.85}$$

Using equations (5.83) and (5.84), we compute

$$\begin{aligned} \omega_{n+3}^i(X_j) &= \langle dY_{n+3}(X_j), Y_i \rangle = d\xi(X_j) \cdot df(X_i) \\ &= -df(AX_j) \cdot df(X_i) = -A_{ij}, \end{aligned} \tag{5.86}$$

where $[A_{ij}]$ is the Euclidean shape operator (second fundamental form) of f. Now by equations (5.81) and (5.85), we have

$$\omega_{n+3}^i(X_j) = \sum h_{ik}\omega_1^k(X_j) = h_{ij}, \tag{5.87}$$

and so $h_{ij} = -A_{ij}$.

Suppose now that λ is an arbitrary Legendre submanifold, and $\{Y_a\}$ is a Lie frame on U such that Y_1 and Y_{n+3} satisfy equations (5.78) and (5.79), respectively. Since the matrix $[h_{ij}]$ is symmetric, we can diagonalize it at any given point $x \in U$ by a change of frame of the form

$$Y_i^* = \sum C_i^j Y_j, \quad 3 \leq i \leq n+1,$$

where $[C_i^j]$ is an $(n-1) \times (n-1)$ orthogonal matrix. In the new frame, equation (5.81) has the following form at x,

$$\omega_{n+3}^i = -\mu_i \omega_1^i, \quad 3 \leq i \leq n+1. \tag{5.88}$$

These μ_i determine the curvature spheres of λ at x. Specifically, given any point $x \in U$, let

$$\{X_3, \ldots, X_{n+1}\}$$

be the dual basis to $\{\omega_1^3, \ldots, \omega_1^{n+1}\}$ in the tangent space T_xM^{n-1}. Then using equation (5.88), we compute the differential of $\mu_i Y_1 + Y_{n+3}$ on X_i to be

$$\begin{aligned} d(\mu_i Y_1 + Y_{n+3})(X_i) &= d\mu_i(X_i)Y_1 + (\mu_i dY_1 + dY_{n+3})(X_i) \\ &\equiv \sum (\mu_i \omega_1^j(X_i) + \omega_{n+3}^j(X_i))Y_j = (\mu_i \omega_1^i(X_i) + \omega_{n+3}^i(X_i))Y_i \\ &= (\mu_i - \mu_i)Y_i = 0, \quad \mathrm{mod}\{Y_1, Y_{n+3}\}. \end{aligned} \tag{5.89}$$

Hence, the curvature spheres of λ at x are precisely

$$K_i = \mu_i Y_1 + Y_{n+3}, \quad 3 \leq i \leq n+1, \tag{5.90}$$

and $X_3, \ldots, X_{n+1}$ are the principal vectors at x. In the case where Y_1 and Y_{n+3} have the form in equation (5.83), the μ_i are just the principal curvatures of the immersion f at the point x, as in Theorem 4.12 on page 218.

Principal Lie frames

Assume now that the number g of distinct curvature spheres is constant on the neighborhood U. Then each distinct curvature sphere has constant multiplicity on U, and so its corresponding curvature sphere map is smooth on U. Furthermore, the principal vector fields $X_3, \ldots, X_{n+1}$ can be chosen smoothly on U (see, for example, Nomizu [402], Reckziegel [457, 458], or Singley [486]). This leads to a smooth choice of frame vectors $Y_3, \ldots, Y_{n+1}$ on U via the formula,

$$Y_i = dY_1(X_i), \quad 3 \leq i \leq n+1.$$

As in equation (5.85), this means that $\{\omega_1^3, \ldots, \omega_1^{n+1}\}$ is the dual basis to $\{X_3, \ldots, X_{n+1}\}$. Equation (5.88) is then satisfied at every point of U.

This frame $\{Y_a\}$ is an example of a principal frame. In general, a Lie frame $\{Z_a\}$ on U is said to be a *principal Lie frame* if there exist smooth functions α_i and β_i on U, which are never simultaneously zero, such that the Maurer–Cartan forms $\{\theta_a^b\}$ for the frame satisfy the equations,

$$\alpha_i \theta_1^i + \beta_i \theta_{n+3}^i = 0, \quad 3 \leq i \leq n+1. \tag{5.91}$$

Note that θ_1^i and θ_{n+3}^i cannot both vanish at a point x in U. To see this, take a Lie frame $\{W_a\}$ on U with

$$W_i = Z_i, \quad 3 \leq i \leq n+1,$$

such that $W_1 = \alpha Z_1 + \beta Z_{n+3}$ is not a curvature sphere at x. Then the Maurer–Cartan form ϕ_1^i for this frame satisfies the equation

$$\phi_1^i = \langle dW_1, W_i \rangle = \langle \alpha Z_1 + \beta Z_{n+3}, Z_i \rangle = \alpha \theta_1^i + \beta \theta_{n+3}^i.$$

Since W_1 is not a curvature sphere, it follows that $\phi_1^i \neq 0$, and thus it is not possible for θ_1^i and θ_{n+3}^i to both equal zero.

We next adapt the choice of frame to study a given curvature sphere map and then consider the impact of the Dupin condition on that map. Suppose that $\{Y_a\}$ is a principal frame on U satisfying equations (5.78) and (5.79) and that the curvature spheres are given by equation (5.90). In particular, suppose that

$$K = \mu Y_1 + Y_{n+3}$$

is a curvature sphere of multiplicity m on U. As noted above, the function μ is smooth on U, and we can re-order the frame vectors $Y_3, \ldots, Y_{n+1}$ so that

$$\mu = \mu_3 = \cdots = \mu_{m+2} \tag{5.92}$$

on U. The function μ does not take the value 0 or ∞ on U, since Y_1 and Y_{n+3} are not curvature spheres at any point of U.

Next we find a frame $\{Y_a^*\}$ with the property that $Y_1^* = K$ is a curvature sphere of multiplicity m. To accomplish this, let

$$\begin{aligned} Y_1^* &= \mu Y_1 + Y_{n+3}, \quad Y_2^* = (1/\mu) Y_2, \\ Y_{n+2}^* &= Y_{n+2} - (1/\mu) Y_2, \quad Y_{n+3}^* = Y_{n+3}, \quad Y_i^* = Y_i, \quad 3 \leq i \leq n+1. \end{aligned} \tag{5.93}$$

Let θ_a^b denote the Maurer–Cartan forms for this frame. Note that

$$dY_1^* = d(\mu Y_1 + Y_{n+3}) = (d\mu) Y_1 + \mu \, dY_1 + dY_{n+3} = \sum \theta_1^a Y_a^*. \tag{5.94}$$

Using equation (5.88), we see that the coefficient of $Y_i^* = Y_i$ in the above equation (5.94) is

$$\theta_1^i = \mu \omega_1^i + \omega_{n+3}^i = (\mu - \mu_i) \, \omega_1^i, \quad 3 \leq i \leq n+1. \tag{5.95}$$

This and equation (5.92) show that

$$\theta_1^r = 0, \quad 3 \leq r \leq m+2. \tag{5.96}$$

Equation (5.96) characterizes the condition that Y_1^* is a curvature sphere of constant multiplicity m on U.

The Dupin condition in terms of moving frames

We now consider the condition that the Legendre submanifold is proper Dupin in terms of our moving frame. In particular, suppose that the curvature sphere map $K = Y_1^*$ is constant along each leaf of its corresponding principal foliation. As noted in Corollary 4.9 on page 216, this is automatic if the multiplicity m of K is greater than one. We denote the corresponding principal foliation by T_1 and choose smooth vector fields

$$\{X_3, \ldots, X_{m+2}\}$$

on U that span T_1. The condition that Y_1^* is constant along each leaf of its principal foliation is given by

$$dY_1^*(X_r) \equiv 0, \quad \bmod Y_1^*, \quad 3 \leq r \leq m+2. \tag{5.97}$$

On the other hand, from equations (5.85) and (5.95), we have

$$dY_1^*(X_r) = \theta_1^1(X_r)Y_1 + \theta_1^{n+3}(X_r)Y_{n+3}, \quad 3 \leq r \leq m+2. \tag{5.98}$$

Comparing equations (5.97) and (5.98), we see that

$$\theta_1^{n+3}(X_r) = 0, \quad 3 \leq r \leq m+2. \tag{5.99}$$

We can make one more change of frame so that in the new frame the Maurer–Cartan form $\alpha_1^{n+3} = 0$ by the following procedure. In terms of the basis $\{\omega_1^3, \ldots, \omega_1^{n+1}\}$, we can write θ_1^{n+3} as

$$\theta_1^{n+3} = \sum s_i \omega_1^i, \tag{5.100}$$

for smooth functions s_i on U. From equation (5.99), we see that this reduces to

$$\theta_1^{n+3} = \sum_{t=m+3}^{n+1} s_t \omega_1^t. \tag{5.101}$$

For $m+3 \leq t \leq n+1$, we use equations (5.85), (5.95), and (5.101) to compute

$$\begin{aligned} dY_1^*(X_t) &= \theta_1^1(X_t)Y_1^* + \theta_1^t(X_t)Y_t + \theta_1^{n+3}(X_t)Y_{n+3} \\ &= \theta_1^1(X_t)Y_1^* + (\mu - \mu_t)Y_t + s_t Y_{n+3} \\ &= \theta_1^1(X_t)Y_1^* + (\mu - \mu_t)(Y_t + (s_t/(\mu - \mu_t))Y_{n+3}. \end{aligned} \tag{5.102}$$

We now make the change of Lie frame,

$$\begin{aligned} Z_1 &= Y_1^*, \quad Z_2 = Y_2^*, \quad Z_{n+3} = Y_{n+3}^*, \quad Z_r = Y_r^* = Y_r, \quad 3 \leq r \leq m+2, \\ Z_t &= Y_t + (s_t/(\mu - \mu_t))Y_{n+3}, \quad m+3 \leq t \leq n+1, \\ Z_{n+2} &= -\sum_t (s_t/(\mu - \mu_t))Y_t + Y_{n+2} - (1/2)\sum_t (s_t/(\mu - \mu_t))^2 Y_{n+3}. \end{aligned} \tag{5.103}$$

Let α_a^b be the Maurer–Cartan forms for this new frame. The equation,

$$\alpha_1^r = \langle dZ_1, Z_r \rangle = \langle dY_1^*, Y_r^* \rangle = \theta_1^r = 0, \quad 3 \leq r \leq m+2, \tag{5.104}$$

is still valid. Further, since $Z_1 = Y_1^*$, the Dupin condition (5.97) still yields

$$\alpha_1^{n+3}(X_r) = 0, \quad 3 \leq r \leq m+2. \tag{5.105}$$

Finally, for $m+3 \leq t \leq n+1$, equations (5.102) and (5.103) give

$$\alpha_1^{n+3}(X_t) = \langle dZ_1(X_t), Z_{n+2} \rangle = \langle \theta_1^1(X_t)Z_1 + (\mu - \mu_t)Z_t, Z_{n+2} \rangle = 0. \tag{5.106}$$

Thus, we have $\alpha_1^{n+3} = 0$. We summarize these results in the following theorem.

Theorem 5.21. *Let $\lambda : M^{n-1} \to \Lambda^{2n-1}$ be a Legendre submanifold. Suppose that K is a curvature sphere of multiplicity m on an open subset U of M^{n-1} that is constant along each leaf of its principal foliation. Then locally on U, there exists a Lie frame $\{Y_1, \ldots, Y_{n+3}\}$ with $Y_1 = K$, such that the Maurer–Cartan forms satisfy the equations*

$$\omega_1^r = 0, \quad 3 \leq r \leq m+2, \quad \omega_1^{n+3} = 0. \tag{5.107}$$

5.5 Cyclides of Dupin

In a book published in 1822, Dupin [143] defined a cyclide to be a surface M in $\mathbf{R}^3$ that is the envelope of the family of spheres tangent to three fixed spheres in $\mathbf{R}^3$. This is equivalent to requiring that M have two distinct principal curvatures at each point, and that both sheets of the focal set of M degenerate into curves, which are fact a pair of focal conics (see, for example, [95, pp. 176–178] for a description of the focal sets of the cyclides of Dupin). Thus M is a proper Dupin hypersurface in modern terminology by Theorem 2.14 on page 28, and M is the envelope of the family of curvature spheres centered along either of the two focal curves. The three fixed spheres in Dupin's definition can be chosen to be three spheres from either family of curvature spheres.

The cyclides of Dupin were studied intensively by many leading mathematicians in the nineteenth century, including Liouville [331], Cayley [69], and Maxwell [356], whose paper contains stereoscopic figures of the various types of cyclides. A good account of the history of the cyclides in the nineteenth century is given by Lilienthal [328] (see also Klein [281, pp. 56–58], Darboux [125, vol. 2, pp. 267–269], Blaschke [42, p. 238], Eisenhart [146, pp. 312–314], Hilbert and Cohn-Vossen [199, pp. 217–219], Fladt and Baur [161, pp. 354–379], and Cecil and Ryan [92, 95, pp. 151–166]). The physicist, Louis Michel, also pointed out to us that the focal conics and the cyclides are prominent in the 1923 paper of G. Friedel [163] on the structure of crystals.

The cyclides of Dupin reappeared in a modern context in the paper of Banchoff [20] on the spherical two-piece property, and then they were studied extensively in the many papers on Dupin hypersurfaces mentioned in the introduction. Pinkall's paper [446] describing the higher dimensional cyclides of Dupin in the context of Lie sphere geometry was particularly influential, and it had its roots in Volume 3 of the book of Blaschke [42], which studied surfaces in the context of Lie sphere geometry. See also [77, pp. 148–159] for a Lie sphere geometric account of the cyclides.

The classical cyclides are the only surfaces in $\mathbf{R}^3$ with two principal curvatures at each point such that all lines of curvature in both families are circles or straight lines. This is just the proper Dupin condition. Using exterior differential systems, Ivey [220] showed that any surface in $\mathbf{R}^3$ containing two orthogonal families of circles is a cyclide of Dupin. For further results involving the cyclides of Dupin, see Garnier et al. [166], Druoton et al. [141, 142], and Bartoszek et al. [26].

The cyclides have also appeared in the context of computer graphics in the papers of Degen [127], Pratt [451, 452], Srinivas and Dutta [496–499], Schrott and Odehnal [478], and Jia [231] among others.

We now turn our attention to the higher dimensional cyclides of Dupin. Since the cyclides are most easily classified in the setting of Lie sphere geometry, we give our definition in that context and then discuss various characterizations of the cyclides in $\mathbf{R}^n$ and S^n.

Cyclides of Dupin of characteristic (p, q)

A proper Dupin submanifold $\lambda : M^{n-1} \to \Lambda^{2n-1}$ with two distinct curvature spheres of respective multiplicities p and q at each point is called a *cyclide of Dupin of characteristic* (p, q). We will prove that any connected cyclide of Dupin of characteristic (p, q) is Lie equivalent to the Legendre lift of an open subset of a standard product of two spheres,

$$S^q(1/\sqrt{2}) \times S^p(1/\sqrt{2}) \subset S^n \subset \mathbf{R}^{q+1} \times \mathbf{R}^{p+1} = \mathbf{R}^{n+1}, \tag{5.108}$$

where p and q are positive integers such that $p+q=n-1$. Thus any two connected cyclides of Dupin of the same characteristic are locally Lie equivalent.

As discussed in Subsection 3.8.2, the product $S^q(1/\sqrt{2}) \times S^p(1/\sqrt{2})$ is an isoparametric hypersurface in S^n with two distinct principal curvatures having multiplicities $m_1 = p$ and $m_2 = q$. Furthermore, every isoparametric hypersurface in S^n with two principal curvatures of multiplicities p and q is Lie equivalent to $S^q(1/\sqrt{2}) \times S^p(1/\sqrt{2})$, since it is congruent to a parallel hypersurface of $S^q(1/\sqrt{2}) \times S^p(1/\sqrt{2})$.

Although $S^q(1/\sqrt{2}) \times S^p(1/\sqrt{2})$ is a good model for the cyclides, it is often easier to work with the two focal submanifolds $S^q(1) \times \{0\}$ and $\{0\} \times S^p(1)$ in proving classification results. The Legendre lifts of these two focal submanifolds are Lie equivalent to the Legendre lift of $S^q(1/\sqrt{2}) \times S^p(1/\sqrt{2})$, since they are parallel submanifolds of the Legendre lift of $S^q(1/\sqrt{2}) \times S^p(1/\sqrt{2})$. In fact, the hypersurface $S^q(1/\sqrt{2}) \times S^p(1/\sqrt{2})$ is a tube of radius $\pi/4$ in S^n over either of its two focal submanifolds.

We now describe our standard model of a cyclide of characteristic (p,q) in detail in the context of Lie sphere geometry, as in Pinkall's paper [446] (see also [77, p. 149]). Let $\{e_1, \ldots, e_{n+3}\}$ be the standard orthonormal basis for $\mathbf{R}^{n+3}_2$. Then S^n is the unit sphere in the Euclidean space $\mathbf{R}^{n+1}$ spanned by $\{e_2, \ldots, e_{n+2}\}$. Let

$$\Omega = \text{Span}\,\{e_1, \ldots, e_{q+2}\}, \quad \Omega^{\perp} = \text{Span}\,\{e_{q+3}, \ldots, e_{n+3}\}. \tag{5.109}$$

These spaces have signatures $(q+1,1)$ and $(p+1,1)$, respectively. The intersection $\Omega \cap Q^{n+1}$ is the quadric given in homogeneous coordinates by

$$x_1^2 = x_2^2 + \cdots + x_{q+2}^2, \quad x_{q+3} = \cdots = x_{n+3} = 0.$$

This set is diffeomorphic to the unit sphere S^q in

$$\mathbf{R}^{q+1} = \text{Span}\,\{e_2, \ldots, e_{q+2}\},$$

by the diffeomorphism $\phi : S^q \to \Omega \cap Q^{n+1}$, defined by $\phi(v) = [e_1 + v]$. Similarly, the quadric $\Omega^{\perp} \cap Q^{n+1}$ is diffeomorphic to the unit sphere S^p in

$$\mathbf{R}^{p+1} = \text{Span}\,\{e_{q+3}, \ldots, e_{n+2}\}$$

by the diffeomorphism $\psi : S^p \to \Omega^{\perp} \cap Q^{n+1}$, defined by $\psi(u) = [u + e_{n+3}]$. The model that we will use for the cyclides in Lie sphere geometry is the Legendre submanifold $\lambda : S^p \times S^q \to \Lambda^{2n-1}$ defined by

$$\lambda(u,v) = [k_1, k_2], \text{with } [k_1(u,v)] = [\phi(v)], \quad [k_2(u,v)] = [\psi(u)]. \tag{5.110}$$

It is easy to check that the Legendre Conditions (1)–(3) of Theorem 4.5 on page 208 are satisfied by the pair $\{k_1, k_2\}$. To find the curvature spheres of λ, we decompose the tangent space to $S^p \times S^q$ at a point (u,v) as

$$T_{(u,v)}S^p \times S^q = T_u S^p \times T_v S^q.$$

Then $dk_1(X,0) = 0$ for all $X \in T_u S^p$, and $dk_2(Y) = 0$ for all Y in $T_v S^q$. (Here we use the notation dk_1 instead of $(k_1)_*$ for the differential of k_1 to be consistent with the notation used in Chapter 4.) Thus, $[k_1]$ and $[k_2]$ are curvature spheres of λ with respective multiplicities p and q. Furthermore, the image of $[k_1]$ lies in the quadric $\Omega \cap Q^{n+1}$, and the image of $[k_2]$ is contained in the quadric $\Omega^\perp \cap Q^{n+1}$. The point sphere map of λ is $[k_1]$, and thus λ is the Legendre lift of the focal submanifold $S^q \times \{0\} \subset S^n$, considered as a submanifold of codimension $p+1$ in S^n.

Lie geometric classification of the cyclides of Dupin

We now prove Pinkall's [446] classification of proper Dupin submanifolds with two distinct curvature spheres at each point (see also [77, pp. 149–151] for an exposition of Pinkall's proof). Pinkall's proof depends on establishing the existence of a local principal coordinate system. This can always be done in the case of $g = 2$ curvature spheres, but not necessarily if $g > 2$. In fact, if M is an isoparametric hypersurface in S^n with more than two distinct principal curvatures, then there cannot exist a local principal coordinate system on M (see Theorem 5.11 on page 249).

Here we give a different proof of Pinkall's theorem using the method of moving frames, following the paper of Cecil-Chern [80]. This approach generalizes to the case of $g > 2$ curvature spheres, as we will see in later sections of this chapter.

Theorem 5.22. (a) *Every connected cyclide of Dupin is contained in a unique compact, connected cyclide of Dupin.*

(b) *Any two cyclides of Dupin of the same characteristic (p,q) are locally Lie equivalent, each being Lie equivalent to an open subset of a standard product of two spheres*

$$S^q(1/\sqrt{2}) \times S^p(1/\sqrt{2}) \subset S^n \subset \mathbf{R}^{q+1} \times \mathbf{R}^{p+1} = \mathbf{R}^{n+1}, \tag{5.111}$$

where $p + q = n - 1$.

Proof. Suppose that $\mu : M^{n-1} \to \Lambda^{2n-1}$ is a connected proper Dupin submanifold with two distinct curvature spheres of multiplicities p and q at each point. As we showed in Theorem 5.21, on any local neighborhood U in M^{n-1}, we can find a local Lie frame, which we now denote by Y_A, whose Maurer–Cartan forms ω_A^B satisfy

$$\omega_1^a = 0, \quad 3 \leq a \leq p+2, \quad \omega_1^{n+3} = 0. \tag{5.112}$$

In this frame, Y_1 is a curvature sphere map of multiplicity p from U to Q^{n+1}. By the hypotheses of Theorem 5.22, there is one other curvature sphere of multiplicity $q = n-1-p$ at each point of M^{n-1}. By repeating the procedure used in constructing the frame Y_A in Theorem 5.21, we can construct a new frame $\overline{Y}_A$ which has $\overline{Y}_{n+3}$ as the other curvature sphere map $sY_1 + Y_{n+3}$, where s is a smooth function on U. The principal space corresponding to this curvature sphere $\overline{Y}_{n+3} = sY_1 + Y_{n+3}$ is the span of the vectors $X_{p+3}, \ldots, X_{n+1}$ in the notation of equation (5.106). The fact that $\overline{Y}_{n+3}$ is a curvature sphere map yields

$$\overline{\omega}^b_{n+3} = 0, \quad p+3 \leq b \leq n+1, \tag{5.113}$$

as in equation (5.99). The Dupin condition analogous to equation (5.97) is

$$d\overline{Y}^*_{n+3}(X_b) \equiv 0, \quad \bmod \overline{Y}_{n+3}, \quad p+3 \leq b \leq n+1. \tag{5.114}$$

This eventually leads to

$$\overline{\omega}^1_{n+3} = 0. \tag{5.115}$$

One can check that this change of frame does not affect the condition (5.112). So we now drop the bars and call this last frame Y_A with Maurer–Cartan forms ω^B_A satisfying,

$$\begin{aligned} \omega^a_1 &= 0, \quad 3 \leq a \leq p+2, \\ \omega^b_{n+3} &= 0, \quad p+3 \leq b \leq n+1, \\ \omega^{n+3}_1 &= 0, \quad \omega^1_{n+3} = 0. \end{aligned} \tag{5.116}$$

Furthermore, the following forms are easily shown to be a basis for the cotangent space at each point of U,

$$\{\omega^3_{n+3}, \ldots, \omega^{p+2}_{n+3}, \omega^{p+3}_1, \ldots, \omega^{n+1}_1\}. \tag{5.117}$$

We first take the exterior derivative of the equations $\omega^a_1 = 0$ and $\omega^b_{n+3} = 0$ in equation (5.116). Using the skew-symmetry of the matrix in (5.75) along with equations (5.76) and (5.116), we obtain

$$0 = \omega^{p+3}_1 \wedge \omega^a_{p+3} + \cdots + \omega^{n+1}_1 \wedge \omega^a_{n+1}, \quad 3 \leq a \leq p+2, \tag{5.118}$$

$$0 = \omega^3_{n+3} \wedge \omega^b_3 + \cdots + \omega^{p+2}_{n+3} \wedge \omega^b_{p+2}, \quad p+3 \leq b \leq n+1. \tag{5.119}$$

We now show that equations (5.118) and (5.119) imply that

$$\omega^a_b = 0, \quad 3 \leq a \leq p+2, \quad p+3 \leq b \leq n+1. \tag{5.120}$$

To see this, note that since $\omega^a_b = -\omega^b_a$, each of the terms ω^a_b occurs in exactly one of the equations (5.118) and in exactly one of the equations (5.119). Equation (5.118) involves the basis forms $\omega^{p+3}_1, \ldots, \omega^{n+1}_1$, while equation (5.119) involves the basis forms $\omega^3_{n+3}, \dots, \omega^{p+2}_{n+3}$. We now show how to handle the form ω^3_{p+3}, and the others are treated in a similar way. The equations from (5.118) and (5.119), respectively, involving $\omega^3_{p+3} = -\omega^{p+3}_3$ are

$$0 = \omega^{p+3}_1 \wedge \omega^3_{p+3} + \omega^{p+4}_1 \wedge \omega^3_{p+4} + \cdots + \omega^{n+1}_1 \wedge \omega^3_{n+1}, \tag{5.121}$$

$$0 = \omega^3_{n+3} \wedge \omega^{p+3}_3 + \omega^4_{n+3} \wedge \omega^{p+3}_4 + \cdots + \omega^{p+2}_{n+3} \wedge \omega^{p+3}_{p+2}. \tag{5.122}$$

We take the wedge product of (5.121) with $\omega^{p+4}_1 \wedge \cdots \wedge \omega^{n+1}_1$ and get

$$0 = \omega^3_{p+3} \wedge (\omega^{p+3}_1 \wedge \cdots \wedge \omega^{n+1}_1),$$

which implies that ω_{p+3} is in the span of $\{\omega^{p+3}_1, \dots, \omega^{n+1}_1\}$. On the other hand, taking the wedge product of equation (5.122) with $\omega^4_{n+3} \wedge \cdots \wedge \omega^{p+2}_{n+3}$ yields

$$0 = \omega^3_{p+3} \wedge (\omega^3_{n+3} \wedge \cdots \wedge \omega^{p+2}_{n+3}),$$

and so ω^3_{p+3} is in the span of $\{\omega^3_{n+3}, \dots, \omega^{p+2}_{n+3}\}$. Thus, we have $\omega^3_{p+3} = 0$, as desired.

We next differentiate $\omega^{n+3}_1 = 0$ and use the skew-symmetry of the matrix in equation (5.75), the equation $\omega^{n+2}_1 = 0$, and equation (5.116) to obtain

$$0 = d\omega^{n+3}_1 = \omega^{p+3}_1 \wedge \omega^{n+3}_{p+3} + \cdots + \omega^{n+1}_1 \wedge \omega^{n+3}_{n+1}. \tag{5.123}$$

This implies that

$$\omega^{n+3}_b \in \text{Span}\ \{\omega^{p+3}_1, \dots, \omega^{n+1}_1\}, \quad p+3 \leq b \leq n+1. \tag{5.124}$$

Similarly, differentiation of the equation $\omega^1_{n+3} = 0$ yields

$$0 = d\omega^1_{n+3} = \omega^3_{n+3} \wedge \omega^1_3 + \cdots + \omega^{p+2}_{n+3} \wedge \omega^1_{p+2}, \tag{5.125}$$

which implies that

$$\omega^a_1 \in \text{Span}\ \{\omega^3_{n+3}, \dots, \omega^{p+2}_{n+3}\}, \quad 3 \leq a \leq p+2. \tag{5.126}$$

We next differentiate equation (5.120). Using the skew-symmetry of the matrix in equation (5.75), and also equations (5.116) and (5.120), we see that all terms drop out except the following,

$$\begin{aligned} 0 = d\omega^a_b &= \omega^2_b \wedge \omega^a_2 + \omega^{n+3}_b \wedge \omega^a_{n+3} \\ &= -(\omega^1_a \wedge \omega^b_1) + \omega^{n+3}_b \wedge \omega^a_{n+3}. \end{aligned} \tag{5.127}$$

Thus, we have

$$\omega_a^1 \wedge \omega_1^b = \omega_b^{n+3} \wedge \omega_{n+3}^a, \quad 3 \leq a \leq p+2, \quad p+3 \leq b \leq n+1. \tag{5.128}$$

We next show that equation (5.128) implies that there exists a function α on U such that

$$\begin{aligned} \omega_a^1 &= \alpha\, \omega_{n+3}^a, \quad 3 \leq a \leq p+2 \\ \omega_b^{n+3} &= -\alpha\, \omega_1^b, \quad p+3 \leq b \leq n+1. \end{aligned} \tag{5.129}$$

To see this, note that for any a, $3 \leq a \leq p+2$, equation (5.126) gives

$$\omega_a^1 = c_3 \omega_{n+3}^3 + \cdots + c_{p+2} \omega_{n+3}^{p+2}, \tag{5.130}$$

for some coefficient functions $c_1, \ldots, c_{p+2}$. Similarly, for any b, $p+3 \leq b \leq n+1$, equation (5.124) gives

$$\omega_b^{n+3} = d_{p+3} \omega_1^{p+3} + \cdots + d_{n+1} \omega_1^{n+1}, \tag{5.131}$$

for some coefficient functions $d_{p+3}, \ldots, d_{n+1}$. Thus, we have

$$\omega_a^1 \wedge \omega_1^b = c_3 \omega_{n+3}^3 \wedge \omega_1^b + \cdots + c_a \omega_{n+3}^a \wedge \omega_1^b + \cdots + c_{p+2} \omega_{n+3}^{p+2} \wedge \omega_1^b, \tag{5.132}$$

$$\omega_b^{n+3} \wedge \omega_{n+3}^a = d_{p+3} \omega_1^{p+3} \wedge \omega_{n+3}^a + \cdots + d_b \omega_1^b \wedge \omega_{n+3}^a + \cdots + d_{n+1} \omega_1^{n+1} \wedge \omega_{n+3}^a. \tag{5.133}$$

From equation (5.128) we know that the right-hand sides of these equations are equal, but these expressions contain no common terms from the basis of 2-forms except those involving $\omega_{n+3}^a \wedge \omega_1^b$. Thus, all of the coefficients except c_a and d_b are zero, and we have

$$c_a \omega_{n+3}^a \wedge \omega_1^b = d_b \omega_1^b \wedge \omega_{n+3}^a = (-d_b) \omega_{n+3}^a \wedge \omega_1^b, \tag{5.134}$$

and we conclude that $c_a = -d_b$. Thus, we have shown that equations (5.130) and (5.131) reduce to

$$\omega_a^1 = c_a \omega_{n+3}^a, \quad \omega_b^{n+3} = d_b \omega_1^b, \tag{5.135}$$

with $d_b = -c_a$. This procedure works for any choice of a and b in the appropriate ranges. By holding a fixed and varying b, we see that all of the quantities d_b are equal to each other and to $-c_a$. Similarly, all the quantities c_a are the same, and thus equation (5.129) holds with α equal to the common value of c_a.

We now consider the expression in equation (5.74) for dY_a, $3 \leq a \leq p+2$. We omit the terms that vanish because of the skew-symmetry of the matrix in (5.75), and equations (5.116) and (5.120). We then have

$$dY_a = \omega_a^1 Y_1 + \omega_a^3 Y_3 + \cdots + \omega_a^{p+2} Y_{p+2} + \omega_a^{n+2} Y_{n+2} + \omega_a^{n+3} Y_{n+3}. \tag{5.136}$$

Using equation (5.129) and the skew-symmetry relation $\omega_a^{n+2} = -\omega_{n+3}^a$, this becomes

$$dY_a = \omega_{n+3}^a (\alpha Y_1 - Y_{n+2}) + \omega_a^3 Y_3 + \cdots + \omega_a^{p+2} Y_{p+2} + \omega_a^{n+3} Y_{n+3}. \tag{5.137}$$

Similarly, for $p+3 \leq b \leq n+1$, we get

$$dY_b = \omega_b^1 Y_1 + \omega_b^2 (Y_2 + \alpha Y_{n+3}) + \omega_b^{p+3} Y_{p+3} + \cdots + \omega_b^{n+1} Y_{n+1}. \tag{5.138}$$

We make the change of frame,

$$\begin{aligned} Y_2^* &= Y_2 + \alpha Y_{n+3}, \quad Y_{n+2}^* = Y_{n+2} - \alpha Y_1, \\ Y_B^* &= Y_B, \quad B \neq 2,\ n+2. \end{aligned} \tag{5.139}$$

We now drop the asterisks but use the new frame. From equations (5.137) and (5.138), we see that in this new frame, we have

$$dY_a = \omega_{n+3}^a (-Y_{n+2}) + \omega_a^3 Y_3 + \cdots + \omega_a^{p+2} Y_{p+2} + \omega_a^{n+3} Y_{n+3}, \tag{5.140}$$

$$dY_b = \omega_b^1 Y_1 + \omega_b^2 Y_2 + \omega_b^{p+3} Y_{p+3} + \cdots + \omega_b^{n+1} Y_{n+1}. \tag{5.141}$$

That is, in this new frame, we have

$$\omega_a^1 = 0, \quad 3 \leq a \leq p+2, \tag{5.142}$$

$$\omega_b^{n+3} = 0, \quad p+3 \leq b \leq n+1. \tag{5.143}$$

We next want to show that the space

$$E = \text{Span}\,\{Y_1, Y_2, Y_{p+3}, \ldots, Y_{n+1}\} \tag{5.144}$$

and its orthogonal complement,

$$E^{\perp} = \text{Span}\,\{Y_3, \ldots, Y_{p+2}, Y_{n+2}, Y_{n+3}\} \tag{5.145}$$

are both invariant under exterior differentiation d, and so they are constant spaces on U.

Invariance of the space E under exterior differentiation

First for the space E, we have that $dY_b \in E$ for $p+3 \leq b \leq n+1$ by equation (5.141). Furthermore, the skew-symmetry of the matrix in (5.75), the equation $\omega_1^{n+2} = 0$, and equation (5.116) imply that

$$dY_1 = \omega_1^1 Y_1 + \omega_1^{p+3} Y_{p+3} + \cdots + \omega_1^{n+1} Y_{n+1}, \tag{5.146}$$

which is in E. Thus, it only remains to show that dY_2 is in E. To show this, we differentiate equation (5.142). As before, we omit terms which are zero because of the skew-symmetry of the matrix in (5.75), or because of equations (5.116), (5.120), and (5.142). We see that the Maurer–Cartan equation for $d\omega_a^1$ reduces to

$$0 = d\omega_a^1 = \omega_a^{n+2} \wedge \omega_{n+2}^1 = -\omega_{n+3}^a \wedge \omega_{n+2}^1 = \omega_{n+3}^a \wedge \omega_2^{n+3}, \tag{5.147}$$

for $3 \leq a \leq p+2$. Similarly, by differentiating equation (5.143), we find that

$$0 = d\omega_b^{n+3} = \omega_b^2 \wedge \omega_2^{n+3} = -\omega_1^b \wedge \omega_2^{n+3}, \quad p+3 \leq b \leq n+1. \tag{5.148}$$

From this and equation (5.147), we see that the wedge product of ω_2^{n+3} with every form in the basis in equation (5.117) is zero, and hence $\omega_2^{n+3} = 0$. Using this and the fact that $\omega_2^{n+2} = -\omega_{n+3}^1 = 0$, and that by the skew-symmetry relations (5.75) and equation (5.142) we have

$$\omega_2^a = -\omega_a^1 = 0, \quad 3 \leq a \leq p+2, \tag{5.149}$$

and we get

$$dY_2 = \omega_2^2 Y_2 + \omega_2^{p+3} Y_{p+3} + \cdots + \omega_2^{n+1} Y_{n+1}, \tag{5.150}$$

which is in E. Thus the space E is invariant under exterior differentiation d, and so E is a fixed subspace of $\mathbf{RP}^{n+2}$, independent of the choice of point in U. Obviously, the orthogonal complement $E^\perp$ defined in equation (5.145) is also a fixed subspace of $\mathbf{RP}^{n+2}$ on U.

Note that E has signature $(q+1, 1)$ as a vector subspace of $\mathbf{R}_2^{n+3}$, and $E^\perp$ has signature $(p+1, 1)$. Take an orthornormal basis $\{w_1, \ldots, w_{n+3}\}$ of $\mathbf{R}_2^{n+3}$ with w_1 and w_{n+3} timelike such that

$$E = \text{Span}\,\{w_1, \ldots, w_{q+2}\}, \quad E^\perp = \text{Span}\,\{w_{q+3}, \ldots, w_{n+3}\}. \tag{5.151}$$

Then $E \cap Q^{n+1}$ is given in homogeneous coordinates $(x_1, \ldots, x_{n+3})$ with respect to this basis by

$$x_1^2 = x_2^2 + \cdots + x_{q+2}^2, \quad x_{q+3} = \cdots = x_{n+3} = 0. \tag{5.152}$$

This quadric is diffeomorphic to the unit sphere S^q in the span $\mathbf{R}^{q+1}$ of the spacelike vectors $w_2, \ldots, w_{q+2}$ with the diffeomorphism $\gamma : S^q \to E \cap Q^{n+1}$ given by

$$\gamma(v) = [w_1 + v], \quad v \in S^q. \tag{5.153}$$

Similarly $E^\perp \cap Q^{n+1}$ is the quadric given in homogeneous coordinates by

$$x_{n+3}^2 = x_{q+3}^2 + \cdots + x_{n+2}^2, \quad x_1 = \cdots = x_{q+2} = 0. \tag{5.154}$$

This space $E^\perp \cap Q^{n+1}$ is diffeomorphic to the unit sphere S^p in the span $\mathbf{R}^{p+1}$ of the spacelike vectors $w_{q+3}, \ldots, w_{n+2}$ with the diffeomorphism

$$\delta : S^p \to E^\perp \cap Q^{n+1}$$

given by

$$\delta(u) = [u + w_{n+3}], \quad u \in S^p. \tag{5.155}$$

The image of the curvature sphere map Y_1 of multiplicity p is contained in the q-dimensional quadric $E \cap Q^{n+1}$ given by equation (5.152), which is diffeomorphic to S^q. The map Y_1 is constant on each leaf of its principal foliation T_1, and so Y_1 factors through an immersion of the q-dimensional space of leaves U/T_1 into the q-dimensional quadric $E \cap Q^{n+1}$. Hence, the image of Y_1 is an open subset of this quadric, and each leaf of T_1 corresponds to a point of v of the quadric.

Similarly, the curvature sphere map Y_{n+3} of multiplicity q factors through an immersion of its p-dimensional space of leaves U/T_2 onto an open subset of the p-dimensional quadric $E^\perp \cap Q^{n+1}$ given by equation (5.154), and each leaf of T_2 corresponds to a point of u of that quadric.

From this it is clear that the restriction of the Legendre map μ to the neighborhood $U \subset M$ is contained in the compact, connected cyclide

$$\nu : S^p \times S^q \to \Lambda^{2n-1}$$

defined by

$$\nu(u, v) = [k_1(u, v), k_2(u, v)], \quad (u, v) \in S^p \times S^q, \tag{5.156}$$

where

$$k_1(u, v) = \gamma(v), \quad k_2(u, v) = \delta(u), \tag{5.157}$$

for the maps γ and δ defined above. By a standard connectedness argument, the Legendre map $\mu : M \to \Lambda^{2n-1}$ is also the restriction of ν to an open subset of $S^p \times S^q$. This proves part (a) of the theorem.

Geometrically, the image of ν consists of all lines joining a point on the quadric in equation (5.152) to a point on the quadric in equation (5.154). Thus any choice of $(q+1)$-plane E in $\mathbf{RP}^{n+2}$ with signature $(q+1,1)$ and corresponding orthogonal complement $E^{\perp}$ with signature $(p+1,1)$ determines a unique compact, connected cyclide of characteristic (p,q) and vice versa. The local Lie equivalence of any two cyclides of the same characteristic is then clear.

Our standard model is the case where E is the space Ω in equation (5.109). Part (b) of the theorem then follows since our standard model is Lie equivalent to the Legendre lift of the standard product of two spheres,

$$S^q(1/\sqrt{2}) \times S^p(1/\sqrt{2}) \subset S^n \subset \mathbf{R}^{q+1} \times \mathbf{R}^{p+1} = \mathbf{R}^{n+1}, \tag{5.158}$$

where $p+q=n-1$, as discussed before the statement of the theorem. □

Möbius geometric classification of the cyclides of Dupin

Theorem 5.22 is a classification of proper Dupin submanifolds with two distinct curvature spheres in the context of Lie sphere geometry. It is also useful to have a Möbius geometric description of proper Dupin hypersurfaces $M^{n-1} \subset \mathbf{R}^n$ with two distinct principal curvatures at each point. This is analogous to the classical characterizations of the cyclides of Dupin in $\mathbf{R}^3$ obtained in the nineteenth century (see, for example, [95, pp. 151–166]). The following Möbius geometric theorem can be proved as a consequence of the Lie sphere geometric Theorem 5.22 above. The proof of this theorem was first given in [74]. The treatment here is taken from the book [77, pp. 151–159].

Theorem 5.23. (a) *Every connected cyclide of Dupin $M^{n-1} \subset \mathbf{R}^n$ of characteristic (p,q) is Möbius equivalent to an open subset of a hypersurface of revolution obtained by revolving a q-sphere $S^q \subset \mathbf{R}^{q+1} \subset \mathbf{R}^n$ about an axis $\mathbf{R}^q \subset \mathbf{R}^{q+1}$ or a p-sphere $S^p \subset \mathbf{R}^{p+1} \subset \mathbf{R}^n$ about an axis $\mathbf{R}^p \subset \mathbf{R}^{p+1}$.*

(b) *Two hypersurfaces obtained by revolving a q-sphere $S^q \subset \mathbf{R}^{q+1} \subset \mathbf{R}^n$ about an axis of revolution $\mathbf{R}^q \subset \mathbf{R}^{q+1}$ are Möbius equivalent if and only if they have the same value of $\rho = |r|/a$, where r is the signed radius of the profile sphere S^q and $a > 0$ is the distance from the center of S^q to the axis of revolution.*

Proof. This theorem follows from Theorem 5.22 by a consideration of Möbius geometry as a subgeometry of Lie sphere geometry. By Theorem 5.22, it suffices to classify compact, connected cyclides up to Möbius equivalence, since every connected cyclide is contained in a unique compact, connected cyclide. Consider a compact, connected cyclide

$$\lambda : S^p \times S^q \to \Lambda^{2n-1}, \quad p+q=n-1,$$

of characteristic (p,q). As shown in the proof of Theorem 5.22, there is a linear space E of $\mathbf{RP}^{n+2}$ with signature $(q+1,1)$ given in equation (5.151) such that the two curvature sphere maps,

$$[k_1] : S^q \to E \cap Q^{n+1}, \quad [k_2] : S^p \to E^{\perp} \cap Q^{n+1},$$

are diffeomorphisms.

Recall that Möbius transformations are precisely those Lie sphere transformations A satisfying $A[e_{n+3}] = [e_{n+3}]$, where $\{e_1, \ldots, e_{n+3}\}$ is the standard orthonormal basis for $\mathbf{R}^{n+3}_2$. We can decompose e_{n+3} as

$$e_{n+3} = \alpha + \beta, \quad \alpha \in E, \quad \beta \in E^{\perp}. \tag{5.159}$$

Since $\langle \alpha, \beta \rangle = 0$, we have

$$-1 = \langle e_{n+3}, e_{n+3} \rangle = \langle \alpha, \alpha \rangle + \langle \beta, \beta \rangle.$$

Thus at least one of the two vectors α, β is timelike.

We first consider the case where β is timelike. Let Z be the orthogonal complement of β in $E^{\perp}$. Then Z is a $(p+1)$-dimensional vector space on which the restriction of $\langle\ ,\ \rangle$ has signature $(p+1, 0)$. Since $Z \subset e_{n+3}^{\perp}$, there is a Möbius transformation A such that

$$A(Z) = S = \text{Span}\,\{e_{q+3}, \ldots, e_{n+2}\}.$$

The curvature sphere map $[Ak_1]$ of the Dupin submanifold $A\lambda$ is a q-dimensional submanifold in the space $S^{\perp} \cap Q^{n+1}$. By equation (4.23) on page 195, this means that these spheres all have their centers in the space

$$\mathbf{R}^q = \text{Span}\,\{e_3, \ldots, e_{q+2}\}.$$

Note that

$$\mathbf{R}^q \subset \mathbf{R}^{q+1} = \text{Span}\,\{e_3, \ldots, e_{q+3}\} \subset \mathbf{R}^n = \text{Span}\,\{e_3, \ldots, e_{n+2}\}.$$

This implies that the Dupin submanifold $A\lambda$ is a hypersurface of revolution in $\mathbf{R}^n$ obtained by revolving a q-dimensional profile submanifold in $\mathbf{R}^{q+1}$ about the axis $\mathbf{R}^q$ (see the proof of Theorem 5.11 in [77, pp. 142–143]). Since $A\lambda$ has two distinct curvature spheres, the profile submanifold has only one curvature sphere. Thus, it is a totally umbilical submanifold of $\mathbf{R}^{q+1}$.

We can distinguish four cases based on the nature of the vector α in equation (5.159). These correspond to different singularity sets of the Euclidean projection of $A\lambda$. Such singularities correspond exactly with the singularities of the Euclidean projection of λ, since the Möbius transformation A preserves the rank of the Euclidean projection. Since we have assumed that β is timelike, we know that for all $u \in S^p$,

$$\langle k_2(u), e_{n+3} \rangle = \langle k_2(u), \alpha + \beta \rangle = \langle k_2(u), \beta \rangle \neq 0,$$

because the orthogonal complement of β in $E^{\perp}$ is spacelike. Thus, the curvature sphere $[Ak_2]$ is never a point sphere. However, it is possible for $[Ak_1]$ to be a point sphere. We now consider the four cases determined by the nature of α.

Case 1: $\alpha = 0$

In this case, the curvature sphere $[Ak_1]$ is a point sphere for every point in $S^p \times S^q$. The image of the Euclidean projection of $A\lambda$ is precisely the axis $\mathbf{R}^q$. The cyclide $A\lambda$ is the Legendre lift of $\mathbf{R}^q$ as a submanifold of codimension $p+1$ in $\mathbf{R}^n$. This is, in fact, the standard model given in equation (5.110). In this case, the Euclidean projection is not an immersion, and so this case does not yield any of the embedded hypersurfaces classified in part (a) of the theorem.

In the remaining cases, we can always arrange that the totally umbilic profile submanifold is a q-sphere and not a q-plane by first inverting $\mathbf{R}^{q+1}$ in a sphere centered at a point on the axis $\mathbf{R}^q$ which is not on the profile submanifold, if necessary. This type of inversion preserves the axis of revolution $\mathbf{R}^q$. Then, by a Euclidean translation, if necessary, we can arrange that the center of the profile sphere is a point $(0, a)$ on the x_{q+3}-axis ℓ in $\mathbf{R}^{q+1}$, as in Figure 5.1. We know that the center of the profile sphere cannot lie on the axis of revolution $\mathbf{R}^q$; otherwise, the hypersurface of revolution would be an $(n-1)$-sphere and not a cyclide of Dupin. Thus, we may take $a > 0$.

The map $[Ak_1]$ is the curvature sphere map that results from the surface of revolution construction. The other curvature sphere of $A\lambda$ corresponds exactly to the curvature sphere of the profile sphere, i.e., to the profile sphere itself. Therefore,

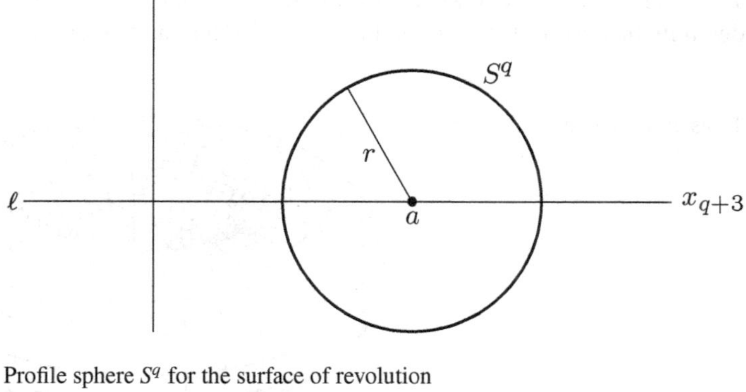

Fig. 5.1 Profile sphere S^q for the surface of revolution

the signed radius r of the profile sphere is equal to the signed radius of the curvature sphere $[Ak_2]$. Since $[Ak_2]$ is never a point sphere, we conclude that $r \neq 0$. From now on, we will identify the profile sphere with the second factor S^q in the domain of λ. We now consider the remaining cases based on whether the nonzero vector α is timelike, lightlike, or spacelike.

Case 2: α is timelike

Since the orthogonal complement of α in E is spacelike, we have for all $v \in S^q$,

$$\langle k_1(v), e_{n+3} \rangle = \langle k_1(v), \alpha \rangle \neq 0,$$

This implies that the Euclidean projection of $A\lambda$ is an immersion at all points. This corresponds to the case $|r| < a$, when the profile sphere is disjoint from the axis of revolution. Note that by interchanging the roles of α and β, we can find a Möbius transformation that takes λ to the Legendre submanifold obtained by revolving a p-sphere around an axis $\mathbf{R}^p \subset \mathbf{R}^{p+1} \subset \mathbf{R}^n$.

We now describe this case in the classical situation of surfaces in $\mathbf{R}^3$. Then the Euclidean projection of $A\lambda$ is a torus of revolution (see Figure 5.2).

It then follows that the Euclidean projection of λ itself is a *ring cyclide* (see Figure 5.3) if the Möbius projection of λ does not contain the improper point, or a *parabolic ring cyclide* (see Figure 5.4) if the Möbius projection of λ does contain the improper point. In either case, the focal set in $\mathbf{R}^3$ consists of a pair of *focal conics*, as we describe in the following remark.

Remark 5.24 (Focal sets of the cyclides in $\mathbf{R}^n$*).* For a ring cyclide in $\mathbf{R}^3$, the focal set consists of an ellipse and a hyperbola in mutually orthogonal planes such that the vertices of the ellipse are the foci of the hyperbola and vice versa. For a parabolic ring cyclide in $\mathbf{R}^3$, the focal set consists of two parabolas in orthogonal planes such that the vertex of each is the focus of the other. For a torus of revolution in $\mathbf{R}^3$, the focal set consists of the core circle and the axis of revolution covered twice. This is a special case of a pair of focal conics. These classical cyclides of Dupin are discussed in more detail in the book of Cecil–Ryan [95, pp. 151–166], and other references are

Fig. 5.2 Torus of revolution

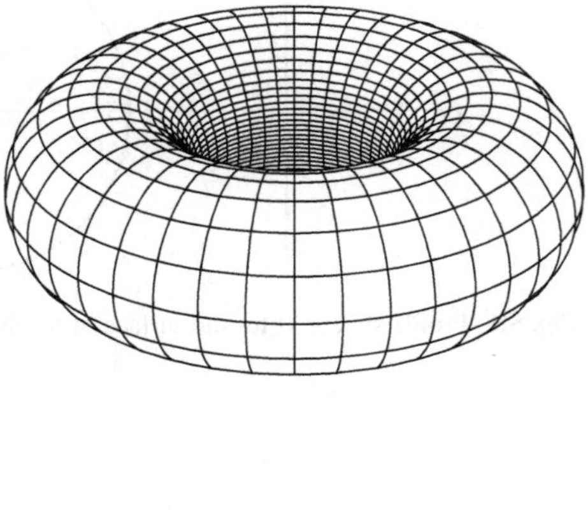

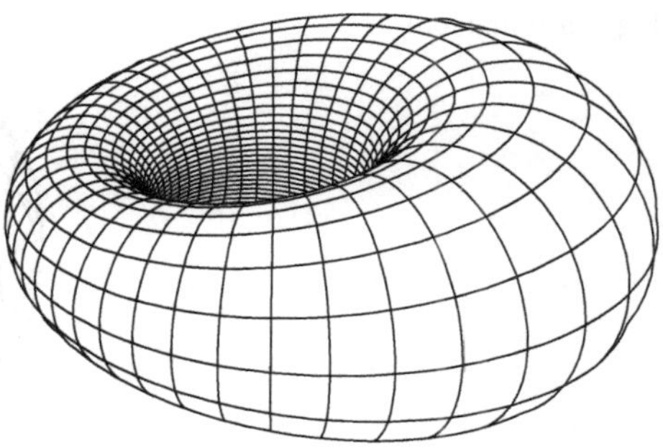

Fig. 5.3 Ring cyclide

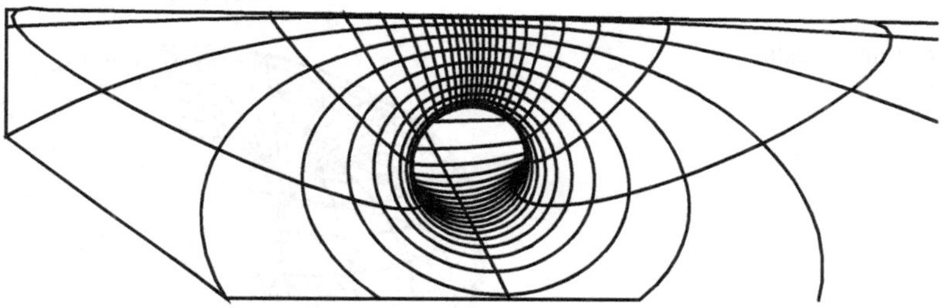

Fig. 5.4 Parabolic ring cyclide

given there. For the higher dimensional cyclides in $\mathbf{R}^n$, the focal set in $\mathbf{R}^n$ consists of a pair of focal quadrics defined in an analogous way. This is shown in detail in [95, pp. 176–178].

Case 3: α is lightlike, but not zero

In this case, there is exactly one $v \in S^q$ such that

$$\langle k_1(v), e_{n+3} \rangle = \langle k_1(v), \alpha \rangle = 0. \tag{5.160}$$

Geometrically, this corresponds to the case $|r| = a$, where the profile sphere intersects the axis in one point. Thus, $S^p \times \{v\}$ is the set of points in $S^p \times S^q$ where the Euclidean projection is singular.

We now describe this case for the classical situation of surfaces in $\mathbf{R}^3$. Then the Euclidean projection of $A\lambda$ is a *limit torus* (see Figure 5.5), and the Euclidean projection of λ itself is a *limit spindle cyclide* (see Figure 5.6) or a *limit horn cyclide* (see Figure 5.7), if the Möbius projection of λ does not contain the improper point.

On the other hand, if the Möbius projection of λ does contain the improper point, then the Euclidean projection of λ is either a *limit parabolic horn cyclide* (see Figure 5.8) or a circular cylinder (in the case where the singularity is at the improper point). For all of these surfaces except the cylinder, the focal set in $\mathbf{R}^3$

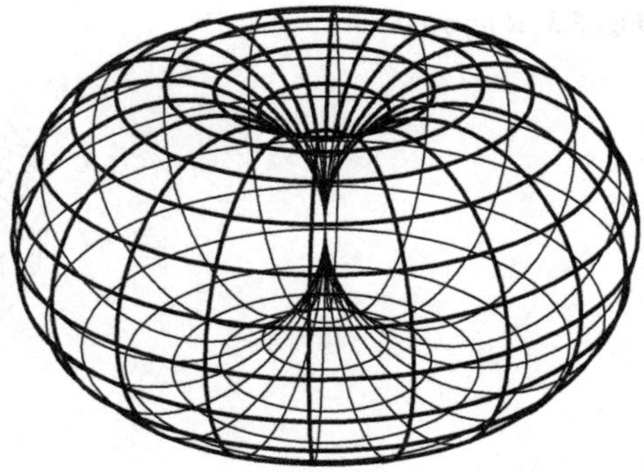

Fig. 5.5 Limit torus

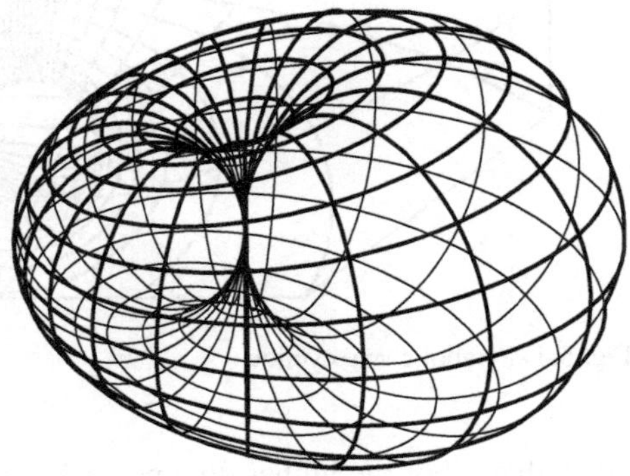

Fig. 5.6 Limit spindle cyclide

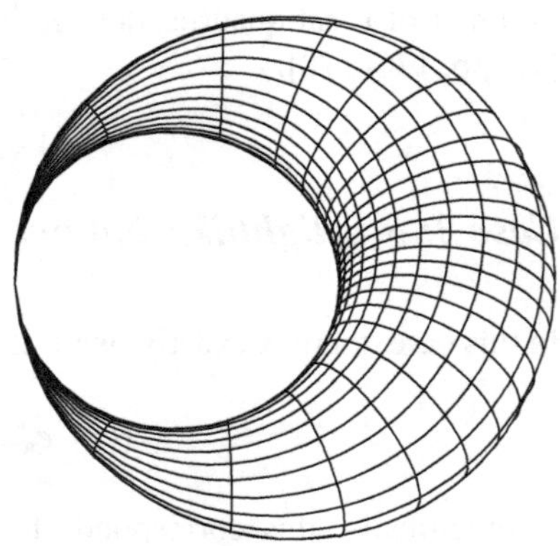

Fig. 5.7 Limit horn cyclide

consists of a pair of focal conics, as in the previous case. To see this, note that each of these surfaces is a parallel surface to a cyclide without singularities, and so it has the same focal set as that cyclide.

For the cylinder, the Euclidean focal set consists only of the axis of revolution, since one of the principal curvatures is identically zero, and so the corresponding focal points are all at infinity. In Lie sphere geometry, both curvature sphere maps are plane curves on the Lie quadric, as shown in the proof of Theorem 5.22.

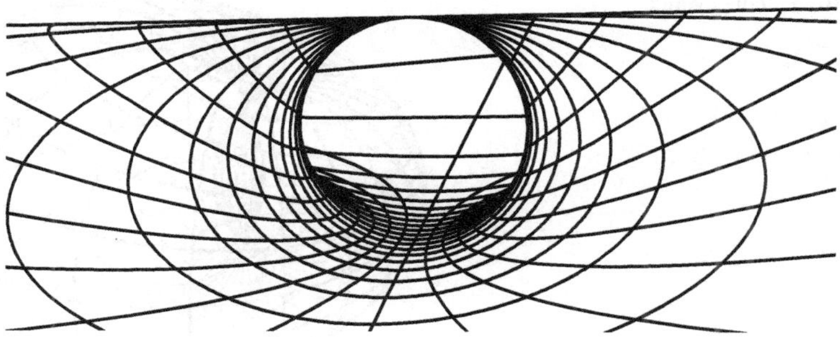

Fig. 5.8 Limit parabolic horn cyclide

Fig. 5.9 Spindle torus

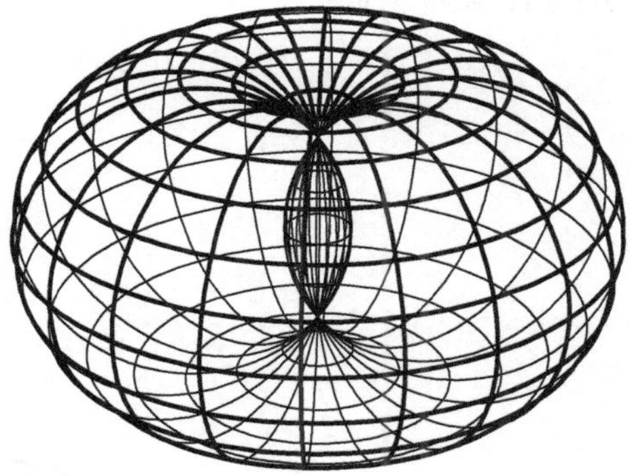

Case 4: α is spacelike

In this case, the condition (5.160) is satisfied by points v in a $(q-1)$-sphere $S^{q-1} \subset S^q$. For points in $S^p \times S^{q-1}$, the point sphere map is a curvature sphere, and thus the Euclidean projection is singular. Geometrically, this is the case $|r| > a$, and so the profile sphere intersects the axis $\mathbf{R}^q$ in a $(q-1)$-sphere.

In the classical situation of surfaces in $\mathbf{R}^3$, the Euclidean projection of $A\lambda$ is a *spindle torus* (see Figure 5.9). The Euclidean projection of λ itself is a *spindle cyclide* (see Figure 5.10) or a *horn cyclide* (see Figure 5.11), if the Möbius projection does not contain the improper point.

On the other hand, if the Möbius projection of λ contains the improper point, then the Euclidean projection of λ is either a *parabolic horn cyclide* (see Figure 5.12) or circular cone (in the case where one of the singularities is at the improper point). For all of these surfaces except the cone, the focal set in $\mathbf{R}^3$ consists of a pair of focal conics, since each of these surfaces is a parallel surface to a cyclide without singularities. For the cone, the Euclidean focal set consists of only the axis of revolution (minus the origin), since one principal curvature is identically zero.

In the four cases above, we assumed that the vector β is timelike. There are also four cases to handle under the assumption that α is timelike. In those cases,

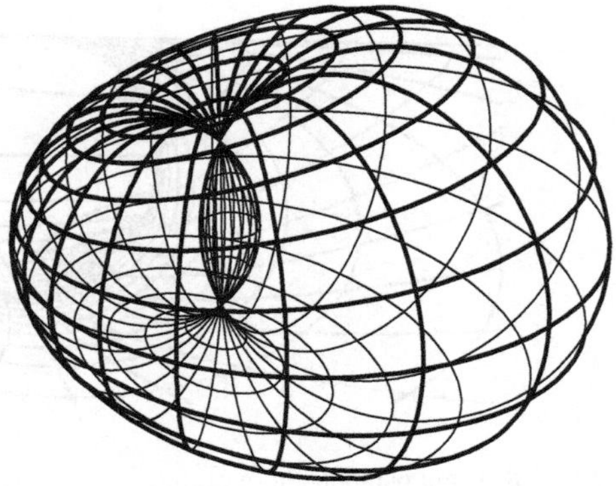

Fig. 5.10 Spindle cyclide

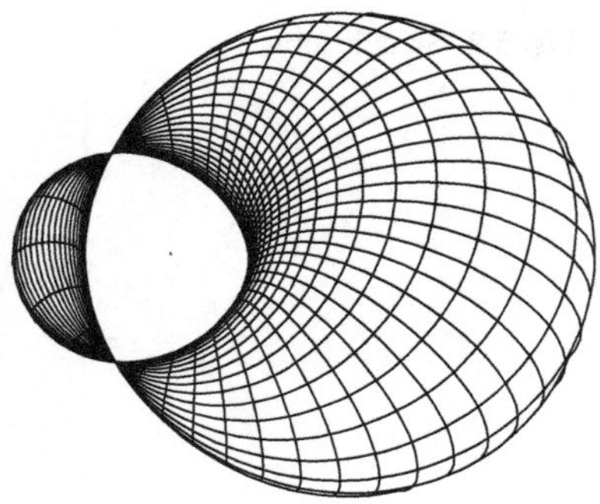

Fig. 5.11 Horn cyclide

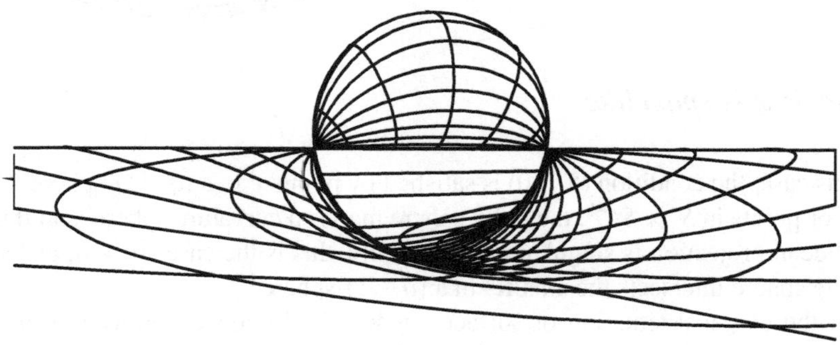

Fig. 5.12 Parabolic horn cyclide

the axis of revolution is a subspace $\mathbf{R}^p \subset \mathbf{R}^{p+1}$, and the profile submanifold is a p-sphere. The roles of p and q in determining the dimension of the singularity set of the Euclidean projection are then reversed from the four cases above. Thus, if $p \neq q$, then only a ring cyclide can be represented as a hypersurface of revolution of both a q-sphere and a p-sphere. This completes the proof of part (a).

Proof of part (b)

Next we turn to the proof of part (b). By part (a), we may assume that the profile sphere S^q of the hypersurface of revolution has center $(0, a)$ with $a > 0$ on the x_{q+3}-axis ℓ. Möbius classification clearly does not depend on the sign of the radius of S^q, since the two hypersurfaces of revolution obtained by revolving spheres with the same center and opposite radii differ only by the change of orientation transformation Γ.

We now show that the ratio $\rho = |r|/a$ is invariant under the subgroup of Möbius transformations of the profile space $\mathbf{R}^{q+1}$ which take one such hypersurface of revolution to another. First, note that symmetry implies that a transformation T in this subgroup maps the axis of revolution $\mathbf{R}^q$ to itself and the axis of symmetry ℓ to itself. Since $\mathbf{R}^q$ and ℓ intersect only at 0 and the improper point ∞, the transformation T maps the set $\{0, \infty\}$ to itself. If T maps 0 to ∞, then the composition ΦT, where Φ is an inversion in a sphere centered at 0, is a member of the subgroup of transformations that map ∞ to ∞ and map 0 to 0. By Theorem 3.16 of [77, p. 47], such a Möbius transformation is a similarity transformation, and so it is the composition of a central dilatation D and a linear isometry Ψ. Therefore, $T = \Phi D \Psi$, and each of the transformations on the right of this equation preserves the ratio ρ. The invariant ρ is the only one needed for Möbius classification, since any two profile spheres with the same value of ρ can be mapped to one another by a central dilatation. □

Remark 5.25. To obtain a collection of hypersurfaces containing one representative from each Möbius equivalence class, we fix $a = 1$ and allow r to vary, $0 < r < \infty$. This results in a family of parallel hypersurfaces of revolution. Note that taking a negative signed radius s for the profile sphere yields a parallel hypersurface that differs only in orientation from the hypersurface corresponding to $r = -s$. Finally, taking $r = 0$ also gives a parallel submanifold in the family, but the Euclidean projection degenerates to a sphere S^p. This is the case $\beta = 0, \alpha = e_{n+3}$ in the proof above, where the point sphere map equals the curvature sphere $[k_2]$ at every point.

Complete cyclides of Dupin in $\mathbf{R}^n$

From Theorem 5.23, we can derive a classification of complete proper Dupin hypersurfaces in $\mathbf{R}^n$ with $g = 2$ principal curvatures (see Theorem 5.26 below). This classification was proven in full in the book [95, pp. 168–179] without using Lie sphere geometry. However, that proof uses the assumption of completeness, which is not required in the local Lie-geometric Theorem 5.23. In the treatment given in [95, pp. 176–178], we also showed that the focal submanifolds of a complete cyclide of Dupin in $\mathbf{R}^n$ are always a pair of focal quadrics.

In the following theorem, we will use the classical terminology to give names to the hypersurfaces of revolution in Theorem 5.23 and their images under Möbius transformations.

Theorem 5.26. *Let $M^{n-1} \subset \mathbf{R}^n$ be a connected, complete cyclide of Dupin of characteristic (p,q).*

(a) *If M^{n-1} is compact, then it is a ring cyclide diffeomorphic to $S^p \times S^q$.*
(b) *If M^{n-1} is not compact, then it is a spherical cylinder $S^p \times \mathbf{R}^{n-1-p}$ or a parabolic ring cyclide.*

Proof. In Theorem 5.23, we showed that every connected cyclide of Dupin $M^{n-1} \subset \mathbf{R}^n$ of characteristic (p,q) is Möbius equivalent to an open subset of a hypersurface of revolution obtained by revolving a q-sphere $S^q \subset \mathbf{R}^{q+1} \subset \mathbf{R}^n$ about an axis $\mathbf{R}^q \subset \mathbf{R}^{q+1}$ or a p-sphere $S^p \subset \mathbf{R}^{p+1} \subset \mathbf{R}^n$ about an axis $\mathbf{R}^p \subset \mathbf{R}^{p+1}$. Thus, we need to determine which Möbius images of open subsets of these hypersurfaces of revolution are complete.

We consider the setup depicted in Figure 5.1 in the proof of Theorem 5.23. If the profile sphere S^q does not intersect the axis of revolution $\mathbf{R}^q$, the hypersurface of revolution is a torus of revolution diffeomorphic to $S^p \times S^q$. Such a hypersurface is also referred to as a *round cyclide* (see Figure 5.2). The image $M \subset \mathbf{R}^n$ of a round cyclide C under a Möbius transformation is called a *ring cyclide* (see Figure 5.3) if M is compact, and this is the only case that results in a compact hypersurface.

On the other hand, if the image $M \subset \mathbf{R}^n$ is not compact, i.e., if the Möbius transformation maps one point of the round cyclide C to the improper point P at infinity, then M is called a *parabolic ring cyclide* (see Figure 5.4), which is also a complete hypersurface in $\mathbf{R}^n$.

We now examine the cases where the profile sphere S^q intersects the axis of revolution $\mathbf{R}^q$. Let O be the point of intersection of the line ℓ with the axis $\mathbf{R}^q$ in Figure 5.1. If the profile sphere S^q in Figure 5.1 intersects the axis in the one point O, then the hypersurface of revolution W is called a *limit torus* (see Figure 5.5). Of course, W has a singularity at O and no open subset of W is a complete hypersurface in $\mathbf{R}^n$. If T is a Möbius transformation that maps O to a proper point Q in $\mathbf{R}^n$, then $T(W)$ has a singularity at Q, and no open subset of $T(W)$ is a complete hypersurface in $\mathbf{R}^n$.

On the other hand, if T is a Möbius transformation that maps O to the improper point P, then the image $M = T(W - \{O\})$ in $\mathbf{R}^n$ is a spherical cylinder $S^p \times \mathbf{R}^q$, where $q = n - 1 - p$. To see this, note that if ϕ is an inversion of $\mathbf{R}^n \cup \{P\}$ in a sphere Σ centered at O, then by symmetry, the copies of the profile sphere S^q obtained as one rotates S^q about the axis $\mathbf{R}^q$ are all mapped by ϕ to q-planes parallel to the axis $\mathbf{R}^q$, since the point O is mapped to the improper point P. Thus, $\phi(W - \{O\})$ is a spherical cylinder $S^p \times \mathbf{R}^q$.

Note that $\phi(O) = P$ and $\phi(P) = O$. Thus, since $T(O) = P$, the Möbius transformation $T \circ \phi$ maps the improper point P to itself. This means that $T \circ \phi$ is both a Laguerre transformation (a Lie sphere transformation that maps P to itself) and a Möbius transformation. Therefore $T \circ \phi$ is a similarity transformation S of $\mathbf{R}^n$

(see Theorem 3.16 of [77, p. 47]). Since ϕ is its own inverse, we can multiply the equation $T \circ \phi = S$ on the right by ϕ to get $T = S \circ \phi$. Since ϕ maps $W - \{O\}$ to a spherical cylinder, and S maps a spherical cylinder to another spherical cylinder, we see that $M = T(W - \{O\})$ is a spherical cylinder, which is a complete hypersurface in $\mathbf{R}^n$.

Finally, if the profile sphere S^q in Figure 5.1 intersects the axis $\mathbf{R}^q$ in more than one point, the resulting hypersurface of revolution is called a *spindle torus* (see Figure 5.9), which has more than one singularity in $\mathbf{R}^n$. The image of a spindle torus under a Möbius transformation T always has a singularity in $\mathbf{R}^n$, so no open subset of $T(W)$ is a complete hypersurface in $\mathbf{R}^n$. Thus, we have handled all the possible cases in Theorem 5.23 and the proof is finished. □

Remark 5.27 (Cyclides in discrete differential geometry). The cyclides of Dupin and their characterization in the setting of Lie sphere geometry play an important role in a recent paper by A. Bobenko and E. Huhnen-Venedy [43]. In that paper, the authors study *cyclidic nets*, which are discrete analogues of surfaces parametrized by lines of curvature (2-dimensional case), and of triply orthogonal coordinate systems (3-dimensional case). Specifically, a 2-dimensional cyclidic net in $\mathbf{R}^3$ is constructed from *cyclidic patches*, which are obtained by restricting an oriented principal coordinate parametrization of a cyclide of Dupin to a closed rectangle. The lines of curvature of the cyclidic patches in a 2-dimensional cyclidic net form a net of C^1-curves composed of circular arcs, which can be considered to be the lines of curvature of the cyclidic net. As the authors note, discretization of surfaces parametrized by lines of curvature is an important area of current research in discrete differential geometry. Some of the most notable discretizations so far are circular nets, conical nets, and contact element nets. The authors show that all of these types of discretizations are cyclidic nets having certain special properties.

5.6 Local Classifications in the Case $\mathbf{g} = \mathbf{3}$

In this section, we discuss local classifications of proper Dupin hypersurfaces with $g = 3$ principal curvatures. The first case is that of a Dupin hypersurface $M^3 \subset \mathbf{R}^4$ with three principal curvatures of multiplicity one. In his dissertation, Pinkall [442, 445] gave a local classification of such Dupin hypersurfaces up to Lie equivalence. This is a fundamental case, and it is the first case where Lie invariants are needed in the classification.

In Pinkall's local classification, he found one Lie invariant (ρ in our treatment below) that completely determines whether or not the Legendre lift λ of the Dupin hypersurface is reducible. If $\rho \neq 0$, then λ is irreducible. Pinkall proved that the Legendre lifts of any two irreducible proper Dupin hypersurfaces with $g = 3$ in $\mathbf{R}^4$ are locally Lie equivalent, each being Lie equivalent to an open subset of Cartan's isoparametric hypersurface in S^4 (see Subsection 3.8.3 on page 151). If $\rho = 0$, then λ is reducible, and Pinkall showed that there is a 1-parameter family of Lie equivalence classes of reducible proper Dupin hypersurfaces with $g = 3$ in $\mathbf{R}^4$ (see [445, p. 111]).

Here we give an exposition of the portion of Pinkall's work concerning the irreducible case, following the paper of Cecil and Chern [80] and using the method of moving frames. (See also the book [70, pp. 168–188] for a similar treatment.) After that we will discuss local classifications of higher dimensional proper Dupin hypersurfaces with $g = 3$ due to Niebergall [393, 394], and Cecil and Jensen [85], although we will not give complete proofs here.

Let $\lambda : M^3 \to \Lambda^7$ be a proper Dupin submanifold with three curvature spheres at each point. Note that the Legendre lift of any proper Dupin hypersurface in $\mathbf{R}^4$ with three distinct principal curvatures is such a map λ.

On a local neighborhood U in M we take a Lie frame Y_A such that for each $x \in M$, the line $\lambda(x) = [Y_1(x), Y_7(x)]$. Using Theorem 5.21 on page 263 as we did in the $g = 2$ case, we can arrange that $[Y_1]$ and $[Y_7]$ are curvature sphere maps, and that the Maurer–Cartan forms satisfy

$$\omega_1^3 = \omega_1^7 = 0, \quad \omega_7^4 = \omega_7^1 = 0. \tag{5.161}$$

In this frame the third curvature sphere has the form $\alpha Y_1 + \beta Y_7$ for some smooth nonvanishing functions α and β on U. If we make a change of Lie frame of the form,

$$Y_1^* = \alpha Y_1, \quad Y_2^* = (1/\alpha) Y_2, \quad Y_7^* = \beta Y_7, \quad Y_6^* = (1/\beta) Y_6, \tag{5.162}$$

then Y_1^* and Y_7^* still represent the first two curvature sphere maps, and $Y_1^* + Y_7^*$ represents the third curvature sphere at each point of U. We drop the asterisks and use this frame. Then by using the method of proof of Theorem 5.21, we can find a new Lie frame whose Maurer–Cartan forms satisfy

$$\omega_1^5 + \omega_7^5 = 0, \quad \omega_1^1 - \omega_7^7 = 0, \tag{5.163}$$

as well as equation (5.161). Such a frame is called a *second order frame* in the terminology of Cecil and Jensen [85, p. 138]. Conditions (5.161) and (5.163) completely determine the frame vectors Y_3, Y_4 and Y_5, while Y_1 and Y_7 are determined up to a transformation of the form,

$$Y_1^* = \tau Y_1, \quad Y_7^* = \tau Y_7, \tag{5.164}$$

for some smooth nonvanishing function τ on U.

Each of the three curvature sphere maps Y_1, Y_7 and $Y_1 + Y_7$ is constant along the leaves of its corresponding principal foliation. Thus, each curvature sphere map factors through an immersion of the corresponding 2-dimensional space of leaves of its principal foliation into the Lie quadric Q^5. In terms of moving frames, this implies that the forms $\omega_1^4, \omega_1^5, \omega_7^3$ are linearly independent on the open set U, i.e.,

$$\omega_1^4 \wedge \omega_1^5 \wedge \omega_7^3 \neq 0. \tag{5.165}$$

This can also be seen by expressing the forms above in terms of a Lie frame $\{Z_1,\ldots,Z_{n+3}\}$ whose Maurer–Cartan forms satisfy the regularity condition (5.78) and using the fact that each curvature sphere has multiplicity one. For simplicity, we will also use the notation,

$$\theta_1 = \omega_1^4, \quad \theta_2 = \omega_1^5, \quad \theta_3 = \omega_7^3. \tag{5.166}$$

Analytically, the Dupin conditions are three partial differential equations, and we are treating an over-determined system. The method of moving frames reduces the handling of its integrability conditions to a straightforward algebraic problem, namely, that of repeated exterior differentiations.

Computing exterior derivatives

We begin by computing the exterior derivatives of the equations,

$$\omega_1^3 = 0, \quad \omega_7^4 = 0, \quad \omega_1^5 + \omega_7^5 = 0. \tag{5.167}$$

These equations come from the fact that Y_1, Y_7 and $Y_1 + Y_7$ are curvature spheres. Using the skew-symmetry of the matrix in equation (5.75), as well as the relations (5.161) and (5.163), the exterior derivatives of the three equations in (5.167) yield

$$\begin{aligned} 0 &= \omega_1^4 \wedge \omega_3^4 + \omega_1^5 \wedge \omega_3^5, \\ 0 &= \omega_1^5 \wedge \omega_4^5 + \omega_7^3 \wedge \omega_3^4, \\ 0 &= \omega_1^4 \wedge \omega_4^5 + \omega_7^3 \wedge \omega_3^5. \end{aligned} \tag{5.168}$$

If we take the wedge product of the first of these equations with ω_1^4, we conclude that ω_3^5 is in the span of ω_1^4 and ω_1^5. On the other hand, taking the wedge product of the third equation with ω_1^4 yields that ω_3^5 is in the span of ω_1^4 and ω_7^3. Consequently, $\omega_3^5 = \rho\omega_1^4$, for some smooth function ρ. Similarly, there exist smooth functions α and β such that $\omega_3^4 = \alpha\omega_1^5$ and $\omega_4^5 = \beta\omega_7^3$. Then, if we substitute these results into equation (5.168), we get that $\rho = \alpha = \beta$, and hence we have

$$\omega_3^5 = \rho\omega_1^4, \quad \omega_3^4 = \rho\omega_1^5, \quad \omega_4^5 = \rho\omega_7^3. \tag{5.169}$$

Next we differentiate the three equations that come from the Dupin conditions,

$$\omega_1^7 = 0, \quad \omega_7^1 = 0, \quad \omega_1^1 - \omega_7^7 = 0. \tag{5.170}$$

As above, use of the skew-symmetry relations in equation (5.75) and equations (5.161) and (5.163) yields the existence of smooth functions a, b, c, p, q, r, s, t, u such that the following relations hold:

$$\begin{aligned}\omega_4^7 &= -\omega_6^4 = a\omega_1^4 + b\omega_1^5,\\ \omega_5^7 &= -\omega_6^5 = b\omega_1^4 + c\omega_1^5;\end{aligned} \tag{5.171}$$

$$\begin{aligned}\omega_3^1 &= -\omega_2^3 = p\omega_7^3 - q\omega_1^5,\\ \omega_5^1 &= -\omega_2^5 = q\omega_7^3 - r\omega_1^5;\end{aligned} \tag{5.172}$$

$$\begin{aligned}\omega_4^1 &= -\omega_2^4 = b\omega_1^5 + s\omega_1^4 + t\omega_7^3,\\ \omega_6^3 &= -\omega_3^7 = q\omega_1^5 + t\omega_1^4 + u\omega_7^3.\end{aligned} \tag{5.173}$$

We next take the exterior derivatives of the three basis forms ω_1^4, ω_1^5, and ω_7^3. Using the relations that we have derived so far, we obtain from the Maurer–Cartan equation (5.76),

$$d\omega_1^4 = \omega_1^1 \wedge \omega_1^4 + \omega_1^5 \wedge \omega_5^4 = \omega_1^1 \wedge \omega_1^4 - \rho\omega_1^5 \wedge \omega_7^3. \tag{5.174}$$

We can obtain similar equations for $d\omega_1^5$ and $d\omega_7^3$. If we write these expressions in terms of the forms θ_1, θ_2 and θ_3 defined in equation (5.166), we get

$$\begin{aligned}d\theta_1 &= \omega_1^1 \wedge \theta_1 - \rho\,\theta_2 \wedge \theta_3,\\ d\theta_2 &= \omega_1^1 \wedge \theta_2 - \rho\,\theta_3 \wedge \theta_1,\\ d\theta_3 &= \omega_1^1 \wedge \theta_3 - \rho\,\theta_1 \wedge \theta_2.\end{aligned} \tag{5.175}$$

We next differentiate equation (5.169). We have $\omega_3^4 = \rho\omega_1^5$. On the one hand,

$$d\omega_3^4 = \rho\, d\omega_1^5 + d\rho \wedge \omega_1^5.$$

Using the second equation in (5.175) with $\omega_1^5 = \theta_2$, this becomes

$$d\omega_3^4 = \rho\omega_1^1 \wedge \omega_1^5 - \rho^2\omega_7^3 \wedge \omega_1^4 + d\rho \wedge \omega_1^5.$$

On the other hand, we can compute $d\omega_3^4$ from the Maurer–Cartan equation (5.76) and use the relationships that we have derived to find

$$d\omega_3^4 = (-p - \rho^2 - a)(\omega_1^4 \wedge \omega_7^3) - q\omega_1^5 \wedge \omega_1^4 + b\omega_7^3 \wedge \omega_1^5.$$

Equating these two expressions for $d\omega_3^4$ yields

$$(-p - a - 2\rho^2)\,\omega_1^4 \wedge \omega_7^3 = (d\rho + \rho\omega_1^1 - q\omega_1^4 - b\omega_7^3) \wedge \omega_1^5. \tag{5.176}$$

Due to the linear independence of the forms $\{\omega_1^4, \omega_1^5, \omega_7^3\}$, both sides of the equation above vanish. Thus, we conclude that

$$2\rho^2 = -a - p, \tag{5.177}$$

and that $d\rho + \rho\omega_1^1 - q\omega_1^4 - b\omega_7^3$ is a multiple of ω_1^5. Similarly, differentiation of the equation $\omega_4^5 = \rho\omega_7^3$ yields the following analogue of equation (5.176),

$$(s - a - r + 2\rho^2)\, \omega_1^4 \wedge \omega_1^5 = (d\rho + \rho\omega_1^1 + t\omega_1^5 - q\omega_1^4) \wedge \omega_7^3, \tag{5.178}$$

and differentiation of $\omega_3^5 = \rho\omega_1^4$ yields

$$(c + p + u - 2\rho^2)\, \omega_1^5 \wedge \omega_7^3 = (-d\rho - \rho\omega_1^1 - t\omega_1^5 + b\omega_7^3) \wedge \omega_1^4. \tag{5.179}$$

In each of the equations (5.176), (5.178), (5.179), both sides of the equation vanish. From the vanishing of the left sides of the equations, we get the fundamental relationship

$$2\rho^2 = -a - p = a + r - s = c + p + u. \tag{5.180}$$

Furthermore, from the vanishing of the right sides of the three equations (5.176), (5.178), (5.179), we can determine after some algebra that

$$d\rho + \rho\omega_1^1 = q\omega_1^4 - t\omega_1^5 + b\omega_7^3. \tag{5.181}$$

This equation shows the importance of ρ.

Covariant derivatives

Following the notation introduced in equation (5.166), we write equation (5.181) as

$$d\rho + \rho\omega_1^1 = \rho_1\theta_1 + \rho_2\theta_2 + \rho_3\theta_3, \tag{5.182}$$

where

$$\rho_1 = q, \quad \rho_2 = -t, \quad \rho_3 = b, \tag{5.183}$$

are the covariant derivatives of ρ.

Using the Maurer–Cartan equations, we can compute

$$\begin{aligned} d\omega_1^1 &= \omega_1^4 \wedge \omega_4^1 + \omega_1^5 \wedge \omega_5^1 \\ &= \omega_1^4 \wedge (b\omega_1^5 + t\omega_7^3) + \omega_1^5 \wedge (q\omega_7^3 - r\omega_1^5) \\ &= b\omega_1^4 \wedge \omega_1^5 + q\omega_1^5 \wedge \omega_7^3 - t\omega_7^3 \wedge \omega_1^4. \end{aligned} \tag{5.184}$$

Using equations (5.166) and (5.183), this can be rewritten as

$$d\omega_1^1 = \rho_3\, \theta_1 \wedge \theta_2 + \rho_1\, \theta_2 \wedge \theta_3 + \rho_2\, \theta_3 \wedge \theta_1. \tag{5.185}$$

The key idea now is to express everything in terms of ρ and its successive covariant derivatives. Ultimately, this leads to the solution of the problem.

We first derive a general form for these covariant derivatives. Suppose that σ is a smooth function which satisfies a relation of the form

$$d\sigma + m\sigma\omega_1^1 = \sigma_1\, \theta_1 + \sigma_2\, \theta_2 + \sigma_3\, \theta_3, \tag{5.186}$$

for some integer m. (Note that equation (5.181) is such a relationship for the function ρ with $m = 1$.) By taking the exterior derivative of equation (5.186) and using equations (5.175) and (5.185) to express both sides in terms of the standard basis of 2-forms $\theta_1 \wedge \theta_2$, $\theta_2 \wedge \theta_3$ and $\theta_3 \wedge \theta_1$, one finds that the functions $\sigma_1, \sigma_2, \sigma_3$ satisfy equations of the form

$$d\sigma_\alpha + (m+1)\sigma_\alpha\omega_1^1 = \sigma_{\alpha 1}\, \theta_1 + \sigma_{\alpha 2}\, \theta_2 + \sigma_{\alpha 3}\, \theta_3, \quad \alpha = 1, 2, 3, \tag{5.187}$$

where the coefficient functions $\sigma_{\alpha\beta}$ satisfy the commutation relations,

$$\begin{aligned} \sigma_{12} - \sigma_{21} &= -\rho\sigma_3 - m\sigma\rho_3, \\ \sigma_{23} - \sigma_{32} &= -\rho\sigma_1 - m\sigma\rho_1, \\ \sigma_{31} - \sigma_{13} &= -\rho\sigma_2 - m\sigma\rho_2. \end{aligned} \tag{5.188}$$

In particular, from equation (5.182), we have the following commutation relations on ρ_1, ρ_2, ρ_3:

$$\begin{aligned} \rho_{12} - \rho_{21} &= -2\rho\rho_3 \\ \rho_{23} - \rho_{32} &= -2\rho\rho_1 \\ \rho_{31} - \rho_{13} &= -2\rho\rho_2. \end{aligned} \tag{5.189}$$

We next take the exterior derivative of equations (5.171)–(5.173). We first differentiate the equation

$$\omega_4^7 = a\omega_1^4 + b\omega_1^5. \tag{5.190}$$

On the one hand, if we write the Maurer–Cartan equation (5.76) for $d\omega_4^7$ and omit those terms that have already been shown to vanish, we get

$$\begin{aligned} d\omega_4^7 &= \omega_4^2 \wedge \omega_2^7 + \omega_4^3 \wedge \omega_3^7 + \omega_4^5 \wedge \omega_5^7 + \omega_4^7 \wedge \omega_7^7 \\ &= -\omega_1^4 \wedge \omega_2^7 + (-\rho)\omega_1^5 \wedge (-q\omega_1^5 - t\omega_1^4 - u\omega_7^3) \\ &\quad + \rho\omega_7^3 \wedge (b\omega_1^4 + c\omega_1^5) + (a\omega_1^4 + b\omega_1^5) \wedge \omega_1^1. \end{aligned} \tag{5.191}$$

On the other hand, differentiation of the right side of equation (5.190) yields

$$\begin{aligned} d\omega_4^7 &= da \wedge \omega_1^4 + ad\omega_1^4 + db \wedge \omega_1^5 + bd\omega_1^5 \\ &= da \wedge \omega_1^4 + a(\omega_1^1 \wedge \omega_1^4 - \rho\omega_1^5 \wedge \omega_7^3) \\ &+ db \wedge \omega_1^5 + b(\omega_1^1 \wedge \omega_1^5 - \rho\omega_1^4 \wedge \omega_7^3). \end{aligned} \tag{5.192}$$

Equating (5.191) and (5.192), we find

$$\begin{aligned} &(da + 2a\omega_1^1 - 2b\rho\omega_7^3 - \omega_2^7) \wedge \omega_1^4 \\ &\quad + (db + 2b\omega_1^1 + (a + u - c)\rho\omega_7^3) \wedge \omega_1^5 + \rho t\omega_1^4 \wedge \omega_1^5 = 0. \end{aligned} \tag{5.193}$$

Since $b = \rho_3$, it follows from (5.181) and (5.187) that

$$db + 2b\omega_1^1 = d\rho_3 + 2\rho_3\omega_1^1 = \rho_{31}\,\theta_1 + \rho_{32}\,\theta_2 + \rho_{33}\,\theta_3. \tag{5.194}$$

By examining the coefficient of $\omega_1^5 \wedge \omega_7^3 = \theta_2 \wedge \theta_3$ in equation (5.193) and using equation (5.194), we find

$$\rho_{33} = \rho(c - a - u). \tag{5.195}$$

Furthermore, the remaining terms in equation (5.193) are

$$\begin{aligned} &(da + 2a\omega_1^1 - \omega_2^7 - 2\rho b\omega_7^3 - (\rho t + \rho_{31})\omega_1^5) \wedge \omega_1^4 \\ &\quad + \text{terms involving } \omega_1^5 \text{ and } \omega_7^3 \text{ only.} \end{aligned} \tag{5.196}$$

Thus, the coefficient in parentheses is a multiple of ω_1^4, call it $\bar{a}\omega_1^4$. We can write this using (5.166) and (5.183) as

$$da + 2a\omega_1^1 = \omega_2^7 + \bar{a}\theta_1 + (\rho_{31} - \rho\rho_2)\theta_2 + 2\rho\rho_3\theta_3. \tag{5.197}$$

In a similar manner, if we differentiate the equation

$$\omega_5^7 = b\omega_1^4 + c\omega_1^5,$$

we obtain

$$dc + 2c\omega_1^1 = \omega_2^7 + (\rho_{32} + \rho\rho_1)\theta_1 + \bar{c}\theta_2 - 2\rho\rho_3\theta_3. \tag{5.198}$$

Thus, from the two equations in (5.171), we have obtained equations (5.195), (5.197), and (5.198). In a completely analogous manner, we can differentiate the two equations in (5.172) to obtain

$$\rho_{11} = \rho(s + r - p), \tag{5.199}$$

$$dp + 2p\omega_1^1 = -\omega_2^7 + 2\rho\rho_1\theta_1 + (-\rho_{13} - \rho\rho_2)\theta_2 + \bar{p}\theta_3, \tag{5.200}$$

$$dr + 2r\omega_1^1 = -\omega_2^7 - 2\rho\rho_1\theta_1 + \bar{r}\theta_2 + (-\rho_{12} + \rho\rho_3)\theta_3. \tag{5.201}$$

Similarly, differentiation of equation (5.173) yields

$$\rho_{22} + \rho_{33} = \rho(p - r - s), \tag{5.202}$$

$$ds + 2s\omega_1^1 = \bar{s}\theta_1 + (\rho_{31} + \rho\rho_2)\theta_2 + (-\rho_{21} + \rho\rho_3)\theta_3, \tag{5.203}$$

$$du + 2u\omega_1^1 = (-\rho_{23} - \rho\rho_1)\theta_1 + (\rho_{13} - \rho\rho_2)\theta_2 + \bar{u}\theta_3. \tag{5.204}$$

In these equations, the coefficients $\bar{a}$, $\bar{c}$, $\bar{p}$, $\bar{r}$, $\bar{s}$, $\bar{u}$ remain undetermined. However, by differentiating equation (5.180) and using the appropriate equations from above, one can show that

$$\begin{aligned} \bar{a} &= -6\rho\rho_1, \quad \bar{c} = 6\rho\rho_2, \\ \bar{p} &= -6\rho\rho_3, \quad \bar{r} = 6\rho\rho_2, \\ \bar{s} &= -12\rho\rho_1, \quad \bar{u} = 12\rho\rho_3. \end{aligned} \tag{5.205}$$

From equations (5.195), (5.199), (5.202), and (5.180), we can easily compute that

$$\rho_{11} + \rho_{22} + \rho_{33} = 0. \tag{5.206}$$

Using equation (5.205), equations (5.203) and (5.204) can be rewritten as

$$ds + 2s\omega_1^1 = -12\rho\rho_1\theta_1 + (\rho_{31} + \rho\rho_2)\theta_2 + (-\rho_{21} + \rho\rho_3)\theta_3, \tag{5.207}$$

$$du + 2u\omega_1^1 = (-\rho_{23} - \rho\rho_1)\theta_1 + (\rho_{13} - \rho\rho_2)\theta_2 + 12\rho\rho_3\theta_3. \tag{5.208}$$

Fundamental equations

By taking the exterior derivatives of these two equations and making use of equation (5.206) and of the commutation relations in equation (5.188) for ρ and its various derivatives, one can ultimately show after a lengthy calculation that the following fundamental equations hold:

$$\begin{aligned} \rho\rho_{12} + \rho_1\rho_2 + \rho^2\rho_3 &= 0, \\ \rho\rho_{21} + \rho_1\rho_2 - \rho^2\rho_3 &= 0, \\ \rho\rho_{23} + \rho_2\rho_3 + \rho^2\rho_1 &= 0, \end{aligned} \tag{5.209}$$

$$\rho\rho_{32} + \rho_2\rho_3 - \rho^2\rho_1 = 0,$$
$$\rho\rho_{31} + \rho_3\rho_1 + \rho^2\rho_2 = 0,$$
$$\rho\rho_{13} + \rho_3\rho_1 - \rho^2\rho_2 = 0.$$

We now briefly outline the details of this calculation. By equation (5.207), we have

$$s_1 = -12\rho\rho_1, \quad s_2 = \rho_{31} + \rho\rho_2, \quad s_3 = \rho\rho_3 - \rho_{21}. \tag{5.210}$$

The commutation relation (5.188) for s with $m = 2$ gives

$$s_{12} - s_{21} = -2s\rho_3 - \rho s_3 = -2s\rho_3 - \rho(\rho\rho_3 - \rho_{21}). \tag{5.211}$$

On the other hand, we can directly compute by taking covariant derivatives of equation (5.211) that

$$s_{12} - s_{21} = -12\rho\rho_{12} - 12\rho_2\rho_1 - (\rho_{311} + \rho_1\rho_2 + \rho\rho_{21}). \tag{5.212}$$

The main problem now is to get the covariant derivative ρ_{311} into a form involving ρ and its first and second covariant derivatives. By taking the covariant derivative of the third equation in (5.189), we find

$$\rho_{311} - \rho_{131} = -2\rho_1\rho_2 - 2\rho\rho_{21}. \tag{5.213}$$

Then using the commutation relation,

$$\rho_{131} = \rho_{113} - 2\rho_1\rho_2 - \rho\rho_{12},$$

we get from equation (5.213) that

$$\rho_{311} = \rho_{113} - 4\rho_1\rho_2 - \rho\rho_{12} - 2\rho\rho_{21}. \tag{5.214}$$

Taking the covariant derivative of the equation,

$$\rho_{11} = \rho(s + r - p),$$

and substituting the expression obtained for ρ_{113} into equation (5.214), we get

$$\rho_{311} = \rho_3(s + r - p) - 3\rho\rho_{21} - 2\rho\rho_{12} + 8\rho^2\rho_3 - 4\rho_1\rho_2. \tag{5.215}$$

If we substitute this expression for ρ_{311} into equation (5.212) and then equate the right sides of equations (5.211) and (5.212), we obtain the first equation in (5.209). The cyclic permutations are obtained in a similar way from $s_{23} - s_{32}$, and so on.

Our frame attached to the line $[Y_1, Y_7]$ is still not completely determined, namely, the following change is allowable:

$$Y_1^* = \tau Y_1, \quad Y_2^* = (1/\tau)Y_2 + \mu Y_7, \qquad (5.216)$$
$$Y_7^* = \tau Y_7, \quad Y_6^* = (1/\tau)Y_6 - \mu Y_1.$$

Under this change of frame, we have

$$\begin{aligned} \omega_1^{4*} &= \tau\omega_1^4, \quad \omega_1^{5*} = \tau\omega_1^5, \quad \omega_7^{3*} = \tau\omega_7^3, \\ \omega_4^{7*} &= (1/\tau)\omega_4^7 + \mu\omega_1^4, \\ \omega_3^{1*} &= (1/\tau)\omega_3^1 - \mu\omega_7^3. \end{aligned} \qquad (5.217)$$

Since Y_3, Y_4, Y_5 are completely determined, we have under this change,

$$\begin{aligned} \omega_1^{4*} &= \tau\omega_1^4, \quad \omega_1^{5*} = \tau\omega_1^5, \quad \omega_7^{3*} = \tau\omega_7^3, \\ \omega_4^{7*} &= (1/\tau)\omega_4^7 + \mu\omega_1^4, \\ \omega_3^{1*} &= (1/\tau)\omega_3^1 - \mu\omega_7^3, \end{aligned} \qquad (5.218)$$

which implies that

$$a^* = \tau^{-2}a + \tau^{-1}\mu, \quad p^* = \tau^{-2}p - \tau^{-1}\mu.$$

Thus, by taking $\mu = (p-a)/2\tau$, we can arrange that $a^* = p^*$. We now make this change of frame and drop the asterisks. In this new frame, we have

$$a = p = -\rho^2, \quad r = 3\rho^2 + s, \quad c = 3\rho^2 - \mu. \qquad (5.219)$$

Using the fact that $a = p$, we can subtract equation (5.200) from equation (5.197) and get that

$$\omega_2^7 = 4\rho\rho_1\theta_1 - ((\rho_{31} + \rho_{13})/2)\theta_2 - 4\rho\rho_3\theta_3. \qquad (5.220)$$

Now through equations (5.196)–(5.201), the covariant derivatives of the functions a, c, p, and r are expressed in terms of ρ and its derivatives. We are now ready to proceed to the main results. Ultimately, we show that it is possible to choose a frame in which ρ is constant. Thus, the classification naturally splits into two cases, $\rho = 0$ and $\rho \neq 0$. We handle the two cases separately.

Case 1: $\rho \neq 0$ (the irreducible case)

Assume that the function ρ is never zero on the open set U on which the frame $\{Y_a\}$ is defined. The key step in getting ρ to be constant is the following lemma due to Pinkall [445, p. 108], where his function c is the negative of our function ρ. The formulation of the proof here using the method of moving frames was given in Cecil-Chern [80, p. 33]. The crucial point here is that since $\rho \neq 0$, the fundamental equations (5.209) allow us to express all of the second covariant derivatives ρ_{ij} in terms of ρ and its first derivatives.

Lemma 5.28. *Suppose that the function ρ never vanishes on the open set U on which the frame $\{Y_a\}$ is defined. Then its covariant derivatives satisfy $\rho_1 = \rho_2 = \rho_3 = 0$ at every point of U.*

Proof. First, note that if ρ_3 vanishes identically, then the equations (5.209) and the assumption that $\rho \neq 0$ imply that ρ_1 and ρ_2 also vanish identically. We now complete the proof by showing that ρ_3 vanishes everywhere on U. This is accomplished by considering the expression $s_{12} - s_{21}$. By the commutation relations (5.209), we have

$$s_{12} - s_{21} = -2s\rho_3 - \rho s_3.$$

By equations (5.209)–(5.210), we have

$$\rho s_3 = \rho^2\rho_3 - \rho\rho_{21} = \rho_1\rho_2,$$

and so

$$s_{12} - s_{21} = -2s\rho_3 - \rho_1\rho_2. \tag{5.221}$$

On the other hand, we can compute s_{12} directly by differentiating the equation

$$s_1 = -12\rho\rho_1.$$

Then using the expression for ρ_{12} obtained from equation (5.209), we get

$$\begin{aligned} s_{12} &= -12\rho_2\rho_1 - 12\rho\rho_{12} = -12(\rho_2\rho_1 + \rho\rho_{12}) \\ &= -12(\rho_2\rho_1 + (-\rho_2\rho_1 - \rho^2\rho_3)) = 12\rho^2\rho_3. \end{aligned} \tag{5.222}$$

Next we have from equation (5.210) that $s_2 = \rho_{31} + \rho\rho_2$. Using equation (5.209), we can write

$$\rho_{31} = -\rho_3\rho_1\rho^{-1} - \rho\rho_2,$$

and thus,

$$s_2 = -\rho_3\rho_1/\rho. \tag{5.223}$$

Then we compute

$$s_{21} = -(\rho(\rho_3\rho_{11} + \rho_{31}\rho_1) - \rho_3\rho_1^2)/\rho^2.$$

Using equation (5.199) for ρ_{11} and (5.209) to get ρ_{31}, this becomes

$$s_{21} = -\rho_3(s + r - p) + 2\rho_3\rho_1^2\rho^{-2} + \rho_1\rho_2. \tag{5.224}$$

Now equate the expression in equation (5.221) for $s_{12} - s_{21}$ with the expression obtained by subtracting equation (5.224) from equation (5.222) to get

$$-2s\rho_3 - \rho_1\rho_2 = 12\rho^2\rho_3 + \rho_3(s + r - p) - 2\rho_3\rho_1^2\rho^{-2} - \rho_1\rho_2.$$

This can be rewritten as

$$0 = \rho_3(12\rho^2 + 3s + r - p - 2\rho_1^2\rho^{-2}). \tag{5.225}$$

Using the expressions in (5.219) for r and p, we see that

$$3s + r - p = 4s + 4\rho^2,$$

and so equation (5.225) can be rewritten as

$$0 = \rho_3(16\rho^2 + 4s - 2\rho_1^2\rho^{-2}). \tag{5.226}$$

Suppose that $\rho_3 \neq 0$ at some point x of U. Then ρ_3 does not vanish on some neighborhood V of x. By equation (5.226), we have

$$16\rho^2 + 4s - 2\rho_1^2\rho^{-2} = 0 \tag{5.227}$$

on V. We now take the θ_2-covariant derivative of equation (5.227) and obtain

$$32\rho\rho_2 + 4s_2 - 4\rho_1\rho_{12}\rho^{-2} + 4\rho_1^2\rho_2\rho^{-3} = 0. \tag{5.228}$$

We now substitute the expression (5.223) for s_2 and the formula

$$\rho_{12} = -\rho_1\rho_2\rho^{-1} - \rho\rho_3$$

obtained from equation (5.209) into equation (5.228). After some algebra, equation (5.228) reduces to

$$\rho_2(32\rho^4 + 8\rho_1^2) = 0.$$

Since $\rho \neq 0$, this implies that $\rho_2 = 0$ on V. But then, the left side of the equation below, obtained from (5.209),

$$\rho\rho_{21} + \rho_1\rho_2 = \rho^2\rho_3,$$

vanishes on V. Since $\rho \neq 0$, we conclude that $\rho_3 = 0$ on V, a contradiction to our assumption. Hence, ρ_3 vanishes identically on the set U, and the lemma is proved. □

We now continue with the case $\rho \neq 0$. According to Lemma 5.28, all the covariant derivatives of ρ are zero, and our formulas simplify greatly. Equations (5.195) and (5.199) give

$$c - a - u = 0, \quad s + r - p = 0.$$

These combined with equation (5.219) give

$$c = r = \rho^2, \quad u = -s = 2\rho^2. \tag{5.229}$$

By equation (5.220), we have $\omega_2^7 = 0$. So the differentials of the frame vectors can now be written as

$$\begin{aligned}
dY_1 - \omega_1^1 Y_1 &= \omega_1^4 Y_4 + \omega_1^5 Y_5, \\
dY_7 - \omega_1^1 Y_7 &= \omega_7^3 Y_3 - \omega_1^5 Y_5, \\
dY_2 + \omega_1^1 Y_2 &= \rho^2(\omega_7^3 Y_3 + 2\omega_1^4 Y_4 + \omega_1^5 Y_5), \\
dY_6 + \omega_1^1 Y_6 &= \rho^2(2\omega_7^3 Y_3 + \omega_1^4 Y_4 - \omega_1^5 Y_5), \\
dY_3 &= \omega_7^3 Z_3 + \rho(\omega_1^5 Y_4 + \omega_1^4 Y_5), \\
dY_4 &= -\omega_1^4 Z_4 + \rho(-\omega_1^5 Y_3 + \omega_7^3 Y_5), \\
dY_5 &= \omega_1^5 Z_5 + \rho(-\omega_1^4 Y_3 - \omega_7^3 Y_4),
\end{aligned} \tag{5.230}$$

where

$$\begin{aligned}
Z_3 &= -Y_6 + \rho^2(-Y_1 - 2Y_7), \\
Z_4 &= Y_2 + \rho^2(2Y_1 + Y_7), \\
Z_5 &= -Y_2 + Y_6 + \rho^2(-Y_1 + Y_7).
\end{aligned} \tag{5.231}$$

From this we notice that

$$Z_3 + Z_4 + Z_5 = 0, \tag{5.232}$$

so that the points Z_3, Z_4, and Z_5 lie on a line in projective space $\mathbf{RP}^6$. From equations (5.182), (5.185) and the lemma above, we see that

$$d\rho + \rho\omega_1^1 = 0, \quad d\omega_1^1 = 0. \tag{5.233}$$

We now make a change of frame of the form

$$\begin{aligned} Y_1^* &= \rho Y_1, \quad Y_2^* = (1/\rho) Y_2, \\ Y_7^* &= \rho Y_7, \quad Y_6^* = (1/\rho) Y_6, \\ Y_i^* &= Y_i, \quad 3 \le i \le 5. \end{aligned} \tag{5.234}$$

Then set

$$\begin{aligned} Z_i^* &= (1/\rho) Z_i, \quad 3 \le i \le 5, \\ \omega_1^{4*} &= \rho \omega_1^4, \quad \omega_1^{5*} = \rho \omega_1^5, \quad \omega_7^{3*} = \rho \omega_7^3. \end{aligned} \tag{5.235}$$

The effect of this change is to make $\rho^* = 1$ and $\omega_1^{1*} = 0$, for we can compute the following differentials of the frame vectors:

$$\begin{aligned} dY_1^* &= \omega_1^{4*} Y_4 + \omega_1^{5*} Y_5, \\ dY_7^* &= \omega_7^{3*} Y_3 - \omega_1^{5*} Y_5, \\ dY_2^* &= \omega_7^{3*} Y_3 + 2\omega_1^{4*} Y_4 + \omega_1^{5*} Y_5, \\ dY_6^* &= 2\omega_7^{3*} Y_3 + \omega_1^{4*} Y_4 - \omega_1^{5*} Y_5, \\ dY_3 &= \omega_7^{3*} Z_3^* + \omega_1^{5*} Y_4 + \omega_1^{4*} Y_5, \\ dY_4 &= -\omega_1^{4*} Z_4^* - \omega_1^{5*} Y_3 + \omega_7^{3*} Y_5, \\ dY_5 &= \omega_1^{5*} Z_5^* - \omega_1^{4*} Y_3 - \omega_7^{3*} Y_4, \end{aligned} \tag{5.236}$$

with

$$\begin{aligned} dZ_3^* &= 2(-2\omega_7^{3*} Y_3 - \omega_1^{4*} Y_4 + \omega_1^{5*} Y_5), \\ dZ_4^* &= 2(\omega_7^{3*} Y_3 + 2\omega_1^{4*} Y_4 + \omega_1^{5*} Y_5), \\ dZ_5^* &= 2(\omega_7^{3*} Y_3 - \omega_1^{4*} Y_4 - 2\omega_1^{5*} Y_5), \end{aligned} \tag{5.237}$$

and

$$\begin{aligned} d\omega_1^{4*} &= -\omega_1^{5*} \wedge \omega_7^{3*}, \quad \text{i.e.,} \quad d\theta_1^* = -\theta_2^* \wedge \theta_3^*, \\ d\omega_1^{5*} &= -\omega_7^{3*} \wedge \omega_1^{4*}, \quad \text{i.e.,} \quad d\theta_2^* = -\theta_3^* \wedge \theta_1^*, \\ d\omega_7^{3*} &= -\omega_1^{4*} \wedge \omega_1^{5*}, \quad \text{i.e.,} \quad d\theta_3^* = -\theta_1^* \wedge \theta_2^*. \end{aligned} \tag{5.238}$$

Comparing the last equation with (5.175), we see that $\omega_1^{1*} = 0$ and $\rho^* = 1$. This is the final frame needed in the case $\rho \neq 0$, so we drop the asterisks once more.

Classification in the irreducible case

We can now prove Pinkall's [442, 445] classification for the case $\rho \neq 0$. As with the cyclides of Dupin, there is only one model up to Lie equivalence. This model is Cartan's isoparametric hypersurface with three principal curvatures in S^4. Cartan's hypersurface is a tube over each of its two focal submanifolds in S^4, both of which are Veronese surfaces. (See Subsection 3.8.3 on page 151 for more detail.)

Theorem 5.29. (a) *Every connected Dupin proper submanifold*

$$\lambda : M^3 \to \Lambda^7$$

with three distinct curvature spheres and $\rho \neq 0$ is contained in a unique compact, connected proper Dupin submanifold with $\rho \neq 0$.

(b) *Any two proper Dupin submanifolds with $\rho \neq 0$ are locally Lie equivalent, each being Lie equivalent to an open subset of Cartan's isoparametric hypersurface in S^4.*

Proof. Let $\{Y_a\}$ be the Lie frame just constructed on a connected open subset $U \subset M^3$ satisfying,

$$\omega_1^1 = 0, \quad \rho = 1. \tag{5.239}$$

Then the derivatives of the frame vectors satisfy the system of equations (5.236), where we again drop the asterisks. The three curvature sphere maps on U are Y_1, Y_7, and $Y_1 + Y_7$. Let

$$W_1 = -Y_1 + Y_6 - 2Y_7, \quad W_2 = -2Y_1 + Y_2 - Y_7. \tag{5.240}$$

Then from equation (5.236), we find that

$$dW_1 = dW_2 = 0.$$

Hence W_1 and W_2 are constant maps. Furthermore, since

$$\langle W_1, W_1 \rangle = \langle W_2, W_2 \rangle = -4, \quad \langle W_1, W_2 \rangle = -2,$$

the line $[W_1, W_2]$ is timelike. Finally, the equations,

$$\langle Y_1, W_1 \rangle = 0, \quad \langle Y_7, W_2 \rangle = 0, \quad \langle Y_1 + Y_7, W_1 - W_2 \rangle = 0, \tag{5.241}$$

imply that the restriction of λ to U is Lie equivalent to an open subset of an isoparametric hypersurface in S^4 by Theorem 4.16 (page 221), since the three curvature sphere maps are orthogonal to three points on a timelike line in $\mathbf{RP}^6$.

If $\{\tilde{Y}_a\}$ is a Lie frame defined on an open subset $\tilde{U} \subset M^3$ by the same construction as $\{Y_a\}$, and $U \cap \tilde{U}$ is nonempty, then the uniqueness of the construction implies that at points of $U \cap \tilde{U}$ the curvature spheres satisfy

$$\tilde{Y}_1 = Y_1, \quad \tilde{Y}_7 = Y_7, \quad \tilde{Y}_1 + \tilde{Y}_7 = Y_1 + Y_7,$$

and the points $\tilde{W}_1 = W_1$, $\tilde{W}_2 = W_2$. Thus, the timelike line $[W_1, W_2]$ and the points W_1 and W_2 on it satisfying equation (5.241) are the same on the set $\tilde{U}$ as they are on U, and hence they are the same on all of the connected manifold M^3. Therefore, the whole Dupin submanifold $\lambda : M^3 \to \Lambda^7$ is Lie equivalent to an open subset of an isoparametric hypersurface in S^4. Since any connected open subset of an isoparametric hypersurface is contained in a unique compact, connected isoparametric hypersurface by Theorem 3.44 (page 131), part (a) is proved. Furthermore, because all isoparametric hypersurfaces in S^4 are locally Lie equivalent by a result of Cartan [54], part (b) is also true. □

Remark 5.30. The proof of Theorem 5.29 above relies on Cartan's classification of isoparametric hypersurfaces with three principal curvatures in S^4 for its completion. However, without needing to invoke Cartan's classification, Cecil and Chern [80] (see also [77, pp. 182–186]) proved directly that a connected Dupin proper submanifold $\lambda : U^3 \to \Lambda^7$ with three distinct curvature spheres and $\rho \neq 0$ is Lie equivalent to an open subset of the Legendre lift of a Veronese surface V^2, considered as a submanifold of codimension two in S^4. Thus, λ is also Lie equivalent to the Legendre lift of an open subset of a tube over the Veronese surface in S^4, i.e., to an open subset of Cartan's isoparametric hypersurface in S^4.

Case 2: $\rho = 0$ (the reducible case)

We now consider the case where ρ is identically zero. It turns out that all such Dupin submanifolds are reducible to cyclides of Dupin in $\mathbf{R}^3$. We return to the frame that we used prior to the assumption that $\rho \neq 0$. Thus, only those relations through equation (5.220) are valid.

If ρ is identically zero, then by equation (5.182) all of its covariant derivatives are also equal to zero. From equations (5.183) and (5.219), we see that the functions defined in equations (5.171)–(5.173) satisfy the equations

$$q = t = b = 0, \quad a = p = 0, \quad r = s, \quad c = -u.$$

Thus, from equation (5.220) we have $\omega_2^7 = 0$. From these and the other relations among the Maurer–Cartan forms which we have derived, we see that the differentials of the frame vectors can be written as

$$dY_1 - \omega_1^1 Y_1 = \omega_1^4 Y_4 + \omega_1^5 Y_5,$$

$$\begin{aligned}
dY_7 - \omega_1^1 Y_7 &= \omega_7^3 Y_3 - \omega_1^5 Y_5, \\
dY_2 + \omega_1^1 Y_2 &= s(-\omega_1^4 Y_4 + \omega_1^5 Y_5), \\
dY_6 + \omega_1^1 Y_6 &= u(\omega_7^3 Y_3 + \omega_1^5 Y_5), \\
dY_3 &= \omega_7^3(-Y_6 + uY_7), \\
dY_4 &= \omega_1^4(sY_1 - Y_2), \\
dY_5 &= \omega_1^5(-sY_1 - Y_2 + Y_6 - uY_7).
\end{aligned} \tag{5.242}$$

Note also that from equations (5.203) and (5.204), we have

$$ds + 2s\omega_1^1 = 0, \quad du + 2u\omega_1^1 = 0, \tag{5.243}$$

and from (5.175) that

$$d\theta_i = \omega_1^1 \wedge \theta_i, \quad i = 1, 2, 3. \tag{5.244}$$

From equation (5.185), we have $d\omega_1^1 = 0$. Hence on any local disk neighborhood U in M, we have

$$\omega_1^1 = d\sigma, \tag{5.245}$$

for some smooth scalar function σ on U. We next consider a change of frame of the form,

$$\begin{aligned}
Y_1^* &= e^{-\sigma} Y_1, \quad Y_7^* = e^{-\sigma} Y_7, \\
Y_2^* &= e^{\sigma} Y_2, \quad Y_6^* = e^{\sigma} Y_6, \\
Y_i^* &= Y_i, \quad 3 \leq i \leq 5.
\end{aligned} \tag{5.246}$$

The effect of this change is to make $\omega_1^{1*} = 0$ while keeping $\rho^* = 0$. If we set

$$\omega_1^{4*} = e^{-\sigma} \omega_1^4, \quad \omega_1^{5*} = e^{-\sigma} \omega_1^5, \quad \omega_7^{3*} = e^{-\sigma} \omega_7^3,$$

then we can compute from equation (5.242) that

$$\begin{aligned}
dY_1^* &= \omega_1^{4*} Y_4 + \omega_1^{5*} Y_5, \\
dY_7^* &= \omega_7^{3*} Y_3 - \omega_1^{5*} Y_5, \\
dY_2^* &= s^*(-\omega_1^{4*} Y_4 + \omega_1^{5*} Y_5) \\
dY_6^* &= u^*(\omega_7^{3*} Y_3 + \omega_1^{5*} Y_5), \\
dY_3 &= \omega_7^{3*} Z_3^*,
\end{aligned}$$

$$\begin{aligned} dY_4 &= \omega_1^{4*} Z_4^*, \\ dY_5 &= \omega_1^{5*} Z_5^*, \end{aligned} \tag{5.247}$$

where

$$\begin{aligned} Z_3^* &= -Y_6^* - u^* Y_7^*, \\ Z_4^* &= s^* Y_1^* - Y_2^*, \\ Z_5^* &= -s^* Y_1^* - Y_2^* + Y_6^* - u^* Y_7^*, \end{aligned} \tag{5.248}$$

and

$$s^* = se^{2\sigma}, \quad u^* = ue^{2\sigma}. \tag{5.249}$$

Using equations (5.243) and (5.249), we can then compute that

$$ds^* = 0, \quad du^* = 0, \tag{5.250}$$

i.e., s^* and u^* are constant functions on the local neighborhood U.

The frame in equation (5.246) is our final frame, and we drop the asterisks in further references to equations (5.246)–(5.250). Since the functions s and u are now constant, we can compute from equation (5.248) that

$$\begin{aligned} dZ_3 &= -2u\omega_7^3 Y_3, \\ dZ_4 &= 2s\omega_1^4 Y_4, \\ dZ_5 &= 2(u - s)\omega_1^5 Y_5. \end{aligned} \tag{5.251}$$

From this we see that the following 4-dimensional subspaces of $\mathbf{RP}^6$,

$$\begin{aligned} &\text{Span}\,\{Y_1, Y_4, Y_5, Z_4, Z_5\}, \\ &\text{Span}\,\{Y_7, Y_3, Y_5, Z_3, Z_5\}, \\ &\text{Span}\,\{Y_1 + Y_7, Y_3, Y_4, Z_3, Z_4\}, \end{aligned} \tag{5.252}$$

are invariant under exterior differentiation, and hence they are constant. Thus, each of the three curvature sphere maps, Y_1, Y_7, and $Y_1 + Y_7$ is contained in a 4-dimensional subspace of $\mathbf{RP}^6$. One can easily show that each of the subspaces in equation (5.252) has signature $(4, 1)$. Thus by Theorem 5.14, our Dupin submanifold λ on U is Lie equivalent to an open subset of a tube over a cyclide of Dupin in $\mathbf{R}^3$ in three different ways. Hence, we have the following result due to Pinkall [442, 445].

Theorem 5.31. *A connected Dupin submanifold $\lambda : M^3 \to \Lambda^7$ with $\rho = 0$ is reducible. It is locally Lie equivalent to a tube over a cyclide of Dupin in $\mathbf{R}^3 \subset \mathbf{R}^4$.*

Pinkall [445, p. 111] proceeds to classify Dupin submanifolds with $\rho = 0$ up to Lie equivalence. We will not do that here. The reader can follow Pinkall's proof using the fact that his constants α and β are our constants s and $-u$, respectively.

Classification in the irreducible case for higher dimensions

We now turn to some generalizations of this approach to higher-dimensional Dupin submanifolds. After Pinkall's result, Niebergall [393] next proved that every connected proper Dupin hypersurface in $\mathbf{R}^5$ with three principal curvatures is reducible. Next Cecil and Jensen [85] proved the following theorem.

Theorem 5.32. *If M^{n-1} is a connected irreducible proper Dupin hypersurface in S^n with three distinct principal curvatures of multiplicities m_1, m_2, m_3, then $m_1 = m_2 = m_3 = m$, and M^{n-1} is Lie equivalent to an isoparametric hypersurface in S^n.*

It then follows from Cartan's classification of isoparametric hypersurfaces with $g = 3$ (see Subsection 3.8.3, page 151) that $m = 1, 2, 4$ or 8. Note that in the original paper [85], Theorem 5.32 was proven under the assumption that M^{n-1} is locally irreducible, i.e., that M^{n-1} does not contain any reducible open subset. However, as noted in Proposition 5.18 on page 255, local irreducibility has now been shown to be equivalent to irreducibility.

We will briefly describe the approach of the paper of Cecil and Jensen [85]. The reader is referred to the paper itself for the details. Let $\lambda : M^{n-1} \to \Lambda^{2n-1}$ be a connected proper Dupin submanifold with three curvature spheres at each point. As in the proof of Pinkall's theorem above, Cecil and Jensen construct a Lie frame $\{Y_a\}$ on a connected open subset U of M^{n-1} so that the three curvature spheres are Y_1, Y_{n+3} and $Y_1 + Y_{n+3}$ with respective multiplicities m_1, m_2, and m_3. Corresponding to the one function ρ in the case above, there are $m_1 m_2 m_3$ functions F^α_{pa}, where

$$\begin{aligned} 1 &\leq a \leq m_1, \\ m_1 + 1 &\leq p \leq m_1 + m_2, \\ m_1 + m_2 + 1 &\leq \alpha \leq m_1 + m_2 + m_3 = n - 1. \end{aligned} \tag{5.253}$$

Corresponding to the case $\rho = 0$ above, Cecil and Jensen show that if there exists a fixed index, say a, such that

$$F^\alpha_{pa} = 0, \quad \text{for all } p, \alpha, \tag{5.254}$$

then the restriction of λ to the open set U is reducible. Thus, by Proposition 5.18, λ is reducible on all of M^{n-1}. Next they show that if the multiplicities are not all equal, then there exists some index a such that equation (5.254) holds, and thus λ is reducible.

Finally, Cecil and Jensen consider the case where all the multiplicities have the same value m. As in the proof of Pinkall's theorem above, they show that if λ is irreducible, then the three curvature sphere maps Y_1, Y_{n+3} and $Y_1 + Y_{n+3}$ are orthogonal to three points on a timelike line in $\mathbf{RP}^{n+2}$, and thus λ is Lie equivalent to the Legendre lift of an isoparametric hypersurface in S^n by Theorem 4.16 on page 221. The classification of such Dupin hypersurfaces then follows from Cartan's [54] classification of isoparametric hypersurfaces with three principal curvatures.

Remark 5.33. The indexing in the paper of Cecil and Jensen [85] is slightly different than that used here, although the approach is very similar. In that paper, the Lie frame is taken so that $\lambda = [Y_0, Y_1]$ rather than $\lambda = [Y_1, Y_{n+3}]$, and the three curvature spheres are Y_0, Y_1 and $Y_0 + Y_1$, rather than Y_1, Y_{n+3} and $Y_1 + Y_{n+3}$. This causes a slight change in all of the indices, and it is simpler to begin reading the paper [85] itself from the beginning, rather than attempting to find a way to transform the indices from our treatment here to the notation in that paper.

An open problem is the classification of reducible Dupin hypersurfaces of arbitrary dimension with three principal curvatures up to Lie equivalence. As noted above, Pinkall [445] found such a classification in the case of $M^3 \subset \mathbf{R}^4$. It may be possible to generalize Pinkall's result to higher dimensions using the approach of [85].

Remark 5.34 (Examples with principal coordinate systems and $g = 3$). As noted in Section 5.2, Otsuki [420, p. 17] gave examples of minimal hypersurfaces in the sphere with three distinct *non-simple* (having constant multiplicity greater than one) principal curvatures. Since all of the multiplicities are greater than one, these hypersurfaces are proper Dupin by Theorem 2.10. Otsuki's examples have the property that each orthogonal complement $T_i^{\perp}$ of a principal foliation is integrable, and thus by Theorem 5.13, each point of the hypersurface M has a principal coordinate neighborhood. Otsuki showed that his examples are not isoparametric and cannot be complete. In fact, Miyaoka [362] showed that if M is a complete hypersurface with constant mean curvature and three non-simple principal curvatures in a real space form $\tilde{M}^{n+1}(c)$ with constant sectional curvature $c \geq 0$, then $c > 0$ and M is isoparametric.

5.7 Local Classifications in the Case g = 4

In this section, we discuss local classification results for proper Dupin submanifolds with $g = 4$ curvature spheres. In these results, a key invariant is the Lie curvature Ψ (see Section 4.5, page 218), which is the cross-ratio of the four curvature spheres (or principal curvatures).

Let $\lambda : M^{n-1} \to \Lambda^{2n-1}$ be a connected proper Dupin submanifold with $g = 4$ curvature spheres at each point. Then in a manner similar to the construction above in the case $g = 3$, one can construct a Lie frame $\{Y_1, \ldots, Y_{n+3}\}$ in $\mathbf{R}_2^{n+3}$ (see Cecil and Jensen [86]) in which the four curvature spheres are Y_1, Y_{n+3}, $Y_1 + Y_{n+3}$ and $Y_1 + \Psi Y_{n+3}$, where Ψ is the Lie curvature of λ.

Remark 5.35. In the paper [86], the indexing is different, and the curvature spheres are Y_0, Y_1, $Y_0 + Y_1$ and $Y_0 + rY_1$, where r denotes the Lie curvature. This is similar to the situation described in Remark 5.33 above.

Denote the multiplicities of these curvature spheres by m_1, m_2, m_3, and m_4, respectively. Corresponding to the one function ρ in the case $g = 3$ above, there are four sets of functions that are crucial in the proof for $g = 4$,

$$F^{\alpha}_{pa}, F^{\mu}_{pa}, F^{\mu}_{\alpha a}, F^{\mu}_{\alpha p}, \tag{5.255}$$

where

$$\begin{aligned} 1 \leq a &\leq m_1, \\ m_1 + 1 \leq p &\leq m_1 + m_2, \\ m_1 + m_2 + 1 \leq \alpha &\leq m_1 + m_2 + m_3. \\ m_1 + m_2 + m_3 + 1 \leq \mu &\leq m_1 + m_2 + m_3 + m_4 = n - 1. \end{aligned} \tag{5.256}$$

As noted after Theorem 3.63 on page 143, Thorbergsson [533] showed that for a compact proper Dupin hypersurface in S^n with four principal curvatures, the multiplicities of the principal curvatures satisfy $m_1 = m_3$, $m_2 = m_4$, when the principal curvatures are appropriately ordered (see also Stolz [502] for more restrictions on the multiplicities). Thus, in the papers of Cecil and Jensen [86] and Cecil, Chi, and Jensen [82], such an assumption is placed on the multiplicities. It is also assumed in [82] that the Lie curvature $\Psi = 1/2$, since that is true for an isoparametric hypersurface with four principal curvatures (when the principal curvatures are listed in ascending order as in equation (4.94) on page 222).

In [86, pp. 3–4], Cecil and Jensen conjectured that an irreducible connected proper Dupin hypersurface in S^n with four principal curvatures having multiplicities satisfying $m_1 = m_3, m_2 = m_4$, and constant Lie curvature Ψ is Lie equivalent to an open subset of an isoparametric hypersurface in S^n.

In that paper [86], the conjecture was verified in the case where all the multiplicities are equal to one (see also Niebergall [394], who obtained the same conclusion under additional assumptions). In the paper of Cecil, Chi, and Jensen [82] mentioned above, the conjecture was proven to be true if $m_1 = m_3 \geq 1$, and $m_2 = m_4 = 1$, and the Lie curvature is assumed to satisfy $\Psi = 1/2$ by proving Theorem 5.36 below. The conjecture in its full generality is still an open problem, since the conjecture does not assume that $m_2 = m_4 = 1$ nor that the constant value of Ψ is $1/2$.

Theorem 5.36. *Let M be an irreducible connected proper Dupin hypersurface in S^n with four principal curvatures having multiplicities $m_1 = m_3$, $m_2 = m_4 = 1$, and constant Lie curvature $\Psi = 1/2$. Then M is Lie equivalent to an open subset of an isoparametric hypersurface.*

An important step in proving this theorem is that under the assumptions on the multiplicities and the Lie curvature given in the theorem, the corresponding Dupin submanifold λ is reducible if there exists some fixed index, say a, such that

$$F^{\alpha}_{pa} = F^{\mu}_{pa} = F^{\mu}_{\alpha a} = 0, \quad \text{for all } p, \alpha, \mu. \tag{5.257}$$

Thus, if λ is irreducible, no such index exists. In that case, it was shown after a lengthy argument that λ is Lie equivalent to the Legendre lift of an isoparametric hypersurface in S^n by invoking Theorem 4.16 (page 221).

The following example [73] (see also [77, pp. 80–82]) shows that the hypothesis of irreducibility is necessary in Theorem 5.36. This example is a noncompact proper Dupin submanifold with $g = 4$ distinct principal curvatures and constant Lie curvature $\Psi = 1/2$, which is not Lie equivalent to an isoparametric hypersurface with four principal curvatures in S^n. Furthermore, one can arrange that all of the principal curvatures have the same multiplicity by an appropriate choice of m in the example. This example is reducible, and it cannot be made compact while preserving the property that the number g of distinct curvatures spheres equals four at each point.

A reducible example with constant Lie curvature $\Psi = 1/2$

Example 5.37 (A reducible example with Lie curvature $\Psi = 1/2$). Let $V \subset S^{n-m}$ be an embedded Dupin hypersurface in S^{n-m} with field of unit normals ξ, such that V has three distinct principal curvatures,

$$\mu_1 < \mu_2 < \mu_3,$$

at each point. Embed S^{n-m} as a totally geodesic submanifold of S^n, and let B^{n-1} be the unit normal bundle of the submanifold $V \subset S^n$. Let

$$\lambda : B^{n-1} \to \Lambda^{2n-1}$$

be the Legendre lift of the submanifold V in S^n. Any unit normal η to V at a point $x \in V$ can be written in the form

$$\eta = \cos\theta\ \xi(x) + \sin\theta\ \zeta,$$

where ζ is a unit normal to S^{n-m} in S^n. Since the shape operator $A_\zeta = 0$, we have

$$A_\eta = \cos\theta\ A_\xi.$$

Thus the principal curvatures of A_η are

$$\kappa_i = \cos\theta\ \mu_i, \quad 1 \le i \le 3. \tag{5.258}$$

If $\langle \eta, \xi \rangle = \cos\theta \neq 0$, then A_η has three distinct principal curvatures. However, if $\langle \eta, \xi \rangle = 0$, then $A_\eta = 0$.

Let U be the open subset of B^{n-1} on which $\cos\theta > 0$, and let α denote the restriction of λ to U. By Theorem 4.13 on page 219, the Legendre submanifold α has four distinct curvature spheres at each point of U. The fourth principal curvature $\kappa_4 = \infty$ has multiplicity m, as in Theorem 4.13.

Since V is proper Dupin in S^{n-m}, it is easy to show that α is proper Dupin (this is similar to the calculations in the tube construction in Section 5.1, see also Section 5.2 of [77]). Furthermore, since $\kappa_4 = \infty$, the Lie curvature Ψ of α at a point (x, η) of U equals the Möbius curvature $\Phi(\kappa_1, \kappa_2, \kappa_3)$. Using equation (5.258), we compute

$$\Psi = \Phi(\kappa_1, \kappa_2, \kappa_3) = \frac{\kappa_1 - \kappa_2}{\kappa_1 - \kappa_3} = \frac{\mu_1 - \mu_2}{\mu_1 - \mu_3} = \Phi(\mu_1, \mu_2, \mu_3). \tag{5.259}$$

Now suppose that V is a minimal isoparametric hypersurface in S^{n-m} with three distinct principal curvatures of multiplicity m for $m = 1, 2, 4$ or 8 (see Subsection 3.8.3, page 151). By Münzner's formula (see Theorem 3.26 on page 108), these principal curvatures have the values,

$$\mu_1 = -\sqrt{3}, \quad \mu_2 = 0, \quad \mu_3 = \sqrt{3}.$$

On the open subset U of B^{n-1} described above, the Lie curvature of α has the constant value $1/2$ by equation (5.259). To construct a reducible proper Dupin hypersurface in S^n with four principal curvatures of multiplicity m and constant Lie curvature $\Psi = 1/2$ in S^n, we simply take the open subset $f_t(U)$ of the tube of radius t around V in S^n.

To see that this example is not Lie equivalent to the Legendre lift of an isoparametric hypersurface in S^n with four distinct principal curvatures, note that the point sphere map $[Y_1]$ of α is a curvature sphere of multiplicity m which lies in the linear subspace of codimension $m + 1$ in $\mathbf{RP}^{n+2}$ orthogonal to the space spanned by e_{n+3} and by those vectors ζ normal to S^{n-m} in S^n. Thus, α is reducible by Theorem 5.15 on page 253.

On the other hand, the Legendre lift of a compact isoparametric hypersurface with four distinct principal curvatures is irreducible by Theorem 5.19 on page 255, and it cannot contain a reducible open subset by Proposition 5.18 (page 255). Since reducibility is a Lie invariant property, α cannot be Lie equivalent to the Legendre lift of an isoparametric hypersurface with four principal curvatures.

The fact that the number of distinct principal curvatures of A_η is not constant as η varies over the unit normal bundle B^{n-1} illustrates why α cannot be extended to a compact proper Dupin submanifold with $g = 4$.

With regard to Theorem 4.16, α comes as close as possible to satisfying the requirements for being Lie equivalent to an isoparametric hypersurface without actually fulfilling them. The principal curvatures $\kappa_2 = 0$ and $\kappa_4 = \infty$ are constant on U. If a third principal curvature were also constant, then the constancy of Ψ would imply that all four principal curvatures were constant, and α would be the Legendre lift of an isoparametric hypersurface.

Using this same method, it is easy to construct noncompact proper Dupin hypersurfaces in S^n with $g = 4$ and $\Psi = c$, for any constant $0 < c < 1$. If $V \subset S^{n-m}$ is an isoparametric hypersurface with three distinct principal curvatures, then Münzner's formula in Theorem 3.26 on page 108 implies that these principal curvatures have the values,

$$\mu_1 = \cot(\theta + \frac{2\pi}{3}), \quad \mu_2 = \cot(\theta + \frac{\pi}{3}), \quad \mu_3 = \cot\theta, \quad 0 < \theta < \frac{\pi}{3}. \qquad (5.260)$$

Furthermore, any value of θ in $(0, \pi/3)$ can be realized by some hypersurface in a parallel family of isoparametric hypersurfaces.

As above, consider $V \subset S^{n-m} \subset S^n$. A direct calculation using equations (5.259) and (5.260) shows that the Lie curvature Ψ of the Legendre submanifold α constructed as above from V satisfies

$$\Psi = \Phi(\kappa_1, \kappa_2, \kappa_3) = \frac{\kappa_1 - \kappa_2}{\kappa_1 - \kappa_3} = \frac{\mu_1 - \mu_2}{\mu_1 - \mu_3} = \frac{1}{2} + \frac{\sqrt{3}}{2} \tan\left(\theta - \frac{\pi}{6}\right),$$

on the set U. This Lie curvature can assume any value c in the interval $(0, 1)$ by an appropriate choice of θ in $(0, \pi/3)$. The open subset $f_t(U)$ of the tube of radius t around V in S^n is a reducible proper Dupin hypersurface with $g = 4$ and $\Psi = \Phi = c$.

Remark 5.38 (Examples with principal coordinate systems and $g = 4$). As noted in Remark 5.12 on page 250, using the theory of higher-dimensional Laplace invariants due to Kamran and Tenenblat [234], Riveros and Tenenblat [463, 464] gave a local classification of proper Dupin hypersurfaces M^4 in $\mathbf{R}^5$ with four distinct principal curvatures which are parametrized by lines of curvatures. (See also Riveros [460].) Then Riveros, Rodrigues, and Tenenblat [462] proved that a proper Dupin hypersurface $M^n \subset \mathbf{R}^{n+1}$, $n \geq 4$, with n distinct principal curvatures and constant Möbius curvatures cannot be parametrized by lines of curvature. They also showed that up to Möbius transformations, there is a unique proper Dupin hypersurface $M^3 \subset \mathbf{R}^4$ with three principal curvatures and constant Möbius curvature that is parametrized by lines of curvature. This M^3 is a cone in $\mathbf{R}^4$ over a standard flat torus in the unit sphere $S^3 \subset \mathbf{R}^4$. In a recent paper, Ferro, Rodrigues, and Tenenblat [158] constructed examples of proper Dupin hypersurfaces in $\mathbf{R}^5$ parametrized by lines of curvature having four distinct principal curvatures and nonconstant Lie curvature.

Submanifolds in Möbius and Laguerre geometries

Finally, we turn to a discussion of some results on submanifolds in the geometries of Möbius and Laguerre that are related to the study of Dupin hypersurfaces (see also the survey paper [78]).

First, C.-P. Wang [544–546] studied the Möbius geometry of submanifolds in S^n in a series of papers. Using the method of moving frames, Wang found a complete set of Möbius invariants for surfaces in $\mathbf{R}^3$ without umbilic points [544] and for hypersurfaces in $\mathbf{R}^4$ with three distinct principal curvatures at each point [545]. Then in [546], Wang defined a Möbius invariant metric g and second fundamental form B for submanifolds in S^n. He then proved that for hypersurfaces in S^n with $n \geq 4$, the pair (g, B) forms a complete Möbius invariant system which determines the hypersurface up to Möbius transformations.

In [318], H. Li, Lui, Wang, and Zhao introduced the concept of a Möbius isoparametric hypersurface in a sphere S^n. They showed that an isoparametric hypersurface in S^n is automatically Möbius isoparametric, whereas a Möbius isoparametric hypersurface is proper Dupin. Later Rodrigues and Tenenblat [465] showed that if $M \subset S^n$ is a hypersurface with a constant number g of distinct principal curvatures at each point, where $g \geq 3$, then M is Möbius isoparametric if and only if M is Dupin with constant Möbius curvatures.

Recently significant progress has been made in the classification of Möbius isoparametric hypersurfaces. First, H. Li, Lui, Wang, and Zhao [318] showed that a connected Möbius isoparametric hypersurface in S^n with two distinct principal curvatures is Möbius equivalent to an open subset of one of the following three types of hypersurfaces in S^n:

(a) a standard product of spheres $S^p(r) \times S^{n-1-p}(s) \subset S^n$, $r^2 + s^2 = 1$,
(b) the image under inverse stereographic projection from $\mathbf{R}^n \to S^n - \{P\}$ of a standard spherical cylinder $S^p(1) \times \mathbf{R}^{n-1-p} \subset \mathbf{R}^n$,
(c) the image under hyperbolic stereographic projection from $H^n \to S^n$ of a standard product $S^p(r) \times H^{n-1-p}(\sqrt{1+r^2}) \subset H^n$.

Later Hu and H. Li [205] classified Möbius isoparametric hypersurfaces in S^4, Hu, H. Li and Wang [206] classified Möbius isoparametric hypersurfaces in S^5, and Hu and Zhai [209] classified those in S^6.

Hu and D. Li [204] studied Möbius isoparametric hypersurfaces with three distinct principal curvatures in S^n and found a complete classification of such hypersurfaces in S^6. Later Hu and Zhai [210] gave a complete classification of Möbius isoparametric hypersurfaces in S^n with three distinct principal curvatures. In a related paper, Hu and Tian [208] studied the relationship between the vanishing of the Möbius form and Möbius isoparametric hypersurfaces.

Remark 5.39 (Laguerre isoparametric hypersurfaces). Recently Y.P. Song [493] studied Laguerre isoparametric hypersurfaces in $\mathbf{R}^n$. These are umbilic free oriented hypersurfaces with nonzero principal curvatures for which the Laguerre 1-form vanishes and the Laguerre shape operator has constant eigenvalues. Song gave a classification of Laguerre isoparametric hypersurfaces in $\mathbf{R}^n, n > 3$, with two distinct nonzero principal curvatures up to Laguerre transformations. The proof relies on the theory of Laguerre embeddings introduced by T. Li and C.P. Wang [320]. (See also Song-Wang [494].) In related results, T.-Z. Li and H.-F. Sun [319] classified Laguerre isoparametric hypersurfaces in $\mathbf{R}^4$. (See also the related papers of Cezana and Tenenblat [97], and Musso and Nicolodi [384, 385].)

Hu, X.X. Li and Zhai [207], and X.X Li and Y.J. Peng [322] studied Blaschke isoparametric hypersurfaces in the unit sphere with three distinct Blaschke eigenvalues. They used the method of moving frames and made extensive use of the algebraic techniques developed by Cecil and Jensen [85] in their classification of irreducible proper Dupin hypersurfaces in spheres with three principal curvatures. X.X. Li and F.Y. Zhang [323] classified Blaschke isoparametric hypersurfaces in S^5 up to Möbius equivalence. See also the note of Li and Wang [321] on Blaschke isoparametric hypersurfaces. In related papers, Shu and Su [483] and Shu and Li [482] obtained some classification results for para-Blaschke isoparametric hypersurfaces in the unit sphere.

5.8 Compact Proper Dupin Hypersurfaces

As we saw in Theorem 2.25 (page 34) due to Pinkall [446], given positive integers $m_1, \ldots, m_g$ with $m_1 + \cdots + m_g = n - 1$, there exists a proper Dupin hypersurface in $\mathbf{R}^n$ with g distinct principal curvatures having respective multiplicities $m_1, \ldots, m_g$. Pinkall constructed the examples needed to prove this theorem using his three standard constructions (cylinder, tube, and surface of revolution), so these examples are all reducible proper Dupin hypersurfaces.

For $g \geq 3$, Pinkall's examples are not compact, and in fact, compact proper Dupin hypersurfaces are far more rare. As noted in Theorem 3.63 on page 143, Thorbergsson [533] proved that if $M^{n-1} \subset S^n$ (or $\mathbf{R}^n$) is a compact, connected proper Dupin hypersurface, then the number g of distinct principal curvatures of M^n is $1, 2, 3, 4$ or 6, the same restriction as for an isoparametric hypersurface in a sphere. Furthermore, the restrictions on the multiplicities of the principal curvatures of isoparametric hypersurfaces are still valid for compact proper Dupin hypersurfaces in the sphere (see Remark 3.51 on page 136, and Stolz [502] for $g = 4$, Grove-Halperin [184] for $g = 6$). Grove and Halperin [184] gave a list of the integral homology of all compact proper Dupin hypersurfaces, and Fang [151] has results on the topology of compact proper Dupin hypersurfaces with $g = 6$ principal curvatures.

We also know from Theorem 5.19 (page 255) due to Cecil, Chi, and Jensen [82] that if M^{n-1} is a compact, connected proper Dupin hypersurface embedded in $\mathbf{R}^n$ with $g \geq 3$ distinct principal curvatures, then M^{n-1} is irreducible as a Dupin hypersurface. That is, the Legendre lift of M^{n-1} is irreducible.

Compact proper Dupin hypersurfaces in S^n have been classified in the cases $g = 1, 2$, and 3. In each case, M^{n-1} is Lie equivalent to an isoparametric hypersurface. The case $g = 1$ is simply the case of umbilic hypersurfaces, and M^{n-1} is a great or small hypersphere in S^n. In the case $g = 2$, Cecil and Ryan [89] showed that M^{n-1} is a cyclide of Dupin (see Section 5.5), and thus it is Möbius equivalent to a standard product of spheres

$$S^p(r) \times S^{n-1-p}(s) \subset S^n(1) \subset \mathbf{R}^{n+1}, \quad r^2 + s^2 = 1.$$

In the case $g = 3$, Miyaoka [363] proved that M^{n-1} is Lie equivalent to an isoparametric hypersurface (see also Cecil-Chi-Jensen [82] for a different proof). Earlier, Cartan [54] had shown that an isoparametric hypersurface with $g = 3$ principal curvatures is a tube over a standard embedding of a projective plane $\mathbf{FP}^2$, for $\mathbf{F} = \mathbf{R}, \mathbf{C}, \mathbf{H}$ (quaternions) or $\mathbf{O}$ (Cayley numbers), in S^4, S^7, S^{13}, and S^{25}, respectively. For $\mathbf{F} = \mathbf{R}$, a standard embedding is a spherical Veronese surface (see Subsection 3.8.3, page 151).

All of these results led to the widely held conjecture (see Cecil–Ryan [95, p. 184]) that every compact, connected proper Dupin hypersurface embedded in S^n is Lie equivalent to an isoparametric hypersurface. All attempts to verify this conjecture in the cases $g = 4$ and 6 were unsuccessful. Finally, in 1988, Pinkall and Thorbergsson [448] and Miyaoka and Ozawa [377] gave two different methods for producing counterexamples to the conjecture with $g = 4$ principal curvatures. The method of Miyaoka and Ozawa also yields counterexamples to the conjecture in the case $g = 6$.

The key ingredient in the construction of these counterexamples to the conjecture is the Lie curvature, i.e., the cross-ratio of the principal curvatures taken four at a time (see Section 4.5 on page 218). Pinkall and Thorbergsson proved that their examples with $g = 4$ are not Lie equivalent to an isoparametric hypersurface by showing that the Lie curvature does not have the constant value $\Psi = 1/2$, as required for a submanifold that is Lie equivalent to an isoparametric hypersurface. In fact, we will show below that the Lie curvature is not constant on their examples.

Miyaoka and Ozawa showed that the Lie curvatures are not constant on their examples with $g = 4$ or $g = 6$, and so these examples cannot be Lie equivalent to an isoparametric hypersurface. In this section, we will present both of these constructions. Our presentation is based on the papers of Pinkall–Thorbergsson [448] and Miyaoka–Ozawa [377], and we follow the treatment of these examples given in the book [77, pp. 112–123] closely.

Pinkall–Thorbergsson examples

The construction of Pinkall and Thorbergsson begins with the Clifford–Stiefel manifold of Clifford orthogonal 2-frames of length $1/\sqrt{2}$ in $\mathbf{R}^l$. This is the focal submanifold M_+ of a family of FKM-type isoparametric hypersurfaces with four principal curvatures. We now recall some details of this construction given in Section 3.9 (see page 162).

Ferus, Karcher, and Münzner [160] begin with a representation of the Clifford algebra C_{m-1} on $\mathbf{R}^l$ determined by a set of orthogonal, skew-symmetric $l \times l$ matrices $E_1, \ldots, E_{m-1}$ that satisfy the equations

$$E_i^2 = -I, \quad E_iE_j = -E_jE_i, \quad i \neq j, \quad 1 \leq i,j \leq m. \tag{5.261}$$

Two vectors u and v in $\mathbf{R}^l$ are said to be Clifford orthogonal if

$$\langle u, v\rangle = \langle E_1 u, v\rangle = \cdots = \langle E_{m-1} u, v\rangle = 0, \tag{5.262}$$

where $\langle\ ,\ \rangle$ is the usual Euclidean inner product on $\mathbf{R}^l$. As shown in equation (3.226), the focal submanifold M_+ of the corresponding isoparametric family is given by

$$M_+ = \left\{(u, v) \in S^{2l-1} \mid |u| = |v| = \frac{1}{\sqrt{2}}, \langle u, v\rangle = 0, \langle E_i u, v\rangle = 0, 1 \leq i \leq m-1\right\}. \tag{5.263}$$

Thus, $M_+ = V_2(C_{m-1})$, the Clifford–Stiefel manifold of Clifford orthogonal 2-frames of length $1/\sqrt{2}$ in $\mathbf{R}^l$. The submanifold M_+ has codimension $m+1$ in the sphere S^{2l-1}.

In Theorem 3.74 on page 176, we showed that for any unit normal ξ at any point $(u, v) \in M_+$, the shape operator A_ξ has three distinct principal curvatures

$$\kappa_1 = -1, \quad \kappa_2 = 0, \quad \kappa_3 = 1, \tag{5.264}$$

with respective multiplicities $l-m-1, m, l-m-1$.

The submanifold M_+ of codimension $m+1$ in S^{2l-1} has a Legendre lift defined on the unit normal bundle $B(M_+)$ of M_+ in S^{2l-1}. As in Theorem 4.13 on page 219, this Legendre lift has a fourth principal curvature $\kappa_4 = \infty$ of multiplicity m at each point of $B(M_+)$. Since $\kappa_4 = \infty$, the Lie curvature Ψ at any point of $B(M_+)$ equals the Möbius curvature Φ, as in equation (4.96) on page 223, i.e.,

$$\Psi = \Phi = \frac{\kappa_1 - \kappa_2}{\kappa_1 - \kappa_3} = \frac{-1-0}{-1-1} = \frac{1}{2}. \tag{5.265}$$

Since all four principal curvatures are constant on $B(M_+)$, a tube M_t of spherical radius t, where $0 < t < \pi$ and $t \notin \{\frac{\pi}{4}, \frac{\pi}{2}, \frac{3\pi}{4}\}$, over M_+ is an isoparametric hypersurface with four distinct principal curvatures, as in Corollary 3.75 on page 177. Note that Münzner [381] proved that if M is any isoparametric hypersurface in S^n with four principal curvatures, then the Lie curvature $\Psi = 1/2$ on all of M, as in equation (4.96).

The construction of Pinkall and Thorbergsson now proceeds as follows. Given positive real numbers α and β with

$$\alpha^2 + \beta^2 = 1, \quad \alpha \neq \frac{1}{\sqrt{2}}, \quad \beta \neq \frac{1}{\sqrt{2}}, \tag{5.266}$$

let

$$T_{\alpha,\beta} : \mathbf{R}^{2l} \to \mathbf{R}^{2l},$$

be the linear map defined by

$$T_{\alpha,\beta}(u,v) = \sqrt{2}\,(\alpha u, \beta v). \tag{5.267}$$

Then for $(u,v) \in M_+$, we have

$$|T_{\alpha,\beta}(u,v)|^2 = 2(\alpha^2\langle u,u\rangle + \beta^2\langle v,v\rangle) = 2\left(\frac{\alpha^2}{2} + \frac{\beta^2}{2}\right) = 1,$$

and thus the image $V_2^{\alpha,\beta} = T_{\alpha,\beta}(M_+)$ is a submanifold of S^{2l-1} of codimension $m+1$ also.

Our first goal is to show that $V_2^{\alpha,\beta}$ is proper Dupin, that is, its Legendre lift is a proper Dupin submanifold. Here we use the notion of curvature surfaces of a submanifold of codimension greater than one defined by Reckziegel [458] (see Remark 2.21 on page 32). Specifically, suppose that $V \subset S^n$ is a submanifold of codimension greater than one, and let $B(V)$ denote its unit normal bundle. A connected submanifold $S \subset V$ is called a curvature surface if there exists a parallel section $\eta : S \to B(V)$ such that for each $x \in S$, the tangent space T_xS is equal to some eigenspace of $A_{\eta(x)}$. The corresponding principal curvature function $\kappa : S \to \mathbf{R}$ is then a smooth function on S. Reckziegel showed that if a principal curvature κ has constant multiplicity μ on $B(V)$ and is constant along each of its curvature surfaces, then each of its curvature surfaces is an open subset of a μ-dimensional metric sphere in S^n. Since our particular submanifold M_+ is compact, all of the curvature surfaces of the principal curvatures κ_1, κ_2 and κ_3 given in equation (5.264) are spheres of the appropriate dimensions in S^{2l-1}.

We now show that the Legendre lift of $V_2^{\alpha,\beta}$ is proper Dupin with four smooth principal curvature functions

$$\lambda_1 < \lambda_2 < \lambda_3 < \lambda_4, \tag{5.268}$$

defined on the unit normal bundle $B(V_2^{\alpha,\beta})$ of $V_2^{\alpha,\beta}$, as in Theorem 4.13 on page 219.

Since $V_2^{\alpha,\beta}$ has codimension $m+1$, the principal curvature $\lambda_4 = \infty$ has multiplicity m and is constant along its curvature surfaces. To complete the proof that $V_2^{\alpha,\beta}$ is proper Dupin, we establish a bijective correspondence between the other curvature surfaces of M_+ and those of $V_2^{\alpha,\beta}$. Let S be any curvature surface of M_+. Since M_+ is compact and proper Dupin, S is a μ-dimensional sphere, where μ is the multiplicity of the corresponding principal curvature of M_+. Along the curvature surface S, the corresponding curvature sphere Σ is constant. Note that Σ is a hypersphere obtained by intersecting S^{2l-1} with a hyperplane π that is tangent to M_+ along S. The image $T_{\alpha,\beta}(\pi)$ is a hyperplane that is tangent to $V_2^{\alpha,\beta}$ along the μ-dimensional sphere $T_{\alpha,\beta}(S)$. Since the hypersphere $T_{\alpha,\beta}(\pi) \cap S^{2l-1}$ is tangent to $V_2^{\alpha,\beta}$ along $T_{\alpha,\beta}(S)$, it is a curvature sphere of $V_2^{\alpha,\beta}$ with multiplicity μ, and $T_{\alpha,\beta}(S)$ is the corresponding curvature surface. Thus, we have a bijective correspondence between the curvature surfaces of M_+ and those of $V_2^{\alpha,\beta}$, and the Dupin condition is clearly satisfied on $V_2^{\alpha,\beta}$. Therefore, $V_2^{\alpha,\beta}$ is a proper Dupin submanifold with four principal curvatures, including $\lambda_4 = \infty$.

Computing the Lie curvature

With the principal curvature functions defined as in equation (5.268) and using the fact that $\lambda_4 = \infty$, there is a unique Lie curvature function Ψ defined on $B(V_2^{\alpha,\beta})$ by

$$\Psi = \Phi = \frac{\lambda_1 - \lambda_2}{\lambda_1 - \lambda_3}, \tag{5.269}$$

where Φ is the Möbius curvature.

We next show that the Legendre lift of $V_2^{\alpha,\beta}$ is not Lie equivalent to the Legendre lift of an isoparametric hypersurface in S^{2l-1} by showing that the Lie curvature Ψ does not equal $1/2$ at some points of the unit normal bundle $B(V_2^{\alpha,\beta})$, as required for the Legendre lift of an isoparametric hypersurface. Moreover, we will show that the Lie curvature is not constant on $B(V_2^{\alpha,\beta})$.

To compute the functions $\lambda_1 < \lambda_2 < \lambda_3$, we first note that

$$V_2^{\alpha,\beta} \subset f^{-1}(0) \cap g^{-1}(0),$$

where f and g are the real-valued functions defined on S^{2l-1} by

$$f(u,v) = \langle \frac{-\beta}{2\alpha} u, u\rangle + \langle \frac{\alpha}{2\beta} v, v\rangle, \quad g(u,v) = -\langle u, v\rangle. \tag{5.270}$$

Thus, the gradients,

$$\xi = \left(\frac{-\beta}{\alpha} u, \frac{\alpha}{\beta} v\right), \quad \eta = (-v, -u),$$

of f and g are two unit normal vector fields on $V_2^{\alpha,\beta}$. Note that by Theorem 3.72 on page 170, we have $l > m + 1$ for the FKM-hypersurfaces, so we can choose $x, y \in \mathbf{R}^l$ such that

$$|x| = \alpha, \quad \langle x, u\rangle = 0, \quad \langle x, v\rangle = 0, \quad \langle x, E_i v\rangle = 0, \quad 1 \le i \le m-1,$$
$$|y| = \beta, \quad \langle y, u\rangle = 0, \quad \langle y, v\rangle = 0, \quad \langle y, E_i u\rangle = 0, \quad 1 \le i \le m-1.$$

We define three curves,

$$\gamma(t) = (\cos t\, u + \sin t\, x, v), \quad \delta(t) = (u, \cos t\, v + \sin t\, y),$$
$$\varepsilon(t) = (\cos t\, u + \frac{\alpha}{\beta} \sin t\, v, -\frac{\beta}{\alpha} \sin t\, u + \cos t\, v). \tag{5.271}$$

It is straightforward to check that each of these curves lies on $V_2^{\alpha,\beta}$ and goes through the point (u, v) when $t = 0$. Along γ, the normal vector ξ is given by

$$\xi(t) = \left(-\frac{\beta}{\alpha}(\cos t\, u + \sin t\, x), \frac{\alpha}{\beta} v \right).$$

Thus, the initial velocity vector to $\xi(t)$ satisfies

$$\overrightarrow{\xi}(0) = \left(-\frac{\beta}{\alpha} x, 0 \right) = -\frac{\beta}{\alpha} \overrightarrow{\gamma}(0).$$

So $X = (x, 0) = \overrightarrow{\gamma}(0)$ is a principal vector of A_ξ at (u, v) with corresponding principal curvature β/α.

Similarly, $Y = (0, y) = \overrightarrow{\delta}(0)$ is a principal vector of A_ξ at (u, v) with corresponding principal curvature $-\alpha/\beta$. Finally, along the curve ε, we have

$$\xi(t) = \left(-\frac{\beta}{\alpha} \left(\cos t\, u + \frac{\alpha}{\beta} \sin t\, v \right), \frac{\alpha}{\beta} \left(-\frac{\beta}{\alpha} \sin t\, u + \cos t\, v \right) \right).$$

Then $\overrightarrow{\xi}(0) = (-v, -u) = \eta$, which is normal to $V_2^{\alpha,\beta}$ at (u, v). Thus, we have $A_\xi Z = 0$, for $Z = \overrightarrow{\varepsilon}(0)$, and Z is a principal vector with corresponding principal curvature zero. Therefore, at the point $\xi(u, v)$ in $B(V_2^{\alpha,\beta})$, there are four principal curvatures written in ascending order as in equation (5.268) (recall that α and β are positive),

$$\lambda_1 = -\frac{\alpha}{\beta}, \quad \lambda_2 = 0, \quad \lambda_3 = \frac{\beta}{\alpha}, \quad \lambda_4 = \infty. \tag{5.272}$$

At this point, the Lie curvature Ψ is

$$\Psi = \Phi = \frac{\lambda_1 - \lambda_2}{\lambda_1 - \lambda_3} = \frac{-\alpha/\beta}{(-\alpha/\beta - \beta/\alpha)} = \alpha^2. \tag{5.273}$$

Since $\alpha^2 \neq 1/2$, the Legendre lift of $V_2^{\alpha,\beta}$ is not Lie equivalent to an isoparametric hypersurface. To obtain a compact proper Dupin hypersurface in S^{2l-1} with four principal curvatures that is not Lie equivalent to an isoparametric hypersurface, one simply takes a tube M over $V_2^{\alpha,\beta}$ in S^{2l-1} of sufficiently small radius so that the tube is an embedded hypersurface.

Since $A_{-\xi} = -A_\xi$, the principal curvatures at the point $-\xi(u,v)$ in $B(V_2^{\alpha,\beta})$ are the negatives of those given in equation (5.272). Thus, since the smooth principal curvature functions $\lambda_1, \lambda_2, \lambda_3$ defined on $B(V_2^{\alpha,\beta})$ by equation (5.268) satisfy $\lambda_1 < \lambda_2 < \lambda_3$, we have

$$\lambda_1 = -\frac{\beta}{\alpha}, \quad \lambda_2 = 0, \quad \lambda_3 = \frac{\alpha}{\beta}, \quad \lambda_4 = \infty. \tag{5.274}$$

at this point $-\xi(u,v)$ in $B(V_2^{\alpha,\beta})$. Thus, at this point $-\xi(u,v)$, we have the Lie curvature,

$$\Psi = \Phi = \frac{\lambda_1 - \lambda_2}{\lambda_1 - \lambda_3} = \frac{-\beta/\alpha}{(-\beta/\alpha - \alpha/\beta)} = \beta^2. \tag{5.275}$$

Since $\beta^2 \neq \alpha^2$, the Lie curvature Ψ is not constant on $B(V_2^{\alpha,\beta})$.

In a related result, Miyaoka [365, Corollary 8.3, p. 252] proved that if the Lie curvature Ψ is constant on a compact, connected proper Dupin hypersurface with four principal curvatures, then, in fact, $\Psi = 1/2$ on the hypersurface.

Miyaoka–Ozawa examples

We next handle the counterexamples due to Miyaoka and Ozawa [377] to the conjecture that every compact, connected proper Dupin hypersurface embedded in S^n is Lie equivalent to an isoparametric hypersurface. The construction of Miyaoka and Ozawa uses the Hopf fibration of S^7 over S^4. Let $\mathbf{R}^8 = \mathbf{H} \times \mathbf{H}$, where $\mathbf{H}$ is the skew field of quaternions. The Hopf fibration of the unit sphere S^7 in $\mathbf{R}^8$ over S^4 is given by

$$h(u,v) = (2u\bar{v}, |u|^2 - |v|^2), \quad u, v \in \mathbf{H}, \tag{5.276}$$

where $\bar{v}$ is the conjugate of v in $\mathbf{H}$. One can easily compute that the image of h lies in the unit sphere S^4 in the Euclidean space $\mathbf{R}^5 = \mathbf{H} \times \mathbf{R}$.

We first recall some important facts about the Hopf fibration and the inverse image of certain types of subsets of S^4 under h. Suppose $(w,t) \in S^4$, with $t \neq 1$, that is, (w,t) is not the point $(0,1)$. We first find the inverse image of (w,t) under h. Suppose that

$$2u\bar{v} = w, \quad |u|^2 - |v|^2 = t. \tag{5.277}$$

Multiplying the first equation in (5.277) by v on the right, we obtain

$$2u|v|^2 = wv, \quad 2u = \frac{w}{|v|}\frac{v}{|v|}. \tag{5.278}$$

Since $|u|^2 + |v|^2 = 1$, the second equation in (5.277) yields

$$|v|^2 = (1-t)/2. \tag{5.279}$$

If we write $z = v/|v|$, then $z \in S^3$, the unit sphere in $\mathbf{H} = \mathbf{R}^4$. Then equations (5.278) and (5.279) give

$$u = \frac{wz}{\sqrt{2(1-t)}}, \quad v = \sqrt{\frac{1-t}{2}}\, z, \quad z \in S^3. \tag{5.280}$$

Thus, if U is the open set $S^4 - \{(0,1)\}$, then $h^{-1}(U)$ is diffeomorphic to $U \times S^3$ by the formula (5.280). The second equation in (5.277) also shows that $h^{-1}(\{(0,1)\})$ is just the 3-sphere in S^7 determined by the equation $v = 0$.

We can find a similar local trivialization containing these points with $v = 0$ by beginning the process above with multiplication of equation (5.277) by $\bar{u}$ on the left, rather than by v on the right. As a consequence of this local triviality, if M is an embedded submanifold in S^4 which does not equal all of S^4, then $h^{-1}(M)$ is diffeomorphic to $M \times S^3$. Finally, recall that the Euclidean inner product $\langle\, ,\, \rangle$ on the space $\mathbf{R}^8 = \mathbf{H} \times \mathbf{H}$ is given by

$$\langle (a,b),(u,v)\rangle = \Re(\bar{a}u + \bar{b}v), \tag{5.281}$$

where $\Re w$ denotes the real part of the quaternion w.

The counterexamples to the conjecture due to Miyaoka and Ozawa all arise as inverse images under h of proper Dupin hypersurfaces in S^4. The proof that these examples are proper Dupin is accomplished by first showing that they are taut, and thus they are Dupin (but not necessarily proper Dupin) by Theorem 2.83 on page 68. Then a separate argument is used to prove that they are in fact proper Dupin.

The Hopf fibration and tautness

We begin with a result about the Hopf fibration and tautness.

Theorem 5.40. *Let M be a compact, connected submanifold of S^4. If M is taut in S^4, then $h^{-1}(M)$ is taut in S^7.*

Proof. Since both M and $h^{-1}(M)$ lie in spheres, tautness is equivalent to tightness for these hypersurfaces by Theorem 2.69 on page 61. We write linear height functions in $\mathbf{R}^8$ in the form

$$f_{ab}(u,v) = \Re(au+bv) = \langle(\bar{a},\bar{b}),(u,v)\rangle, \quad (a,b) \in S^7. \tag{5.282}$$

This is the height function in the direction $(\bar{a},\bar{b})$. We want to determine when the point (u,v) is a critical point of f_{ab}. Without loss of generality, we may assume that (u,v) lies in a local trivialization of the form (5.280) when making local calculations. Let $x=(w,t)$ be a point of $M \subset S^4$, and let

$$(x,z) = (w,t,z)$$

be a point in the fiber $h^{-1}(x)$. The tangent space to $h^{-1}(M)$ at (x,z) can be decomposed as $T_xM \times T_zS^3$. We first locate the critical points of the restriction of f_{ab} to the fiber through (x,z). By equations (5.280) and (5.282), we have

$$\begin{aligned} f_{ab}(w,t,z) &= \Re\left(\frac{awz}{\sqrt{2(1-t)}} + bz\sqrt{(1-t)/2}\right) \\ &= \Re(\alpha(w,t)z) = \langle\alpha(w,t),\bar{z}\rangle, \end{aligned} \tag{5.283}$$

where

$$\alpha(w,t) = \frac{aw}{\sqrt{2(1-t)}} + b\sqrt{(1-t)/2}.$$

This defines the map α from S^4 to $\mathbf{H}$. If Z is any tangent vector to S^3 at z, we write Zf_{ab} for the derivative of f_{ab} in the direction $(0,Z)$. Then

$$Zf_{ab} = \langle\alpha(w,t),\bar{Z}\rangle \tag{5.284}$$

at (x,z). Now there are two cases to consider. First, if $\alpha(w,t) \neq 0$, then in order to have $Zf_{ab}=0$ for all $Z \in T_zS^3$, we must have

$$\bar{z} = \pm\frac{\alpha(w,t)}{|\alpha(w,t)|}. \tag{5.285}$$

So the restriction of f_{ab} to the fiber has exactly two critical points with corresponding values

$$\pm\,|\alpha(w,t)|. \tag{5.286}$$

The second case is when $\alpha(w,t)=0$. Then the restriction of f_{ab} to the fiber is identically zero by equation (5.283). In both cases the function,

$$g_{ab}(w,t) = |\alpha(w,t)|^2,$$

satisfies the equation

$$g_{ab}(w,t) = f_{ab}^2(w,t,z),$$

at the critical point. The key in relating this fact to information about the submanifold M is to note that

$$\begin{aligned} g_{ab}(w,t) = |\alpha(w,t)|^2 &= \frac{1}{2}\Re\left\{2a\bar{b}w + (|a|^2 - |b|^2)t\right\} + \frac{1}{2}(|a|^2 + |b|^2) \\ &= \frac{1}{2} + \frac{1}{2}\langle (w,t), (2\bar{a}b, |a|^2 - |b|^2)\rangle \qquad (5.287) \\ &= \frac{1}{2} + \frac{1}{2}\ell_{ab}(w,t), \end{aligned}$$

where ℓ_{ab} is the linear height function on $\mathbf{R}^5$ in the direction

$$(2\bar{a}b, |a|^2 - |b|^2) = h(\bar{a}, \bar{b}).$$

This shows that $g_{ab}(w,t) = 0$ if and only if $(w,t) = -h(\bar{a},\bar{b})$. Thus, if $-h(\bar{a},\bar{b})$ is not in M, the restriction of f_{ab} to each fiber has exactly two critical points of the form (x,z), with z as in equation (5.285). For $X \in T_xM$, we write Xf_{ab} for the derivative of f_{ab} in the direction $(X,0)$. At the two critical points, we have

$$Xf_{ab} = \langle d\alpha(X), \bar{z}\rangle, \qquad (5.288)$$

$$Xg_{ab} = 2\langle d\alpha(X), \alpha(x)\rangle = \pm 2|\alpha(x)|\langle d\alpha(X), \bar{z}\rangle = \pm 2|\alpha(X)|Xf_{ab}. \qquad (5.289)$$

Thus (x,z) is a critical point of f_{ab} if and only if x is a critical point of g_{ab}. By equation (5.287), this happens precisely when x is a critical point of ℓ_{ab}.

We conclude that if $-h(\bar{a},\bar{b})$ is not in M, then f_{ab} has two critical points for every critical point of ℓ_{ab} on M. The set of points (a,b) in S^7 such that $-h(\bar{a},\bar{b})$ belongs to M has measure zero. If (a,b) is not in this set, then f_{ab} has twice as many critical points as the height function ℓ_{ab} on M. Since M is taut, every nondegenerate height function ℓ_{ab} has $\beta(M,\mathbf{Z}_2)$ critical points on M, where $\beta(M,\mathbf{Z}_2)$ is the sum of the $\mathbf{Z}_2$-Betti numbers of M. Thus, except for a set of measure zero, every height function f_{ab} has $2\beta(M,\mathbf{Z}_2)$ critical points on $h^{-1}(M)$. Since $h^{-1}(M)$ is diffeomorphic to $M \times S^3$, we have

$$\beta(h^{-1}(M),\mathbf{Z}_2) = \beta(M \times S^3,\mathbf{Z}_2) = 2\beta(M,\mathbf{Z}_2).$$

Thus, $h^{-1}(M)$ is taut in S^7. □

The Hopf fibration and Dupin submanifolds

We next use Theorem 5.40 to show that the inverse image under h of a compact proper Dupin submanifold in S^4 is proper Dupin. The key idea here is due to Ozawa [421], who proved that a taut submanifold $M \subset S^n$ is proper Dupin if and only if every connected component of a critical set of a linear height function on M is a point or is homeomorphic to a sphere of some dimension k. (See also Hebda [194].)

Theorem 5.41. *Let M be a compact, connected proper Dupin submanifold embedded in S^4. Then $h^{-1}(M)$ is a proper Dupin submanifold in S^7.*

Proof. As noted in Theorem 2.80 on page 66, Thorbergsson [533] proved that a compact proper Dupin hypersurface embedded in S^n is taut, and Pinkall [447] (see Theorem 2.82) extended this result to the case where M has codimension greater than one and the number of distinct principal curvatures is constant on the unit normal bundle $B(M)$. Thus, our M is taut in S^4, and therefore $h^{-1}(M)$ is taut in S^7 by Theorem 5.40. To show that $h^{-1}(M)$ is proper Dupin, we need to show that each connected component of a critical set of a height function f_{ab} on $h^{-1}(M)$ is a point or is homeomorphic to a sphere.

We use the same notation as in the proof of Theorem 5.40. Suppose that (x, z) is a critical point of f_{ab}. For $X \in T_xM$, we compute from equation (5.283) that

$$Xf_{ab} = \langle d\alpha(X), \bar{z} \rangle. \tag{5.290}$$

From (5.289), we see that Xg_{ab} also equals zero, and the argument again splits into two cases, depending on whether or not $g_{ab}(x)$ is zero. If $g_{ab}(x)$ is nonzero, then there are two critical points of f_{ab} on the fiber $h^{-1}(x)$. Thus a component in $h^{-1}(M)$ of the critical set of f_{ab} through (x, z) is homeomorphic to the corresponding component of the critical set containing x of the linear function ℓ_{ab} on M. Since M is proper Dupin, such a component is a point or a sphere.

The second case is when $g_{ab}(x) = f_{ab}^2(x, z) = 0$. As we have seen, this happens only if $x = -h(\bar{a}, \bar{b})$. In that case, x is an isolated absolute minimum of the height function ℓ_{ab}. Thus, the corresponding component of the critical set of f_{ab} through (x, z) lies in the fiber $h^{-1}(x)$, which is diffeomorphic to S^3. From equation (5.290), we see that this component of the critical set consists of those points (x, y) in the fiber such that $\bar{y}$ is orthogonal to $d\alpha(X)$, for all $X \in T_xM$. We know that

$$g_{ab}(x) = \frac{1}{2} + \frac{1}{2}\ell_{ab}(x), \tag{5.291}$$

and x is an isolated critical point of ℓ_{ab} on M. The tautness of M and the results of Ozawa [421] imply that x is a nondegenerate critical point of ℓ_{ab}, since the component of the critical set of a height function containing a degenerate critical point is a sphere of dimension greater than zero. By equation (5.291), x is also a nondegenerate critical point of g_{ab}, and so the Hessian $H(X, Y)$ of g_{ab} is nondegenerate at x. Since $\alpha(x) = 0$, we compute that for X and Y in T_xM,

$$H(X,Y) = 2\langle d\alpha(X), d\alpha(Y)\rangle.$$

Hence, $d\alpha$ is nondegenerate at x, and the rank of $d\alpha$ is the dimension of M. From this it follows that the component of the critical set of f_{ab} through (x,z) is a sphere in $h^{-1}(x)$ of dimension $(3 - \dim M)$. Therefore, we have shown that every component of the critical set of a linear height function f_{ab} on $h^{-1}(M)$ is homeomorphic to a point or a sphere. Thus, $h^{-1}(M)$ is proper Dupin. □

We next relate the principal curvatures of $h^{-1}(M)$ to those of M.

Theorem 5.42. *Let M be a compact, connected proper Dupin hypersurface embedded in S^4 with g principal curvatures. Then the proper Dupin hypersurface $h^{-1}(M)$ in S^7 has $2g$ principal curvatures. Each principal curvature,*

$$\lambda = \cot\theta, \quad 0 < \theta < \pi,$$

of M at a point $x \in M$ yields two principal curvatures of $h^{-1}(M)$ at points in $h^{-1}(x)$ with values

$$\lambda^+ = \cot(\theta/2), \quad \lambda^- = \cot((\theta+\pi)/2).$$

Proof. By Theorem 2.1 on page 11, a principal curvature $\lambda = \cot\theta$ of a hypersurface M at x corresponds to a focal point at oriented distance θ along the normal geodesic to M at x. A point (x,z) in $h^{-1}(M)$ is a critical point of f_{ab} if and only if $(\bar{a},\bar{b})$ lies along the normal geodesic to $h^{-1}(M)$ at (x,z). The critical point is degenerate if and only if $(\bar{a},\bar{b})$ is a focal point of $h^{-1}(M)$ at (x,z). Note further that (x,z) is a degenerate critical point of f_{ab} if and only if x is a degenerate critical point of ℓ_{ab}. This follows from the fact that both embeddings are taut, and the dimensions of the components of the critical sets agree by Theorem 5.41. The latter claim holds even when $x = -h(\bar{a},\bar{b})$, since the fact that M has dimension three implies that the critical point (x,z) of f_{ab} is isolated. Thus, $(\bar{a},\bar{b})$ is a focal point of $h^{-1}(M)$ if and only if $h(\bar{a},\bar{b})$ is a focal point of M.

Suppose now that $(\bar{a},\bar{b})$ lies along the normal geodesic to $h^{-1}(M)$ at (x,z) and that $f_{ab}(x,z) = \cos\phi$. Then by equation (5.287),

$$g_{ab}(x) = \frac{1}{2} + \frac{1}{2}\ell_{ab}(x) = \frac{1}{2} + \frac{1}{2}\cos\theta,$$

where θ is the distance from $h(\bar{a},\bar{b})$ to x. Since (x,z) is a critical point of f_{ab}, we have $g_{ab}(x) = f_{ab}^2(x,z)$. Thus,

$$\frac{1}{2} + \frac{1}{2}\cos\theta = \cos^2\phi = \frac{1}{2} + \frac{1}{2}\cos 2\phi,$$

and so $\cos\theta = \cos 2\phi$. This means that under the map h, the normal geodesic to $h^{-1}(M)$ at (x,z) double covers the normal geodesic to M at x, since the points corresponding to the values $\phi = \theta/2$ and $\phi = (\theta + \pi)/2$ are mapped to the same point by h. In particular, a focal point corresponding to a principal curvature $\lambda = \cot\theta$ on the normal geodesic to M at x gives rise to two focal points on the normal geodesic to $h^{-1}(M)$ at (x, z) with corresponding principal curvatures

$$\lambda^+ = \cot(\theta/2), \quad \lambda^- = \cot((\theta + \pi)/2),$$

thus completing the proof of the theorem. □

We now construct the examples of Miyaoka and Ozawa. As noted above, a compact proper Dupin hypersurface M in S^4 with two principal curvatures is a cyclide of Dupin, which is the image under a Möbius transformation of S^4 of a standard product of spheres,

$$S^1(r) \times S^2(s) \subset S^4(1) \subset \mathbf{R}^5, \quad r^2 + s^2 = 1.$$

A conformal, non-isometric image of an isoparametric cyclide does not have constant principal curvatures. Similarly, a compact, connected proper Dupin hypersurface in S^4 with three principal curvatures is Lie equivalent to an isoparametric hypersurface in S^4 with three principal curvatures, but it can have three nonconstant principal curvature functions.

Corollary 5.43. *Let M be a compact, connected proper Dupin hypersurface embedded in S^4 with g nonconstant principal curvatures, where $g = 2$ or 3. Then $h^{-1}(M)$ is a compact, connected proper Dupin hypersurface in S^7 with $2g$ principal curvatures that is not Lie equivalent to an isoparametric hypersurface in S^7.*

Proof. Suppose that $\lambda = \cot\theta$ and $\mu = \cot\alpha$ are two distinct nonconstant principal curvature functions on M. (In the case $g = 3$, just use any two of the principal curvature functions.) Let

$$\lambda^+ = \cot(\theta/2), \quad \lambda^- = \cot((\theta + \pi)/2),$$
$$\mu^+ = \cot(\alpha/2), \quad \mu^- = \cot((\alpha + \pi)/2),$$

be the four distinct principal curvature functions on $h^{-1}(M)$ induced from λ and μ. Then the Lie curvature

$$\Psi = \frac{(\lambda^+ - \lambda^-)(\mu^+ - \mu^-)}{(\lambda^+ - \mu^-)(\mu^+ - \lambda^-)} = \frac{2}{1 + \cos(\theta - \alpha)},$$

is not constant on $h^{-1}(M)$, and therefore $h^{-1}(M)$ is not Lie equivalent to an isoparametric hypersurface in S^7. □

Miyaoka and Ozawa note that certain parts of their construction are also valid if **H** is replaced by the Cayley numbers or a more general Clifford algebra. See their paper [377] for more detail on this point.

Compact proper Dupin hypersurfaces with constant Lie curvatures

In these examples due to Pinkall–Thorbergsson and Miyaoka–Ozawa, the Lie curvatures are not constant, and so these hypersurfaces cannot be Lie equivalent to an isoparametric hypersurface. This leaves open the possibility that a compact proper Dupin hypersurface with 4 or 6 principal curvatures and constant Lie curvatures must be Lie equivalent to an isoparametric hypersurface. In the case $g = 4$, Miyaoka [365] showed that this is true if the hypersurface also satisfies some additional assumptions on the intersections of the leaves of the various principal foliations. In the same paper, Miyaoka also proved that if the Lie curvature of compact proper Dupin hypersurface with $g = 4$ is constant, then it has the value $1/2$.

Later Cecil, Chi, and Jensen [83] formulated the following conjecture which remains as an open problem, although partial results have been obtained.

Conjecture 5.44 (Compact proper Dupin hypersurfaces with constant Lie curvatures). Every compact, connected proper Dupin hypersurface in S^n with $g = 4$ or $g = 6$ principal curvatures and constant Lie curvatures is Lie equivalent to an isoparametric hypersurface.

In [85], Cecil and Jensen proved that conjecture is true for a compact proper Dupin hypersurface with four principal curvatures of multiplicity one. Then Cecil, Chi, and Jensen [82] verified the conjecture in the case where the multiplicities satisfy $m_1 = m_3 \geq 1, m_2 = m_4 = 1$ to obtain the following theorem.

Theorem 5.45. *Let M be a compact, connected proper Dupin hypersurface in S^n with four principal curvatures having multiplicities $m_1 = m_3 \geq 1$, $m_2 = m_4 = 1$, and constant Lie curvature. Then M is Lie equivalent to an isoparametric hypersurface.*

Note that since the multiplicities of a compact, connected proper Dupin hypersurface with four principal curvatures, satisfy the conditions $m_1 = m_3$ and $m_2 = m_4$ when the principal curvatures are appropriately ordered. This means that the full conjecture for $g = 4$ would be proven if the assumption that the value of $m_2 = m_4$ is equal to one could be eliminated from the theorem above.

Cecil, Chi, and Jensen proved Theorem 5.45 as a consequence of the local classification (Theorem 5.36) of irreducible proper Dupin hypersurfaces with four principal curvatures having the given multiplicities and constant Lie curvature. The fact that the constant Lie curvature must equal $1/2$ in the compact case is due to

Miyaoka [365], as mentioned above. The proof of Theorem 5.36 involves some complicated calculations, which become even more elaborate if the assumption that $m_2 = m_4 = 1$ is dropped. Even so, this approach to proving Conjecture 5.44 could possibly be successful with some additional insight regarding the structure of the calculations involved.

In the case $g = 6$, we do not know of any results beyond those of Miyaoka [366], who showed that Conjecture 5.44 is true if the hypersurface satisfies some additional assumptions on the intersections of the leaves of the various principal foliations. An approach similar to that used by Cecil, Chi, and Jensen in [82] for the $g = 4$ case is plausible, but the calculations involved would be very complicated, unless some new algebraic insight is found to simplify the situation.

Remark 5.46 (Dupin hypersurfaces with constant scalar curvature). S. de Almeida and A. Brasil [11] proved that if M^n, $n \leq 4$, is a compact proper Dupin hypersurface in S^{n+1} with constant mean curvature and constant scalar curvature $s \geq 0$, then M^n is isoparametric. In the case $n = 3$, the conclusion holds even without the assumption that M^n is Dupin. This was shown in an earlier paper by S. de Almeida and F. Brito [12]. In another paper, X.M. Wang [551] also found sufficient conditions on the principal curvatures for a proper Dupin hypersurface with constant mean curvature to be isoparametric.

5.9 Taut Embeddings of Surfaces

In this section, we give a complete classification of taut (2-dimensional) surfaces in Euclidean spaces. Many of the results that we need have been covered already, and we begin by recalling them here.

If M^2 is a compact, connected surface, and $f : M^2 \to \mathbf{R}^m$ is a substantial taut embedding, then $m \leq 5$ by Theorem 2.76 on page 64. Furthermore, $m = 5$, then $M^2 = \mathbf{RP}^2$, and $f(M^2)$ is a spherical Veronese surface contained in a metric sphere $S^4 \subset \mathbf{R}^5$. Next if $f : M^2 \to \mathbf{R}^4$ is a substantial, non-spherical taut embedding of a compact surface, then $f(M^2)$ is the image under stereographic projection $\tau : S^4 - \{P\} \to \mathbf{R}^4, P \in S^4$, of a spherical Veronese surface in $S^4 \subset \mathbf{R}^5$ by Corollary 2.72 on page 62.

Finally, if $f : M^2 \to S^3 \subset \mathbf{R}^4$ is a substantial, spherical taut embedding, and $\tau : S^3 - \{P\} \to \mathbf{R}^3$ is stereographic projection with pole $P \in S^3$, then $\tau \circ f : M^2 \to \mathbf{R}^3$ is a taut embedding, and conversely any taut spherical embedding of a compact, connected surface M^2 into $S^3 \subset \mathbf{R}^4$ is the image under inverse stereographic projection of a taut compact, connected surface in $\mathbf{R}^3$. Thus, the classification of taut compact, connected surfaces in Euclidean spaces is reduced to the classification of such surfaces in $\mathbf{R}^3$.

If $g : M^2 \to \mathbf{R}^m$ is a substantial, taut embedding of a non-compact surface M^2, then $m \leq 4$ by Theorem 2.77 on page 65, and if $m = 4$, then $g = \tau \circ f$, where $f : \mathbf{RP}^2 \to S^4 \subset \mathbf{R}^5$ is a spherical Veronese surface, and $\tau : S^4 - \{P\} \to \mathbf{R}^4$

is stereographic projection with pole $P \in f(\mathbf{RP}^2)$. Thus, in the non-compact case also, the classification of taut embeddings is reduced to the classification of taut non-compact surfaces in $\mathbf{R}^3$.

The classification of taut embeddings of surfaces $f : M^2 \to \mathbf{R}^3$ follows fairly quickly from known results. If $f(M^2)$ has even one umbilic point, then it is totally umbilic by Corollaries 2.66 and 2.68 (see page 60). Thus, $f(M^2)$ is a hyperplane or metric sphere in $\mathbf{R}^3$.

If $f(M^2)$ has no umbilic points, then there are two distinct principal curvatures at each point of $f(M^2)$. Then since taut implies Dupin by Theorem 2.83 on page 68, we have that $f(M^2)$ is a proper Dupin surface in $\mathbf{R}^3$ with $g = 2$ principal curvatures at each point. Since a taut surface is complete, we get that $f(M^2)$ is a ring cyclide, a parabolic ring cyclide or a circular cylinder by Theorem 5.26 on page 282.

This proof relies on Theorem 2.83 that taut implies Dupin. We can actually get the classification of taut surfaces in $\mathbf{R}^3$ from Lemma 5.47 below, which is more elementary than Theorem 2.83. We will formulate the lemma for hypersurfaces of arbitrary dimension, since it will be useful in the next section on higher-dimensional Dupin hypersurfaces.

Consequences of tautness

We first make some preliminary remarks before stating the lemma. Let $f : M^n \to \mathbf{R}^{n+1}$ be a taut hypersurface. Tautness implies that f is a proper embedding, and therefore $f(M^n)$ is a closed subset of $\mathbf{R}^{n+1}$. Thus, $f(M^n)$ is orientable (see Samelson [471]), and we can take ξ to be a field of unit normals on $f(M^n)$. Thus, there are globally defined continuous principal curvature functions

$$\lambda_1 \geq \lambda_2 \geq \cdots \geq \lambda_n, \tag{5.292}$$

on M^n determined by the field of unit normals ξ (see Ryan [468, p. 371]).

Tautness implies that if L_q is any nondegenerate Euclidean distance function on M^n and μ_k denotes the number of critical points of L_q of index k, then $\mu_0 = 1$, and $\mu_n = 1$ or 0, depending on whether or not M^n is compact. We now show that these conditions on μ_0 and μ_n imply that if p is the focal point of (M^n, x) corresponding to the largest or smallest principal curvature at $x \in M^n$, then L_p has an absolute minimum or absolute maximum at x.

Let x be a point of M^n. If $\lambda_1(x) > 0$, then the focal point p determined by $\lambda_1(x)$ is the first focal point on the normal ray to $f(M^n)$ at $f(x)$ in the direction $\xi(x)$, and so L_p has an absolute minimum at x by Theorem 2.64 (a) on page 59.

If $\lambda_1(x) < 0$, then the corresponding focal point p (which lies on the ray in the direction $-\xi(x)$) has the property that the sum of the multiplicities of the focal points of (M^n, x) on the closed segment from $f(x)$ to p is n. Thus, if q is a point beyond p on the normal ray in the direction $-\xi(x)$, then L_q has a nondegenerate maximum at x.

By Corollary 2.52 (b) on page 53, there is a nondegenerate distance function L_z that has a nondegenerate maximum at some point near x. Thus, we have $\mu_n(L_z) \geq 1$. Using the assumption that $\mu_n = 1$ or 0, depending on whether or not M^n is compact, we get that $\mu_n(L_z) = 1$, and M^n is compact. Then by Theorem 2.64 (b), the distance function L_p determined by the focal point p has an absolute maximum at x. Similar statements can be made about the smallest principal curvature λ_n, which is just the negative of the largest principal curvature for the unit normal field $-\xi$.

In summary, by Theorem 2.64, the conditions $\mu_0 = 1$, and $\mu_n = 1$ or 0, depending on whether or not M^n is compact, imply that if p is the focal point of (M^n, x) corresponding to the largest or smallest principal curvature at x, then L_p has an absolute minimum or absolute maximum at x. This is the key idea in the proof of the following lemma, which is due to Banchoff [20] and Cecil [72], and we follow the presentation in the book [95, pp. 191–193].

Lemma 5.47. *Let $M \subset \mathbf{R}^{n+1}$ be a properly embedded hypersurface. Suppose that for every nondegenerate Euclidean distance function L_p, we have $\mu_0 = 1$, and $\mu_n = 1$ or 0, depending on whether or not M^n is compact. Let λ be the largest or smallest principal curvature function on M. If λ has constant multiplicity 1 on some open set $U \subset M$, then λ is constant along its lines of curvature in U.*

Proof. Let γ be an arbitrary (connected) line of curvature of the principal curvature λ in the open set U. If λ is identically equal to zero on γ, then λ is constant along γ as needed. Suppose $\lambda(x_0) \neq 0$ for some point x_0 on γ. Since λ is a continuous function, the set

$$B = \{x \in \gamma \mid \lambda(x) = \lambda(x_0)\}$$

is closed in γ. We will now show that B is also open in γ, and thus it is all of γ by connectedness.

Let x be an arbitrary point in B, and let $W \subset U$ be a neighborhood of x on which λ is nonzero. Let X be a unit vector field in the principal foliation T_λ on W. We will show that $X\lambda = 0$ on W, and so λ is constant along the line of curvature γ in W, as needed.

Let y be a point in W, and let β be the normal section of M at y obtained by intersecting W with the plane spanned by $X(y)$ and the hypersurface normal $\xi(y)$. Parametrize β by arc-length so that $\beta(0) = y$ and the initial tangent vector $\overrightarrow{\beta}(0) = X(y)$. Let $\kappa(s)$ denote the curvature function of β, and let $\lambda(s) = \lambda(\beta(s))$. We now want to show that $\kappa(0) = \lambda(0)$ and $\kappa'(0) = \lambda'(0)$, where the prime denotes differentiation with respect to s.

The normal curvature $k_n(s)$ at the point $\beta(s)$ in the direction $\overrightarrow{\beta}(s)$ is given by the formula

$$k_n(s) = \langle A(\overrightarrow{\beta}(s)), \overrightarrow{\beta}(s)\rangle, \tag{5.293}$$

where A is the shape operator of M. By Meusnier's theorem, we have

$$k_n(s) = \kappa(s)\cos\phi(s), \tag{5.294}$$

where $\phi(s)$ is the angle between the principal normal to the curve β and the hypersurface normal ξ at the point $\beta(s)$. Since $\xi(y)$ is the principal normal to the curve β at $y = \beta(0)$, we have $\phi(0) = 0$, and thus $k_n(0) = \kappa(0)$. Since

$$k_n(0) = \langle A(\vec{\beta}(0)), \vec{\beta}(0)\rangle = \langle AX, X\rangle = \langle \lambda(0)X, X\rangle = \lambda(0),$$

we have $\kappa(0) = k_n(0) = \lambda(0)$. Furthermore, by differentiating equation (5.294) and evaluating at $s = 0$, we get $k_n'(0) = \kappa'(0)$. We now want to show that $k_n'(0) = \lambda'(0)$.

Decompose $\vec{\beta}(s)$ into its components in T_λ and $T_\lambda^\perp$, i.e.,

$$\vec{\beta}(s) = a(s)X(\beta(s)) + b(s)Y(\beta(s)), \tag{5.295}$$

where X is the unit vector field in T_λ and Y is a unit vector field along β in $T_\lambda^\perp$. Then X and Y are smooth vector fields, and a and b are smooth functions along the curve β. Furthermore, we have $a(0) = 1$, $b(0) = 0$ and $a'(0) = 0$, since 1 is the maximum value that the function a can attain along β. Applying A to equation (5.295), we get the following equation along the curve β,

$$A\vec{\beta} = a\lambda X + bAY, \tag{5.296}$$

where AY is also orthogonal to T_λ, since $T_\lambda^\perp$ is invariant under A. Thus, we get

$$k_n(s) = \langle A\vec{\beta}, \vec{\beta}\rangle = a^2\lambda + b^2\langle Y, AY\rangle. \tag{5.297}$$

Differentiating this equation and evaluating at $s = 0$, we get $k_n'(0) = \lambda'(0)$. Since we already have $k_n'(0) = \kappa'(0)$, we conclude that $\kappa'(0) = \lambda'(0)$.

We now complete the proof of the lemma as follows. Let p be the focal point

$$p = y + \frac{1}{\lambda(y)}\xi(y).$$

Let C be the osculating circle to the plane curve β at y, that is, the circle through y centered at p. Using Taylor's formula and the Frenet equations, it is easy to show (see, for example, Goetz [175, p. 84]) that β crosses C unless $\kappa'(0) = 0$. Thus, if $\kappa'(0) \neq 0$, the function L_p does not have an extreme value at y.

However, the hypotheses of the lemma imply that M is 0-taut, and if it is compact, it has the STPP. Therefore, M satisfies the hypotheses of Theorem 2.64 on page 59, and by that theorem L_p has an absolute minimum or absolute maximum at y, since λ is the largest or smallest principal curvature, as explained above the statement of the lemma.

Thus, the assumption that $\kappa'(0) \neq 0$ leads to a contradiction, and so we have

$$X\lambda = \lambda'(0) = \kappa'(0) = 0,$$

at every point $y \in W$. This implies that λ is constant along the line of curvature γ in W, and so the set B is open. Since B is both open and closed, it is all of γ. Therefore, λ is constant along the line of curvature γ in U. □

Classifications of taut surfaces in Euclidean space

From Lemma 5.47 and Theorem 5.26, we obtain the following classification of taut surfaces in $\mathbf{R}^3$ due to Banchoff [20] for the compact case and Cecil [72] for the non-compact case.

Theorem 5.48. *Let $M \subset \mathbf{R}^3$ be a taut connected surface.*

(a) *If M is compact, then M is a metric sphere or a ring cyclide.*
(b) *If M is not compact, then M is a plane, a circular cylinder, or a parabolic ring cyclide.*

Proof. (a) Suppose that M is compact. If M has one umbilic point, then it is a metric sphere by Corollary 2.68 on page 60. If M has no umbilic points, then M has two distinct principal curvatures at each point, one of which is the largest principal curvature, and the other is the smallest principal curvature. Tautness implies that $\mu_0 = 1$, and $\mu_2 = 1$ for every nondegenerate Euclidean distance function L_p. Thus by Lemma 5.47, each principal curvature is constant along each of its corresponding lines of curvature, that is, M is proper Dupin. Therefore, by Theorem 5.26, M is a ring cyclide.

(b) Suppose that M is not compact. If M has one umbilic point, then it is a plane in $\mathbf{R}^3$ by Corollary 2.66 on page 60. If M has no umbilic points, then M has two distinct principal curvatures at each point, one of which is the largest principal curvature, and the other is the smallest principal curvature. Tautness implies M is complete and that that $\mu_0 = 1$, and $\mu_2 = 0$ for every nondegenerate Euclidean distance function L_p. Thus by Lemma 5.47, each principal curvature is constant along each of its corresponding lines of curvature, that is, M is proper Dupin. Then by the classification of complete non-compact proper Dupin hypersurfaces with two principal curvatures in Theorem 5.26, M is a circular cylinder, or a parabolic ring cyclide. □

Theorem 5.48 and the arguments given at the beginning of this section yield the following classification of taut surfaces in Euclidean spaces.

Theorem 5.49. *Let M be a taut connected surface substantially embedded in $\mathbf{R}^m$.*

(a) *If M is compact, then M is a metric sphere or a ring cyclide in* $\mathbf{R}^3$*, a spherical Veronese surface in* $\mathbf{R}^5$*, or a compact surface in* $\mathbf{R}^4$ *related to one of these by stereographic or inverse stereographic projection.*
(b) *If M is not compact, then M is a plane, a circular cylinder, or a parabolic ring cyclide in* $\mathbf{R}^3$*, or it is the image in* $\mathbf{R}^4$ *of a punctured spherical Veronese surface in* $\mathbf{R}^5$ *under stereographic projection.*

Conversely, all of the surfaces listed in (a) *and* (b) *are taut.*

Remark 5.50 (Taut subsets of $\mathbf{R}^2$ *and* $\mathbf{R}^3$*).* Recall from Section 2.7 that a subset $X \subset \mathbf{R}^m$ is taut if the inclusion map f is a taut map, i.e., for every closed ball, complement of an open ball, and closed half-space Ω in $\mathbf{R}^m$, the induced homomorphism $H_*(f^{-1}\Omega) \to H_*(X)$ in Čech homology with $\mathbf{Z}_2$ coefficients is injective. An important type of subset $\mathbf{R}^2$ is the set X obtained by deleting $k \leq \infty$ disjoint open round disks from a closed round disk D. Such a set is called a *Swiss cheese*, and if the round interiors are everywhere dense in D, then X is called a *limit Swiss cheese*. Banchoff [20] showed that a compact, connected $X \subset \mathbf{R}^2$ is 0-taut if and only if X is a point, a circle, or a Swiss cheese. Kuiper [304] then showed that $X \subset \mathbf{R}^2$ is taut if and only if it is a point, a circle or a limit Swiss cheese. This follows from his lemma which states that if $X \subset S^m$ is taut and has an interior point, then $X = S^m$. The determination of taut subsets of $\mathbf{R}^3$ is much more difficult, but Kuiper [304] proved the following theorem.

Theorem 5.51. *A taut compact, connected ANR (absolute neighborhood retract) subset* $X \subset \mathbf{R}^3$ *is a point, a circle, a round 2-sphere or a ring cyclide of Dupin.*

5.10 Classifications of Taut Submanifolds

In this section, we present several classification results for taut submanifolds in detail, and then survey other known results on taut submanifolds. We will follow the presentations in the book [95, pp. 197–207] and the article [76]. Results have been obtained for manifolds with relatively simple homology, but general classifications of taut submanifolds are rare.

We first recall some known classifications that we have proven already. We have just classified taut embeddings of 2-dimensional manifolds in Theorem 5.49 in the previous section. Of course, the manifolds with the simplest homology are spheres, and we have Theorem 2.73 on page 63, which states that if $f : S^n \to \mathbf{R}^m$ is a substantial taut immersion, then $m = n + 1$, and f embeds S^n as a metric hypersphere. This is due to Banchoff [20] for $n = 2$ and Carter and West [61] for arbitrary dimensions, and it was also proven independently by Nomizu and Rodriguez [405]. The proof of Nomizu and Rodriguez also yields the slightly more general Theorem 2.74 on page 63, which we restate here for the sake of completeness.

Theorem 5.52. *Let $M^n, n \geq 2$, be a connected, complete Riemannian manifold isometrically immersed in $\mathbf{R}^m$. If every nondegenerate distance function L_p has index 0 or n at each of its critical points, then M^n is embedded as a totally geodesic n-plane or a metric n-sphere $S^n \subset \mathbf{R}^{n+1} \subset \mathbf{R}^m$.*

In Remark 2.60 on page 56, we saw that a spherical cylinder is taut. We next prove the related classification result due to Carter and West [61].

Theorem 5.53. *Let $f : M \to \mathbf{R}^{n+1}$ be a taut embedding of a connected, noncompact n-manifold with $H_k(M, \mathbf{Z}_2) = \mathbf{Z}_2$ for some k, $0 < k < n$, and $H_i(M, \mathbf{Z}_2) = 0$ for $i \neq 0, k$. Then M is diffeomorphic to $S^k \times \mathbf{R}^{n-k}$ and f is a standard product embedding of $S^k \times \mathbf{R}^{n-k}$ as a spherical cylinder.*

An important step in the proof of this theorem is to prove that the complement of the focal set of $f(M)$ in $\mathbf{R}^{n+1}$ is connected. In this case, this follows fairly quickly from Lemma 5.54 below. Later we will prove a more general result (Theorem 5.64) concerning the connectedness of the complement of the focal set of taut submanifold.

We first recall some terminology. Let $f : M \to \mathbf{R}^m$ be an embedding of a connected manifold, and let Γ denote the set of critical points of the normal exponential map $E : NM \to \mathbf{R}^m$. The focal set of $f(M)$ is the set $E(\Gamma) \subset \mathbf{R}^m$, i.e., the set of critical values of the normal exponential map E.

Lemma 5.54. *Let $f : M \to \mathbf{R}^m$ be a taut embedding of a connected manifold. Suppose that for each $x \in M$ there is at most one focal point of (M, x) on each normal ray to $f(M)$ at $f(x)$. Then $f(M)$ is disjoint from the focal set $E(\Gamma)$, and the complement of $E(\Gamma)$ in $\mathbf{R}^m$ is path-connected.*

Proof. Suppose that p is a point in $\mathbf{R}^m$ and that L_p has a nondegenerate minimum at $x \in M$. Then by Theorem 2.64 (a) on page 59, L_p has a strict absolute minimum on M at x. Now suppose that p is also a focal point of M, and thus that $p = E(y, \eta)$, where $(y, \eta) \in \Gamma$. By assumption, p is the only point of (M, y) on the normal ray to $f(M)$ at $f(y)$ in the direction η. Applying Theorem 2.64 (a) again, we get that L_p has an absolute minimum at y. This contradicts the fact that L_p has a strict absolute minimum at x. In particular, this applies when $p = f(x)$, and so no point in $f(M)$ lies in the focal set $E(\Gamma)$.

Now suppose that p and q are two points in the complement of the focal set $E(\Gamma)$ in $\mathbf{R}^m$. Then L_p and L_q are nondegenerate functions on M. By tautness, the nondegenerate function L_p has a unique strict absolute minimum at some point $x \in M$. Similarly, L_q has a unique strict absolute minimum at some point $y \in M$. If $z \in \mathbf{R}^m$ is any point on the closed line segment from p to $f(x)$, then the function L_z has a nondegenerate minimum at x, and so by the preceding paragraph, z is not in the focal set of $f(M)$. Similarly, every point on the closed line segment from q to $f(y)$ is in the complement of the focal set. We can now construct a path in the complement of the focal $E(\Gamma)$ from p to q. First traverse the line segment from p to $f(x)$, then follow a path in $f(M)$ from $f(x)$ to $f(y)$, and then traverse the line segment from $f(y)$ to q. Thus, the complement of the focal set is path-connected. □

We now use the lemma to complete the proof of Theorem 5.53 due to Carter and West [61].

Proof (of Theorem 5.53). By the assumption on the homology of M, tautness implies that every nondegenerate distance function L_p has exactly two critical points on M, one of index 0 and one of index k. Thus by Corollary 2.52 on page 53, if p is a focal point of $f(M)$, but L_p has a nondegenerate critical point of index j, then j equals 0 or k. Hence, there is at most one focal point of $f(M)$ along each normal ray, and that focal point has multiplicity k. Therefore, by Lemma 5.54, the complement Σ of the focal set is path-connected. The restriction of the normal exponential map E to the set $E^{-1}(\Sigma)$ is a double covering of Σ, since E is an immersion on $NM - \Gamma$, and since for each $p \in \Sigma$, the function L_p has exactly two critical points.

Since a taut embedding is a proper map, the image $f(M)$ is a closed subset of $\mathbf{R}^{n+1}$, and thus it is orientable, and we take a field of unit normals ξ on $f(M)$. By identifying $(x, t\xi) \in NM$ with (x, t), the normal bundle is diffeomorphic to $M \times \mathbf{R}$. Let

$$N^+ = \{(x,t) \mid t > 0\}, \quad N^- = \{(x,t) \mid t < 0\} \tag{5.298}$$

in $M \times \mathbf{R}$, and let $M_0 = M \times \{0\}$, the zero-section of NM.

For each $(x,t) \in E^{-1}(\Sigma)$, let $p = E(x,t)$. Then L_p has a nondegenerate critical point at x whose index will be denoted by $I(x,t)$. This index function is continuous and thus it is constant on any path component of $E^{-1}(\Sigma)$. Since Σ is connected, the set $E^{-1}(\Sigma)$ has just two components V_0 and V_k, on which the index function is equal to 0 and k, respectively. Since $M_0 \cap E^{-1}(\Sigma)$ lies in V_0, the component V_k lies entirely in either N_+ or N_-. By considering the unit normal field $-\xi$ instead of ξ, if necessary, we can assume that V_k lies in M_+. Then for each $x \in M^n$, there is at most one focal point of (M^n, x), and it lies on the normal ray in the direction ξ determined by N_+. On the other hand, there must be at least one focal point on each normal line, otherwise by Corollary 2.66 on page 60, tautness implies that $f(M)$ lies in a hyperplane in $\mathbf{R}^{n+1}$, which is clearly impossible, since f is not totally geodesic. Thus, there is exactly one focal point of multiplicity k on each normal line to $f(M)$, and so there are exactly two distinct principal curvatures at each point: a nonzero principal curvature λ of multiplicity k, and a second principal curvature $\mu = 0$ of multiplicity $n-k$.

We complete the proof by showing that $f(M)$ is proper Dupin. If $k > 1$, then λ is constant along each leaf of its principal foliation by Theorem 2.10 on page 24. If $k = 1$, then since λ is the largest or smallest principal curvature function on M, tautness implies that λ is constant along its lines of curvature by Lemma 5.47. Thus, $f(M)$ is a non-compact proper Dupin hypersurface with two principal curvatures at each point, one of which is identically zero, and it is complete since f is a proper embedding. Thus, by the classification of complete proper Dupin hypersurfaces with $g = 2$ in Theorem 5.26, $f(M)$ is a spherical cylinder $S^k \times \mathbf{R}^{n-k}$ in $\mathbf{R}^{n+1}$. □

We next prove a similar classification result which also characterizes spherical cylinders.

Theorem 5.55. *Let $f : M^n \to \mathbf{R}^{n+q}$ be a substantial taut embedding of a connected, non-compact n-manifold whose $\mathbf{Z}_2$-Betti numbers satisfy $\beta_k(M) = j > 0$ for some k with $\frac{n}{2} < k < n$, and $\beta_i(M) = 0$ for $i \neq 0, k$. Then $q = 1, j = 1$, and f embeds M as a spherical cylinder $S^k \times \mathbf{R}^{n-k} \subset \mathbf{R}^{n+1}$.*

Proof. As shown in the proof of the previous theorem, the index of any nondegenerate critical point of any distance function L_p is 0 or k. Thus, there can be at most one focal point on every normal ray to $f(M)$, and it has multiplicity k. Furthermore, since $2k > n$, and the sum of the multiplicities of the focal points on any normal line is at most n, there can be at most one focal point on any normal line to $f(M)$. On the other hand, since f is taut and substantial, there is at least one focal point on every normal line by Corollary 2.66 on page 60. Thus there is exactly one focal point on each normal line, and as in the previous proof, we conclude that for each unit normal vector ξ to $f(M)$, the shape operator A_ξ has exactly two eigenvalues, λ of multiplicity k which is never zero, and μ of multiplicity $n-k$ which is identically zero.

If the codimension q is greater than one, then the unit normal bundle BM is connected and there exists a continuous path in BM from a unit normal ξ to its negative $-\xi$. Since the principal curvature function λ is never zero, it has the same sign at ξ as it does at $-\xi$. On the other hand, $A_{-\xi} = -A_\xi$, and so the nonzero principal curvature λ of $A_{-\xi}$ has the opposite sign of the nonzero principal curvature of A_ξ, a contradiction. Thus, the codimension q equals one, and we have shown that $f(M)$ is a taut, connected non-compact hypersurface in $\mathbf{R}^{n+1}$ with two principal curvatures at each point, one of which is identically zero. Then, as in the previous proof, tautness implies that $f(M)$ is proper Dupin, and by Theorem 5.26, we get that $f(M)$ is a spherical cylinder $S^k \times \mathbf{R}^{n-k}$ in $\mathbf{R}^{n+1}$. □

Remark 5.56. The taut substantial embedding of the Möbius band $M^2 = \mathbf{RP}^2 - \{P\}$ into $\mathbf{R}^4$ obtained from a punctured Veronese surface $V^2 - \{P\} \subset S^4$ by stereographic projection $\tau : S^4 - \{P\} \to \mathbf{R}^4$ with pole P on V^2 shows that the hypothesis $\frac{n}{2} < k$ in Theorem 5.55 and the hypothesis that the codimension is one in Theorem 5.53 are both necessary.

Taut embeddings and the cyclides of Dupin

We now begin a sequence of classification results that include taut hypersurfaces, in particular, the higher-dimensional cyclides of Dupin. The first result is a generalization of the classification of taut surfaces in $\mathbf{R}^3$ given in Theorem 5.48. In 1971, Carter and West [61, pp. 712–714] showed that if M^{2k} is a taut compact $(k-1)$-connected hypersurface in $\mathbf{R}^{2k+1}$, then $H_k(M, \mathbf{Z})$ is either 0 or $\mathbf{Z} \oplus \mathbf{Z}$. In 1978, Cecil and Ryan [89] proved that if $M^n \subset \mathbf{R}^{n+1}$ is a taut compact hypersurface with the same integral homology as $S^k \times S^{n-k}$, $1 \leq k \leq \frac{n}{2}$, then M is a ring cyclide. Using tightness arguments, C.S. Chen [101] (see also [99, 100]) independently proved a similar result. Chen's theorem differs in that he only assumes that M is tight and lies in an ovaloid in $\mathbf{R}^{n+2}$. However, he excludes the case where $k/(n-k)$ is equal to 2 or $1/2$.

We first handle the case $k = n - k$, which is simpler than the case $k \neq n - k$. The proof originally given in [89, p. 184] actually works under the following weaker hypothesis which include non-compact hypersurfaces as well as compact ones (see also [95, pp. 200–201]).

Theorem 5.57. *Let $M^{2k} \subset \mathbf{R}^{2k+1}$ be a taut connected hypersurface such that $H_i(M, \mathbf{Z}_2) = 0$ for i not equal to 0, k, or $2k$.*

(a) *If M is compact, then M is a metric sphere or a ring cyclide.*
(b) *If M is not compact, then M is a hyperplane, a spherical cylinder, or a parabolic ring cyclide.*

Proof. For $k = 1$, there is no restriction on the homology, and the theorem follows from the classification of taut surfaces in $\mathbf{R}^3$ given in Theorem 5.48. So we now assume that $k > 1$.

Tautness implies that M is properly embedded, and so it is a closed subset of $\mathbf{R}^{n+1}$. Thus, M is orientable (see Samelson [471]). Let ξ be a field of unit normals on M, and we consider the principal curvature functions on M determined by the shape operator A_ξ. If the taut hypersurface M has one umbilic point, then it is totally umbilic by Corollaries 2.66 and 2.68 (see page 60), and then it is a hyperplane or a metric sphere.

Assume now that M has no umbilic points. Let x be an arbitrary point in M. There is at least one focal point of (M, x) along the normal line to M at x, otherwise M would have a planar (umbilic) point at x. Let p be the first focal point on a normal ray to M at x. Let q be a point beyond p on that normal ray, but before the second focal point on that normal ray (if such a second focal point exists). By the Index Theorem (Theorem 2.51 on page 53), the function L_q has a nondegenerate critical point of index $j > 0$ at x, where j is the multiplicity of the focal point p.

The function L_q may or may not be a Morse function, but by Corollary 2.52 on page 53, there is a Morse function L_z that has a nondegenerate critical point $y \in M$ of index j. Tautness and the assumption on the homology of M imply that this index j is 0, k or $2k$. The number j is greater than zero, and it cannot equal $2k$, for then p would be a focal point of multiplicity $2k$, and thus x would be an umbilic point. Therefore, the value of j is k. Furthermore, if there is a second focal point on this normal ray, it also has multiplicity k. Otherwise it would give rise to a distance function having a critical point at x with index between k and $2k$, contradicting tautness. Thus, every focal point of (M, x) has multiplicity k.

Given the correspondence between focal points and principal curvatures in Theorem 2.1 on page 11, we have shown that at every point $x \in M$, there are two distinct principal curvatures, each of which has multiplicity k. Since $k > 1$, these principal curvatures are constant along the leaves of their corresponding principal foliations by Theorem 2.10 on page 24. Thus, M is a complete proper Dupin hypersurface with $g = 2$ principal curvatures at each point, and the theorem now follows from the classification of such hypersurfaces in Theorem 5.26. □

Taut embeddings of highly connected manifolds

Remark 5.58 (Taut embeddings of highly connected manifolds). The hypotheses of Theorem 5.57 imply that M^{2k} is $(k-1)$-connected. To see this, note that tautness and the assumptions on the homology of M^{2k} imply that every nondegenerate L_p function has only critical points of index 0, k, and $2k$. By applying basic Morse theory to any choice of nondegenerate L_p function, we get that M^{2k} has the homotopy type of a CW-complex with cells of dimension 0, k, and $2k$ only. Thus, the $(k-1)$-skeleton of M^{2k} is just the 0-skeleton, and so M^{2k} is $(k-1)$-connected (see, for example, [355, p. 297]).

Thorbergsson [534] generalized Theorem 5.57 to higher codimension by proving that if M^{2k} is a compact $(k-1)$-connected (not k-connected) taut submanifold of $\mathbf{R}^m$ that is substantial and non-spherical, then one of the following holds:

(a) $m = 2k+1$, and M^{2k} is a cyclide of Dupin diffeomorphic to $S^k \times S^k$,
(b) $m = 3k+1$, and M^{2k} is diffeomorphic to one on the projective planes $\mathbf{RP}^2$, $\mathbf{CP}^2$, $\mathbf{HP}^2$, $\mathbf{OP}^2$, for $k = 1, 2, 4, 8$, respectively.

Of course, the standard embeddings of these projective planes (see Section 2.9, page 74) are substantial taut embeddings into $\mathbf{R}^{3k+2}$, but they are spherical. One obtains a substantial, non-spherical taut embedding of the appropriate projective plane into $\mathbf{R}^{3k+1}$ by applying stereographic projection to the image of a standard embedding. Thorbergsson [534] has similar results for noncompact taut submanifolds in $\mathbf{R}^m$, and for taut, compact $(k-1)$-connected hypersurfaces in hyperbolic space H^{2k+1}.

Kuiper [302, p. 231] and [303, p. 133] proved that a tight immersion of a $(k-1)$-connected M^{2k} satisfies rather stringent conditions, and his results were further refined by Thorbergsson [535]. Hebda [193] showed that a connected sum of arbitrarily many copies of $S^k \times S^k$ can be realized as a tight hypersurface in $\mathbf{R}^{2k+1}$. Hebda's examples are counterexamples to a conjecture of Kuiper [303, p. 116] that every smooth $(k-1)$-connected $2k$-dimensional manifold $k \geq 2$ which admits a smooth tight immersion into Euclidean space is diffeomorphic to one of the following manifolds: $\mathbf{CP}^2$, $\mathbf{HP}^2$, $\mathbf{OP}^2$ or $S^k \times S^k$. By (a) above, we see that Hebda's examples are not taut, as Hebda showed in his paper.

Taut embeddings of $S^k \times S^{n-k}$ for $k \neq n-k$

We now prove the analogue of part (a) of Theorem 5.57 in the case $k \neq n-k$. This proof is substantially more difficult than in the case $k = n-k$ given in Theorem 5.57. The theorem below was first proven in the paper of Cecil and Ryan [89], and we follow the presentation given in the book [95, pp. 202–207].

Theorem 5.59. *Let $M^n \subset \mathbf{R}^{n+1}$ be a taut connected hypersurface with the same $\mathbf{Z}_2$-homology as $S^k \times S^{n-k}$ for $k \neq n-k$. Then M is a ring cyclide.*

Proof. For the sake of definiteness, we assume that $k < n - k$. By the assumption on the homology of M, every nondegenerate distance function L_p has four critical points with respective indices $0, k, n - k$ and n. As in the proof of Theorem 5.57, tautness implies that M is orientable, and we take ξ to be a field of unit normals on M and consider the principal curvature functions on M determined by the shape operator A_ξ.

Let x be an arbitrary point in M. We want to try to determine the number of distinct principal curvatures at x. This number is greater than one, otherwise x is an umbilic point, and we know from Corollary 2.68 on page 60 that if a taut compact, connected hypersurface has one umbilic point, then it is a metric sphere. That is impossible by the assumption on the homology of M.

Next we use the Index Theorem (Theorem 2.51 on page 53) to prove that the number of distinct principal curvatures at x is at most three. Let p the first focal point on a normal ray to M at x, and let q be a point beyond p on that normal ray, but before the second focal point on that normal ray (if such a second focal point exists). The Index Theorem implies that the function L_q has a nondegenerate critical point of index $j > 0$ at x, where j is the multiplicity of the focal point p. Then by Corollary 2.52 on page 53, there is a Morse function L_z that has a nondegenerate critical point $y \in M$ of the same index j. Tautness and the assumption on the homology of M imply that this index j is k, $n - k$ or n. The value n is impossible, since then the focal point p would have multiplicity n, and x would be an umbilic point. Thus, the value of j is k or $n - k$.

If multiplicity of p is k, then there could exist two other focal points of (M, x) on the normal line to M at x. In that case, the multiplicities of those two focal points are $n - 2k$ and k by the Index Theorem and the fact that the index of any nondegenerate critical point of any distance function is $0, k, n-k$ or n. Then the sum of the multiplicities of the focal points is n, and so there are exactly three distinct principal curvatures at x with respective multiplicities k, $n - 2k$ and k. There could also be less than three focal points on the normal to M at x. In that case, there are at most three distinct principal curvatures at x, counting the possibility that one principal curvature could equal zero and not give rise to a focal point. Again if there are three distinct principal curvatures, they have respective multiplicities k, $n - 2k$ and k.

If the first focal point p on a normal ray has multiplicity $n - k$, then there can be at most one other focal point of (M, x) on the normal line to M at x, and it has multiplicity k. In that case, there can be at most two distinct principal curvatures at x. The following three lemmas show that M has two principal curvatures at each point and is a ring cyclide. □

Number of distinct principal curvatures

In summary, we have shown that the number of distinct principal curvatures at a given point $x \in M$ is either two or three. Furthermore, if there are three principal curvatures, they have respective multiplicities k, $n-2k$ and k. Let U be the subset of M on which there are three distinct principal curvatures. Then U is an open subset of M, and on U there are three distinct smooth principal curvature functions

$$\lambda_1 > \lambda_2 > \lambda_3, \tag{5.299}$$

which have respective multiplicities k, $n-2k$ and k. We now begin a sequence of lemmas which eventually lead to the conclusion that U is the empty set, and thus there are exactly two distinct principal curvatures at each point of M. In all of these lemmas, we assume the hypotheses of Theorem 5.59.

Lemma 5.60. *The number of distinct principal curvatures is a constant which is either 2 or 3.*

Proof. We now assume that the set U defined above is non-empty and is not all of M, and then derive a contradiction. Since M is an embedded compact, connected hypersurface in $\mathbf{R}^{n+1}$, the concept of inner and outer normal ray is well defined. Let x be any point on the boundary of the set U in M. Since U is open, and since M has no umbilic points, there are exactly two distinct principal curvatures at x, and they have multiplicities k and $n-k$. By Corollary 2.65 on page 59, there is a focal point of M on every inner normal ray. By inverting M in a sphere, if necessary, we can assume that the first focal point on the inner normal ray to M at x has multiplicity $n-k$. Of course, inversion in a sphere preserves tautness by Theorem 2.70 on page 61, and it preserves the multiplicities of the principal curvatures by Theorem 2.6 on page 20. We will complete the proof of the lemma by constructing a nondegenerate distance function that has at least five critical points, contradicting tautness.

Let ξ be the globally defined unit inner normal field. The first focal point on the inner normal ray to M at x is $p = x + \rho\xi$, where $\rho = 1/\lambda$, and λ is the largest principal curvature at x. By our construction in the preceding paragraph, the focal point p has multiplicity $n-k$. By Theorem 2.64 on page 59, the function L_p has an absolute minimum value $\alpha = \rho^2$ at x. Since the segment from x to p cannot intersect M, the point p lies inside M (i.e., in the bounded component of the complement of M).

Let $y \in M$ be a point where L_p has an absolute maximum on M. Since M is not a metric sphere, we know that $L_p(x) < L_p(y)$. The function L_p has a maximum at y, and so the sum of the multiplicities of the focal points of (M, y) on the closed segment from y to p is n. Since we also know that there is at least one focal point on any inner normal ray, we conclude that p lies on the inner normal ray to M at y. Let q be the first focal point of (M, y) on the inner normal ray to M at y. Then $q \neq p$, since L_q has an absolute minimum at y by Theorem 2.64, while L_p has an absolute maximum at y, and the minimum and maximum values of L_p cannot be the same, since M is not a metric sphere. Furthermore, the point q also lies inside M, since the segment from y to q cannot intersect M.

Let $\gamma = L_q(y)$, the absolute minimum value of L_q on M, and recall that $\alpha = L_p(x)$ is the absolute minimum value of L_p on M. Using our notation from critical point theory, let

$$M_\alpha(L_p) = \{z \in M \mid L_p(z) \leq \alpha\}, \tag{5.300}$$

and let $M_\gamma(L_q)$ be defined in a similar way. We first show that $M_\alpha(L_p)$ and $M_\gamma(L_q)$ are disjoint.

Suppose that z is a point in the intersection of $M_\alpha(L_p)$ and $M_\gamma(L_q)$. Then both L_p and L_q have an absolute minimum at z. Since p and q are both inside M, and the closed segments $[z, p]$ and $[z, q]$ contain no points of M other than z, these segments both lie on the inner normal ray to M at z. Furthermore, the normal lines to M at z and y coincide, since they both contain the points p and q.

We know that $z \neq y$, since L_p does not have a minimum at y. If $z = x$, then p is the first focal point on the inner normal ray to M at z. If $z \neq x$, then L_p does not have a strict absolute minimum at z, since $L_p(x) = L_p(z)$. By Theorem 2.64, L_p has a degenerate minimum at z, and again we conclude that p is the first focal point on the inner normal to M at z. Thus, in either case, q lies beyond the first focal point p on the inner normal ray to M at z, and so L_q cannot have an absolute minimum at z, contradicting the assumption that z is in $M_\gamma(L_q)$.

Since $M_\alpha(L_p)$ and $M_\gamma(L_q)$ are compact, there exists $\varepsilon > 0$ such that $M_{\alpha+\varepsilon}(L_p)$ and $M_{\gamma+\varepsilon}(L_q)$ are also disjoint. Since the focal set of M has measure zero in $\mathbf{R}^{n+1}$, there exists a point $p' \in \mathbf{R}^{n+1}$ and $r > 0$ such that $L_{p'}$ is a nondegenerate function on M, and r is a non-critical value of $L_{p'}$ satisfying

$$M_\alpha(L_p) \subset M_r(L_{p'}) \subset M_{\alpha+\varepsilon}(L_p). \tag{5.301}$$

Since x is on the boundary of U, x has an arbitrarily close neighbor w at which there are three distinct principal curvatures. By considering the focal points on the inner normal to M at w and using Corollary 2.52 on page 53, one can produce points a and b in $\mathbf{R}^{n+1}$ arbitrarily near to p, such that L_a and L_b are Morse functions with respective critical points u and v with indices k and $n - k$, arbitrarily close to x.

Let $\sigma = L_a(u)$ and $\delta = L_b(v)$. The points a and b can be chosen so that $M_\sigma(L_a)$ and $M_\delta(L_b)$ are both contained in $M_r(L_{p'})$. The function L_a has at least two critical points in $M_\sigma(L_a)$, a minimum and a critical point of index k at u. By tautness and the homology of M, we conclude that the $\mathbf{Z}_2$-Betti numbers of $M_\sigma(L_a)$ satisfy

$$\beta_0(M_\sigma(L_a)) = 1, \quad \beta_k(M_\sigma(L_a)) = 1. \tag{5.302}$$

Similarly, we can show that

$$\beta_0(M_\delta(L_b)) = 1, \quad \beta_{n-k}(M_\delta(L_b)) = 1. \tag{5.303}$$

We have the inclusion maps,

$$M_\sigma(L_a) \xrightarrow{i} M_r(L_{p'}) \xrightarrow{j} M. \tag{5.304}$$

By tautness, the induced maps j_* and $(j \circ i)_*$ on homology are injective, and so i_* is also injective. Thus, we get $\beta_k(M_r(L_{p'})) = 1$. By applying the same argument to $M_\delta(L_b)$, we get $\beta_{n-k}(M_r(L_{p'})) = 1$. Since $L_{p'}$ must also have a minimum in $M_r(L_{p'})$, tautness implies that $L_{p'}$ has at least three critical points in $M_r(L_{p'})$ with respective indices $0, k, n-k$.

On the other hand, we know that the sets $M_r(L_{p'})$ and $M_{\gamma+\varepsilon}(L_q)$ are disjoint. Let q' be a point just beyond q on the normal line to M at y and s a real number such that $M_s(L_{q'}) \subset M_{\gamma+\varepsilon}(L_q)$. Then the function $L_{q'}$ has a nondegenerate critical point of index h at y, where h is the multiplicity of the focal point q of (M, y). Thus, h is either k or $n-k$. By tautness, we get $\beta_h(M_s(L_{q'})) = 1$.

Note that $-L_{p'}$ is a perfect Morse function, since $L_{p'}$ is a perfect Morse function, and so the injectivity condition on homology holds for sublevel sets of $-L_{p'}$ as well. Let

$$M_{-r}(-L_{p'}) = \{z \in M \mid -L_{p'}(z) \leq -r\} = \{z \in M \mid L_{p'}(z) \geq r\}. \tag{5.305}$$

Since $M_s(L_{q'}) \subset M_{\gamma+\varepsilon}(L_q)$, and $M_{\gamma+\varepsilon}(L_q)$ is disjoint from $M_r(L_{p'})$, we get that

$$M_s(L_{q'}) \subset M_{-r}(-L_p) \subset M. \tag{5.306}$$

As before, tautness and the fact that $\beta_h(M_s(L_{q'})) = 1$ imply that

$$\beta_h(M_{-r}(-L_{p'})) = 1,$$

and so the perfect Morse function $-L_{p'}$ has critical points of index 0 and h in $M_{-r}(-L_{p'})$. Thus, $L_{p'}$ has critical points of index n and $n-h$ in $M_{-r}(-L_{p'})$. These two critical points are distinct from the three critical points found earlier in $M_r(L_{p'})$, since r is not a critical value of $L_{p'}$. Thus $L_{p'}$ has at least five critical points on M, contradicting tautness.

This contradiction results from the assumption that the set U is a non-empty proper subset of M. We conclude that U is either empty or else all of M, that is, there are either two distinct principal curvatures at each point of M, or else three distinct principal curvatures at each point of M. □

The complement of the focal set is connected

Using Lemma 5.60 and the fact that a taut embedding is Dupin (see Theorem 2.83 on page 68), we can prove the next lemma.

Lemma 5.61. *Each sheet of the focal set of M is an immersed submanifold of codimension greater than one. Thus, the complement of the focal set in $\mathbf{R}^{n+1}$ is path-connected.*

Proof. By Lemma 5.60, the number of distinct principal curvatures is constant on M. Thus, any given principal curvature function λ has constant multiplicity ν, and both λ and its corresponding ν-dimensional principal foliation are smooth. If $\nu > 1$, then the corresponding sheet of the focal set $f_\lambda(M)$ is an immersed submanifold of dimension $n - \nu$ (codimension $\nu + 1$) by Theorem 2.12 on page 26. If $\nu = 1$, then since a taut embedding is Dupin by Theorem 2.83, λ is constant along its lines of curvature. Thus $f_\lambda(M)$ is an immersed submanifold of dimension $n-1$ (codimension 2) by Theorem 2.14 on page 28.

Next we show that the complement $\Sigma = \mathbf{R}^{n+1} - E(\Gamma)$ of the focal set $E(\Gamma)$ is path-connected. We have shown that the focal set $E(\Gamma)$ is the union of the images of at most three immersions $g_i : M_i \to \mathbf{R}^{n+1}$, where $\dim M_i = n - \nu_i$, for ν_i equal to the multiplicity of the principal curvature λ_i. Thus, these immersions all have codimension greater than one.

Let p and q be any two points in Σ. Cover each M_i by a countable number of compact $(n - \nu_i)$-dimensional disks D_{ij} such that the restriction of g_i to each D_{ij} is an embedding. Transversality (see, for example, Hirsch [200, p. 74]) implies that there is a path from p to q which is disjoint from $g_i(D_{ij})$ for all i, j, and so Σ is path-connected. □

We now complete the proof of Theorem 5.59 following the approach used by Carter and West [61] in proving Theorem 5.53.

Lemma 5.62. *There exist exactly two principal curvatures on M and they are constant along the leaves of their corresponding principal foliations. Hence, M is a ring cyclide.*

Proof. Since M is taut, it is Dupin by Theorem 2.83, and so M is a compact proper Dupin hypersurface in $\mathbf{R}^{n+1}$ with $g = 2$ or $g = 3$ principal curvatures at each point. We now show that $g = 2$ and not 3. The normal bundle NM is diffeomorphic to $M \times \mathbf{R}$, and we can write $NM = N^+ \cup N^- \cup N_0$, as in the proof of Theorem 5.53.

Let $\Sigma = \mathbf{R}^{n+1} - E(\Gamma)$ be the complement of the focal set in $\mathbf{R}^{n+1}$, and let $V = E^{-1}(\Sigma)$. Then $E : V \to \Sigma$ is a four-fold covering map with $E^{-1}(p)$ consisting of four points in V corresponding to the four critical points of L_p. As in Theorem 5.53, the index function is locally constant on V, and it induces the decomposition of V into the disjoint union of open sets,

$$V = V_0 \cup V_k \cup V_{n-k} \cup V_n. \tag{5.307}$$

Since Σ is connected by Lemma 5.61, each V_i is connected. Since the intersection of the zero-section M_0 with V is contained in V_0, the connected set V_k lies entirely in either N^+ or N^-. Suppose now that there are three distinct principal curvatures $\lambda_1 > \lambda_2 > \lambda_3$ on M with respective multiplicities $k, n-2k, k$. The homology of M is different than that of a sphere S^n, and thus we know that M is not a convex hypersurface in $\mathbf{R}^{n+1}$. Thus, there exists a point $x \in M$ such that there are focal points of (M, x) on both the inner and outer normal rays to M at x.

Since M has three principal curvatures at each point, the first focal point on each normal ray has multiplicity k. Suppose the outer normal ray at x is equal to the set of points of the form $E(x,t)$ for $t > 0$. If we take a point $p = E(x,t)$, with $t > 0$, beyond the first focal point of (M,x) but before the second focal point of (M,x) on this outer normal ray, then L_p has a nondegenerate critical point of index k at x. By Corollary 2.52 on page 53, there is a point $q = E(y,s)$ with $s > 0$ near p such that L_q is a nondegenerate function having a critical point y of index k near to x. Thus, the point (y,s) is in the set $V_k \cap N^+$. Similarly, by considering the inner normal ray at x, we can produce a point in $V_k \cap N^-$. Thus, the set V_k is not connected, a contradiction. We conclude that the number of distinct principal curvatures of M is two. Thus, M is a connected, compact proper Dupin hypersurface in $\mathbf{R}^{n+1}$ with $g = 2$ principal curvatures, and so it is a ring cyclide by Theorem 5.26. This completes the proof of Theorem 5.59. □

Taut conformally flat hypersurfaces

One can get a similar characterization of taut conformally flat hypersurfaces in $\mathbf{R}^{n+1}$ as follows. Recall that a Riemannian manifold (M,g) with Riemannian metric g is *conformally flat* if every point of M has a neighborhood that is conformally equivalent to an open subset in Euclidean space $\mathbf{R}^n$.

Schouten [477] proved that if M^n, $n \geq 4$, is an immersed hypersurface in $\mathbf{R}^{n+1}$, then M^n is conformally flat in the induced metric if and only if at least $n-1$ of the principal curvatures coincide at each point of M^n. This characterization does not hold if $n = 3$, and Lancaster [312, p. 6] gave an example of a conformally flat hypersurface M^3 in $\mathbf{R}^4$ with three distinct principal curvatures. Using Schouten's result, Theorem 5.59, and some basic results for taut immersions, Cecil and Ryan [92] proved the following theorem.

Theorem 5.63. *Let M^n, $n \geq 4$, be a connected manifold tautly embedded in $\mathbf{R}^{n+1}$. Then M is conformally flat in the induced metric if and only if it is one of the following:*

(a) *a hyperplane or metric hypersphere;*
(b) *a cylinder over a circle or an $(n-1)$-sphere;*
(c) *a ring cyclide diffeomorphic to $S^1 \times S^{n-1}$;*
(d) *a parabolic ring cyclide diffeomorphic to $(S^1 \times S^{n-1}) - \{p\}$.*

Proof. Since M is tautly embedded, it is a complete hypersurface in $\mathbf{R}^{n+1}$. By Corollaries 2.66 and 2.68, tautness implies that if M has one umbilic point, then M is totally umbilic, and thus it is a hyperplane or a metric hypersphere. Otherwise, M has at least two distinct principal curvatures at each point, and then by Schouten's theorem, M has exactly two distinct principal curvatures at each point, λ of multiplicity $n-1$, and μ of multiplicity one. Since taut implies Dupin by Theorem 2.83, M is a complete proper Dupin hypersurface with two principal curvatures at each point, and the theorem follows from Theorem 5.26. □

Taut implies connectedness of the complement of the focal set

The following generalization of Lemma 5.61 is of independent interest. This was proven by Cecil, Chi, and Jensen [84, p. 237]. For simplicity, we will prove the theorem for submanifolds of the sphere S^n, since then each principal curvature of each shape operator A_ξ gives rise to a focal point (actually two antipodal focal points) of M in S^n.

Theorem 5.64. *Let $M \subset S^n$ be a connected taut submanifold of S^n. Then the complement of the focal set of M in S^n is connected.*

Proof. We first handle the case where $M \subset S^n$ is a connected taut hypersurface. Then M is compact and orientable, and we let ξ be a field of unit normal vectors on M. The normal bundle NM is diffeomorphic to $M\times\mathbf{R}$, and the normal exponential map $E : M \times \mathbf{R} \to S^n$ is defined by

$$E(x,t) = \cos t\, x + \sin t\, \xi. \tag{5.308}$$

A point $p = E(x,t)$ is a focal point of (M,x) of multiplicity $m > 0$ if the nullity of the derivative map E_* is equal to m at (x,t). Let $\Gamma \subset M \times \mathbf{R}$ denote the set of critical points of E. Then the focal set of M is the set $E(\Gamma) \subset S^n$ of critical values of E. Let Σ denote the complement of the focal set $E(\Gamma)$ in S^n.

In the proof of this theorem, we will use Federer's version of Sard's Theorem [154, p. 316] which implies that the image of the set of critical points of a smooth function $f : \mathbf{R}^k \to \mathbf{R}^l$ at which the rank of the derivative is less than or equal to ν is of $\mathcal{H}^\nu$-measure 0, where $\mathcal{H}^\nu$ denotes the Hausdorff ν-dimensional measure.

By Theorem 2.1 on page 11, each principal curvature λ of M at x gives rise to two antipodal focal points, $p = E(x,t)$ and $-p = E(x, t-\pi)$, where $\lambda = \cot t$ for $0 < t < \pi$. Thus, the set of focal points of (M,x) is antipodally symmetric along the normal geodesic to M at x. Label the principal curvature functions on M as

$$\lambda_1 \geq \lambda_2 \geq \cdots \geq \lambda_{n-1}. \tag{5.309}$$

Then each λ_i is a continuous function (see Ryan [468, p. 371]). Furthermore, if a continuous principal curvature function λ has constant multiplicity m on M, then λ is a smooth function, and its m-dimensional distribution T_λ of principal vectors is a smooth foliation by Theorem 2.10. Reckziegel [457, 458] and Singley [486], independently, showed that for each i, there is an open dense subset of M on which the principal curvature function λ_i has locally constant multiplicity and is therefore smooth.

Choose t_i, $0 < t_i < \pi$, such that $\lambda_i = \cot t_i$, then the corresponding focal map,

$$f_i(x) = \cos t_i\, x + \sin t_i\, \xi, \tag{5.310}$$

is continuous on M and smooth on a dense open subset of M.

For each i, let O_i be the open subset of M on which λ_i has multiplicity 1. (The set O_i could be empty.) If non-empty, then O_i consists of countably many open components O_{ij}, for $j = 1, 2, \ldots$, such that the restriction of f_i to O_{ij} is an embedded submanifold of dimension $n-2$. This follows from Theorem 2.14, since taut implies Dupin (Theorem 2.83), and Dupin implies that λ_i is constant along its lines of curvature in O_i.

Let Z_i be the complement of O_i in M. (The set Z_i could be empty.) At each point x of Z_i, the principal curvature λ_i has multiplicity at least 2, and so the normal exponential map E has rank $\leq n-2$ at the point (x, t_i). Thus, the focal point $f_i(x)$ lies in the singular value set K of points for which the derivative of E has rank $\leq n-2$.

Thus the entire focal set $E(\Gamma)$ is composed of the countably many embedded submanifolds $f_i(O_{ij})$ of dimension $n-2$, their antipodal sets, and the set K which has Hausdorff $(n-2)$-measure zero by the theorem of Federer mentioned above. Therefore, the Hausdorff $(n-1)$-measure of the whole focal set $E(\Gamma)$ is zero, which implies that the complement Σ of $E(\Gamma)$ in S^n is connected (see Schoen and Yau [476, p. 269]).

Next suppose that $M \subset S^n$ is a connected taut submanifold of codimension greater than one. Then M must be compact. Let M_ε be a tube over M of sufficiently small radius ε so that M_ε is an embedded hypersurface in S^n. By Theorem 2.81, M_ε is taut, and so by the argument above, the complement of the focal set of M_ε in S^n is connected. By Theorem 2.3, the focal set of the tube M_ε consists of the union of the focal set of M with M itself. Thus, the complement of the focal set of M is also connected. □

Taut embeddings of 3-manifolds

Concerning taut embeddings of 3-manifolds, Pinkall and Thorbergsson [449] have proven the following result.

Theorem 5.65. *A compact taut 3-dimensional submanifold in Euclidean space is diffeomorphic to one of the following seven manifolds:*
S^3, $\mathbf{RP}^3$, *the quaternion space* $S^3/\{\pm 1, \pm i, \pm j, \pm k\}$, *the 3-torus* T^3, $S^1 \times S^2$, $S^1 \times \mathbf{RP}^2$, $S^1 \times_h S^2$, *where h denotes an orientation reversing diffeomorphism of* S^2. *Furthermore, all of these manifolds admit taut embeddings.*

Pinkall and Thorbergsson gave more detail about these embeddings as follows. Since tautness is invariant under stereographic projection (see Theorem 2.70 on page 61), they classified *spherically substantial* taut embeddings, i.e., those which do not lie in any hypersphere. In the description below, the codimension means the spherically substantial codimension.

A taut embedding of S^3 is a metric hypersphere, as shown in Theorem 2.73 on page 63. Real projective space $\mathbf{RP}^3$ can be tautly embedded with codimension 2 as the Stiefel manifold $V_{3,2} \subset S^5 \subset \mathbf{R}^6$ (see Subsection 3.8.4, page 155), and with codimension 5 as $SO(3)$ in the unit sphere in the space of 3×3 matrices. It is not known whether the codimensions 3 and 4 are possible.

The quaternion space is embedded as Cartan's isoparametric hypersurface in S^4 (see Subsection 3.8.3), where it is unique up to Lie equivalence, and no other codimensions are possible. The 3-torus can be tautly embedded with codimension one as a tube in $\mathbf{R}^4$ around a torus of revolution $T^2 \subset \mathbf{R}^3 \subset \mathbf{R}^4$, and with codimension 2 as $T^2 \times S^1 \subset \mathbf{R}^5$.

The space $S^1 \times S^2$ can be tautly embedded with codimension 1 as a cyclide of Dupin (see Theorem 5.26 on page 282), and no other codimension is possible. The manifold $S^1 \times \mathbf{RP}^2$ can be tautly embedded with codimension 3 as the product of a metric circle and a Veronese surface. It can be tautly embedded with codimension 2 as a rotational submanifold with profile submanifold $\mathbf{RP}^2$, and the only codimensions possible are 2 and 3. Finally, $S^1 \times_h S^2$ can be tautly embedded with codimension 2 as the "complexified unit sphere"

$$\{e^{i\theta}x \mid \theta \in \mathbf{R}, x \in S^2 \subset \mathbf{R}^3\} \subset S^5 \subset \mathbf{C}^3. \tag{5.311}$$

This is one of the focal submanifolds of a homogeneous family of isoparametric hypersurfaces with four principal curvatures in S^5, the other being a Stiefel manifold $V_{2,3}$ (see Subsection 3.8.4, page 155). No other codimensions are possible for a taut embedding of $S^1 \times_h S^2$.

Other results on taut submanifolds

Remark 5.66 (Taut embeddings of 4-manifolds). In a nice survey article on taut submanifolds, Gorodski [177] obtained some partial results on taut embeddings of 4-manifolds into spheres. Let M be a compact, connected smooth 4-dimensional taut submanifold of a sphere S^n for some n. Gorodski showed that if M has vanishing first Betti number, then M is diffeomorphic to S^4, $S^2 \times S^2$ or $\mathbf{CP}^2$, and if M has vanishing second Betti number, then M is diffeomorphic to S^4 or $S^1 \times S^3$.

Remark 5.67 (Taut embeddings of homogeneous spaces). Many important examples of taut embeddings are homogeneous spaces, e.g., principal orbits of isotropy representations of symmetric spaces. Thorbergsson [536] found some necessary topological conditions for the existence of a taut embedding which enabled him to prove that certain homogeneous spaces do not admit taut embeddings. Similarly, Hebda [194] found certain necessary cohomological conditions for the existence of a taut embedding, and he used these results to give examples of manifolds which cannot be tautly embedded. In the case where M is a compact homogeneous submanifold substantially embedded in Euclidean space with flat normal bundle, Olmos [410] showed that the following statements are equivalent:

(a) M is taut;
(b) M is Dupin;

(c) M has constant principal curvatures;
(d) M is an orbit of the isotropy representation of a symmetric space;
(e) the first normal space of M coincides with the normal space.

Remark 5.68 (Taut representations). Gorodski and Thorbergsson [180, 181] studied *taut representations*, i.e., representations of compact Lie groups all of whose orbits are tautly embedded. Bott and Samelson [49] proved that isotropy representations of symmetric spaces (also called s-representations) are taut. For a long time, the s-representations were the only known examples of taut representations, but in the paper [181], Gorodski and Thorbergsson classified taut irreducible representations of compact Lie groups. Their classification includes three families of representations that are not s-representations, thereby supplying many new examples of tautly embedded homogeneous spaces. In a subsequent paper, Gorodski [178] gave a complete classification of all taut representations of compact simple Lie groups.

In related work, the class of polar representations was introduced by Dadok and Kac [123] in 1985. In that same year, Dadok [122] proved that a polar representation of a compact Lie group has the same orbits as the isotropy representation of a Riemannian symmetric space. More recently, Geatti and Gorodski [172] extended this theory by showing that a polar orthogonal representation of a connected real reductive algebraic group has the same closed orbits as the isotropy representation of a semi-Riemannian symmetric space.

In a related area, a proper isometric action of a Lie group G on a Riemannian manifold M is called *polar* if there exists a connected, complete submanifold Σ (called a *section*) that meets all orbits of G orthogonally. A basic result is that a section Σ is a totally geodesic submanifold of M. Biliotti and Gorodsky [40] proved that the orbits of a polar action of a compact Lie group on a compact rank one symmetric space are $\mathbf{Z}_2$-taut.

Remark 5.69 (Cylindrically taut immersions). Carter, Mansour, and West [59, 65] introduced a notion of *k-cylindrical taut* immersion $f : M \to \mathbf{R}^n$ by using distance functions from k-planes in $\mathbf{R}^n$ (see also Carter and Şentürk [60], and Carter and West [65]). For $k = 0$, this is equivalent to tautness, and for $k = n - 1$ it is equivalent to tightness. This theory turns out to closely related to the theory of convex sets and many of the results concern embeddings of spheres. (See also Wegner [552] for more on cylindrical distance functions.)

Chapter 6
Real Hypersurfaces in Complex Space Forms

The study of real hypersurfaces in complex projective space $\mathbf{CP}^n$ and complex hyperbolic space $\mathbf{CH}^n$ began at approximately the same time as Münzner's work on isoparametric hypersurfaces in spheres. A key early work was Takagi's classification [507] in 1973 of homogeneous real hypersurfaces in $\mathbf{CP}^n$. These hypersurfaces necessarily have constant principal curvatures, and they serve as model spaces for many subsequent classification theorems. Later Montiel [378] provided a similar list of standard examples in complex hyperbolic space $\mathbf{CH}^n$. In this chapter, we describe these examples of Takagi and Montiel in detail, and later we prove many important classification results involving them.

Each of these ambient spaces can be endowed with a well-known Riemannian metric. For $\mathbf{CP}^n$ this is a Fubini-Study metric of constant positive holomorphic sectional curvature, and for $\mathbf{CH}^n$ it is a Bergman metric of constant negative holomorphic sectional curvature. While these spaces may be regarded as the next simplest ambient spaces after the real space forms, their geometry is significantly different in fundamental ways. For example, there are no totally umbilic real hypersurfaces in either of these spaces, and the geodesic spheres in these spaces do not have constant sectional curvature. There are also no Einstein hypersurfaces in either of these spaces.

Let M be a real hypersurface in $\mathbf{CP}^n$ or $\mathbf{CH}^n$ with field of unit normals ξ. The *structure vector* on M is defined by $W = -J\xi$, where J is the complex structure of the ambient space. In early papers on this subject, computations involving the shape operator and focal submanifolds of M were found to be much simpler in the case where W is a principal vector field on M. Furthermore, a tube of constant radius over a complex submanifold in $\mathbf{CP}^n$ or $\mathbf{CH}^n$ always has this property. Eventually, hypersurfaces for which W is a principal vector field were given the name *Hopf hypersurfaces*, and the study of Hopf hypersurfaces has become a major part of the theory. An important first result is that the *Hopf principal curvature* α corresponding to the principal vector field W is always constant.

T.E. Cecil, P.J. Ryan, *Geometry of Hypersurfaces*, Springer Monographs in Mathematics, DOI 10.1007/978-1-4939-3246-7_6

The homogeneous hypersurfaces in $\mathbf{CP}^n$ or $\mathbf{CH}^n$ on the lists of Takagi and Montiel are all Hopf hypersurfaces with constant principal curvatures, as we show in this chapter. The principal curvatures of these hypersurfaces are most easily computed by considering the hypersurface as a tube over one of its focal submanifolds. Here we develop formulas for the principal curvatures of a tube similar to those found for tubes over submanifolds of real space forms in Theorem 2.2 (page 17).

A major theorem due to Kimura [270] in $\mathbf{CP}^n$ (see Theorem 8.13) and Berndt [27] in $\mathbf{CH}^n$ (see Theorem 8.12) states that a Hopf hypersurface with constant principal curvatures is one of the hypersurfaces on the list of Takagi for $\mathbf{CP}^n$ and Montiel for $\mathbf{CH}^n$. The proofs of these classifications require an analogue of Cartan's formula involving the principal curvatures (see Theorem 8.6), as well as a detailed study of the focal submanifold determined by the Hopf principal curvature α. This focal submanifold (if non-empty) is necessarily a complex submanifold with certain special properties, and these will be studied in Chapter 7.

Certain important classes of hypersurfaces can also be defined by conditions on the *holomorphic distribution* $\mathcal{W}^\perp$ orthogonal to W. For example, a hypersurface is said to be *pseudo-Einstein* if the Ricci tensor acts as a multiple of the identity on both $\mathcal{W} = \text{span } W$ and $\mathcal{W}^\perp$. This is the appropriate generalization of the Einstein condition for hypersurfaces of complex space forms. We will study this condition as well as several other natural conditions on the shape operator, curvature tensor, and Ricci tensor of a real hypersurface M in detail in Chapter 8.

Remark 6.1. As we continue our study of hypersurfaces in a new context, that of complex space forms, we introduce a few changes in terminology.

- Rather than *metric sphere*, we use the equivalent term *geodesic sphere* which is more commonly used in complex space forms.
- The terms *manifold*, *hypersurface*, and *submanifold* will include the attribute *connected* unless otherwise noted.
- When dealing with real space forms, we chose, for convenience, to work with ambient spaces having constant curvature ± 1 since this simplification has no significant effect on our results and any quantitative information arising can be easily scaled. From now on, however, we will need to deal, for example, with hypersurfaces in spheres of different radii. If a hypersurface in a sphere of radius 1 has a principal curvature $\cot\theta$, the corresponding hypersurface in a sphere of radius r will have principal curvature $\frac{1}{r}\cot\theta$. We will often make this translation in what follows, especially when using the results on isoparametric hypersurfaces from Chapter 3.
- The formulation used for Clifford algebras in Section 7.5 differs slightly from that used in Section 3.9 and thus should be regarded as independent. Background material on Clifford algebras may be found in Pressley and Segal [453].

6.1 Complex Space Forms

In this section we construct the standard models of the nonflat complex space forms, complex projective space $\mathbf{CP}^n$ and complex hyperbolic space $\mathbf{CH}^n$. We assume that $n \geq 2$ throughout.

Complex projective space

We first construct complex projective space. For $z = (z_0, \ldots, z_n)$ and $w = (w_0, \ldots, w_n)$ in $\mathbf{C}^{n+1}$, write

$$F(z, w) = \sum_{k=0}^{n} z_k \bar{w}_k$$

and let $\langle z, w \rangle = \Re F(z, w)$, the real part of $F(z, w)$. The sphere $S^{2n+1}(r)$ of radius r is defined by

$$S^{2n+1}(r) = \{z \in \mathbf{C}^{n+1} \mid \langle z, z \rangle = r^2\}.$$

We may identify $\mathbf{C}^{n+1}$ with $\mathbf{R}^{2n+2}$, defining $u, v \in \mathbf{R}^{2n+2}$ by

$$\begin{aligned} z_\ell &= u_{2\ell} + u_{2\ell+1}\, i \\ w_\ell &= v_{2\ell} + v_{2\ell+1}\, i \end{aligned} \tag{6.1}$$

for $0 \leq \ell \leq n$. Then

$$\langle z, w \rangle = \langle u, v \rangle = \sum_{\ell=0}^{2n+1} u_\ell v_\ell$$

is the usual inner product on $\mathbf{R}^{2n+2}$. We will use $\langle z, w \rangle$ and $\langle u, v \rangle$ interchangeably. When desired, we can work exclusively in real terms by introducing the complex structure J for multiplication by the complex number i. Note that for $z \in S^{2n+1}(r)$,

$$T_z S^{2n+1}(r) = \{w \in \mathbf{R}^{2n+2} \mid \langle z, w \rangle = 0\}.$$

The restriction of $\langle\ ,\ \rangle$ to $S^{2n+1}(r)$ is a Riemannian metric whose Levi-Civita connection $\tilde{\nabla}$ satisfies

$$D_X Y = \tilde{\nabla}_X Y - \langle X, Y \rangle \frac{z}{r^2}$$

for X, Y tangent to $S^{2n+1}(r)$ at z, where D is the Levi-Civita connection of $\mathbf{R}^{2n+2}$. The usual calculations of the Gauss equation show that the curvature tensor $\tilde{R}$ of $\tilde{\nabla}$ satisfies

$$\tilde{R}(X,Y) = \frac{1}{r^2} X \wedge Y, \tag{6.2}$$

where $X \wedge Y$ denotes the linear transformation satisfying

$$(X \wedge Y)Z = \langle Y, Z\rangle X - \langle X, Z\rangle Y. \tag{6.3}$$

Let $\mathcal{V}$ be the span of $\{Jz\}$ and write down the orthogonal decomposition into so-called vertical and horizontal subspaces,

$$T_z S^{2n+1}(r) = \mathcal{V} \oplus \mathcal{V}^{\perp}.$$

The complex projective space $\mathbf{CP}^n$ is the set of complex 1-dimensional subspaces of $\mathbf{C}^{n+1}$. The canonical projection is

$$\pi : S^{2n+1}(r) \to \mathbf{CP}^n$$

with fiber S^1.

Complex hyperbolic space

Next, we introduce the complex hyperbolic space $\mathbf{CH}^n$. The construction is parallel to that of $\mathbf{CP}^n$ with some important differences. For z, w in $\mathbf{C}^{n+1}$, write

$$F(z,w) = -z_0\bar{w}_0 + \sum_{k=1}^{n} z_k \bar{w}_k$$

and let $\langle z, w\rangle = \Re F(z,w)$. Using the same identification of $\mathbf{C}^{n+1}$ with $\mathbf{R}^{2n+2}$ we get

$$\langle z, w\rangle = \langle u, v\rangle = -\sum_{\ell=0}^{1} u_\ell v_\ell + \sum_{\ell=2}^{2n+1} u_\ell v_\ell.$$

Set

$$\mathbf{H} = \{z \in \mathbf{C}^{n+1} \mid \langle z, z\rangle = -r^2\}.$$

The restriction of $\langle\ ,\ \rangle$ to $\mathbf{H}$ makes it into the anti-de Sitter space $H_1^{2n+1}(r)$, a semi-Riemannian space form of constant curvature $-1/r^2$. It is a Lorentz space as its index is 1 (see O'Neill [412, p. 110]). Its tangent space is given by

$$T_z\mathbf{H} = \{w \in \mathbf{C}^{n+1} \mid \langle z, w\rangle = 0\},$$

and its Levi-Civita connection $\tilde{\nabla}$ satisfies

$$D_X Y = \tilde{\nabla}_X Y + \langle X, Y\rangle \frac{z}{r^2}$$

at z. The Gauss equation takes the form

$$\tilde{R}(X, Y) = -\frac{1}{r^2}\, X \wedge Y. \tag{6.4}$$

Again we get an orthogonal decomposition

$$T_z\mathbf{H} = \mathcal{V} \oplus \mathcal{V}^{\perp}.$$

Denote by $\mathbf{CH}^n$ the image of $\mathbf{H}$ by the canonical projection π to complex projective space,

$$\pi : \mathbf{H} \to \mathbf{CH}^n \subset \mathbf{CP}^n.$$

Thus, topologically, $\mathbf{CH}^n$ is an open subset of $\mathbf{CP}^n$. However, as Riemannian manifolds, they have quite different structures.

The Complex Space Forms $\mathbf{CP}^n$ *and* $\mathbf{CH}^n$

From here on we make a uniform exposition covering both $\mathbf{CP}^n$ and $\mathbf{CH}^n$. When convenient, we make use of the letter ϵ to distinguish the two cases. It will serve as the sign of the constant holomorphic sectional curvature $4c = 4\epsilon/r^2$. For example, equations (6.2) and (6.4) could be written as

$$\tilde{R}(X, Y) = \frac{\epsilon}{r^2}\, X \wedge Y.$$

We also use $\tilde{M}$ to stand for either $\mathbf{CP}^n$ or $\mathbf{CH}^n$ and $\tilde{M}'$ for $S^{2n+1}(r)$ or $\mathbf{H}$.

Note that $\pi_*\mathcal{V} = 0$ but that π_* is an isomorphism on $\mathcal{V}^{\perp}$. Let z be any point of $\tilde{M}'$. For $X \in T_{\pi z}\tilde{M}$, let X^L be the vector in $\mathcal{V}_z^{\perp}$ that projects to X. The vector X^L is called the *horizontal lift* of X to z. Define a Riemannian metric on $\tilde{M}$ by $\langle X, Y\rangle = \langle X^L, Y^L\rangle$. It is well defined since the metric on $\tilde{M}'$ is invariant by the fiber S^1. Since $\mathcal{V}^{\perp}$ is invariant by J, the manifold $\tilde{M}$ can be assigned a complex structure which we (by abuse of notation) also denote by J by setting $JX = \pi_*(JX^L)$. The reader can easily distinguish by context. Specifically, we define for $X \in T_z\tilde{M}$,

$$JX = \pi_*(JX^L).$$

It is easy to check that $\langle \, , \, \rangle$ is Hermitian with respect to J and its Levi-Civita connection $\tilde{\nabla}$ satisfies

$$\tilde{\nabla}_X Y = \pi_*(\tilde{\nabla}_{X^L} Y^L).$$

We also note that on $\tilde{M}'$

$$\tilde{\nabla}_{X^L} V = \tilde{\nabla}_V X^L = JX^L = (JX)^L \tag{6.5}$$

for $V = Jz \in \mathcal{V}$, while

$$\tilde{\nabla}_V V = 0.$$

See O'Neill [411] for background on Riemannian submersions.

The curvature tensor of $\tilde{M}$ follows from the relationship between the respective Levi-Civita connections on $\tilde{M}$ and $\tilde{M}'$.

Theorem 6.2. *The curvature tensor $\tilde{R}$ of $\tilde{M}$ satisfies*

$$\tilde{R}(X,Y)Z = \frac{\epsilon}{r^2}(X \wedge Y + JX \wedge JY + 2\langle X, JY\rangle J)Z. \tag{6.6}$$

In particular, it follows from equation (6.6) that every holomorphic 2-plane (i.e., one with a basis of the form $\{X, JX\}$) has sectional curvature $4\epsilon/r^2$ which we can write as $4c$. Such a space is said to have *constant holomorphic sectional curvature*. Note also that $\tilde{\nabla} J = 0$ on $\tilde{M}$ so that our metrics are Kähler. These metrics are traditionally known as the *Fubini-Study metric* on $\mathbf{CP}^n$ and the *Bergman metric* on $\mathbf{CH}^n$. A detailed discussion of these metrics may be found in Chapter IX of Kobayashi and Nomizu [283].

6.2 Real Hypersurfaces

Now take any space $\tilde{M}$ of constant holomorphic curvature $4c$ with real dimension $2n$ and Levi-Civita connection $\tilde{\nabla}$. For an immersed manifold $f : M^{2n-1} \to \tilde{M}$, the Levi-Civita connection ∇ of the induced metric and the shape operator A of the immersion are characterized, respectively, by

$$\tilde{\nabla}_X Y = \nabla_X Y + \langle AX, Y\rangle \xi,$$

and

$$\tilde{\nabla}_X \xi = -AX,$$

where ξ is a local choice of unit normal. We omit mention of the immersion f in these equations for brevity of notation. We define the *structure vector* $W = -J\xi$ on M. Also, we get the $(1,1)$ tensor field φ on M by projection of J. Specifically, for all tangent vectors X, we define

$$\varphi X = JX - \langle JX, \xi \rangle \xi = JX - \langle X, W \rangle \xi.$$

Let $\mathcal{W}$ be the span of W. Then φ preserves $\mathcal{W}$ and $\mathcal{W}^{\perp}$. In fact, $\varphi W = 0$ and $\varphi^2 X = -X$ for $X \in \mathcal{W}^{\perp}$. The distribution $\mathcal{W}^{\perp}$ is called the *holomorphic distribution.*

The relationship between $\tilde{\nabla}$ and ∇ gives rise to the Gauss and Codazzi equations given, respectively, by

$$\tilde{R}(X,Y) = AX \wedge AY + c\,(X \wedge Y + \varphi X \wedge \varphi Y + 2\langle X, \varphi Y \rangle \varphi)\,, \tag{6.7}$$

$$(\nabla_X A)Y - (\nabla_Y A)X = c(\langle X, W \rangle \varphi Y - \langle Y, W \rangle \varphi X + 2\langle X, \varphi Y \rangle W). \tag{6.8}$$

Proposition 6.3.

$$\langle (\nabla_X A)Y - (\nabla_Y A)X, W \rangle = 2c\,\langle X, \varphi Y \rangle$$
$$\langle (\nabla_X A)W, W \rangle = \langle (\nabla_W A)X, W \rangle = \langle (\nabla_W A)W, X \rangle.$$

Proof. The first equation follows by taking the inner product of the Codazzi equation with W, and the second follows by letting $Y = W$. □

Proofs of the following two basic propositions may be found in Niebergall and Ryan [399, pp. 239–240].

Proposition 6.4.

$$\nabla_X W = \varphi AX.$$
$$(\nabla_X \varphi)Y = \langle Y, W \rangle AX - \langle AX, Y \rangle W.$$

Proposition 6.5. *If $c \neq 0$, then ∇W cannot be identically zero. Equivalently, φA cannot be identically zero.*

We next show that there are no totally umbilic hypersurfaces in $\mathbf{CP}^n$ or $\mathbf{CH}^n$. This was first shown by Tashiro and Tachibana [523] in 1963. We also find that the shape operator cannot be parallel.

Theorem 6.6. *Let M^{2n-1}, where $n \geq 2$, be a hypersurface in a complex space form of constant holomorphic sectional curvature $4c \neq 0$. Then the shape operator A cannot be parallel. Also, no identity of the form $A = \lambda I$ can hold, even with λ nonconstant. In particular, totally umbilic hypersurfaces cannot occur.*

Proof. Suppose first that $A = \lambda I$. Then the Codazzi equation (6.8) becomes

$$(X\lambda)Y - (Y\lambda)X = c(\langle X, W\rangle\varphi Y - \langle Y, W\rangle\varphi X + 2\langle X, \varphi Y\rangle W). \tag{6.9}$$

If we take $Y = W$ in this equation, it reduces to

$$(X\lambda)Y - (Y\lambda)X = -c\varphi X. \tag{6.10}$$

For $X \neq 0$ orthogonal to W, the set $\{X, \varphi X, W\}$ is linearly independent, and so $c = 0$, contradicting the hypothesis. Now suppose that $\nabla A = 0$. Taking $X \neq 0$ orthogonal to W and $Y = W$ in the Codazzi equation yields $-c\varphi X = 0$, another contradiction. □

Remark 6.7. We note that this does not rule out the possibility of umbilic points. However, they cannot be so numerous as to form an open set.

The Ricci tensor S of type $(1, 1)$ is defined by the equation

$$\langle SX, Y\rangle = \text{trace}\,\{Z \mapsto R(Z, X)Y\}. \tag{6.11}$$

Using the Gauss equation, we compute the Ricci tensor S to be given by

$$SX = (2n + 1)cX - 3c\langle X, W\rangle W + (\text{trace}\,A)AX - A^2X. \tag{6.12}$$

The trace of the shape operator A is denoted by $\mathbf{m}$ and we reserve the symbol $\mathbf{m}$ for this purpose throughout. It is, of course, closely related to the mean curvature $\mathbf{m}/(2n-1)$.

A hypersurface is said to be *pseudo-Einstein* if the Ricci tensor acts as a multiple of the identity on both $\mathcal{W}$ and $\mathcal{W}^\perp$. Thus, M is pseudo-Einstein if there exist functions ρ and σ such that

$$SX = \rho X + \sigma\langle X, W\rangle W$$

for all tangent vectors X. Although it is traditional to require that ρ and σ be constant (see, for example, Kon [289]), we will not do this since it follows from the classification (see Theorems 8.63 and 8.64). In fact, it is easy to see that even the smoothness of ρ and σ need not be assumed. Note that if σ is identically zero, we have the familiar Einstein condition. However, it will turn out that Einstein hypersurfaces cannot occur in $\mathbf{CP}^n$ or $\mathbf{CH}^n$ (see Theorem 8.69).

6.3 Examples in $\mathbf{CP}^n$ (Takagi's list)

In this section, we introduce a list of examples that first appeared in Takagi's [507] classification of homogeneous hypersurfaces in $\mathbf{CP}^n$ – hence the designation "Takagi's list." The principal curvatures listed below can easily be computed from the formulas for the shape operator of a tube over a submanifold given later in Subsection 6.7.1, and several of them are computed there. Here we simply list the principal curvatures without giving the calculations.

Let M be a totally geodesic $\mathbf{CP}^k$ in $\mathbf{CP}^n$. For $0 < u < \frac{\pi}{2}$, the tube of radius ru over M is a hypersurface in $\mathbf{CP}^n$. Such tubes fall into two categories, Type A_1 and Type A_2. We present them in the following order to simplify the exposition.

Type A_2
If $1 \leq k \leq n-2$, then the tube has the following principal curvatures:

- $\alpha = \frac{2}{r}\cot 2u$ of multiplicity 1;
- $\lambda = \frac{1}{r}\cot u$ of multiplicity 2ℓ;
- $\mu = -\frac{1}{r}\tan u$ of multiplicity $2k$,

where $k + \ell = n - 1$.

Type A_1
If $k = n - 1$, the tube has the following principal curvatures:

- $\alpha = \frac{2}{r}\cot 2u$ of multiplicity 1;
- $\mu = -\frac{1}{r}\tan u$ of multiplicity $2k = 2n - 2$.

The Type A_1 hypersurfaces are tubes over complex projective hyperplanes. They are also geodesic spheres. For example, the geodesic sphere centered at πe_0 with radius $r(\frac{\pi}{2} - u)$ coincides with the tube of radius ru over the totally geodesic $\mathbf{CP}^{n-1} = \pi\{z \mid z_0 = 0\}$. In fact, if we abuse notation slightly and set $k = 0$ in the prescription for Type A_2 hypersurfaces, we get

- $\alpha = \frac{2}{r}\cot 2u$ of multiplicity 1;
- $\lambda = \frac{1}{r}\cot u$ of multiplicity $2n - 2$

which, upon substitution of $\frac{\pi}{2} - u$ for u, would give the configuration of principal curvatures derived for the Type A_1 case (with a change of sign).

Type B
Let M be a totally geodesic real projective space $\mathbf{RP}^n$ in $\mathbf{CP}^n$. This can be obtained, for example, by setting the imaginary part of all coordinates to zero, i.e.,

$$M = \pi\{z \in S^{2n+1}(r) \mid \Im z = 0\} \subset \mathbf{CP}^n, \tag{6.13}$$

where $\Im z$ denotes the imaginary part of z. For $0 < u < \frac{\pi}{4}$, the tube of radius ru over M is a hypersurface in $\mathbf{CP}^n$. Such a hypersurface is said to be of Type B. Its principal curvatures are as follows:

- $\alpha = \frac{2}{r} \tan 2u$ of multiplicity 1;
- $\lambda = -\frac{1}{r} \cot u$ of multiplicity $n-1$;
- $\mu = \frac{1}{r} \tan u$ of multiplicity $n-1$.

This hypersurface coincides with the tube of radius $r(\frac{\pi}{4} - u)$ over the complex quadric Q^{n-1} and a description from this point of view is given in Subsection 6.7.1.

The Type B hypersurfaces are the images under π of the family of isoparametric hypersurfaces in the sphere $S^{2n+1}(r)$ with four principal curvatures discussed in Subsection 3.8.4 (page 155).

Type C
These are tubes over the Segre embedding of $\mathbf{CP}^1 \times \mathbf{CP}^{m-1}$ in $\mathbf{CP}^n$. The specific construction will be given in Section 7.3 (page 400).

Type D
These are tubes over the Plücker embedding into $\mathbf{CP}^9$ of the Grassmann manifold of complex 2-planes in $\mathbf{C}^5$. The specific construction will be given in Section 7.4 (page 405).

Type E
These are tubes over the half-spin embedding of $SO(10)/U(5)$ in $\mathbf{CP}^{15}$. The specific construction will be given in Section 7.5 (page 412).

The principal curvatures of the hypersurfaces of types C, D, and E may be found in [399, p. 261]. See also Proposition 8.14 and the ambient discussion.

6.4 Examples in $\mathbf{CH}^n$ (Montiel's list)

The hypersurfaces that we introduce in this section first appeared in Montiel's paper [378] – hence the designation "Montiel's list." The principal curvatures listed below can easily be computed from the formulas for the shape operator of a tube over a submanifold given later in Subsection 6.7.1.

Let M be a totally geodesic $\mathbf{CH}^k$ in $\mathbf{CH}^n$. For $u > 0$ the tube of radius ru over M is a hypersurface in $\mathbf{CH}^n$. Such tubes fall into two categories, Type A_1 and Type A_2.

Type A_2
If $1 \leq k \leq n-2$, the tube has the following principal curvatures:

- $\alpha = \frac{2}{r} \coth 2u$ of multiplicity 1;
- $\lambda = \frac{1}{r} \coth u$ of multiplicity 2ℓ;
- $\mu = \frac{1}{r} \tanh u$ of multiplicity $2k$.

where $k + \ell = n - 1$.

Type A_1

If $k = n - 1$, the tube has the following principal curvatures:

- $\alpha = \frac{2}{r}\coth 2u$ of multiplicity 1;
- $\mu = \frac{1}{r}\tanh u$ of multiplicity $2k = 2n - 2$.

These hypersurfaces are tubes over complex hyperbolic hyperplanes. The geodesic spheres in $\mathbf{CH}^n$ (unlike those in $\mathbf{CP}^n$) form a distinct class of hypersurfaces, formally corresponding to the $k = 0$ case. The geodesic spheres of radius ru in $\mathbf{CH}^n$ have principal curvatures

- $\alpha = \frac{2}{r}\coth 2u$ of multiplicity 1;
- $\lambda = \frac{1}{r}\coth u$ of multiplicity $2n - 2$,

and they are also designated as Type A_1 hypersurfaces.

Type A_0

These are the horospheres in $\mathbf{CH}^n$. The principal curvatures are as follows:

- $\alpha = \frac{2}{r}$ of multiplicity 1;
- $\lambda = \frac{1}{r}$ of multiplicity $2n - 2$.

Type B

Let M be a totally geodesic real hyperbolic space $\mathbf{RH}^n$ in $\mathbf{CH}^n$. In a similar fashion to the $\mathbf{CP}^n$ case, we take

$$M = \pi(\{z \in \mathbf{H} \mid \Im z = 0\}) \subset \mathbf{CH}^n. \tag{6.14}$$

For $u > 0$, the tube of radius ru over M is a hypersurface in $\mathbf{CH}^n$. Its principal curvatures are as follows:

- $\alpha = \frac{2}{r}\tanh 2u$ of multiplicity 1 (except when $\coth u = 2\tanh 2u$);
- $\lambda = \frac{1}{r}\coth u$ of multiplicity $n - 1$ (except when $\coth u = 2\tanh 2u$);
- $\mu = \frac{1}{r}\tanh u$ of multiplicity $n - 1$.

Such a hypersurface is said to be of Type B. In the special case noted, λ and α have the same value and so the common multiplicity is n.

Remark 6.8. When introducing the complex space forms $\mathbf{CP}^n$ and $\mathbf{CH}^n$ earlier in this chapter, we specified that $n \geq 2$. More specifically, we will assume this when these spaces are used as ambient spaces. However, our definitions work equally well for $n = 1$ and both $\mathbf{CP}^1$ and $\mathbf{CH}^1$ can occur as submanifolds, as they have in the construction of Type A_1 and Type A_2 hypersurfaces. Note that the 1-dimensional complex space forms are surfaces of constant curvature $4c$ since all 2-planes are holomorphic planes. Thus they are, in fact, isometric to spheres or real hyperbolic planes depending on the sign of c.

6.5 Properties of Type A, B, C, D, and E Hypersurfaces

For brevity of notation, hypersurfaces of types A_0, A_1, and A_2 are all referred to as Type A hypersurfaces. An immediate observation is that all the hypersurfaces we have introduced so far have constant principal curvatures. It will follow from our construction later on that the structure vector $W = -J\xi$ is a principal vector for the principal curvature α, i.e., they are Hopf hypersurfaces (see definition in Section 6.6). There is a useful criterion, following directly from (6.12), for such a hypersurface to be pseudo-Einstein.

Proposition 6.9. *A Hopf hypersurface is pseudo-Einstein if and only if*

$$\lambda + \mu = \mathbf{m}$$

whenever λ and μ are distinct principal curvatures corresponding to principal vectors in $\mathcal{W}^\perp$.

Proposition 6.9 includes the possibility that no such distinct pair $\{\lambda, \mu\}$ exists, (i.e., A acts as a scalar multiple of the identity on $\mathcal{W}^\perp$), in which case the hypersurface in question is pseudo-Einstein. We also have

Proposition 6.10.
1. *For Type A_2 and B hypersurfaces $\lambda\mu + c = 0$.*
2. *For Type A_2 hypersurfaces, $\lambda + \mu = \alpha$.*
3. *For Type B hypersurfaces, $(\lambda + \mu)\alpha + 4c = 0$.*
4. *Type A_1 hypersurfaces are pseudo-Einstein with*

$$SX = \rho_\nu X + (\rho_\alpha - \rho_\nu)\langle X, W\rangle W,$$

where

$$\begin{aligned} \rho_\nu &= 2nc + 2(n-1)\nu^2, \\ \rho_\alpha &= 2(n-1)\nu^2, \end{aligned} \tag{6.15}$$

and ν is the principal curvature of multiplicity $2n-2$.

5. *Type A_0 hypersurfaces are pseudo-Einstein with*

$$SX = \rho_\lambda X + (\rho_\alpha - \rho_\lambda)\langle X, W\rangle W,$$

where

$$\begin{aligned} \rho_\lambda &= 2c, \\ \rho_\alpha &= -2(n-1)c. \end{aligned} \tag{6.16}$$

(Note that these values are limits as $u \to \infty$ of the respective values for the corresponding Type A_1 hypersurfaces.)

6. *The principal vectors of Type* A_2 *hypersurfaces are also eigenvectors of the Ricci tensor* S *with corresponding eigenvalues*

$$\begin{aligned} \rho_\lambda &= 2(\ell + 1)c + 2\ell\lambda^2, \\ \rho_\mu &= 2(k + 1)c + 2k\mu^2, \\ \rho_\alpha &= 2\ell\lambda^2 + 2k\mu^2. \end{aligned} \tag{6.17}$$

7. *A Type* A_2 *hypersurface in* $\mathbf{CP}^n$ *is pseudo-Einstein if and only if its radius as a tube over* $\mathbf{CP}^k$ *is* ru *where*

$$\cot^2 u = \frac{k}{\ell}.$$

In this case, $\rho_\lambda = \rho_\mu = 2nc$*, while* $\rho_\alpha = 2(n-1)c$*. No Type* A_2 *hypersurface in* $\mathbf{CH}^n$ *is pseudo-Einstein.*

8. *The principal vectors of Type B hypersurfaces are also eigenvectors of the Ricci tensor* S *with corresponding eigenvalues*

$$\begin{aligned} \rho_\alpha &= -2(n-1)c, \\ \rho_\lambda &= (n+2)c + (n-2)\lambda^2 + \alpha\lambda, \\ \rho_\mu &= (n+2)c + (n-2)\mu^2 + \alpha\mu. \end{aligned} \tag{6.18}$$

9. *A Type B hypersurface in* $\mathbf{CP}^n$ *is pseudo-Einstein if and only if its radius as a tube over* $\mathbf{RP}^n$ *is* ru *where*

$$\tan^2 2u = n - 2.$$

In this case, $\rho_\lambda = \rho_\mu = 2nc$*, while* $\rho_\alpha = -2(n-1)c$*. No Type B hypersurface in* $\mathbf{CH}^n$ *is pseudo-Einstein.*

10. *The hypersurfaces discussed in this section are not Einstein.*

Lifts of hypersurfaces in $\tilde{M}$ *to* $\tilde{M}'$

We now show how to relate the shape operators of hypersurfaces in complex space forms to shape operators in the more familiar setting of hypersurfaces in real space forms. Detailed proofs of the assertions in this section may be found in Niebergall and Ryan [399, pp. 240–242].

Once again, we let $\tilde{M}$ represent $\mathbf{CP}^n$ or $\mathbf{CH}^n$ and $\tilde{M}'$ represent $S^{2n+1}(r)$ or $\mathbf{H}$ respectively, with the canonical projection

$$\pi : \tilde{M}' \to \tilde{M}.$$

Now consider a hypersurface M in $\tilde{M}$. Then $M' = \pi^{-1}M$ is an S^1-invariant hypersurface in $\tilde{M}'$ (Lorentzian in the case $\tilde{M}' = \mathbf{H}$). If ξ is a unit normal for M, then $\xi' = \xi^L$ is a unit normal for M'. The induced connection ∇' and the shape operator A' for M' satisfy

$$\tilde{\nabla}_X Y = \nabla'_X Y + \langle A'X, Y\rangle \xi',$$
$$\tilde{\nabla}_X \xi' = -A'X,$$

and the more familiar form of the Codazzi equation

$$(\nabla'_X A')Y = (\nabla'_Y A')X$$

holds. There is also a Gauss equation, but we will not have occasion to use it. In addition, $W^L = U = -J\xi'$, where $W = -J\xi$ is the structure vector introduced earlier.

Lemma 6.11. *For X and Y tangent to $\tilde{M}$,*

$$(\tilde{\nabla}_X Y)^L = \tilde{\nabla}_{X^L} Y^L + \langle JX^L, Y^L\rangle \frac{\epsilon}{r^2} V.$$

Recalling that $c = \epsilon/r^2$ and computing the shape operator A' of M', we have

Lemma 6.12. • $\tilde{\nabla}_V \xi' = J\xi' = -U$, *so* $A'V = U$.
- *For X tangent to M,* $(AX)^L = A'X^L - \langle X^L, U\rangle cV$.
- *In particular,* $(AW)^L = A'U - cV$.
- *If* $AW = \alpha W$, *then* $A'U = \alpha U + cV$.

We now look at the relationship between the covariant derivatives of the respective shape operators of M and M'.

Theorem 6.13. *Let M^{2n-1}, where $n \geq 2$, be a real hypersurface in a complex space form of constant holomorphic sectional curvature $4c \neq 0$. Then the shape operator A' of $M' = \pi^{-1}M$ satisfies*

$$\pi_*((\nabla'_{X^L} A')Y^L) = (\nabla_X A)Y + c(\langle \varphi X, Y\rangle W + \langle Y, W\rangle \varphi X)$$

for all X, Y tangent to M.

Clearly this, together with the Codazzi equation for M' in $\tilde{M}'$, leads to the Codazzi equation (6.8) for M in $\tilde{M}$. We also look at the vertical component of the covariant derivative of A' and observe the following nice relationship.

Proposition 6.14. *Under the hypothesis of Theorem 6.13,*

$$\langle (\nabla'_{X^L} A')Y^L, V\rangle = \langle (\varphi A - A\varphi)X, Y\rangle.$$

Therefore $(\nabla'_{X^L} A')Y^L$ is horizontal for all X and Y if and only if φ and A commute.

Using Theorem 6.13, we can strengthen the result that the shape operator cannot be parallel. In fact, its covariant derivative cannot vanish even at one point. Specifically we have

Theorem 6.15. *Let M^{2n-1}, where $n \geq 2$, be a real hypersurface in a complex space form of constant holomorphic sectional curvature $4c \neq 0$. Then the shape operator A satisfies*

$$|\nabla A|^2 \geq 4c^2(n-1).$$

Equality holds if and only if

$$(\nabla_X A)Y = -c(\langle \varphi X, Y\rangle W + \langle Y, W\rangle \varphi X)$$

for all X and Y.

The proof of this theorem is given in detail in [399, p. 243]. In a later section, we will show an even stronger result. Specifically, equality holds if and only if M is an open subset of a Type A hypersurface.

6.6 Basic Results on Hopf Hypersurfaces

A hypersurface in a complex space form is said to be a *Hopf hypersurface* if $W = -J\xi$ is a principal vector. If $AW = \alpha W$, then α is called the *Hopf principal curvature*. We shall see in Chapter 8 that the hypersurfaces in the Takagi/Montiel lists are Hopf. Further, we shall see that all hypersurfaces that are tubes over complex submanifolds are Hopf.

The following fact is fundamental.

Theorem 6.16. *Let M^{2n-1}, where $n \geq 2$, be a Hopf hypersurface in $\mathbf{CP}^n$ or $\mathbf{CH}^n$. Then the Hopf principal curvature α is constant.*

The proof of this is easy for $\mathbf{CP}^n$ but rather difficult for $\mathbf{CH}^n$. See [399, pp. 244–252] for a complete proof. We also have the following useful result.

Theorem 6.17. *Let M^{2n-1}, where $n \geq 2$, be a Hopf hypersurface in $\mathbf{CP}^n$ or $\mathbf{CH}^n$. Then*

1. $A\varphi A - \frac{\alpha}{2}(A\varphi + \varphi A) - c\varphi = 0.$
2. *If $X \in \mathcal{W}^\perp$ and $AX = \lambda X$, then*

$$\left(\lambda - \frac{\alpha}{2}\right) A\varphi X = \left(\frac{\lambda\alpha}{2} + c\right) \varphi X.$$

3. If $0 \neq X \in \mathcal{W}^{\perp}$ *satisfies* $AX = \lambda X$ *and* $A\varphi X = \mu\varphi X$*, then*

$$\lambda\mu = \frac{\lambda+\mu}{2}\alpha + c.$$

If T_λ *is* φ*-invariant, then* $\lambda^2 = \alpha\lambda + c$*. (Here* T_λ *denotes the space of principal vectors for a principal curvature* λ*.)*

The following criteria for principal curvatures and principal vectors are equivalent to Theorem 6.17.

Corollary 6.18. *Let* M^{2n-1}*, where* $n \geq 2$*, be a Hopf hypersurface in* $\mathbf{CP}^n$ *or* $\mathbf{CH}^n$*. Then for each* $p \in M$*,* $\mathcal{W}_p^{\perp}$ *has an orthonormal basis consisting of* $n-1$ *pairs* $\{X, \varphi X\}$ *of principal vectors with corresponding respective principal curvatures* λ *and* μ *such that exactly one of the following holds:*

1. $\lambda \neq \mu$*, neither* λ *nor* μ *is equal to* $\frac{\alpha}{2}$ *and* $\lambda\mu = \frac{\lambda+\mu}{2}\alpha + c$*;*
2. $\lambda = \mu \neq \frac{\alpha}{2}$ *and* $\lambda^2 = \alpha\lambda + c$*;*
3. $\lambda \neq \mu = \frac{\alpha}{2}$ *and* $\lambda\mu = \frac{\lambda+\mu}{2}\alpha + c$*;*
4. $\lambda = \mu = \frac{\alpha}{2}$ *and* $\lambda^2 = \alpha\lambda + c$*.*

Clearly, *3.* and *4.* cannot occur unless $\alpha^2 + 4c = 0$. In particular, they cannot occur when the ambient space is $\mathbf{CP}^n$. On the other hand, if a Hopf hypersurface M in $\mathbf{CH}^n$ happens to satisfy $\alpha^2 + 4c = 0$, then *1.* and *2.* cannot occur and the only possibilities are *3.* and *4.* In this case, the condition $\lambda\mu = \frac{\lambda+\mu}{2}\alpha + c$ does not place any restriction on λ but implies that $\mu = \frac{\alpha}{2}$.

We can now refine Theorem 6.15 as follows:

Theorem 6.19. *Let* M^{2n-1}*, where* $n \geq 2$*, be a real hypersurface in* $\mathbf{CP}^n$ *or* $\mathbf{CH}^n$*. Then* $\varphi A = A\varphi$ *if and only if*

- *M is a Hopf hypersurface with constant principal curvatures, and*
- *the number of distinct principal curvatures is 2 or 3, and*
- *the principal subspaces are* φ*-invariant.*

That is, the principal curvature configuration matches that of a Type A hypersurface.

Later, (see Theorem 8.37) we will show that such an M is actually an open subset of a Type A hypersurface.

Proof. First note that the "if" part of this theorem is almost trivial. Since $AW = \alpha W$, we have $\varphi AW = \varphi\alpha W = \alpha\varphi W = 0 = A\varphi W$. Also, if $AX = \lambda X$ for $X \in \mathcal{W}^{\perp}$, then $A\varphi X = \lambda\varphi X = \varphi\lambda X = \varphi AX$. Thus, $A\varphi - \varphi A$ vanishes on a basis of the tangent space, and therefore it vanishes identically.

Now assume that $\varphi A = A\varphi$. First note that $\varphi AW = A\varphi W = 0$. Thus M is a Hopf hypersurface. Now suppose that λ is a principal curvature with associated principal vector $X \in \mathcal{W}^{\perp}$. Then $A\varphi X = \varphi AX = \lambda\varphi X$. Applying Theorem 6.17, we find that T_λ is φ-invariant and $\lambda^2 = \alpha\lambda + c$. Noting that the Hopf principal curvature α is constant, the same is also true about the two possible roots of the quadratic equation. The number of distinct principal curvatures is therefore either 2 or 3. Note that the principal space of α, namely $\mathcal{W}$, is also φ-invariant. This completes the proof. □

6.7 Parallel Hypersurfaces, Focal Sets, and Tubes

Let M be a submanifold of a complex space form $\tilde{M}$ (either $\mathbf{CP}^n$ or $\mathbf{CH}^n$) of (real) codimension κ. As in Section 2.2 for submanifolds of real space forms, the normal exponential map $E : NM \to \tilde{M}$ is smooth and nonsingular at points on the 0-section of NM. Thus, in trying to locate the critical values of E, we restrict our attention to points NM that are not in the 0-section.

As in Section 2.2 (page 11) for submanifolds of real space forms, the normal exponential map can be formulated as follows:

1. Choose a point $p \in M$.
2. Let $U \subset M$ be a normal coordinate neighborhood centered at p.
3. Choose an orthonormal basis $\{\xi_j\}_{1 \leq j \leq \kappa}$ for the normal space at p.
4. Extend this basis to an orthonormal frame in U by parallel translation (with respect to the normal connection) along radial geodesics beginning at p.

We parametrize the normal bundle (minus the 0-section) locally, defining

$$\Psi : (0, \infty) \times S^{\kappa-1} \times U \to NM \tag{6.19}$$

by

$$\Psi(\mu, a, x) = \mu \sum_{j=1}^{\kappa} a_j \xi_j(x), \tag{6.20}$$

where the vector $\Psi(\mu, a, x)$ is normal to M at the point $x \in U$. Then

$$(E \circ \Psi)(\mu, a, x)$$

is the point of $\tilde{M}$ reached by traveling a distance μ along the geodesic beginning at x and having direction $\sum_{j=1}^{\kappa} a_j \xi_j(x)$.

Let $\tilde{M}'$ be the set of "length r" vectors in $\mathbf{C}^{n+1}$ with respect to the metric that defines $\tilde{M}$, that is, $\tilde{M}'$ is $S^{2n+1}(r)$ or $H_1^{2n+1}(r)$. We have the natural projection π : $\tilde{M}' \to \tilde{M}$ with fiber S^1 and denote $\pi^{-1}M \subset \tilde{M}'$ by M'. For any $x \in U$, let w be a point in M' such that $\pi w = x$. Each tangent vector X to $\tilde{M}$ at x has a unique horizontal lift X^L, tangent to $\tilde{M}'$ at w. (In fact, a vector field on $\tilde{M}$ can be lifted uniquely to a horizontal vector field on $\tilde{M}'$.)

Focal points occur when the differential of the normal exponential map E is singular, i.e., when its rank is less than $2n$. A typical point in the normal bundle (not in the 0-section) may be regarded as $\mu\xi_1(p)$ using the setup described above. Evaluating E_* at this point is equivalent to evaluating the differential $(E \circ \Psi)_*$ at the point (μ, ϵ_1, p), where $\{\epsilon_j\}$, $1 \leq j \leq \kappa$, is the standard basis of $\mathbf{R}^\kappa$. Now, writing $u = \mu/r$, we have

$$(E \circ \Psi)(\mu, a, x) = \pi z, \tag{6.21}$$

where

$$z = \cos u \, w + r \sin u \, \xi_1^L(w), \tag{6.22}$$

and

$$z = \cosh u \, w + r \sinh u \, \xi_1^L(w), \tag{6.23}$$

in the $\mathbf{CP}^n$ and $\mathbf{CH}^n$ cases, respectively. Here, $\xi_1^L(w)$ is the horizontal lift of $\xi_1(x)$ to w. Locally, z is determined by a single choice of w so that it may be regarded a smooth function of (μ, a, x). Let

$$\zeta = -\sin u \, \frac{w}{r} + \cos u \, \xi_1^L(w) \tag{6.24}$$

respectively,

$$\zeta = \sinh u \, \frac{w}{r} + \cosh u \, \xi_1^L(w). \tag{6.25}$$

Note that both ζ and $i\zeta$ are horizontal at z.

We now express $(E \circ \Psi)_*$ in terms of a basis consisting of $\frac{\partial}{\partial \mu}$ for $(0, \infty)$, $\{\epsilon_j\}$, $2 \leq j \leq \kappa$, for $T_{\epsilon_1} S^{\kappa-1}$, and an orthonormal basis of $T_x M$. The latter is made up of eigenvectors X of the shape operator A_{ξ_1} with corresponding eigenvalue λ. We compute

$$\begin{aligned} (E \circ \Psi)_* \frac{\partial}{\partial \mu} &= (\pi_*)_z \zeta \\ (E \circ \Psi)_* \epsilon_j &= (\pi_*)_z (r \sin u \, \xi_j^L(w)) \\ (E \circ \Psi)_* X &= (\pi_*)_z \big((\cos u \, - r\lambda \sin u) X^L + \sin u \, \langle J\xi_1, X \rangle \frac{iw}{r} \big) \end{aligned} \tag{6.26}$$

for $\mathbf{CP}^n$, and

$$\begin{aligned} (E \circ \Psi)_* \frac{\partial}{\partial \mu} &= (\pi_*)_z \zeta \\ (E \circ \Psi)_* \epsilon_j &= (\pi_*)_z (r \sinh u \, \xi_j^L(w)) \\ (E \circ \Psi)_* X &= (\pi_*)_z \big((\cosh u \, - r\lambda \sinh u) X^L - \sinh u \, \langle J\xi_1, X \rangle \frac{iw}{r} \big) \end{aligned} \tag{6.27}$$

for $\mathbf{CH}^n$. However, the arguments of $(\pi_*)_z$ are not automatically horizontal. There are two special cases for which it is relatively easy to modify the arguments so that they will be horizontal, as we now discuss.

Case 1: when M is a complex submanifold

If M is a complex submanifold, then $J\xi_1$ is also a unit normal vector, so that we may choose $\xi_2 = J\xi_1$. Note that $J\xi_1$ is parallel along geodesics emanating from p. Our equations now become

$$\begin{aligned}(E \circ \Psi)_* \frac{\partial}{\partial \mu} &= (\pi_*)_z \zeta \\ (E \circ \Psi)_* \epsilon_2 &= (\pi_*)_z \left(\frac{r}{2} \sin 2u \; i\zeta \right) \\ (E \circ \Psi)_* \epsilon_j &= (\pi_*)_z (r \sin u \; \xi_j^L(w)) \text{ for } j \geq 3 \\ (E \circ \Psi)_* X &= (\pi_*)_z \big((\cos u \; - r\lambda \sin u) X^L \big)\end{aligned} \tag{6.28}$$

and

$$\begin{aligned}(E \circ \Psi)_* \frac{\partial}{\partial \mu} &= (\pi_*)_z \zeta \\ (E \circ \Psi)_* \epsilon_2 &= (\pi_*)_z \left(\frac{r}{2} \sinh 2u \; i\zeta \right) \\ (E \circ \Psi)_* \epsilon_j &= (\pi_*)_z (r \sinh u \; \xi_j^L(w)) \text{ for } j \geq 3 \\ (E \circ \Psi)_* X &= (\pi_*)_z \big((\cosh u \; - r\lambda \sinh u) X^L \big)\end{aligned} \tag{6.29}$$

in their respective contexts. In the $\mathbf{CP}^n$ case, we may choose θ to satisfy $0 < \theta < \pi$ and $\lambda = \frac{1}{r} \cot \theta$, in which case the last equation in (6.28) may be rewritten as

$$(E \circ \Psi)_* X = \frac{\sin(\theta - u)}{\sin \theta} (\pi_*)_z X^L. \tag{6.30}$$

Case 2: when $J\xi_1$ is an eigenvector for A_{ξ_1}

In the second special case, we write ξ_1 as ξ and A_{ξ_1} as A. We assume that $W = -J\xi$ is a tangent vector to M satisfying $AW = \alpha W$. This generalizes the Hopf condition. Our equations become:

$\mathbf{CP}^n$ case:

$$
\begin{aligned}
(E\circ\Psi)_*\frac{\partial}{\partial\mu} &= (\pi_*)_z\zeta \\
(E\circ\Psi)_*W &= -(\pi_*)_z\big((\cos 2u - \frac{r\alpha}{2}\sin 2u)i\zeta\big) \\
(E\circ\Psi)_*\epsilon_j &= (\pi_*)_z(r\sin u\ \xi_j^L(w)) \text{ for } j\geq 2 \\
(E\circ\Psi)_*X &= (\pi_*)_z\big((\cos u - r\lambda\sin u)X^L\big) \text{ for } X\in\mathcal{W}^\perp
\end{aligned}
\tag{6.31}
$$

where X is any eigenvalue of A with corresponding eigenvalue λ. The arguments for $(\pi_*)_z$ are horizontal tangent vectors to $\tilde{M}'$ at z. All of this is immediate, except for the formula involving W. We verify it as follows:

$$
\begin{aligned}
(\cos u &- r\alpha\sin u)W^L + \sin u\ \langle J\xi, W\rangle\frac{iw}{r} \\
&= i\zeta + (2\cos u - r\alpha\sin u)\ W^L \\
&\equiv i\zeta + (2\cos u - r\alpha\sin u)\big(W^L + \sin u\ \frac{iz}{r}\big) \\
&= i\zeta + (2\cos u - r\alpha\sin u)\big(\cos^2 u\ W^L + \sin u\ \cos u\ \frac{iw}{r}\big) \\
&= i\zeta + (2\cos^2 u - r\alpha\sin u\ \cos u)(-i\zeta) \\
&= -\left(\cos 2u - \frac{r\alpha}{2}\sin 2u\right)\ i\zeta
\end{aligned}
\tag{6.32}
$$

where $\equiv$ represents equivalence with respect to $(\pi_*)_z$.

$\mathbf{CH}^n$ case:

Under analogous conditions on X and λ, we have

$$
\begin{aligned}
(E\circ\Psi)_*\frac{\partial}{\partial\mu} &= (\pi_*)_z\zeta \\
(E\circ\Psi)_*W &= -(\pi_*)_z\big((\cosh 2u - \frac{r\alpha}{2}\sinh 2u)i\zeta\big) \\
(E\circ\Psi)_*\epsilon_j &= (\pi_*)_z(r\sinh u\ \xi_j^L(w)) \text{ for } j\geq 2 \\
(E\circ\Psi)_*X &= (\pi_*)_z\big((\cosh u - r\lambda\sinh u)X^L\big),
\end{aligned}
\tag{6.33}
$$

where again, the arguments of $(\pi_*)_z$ are horizontal and we need only verify the equation involving W, as follows:

$$
\begin{aligned}
(\cosh u &- r\alpha\sinh u)W^L - \sinh u\ \langle J\xi, W\rangle\frac{iw}{r} \\
&= i\zeta + (2\cosh u - r\alpha\sinh u)\ W^L \\
&\equiv i\zeta + (2\cosh u - r\alpha\sinh u)\big(W^L - \sinh u\ \frac{iz}{r}\big)
\end{aligned}
\tag{6.34}
$$

$$= i\zeta + (2\cosh u - r\alpha \sinh u)\big(\cosh^2 u\, W^L - \sinh u\, \cosh u\, \frac{iw}{r}\big)$$

$$= i\zeta + (2\cosh^2 u - r\alpha \sinh u\, \cosh u)(-i\zeta)$$

$$= -(\cosh 2u\, - \frac{r\alpha}{2} \sinh 2u)\; i\zeta$$

where $\equiv$ represents equivalence with respect to $(\pi_*)_z$.

6.7.1 Shape operators of tubes

In this section, we find formulas for shape operators of tubes over submanifolds in complex space forms similar to the formulas in Theorem 2.2 on page 17 for tubes over submanifolds of real space forms.

Fix a value of $\mu > 0$ and the associated value $u = \mu/r$. Define ϕ_u and E_μ by

$$\phi_u(a,x) = (E \circ \Psi)(\mu, a, x) = E_\mu(\Psi(1, a, x)).$$

Then $(\phi_u)_*$ may be read off from the formulas displayed above as follows. Just replace $E \circ \Psi$ by ϕ_u and ignore the $\partial/\partial\mu$ equations in (6.28), (6.29), (6.31), and (6.33).

If μ is chosen so that $E \circ \Psi$ is nonsingular at (μ, ϵ_1, p), then ϕ_u embeds a neighborhood of (ϵ_1, p) into $\tilde{M}$ as a (real) hypersurface M_μ, the tube over M of radius $\mu = ru$. We now compute the shape operator of this hypersurface.

It is easy to check that $\eta = -(\pi_*)_z\zeta$ is a field of unit normals for M_μ. The shape operator A_μ is defined as follows:

$$(\phi_u)_*(A_\mu v) = -\tilde{\nabla}_{(\phi_u)_* v}\,\eta \tag{6.35}$$

for v tangent to M.

Shape operators for tubes over complex submanifolds

In the $\mathbf{CP}^n$ case,

$$A_\mu \epsilon_2 = \frac{2}{r} \cot\, 2u\; \epsilon_2, \tag{6.36}$$

$$A_\mu \epsilon_j = \frac{1}{r} \cot\, u\; \epsilon_j \text{ for } j \geq 3,$$

$$A_\mu X = -\frac{1}{r} \cot(\theta - u) X.$$

There will be a version of the third equation for each eigenvalue of $A = A_{\xi_1}$.

For $\mathbf{CH}^n$, we have the analogous equations

$$\begin{aligned} A_\mu \epsilon_2 &= \frac{2}{r} \coth 2u \; \epsilon_2, \\ A_\mu \epsilon_j &= \frac{1}{r} \coth u \; \epsilon_j \text{ for } j \geq 3, \\ A_\mu X &= \frac{1}{r} \frac{\sinh u - r\lambda \cosh u}{\cosh u - r\lambda \sinh u} X. \end{aligned} \tag{6.37}$$

The last equation breaks down into separate cases depending on the magnitude of $r\lambda$ which can be $\coth\theta$, $\tanh\theta$, 1 or -1. Thus $A_\mu X$ is given by:

$$\begin{aligned} A_\mu X &= -\frac{1}{r} \coth(\theta - u)\, X, \text{ if } AX = \frac{1}{r} \coth\theta\, X, \\ A_\mu X &= -\frac{1}{r} \tanh(\theta - u)\, X, \text{ if } AX = \frac{1}{r} \tanh\theta\, X, \\ A_\mu X &= -\frac{1}{r} X, \text{ if } AX = \frac{1}{r}\, X, \\ A_\mu X &= \frac{1}{r} X, \text{ if } AX = -\frac{1}{r}\, X. \end{aligned} \tag{6.38}$$

Typical examples are the Type A_2 hypersurfaces which are tubes over totally geodesic complex projective (respectively complex hyperbolic) subspaces $\mathbf{CP}^k$ (respectively, $\mathbf{CH}^k$). The codimension of M is $2(n-k)$. In the $\mathbf{CP}^n$ case, $\theta = \pi/2$ and the three principal curvatures of the tube are

$$\frac{2}{r} \cot 2u, \quad \frac{1}{r} \cot u, \quad -\frac{1}{r} \cot\left(\frac{\pi}{2} - u\right) = -\frac{1}{r} \tan\, u,$$

with respective multiplicities 1, 2ℓ, $2k$, where $k + \ell = n - 1$.

In the $\mathbf{CH}^n$ case, we must use the tanh alternative with $\theta = 0$. The three principal curvatures of the tube will be

$$\frac{2}{r} \coth 2u, \quad \frac{1}{r} \coth u, \quad -\frac{1}{r} \tanh(-u) = \frac{1}{r} \tanh u,$$

with respective multiplicities 1, 2ℓ, and $2k$. See [399, pp. 257–260].

Note that $J\eta = -(\pi_*)_z(i\zeta)$ since $-\zeta$ is the horizontal lift of η to z. Thus $(\phi_u)_*\epsilon_2$ is a nonzero multiple of $J\eta$. This shows that M_μ is embedded as a Hopf hypersurface with Hopf principal curvature $\alpha = \frac{2}{r} \cot 2u$ (respectively, $\frac{2}{r} \coth 2u$).

Another example is given by the Type B hypersurfaces in $\mathbf{CP}^n$. In that case, M is the complex quadric, and if ξ is a unit normal, then A_ξ has eigenvalues $\frac{1}{r}\cot\frac{\pi}{4}$ and $\frac{1}{r}\cot\frac{3\pi}{4}$, each of multiplicity $n-1$. Then M_μ has principal curvatures

$$\frac{2}{r}\cot 2u, \quad -\frac{1}{r}\cot\left(\frac{\pi}{4}-u\right), \quad -\frac{1}{r}\cot\left(\frac{3\pi}{4}-u\right),$$

with respective multiplicities $(1, n-1, n-1)$. Another way of describing the same hypersurfaces would be to use a radius of $r(\frac{\pi}{4}-u)$, in which case the principal curvatures reduce to the "simpler" expressions

$$\frac{2}{r}\tan 2u, \quad -\frac{1}{r}\cot u, \quad \frac{1}{r}\tan u.$$

These data agree with [399, p. 260]. See also the alternative description in Section 6.3.

Shape operator for tubes for which $J\xi_1$ is an eigenvector of A_{ξ_1}

For the calculations of this section, we will choose $\eta = \pi_*\zeta$. Then the shape operator of the tube will have the following behavior. As before, we write W for $-J\xi_1$ and A for A_{ξ_1}. We choose α so that $AW = \alpha W$. Then we get separate formulas for $A_\mu W$ and $A_\mu X$, where $X \in \mathcal{W}^\perp$ is an eigenvector of A with corresponding eigenvalue λ. In the $\mathbf{CP}^n$ case, we have

$$\begin{aligned} A_\mu W &= \frac{2}{r}\cot 2(\theta-u)W, \text{ if } \alpha = \frac{2}{r}\cot 2\theta, \\ A_\mu \epsilon_j &= -\frac{1}{r}\cot u\; \epsilon_j, \text{ for } j \geq 2, \\ A_\mu X &= \frac{1}{r}\cot(\theta-u)\, X, \text{ if } \lambda = \frac{1}{r}\cot\theta, \end{aligned} \tag{6.39}$$

while in the $\mathbf{CH}^n$ case, we get

$$\begin{aligned} A_\mu W &= -\frac{2}{r}\frac{\sinh 2u - \frac{r\alpha}{2}\cosh 2u}{\cosh 2u - \frac{r\alpha}{2}\sinh 2u} W, \text{ if } AW = \alpha W, \\ A_\mu \epsilon_j &= -\frac{1}{r}\coth u\; \epsilon_j, \text{ for } j \geq 2, \\ A_\mu X &= -\frac{1}{r}\frac{\sinh u - r\lambda\cosh u}{\cosh u - r\lambda\sinh u} X, \text{ if } AX = \lambda X. \end{aligned} \tag{6.40}$$

If M is a Hopf hypersurface, the shape operator of the tube M_μ is related to that of M in a particularly simple way. First of all, a Hopf hypersurface satisfies the conditions of this section. The codimension κ is 1, so that $\xi = \xi_1$ is a field of unit normals with $A_\xi X = \alpha X$. The principal vectors for M_μ are the same as those for M. In the $\mathbf{CP}^n$ case, with $\alpha = \frac{2}{r}\cot 2\theta$ where $0 < \theta < \frac{\pi}{2}$, M_μ is a Hopf hypersurface with Hopf principal curvature $\frac{2}{r}\cot 2(\theta - u)$. Further, for each principal curvature λ, we can find θ with $0 < \theta < \pi$, so that $\lambda = \frac{1}{r}\cot\theta$, and M_μ has corresponding principal curvature $\frac{1}{r}\cot(\theta - u)$.

For the $\mathbf{CH}^n$ case, there are three possibilities, depending on the magnitudes of α and λ. We can check that $A_\mu W$ satisfies:

$$A_\mu W = \frac{2}{r}\coth 2(\theta - u)\ W, \text{ if } AW = \frac{2}{r}\coth 2\theta\ W, \tag{6.41}$$

$$A_\mu W = \frac{2}{r}\tanh 2(\theta - u)\ W, \text{ if } AW = \frac{2}{r}\tanh 2\theta\ W,$$

$$A_\mu W = \frac{2}{r} W, \text{ if } AW = \frac{2}{r}\ W,$$

$$A_\mu W = -\frac{2}{r} W, \text{ if } AW = -\frac{2}{r}\ W.$$

In particular, we notice that all tubes over a given horosphere have the same principal curvatures. It turns out that they form a parallel family of congruent horospheres. Montiel called the horosphere a "self-tube."

The Type B hypersurfaces also arise as examples of this construction. In the $\mathbf{CP}^n$ case, let M be the totally geodesic real projective space $\mathbf{RP}^n$. The real dimension is n, so the codimension is $\kappa = n$ also. For a choice of ξ_1 and $W = -J\xi_1$, we have $AW = 0$, so that $\alpha = 2\cot 2\theta/r$ with $\theta = \pi/4$. For $X \in \mathcal{W}^\perp$, we also have $AX = 0$, so that $\lambda = \cot\theta/r$ with $\theta = \pi/2$. Thus M_μ has principal curvatures

$$\frac{2}{r}\cot 2\left(\frac{\pi}{4} - u\right), \quad -\frac{1}{r}\cot u, \quad \frac{1}{r}\cot\left(\frac{\pi}{2} - u\right),$$

with respective multiplicities $(1, n-1, n-1)$. These principal curvatures can also be written as

$$\frac{2}{r}\tan 2u, \quad -\frac{1}{r}\cot u, \quad \frac{1}{r}\tan u.$$

This is consistent with the observation that the tube of radius ru over $\mathbf{RP}^n$ is also the tube of radius $r(\frac{\pi}{4} - u)$ over the complex quadric.

In the $\mathbf{CH}^n$ case, let M be the totally geodesic real hyperbolic space $\mathbf{RH}^n$. Noting that α and λ are both zero, we can see immediately that the principal curvatures of the tube are

$$-\frac{2}{r}\tanh 2u, \quad -\frac{1}{r}\coth u, \quad -\frac{1}{r}\tanh u,$$

with respective multiplicities $(1, n-1, n-1)$. The negative sign could be regarded as coming from the fact that

$$\tanh 2(\theta - u) = -\tanh 2u, \text{ and } \tanh(\theta - u) = -\tanh u, \text{ when } \theta = 0.$$

These principal curvatures differ in sign from those in [399, p. 259].

6.7.2 *Geometry of focal sets*

When the rank of ϕ_u is less than $2n-1$, its range consists of focal points. Assume that this rank is a constant m in a neighborhood of the point (ϵ_1, p). Then ϕ_u embeds this neighborhood as an m-dimensional focal submanifold of $\tilde{M}$. As before, $\eta = \pi_* \zeta$ is a unit normal to the focal submanifold at the point $\phi_u(\epsilon_1, x) = \pi z$, and we can use our previous calculations to study the focal submanifold in two important cases.

Case 1: when M is a complex submanifold

We can see from equation (6.28) that the following focal point behavior occurs in the $\mathbf{CP}^n$ case.

- If 0 is not an eigenvalue of A_{ξ_1} at p and $u = \frac{\pi}{2}$, then ϕ_u has rank $2n-2$ near p and ϕ_u maps a neighborhood of (ϵ_1, p) onto a focal submanifold of codimension 2.
- If 0 is an eigenvalue of constant multiplicity k and $u = \frac{\pi}{2}$, then ϕ_u has rank $2n-2-k$ near p and ϕ_u maps a neighborhood of (ϵ_1, p) onto a focal submanifold of codimension $k+2$.
- If $\frac{1}{r}\cot u \neq 0$ is an eigenvalue of constant multiplicity k, then ϕ_u has rank $2n-1-k$ near p and ϕ_u maps a neighborhood of (ϵ_1, p) onto a focal submanifold of codimension $k+1$.

Looking at equation (6.29) for the $\mathbf{CH}^n$ case, we see that the situation is simpler.

- If $\frac{1}{r}\coth u$ is an eigenvalue of constant multiplicity k, then ϕ_u has rank $2n-1-k$ near p and ϕ_u maps a neighborhood of (ϵ_1, p) onto a focal submanifold of codimension $k+1$.

In all cases, η is a unit normal for the focal submanifold. For the first two $\mathbf{CP}^n$ possibilities (where $u = \frac{\pi}{2}$), A_η can be read off from equation (6.36) by merely discarding the equation involving $\cot 2u$ and the equation for which $\theta = u$, if it occurs. For third $\mathbf{CP}^n$ case, we discard the equation for which $\theta = u$. For $\mathbf{CH}^n$, we use equation (6.37) and discard the term for which $r\lambda = \coth u$. Specifically, we have

$$A_\eta \epsilon_2 = \frac{2}{r} \coth 2u\, \epsilon_2, \tag{6.42}$$

$$A_\eta \epsilon_j = \frac{1}{r} \coth u\, \epsilon_j \text{ for } j \geq 3,$$

$$A_\eta X = \frac{1}{r} \frac{\sinh u - r\lambda \cosh u}{\cosh u - r\lambda \sinh u} X, \text{ if } AX = \lambda X,$$

the third equation holding for eigenvalues λ of A for which $r\lambda \neq \coth u$.

Case 2: when $J\xi_1$ is an eigenvector of A_{ξ_1}

The values of u that give rise to focal points are those for which $r\alpha = 2\cot 2u$ and $r\lambda = \cot u$ (resp. $r\alpha = 2\coth 2u$ and $r\lambda = \coth u$) for the $\mathbf{CP}^n$ and $\mathbf{CH}^n$ cases, respectively. If the eigenvalues of A maintain constant multiplicities, then these ϕ_u will map locally to focal submanifolds for these values of u. The shape operator A_η can be read off from the corresponding tube equations, discarding equations involving expressions that are undefined at the relevant value of u.

In the $\mathbf{CH}^n$ case, suppose first that either $|r\alpha| > 2$ and u is such that

$$\alpha \neq \frac{2}{r} \coth 2u$$

or that $|r\alpha| \leq 2$. Then $(\phi_u)_* W \neq 0$. For a focal point, we need a unit tangent vector $Y \in \mathcal{W}^\perp$ at x and a number ν such that $AY = \nu Y$ and $r\nu = \coth u$. Provided that ν is a principal curvature of constant multiplicity, we will have a focal submanifold and the A_η can be read off from equation (6.40), discarding the equation for which $\lambda = \nu$.

If $|r\alpha| > 2$ and $\alpha = \frac{2}{r} \coth 2u$, there are basically two possibilities. One is that $\nu = \frac{1}{r} \coth u$ is also an eigenvalue of A (with constant multiplicity). Then shape operator A_η can be read off from equation (6.40) discarding both the W equation and the equation for which $\lambda = \nu$. The second is that $\frac{1}{r} \coth u$ is not an eigenvalue. Then dimension of the focal submanifold is greater and the only equation that should be discarded is that involving W.

Note that in all cases, existence of a focal submanifold requires that certain multiplicities be constant, to ensure that ϕ_u has constant rank. However, other principal curvatures need not have constant multiplicities. The shape operator formulas for the focal submanifold will still be correct.

A special case of the situation we have been discussing is that of a Hopf hypersurface. We carry out a detailed study of the focal set behavior of Hopf hypersurfaces in the next section.

6.8 Structure Theorem for Hopf Hypersurfaces

We now have the machinery to work out the fundamental structure of Hopf hypersurfaces. We have seen that tubes over complex submanifolds are Hopf. We will show in this section that typically Hopf hypersurfaces in $\mathbf{CP}^n$ are tubes over complex submanifolds and that these submanifolds are focal submanifolds. In $\mathbf{CH}^n$, focal submanifolds may not always exist but similar structure theorems hold. These results were first proved by Cecil and Ryan [94] in $\mathbf{CP}^n$, and Montiel [378] in $\mathbf{CH}^n$.

We start by briefly revisiting the previous three sections in the codimension one context. Recall that U is a normal coordinate neighborhood. For a hypersurface with a specified choice of unit normal, we simplify the function ϕ_u, considering instead the function

$$f_u : U \to \tilde{M}$$

defined by

$$f_u(x) = \phi_u(\epsilon_1, x).$$

Thus, $f_u(x)$ is the point reached by traveling along the (unique) normal geodesic through x a distance ru in the direction determined by the choice of unit normal. Choosing $w, z \in \tilde{M}'$ so that $\pi w = x$ and $\pi z = f_u(x)$, we can read off the differential of f_u from equations (6.26) and (6.27) to get

$$(f_u)_*X = (\pi_*)_z\big((\cos u \ - r\lambda \sin u)X^L + \sin u \ \langle J\xi, X\rangle \frac{iw}{r}\big) \tag{6.43}$$

$$(f_u)_*X = (\pi_*)_z\big((\cosh u \ - r\lambda \sinh u)X^L - \sinh u \ \langle J\xi, X\rangle \frac{iw}{r}\big)$$

for $\mathbf{CP}^n$ and $\mathbf{CH}^n$, respectively. (Here X is a principal vector with principal curvature λ.) When the range of f_u consists of focal points, we refer to it as a "focal map." When a focal map has constant rank, its range is a focal submanifold. We have the following theorem.

Theorem 6.20. *Let M^{2n-1}, where $n \geq 2$, be a Hopf hypersurface in $\mathbf{CP}^n$ or $\mathbf{CH}^n$. Suppose that f_u has constant rank m, where $0 < m < 2n-1$ in a neighborhood of a point $x \in M$. Then f_u maps a neighborhood U of x onto an m-dimensional focal submanifold V. Furthermore, U lies on the tube of radius ru over V.*

Proof. The "constant rank theorem" (see, for example, Conlon [120, p. 39]) implies that f_u maps a neighborhood U of x onto an m-dimensional submanifold V of $\tilde{M}$. Define a mapping

$$\eta : U \to BV,$$

where BV is the unit normal bundle of V , by

$$\eta = \pi_* \zeta.$$

In other words, for each $x \in U$,

$$\eta(x) = (\pi_*)_z \zeta = (\pi_*)_z(-\sin u \, \frac{w}{r} + \cos u \, \xi^L(w)).$$

Since $\{X^L, iw, w, \xi^L\}$ form a mutually orthogonal set, it is clear that $\eta(x)$ is a unit normal to V at p. Now, $\eta(x)$ is just the velocity vector at the point $p = f_u(x)$ of the normal geodesic to M with initial conditions (x, ξ). On the other hand, our construction for the tube of radius ru over V involves traversing this same geodesic in the opposite direction. Thus

$$E_\mu(-\eta(x)) = x$$

for all x in U, where $E_\mu : BV \to \tilde{M}$ comes from the normal exponential map of V (see the beginning of Subsection 6.7.1). We conclude that U lies on the tube as required. □

Remark 6.21. It is possible for a Hopf hypersurface that f_u is a constant map, and thus has rank $m = 0$. Then the focal set consists of a single point and U lies on a geodesic sphere of radius ru centered at that point. Such hypersurfaces are open subsets of Type A_1 hypersurfaces. Although it is possible to interpret the focal set as a submanifold of dimension 0, we shall not do this. All of our manifolds have dimension at least 1.

The focal submanifolds of a Hopf hypersurface fall into two categories, depending on the Hopf principal curvature α. In $\mathbf{CP}^n$, at least one focal submanifold is a complex (Kähler) submanifold. All non-complex focal submanifolds are generic in the sense of Yano and Kon [560] (as defined in the paragraph below). In $\mathbf{CH}^n$ it is possible for a Hopf hypersurface to have no focal points at all. Horospheres are examples of this phenomenon.

A submanifold V of a complex space form is said to be *generic* if for every normal vector v, the vector Jv is tangent to V. A submanifold V is said to be *totally real* if for every vector X tangent to V, JX is a normal vector.

We now refine Theorem 6.20 using the relationship between α and u.

Theorem 6.22. *Let M^{2n-1}, where $n \geq 2$, be a Hopf hypersurface in $\mathbf{CP}^n$ with Hopf principal curvature $\alpha = \frac{2}{r}\cot 2u$, $0 < u \leq \frac{\pi}{2}$. Suppose that f_u has constant rank $m > 0$ in a neighborhood of a point $x \in M$. Then m is even and f_u maps a neighborhood U of x onto a complex $\frac{m}{2}$-dimensional submanifold V. Furthermore, $V = f_u U$ is a focal submanifold and U lies on the tube of radius ru over V.*

Proof. We need only show that V is a complex submanifold. Because of the relationship between α and u, we have $(f_u)_* W = 0$. On the other hand, if $X \in \mathcal{W}^\perp$

is principal with principal curvature λ, then $(f_u)_*X$ is a scalar multiple of $(\pi_*)_z X^L$. In order to show that $J\eta$ is normal to $f_u U$ at p, it will be sufficient to check that $\langle(\pi_*)_z X^L, J\eta\rangle = 0$ for all such X. Now

$$\begin{aligned}\langle(\pi_*)_z X^L, J\eta\rangle &= \langle((\pi_*)_z X^L)^L, i\zeta\rangle = \langle X^L - az - biz, i\zeta\rangle \\ &= \langle X^L, i\zeta\rangle = \langle X^L, -\sin u\,\frac{iw}{r} + \cos u\; i\xi^L(w)\rangle,\end{aligned} \tag{6.44}$$

where a and b are suitable scalars. The terms involving a and b vanish since $i\zeta$ is horizontal at z. Then, since X^L is horizontal at w, the term involving iw must also vanish. We need only check that $\langle X^L, i\xi^L\rangle = 0$. Observe that the difference between $i\xi^L$ and $(J\xi)^L$ is a linear combination of w and iw. Since $X \in \mathcal{W}^\perp$, we have

$$0 = \langle X, J\xi\rangle = \langle X^L, (J\xi)^L\rangle = \langle X^L, i\xi^L\rangle,$$

as required. Now, of course, $\eta(x)$ is only one of the many unit normals to V at p. However, all other normals at p can be expressed as linear combinations of similar terms. To see this, first recall the equation $E_\mu(-\eta(x)) = x$ from the proof of Theorem 6.20. From the fact that U and BV have the same dimension it follows that η is a diffeomorphism of U onto an open subset $\eta(U)$ of BV. The fiber of BV over p is a sphere of dimension $2n - m - 1$. Its intersection with $\eta(U)$ is nonempty and open in the sphere. Therefore, it contains a basis for the normal space of V at p. Each such basis element is of the form $\eta(y)$ for some $y \in f_u^{-1}(p)$ and, as we have seen, $J\eta(y)$ is normal at p. Thus for any normal vector v at p, Jv is a linear combination of unit normals and hence is normal itself. Consequently, J preserves the normal space (and hence the tangent space) to V, and V is a complex submanifold. □

The situation for $\mathbf{CH}^n$ is slightly more complicated. If the Hopf principal curvature satisfies $|r\alpha| > 2$, we may write $\alpha = \frac{2}{r}\coth 2u$. Then U lies on a tube over a complex submanifold, as in the $\mathbf{CP}^n$ case. Specifically, we have

Theorem 6.23. *Let M^{2n-1}, where $n \geq 2$, be a Hopf hypersurface in $\mathbf{CH}^n$ with Hopf principal curvature $\alpha = \frac{2}{r}\coth 2u$, where $u > 0$. Suppose that f_u has constant rank $m > 0$ in a neighborhood of a point $x \in M$. Then m is even and f_u maps a neighborhood U of x onto a complex $\frac{m}{2}$-dimensional submanifold V. Furthermore, V is a focal submanifold and U lies on the tube of radius ru over V.*

The proof is exactly the same as for $\mathbf{CP}^n$ with the trigonometric functions replaced by the appropriate hyperbolic functions.

If $|r\alpha| \leq 2$, on the other hand, one cannot guarantee that focal points exist. In other words, it may happen that f_u embeds a neighborhood U of x onto a parallel hypersurface for all u. However, when focal submanifolds exist, they can be characterized as follows.

Theorem 6.24. *Let M^{2n-1}, where $n \geq 2$, be a Hopf hypersurface in $\mathbf{CP}^n$ (resp. $\mathbf{CH}^n$). Suppose that $0 < u < \frac{\pi}{2}$ (resp. $u > 0$) is such that f_u has constant rank*

$m < 2n - 1$ in a neighborhood of a point $x \in M$. Except in the case $\frac{2}{r}\cot 2u = \alpha$ (resp. $\frac{2}{r}\coth 2u = \alpha$), f_u maps a neighborhood U of x onto an m-dimensional generic submanifold V. Furthermore, V is a focal submanifold and U lies on the tube of radius ru over V.

Proof. We proceed as in the proof of the preceding theorem. The difference is that $(f_u)_* W$ does not vanish, but is a nonzero multiple of $(\pi_*)_z i\zeta$. Now $J\eta(x) = J(\pi_*)_z\zeta = (\pi_*)_z(i\zeta)$ since ζ is horizontal. This exhibits $J\eta(x)$ as a tangent vector to V at p. By the same argument used in the preceding theorem, every normal vector at p is a linear combination of suitable $\eta(y)$ all of which are mapped into the tangent space by J. □

Corollary 6.25. *Let V^m be a generic focal submanifold of a Hopf hypersurface in $\mathbf{CP}^n$ or $\mathbf{CH}^n$. Then $m \geq n$. In other words, the codimension cannot be larger than half the dimension of the ambient space. If $m = n$, then V is totally real.*

Remark 6.26. For complex focal submanifolds there are no corresponding restrictions on the dimension. There are Type A hypersurfaces having complex focal submanifolds of every complex dimension k for $1 \leq k \leq n - 1$.

For a Hopf hypersurface M, the shape operator of a parallel hypersurface M_μ can be read off from equations (6.39) and (6.40),

$$A_\mu W = \frac{2}{r}\cot 2(\theta - u)W, \text{ if } \alpha = \frac{2}{r}\cot 2\theta, \tag{6.45}$$
$$A_\mu X = \frac{1}{r}\cot(\theta - u)\, X, \text{ if } \lambda = \frac{1}{r}\cot\theta, AX = \lambda X,$$

$$A_\mu W = -\frac{2}{r}\,\frac{\sinh 2u - \frac{r\alpha}{2}\cosh 2u}{\cosh 2u - \frac{r\alpha}{2}\sinh 2u}\,W, \text{ if } AW = \alpha W, \tag{6.46}$$
$$A_\mu X = -\frac{1}{r}\,\frac{\sinh u - r\lambda\cosh u}{\cosh u - r\lambda\sinh u}\,X, \text{ if } AX = \lambda X,$$

for $\mathbf{CP}^n$ and $\mathbf{CH}^n$, respectively.

With regard to focal sets, the $\mathbf{CP}^n$ case is simplest. Suppose that the Hopf hypersurface M has principal curvatures $\{\lambda_j, \mu_j\}$ on $\mathcal{W}^\perp$ at a point x. We write

$$\lambda_j = \frac{1}{r}\cot\theta_j$$

for suitable $\theta_j \in (0, \pi)$. For each j we assume that the pair (λ_j, μ_j) satisfies condition *3.* of Theorem 6.17 on page 357. For at most two values of j we may have $\lambda_j = \mu_j$. In these cases, the λ_j can be expressed in terms of α. In fact, if $\alpha = \frac{2}{r}\cot 2\theta$, then the two possible values of λ_j are $\frac{1}{r}\cot\theta$ and

$$\frac{1}{r}\cot\left(\theta - \frac{\pi}{2}\right) = -\frac{1}{r}\tan\theta.$$

Except for this, elements of the list $\{\lambda_j, \mu_j\}$ are distinct. Condition *3.* of Theorem 6.17 also implies that if $\lambda_j = \frac{1}{r}\cot\theta_j$, then $\mu_j = \frac{1}{r}\cot(2\theta - \theta_j)$. We might as well assume that $\theta_j \leq \theta$ for all j.

We first look at a complex focal submanifolds in $\mathbf{CP}^n$. Assume that f_u has constant rank m near x where $\alpha = \frac{2}{r}\cot 2u$. If it happens that $\frac{1}{r}\cot u$ is also a principal curvature at x, the same must hold in a neighborhood U (because of the constant rank assumption). The tangent space at any point of V is spanned by (the image by $(f_u)_*$ of) the principal vectors $X \in \mathcal{W}^\perp$ corresponding to principal curvatures other than $\frac{1}{r}\cot u$. The shape operator A_μ of the focal submanifold V at such a point satisfies

$$A_\mu X = \frac{1}{r}\cot(\theta - u)\, X, \tag{6.47}$$

where $AX = \frac{1}{r}\cot\theta X$. This result can be read off the formulas (6.39) for parallel hypersurfaces M_μ. Even though V is a (smooth) complex submanifold, there is no guarantee that the principal curvatures will be smooth functions near x so that the result is a pointwise one. However, if the multiplicities of the principal curvatures of M remain constant in a neighborhood of x, then one can choose the various X and θ smoothly near x.

If it happens that $\frac{1}{r}\cot u$ is not a principal curvature at x, then f_u has rank $m = 2n-2$ near x and V is a complex hypersurface (i.e., the complex dimension is $n-1$). This happens, for example, with a Type B hypersurface. The tangent space to V is spanned by $\mathcal{W}^\perp$. The shape operator A_μ satisfies the same conditions (6.47).

We now consider generic focal submanifolds. If $\frac{1}{r}\cot u$ is a principal curvature of constant multiplicity k in a neighborhood of x, then the rank of f_u will be $m = 2n-1-k$ there. The tangent space to the focal submanifold V will be (the image by $(f_u)_*$ of) the orthogonal complement of the principal subspace corresponding to the principal curvature $\frac{1}{r}\cot u$. This includes W, so that shape operator A_μ of V satisfies

$$A_\mu W = \frac{2}{r}\cot 2(\tilde{u} - u)\, W \tag{6.48}$$

$$A_\mu X = \frac{1}{r}\cot(\theta - u)\, X$$

where we have written $\alpha = \frac{2}{r}\cot 2\tilde{u}$ and X and θ are as before.

The situation for $\mathbf{CH}^n$ is complicated by the fact that there are three possible forms for the Hopf principal curvature α and three possible forms for each of the other principal curvatures of M. However, the computation of the principal curvatures of the parallel hypersurface M_μ and the focal submanifolds follows a similar pattern, using hyperbolic instead trigonometric functions.

The case of constant principal curvatures

We have presented many examples of Hopf hypersurfaces whose principal curvatures are constant. Most of them are tubes over focal submanifolds and these focal submanifolds exhibit a high degree of symmetry. The following lemma will be useful for our classification theorem.

Lemma 6.27. *Let M^{2n-1}, where $n \geq 2$, be a Hopf hypersurface in* $\mathbf{CP}^n$ *or* $\mathbf{CH}^n$*. Suppose that its principal curvatures are constant and that M_u is a focal submanifold. Let ξ be any unit normal to M_u. Then a number λ is an eigenvalue of the shape operator A_ξ if and only if $-\lambda$ is an eigenvalue with the same multiplicity. In other words, A_ξ and $-A_\xi$ have the same characteristic polynomial. Furthermore, the eigenvalues and multiplicities are independent of the choice of ξ in the unit normal bundle BM_u.*

Proof. Our assumptions guarantee that f_u is of constant rank. For any $x \in M$, the arguments presented in the proofs of the previous three theorems show that f_u maps a neighborhood U of x onto an open subset V of M_u. Also, η maps U diffeomorphically onto an open subset $\eta(U)$ of the unit normal bundle BV. Since the principal curvatures are constant on U, the eigenvalues of the shape operator $A_{\eta(y)}$ are the same for all $y \in U$ (just subtract u from the arguments of the functions used to express the principal curvatures). Thus the characteristic polynomial $\{\xi \mapsto \det(A_\xi - tI)\}$ is a constant function on the open set $\eta(U)$. Since every unit normal to M_u comes from such an x, we see that $\det(A_\xi - tI)$ is locally constant, and hence constant, on BM_u. In particular, A_ξ and $A_{-\xi} = -A_\xi$ have the same principal curvatures and multiplicities. Note also that the values and multiplicities of the eigenvalues of A_ξ are completely determined by the principal curvatures of M and the value of u. They do not change if we vary the unit normal ξ or the point at which it is based. This completes the proof of our lemma. □

6.9 Focal Sets and Tubes Using Jacobi Fields

In the previous sections, we have studied this topic by explicitly parametrizing geodesics on the real space forms $\tilde{M}'$ and exploiting the relationship between the geometry of $\tilde{M}'$ and that of our desired ambient space $\tilde{M}$. We now present a second approach which is more general. In particular, it will be useful when working with more complicated ambient spaces, such as quaternionic space forms. It will also be more adaptable to situations when we wish to emphasize the symmetric space structure of the ambient space.

Vector fields along a curve

Let M^n be a manifold with linear connection ∇ and let p be an arbitrary point of M. If X and Y are vector fields on M, then the value of $\nabla_X Y$ at p is determined by X_p and the value of Y along any curve through p whose velocity vector at p is X_p. Put another way, if x_t is any curve and $Y_t \in T_{x_t}M$ for each t, then (by extending the velocity vector $\overrightarrow{x_t}$ and Y_t to vector fields locally), we may arrive at a unique value for $\nabla_X Y$ at x_t for any particular parameter value t satisfying $\overrightarrow{x_t} \neq 0$. In order to illustrate this dependence, we call Y_t a vector field along x_t and write its covariant derivative as $\nabla_{\overrightarrow{x_t}} Y$ or $\nabla_t Y$. We can also extend this notation to handle curves for which $\overrightarrow{x_t}$ is allowed to vanish, and so determine $\nabla_t Y$, by using local coordinates.

Specifically, let $x : U \to R^n$ define a local coordinate system on a suitable open set $U \subset M$. Express the curve x_t in local coordinates x_t^i. Then $\overrightarrow{x_t}$ is expressed as

$$\sum_{k=1}^{n} \frac{dx^k}{dt} \frac{\partial}{\partial x^k}\bigg|_{x_t}$$

and $\nabla_t Y_t$ as

$$\sum_{k=1}^{n} \left(\frac{dY^k}{dt} + \sum_{i,j=1}^{n} Y^i \frac{dx^j}{dt} \Gamma_{ij}^k \right) \frac{\partial}{\partial x^k}\bigg|_{x_t} \tag{6.49}$$

where Γ_{ij}^k are the usual Christoffel symbols. This expression may be used to define $\nabla_t Y$, even at values of t for which $\overrightarrow{x_t}$ vanishes. The construction is sometimes referred to as the "associated covariant derivative," see Conlon [120, p. 297].

We say that Y_t is *parallel along* x_t if $\nabla_t Y = 0$ for all t. The curve x_t is called a *geodesic* if its velocity vector $\overrightarrow{x_t}$ is parallel along x_t. It is well known that for a given curve x_t, a parallel vector field Y_t along x_t is uniquely determined by specifying its value arbitrarily at one particular t. Furthermore, given a point p in M and a tangent vector v at p, there is (at least locally) a unique geodesic passing through p and having velocity vector v there. These facts follow from standard theory of ordinary differential equations.

The geodesic flow on a complete Riemannian manifold

Let M^n be a complete Riemannian manifold. Completeness means that all geodesics have parameter interval $(-\infty, \infty)$. We use, of course, the Levi-Civita connection of M, the unique torsion-free linear connection that is compatible with the Riemannian metric $\langle\ ,\ \rangle$. Each point $\theta = (p, v)$ in the tangent bundle TM determines a unique geodesic γ_θ. Abbreviating γ_θ as γ, so that γ_t now denotes the point along the geodesic corresponding to parameter value t, we have $\gamma_0 = p$ and $\overrightarrow{\gamma_0} = v$. Define

$$\Phi_t : TM \to TM$$

by

$$\Phi_t\theta = (\gamma_t, \overrightarrow{\gamma_t}),$$

and let G_θ be the initial tangent vector to the curve $t \mapsto \Phi_t\theta$. The vector field G on TM is called the *geodesic spray*. It generates the one-parameter group of transformations $\{\Phi_t\}$ and is complete in the sense of Kobayashi and Nomizu [283, Vol. I, pp. 12–14]. This group of transformations is called the *geodesic flow*. We note that the unit tangent bundle SM is invariant by the geodesic flow. In other words, we can write

$$\Phi_t : SM \to SM$$

using the same formula. This defines G as a vector field on SM. Denote by

$$\Pi : TM \to M$$

the canonical projection map of the bundle TM. For $\theta = (p, v)$ in TM, we note that

$$(\Pi_*)_\theta G_\theta = v \in T_pM.$$

Decomposition of $T_\theta TM$

We will now describe a splitting of $T_\theta TM$ into vertical and horizontal subspaces. The vertical subspace V_θ is the kernel of

$$(\Pi_*)_\theta : T_\theta TM \to T_pM.$$

Note that the vertical subspace depends only on the differentiable structure. The horizontal subspace, however, will depend on the Riemannian metric, or more specifically, the Levi-Civita connection ∇ of that metric. Let

$$z_t = (\alpha_t, Z_t)$$

be a curve in TM with $z_0 = \theta$ and $\overrightarrow{z_0} = \xi \in T_\theta TM$. We define a map

$$K_\theta : T_\theta TM \to T_pM$$

to be the (associated) covariant derivative of Z_t along α_t evaluated at $t = 0$. In other words, we have

$$K_\theta \xi = \nabla_t Z|_{t=0}.$$

The horizontal subspace H_θ at θ is, by definition, the kernel of the map K_θ. The Sasaki metric on *TM* is given by

$$\langle \xi, \eta \rangle = \langle \Pi_* \xi, \Pi_* \eta \rangle + \langle K_\theta \xi, K_\theta \eta \rangle,$$

for ξ, η in $T_\theta TM$. The right side represents evaluation of the Riemannian metric of M on T_pM. Then we have

Proposition 6.28.

- *T_θ is the orthogonal direct sum of V_θ and H_θ;*
- *K_θ is a linear isomorphism of V_θ onto T_pM;*
- *$(\Pi_*)_\theta$ is a linear isomorphism of H_θ onto T_pM.*

Proof of this proposition and further discussion of the decomposition will be provided in the next few sections. A good reference for this material is Paternain [428].

Local coordinates in TM

Given a coordinate map $x : U \to R^n$ for M, there is a natural way to define a coordinate map $z : \Pi^{-1}U \to R^{2n}$ for TM. We write

$$z(p, v) = (x(p), y(p, v)) \in R^n \times R^n = R^{2n},$$

where

$$v = \sum_{i=1}^{n} y^i(p, v) \left.\frac{\partial}{\partial x^i}\right|_p.$$

It is easy to check that

$$(\Pi_*)_\theta \left.\frac{\partial}{\partial z^i}\right|_\theta = \left.\frac{\partial}{\partial x^i}\right|_p \text{ and } (\Pi_*)_\theta \left.\frac{\partial}{\partial z^{n+i}}\right|_\theta = 0,$$

for $1 \leq i \leq n$.

If $\xi \in T_\theta TM$ is expressed in terms of this coordinate system, then ξ is vertical if and only if $\xi^i = 0$ for $1 \leq i \leq n$. The vertical space V_θ is spanned by the last n coordinate vectors. Using the coordinate expression (6.49), we see that K_θ is well defined. Specifically, the coordinate expression for $K_\theta \xi$ in this case is

$$\sum_{k=1}^{n}\left(\frac{dZ^k}{dt}+\sum_{i,j=1}^{n} Z^i\frac{d\alpha^j}{dt}\Gamma_{ij}^k\right)\frac{\partial}{\partial x^k}\bigg|_{\alpha_t} \tag{6.50}$$

evaluated at $t = 0$. The terms in this expression may be interpreted as follows. The $\frac{d\alpha^j}{dt}$ are the first n components of ξ, while the $\frac{dZ^k}{dt}$ are the last n components of ξ. The Z^i are the last n coordinates of θ and the Γ_{ij}^k are evaluated at p. Thus, all elements of the expression for the covariant derivative at $t = 0$ depend only on ξ, even though there are many choices of z_t for which $\overrightarrow{z_0} = \xi$. The kth component of $K_\theta\xi$ is

$$\xi^{n+k} + \sum_{i,j=1}^{n} v^i\xi^j\Gamma_{ij}^k(p)$$

where

$$v = \sum_{i=1}^{n} v^i\frac{\partial}{\partial x^i}\bigg|_p.$$

This also shows that K_θ is linear in ξ. Clearly, the restriction of K_θ to vertical vectors is injective and is thus an isomorphism. On the other hand, equation (6.49) shows that if Z is parallel along α, then $K_\theta\xi = 0$ and so ξ is horizontal.

There is an identification of T_pM with the horizontal subspace of $T_\theta TM$ by means of the *horizontal lift*. Specifically, for each $X \in T_pM$, there is a unique horizontal element of $T_\theta TM$, denoted by X_θ^L, satisfying $\Pi_* X_\theta^L = X$. It may be constructed as follows. Let α be a curve in M with initial tangent vector X. Let $Z_t \in T_{\alpha_t}M$ be defined by parallel translation of v along α. Consider $z_t = (\alpha_t, Z_t)$ as a curve in TM. Set $X_\theta^L = \overrightarrow{z_0}$. The horizontal lift X_θ^L has the X^j as its first n components and the

$$\frac{dZ^k}{dt} = -\sum_{i,j=1}^{n} Z^i\frac{d\alpha^j}{dt}\Gamma_{ij}^k$$

as its last n components. Thus, the kth component of $K_\theta X_\theta^L$ is

$$\sum_{i,j=1}^{n}\big(-v^iX^j\Gamma_{ij}^k(p) + v^iX^j\Gamma_{ij}^k(p)\big) = 0$$

so X_θ^L is horizontal. Since $(\Pi_*)_\theta X_\theta^L = X$, it is the unique horizontal lift. This shows that (the restriction of) $(\Pi_*)_\theta : H_\theta \to T_pM$ is a linear isomorphism and completes the proof of our proposition.

The exponential map

The exponential map

$$\mathrm{Exp} : TM \to TM$$

is defined by $\mathrm{Exp}(\theta) = \Phi_1(\theta)$ and induces for each $p \in M$ the familiar map

$$\exp_p : T_pM \to M$$

by

$$\exp_p(v) = \Pi(\mathrm{Exp}(\theta))$$

where $\theta = (p, v)$. We will be using (for our study of submanifolds) a slightly more general version of the exponential map and we will need to compute its differential. Our calculation will also yield the differential of the usual exponential map, as we point out in Subsection 6.9.2.

6.9.1 *Jacobi fields*

Let $z_s = (\alpha_s, Z_s)$ be a curve in TM with $z_0 = \theta = (p, v)$ and $\overrightarrow{z_0} = \xi$. Let $\tilde{\xi} = (\Phi_t)_* \xi \in T_{\tilde{\theta}}TM$ where $\tilde{\theta} = \Phi_t\theta = (\tilde{p}, \tilde{v})$. Write $\tilde{z} = \Phi_t \circ z = (\tilde{\alpha}, \tilde{Z})$ so that $\tilde{z}_0 = \tilde{\theta}$ and $\overrightarrow{\tilde{z}_0} = \tilde{\xi}$. Now set

$$V(s,t) = \Pi(\Phi_t z_s).$$

V is said to be a variation of the geodesic γ_θ. The initial tangent vector Y_t to the curve $s \mapsto V(s,t)$, also written

$$\left.\frac{\partial V}{\partial s}\right|_{s=0}$$

is a vector field along the geodesic γ_θ. Such a Y_t is said to be a *Jacobi field* along the geodesic. Jacobi fields are characterized by the fact that they satisfy the *Jacobi equation*

$$\nabla^2 Y_t + R(Y_t, \overrightarrow{\gamma_t})\overrightarrow{\gamma_t} = 0,$$

where R is the curvature tensor of the Riemannian metric and γ_θ is abbreviated γ.

The set of Jacobi fields along γ is a $2n$-dimensional vector space. A unique Jacobi field is determined by specifying its value Y_t and its associated covariant derivative

$\nabla_t Y$ at $t = 0$. We also sometimes write Y_t as $Y(t)$ and $\nabla_t Y$ as $Y'(t)$. We denote by $J_\xi(t)$ the unique Jacobi field Y_t along γ_θ such that $Y_0 = (\Pi_*)_\theta \xi$ and $Y'(0) = K_\theta \xi$. Since $(\Pi_*)_\theta H_\theta = K_\theta V_\theta = T_p M$, all possible Jacobi fields along γ_θ are obtained in this way.

Equality of "mixed partials"

Lemma 6.29. *Let M be a Riemannian manifold and U an open connected subset of $\mathbf{R}^2$. A mapping $V : U \to M$ may be considered as a 1-parameter family of curves on M in two ways. If we write $\alpha_s = \beta_t = V(s,t)$, then $\overrightarrow{\beta_t}$ is a vector field along α_s and $\overrightarrow{\alpha_s}$ is a vector field along β_t. Then*

$$\nabla_s \overrightarrow{\beta_t} = \nabla_t \overrightarrow{\alpha_s}$$

for all $(s,t) \in U$.

Proof. If we carry out the calculation in local coordinates (as introduced earlier), the kth component of $\nabla_t \overrightarrow{\alpha_s}$ is easily computed to be

$$\frac{\partial^2 V^k}{\partial t \partial s} + \sum_{i,j=1}^{n} \frac{\partial V^i}{\partial s} \frac{\partial V^j}{\partial t} \Gamma_{ij}^k,$$

where the V^i are the components of V. Because mixed partial derivatives of real-valued functions commute, it is clear that this expression is unchanged when s and t are interchanged. □

Remark 6.30. In this situation, it is standard to write $\frac{\partial V}{\partial s}$ for $\overrightarrow{\alpha_s}$ and $\frac{\partial V}{\partial t}$ for $\overrightarrow{\beta_t}$. By an abuse of notation, the result of the lemma is sometimes written

$$\frac{\partial^2 V}{\partial t \partial s} = \frac{\partial^2 V}{\partial s \partial t}$$

in analogy with the fact from elementary calculus that we have used in the proof, i.e., for real-valued functions, repeated partial derivatives are independent of the order of differentiation used in computing them.

6.9.2 *Differential of the exponential and related maps*

Using the setup of Subsection 6.9.1, we have $\frac{\partial V}{\partial t} = \tilde{Z}_s$ and

$$\left.\frac{\partial V}{\partial s}\right|_{s=0} = \Pi_*((\Phi_t)_* \overrightarrow{z_0}) = \Pi_* \tilde{\xi}.$$

By Lemma 6.29 and the definition of K, we have

$$\nabla_t \frac{\partial V}{\partial s}\bigg|_{s=0} = \nabla_s \frac{\partial V}{\partial t}\bigg|_{s=0} = K_{\tilde{\theta}}\tilde{\xi}.$$

In particular, for $t = 0$, this is $K_\theta \xi$. This means that $\Pi_* \tilde{\xi}$ coincides with the Jacobi field $J_\xi(t)$ and also shows that for all t, $J'_\xi(t) = K_{\tilde{\theta}}\tilde{\xi}$.

This allows us to improve the result of Proposition 6.28 as follows.

Proposition 6.31. *If $\tilde{\xi} = \tilde{\xi}_V + \tilde{\xi}_H$ is the decomposition of $\tilde{\xi} \in T_{\tilde{\theta}}TM$ into vertical and horizontal components, then $K_{\tilde{\theta}}\, \tilde{\xi}_V = J'_\xi(t)$ and $\Pi_* \tilde{\xi}_H = J_\xi(t)$.*

Proof. The first assertion has already been established. From the discussion above, we also have

$$J_\xi(t) = \Pi_* \tilde{\xi} = \Pi_*(\tilde{\xi}_V + \tilde{\xi}_H) = \Pi_* \tilde{\xi}_H,$$

which completes the proof. □

Remark 6.32. The geodesic spray G is a horizontal vector field. In fact, $G_\theta = v^L_\theta$ for each $\theta = (p, v)$ in TM. G is the unique horizontal vector field such that $(\Pi_*)_\theta G_\theta = v \in T_pM$. The associated Jacobi field $J_G(t) = \overrightarrow{\gamma_\theta(t)}$ is the unique Jacobi field with initial conditions $(v, 0)$, that is, $J_G(0) = v$ and $J'_G(0) = 0$. Note that $J'_G(t) = 0$ for all t.

Lemma 6.33. *In the notation of the previous paragraph, $K_\theta X^L_\theta = 0$. Furthermore, for every horizontal $\xi \in T_\theta TM$, $(\Pi_* \xi)^L_\theta = \xi$.*

The $2n$-dimensional space of Jacobi fields along the geodesic γ_θ may be regarded as the direct sum of two n-dimensional subspaces, one consisting of Jacobi fields with initial conditions $(0, \Pi_* \xi)$ with $\xi \in H_\theta$, the other consisting of Jacobi fields with initial conditions $(K_\theta \xi, 0)$ with $\xi \in V_\theta$. The rank of the exponential map $TM \to TM$ at θ is the dimension of the space spanned by these Jacobi fields at $t = 1$.

We now look at the map $\exp_p : T_pM \to M$ which may be broken down as $\Pi \circ \text{Exp} \circ \rho$ where $\rho : T_pM \to TM$ is given by $\rho(v) = (p, v) = \theta$. Take any $X \in T_v(T_pM)$. Consider the curve $s \mapsto v + sX$ whose initial tangent vector is X. Let $z_s = (p, v + sX)$ so that $z_0 = \theta$ and $\overrightarrow{z_0} = (\rho_*)_v X$. In keeping with the notation of the preceding subsection, denote $\overrightarrow{z_0}$ by ξ. Then $((\exp_p)_*)_v X = J_\xi(1)$. In fact, we have

Proposition 6.34. *For $X \in T_v(T_pM)$, $(\exp_p)_* X = J_\xi(1)$, where $J_\xi(t)$ is the unique Jacobi field along γ_t satisfying $J_\xi(0) = 0$ and $J'_\xi(0) = X$. Here, we have used the natural identification of $T_v(T_pM)$ with T_pM itself.*

Proof. In the coordinate system introduced earlier, z_s is expressed as

$$(p^1, p^2, \ldots, p^n, v^1 + sX^1, v^2 + sX^2, \ldots, v^n + sX^n).$$

Then

$$\xi = \overrightarrow{z_0} = \sum_{i=1}^{n} X^i \left.\frac{\partial}{\partial z^{n+i}}\right|_\theta \in T_\theta TM.$$

Then $\xi^{n+k} + v^i \xi^j \Gamma_{ij}^k(p) = X^k$, so that

$$K_\theta \xi = \sum_{i=1}^{n} X^i \left.\frac{\partial}{\partial x^i}\right|_p.$$

In other words, $K_\theta \xi = X$. □

Submanifolds and the normal bundle

Let $\tilde{M}$ be a complete Riemannian manifold and M a submanifold with the induced Riemannian metric. We now discuss the restriction of the exponential map to the normal bundle NM and apply our previous discussion to the study of tubes over M. Take any $\theta = (p, v) \in NM$. Let $z_s = (\alpha_s, Z_s)$ be a curve in NM such that $z_0 = \theta$. For a fixed real number μ, consider the map

$$\phi_\mu : NM \to \tilde{M}$$

defined by $\phi_\mu = \Pi \circ \Phi_\mu$. Then $(\phi_\mu)_* \xi = (\Pi)_*(\Phi_\mu)_* \xi = J_\xi(\mu)$. If ϕ_μ has maximal rank (i.e., the same as the dimension of $\tilde{M}$) at θ, then ϕ_μ maps a neighborhood of $\theta \in NM$ diffeomorphically onto an open neighborhood of $q = \gamma_\mu = \phi_\mu \theta$. The restriction of ϕ_μ to the unit normal bundle BM maps a neighborhood of θ diffeomorphically to a neighborhood of q in the tube M_μ of radius μ (for as long as γ_θ is a minimizing geodesic). The tangent space to the tube at q is the direct sum of two subspaces. The first is spanned by the Jacobi fields $J_\xi(\mu)$ with initial conditions $J_\xi(0) = X \in T_pM$ and $J'_\xi(0) = -A_v X$. The second is spanned by Jacobi fields $J_\xi(\mu)$ with initial conditions $J_\xi(0) = 0$ and $J'_\xi(0) = w$, where w is normal to M but orthogonal to v at p.

To see this, first choose z_s so that $\overrightarrow{\alpha_s}$ is nonvanishing and Z is parallel along α with respect to the normal connection. In other words,

$$\tilde{\nabla}_s Z = -A_{Z_s} \overrightarrow{\alpha_s}.$$

Let $X = \overrightarrow{\alpha_0}$. Then

$$K_\theta \xi = \tilde{\nabla}_s Z\big|_{s=0} = -A_v X,$$
$$\Pi_* \xi = \overrightarrow{\alpha_0} = X.$$

The resulting Jacobi fields J_ξ provide the second summand in the tangent space to the tube at q.

Lemma 6.35. *The unit normal to the tube M_μ at $q = \gamma_\mu$ is $\overrightarrow{\gamma_\mu}$.*

Proof. We show that

$$\langle J_\xi(t), \overrightarrow{\gamma_t} \rangle = 0$$

for all t. For either of the two types of tangent vectors to M_μ at q, we have $|Z_s| = |v| = 1$. Thus, $\overrightarrow{(\gamma_{z_s})_t}$ is a unit vector for all s, t. As we saw in Lemma 6.29,

$$J'_\xi(t) = \tilde{\nabla}_s \frac{\partial}{\partial t}(\Pi(\Phi_t(z_s)))\bigg|_{s=0} = \tilde{\nabla}_s \overrightarrow{(\gamma_{z_s})_t}\bigg|_{s=0},$$

which is, of course, orthogonal to $\overrightarrow{\gamma_t}$, since $\overrightarrow{(\gamma_{z_s})_t}$ is a unit vector. So

$$\langle J'_\xi(t), \overrightarrow{(\gamma_{z_s})_t}\bigg|_{s=0} \rangle = \langle J'_\xi(t), \overrightarrow{\gamma_t} \rangle = 0.$$

Now,

$$\frac{d}{dt}\langle J_\xi(t), \overrightarrow{\gamma_t} \rangle = \langle J'_\xi(t), \overrightarrow{\gamma_t} \rangle + \langle J_\xi(t), \tilde{\nabla}_t \overrightarrow{\gamma_t} \rangle.$$

The first term is zero as we showed in the previous equation. The second term vanishes because γ_t is a geodesic. Thus $\langle J_\xi(t), \overrightarrow{\gamma_t} \rangle$ is constant and equal to its value at $t = 0$, namely $\langle J_\xi(0), v \rangle = 0$. This shows that every vector in $T_q M_\mu$ is orthogonal to $\overrightarrow{\gamma_\mu}$ and thus completes the proof. □

If ϕ_μ has constant (but non-maximal) rank k in a neighborhood of θ, then ϕ_μ maps a neighborhood of θ onto a smooth focal submanifold whose tangent space is spanned by the same Jacobi fields $J_\xi(\mu)$.

6.9.3 *Shape operators of tubes using Jacobi fields*

Having determined the tangent space to the tube M_μ we now compute the shape operator with respect to the unit normal

$$N = -\overrightarrow{(\gamma_\theta)_\mu}$$

at the point $q = \phi_\mu\theta$. Write $\alpha_s = \gamma_t = V(s,t)$. Then $J_\xi(t) = \overrightarrow{\alpha_0}$. The shape operator A with respect to N_q satisfies

$$\tilde{\nabla}_Y N = -AY$$

for $Y \in T_q M_\mu$. Since each such Y is of the form $J_\xi(\mu)$, we consider

$$\begin{aligned} -\tilde{\nabla}_{J_\xi(t)} N &= \tilde{\nabla}_s \overrightarrow{\gamma_t}\Big|_{s=0} = \tilde{\nabla}_t \overrightarrow{\alpha_s}\Big|_{s=0} \\ &= \tilde{\nabla}_t J_\xi(t) = J'_\xi(t) \end{aligned} \tag{6.51}$$

for t near μ. Thus we have the following.

Theorem 6.36. *Let M be a submanifold of a complete Riemannian manifold $\tilde{M}$. For a fixed real number μ, suppose that the map $\phi_\mu : BM \to \tilde{M}$ has maximal rank (one less than the dimension of $\tilde{M}$) at a point $\theta = (p, v)$. Then*

1. *ϕ_μ embeds a neighborhood $\mathcal{U}$ of θ onto a hypersurface $\mathcal{U}_\mu$ lying on a tube of constant radius over M;*
2. *The tangent space to $\mathcal{U}_\mu$ at $q = \phi_\mu(\theta)$ is spanned by Jacobi fields $J_\xi(\mu)$ as described in the preceding subsection.*
3. *The shape operator of $\mathcal{U}_\mu$ at q satisfies $AJ_\xi(\mu) = J'_\xi(\mu)$ for each $J_\xi(\mu)$ in the tangent space.*

Tubes over submanifolds of CP^n and CH^n

Let M be a submanifold of a complex space form $\tilde{M}$ of constant holomorphic sectional curvature $4c = 4\epsilon/r^2$. For $\theta = (p, v)$ in the unit normal bundle, let X be an eigenvector of A_v corresponding to eigenvalue λ. Let γ_t be the (normal) geodesic determined by θ. Let B_t be the parallel vector field along γ_t with $B_0 = X$. We are interested in the shape operators of tubes over M.

Lemma 6.37. *Assume that X is orthogonal to the vector Jv.*

- *If $c > 0$, then $X_t = (\cos u - r\lambda \sin u)B_t$ is a Jacobi field along γ_t with $X_0 = X$ and $X'_0 = -\lambda X$;*
- *If $c < 0$, then $X_t = (\cosh u - r\lambda \sinh u)B_t$ is a Jacobi field along γ_t with $X_0 = X$ and $X'_0 = -\lambda X$,*

where $u = t/r$.

Proof. First it is easy to check that

$$\tilde{\nabla}_t^2 X_t = X''_t = -cX_t.$$

We need to compute $\tilde{R}(X_t, \overrightarrow{\gamma_t})\overrightarrow{\gamma_t}$. We consider the terms of equation (6.6) separately. First, we see that

$$\frac{\epsilon}{r^2}(X_t \wedge U_t)\, U_t = \frac{\epsilon}{r^2}(\langle U_t, U_t\rangle\, X_t - \langle X_t, U_t\rangle\, U_t) = cX_t, \tag{6.52}$$

where U_t is the unit vector $\overrightarrow{\gamma_t}$. Since X_t is a scalar multiple of B_t, the term $\langle B_t, U_t\rangle$ is constant along γ_t (being the inner product of two vector fields that are parallel along γ_t), and $\langle B_0, U_0\rangle = \langle X, v\rangle = 0$. Now,

$$\begin{aligned}\big(JX_t \wedge JU_t + 2\langle X_t, JU_t\rangle\, J\big)\, U_t &= -\langle JX_t, U_t\rangle\, JU_t + 2\langle X_t, JU_t\rangle\, JU_t \\ &= 3\langle X_t, JU_t\rangle\, JU_t. \end{aligned}\tag{6.53}$$

Recalling that X_t is a scalar multiple of B_t and that J, X_t and U_t are parallel along γ_t, we see that $\langle B_t, JU_t\rangle$ is constant along γ_t and hence equal to $\langle B_0, JU_0\rangle = \langle X, Jv\rangle = 0$. This completes the proof. □

Lemma 6.38. *Assume that $X = Jv$.*

- *If $c > 0$, then $X_t = (\cos 2u - \frac{r}{2}\lambda \sin 2u)B_t$ is a Jacobi field along γ_t with $X_0 = X$ and $X_0' = -\lambda X$.*
- *If $c < 0$, then $X_t = (\cosh 2u - \frac{r}{2}\lambda \sinh 2u)B_t$ is a Jacobi field along γ_t with $X_0 = X$ and $X_0' = -\lambda X$,*

where $u = t/r$.

Proof. The proof is similar to that of the previous lemma. First of all, we get

$$\tilde{\nabla}_t^2 X_t = X_t'' = -4cX_t.$$

The first term in the curvature expression is unchanged from equation (6.52). Further, $\langle B_t, JU_t\rangle JU_t$ is parallel along γ_t. Its initial value is $\langle X, Jv\rangle Jv = X$. Thus $\langle B_t, JU_t\rangle JU_t = B_t$ and hence $\langle X_t, JU_t\rangle JU_t = X_t$. Consequently,

$$\tilde{R}(X_t, \overrightarrow{\gamma_t})\overrightarrow{\gamma_t} = cX_t + 3cX_t = 4cX_t, \tag{6.54}$$

and the Jacobi equation is satisfied. □

To complete the tangent space to the tube, we need to look at normals orthogonal to v.

Lemma 6.39. *Assume that (p, w) is in the unit normal bundle BM with $\langle w, v\rangle = \langle w, Jv\rangle = 0$. Let W_t be parallel along γ_t with $W_0 = w$.*

- *If $c > 0$, then $Y_t = (\sin u)W_t$ is a Jacobi field along γ_t with $Y_0 = 0$ and $Y_0' = w/r$.*
- *If $c < 0$, then $Y_t = (\sinh u)W_t$ is a Jacobi field along γ_t with $Y_0 = 0$ and $Y_0' = w/r$,*

where $u = t/r$.

Proof. Following the same procedures as in earlier cases, we find that $Y_t'' = -4cY_t$ and that $\langle W_t, JU_t\rangle JU_t = W_t$, so that $\tilde{R}(Y_t, \vec{\gamma_t})\vec{\gamma_t} = 4cY_t$, and the Jacobi equation is satisfied. □

Remark 6.40. These constructions are also useful in computing the shape operators of parallel hypersurfaces in Section 8.2.

Chapter 7
Complex Submanifolds of $\mathbf{CP}^n$ and $\mathbf{CH}^n$

A primary goal in our study of real hypersurfaces in $\mathbf{CP}^n$ or $\mathbf{CH}^n$ is the classification of Hopf hypersurfaces with constant principal curvatures given in Chapter 8. Since the focal submanifold corresponding to the Hopf principal curvature must be a complex submanifold, certain aspects of the theory of complex submanifolds are important in our study of real hypersurfaces. In particular, a knowledge of the behavior of the principal curvatures of certain well-known complex submanifolds is needed.

In this chapter, we study those aspects of complex submanifolds in $\mathbf{CP}^n$ or $\mathbf{CH}^n$ that are pertinent to our theory. In Section 7.1, we recall some basic terminology from the theory of submanifolds in a Riemannian manifold. We then specialize to the case of a Kähler submanifold M of a complex space form $\tilde{M}$. We prove an important formula of Simons' type, which expresses the Laplacian of the length of the second fundamental form σ in terms of σ and its derivative. Such a formula was introduced by Simons [485] for minimal submanifolds of Riemannian manifolds, and developed by Nomizu and Smyth [406] and many other authors in different contexts. This formula will be needed in Chapter 8 in our classification of Hopf hypersurfaces with constant principal curvatures.

In Sections 7.2–7.5, we study four important embeddings of complex submanifolds in $\mathbf{CP}^n$: the Veronese embedding of $\mathbf{CP}^m$ in $\mathbf{CP}^n$, the Segre embedding of $\mathbf{CP}^h \times \mathbf{CP}^k$ in $\mathbf{CP}^n$, the Plücker embedding of complex Grassmannians in $\mathbf{CP}^n$, and the half-spin embedding of $SO(2d)/U(d)$ in $\mathbf{CP}^n$.

For each of these embeddings we give a detailed analysis of the behavior of the principal curvature functions on the unit normal bundle of the submanifold. Such an analysis is needed to complete the proof of Kimura's [270] classification of Hopf hypersurfaces with constant principal curvatures in $\mathbf{CP}^n$ (Theorem 8.13, page 432).

T.E. Cecil, P.J. Ryan, *Geometry of Hypersurfaces*, Springer Monographs in Mathematics, DOI 10.1007/978-1-4939-3246-7_7

7.1 A Formula of Simons' Type

We begin in a general setting. Let M be a submanifold of a Riemannian manifold $\tilde{M}$. The second fundamental form σ is defined by

$$\sigma(X, Y) = \tilde{\nabla}_X Y - \nabla_X Y \tag{7.1}$$

where X and Y are tangent vector fields on M. Then, corresponding to each normal vector ξ, we have the shape operator A_ξ which satisfies

$$\langle A_\xi X, Y\rangle = \langle \sigma(X, Y), \xi\rangle. \tag{7.2}$$

A formula of Simons' type expresses the Laplacian of the length of the second fundamental form σ in terms of σ and its derivative. The main result in this section is a self-contained derivation of such a formula for Kähler submanifolds of complex space forms. Specifically, we will prove the following theorem. In order to state the theorem succinctly, we introduce the following terminology. For a $2d$-dimensional (real) vector space with complex structure J, a basis $\{v_i\}$ is said to be *J-invariant* if it consists of d pairs of the form $\{v, Jv\}$. Unless otherwise specified, we will assume that $Jv_i = v_{i+d}$ for $1 \leq i \leq d$.

Theorem 7.1. *Let M be a Kähler submanifold of a complex space form $\tilde{M}$ of constant holomorphic curvature $4c$. Assume that the complex dimensions of M and $\tilde{M}$ are m and n respectively, with $p = n - m$ denoting the complex codimension. If $\{\xi_\alpha\}_{\alpha=1}^{2p}$ is a J-invariant orthonormal basis for the normal space with corresponding shape operators $\{A_\alpha\}$, then*

$$\frac{1}{2}\Delta|\sigma|^2 = |\nabla'\sigma|^2 + 2(m+2)c|\sigma|^2 - \Sigma(\text{ trace } (A_\alpha A_\beta))^2 - 2 \text{ trace } (\Sigma A_\alpha^2)^2.$$

In the summations, the indices α and β run from 1 to $2p$.

We recall a few basic concepts from submanifold theory. The companion formula to equation (7.1) is

$$\tilde{\nabla}_X \xi = -A_\xi X + \nabla_X^\perp \xi, \tag{7.3}$$

where the two terms on the right side are, by definition, the tangential and normal components, respectively, of the left side. This formula defines the shape operator A_ξ as a symmetric tensor field of type $(1, 1)$ and the normal connection $\nabla^\perp$ as a connection on the normal bundle. With respect to an orthonormal normal frame $\{\xi_\alpha\}$, the normal connection can be expressed in terms of a set of 1-forms by

$$\nabla_X^\perp \xi_\alpha = \Sigma s_{\alpha\beta}(X)\xi_\beta. \tag{7.4}$$

The normal connection has a natural extension to higher-order objects over M. In particular, we can differentiate σ as follows:

$$(\nabla'_X\sigma)(Y,Z) = \nabla^{\perp}_X(\sigma(Y,Z)) - \sigma(\nabla_X Y,Z) - \sigma(Y,\nabla_X Z) \tag{7.5}$$

which, when expressed in terms of the normal frame, yields

$$(\nabla'_X\sigma)(Y,Z) = \langle(\nabla_X A_\alpha)Y,Z\rangle\ \xi_\alpha + \langle A_\alpha Y,Z\rangle\ s_{\alpha\beta}(X)\xi_\beta. \tag{7.6}$$

The first term is to be summed over all α; the second term over all ordered pairs (α,β). The length of the second fundamental form and its covariant derivative may be expressed as

$$\begin{aligned}|\sigma|^2 &= \Sigma\ |\sigma(e_i,e_j)|^2\\ |\nabla'\sigma|^2 &= \Sigma\ |(\nabla'_{e_i}\sigma)(e_j,e_k)|^2,\end{aligned} \tag{7.7}$$

where $\{e_i\}$ is an orthonormal basis for the tangent space to M. The first summation is over all ordered pairs (i,j) and the second over all ordered triples (i,j,k). We will now relate these to the norms of the shape operators and their covariant derivatives. Note that for any tensor T of type $(1,1)$ we have

$$\begin{aligned}|T|^2 &= \Sigma\ |\langle Te_i,e_j\rangle|^2\\ |\nabla T|^2 &= \Sigma\ |\langle(\nabla_{e_i}T)e_j,e_k\rangle|^2\end{aligned} \tag{7.8}$$

with the same summation indices as in equation (7.7).

Lemma 7.2. *In terms of the notation established above, we have*

$$\begin{aligned}|\sigma|^2 &= \Sigma|A_\alpha|^2 = \Sigma\ \text{trace}\ A_\alpha^2\\ |\nabla'\sigma|^2 &= \Sigma|\nabla A_\alpha|^2 - 2\ \Sigma s_{\alpha\beta}(e_i)\ \text{trace}\ ((\nabla_{e_i}A_\alpha)A_\beta)\\ &\quad -\Sigma s_{\alpha\beta}(e_i)s_{\beta\gamma}(e_i)\ \text{trace}\ (A_\alpha A_\gamma)\end{aligned} \tag{7.9}$$

with the obvious ranges of summation.

For a tensor field T of type $(1,1)$, the second covariant derivative is defined by

$$K(U,V) = \nabla_U\nabla_V T - \nabla_{\nabla_U V}T$$

where U and V are vector fields. In the notation of Kobayashi and Nomizu, [283, Vol. I, p. 124],

$$K(U,V) = \nabla^2 T(;V;U).$$

Also, the curvature operator $R(U,V)$ is expressed in terms of K by

$$K(U,V) - K(V,U) = R(U,V)\cdot T = [R(U,V),T].$$

Following Nomizu and Smyth [406], we define the restricted Laplacian $\Delta' A$ of A by

$$\Delta' A = \Sigma K(e_i,e_i).$$

The following is straightforward to check (see, for example, [406]).

Lemma 7.3. *For any tensor T of type* $(1,1)$, *the Laplacian of its length satisfies*

$$\frac{1}{2}\Delta|T|^2 = \text{trace}((\Delta' T)T) + |\nabla T|^2. \tag{7.10}$$

Gauss, Codazzi, and Ricci equations

The Gauss, Codazzi, and Ricci equations relate the curvature operator of the ambient space $\tilde{M}$ to the curvature tensor of M and to the second fundamental form and shape operators. For X and Y tangent to M, we use equations (7.1) and (7.3) to compute $\tilde{R}(X,Y)Z$ for Z tangent to M and $\tilde{R}(X,Y)\xi$ for ξ normal to M. It is straightforward to compute that

$$\tilde{R}(X,Y)Z = R(X,Y)Z - (A_{\sigma(Y,Z)}X - A_{\sigma(X,Z)}Y) + (\nabla'_X\sigma)(Y,Z) - (\nabla'_Y\sigma)(X,Z).$$

Thus, we have the Gauss equation

$$R(X,Y)Z = A_{\sigma(Y,Z)}X - A_{\sigma(X,Z)}Y + \text{tangential component of } \tilde{R}(X,Y)Z, \tag{7.11}$$

and the Codazzi equation

$$(\nabla'_X\sigma)(Y,Z) - (\nabla'_Y\sigma)(X,Z) = \text{normal component of } \tilde{R}(X,Y)Z. \tag{7.12}$$

On the other hand, if we apply the curvature operator $\tilde{R}(X,Y)$ to a normal vector ξ, we get

$$\begin{aligned}\tilde{R}(X,Y)\xi &= -((\nabla_X A_\xi)Y - (\nabla_Y A_\xi)X) + A_{\nabla^\perp_X\xi}Y - A_{\nabla^\perp_Y\xi}X \\ &\quad - (\sigma(X,A_\xi Y) - \sigma(Y,A_\xi X)) + R^\perp(X,Y)\xi,\end{aligned} \tag{7.13}$$

where $R^\perp$ is the curvature operator of the normal connection. Again separating out the tangential and normal components we get

$$(\nabla_X A_\xi)Y - (\nabla_Y A_\xi)X = A_{\nabla^\perp_X\xi}Y - A_{\nabla^\perp_Y\xi}X - (\text{tang. comp. of } \tilde{R}(X,Y)\xi), \tag{7.14}$$

where "tang. comp." means the tangential component.

Using the fact that $\langle \tilde{R}(X,Y)Z, \xi\rangle = -\langle Z, \tilde{R}(X,Y)\xi\rangle$, it is easy to check that equation (7.14) is just another form of the Codazzi equation (7.12). Specifically, in terms of our orthonormal basis for the normal space, we can write equation (7.14) as

$$\begin{aligned}(\nabla_X A_\alpha)Y - (\nabla_Y A_\alpha)X &= s_{\alpha\beta}(X)A_\beta Y - s_{\alpha\beta}(Y)A_\beta X \\ &\quad -(\text{tang. comp. of } \tilde{R}(X,Y)\xi_\alpha) \qquad (7.15)\end{aligned}$$

with summation over β. Also we have,

$$\sigma(X, A_\xi Y) - \sigma(Y, A_\xi X) = R^\perp(X,Y)\xi - (\text{norm. comp. of } \tilde{R}(X,Y)\xi), \qquad (7.16)$$

which is the Ricci equation. In terms of the normal basis, we have for each α,

$$\begin{aligned}\langle [A_\alpha, A_\beta]X, Y\rangle \xi_\beta &= ((\nabla_X s_{\alpha\beta})Y - (\nabla_Y s_{\alpha\beta})X)\ \xi_\beta \\ &\quad -(s_{\alpha\beta}(X)s_{\beta\gamma}(Y) - s_{\alpha\beta}(Y)s_{\beta\gamma}(X))\ \xi_\gamma \\ &\quad -(\text{norm. comp. of } \tilde{R}(X,Y)\ \xi_\alpha), \qquad (7.17)\end{aligned}$$

with summation of β and γ. In other words, for each fixed pair of indices (α, β), we have

$$\begin{aligned}\langle [A_\alpha, A_\beta]X, Y\rangle &= (\nabla_X s_{\alpha\beta})Y - (\nabla_Y s_{\alpha\beta})X \\ &\quad -\Sigma(s_{\alpha\gamma}(X)s_{\gamma\beta}(Y) - s_{\alpha\gamma}(Y)s_{\gamma\beta}(X)) \\ &\quad -\langle \tilde{R}(X,Y)\xi_\alpha, \xi_\beta\rangle, \qquad (7.18)\end{aligned}$$

with summation over γ.

Kähler submanifolds

We now specialize to the case of Kähler manifolds. The complex structure is denoted by J and it interacts with the shape operators as follows:

Lemma 7.4. *For a Kähler submanifold M of a Kähler manifold $\tilde{M}$, and any unit normal vector ξ, we have*

1. $A_{J\xi} = JA_\xi$,
2. $JA_\xi = -A_\xi J$.

This allows us to verify the following well-known fact.

Corollary 7.5. *A Kähler submanifold M of a Kähler manifold $\tilde{M}$ is minimal, i.e.,* trace $A_\xi = 0$ *for all normal vectors ξ.*

Proof. Fix a point x in M. Choose a J-invariant orthonormal basis $\{e_\ell\}$ for T_xM and a J-invariant orthonormal basis $\{\xi_\alpha\}$ for the normal space at x. Then

$$\begin{aligned}
\text{trace } A_\xi &= \sum_{\ell=1}^{2m} \langle A_\xi e_\ell, e_\ell\rangle \\
&= \sum_{\ell=1}^{m} \left(\langle A_\xi e_\ell, e_\ell\rangle + \langle A_\xi J e_\ell, J e_\ell\rangle\right) \\
&= \sum_{\ell=1}^{m} \langle A_\xi e_\ell, e_\ell\rangle - \langle J A_\xi J e_\ell, e_\ell\rangle \\
&= \sum_{\ell=1}^{m} \langle A_\xi e_\ell, e_\ell\rangle + \langle J^2 A_\xi e_\ell, e_\ell\rangle \\
&= \sum_{\ell=1}^{m} \langle A_\xi e_\ell, e_\ell\rangle - \langle A_\xi e_\ell, e_\ell\rangle = 0.
\end{aligned} \tag{7.19}$$

□

We have used the fact that for a Kähler submanifold of a Kähler manifold both the tangent space and the normal space are J-invariant. When $\tilde{M}$ is a complex space form, we can see using (6.6) that the $\tilde{R}$ term makes no contribution to the Codazzi equation in (7.12) or (7.14)). Thus (7.15) takes the form

$$(\nabla_X A_\alpha)Y - (\nabla_Y A_\alpha)X = s_{\alpha\beta}(X)A_\beta Y - s_{\alpha\beta}(Y)A_\beta X \tag{7.20}$$

with summation over β. Similarly, the Ricci equation becomes

$$\begin{aligned}
\langle [A_\alpha, A_\beta]X, Y\rangle &= (\nabla_X s_{\alpha\beta})Y - (\nabla_Y s_{\alpha\beta})X \\
&\quad - \Sigma(s_{\alpha\gamma}(X)s_{\gamma\beta}(Y) - s_{\alpha\gamma}(Y)s_{\gamma\beta}(X)) - 2c\langle X, JY\rangle\langle J\xi_\alpha, \xi_\beta\rangle
\end{aligned} \tag{7.21}$$

with summation over γ.

We now embark upon the calculation of trace $(\Delta' A \circ A)$, where A is a shape operator. We will suppress the explicit summation signs. Repeated indices indicate summation over the appropriate ranges. Also, where necessary, we extend $\{e_\ell\}$ to a local orthonormal frame near x by parallel translation along geodesics emanating from x, and we do the same for other tangent vectors used in the proofs.

Lemma 7.6. *Let M be a Kähler submanifold of a complex space form $\tilde{M}$ of constant holomorphic curvature* $4c$. *Then*

$$\begin{aligned}
\Delta' A_\alpha &= 2(m+2)cA_\alpha + [A_\beta, [A_\alpha, A_\beta]] - (\text{trace } (A_\alpha A_\beta))A_\beta \\
&\quad + (\nabla_{e_\ell} s_{\alpha\beta})(e_\ell)A_\beta + 2s_{\alpha\beta}(e_\ell)\nabla_{e_\ell} A_\beta - s_{\alpha\beta}(e_\ell)s_{\beta\gamma}(e_\ell)A_\gamma.
\end{aligned} \tag{7.22}$$

Proof. We will begin by deriving rules for permuting the arguments of the second covariant derivative operator K. In particular, we have for the K arising from a shape operator A_α,

$$\begin{aligned} K(U,V)X &= \nabla_U((\nabla_V A_\alpha)X) \\ &= \nabla_U((\nabla_X A_\alpha)V - s_{\alpha\beta}(X)A_\beta V + s_{\alpha\beta}(V)A_\beta X), \end{aligned} \tag{7.23}$$

(evaluated at x) where we have used the Codazzi equation and discarded terms involving first derivatives of the vector fields U, V, and X since

$$\nabla U = \nabla V = \nabla X = 0$$

at x. Continuing in this fashion, we find that the right side is equal to

$$\begin{aligned} &K(U,X)V - (\nabla_U s_{\alpha\beta})(X)A_\beta V + (\nabla_U s_{\alpha\beta})(V)A_\beta X \\ &\quad - s_{\alpha\beta}(X)(\nabla_U A_\beta)V + s_{\alpha\beta}(V)(\nabla_U A_\beta)X. \end{aligned} \tag{7.24}$$

Thus

$$\begin{aligned} K(U,V)X &= K(X,U)V - [R(X,U),A_\alpha]V \\ &\quad - (\nabla_U s_{\alpha\beta})(X)A_\beta V + (\nabla_U s_{\alpha\beta})(V)A_\beta X \\ &\quad - s_{\alpha\beta}(X)(\nabla_U A_\beta)V + s_{\alpha\beta}(V)(\nabla_U A_\beta)X. \end{aligned} \tag{7.25}$$

Now we are interested in computing $(\Delta' A)X = \Sigma K(e_i,e_i)X$. Setting $U = V = e_i$ in the previous equation gives us three terms to evaluate. The first term is

$$\begin{aligned} K(X,e_i)e_i &= \nabla_X((\nabla_{e_i} A_\alpha)e_i) \\ &= (\nabla_X s_{\alpha\beta})(e_i)A_\beta e_i + s_{\alpha\beta}(e_i)(\nabla_X A_\beta)e_i. \end{aligned} \tag{7.26}$$

Here we have used the identity

$$(\nabla_{e_i} A_\alpha)e_i = s_{\alpha\beta}(e_i)A_\beta e_i, \tag{7.27}$$

which follows from the Codazzi equation and the fact that trace $A_\alpha = 0$. The second term can be evaluated using the Gauss equation

$$\begin{aligned} R(X,e_i)A_\alpha e_i &= c(X \wedge e_i + JX \wedge Je_i + 2\langle X, Je_i\rangle J)A_\alpha e_i + (A_\beta X \wedge A_\beta e_i)A_\alpha e_i \\ &= c((\text{trace } A_\alpha)X - A_\alpha X + \text{trace } (A_\alpha J)JX - JA_\alpha JX \\ &\quad + 2JA_\alpha(-JX) + (\text{trace } (A_\alpha A_\beta)A_\beta)X - A_\beta A_\alpha A_\beta X \\ &= -4cA_\alpha X + (\text{trace } (A_\alpha A_\beta))A_\beta X - A_\beta A_\alpha A_\beta X. \end{aligned} \tag{7.28}$$

On the other hand,

$$\begin{aligned} A_\alpha R(X, e_i)e_i &= A_\alpha(c(X \wedge e_i + JX \wedge Je_i + 2\langle X, Je_i\rangle J)e_i + (A_\beta X \wedge A_\beta e_i)e_i) \\ &= c((2m-1)A_\alpha X + A_\alpha X + 2A_\alpha X) \\ &+ A_\alpha((\text{trace } A_\beta)A_\beta X - A_\beta^2 X) \\ &= 2(m+1)cA_\alpha X - A_\alpha A_\beta^2 X, \end{aligned} \tag{7.29}$$

so that

$$[R(X, e_i), A_\alpha]e_i = -2(m+3)cA_\alpha X + (\text{trace } (A_\alpha A_\beta))A_\beta X + [A_\alpha, A_\beta]A_\beta X.$$

The third term is obtained by merely substituting e_i for each of U and V and becomes

$$\begin{aligned} &- (\nabla_{e_i} s_{\alpha\beta})(X)A_\beta e_i + (\nabla_{e_i} s_{\alpha\beta})(e_i)A_\beta X \\ &- s_{\alpha\beta}(X)(\nabla_{e_i} A_\beta)e_i + s_{\alpha\beta}(e_i)(\nabla_{e_i} A_\beta)X. \end{aligned} \tag{7.30}$$

Collecting and slightly rearranging the terms, we get

$$\begin{aligned} (\Delta' A_\alpha)X &= (\nabla_X s_{\alpha\beta})(e_i)A_\beta e_i - (\nabla_{e_i} s_{\alpha\beta})(X)A_\beta e_i \\ &+ s_{\alpha\beta}(e_i)(\nabla_X A_\beta)e_i + (\nabla_{e_i} s_{\alpha\beta})(e_i)A_\beta X \\ &- s_{\alpha\beta}(X)(\nabla_{e_i} A_\beta)e_i + s_{\alpha\beta}(e_i)(\nabla_{e_i} A_\beta)X \\ &+ 2(m+3)cA_\alpha X - (\text{trace } (A_\alpha A_\beta))A_\beta X - [A_\alpha, A_\beta]A_\beta X. \end{aligned} \tag{7.31}$$

Applying equation (7.21) to the first two terms, the Codazzi equation to the 3rd term and equation (7.27) to the 5th term, we get

$$\begin{aligned} (\Delta' A_\alpha)X &= \langle [A_\alpha, A_\beta]X, e_i\rangle A_\beta e_i + \langle \tilde{R}(X, e_i)\xi_\alpha, \xi_\beta\rangle A_\beta e_i \\ &+ \Sigma s_{\alpha\gamma}(X)s_{\gamma\beta}(e_i)A_\beta e_i - \Sigma s_{\alpha\gamma}(e_i)s_{\gamma\beta}(X)A_\beta e_i \\ &+ s_{\alpha\beta}(e_i)((\nabla_{e_i} A_\beta)X + s_{\beta\gamma}(X)A_\gamma e_i - s_{\beta\gamma}(e_i)A_\gamma X) \\ &+ (\nabla_{e_i} s_{\alpha\beta})(e_i)A_\beta X \\ &+ s_{\alpha\beta}(e_i)(\nabla_{e_i} A_\beta)X \\ &- s_{\alpha\beta}(X)s_{\beta\gamma}(e_i)A_\gamma e_i \\ &+ 2(m+3)cA_\alpha X - (\text{trace } (A_\alpha A_\beta))A_\beta X - [A_\alpha, A_\beta]A_\beta X. \end{aligned} \tag{7.32}$$

It is easy to check that $\langle \tilde{R}(X, e_i)\xi_\alpha, \xi_\beta\rangle A_\beta e_i = -2cA_\alpha X$. The key observation is that the only value of β that produces a nonzero summand is the one for which $\xi_\beta = J\xi_\alpha$. Combining some of the 13 terms in (7.32) and noting cancellation of terms 4 and 6, also terms 3 and 10, we get

$$(\Delta' A_\alpha)X = [A_\beta, [A_\alpha, A_\beta]]X + 2(m+2)cA_\alpha X - (\text{trace } (A_\alpha A_\beta))A_\beta X$$
$$- s_{\alpha\beta}(e_i)s_{\beta\gamma}(e_i)A_\gamma X + (\nabla_{e_i} s_{\alpha\beta})(e_i)A_\beta X + 2s_{\alpha\beta}(e_i)(\nabla_{e_i} A_\beta)X.$$

This agrees with equation (7.22) and so we have completed the proof of Lemma 7.6. □

Lemma 7.7. *Let M be a Kähler submanifold of a complex space form $\tilde{M}$ of constant holomorphic curvature $4c$. Then*

$$\begin{aligned}\text{trace } (\Delta' A_\alpha \circ A_\alpha) &= 2(m+2)c \text{ trace } A_\alpha^2 + 2 \text{ trace } ((A_\alpha A_\beta)^2 - A_\alpha^2 A_\beta^2) \\ &\quad - (\text{trace } (A_\alpha A_\beta))^2 + (\nabla_{e_i} s_{\alpha\beta})(e_i) \text{ trace } (A_\beta A_\alpha) \\ &\quad + 2s_{\alpha\beta}(e_i) \text{ trace } ((\nabla_{e_i} A_\beta)A_\alpha) \\ &\quad - s_{\alpha\beta}(e_i)s_{\beta\gamma}(e_i) \text{ trace } (A_\gamma A_\alpha).\end{aligned} \tag{7.33}$$

Proof. Composing equation (7.22) with A_α and taking the trace, we get

$$\begin{aligned}\text{trace } (\Delta' A_\alpha \circ A_\alpha) &= 2(m+2)c \text{ trace } A_\alpha^2 \\ &\quad + \text{trace } (A_\beta A_\alpha A_\beta A_\alpha - A_\beta^2 A_\alpha^2 - A_\alpha A_\beta^2 A_\alpha + A_\beta A_\alpha A_\beta A_\alpha) \\ &\quad - (\text{trace } (A_\alpha A_\beta))^2 + (\nabla_{e_i} s_{\alpha\beta})(e_i) \text{ trace } (A_\beta A_\alpha) \\ &\quad + 2s_{\alpha\beta}(e_i) \text{ trace } ((\nabla_{e_i} A_\beta)A_\alpha) \\ &\quad - s_{\alpha\beta}(e_i)s_{\beta\gamma}(e_i) \text{ trace } (A_\gamma A_\alpha) \\ &= 2(m+2)c \text{ trace } A_\alpha^2 + 2 \text{ trace } ((A_\alpha A_\beta)^2 - A_\alpha^2 A_\beta^2) \\ &\quad - (\text{trace } (A_\alpha A_\beta))^2 + (\nabla_{e_i} s_{\alpha\beta})(e_i) \text{ trace } (A_\alpha A_\beta) \\ &\quad + 2s_{\alpha\beta}(e_i) \text{ trace } ((\nabla_{e_i} A_\beta)A_\alpha) \\ &\quad - s_{\alpha\beta}(e_i)s_{\beta\gamma}(e_i) \text{ trace } (A_\gamma A_\alpha).\end{aligned}$$

Thus we have equation (7.33) which proves Lemma 7.7. □

We can now complete the proof of Theorem 7.1. From Lemmas 7.2 and 7.3, we have

$$\begin{aligned}\frac{1}{2}\Delta|\sigma|^2 &= \Sigma \text{ trace } (\Delta' A_\alpha \circ A_\alpha) + |\nabla'\sigma|^2 \\ &\quad + 2\Sigma s_{\alpha\beta}(e_i) \text{ trace}((\nabla_{e_i} A_\alpha)A_\beta) + \Sigma s_{\alpha\beta}(e_i)s_{\beta\gamma}(e_i) \text{ trace}(A_\alpha A_\gamma) \\ &= \Sigma \text{ trace } (\Delta' A_\alpha \circ A_\alpha) + |\nabla'\sigma|^2 \\ &\quad - 2\Sigma s_{\alpha\beta}(e_i) \text{ trace}((\nabla_{e_i} A_\beta)A_\alpha) + \Sigma s_{\alpha\beta}(e_i)s_{\beta\gamma}(e_i) \text{ trace}(A_\alpha A_\gamma),\end{aligned}$$

where we have interchanged α and β in the next-to-last summation. Substituting for trace $(\Delta' A_\alpha \circ A_\alpha)$ from Lemma 7.7, we get

$$\begin{aligned}\frac{1}{2}\Delta|\sigma|^2 &= 2(m+2)c \text{ trace } (\Sigma A_\alpha^2) + 2 \text{ trace } \Sigma((A_\alpha A_\beta)^2 - A_\alpha^2 A_\beta^2)\\ &- \Sigma(\text{trace } (A_\alpha A_\beta))^2 + \Sigma((\nabla_{e_i} s_{\alpha\beta})(e_i) \text{ trace } (A_\beta A_\alpha))\\ &+ 2\Sigma(s_{\alpha\beta}(e_i) \text{ trace } ((\nabla_{e_i} A_\beta)A_\alpha))\\ &- \Sigma(s_{\alpha\beta}(e_i)s_{\beta\gamma}(e_i) \text{ trace } (A_\gamma A_\alpha)) + |\nabla'\sigma|^2\\ &- 2\Sigma(s_{\alpha\beta}(e_i) \text{ trace}((\nabla_{e_i} A_\beta)A_\alpha)) + \Sigma(s_{\alpha\beta}(e_i)s_{\beta\gamma}(e_i) \text{ trace}(A_\alpha A_\gamma)).\end{aligned}$$

After the obvious cancellations, we notice that $(\nabla_{e_i} s_{\alpha\beta})(e_i)$ trace $(A_\alpha A_\beta)$ sums to zero since trace $(A_\alpha A_\beta)$ is symmetric in $\{\alpha, \beta\}$, while $(\nabla_{e_i} s_{\alpha\beta})(e_i)$ is skew-symmetric. In addition, we note that the $(A_\alpha A_\beta)^2$ term contributes zero to the sum since the $2p$ terms occurring with each A_α can be paired as A_β and JA_β, yielding a summand of $(A_\alpha A_\beta)^2 + (A_\alpha(JA_\beta))^2 = 0$. Also, we may rewrite the $A_\alpha^2 A_\beta^2$ term in terms of one index α. Thus, we get

$$\begin{aligned}\frac{1}{2}\Delta|\sigma|^2 &= 2(m+2)c|\sigma|^2 + |\nabla'\sigma|^2\\ &- 2 \text{ trace } (\Sigma A_\alpha^2)^2 - \Sigma(\text{trace } (A_\alpha A_\beta))^2. \end{aligned} \tag{7.34}$$

This completes the proof of Theorem 7.1.

7.2 The Veronese Embedding of CP^m in CP^n

In the next four sections, we work out the shape operators of some complex submanifolds in **CP**^{n} that are important in our classification of Hopf hypersurfaces with constant principal curvatures.

We begin with the Veronese embedding of **CP**^{m} into **CP**^{n}, where $n = m(m+3)/2$. Thus the complex codimension is $p = m(m+1)/2$. The embedding is constructed in the following way. For $z \in \mathbf{C}^{m+1}$, let $\tilde{f}(z) = zz^T$, where z^T denotes the transpose of the column vector z. Thus $\tilde{f}(z)$ is an $(m+1) \times (m+1)$ symmetric matrix of complex numbers with (j,k) entry equal to $z_j z_k$. We may regard the space of all $(m+1) \times (m+1)$ matrices as $\mathbf{C}^{(m+1)^2}$ and the space of symmetric matrices as $\mathbf{C}^{n+1}$. In fact, if E_{jk} denotes the matrix with 1 in the (j,k) position and 0 elsewhere, then the basis consisting of the $m+1$ matrices $\{E_{jj}\}$ and the $m(m+1)/2$ matrices

$$\frac{1}{\sqrt{2}}(E_{jk} + E_{kj})$$

is orthonormal with respect to the standard Hermitian inner product on $\mathbf{C}^{(m+1)^2}$. This inner product may be written as

$$\langle U, V \rangle = \Re \operatorname{trace} V^* U,$$

where V^* denotes the conjugate transpose of the complex matrix V, and $\Re$ denotes the real part. The mapping $\tilde{f}$ determines an embedding f of $\mathbf{CP}^m$ into $\mathbf{CP}^n$ given by $f(\pi z) = \pi \tilde{f}(z)$. With respect to the usual Hermitian inner product on $\mathbf{C}^{m+1}$, we can check that $\tilde{f}$ maps vectors of length r into vectors of length r^2. The Fubini-Study metric on $\mathbf{CP}^m$, as given in Section 6.1, has constant holomorphic curvature $4c = 4/r^2$. The Fubini-Study metric on $\mathbf{CP}^n$ is constructed in the same way. We write the constant holomorphic curvature of $\mathbf{CP}^n$ as $4c^2 = 4/r^4$.

If we temporarily denote the metric on $\mathbf{CP}^m$ by g, it is easy to check that for vectors v and w tangent to $\mathbf{CP}^m$, we have

$$\langle f_* v, f_* w \rangle = 2r^2 g(v, w),$$

so that the induced metric on $\mathbf{CP}^m$ is $2r^2$ times the original metric. This implies that all sectional curvatures in the induced metric are $1/2r^2$ times the curvatures in the original metric. Thus, in the induced metric, $\mathbf{CP}^m$ has constant holomorphic sectional curvature

$$\frac{1}{2r^2}\, 4c = \frac{2}{r^4}$$

which is half that of the ambient space $\mathbf{CP}^n$.

When $m = 1$ and hence $n = 2$, the Veronese embedding of $\mathbf{CP}^1$ is also an embedding of the complex quadric Q^1. Put another way, the quadric Q^1 is congruent to $\mathbf{CP}^1$ and both are isometric to the 2-sphere S^2 with constant curvature which is half the holomorphic curvature of the ambient space $\mathbf{CP}^2$. We can see this as follows:

Lemma 7.8. *The standard quadric $Q^1 = \{\pi z \mid z_0^2 + z_1^2 + z_2^2 = 0\}$ in $\mathbf{CP}^2$ is related to the quadric $\tilde{Q}^1 = \{\pi z \mid z_2^2 = 2 z_0 z_1\}$ by a holomorphic isometry of $\mathbf{CP}^2$.*

Proof. Consider the unitary matrix

$$\tilde{T} = \begin{bmatrix} \frac{1}{\sqrt{2}} & -\frac{i}{\sqrt{2}} & 0 \\ \frac{1}{\sqrt{2}} & \frac{i}{\sqrt{2}} & 0 \\ 0 & 0 & -i \end{bmatrix} \in SU(3), \tag{7.35}$$

and let T be the holomorphic isometry of $\mathbf{CP}^2$ determined by $\tilde{T}$, that is, $T(\pi z) = \pi(\tilde{T} z)$. It is easy to check that $TQ^1 = \tilde{Q}^1$. □

Now if

$$\tilde{f}(z) = z z^T = \begin{bmatrix} z_0^2 & z_0 z_1 \\ z_0 z_1 & z_1^2 \end{bmatrix} = w_0\, E_{00} + w_1\, E_{11} + w_2\, \frac{1}{\sqrt{2}}(E_{01} + E_{10}),$$

it is straightforward to verify that $w_2^2 = 2w_0w_1$, so that $f(\mathbf{CP}^1) \subset \tilde{Q}^1$. On the other hand, the map is surjective, since for every point q of $\tilde{Q}^1$, either $q = \pi(1, 2a^2, 2a) = \pi f(1, \sqrt{2}\, a)$ or $q = \pi(2a^2, 1, 2a) = \pi f(\sqrt{2}\, a, 1)$ for a suitable complex number a.

Shape operators of the Veronese embedding

Let $p_0 = \pi(r\epsilon_0) \in \mathbf{CP}^m$ and $q_0 = f(p_0) = \pi(r^2 E_{00})$. The tangent space to $f(M)$ at q_0 is spanned by

$$\{\pi_* \tilde{f}_* \epsilon_j, \pi_*(i\tilde{f}_* \epsilon_j)\} = \{\pi_*(E_{0j} + E_{j0}), \pi_*(i(E_{0j} + E_{j0}))\},$$

where $1 \leq j \leq m$. Thus, the normal space at q_0 is spanned by $\{\pi_* E_{jj}, \pi_*(iE_{jj})\}$, $1 \leq j \leq m$, together with $\{\pi_*(E_{jk} + E_{kj}), \pi_*(i(E_{jk} + E_{kj}))\}$, $1 \leq j < k \leq m$.

Let $\xi = \pi_* E_{11}$. We wish to compute A_ξ. In particular, we will compute $A_\xi(\pi_*\epsilon_j)$ for $1 \leq j \leq m$. Let

$$z_t = \cos\frac{t}{r}\, r\epsilon_0 + \sin\frac{t}{r}\, r\epsilon_j. \tag{7.36}$$

This is a curve on $S^{2m+1}(r)$ with $z_0 = r\epsilon_0$ and $\overrightarrow{z_0} = \epsilon_j$ which embeds in $S^{2n+1}(r^2)$ to give

$$\tilde{f}(z_t) = r^2(\cos^2\frac{t}{r} E_{00} + \sin^2\frac{t}{r} E_{jj} + \cos\frac{t}{r}\sin\frac{t}{r}(E_{0j} + E_{j0})) \tag{7.37}$$

and

$$\tilde{f}_* \overrightarrow{z_t} = r(-\sin\frac{2t}{r}(E_{00} - E_{jj}) + \cos\frac{2t}{r}(E_{0j} + E_{j0})). \tag{7.38}$$

We need to extend ξ along $\pi\tilde{f}(z_t)$ so that it remains a unit normal. For $j > 1$ we let

$$\xi_t = E_{11}, \tag{7.39}$$

while for $j = 1$, we let

$$\xi_t = \sin^2\frac{t}{r} E_{00} + \cos^2\frac{t}{r} E_{11} - \cos\frac{t}{r}\sin\frac{t}{r}(E_{01} + E_{10})). \tag{7.40}$$

The space of horizontal tangent vectors to $S^{2m+1}(r)$ at z_t is spanned by

$$\{\overrightarrow{z_t}, i\overrightarrow{z_t}, \epsilon_k, i\epsilon_k\},$$

where k runs from 1 through m, skipping j. Thus, the space of horizontal tangent vectors to $\tilde{M} = \tilde{f}(S^{2m+1}(r))$ at $\tilde{f}(z_t)$ is spanned by

$$\{\tilde{f}_* \overrightarrow{z_t}, i\tilde{f}_* \overrightarrow{z_t}, E_{0k} + E_{k0}, i(E_{0k} + E_{k0})\}.$$

From this, it is straightforward to check that ξ_t is a unit horizontal tangent vector to $S^{2n+1}(r^2)$ that is orthogonal to the tangent space to $\tilde{M}$. Thus, $\pi_* \xi_t$ is an appropriate extension of ξ and can be used in computing $A_\xi \pi_* \epsilon_j$ at p_0. Specifically, we get for $j = 1$,

$$\begin{aligned} \frac{d\xi_t}{dt} &= \frac{1}{r}\Big(2\cos\frac{t}{r}\sin\frac{t}{r}(E_{00} - E_{11}) - (\cos^2\frac{t}{r} - \sin^2\frac{t}{r})(E_{10} + E_{01})\Big) \\ &= \frac{1}{r}\Big(\sin\frac{2t}{r}(E_{00} - E_{11}) - \cos\frac{2t}{r}(E_{10} + E_{01})\Big) \end{aligned}$$

Evaluating this at $t = 0$, gives

$$-\frac{1}{r}(E_{01} + E_{10}) = -\frac{1}{r^2}\tilde{f}_* \overrightarrow{z_0}.$$

If we write $e_j = \pi_* \epsilon_j$ for $1 \leq j \leq m$, we thus get $A_\xi e_1 = \lambda e_1$, where $\lambda = 1/r^2$. When $j \neq 1$, on the other hand, we have

$$\frac{d\xi_t}{dt} = 0,$$

so $A_\xi e_j = 0$ for $j \geq 2$. Of course, using $AJ = -JA$, we also have

$$A_\xi J e_1 = -\lambda J e_1,$$

and $A_\xi J e_j = 0$ for $j \geq 2$. It is easy to see that shape operators of all normals of the form $\pi_* E_{jj}$ will behave similarly. We now turn to normals of the form

$$\pi_* \frac{E_{jk} + E_{kj}}{\sqrt{2}} \tag{7.41}$$

where $k \neq j$. Specifically, let

$$\eta = \pi_* \frac{E_{12} + E_{21}}{\sqrt{2}}.$$

We begin the computation for $A_\eta e_1$ by extending η as follows:

$$\eta_t = -\sin\frac{t}{r}\frac{E_{02} + E_{20}}{\sqrt{2}} + \cos\frac{t}{r}\frac{E_{12} + E_{21}}{\sqrt{2}}. \tag{7.42}$$

Again, we may check that η_t is a unit normal to $\tilde{M}$ along $\tilde{f}(z_t)$. Computing

$$\frac{d\eta_t}{dt} = -\frac{1}{r}\left(\cos\frac{t}{r}\frac{E_{02}+E_{20}}{\sqrt{2}} + \sin\frac{t}{r}\frac{E_{12}+E_{21}}{\sqrt{2}}\right) \tag{7.43}$$

and setting $t = 0$ yields $(-1/r^2)\tilde{f}_*\epsilon_2$, from which we deduce that $A_\eta e_1 = \lambda e_2$ and also that $A_\eta J e_1 = -\lambda J e_2$. A similar calculation gives $A_\eta e_2 = \lambda e_1$ and $A_\eta J e_2 = -\lambda J e_1$. For $j \geq 3$, we choose η_t to be constant and find that $A_\eta e_j = A_\eta J e_j = 0$.

Again, we note that the same calculation applies to all unit normals of the form

$$\pi_* \frac{E_{jk}+E_{kj}}{\sqrt{2}}.$$

Thus we are able to compute the shape operator A_ζ for all unit normal vectors ζ. In fact, for any real symmetric $m \times m$ matrix C with $|C| = 1$, $\zeta = \pi_* C$ is a unit normal to $f(M)$ at q_0. If we write $C = D + E$ as the sum of a diagonal matrix and a matrix with zero diagonal, then the matrix of A_ζ, restricted to the span of $\{e_1, e_2, \ldots, e_m\}$, is

$$A_\zeta = \lambda(D + \sqrt{2}E). \tag{7.44}$$

We then use $A_\zeta \circ J = -J \circ A_\zeta$ to complete our calculation of A_ζ. By appropriate choice of C (in fact, with $E = 0$), we see that the rank of A_ζ can be as low as 2 and as high as $2m$ and that the number of distinct eigenvalues of A_ζ can be as high as m and as low as 2.

7.3 The Segre Embedding of $\mathbf{CP}^h \times \mathbf{CP}^k$ in $\mathbf{CP}^n$

The Segre embedding $f : \mathbf{CP}^h \times \mathbf{CP}^k \to \mathbf{CP}^n$ satisfies $n = h + k + hk$. The complex dimension of the submanifold is $m = h+k$ and the complex codimension is $p = hk$. The construction is similar to that of the Veronese embedding. For $z \in \mathbf{C}^{h+1}$ and $w \in \mathbf{C}^{k+1}$, let $\tilde{f}(z, w) = zw^T$. Thus $\tilde{f}(z, w)$ is a $(h + 1) \times (k + 1)$ matrix of complex numbers with (i,j) entry equal to $z_i w_j$. We may regard the space of all $(h + 1) \times (k + 1)$ matrices as $\mathbf{C}^{n+1}$. The standard Hermitian inner product is given by the same formula,

$$\langle U, V \rangle = \Re \text{ trace } V^* U,$$

and the E_{ij} form an orthonormal basis. The mapping $\tilde{f}$ determines an embedding f of $\mathbf{CP}^h \times \mathbf{CP}^k$ into $\mathbf{CP}^n$ given by $f(\pi z, \pi w) = \pi \tilde{f}(z, w)$. With respect to the respective standard Hermitian inner products on $\mathbf{C}^{h+1}$, $\mathbf{C}^{k+1}$, and $\mathbf{C}^{n+1}$, $\tilde{f}$

maps $S^{2h+1}(r) \times S^{2k+1}(s)$ into $S^{2n+1}(rs)$. The usual Fubini-Study metric on $\mathbf{CP}^h$ (respectively $\mathbf{CP}^k$) has constant holomorphic curvature $4/r^2$ (respectively $4/s^2$). In terms of these parameters, we can write the constant holomorphic curvature of $\mathbf{CP}^n$ as $4/(r^2s^2)$.

When $m = 2$, f embeds $\mathbf{CP}^1 \times \mathbf{CP}^1$ onto a complex quadric $\tilde{Q}^2$ in $\mathbf{CP}^3$. The situation is similar to the Veronese case discussed earlier. Specifically, we have the following lemma.

Lemma 7.9. *The standard quadric* $Q^2 = \{\pi z \mid z_0^2 + z_1^2 + z_2^2 + z_3^2 = 0\}$ *in* $\mathbf{CP}^3$ *is related to the quadric* $\tilde{Q}^2 = \{\pi z \mid z_0 z_1 = z_2 z_3\}$ *by a holomorphic isometry of* $\mathbf{CP}^3$.

Proof. Consider the unitary matrix

$$\tilde{T} = \begin{bmatrix} \frac{1}{\sqrt{2}} & -\frac{i}{\sqrt{2}} & 0 & 0 \\ \frac{1}{\sqrt{2}} & \frac{i}{\sqrt{2}} & 0 & 0 \\ 0 & 0 & -\frac{1}{\sqrt{2}} & \frac{i}{\sqrt{2}} \\ 0 & 0 & \frac{1}{\sqrt{2}} & \frac{i}{\sqrt{2}} \end{bmatrix} \in SU(4), \tag{7.45}$$

determining the holomorphic isometry T of $\mathbf{CP}^3$. It is straightforward to check that $f(\mathbf{CP}^1 \times \mathbf{CP}^1) = \tilde{Q}^2$ and $TQ^2 = \tilde{Q}^2$. □

Returning to the general case, an easy calculation yields the following:

Proposition 7.10. *The metric on* $\mathbf{CP}^h$ *(respectively* $\mathbf{CP}^k$*) induced by the Segre embedding is* s^2 *times (respectively,* r^2 *times) its original metric. The metric on* $\mathbf{CP}^h \times \mathbf{CP}^k$ *induced by the embedding* f *is the product metric of two spaces of constant holomorphic curvature* $4/(r^2s^2)$*, the same as that of the ambient* $\mathbf{CP}^n$.

Shape operators of the Segre embedding

In this section, we adopt the following conventions and notation. Components of vectors and matrices are numbered beginning with 0. The standard unit basis vectors are denoted by ϵ_j that has 1 in position j and zeroes elsewhere. $E_{j\ell}$ denotes the matrix that has 1 in position (j, ℓ) and zeroes elsewhere. The dimensions (number of rows and columns) are determined by the context.

Due to the homogeneity, it is sufficient to compute shape operators at one point $p_0 \in M = \mathbf{CP}^h \times \mathbf{CP}^k$. For convenience, we choose the point $p_0 = \pi(r\epsilon_0, s\epsilon_0)$ where $r\epsilon_0 \in S^{2h+1}(r) \subset \mathbf{C}^{h+1}$ and $s\epsilon_0 \in S^{2k+1}(s) \subset \mathbf{C}^{k+1}$.

Then $f(p_0) = \pi(rsE_{00})$ and $f_* T_{p_0}M$ is spanned by the $h + k$ vectors of the form $\pi_* E_{j0}$ and $\pi_* E_{0\ell}$, where $1 \leq j \leq h$ and $1 \leq \ell \leq k$. The normal space is spanned by the hk vectors $\pi_* E_{j\ell}$. We consider $\xi = \pi_* E_{11}$. For $X \in T_{p_0}M$, we have

$$f_*(-A_\xi X) = \pi_* D_{(f_*X)^L} \xi^L. \tag{7.46}$$

We will need to consider parametrized curves in the space of matrices, and so we introduce the following notation. Here τ and a are real numbers with a positive and j is an integer whose range will be clear from the context. The curve C (actually a 1-parameter subgroup of the orthogonal group in case the matrices are square) is defined by

$$C^j_\tau(a) = \cos\frac{\tau}{a}E_{00} + \sin\frac{\tau}{a}E_{j0} - \sin\frac{\tau}{a}E_{0j} + \cos\frac{\tau}{a}E_{jj}.$$

Now consider

$$\begin{aligned} z^{j\ell}_{t,u} &= C^j_t(r)(rsE_{00})(C^\ell_u(s))^T \\ &= rs(\cos\frac{t}{r}E_{00} + \sin\frac{t}{r}E_{j0})(\cos\frac{u}{s}E_{00} + \sin\frac{u}{s}E_{0\ell}) \\ &= rs\big((\cos\frac{t}{r}\ (\cos\frac{u}{s}E_{00} + \sin\frac{u}{s}E_{0\ell}) + \sin\frac{t}{r}\ (\cos\frac{u}{s}E_{j0} + \sin\frac{u}{s}E_{j\ell})\big). \end{aligned} \tag{7.47}$$

In essence, we have chosen one coordinate direction related to the first factor and one related to the second factor. Then $\pi(z^{j\ell}_{t,u})$ is a 2-parameter family of curves on $f(M)$. We compute

$$\begin{aligned} \frac{dz}{dt} &= s(-\sin\frac{t}{r}E_{00} + \cos\frac{t}{r}E_{j0})(\cos\frac{u}{s}E_{00} + \sin\frac{u}{s}E_{0\ell}) \\ &= s\big(\cos\frac{t}{r}\ (\cos\frac{u}{s}E_{j0} + \sin\frac{u}{s}E_{j\ell}) - \sin\frac{t}{r}\ (\cos\frac{u}{s}E_{00} + \sin\frac{u}{s}E_{0\ell})\big), \end{aligned} \tag{7.48}$$

and

$$\begin{aligned} \frac{dz}{du} &= r(\cos\frac{t}{r}E_{00} + \sin\frac{t}{r}E_{j0})(-\sin\frac{u}{s}E_{00} + \cos\frac{u}{s}E_{0\ell}) \\ &= r\big(\cos\frac{t}{r}\ (-\sin\frac{u}{s}E_{00} + \cos\frac{u}{s}E_{0\ell}) + \sin\frac{t}{r}\ (-\sin\frac{u}{s}E_{j0} + \cos\frac{u}{s}E_{j\ell})\big). \end{aligned} \tag{7.49}$$

Let

$$N^{j\ell}_{t,u} = \cos\frac{t}{r}\ (-\sin\frac{u}{s}E_{j0} + \cos\frac{u}{s}E_{j\ell}) - \sin\frac{t}{r}\ (-\sin\frac{u}{s}E_{00} + \cos\frac{u}{s}E_{0\ell}) \tag{7.50}$$

and note that $N^{j\ell}_{t,u}$ is tangent to the sphere $S^{2n+1}(rs)$ at $z^{j\ell}_{t,u}$ but orthogonal to $\frac{dz}{dt}$ and $\frac{dz}{du}$ there. Also observe that $N^{j\ell}_{t,u}$ is horizontal at $z^{j\ell}_{t,u}$, and note that $N^{j\ell}_{0,0} = E_{j\ell}$. We are thus in a position to compute the shape operators at p, in particular, A_ξ with $N^{11}_{0,0} = \xi^L$.

To this end, consider first $j > 1$. Then the constant vector $N^{11}_{0,0}$ is orthogonal to $z^{j\ell}_{t,u}$, $\frac{dz}{dt}$, and $\frac{dz}{du}$ along the curve $\{t \mapsto z^{j\ell}_{t,0}\}$ (i.e., ξ^L can be extended to be a constant vector along the curve). This leads to the conclusion that $A_\xi(\pi_*\epsilon_j, 0) = 0$. Similarly, for $\ell > 1$, we have $A_\xi(0, \pi_*\epsilon_\ell) = 0$. For $j = 1$, however, we extend ξ using $N^{11}_{t,0}$. This gives the first of the following two equations. The second follows by a similar argument.

$$\frac{dN}{dt} = -\frac{1}{r}E_{0\ell} = -\frac{1}{r^2}\tilde{f}_*(0, \epsilon_\ell)$$
$$\frac{dN}{du} = -\frac{1}{s}E_{j0} = -\frac{1}{s^2}\tilde{f}_*(\epsilon_j, 0). \tag{7.51}$$

In fact, setting $X = (\pi_*\epsilon_1, 0)$ and $U = (0, \pi_*\epsilon_1)$, we have

$$A_\xi X = \frac{1}{r^2}U$$
$$A_\xi U = \frac{1}{s^2}X. \tag{7.52}$$

Even though X and U are unit vectors in the respective natural metrics on $\mathbf{CP}^h$ and $\mathbf{CP}^k$, they are not unit vectors in the metric induced by the embedding into $\mathbf{CP}^n$. Setting $\hat{X} = \frac{1}{s}X$ and $\hat{U} = \frac{1}{r}U$, we get an orthonormal pair satisfying

$$A_\xi \hat{X} = \tfrac{1}{rs}\hat{U}$$
$$A_\xi \hat{U} = \tfrac{1}{rs}\hat{X}. \tag{7.53}$$

With respect to an orthonormal basis whose first four vectors are $\hat{X}$, $\hat{U}$, $J\hat{X}$, and $J\hat{U}$, the upper-left 4×4 block of the matrix of A_ξ is

$$\begin{bmatrix} 0 & \lambda & 0 & 0 \\ \lambda & 0 & 0 & 0 \\ 0 & 0 & 0 & -\lambda \\ 0 & 0 & -\lambda & 0 \end{bmatrix} \tag{7.54}$$

where $\lambda = 1/rs$. All the other entries in the matrix of A_ξ will be zero.

Now consider the unit normal vector $\eta = \pi_* E_{12}$. The preceding calculation with $j = 1$ and $\ell = 2$ will give

$$A_\xi X = \frac{1}{r^2}V$$
$$A_\xi V = \frac{1}{s^2}X, \tag{7.55}$$

where $V = (0, \pi_*\epsilon_2)$. We also set $\hat{V} = \frac{1}{r}V$. With respect to an orthonormal basis whose first six vectors are $\hat{X}$, $\hat{U}$, $\hat{V}$, $J\hat{X}$, $J\hat{U}$, and $J\hat{V}$, the upper-left 6×6 block of the matrix of A_η will look like this:

$$\begin{bmatrix} 0 & 0 & \lambda & 0 & 0 & 0 \\ 0 & 0 & 0 & 0 & 0 & 0 \\ \lambda & 0 & 0 & 0 & 0 & 0 \\ 0 & 0 & 0 & 0 & 0 & -\lambda \\ 0 & 0 & 0 & 0 & 0 & 0 \\ 0 & 0 & 0 & -\lambda & 0 & 0 \end{bmatrix}. \tag{7.56}$$

All other entries in the matrix are zero. Note that both A_ξ and A_η have eigenvalues $(\lambda, -\lambda, 0)$ with respective multiplicities $(2, 2, 2(h+k-2))$. The same holds for $\pi_* E_{j\ell}$ for any particular index choice (j, ℓ). Further, consider any unit normal ζ that is a linear combination of ξ and η. Write $\zeta = a\xi + b\eta$. Then, the upper-left 6×6 block of A_ζ with respect to this same basis is

$$\lambda \begin{bmatrix} 0 & a & b & 0 & 0 & 0 \\ a & 0 & 0 & 0 & 0 & 0 \\ b & 0 & 0 & 0 & 0 & 0 \\ 0 & 0 & 0 & 0 & -a & -b \\ 0 & 0 & 0 & -a & 0 & 0 \\ 0 & 0 & 0 & -b & 0 & 0 \end{bmatrix}. \tag{7.57}$$

A basis of eigenvectors for this matrix can be worked out as follows:

- $(\hat{X} + a\hat{U} + b\hat{V})$ and $(J\hat{X} - a\hat{U} - bJ\hat{V})$ with eigenvalue λ;
- $(\hat{X} - a\hat{U} - b\hat{V})$ and $(J\hat{X} + a\hat{U} + bJ\hat{V})$ with eigenvalue $-\lambda$;
- $-b\hat{U} + a\hat{V}$ and $-bJ\hat{U} + aJ\hat{V}$ with eigenvalue 0,

so that A_ζ has the same eigenvalue behavior as A_ξ and A_η.

We can repeat this process, with $a\hat{U} + b\hat{V}$ playing the role of $\hat{U}$, for the span of

$$\{\pi_* E_{11}, \pi_* E_{12}, \pi_* E_{13}\}$$

and so on, finally determining that every unit normal in the span of

$$\{\pi_* E_{1\ell} \mid 2 \leq \ell \leq k\}$$

has these same eigenvalues and multiplicities.

However, if $h \geq 2$ and $k \geq 2$, let $\xi = \pi_* E_{11}$, $\eta = \pi_* E_{22}$, and $\zeta = (\xi + \eta)/\sqrt{2}$. It is easy to see that with respect to a basis beginning with $\{\hat{X}, \hat{U}, \hat{Y}, \hat{V}, J\hat{X}, J\hat{U}, J\hat{Y}, J\hat{V}\}$, where $Y = (\pi_* \epsilon_2, 0)$, the upper-left 8×8 block of A_ζ is

$$\frac{1}{\sqrt{2}}\begin{bmatrix} 0 & \lambda & 0 & 0 & 0 & 0 & 0 & 0 \\ \lambda & 0 & 0 & 0 & 0 & 0 & 0 & 0 \\ 0 & 0 & 0 & \lambda & 0 & 0 & 0 & 0 \\ 0 & 0 & \lambda & 0 & 0 & 0 & 0 & 0 \\ 0 & 0 & 0 & 0 & 0 & -\lambda & 0 & 0 \\ 0 & 0 & 0 & 0 & -\lambda & 0 & 0 & 0 \\ 0 & 0 & 0 & 0 & 0 & 0 & 0 & -\lambda \\ 0 & 0 & 0 & 0 & 0 & 0 & -\lambda & 0 \end{bmatrix}, \tag{7.58}$$

so that the eigenvalues of A_ζ are $(\lambda, -\lambda, 0)$ with respective multiplicities

$$(4, 4, 2(h+k-4)).$$

In fact, considering linear combinations of ξ and η as above, we get eigenvalues

$$(a\lambda, -a\lambda, b\lambda, -b\lambda, 0)$$

with multiplicities $(2, 2, 2, 2, 2(h+k-4))$ except when $a = b$ or $ab = 0$.

7.4 The Plücker Embedding of Complex Grassmannians

The complex Grassmann manifold of complex h-planes in $\mathbf{C}^{h+k}$ is expressed as a symmetric space

$$G_{h,k}(\mathbf{C}) = \frac{U(h+k)}{U(h) \times U(k)},$$

see [283, Vol. II, pp. 160 and 286], for more details. In case $h = 1$, $G_{1,k}(\mathbf{C})$ is just the complex projective space $\mathbf{CP}^k$. The Plücker embedding realizes $G_{h,k}(\mathbf{C})$ as a Kähler submanifold of a higher-dimensional complex projective space $\mathbf{CP}^n$, specifically the complex projective space over the h-th complex exterior power of $\mathbf{C}^{h+k}$. Thus

$$n = \binom{h+k}{h} - 1.$$

In fact, an h-plane spanned by h linearly independent elements $\mathbf{v}_i$ of $\mathbf{C}^{h+k}$, gives rise to a totally decomposable h-vector $\mathbf{v}_1 \wedge \mathbf{v}_2 \wedge \cdots \wedge \mathbf{v}_h$. This correspondence gives a well-defined bijection from $G_{h,k}(\mathbf{C})$ onto the complex submanifold of totally decomposable h-vectors. This is the Plücker embedding.

When considering the submanifolds $G_{h,k}(\mathbf{C})$, it is sufficient to assume $h \leq k$. This is because of a duality relationship, i.e., $G_{h,k}(\mathbf{C})$ and $G_{k,h}(\mathbf{C})$ are holomorphically congruent as submanifolds of $\mathbf{CP}^n$. It turns out that the $k = 2$ case is particularly interesting for two reasons:

- The complex 2-plane Grassmannians $G_{2,3}(\mathbf{C})$ occur as focal submanifolds of Hopf hypersurfaces in $\mathbf{CP}^9$ (see Theorem 8.26 on page 444).
- Hypersurfaces in $G_{2,k}(\mathbf{C})$ exhibit many of the properties of hypersurfaces in real, complex, and quaternionic space forms. There has been much recent interest in studying such hypersurfaces beginning with J. Berndt and Y.J. Suh [29, 37].

We will restrict our attention to this case in the rest of the section.

The complex 2-plane Grassmannian

We now restrict our attention to the $h = 2$ case. For any m, the 2nd exterior power of $\mathbf{C}^m$ may be identified with the space of complex $m \times m$ skew-symmetric matrices. Here $m = 2 + k$. Consider the space of all complex $m \times m$ matrices with its standard Hermitian inner product as used in Section 7.3 in connection with the Segre embedding. We decompose this $\mathbf{C}^{m^2}$ into a direct sum of two orthogonal subspaces, the symmetric and the skew-symmetric matrices, as in Section 7.2. In terms of dimensions, the breakdown is

$$m^2 = \frac{m(m+1)}{2} + \frac{m(m-1)}{2},$$

so that the dimension n of the ambient projective space will satisfy

$$n + 1 = \frac{m(m-1)}{2} = \frac{(2+k)(1+k)}{2} = \frac{k(k+3)}{2} + 1.$$

Note that the Hermitian inner product, when restricted to the space of skew-symmetric matrices, yields the usual Hermitian inner product on $\mathbf{C}^{n+1}$. Thus, we can form the familiar sphere $S^{2n+1}(r)$ and the complex projective space $\mathbf{CP}^n$ with constant holomorphic sectional curvature $4c = 4/r^2$.

The unitary group $U(2+k) = U(m)$ acts on $\mathbf{C}^{m^2}$ by

$$(B, z) \mapsto BzB^T,$$

where $B \in U(m)$ and $z \in \mathbf{C}^{m^2}$. Clearly, this action preserves symmetry and skew-symmetry. It is also easy to check that it preserves the Hermitian inner product. In fact, we have

$$\text{trace } (BVB^T)^* BUB^T = \text{trace } V^* U. \tag{7.59}$$

For $1 \leq j, \ell \leq 2 + k$, we introduce

$$F_{j\ell} = E_{j\ell} - E_{\ell j},$$

and observe that the $n + 1$ elements,

$$e_{j\ell} = \frac{1}{\sqrt{2}} F_{j\ell}, \text{ where } j < \ell,$$

form an orthonormal set of complex skew-symmetric matrices, and the set

$$\{e_{j\ell}, ie_{j\ell}\}$$

constitute an orthonormal basis for $\mathbf{C}^{n+1}$.

In this context, the Plücker embedding may be described as follows. Let M' be the orbit of $z_0 = r\, e_{12}$ under the action of $U(2 + k)$ and let $M = \pi M'$. The isotropy subgroup is $U(2) \times U(k)$, understood as the set of matrices with the 2×2 upper-left block and the $k \times k$ lower-right block, both unitary. The other entries in such matrices must all be zero. The Cartan decomposition of the Lie algebra $\mathfrak{u}(2 + k)$ is

$$\mathfrak{u}(2 + k) = (\mathfrak{u}(2) + \mathfrak{u}(k)) + \mathfrak{m}$$

where $\mathfrak{m}$ may be identified with the tangent space to $G_{2,k}(\mathbf{C})$ at the origin. Note that $\mathfrak{m}$ is spanned as a real vector space by the $4k$ elements $\{F_{1j}, iF_{1j}, F_{2j}, iF_{2j}\}$ where $3 \leq j \leq k + 2$. For $Z \in \mathfrak{m}$, let

$$\gamma_t = (\exp tZ)\, z_0\, (\exp tZ)^T.$$

Then $\gamma_0 = z_0$ and

$$\overrightarrow{\gamma_0} = [Z, z_0] = \frac{r}{\sqrt{2}} [Z, F_{12}].$$

This means that if

$$f : G_{2,k} \to S^{2n+1}(r)$$

denotes the map we have been describing (so that the Plücker embedding is $\pi \circ f$), its differential at the origin satisfies

$$f_* Z = \frac{r}{\sqrt{2}} [Z, F_{12}].$$

In particular, for $j \geq 3$,

$$f_* F_{1j} = \frac{r}{\sqrt{2}} [F_{1j}, F_{12}] = \frac{r}{\sqrt{2}} F_{2j}.$$

By a similar calculation, $f_* F_{2j} = -\frac{r}{\sqrt{2}} F_{1j}$. Recalling the basis of $\mathfrak{m}$, we see that the $4k$ vectors of the form

$$\{\pi_* F_{1j}, J\pi_* F_{1j}, \pi_* F_{2j}, J\pi_* F_{2j}\},$$

where $3 \leq j \leq k+2$, are mutually orthogonal vectors of length $\sqrt{2}$ spanning the tangent space to M at πz_0. From this, we note that for tangent vectors X, Y to $G_{2,k}$, we have $\langle f_* X, f_* Y\rangle = \frac{r^2}{2}\langle X, Y\rangle$ with respect to the usual Hermitian inner product on the respective spaces of matrices. Also, the normal space to M at πz_0 is spanned by the $k(k-1)/2$ vectors $\{\pi_* F_{j\ell}, J\pi_* F_{j\ell}\}$ where $3 \leq j < \ell \leq k+2$. These vectors also have length $\sqrt{2}$.

Shape operators of the Plücker embedding

Theorem 7.11. *For the Plücker embedding of $G_{2,k}$ into $\mathbf{CP}^n$, where $2n = k(k+3)$, the shape operator with respect to all unit normals at all points has the following properties:*

- *The nonzero eigenvalues are $\lambda = \frac{1}{r}$ and $-\lambda = -\frac{1}{r}$, each of multiplicity 4.*
- *The corresponding eigenspaces T_λ and $T_{-\lambda}$ satisfy $JT_\lambda = T_{-\lambda}$.*
- *If $k \geq 3$, the zero eigenvalue has multiplicity $4(k-2)$. Otherwise, there is no zero eigenvalue.*

Because of homogeneity, it is sufficient to check the shape operator at one point. We choose the point $p_0 = \pi z_0$. Now consider the unit normal

$$\xi = \pi_* \xi_0 \text{ where } \xi_0 = e_{34} = \frac{1}{\sqrt{2}} F_{34}.$$

Also, it is easy to check that $A_\xi X = \pi_* A'_{\xi_0} X^L$ so that it is sufficient to compute the shape operator A'_{ξ_0} for the embedding f of $G_{2,k}$ onto the horizontal submanifold M' of the sphere S^{2n+1}.

To this end, we set

$$\xi_t = (\exp tZ)\, \xi_0\, (\exp tZ)^T$$

where Z is one of the basis vectors of $\mathfrak{m}$.

Lemma 7.12. *ξ_t is a unit normal to M' at γ_t.*

Proof. Clearly, $\langle \xi_t, \gamma_t \rangle = \langle \xi_0, z_0 \rangle = 0$, since F_{12} and F_{34} are orthogonal. Furthermore, in view of equation (7.59), we have $\langle \xi_t, i\gamma_t \rangle = 0$ so that ξ_t is horizontal at γ_t. We compute

$$\overrightarrow{\gamma_t} = (\exp tZ)\ [Z, z_0]\ (\exp tZ)^T,$$

and note that $\langle \xi_t, \overrightarrow{\gamma_t} \rangle = 0$. However, we need also to check that ξ_t is orthogonal to all other tangent vectors to M' at γ_t. Let $\tilde{Z}$ be a basis vector of $\mathfrak{m}$ which is orthogonal to Z and iZ. Let

$$\delta_u = (\exp tZ)(\exp u\tilde{Z})z_0(\exp u\tilde{Z})^T(\exp tZ)^T,$$

which is a curve on M' with initial conditions $\delta_0 = \gamma_t$ and

$$\overrightarrow{\delta_0} = (\exp tZ)\ [\tilde{Z}, z_0]\ (\exp tZ)^T.$$

These initial tangent vectors, together with Z and iZ span the tangent space to M' at γ_t. Since $\langle \xi_t, \overrightarrow{\delta_0} \rangle = \langle \xi_0, [\tilde{Z}, z_0] \rangle = 0$, we see that ξ_t is normal to M' at γ_t. □

Using the fact that the sphere S^{2n+1} is totally umbilic in $\mathbf{C}^{n+1}$, we have for $Z \in \mathfrak{m}$,

$$-f_* A'_{\xi_0} Z = \overrightarrow{\xi_0} = \frac{1}{\sqrt{2}}[Z, F_{34}].$$

In particular, for $Z = F_{1j}$, we have $[Z, F_{34}] = [F_{1j}, F_{34}]$ which is equal to F_{14} if $j = 3$, $-F_{13}$ if $j = 4$, and zero otherwise. Thus, we have

$$-f_* A'_{\xi_0} F_{13} = -\frac{1}{r} f_* F_{24},$$

and similar equations for F_{14}, F_{23}, and F_{24}. This gives us

$$\begin{aligned}
A'_{\xi_0} F_{13} &= \frac{1}{r} F_{24} \\
A'_{\xi_0} F_{14} &= -\frac{1}{r} F_{23} \\
A'_{\xi_0} F_{23} &= -\frac{1}{r} F_{14} \\
A'_{\xi_0} F_{24} &= \frac{1}{r} F_{13} \\
A'_{\xi_0} F_{ij} &= 0 \text{ for } i = 1, 2 \text{ and } j \geq 5.
\end{aligned} \tag{7.60}$$

Interpreting our result on M, we see that the eigenvalue/eigenspace configuration of A_ξ is as follows:

- The eigenvalues are λ, $-\lambda$ and 0, where $\lambda = 1/r$.
- The eigenspace T_λ is spanned by

$$\pi_*(F_{13} + F_{24}),\ \pi_*(F_{14} - F_{23}),\ J\pi_*(F_{13} - F_{24}),\ J\pi_*(F_{14} + F_{23}).$$

- The eigenspace $T_{-\lambda}$ is spanned by

$$\pi_*(F_{13} - F_{24}),\ \pi_*(F_{14} + F_{23}),\ J\pi_*(F_{13} + F_{24}),\ J\pi_*(F_{14} - F_{23}).$$

- The eigenspace T_0 is spanned by the set $\{F_{ij}, JF_{ij}\}$ for $1 \leq i \leq 2$ and $5 \leq j \leq k+2$.

Clearly, the multiplicities of the eigenvalues and their relationship with the complex structure J are as announced in the statement of the theorem.

When $k = 3$, we can apply a similar computation to e_{35} and e_{45}. In the interest of symmetry, we write e_{53} for $-e_{35}$ and let

$$\zeta_0 = a\, e_{34} + b\, e_{45} + c\, e_{53},$$

where $a^2 + b^2 + c^2 = 1$. Then the matrix of A'_{ζ_0} (restricted to the span of F_{13}, F_{14}, F_{15}, F_{23}, F_{24}, F_{25}) takes the form,

$$A = \frac{1}{r}\begin{bmatrix} 0 & 0 & 0 & 0 & a & c \\ 0 & 0 & 0 & -a & 0 & -b \\ 0 & 0 & 0 & -c & b & 0 \\ 0 & -a & -c & 0 & 0 & 0 \\ a & 0 & b & 0 & 0 & 0 \\ c & -b & 0 & 0 & 0 & 0 \end{bmatrix}. \tag{7.61}$$

The square of this matrix is

$$A^2 = \frac{1}{r^2}\begin{bmatrix} a^2+c^2 & -bc & ab & 0 & 0 & 0 \\ -bc & a^2+b^2 & ac & 0 & 0 & 0 \\ ab & ac & b^2+c^2 & 0 & 0 & 0 \\ 0 & 0 & 0 & a^2+c^2 & -bc & ab \\ 0 & 0 & 0 & -bc & a^2+b^2 & ac \\ 0 & 0 & 0 & ab & ac & b^2+c^2 \end{bmatrix}. \tag{7.62}$$

One can check that the $bF_{13}+cF_{14}-aF_{15}$, and $bF_{23}+cF_{24}-aF_{25}$ span the nullspace of A^2. Provided that $a \neq 0$, $aF_{13}+bF_{15}$, $aF_{23}+bF_{25}$, $aF_{14}+cF_{15}$, and $aF_{24}+cF_{25}$ are linearly independent vectors spanning an eigenspace of A^2 with eigenvalue λ^2. This means that the eigenvalues of A are $\{0, \lambda, -\lambda\}$ and the zero eigenvalue has multiplicity 2. Further, noting that A has trace 0, we conclude that λ and $-\lambda$ each have multiplicity 2. If $a = 0$, a different linear combination provides the same result. In fact, $cF_{13} - bF_{14}$, $cF_{23} - bF_{24}$, F_{15}, and F_{25}, will span the λ^2-eigenspace of A^2. Finally, recalling that the shape operator anticommutes with J, we see that each of the eigenvalues discussed above has multiplicity 4 when the shape operator is applied to the full tangent space.

We now examine the case $k \geq 4$. We will show that not all unit normals have shape operators with the same eigenvalue configuration. This turns out to have the implication that tubes over such Grassmannians do not have constant principal curvatures. We will use this later on, in the proof of Theorem 8.26.

Specifically, let $\eta_0 = a\, e_{34} + b\, e_{56}$. Then the matrix of A'_{η_0} (restricted to the span of F_{13}, F_{14}, F_{23}, F_{24}, F_{15}, F_{16}, F_{25}, and F_{26}) takes the form

$$A = \frac{1}{r}\begin{bmatrix} 0 & 0 & 0 & a & 0 & 0 & 0 & 0 \\ 0 & 0 & -a & 0 & 0 & 0 & 0 & 0 \\ 0 & -a & 0 & 0 & 0 & 0 & 0 & 0 \\ a & 0 & 0 & 0 & 0 & 0 & 0 & 0 \\ 0 & 0 & 0 & 0 & 0 & 0 & 0 & b \\ 0 & 0 & 0 & 0 & 0 & 0 & -b & 0 \\ 0 & 0 & 0 & 0 & 0 & -b & 0 & 0 \\ 0 & 0 & 0 & 0 & b & 0 & 0 & 0 \end{bmatrix}. \tag{7.63}$$

The square of this matrix is

$$A^2 = \frac{1}{r^2}\begin{bmatrix} a^2 & 0 & 0 & 0 & 0 & 0 & 0 & 0 \\ 0 & a^2 & 0 & 0 & 0 & 0 & 0 & 0 \\ 0 & 0 & a^2 & 0 & 0 & 0 & 0 & 0 \\ 0 & 0 & 0 & a^2 & 0 & 0 & 0 & 0 \\ 0 & 0 & 0 & 0 & b^2 & 0 & 0 & 0 \\ 0 & 0 & 0 & 0 & 0 & b^2 & 0 & 0 \\ 0 & 0 & 0 & 0 & 0 & 0 & b^2 & 0 \\ 0 & 0 & 0 & 0 & 0 & 0 & 0 & b^2 \end{bmatrix}. \tag{7.64}$$

Also, A'_{η_0} vanishes on the orthogonal complement of the 16-dimensional subspace spanned by $\{F_{ij}, JF_{ij}\}$ where $1 \leq i \leq 2$, $3 \leq j \leq 6$. Arguing as in the previous case, A can be diagonalized with $a\lambda$, $-a\lambda$, $b\lambda$, and $-b\lambda$ each occurring twice along the diagonal. Depending on the choice of a and b, the eigenvalues of A can be $(0, \lambda, -\lambda)$ with respective multiplicities, $(4, 2, 2)$, $(a\lambda, -a\lambda, b\lambda, -b\lambda)$ each of multiplicity 2 (when $a^2 \neq b^2$ and both are nonzero), or, finally, $(\frac{1}{\sqrt{2}}\lambda, -\frac{1}{\sqrt{2}}\lambda)$, each of multiplicity 4. In any case, many different multiplicities are possible.

7.5 The Half-spin Embedding of $SO(2d)/U(d)$ in $\mathbf{CP}^n$

This is the space $SO(2d)/U(d)$ of real dimension $d(d-1)$. It is embedded as an $m = \frac{d(d-1)}{2}$-dimensional complex submanifold M of $\mathbf{CP}^n$ where $n+1 = 2^{d-1}$ using the half-spin embedding. We assume that $d \geq 2$. In the special case where $d = 5$, we have $m = 10$, $n = 15$, and M is a focal submanifold of the Type E hypersurface in $\mathbf{CP}^{15}$.

Construction of the ambient space – Part 1

1. Let V be a real vector space of dimension $2d$. We may regard V as $\mathbf{R}^{2d}$ with the usual inner product $\langle \, , \, \rangle$ and standard orthonormal basis $\{\epsilon_k\}$, where $1 \leq k \leq 2d$.
2. Let J be the complex structure on V satisfying $\epsilon_{d+k} = J\epsilon_k$ for $1 \leq k \leq d$.
3. Form the Clifford algebra $C\ell(V)$ generated by V subject to the relations
$$v_1 v_2 + v_2 v_1 = 2\langle v_1, v_2\rangle.$$
Then $\epsilon_j^2 = 1$ and $\epsilon_k\epsilon_j = -\epsilon_j\epsilon_k$ for $1 \leq j \neq k \leq 2d$. The dimension of this Clifford algebra is 2^{2d}.
4. Let $Pin(V)$ be the group generated by unit vectors of V using the Clifford multiplication, and let $Spin(V)$ be the subgroup of $Pin(V)$ generated by products of an even number of unit vectors.
5. Note that each element of the orthogonal group $O(V)$ is a product of reflections Ω_u where
$$\Omega_u : V \to V$$
is the reflection in the hyperplane through the origin with unit normal u. The action can be expressed in terms of Clifford multiplication by
$$\Omega_u X = -uXu = -uXu^{-1}$$
for $X \in V$. $Pin(V)$ is a double cover of $O(V)$ (and $Spin(V)$ of $SO(V)$) since $\Omega_u = \Omega_{-u}$.
6. Let $V_{\mathbf{C}}$ be the complexification of V and extend J, the inner product $\langle \, , \, \rangle$, and the Clifford multiplication by (complex) linearity. The complex Clifford algebra $C\ell(V_{\mathbf{C}})$ is just the complexification of $C\ell(V)$.
7. The real (resp. complex) Clifford algebras are 2^{2d}-dimensional real (resp. complex) vectors spaces which may be expressed as the direct sum of the $\binom{2d}{j}$-dimensional subspaces $C\ell^j(V)$, (resp., $C\ell^j(V_{\mathbf{C}})$), consisting of elements of pure degree j. For any nonempty subset K of the integers 1 through $2d$, denote by ϵ_K the product (in ascending order) of $\{\epsilon_k \mid k \in K\}$. When K is the empty set,

$\epsilon_K = 1$, by definition. The set of all ϵ_K forms a basis for the Clifford algebra and the cardinality of K corresponds to the degree.

We decompose $V_\mathbf{C} = W \bigoplus \overline{W}$, where W is the $+i$ eigenspace of J and $\overline{W}$ is the $-i$ eigenspace. Each is a complex vector space of dimension d.

Lemma 7.13. *W is an isotropic subspace, i.e., $w^2 = 0$ for all $w \in W$. Similarly, $\overline{W}$ is isotropic.*

Remark 7.14. The Clifford algebras $\mathcal{C}\ell(W)$ and $\mathcal{C}\ell(\overline{W})$ are Clifford subalgebras of $\mathcal{C}\ell(V_\mathbf{C})$. Note that the restriction to W and to $\overline{W}$ of the bilinear form on which Clifford multiplication is based, is identically zero, so that any two elements anticommute. Thus we can identify $\mathcal{C}\ell(W)$ and $\mathcal{C}\ell(\overline{W})$ with the exterior algebras $\bigwedge W$ and $\bigwedge \overline{W}$. In this identification, the Clifford multiplication corresponds to the usual wedge product.

In this context, we use the notations $\bigwedge^j W$ and $\mathcal{C}\ell^j(W)$ interchangeably (and similarly for $\overline{W}$). Write

$$\alpha_k = \frac{1}{\sqrt{2}}(\epsilon_k - iJ\epsilon_k); \quad \bar{\alpha}_k = \frac{1}{\sqrt{2}}(\epsilon_k + iJ\epsilon_k), \tag{7.65}$$

for $1 \leq k \leq d$. The α_k (resp. $\bar{\alpha}_k$) constitute a basis for W (resp. $\overline{W}$), and as before, the α_K and $\bar{\alpha}_K$ are a basis for the respective Clifford (exterior) algebras.

Construction of the ambient space – Part 2

We first note that there is a unique Hermitian inner product on W that makes the d vectors α_k orthonormal. This is obtained by defining

$$(\alpha_j, \alpha_k) = \langle \bar{\alpha}_j, \alpha_k \rangle,$$

for $1 \leq j, k \leq d$. We extend $(\ ,\ )$ to $\mathcal{C}\ell(W)$ so that the canonical basis elements α_K are orthonormal. The Clifford algebra $\mathcal{C}\ell(W)$ may be further broken down, separating terms of even and odd degree. Specifically, let $\bigwedge^+ W$ be the span of all α_K such that $|K|$ is even. This is a complex vector space of dimension 2^{d-1} which may therefore be identified with $\mathbf{C}^{n+1}$. Restricting the Hermitian inner product $(\ ,\ )$ to this $\mathbf{C}^{n+1}$, we may carry out the construction of $\mathbf{CP}^n$ as usual. For example, when $d = 5$, so that $\mathbf{C}^{n+1} = \mathbf{C}^{16}$, we will have the following 16 orthonormal basis elements:

- 1 of degree 0;
- $\alpha_1\alpha_2$; $\alpha_1\alpha_3$; $\alpha_1\alpha_4$; $\alpha_1\alpha_5$; $\alpha_2\alpha_3$; $\alpha_2\alpha_4$; $\alpha_2\alpha_5$; $\alpha_3\alpha_4$; $\alpha_3\alpha_5$; $\alpha_4\alpha_5$; all of degree 2;
- $\alpha_1\alpha_2\alpha_3\alpha_4$; $\alpha_1\alpha_2\alpha_3\alpha_5$; $\alpha_1\alpha_2\alpha_4\alpha_5$; $\alpha_1\alpha_3\alpha_4\alpha_5$; $\alpha_2\alpha_3\alpha_4\alpha_5$; all of degree 4.

The Lie algebras and exponential maps

We recall that the Lie algebra $\mathfrak{o}(2d)$ of $SO(2d)$ is spanned by the elements F_{jk} where $j < k$ (skew-symmetric real $2d \times 2d$ matrices). On the other hand, $\mathcal{C}\ell^2(V)$ may be regarded as a Lie algebra using the operation $[a, b] = ab - ba$. Note that the right side (which uses the Clifford multiplication) indeed has degree 2. Define the map

$$\tau : \mathfrak{o}(2d) \to \mathcal{C}\ell^2(V)$$

on basis elements by $\tau(F_{jk}) = \frac{1}{2}\epsilon_j\epsilon_k$ and extend linearly. Then it is easy to check that τ is a Lie algebra isomorphism. Further, if $\tau(A) = a$, then for all $v \in V$, we have

$$Av = av - va,$$

where the operation on the left is ordinary matrix multiplication, while that on the right is again Clifford multiplication. We need to check that if $v \in \mathcal{C}\ell^1(V)$ and $a \in \mathcal{C}\ell^2(V)$, then $va - av$ is in $\mathcal{C}\ell^1(V)$. Furthermore, we have

$$(\exp A)v = (\exp a)\ v\ (\exp(-a)),$$

where again, the operations on the left are ordinary matrix exponentiation and multiplication, while the exponentiations and multiplications on the right are defined in terms of the Clifford multiplication in the finite-dimensional algebra $\mathcal{C}\ell(V)$. Of course, individual terms on the right need not be of degree 1, but the product must be.

We will need to discuss one-parameter subgroups of $SO(2d)$ in order to analyze its action on $\mathbf{C}^{n+1}$. We begin with reflections as represented in $Pin(V)$. For $1 \leq j \neq k \leq 2d$, let

$$u_t = -\sin\frac{t}{2}\ \epsilon_j + \cos\frac{t}{2}\ \epsilon_k. \tag{7.66}$$

This particular element of $Pin(V)$ determines a reflection in the hyperplane with unit normal u_t. Note that this reflection fixes all basis vectors not in the span of $\{\epsilon_j, \epsilon_k\}$. Now consider the element $u_0u_t \in Spin(V)$. As an element of $\mathcal{C}\ell(V)$, this is

$$\cos\frac{t}{2} + \sin\frac{t}{2}\ \epsilon_j\epsilon_k = \exp\left(\frac{t}{2}\ \epsilon_j\epsilon_k\right).$$

The "action" of SO(2d) on $\mathbf{C}^{n+1}$

We put the word "action" in quotation marks, as it is only defined up to a sign. In fact, it is $Spin(2d)$ that acts in an unambiguous sense. (We use $Spin(2d)$ to mean $Spin(R^{2d})$ here just as $SO(2d)$ is $SO(R^{2d})$.) However, this will be sufficient for our purposes, since ultimately, we are interested in $\mathbf{CP}^n$ rather than on $\mathbf{C}^{n+1}$.

For $v \in V = \mathcal{C}\ell^1(V) \subset \mathcal{C}\ell^1(V_{\mathbf{C}})$, we may write v uniquely as

$$v = v_W + v_{\overline{W}}$$

where $v_W \in W$ and $v_{\overline{W}} \in \overline{W}$. For $\omega \in \bigwedge W$, write

$$v \bullet \omega = \sqrt{2}\, i(v_W \wedge \omega - v_{\overline{W}} \lrcorner \omega),$$

where the notation "$\lrcorner$" denotes the interior product (see, for example, Sternberg [500, p.20]). The action preserves $\bigwedge W$ and raises or lowers the degrees of individual terms by 1. For u, v in $Pin(V)$ so that $a = uv \in Spin(V)$, we have

$$a \bullet \omega = u \bullet (v \bullet \omega)$$

so that the action of $Spin(V)$ preserves the degree parity. In particular, it preserves $\bigwedge^+ W = \mathbf{C}^{n+1}$.

Let $M' \subset S^{2n+1}(r)$ be the orbit of $r \cdot 1$ under $Spin(2d)$ and let $M = \pi M'$. For $X \in \mathfrak{o}(2d)$, consider the map

$$\omega \mapsto (\exp X) \bullet \omega.$$

The differential of this map at $r \cdot 1$ sends X to a tangent vector to M' at the point $(\exp X) \bullet (r \cdot 1)$. We may break down the Lie algebra

$$\mathfrak{o}(2d) = \mathfrak{h} + \mathfrak{m} \tag{7.67}$$

where $\mathfrak{h}$ is a Lie subalgebra, $[\mathfrak{m}, \mathfrak{m}] \subset \mathfrak{h}$, and $[\mathfrak{h}, \mathfrak{m}] \subset \mathfrak{m}$.

Recalling that $\mathfrak{o}(2d)$ has a basis of the form $\{F_{jk}\}$, $1 \leq j < k \leq 2d$, we now choose an alternative basis that is adapted to the action we are studying. For $1 \leq j < k \leq d$, we define

$$\begin{aligned} K_j &= F_{j\,j+d} \\ G^+_{jk} &= F_{jk} + F_{j+d\;k+d}\;; \quad H^+_{jk} = F_{j\;k+d} + F_{k\;j+d} \\ G^-_{jk} &= F_{jk} - F_{j+d\;k+d}\;; \quad H^-_{jk} = F_{j\;k+d} - F_{k\;j+d}. \end{aligned} \tag{7.68}$$

Let $\mathfrak{k}$, $\mathfrak{g}^+$, $\mathfrak{h}^+$, $\mathfrak{g}^-$, and $\mathfrak{h}^-$ be the respective spans of these five sets of matrices. Then set

$$\mathfrak{h} = \mathfrak{k} \oplus \mathfrak{g}^+ \oplus \mathfrak{h}^+\;; \quad \mathfrak{m} = \mathfrak{g}^- \oplus \mathfrak{h}^-; \tag{7.69}$$

One can check that the specified conditions for the decomposition in equation (7.67) are satisfied. To see this more easily, we extend our notation, temporarily replacing the condition $j < k$ by $j \neq k$, and setting

$$G^+_{kj} = -G^+_{jk}; \quad G^-_{kj} = -G^-_{jk}; \quad H^+_{kj} = -H^+_{jk}; \quad H^-_{kj} = -H^-_{jk}.$$

Then, we have the following identities for distinct indices k, j, and ℓ. The Lie brackets involving different configurations of indices (for example, four distinct or only two distinct) all vanish.

$$[G^+_{kj}, G^+_{\ell j}] = G^+_{\ell k}, \quad [H^+_{kj}, H^+_{\ell j}] = G^+_{\ell k}, \quad [G^+_{kj}, H^+_{\ell j}] = H^+_{\ell k},$$
$$[G^+_{kj}, K_j] = H^+_{jk}, \quad [H^+_{kj}, K_j] = G^+_{jk}, \quad [K_k, K_j] = 0,$$

and thus $\mathfrak{h}$ is a Lie subalgebra. Further, we have $[\mathfrak{m}, \mathfrak{m}] \subset \mathfrak{h}$, since

$$[G^-_{kj}, G^-_{\ell j}] = G^+_{\ell k}, \quad [H^-_{kj}, H^-_{\ell j}] = G^+_{\ell k}, \quad [H^-_{kj}, G^-_{\ell j}] = H^+_{\ell k}.$$

Finally, we check that

$$[H^-_{kj}, G^+_{\ell j}] = H^-_{\ell k}, \quad [G^-_{kj}, G^+_{\ell j}] = G^-_{\ell k}, \quad [H^+_{kj}, G^-_{\ell j}] = H^-_{\ell k},$$
$$[H^-_{kj}, H^+_{\ell j}] = G^-_{\ell k}, \quad [G^-_{kj}, K_j] = H^-_{kj}, \quad [H^-_{kj}, K_j] = G^-_{jk},$$

which shows that $[\mathfrak{h}, \mathfrak{m}] \subset \mathfrak{m}$.

We now check that $\mathfrak{m}$ may be identified with the tangent space to M' at $r \cdot 1$. First take $X = G^-_{jk}$. The associated member of $C\ell^1(V)$ is

$$a = \frac{1}{2}\epsilon_j\epsilon_k - \frac{1}{2}\epsilon_{j+d}\epsilon_{k+d}.$$

Then

$$\exp ta = (\exp(\frac{t}{2}\epsilon_j\epsilon_k))\,(\exp(-\frac{t}{2}\epsilon_{j+d}\epsilon_{k+d}))$$

since $\epsilon_j\epsilon_k$ and $\epsilon_{j+d}\epsilon_{k+d}$ commute in $C\ell(V)$. We can write

$$\exp ta = u_0 u_t \tilde{u}_0 \tilde{u}_{-t}$$

where u_t is defined by (7.66) and $\tilde{u}_t$ is the analogous expression involving $\epsilon_{j+d}\epsilon_{k+d}$.

We may rewrite u_t in the form

$$u_t = \frac{1}{2}(u_t - iJu_t) + \frac{1}{2}(u_t + iJu_t)$$

which reduces to

$$u_t = \frac{1}{\sqrt{2}}(-\sin\frac{t}{2}\,\alpha_j + \cos\frac{t}{2}\,\alpha_k) + \frac{1}{\sqrt{2}}(-\sin\frac{t}{2}\,\bar{\alpha}_j + \cos\frac{t}{2}\,\bar{\alpha}_k).$$

Then

$$u_0 = \frac{1}{\sqrt{2}}\alpha_k + \frac{1}{\sqrt{2}}\bar{\alpha}_k.$$

Similarly,

$$\tilde{u}_t = Ju_t = \frac{i}{\sqrt{2}}(-\sin\frac{t}{2}\,\alpha_j + \cos\frac{t}{2}\,\alpha_k) - \frac{i}{\sqrt{2}}(-\sin\frac{t}{2}\,\bar{\alpha}_j + \cos\frac{t}{2}\,\bar{\alpha}_k)$$

and

$$\tilde{u}_0 = \frac{i}{\sqrt{2}}\alpha_k - \frac{i}{\sqrt{2}}\bar{\alpha}_k.$$

These are the expressions we need to compute the action of $\exp ta$ on $\bigwedge^+ W$. In particular, we compute $(\exp ta) \bullet 1$ (and hence $(\exp ta) \bullet (r \cdot 1)$) as follows:

$$\begin{aligned}
\tilde{u}_{-t} \bullet 1 &= \sin\frac{t}{2}\,\alpha_j - \cos\frac{t}{2}\,\alpha_k \\
\tilde{u}_0\tilde{u}_{-t} \bullet 1 &= \cos\frac{t}{2} - \sin\frac{t}{2}\,\alpha_j\alpha_k \\
u_t\tilde{u}_0\tilde{u}_{-t} \bullet 1 &= i(-\sin t\,\alpha_j + \cos t\,\alpha_k) \\
u_0u_t\tilde{u}_0\tilde{u}_{-t} \bullet 1 &= \cos t\, - \sin t\,\alpha_j\alpha_k
\end{aligned} \tag{7.70}$$

so that

$$\exp_* G^-_{jk} = -r\alpha_j\alpha_k.$$

Similarly, by choosing

$$a = \frac{1}{2}\epsilon_j\epsilon_{k+d} - \frac{1}{2}\epsilon_k\epsilon_{j+d}$$

and

$$u_t = -\sin\frac{t}{2}\,\epsilon_j + \cos\frac{t}{2}\,\epsilon_{k+d}$$

we get

$$u_t = \frac{1}{\sqrt{2}}(-\sin\frac{t}{2}\,\alpha_j + i\cos\frac{t}{2}\,\alpha_k) - \frac{1}{\sqrt{2}}(\sin\frac{t}{2}\,\bar{\alpha}_j + i\cos\frac{t}{2}\,\bar{\alpha}_k).$$

Then

$$u_0 = \frac{i}{\sqrt{2}}\alpha_k - \frac{i}{\sqrt{2}}\bar{\alpha}_k,$$

$$\tilde{u}_t = \frac{1}{\sqrt{2}}(-\sin\frac{t}{2}\ \alpha_k + i\cos\frac{t}{2}\ \alpha_j) - \frac{1}{\sqrt{2}}(\sin\frac{t}{2}\ \bar{\alpha}_k + i\cos\frac{t}{2}\ \bar{\alpha}_j)$$

and

$$\tilde{u}_0 = \frac{i}{\sqrt{2}}\alpha_j - \frac{i}{\sqrt{2}}\bar{\alpha}_j,$$

and we compute

$$u_0 u_t \tilde{u}_0 \tilde{u}_{-t} \bullet 1 = \cos t\ - i\sin t\ \alpha_j\alpha_k \tag{7.71}$$

so that

$$\exp_* H_{j\bar{k}} = -ir\alpha_j\alpha_k.$$

The detailed computation for the above is as follows:

$$\begin{aligned}
\tilde{u}_{-t} \bullet 1 &= -\cos\frac{t}{2}\ \alpha_j + i\sin\frac{t}{2}\ \alpha_k \\
\tilde{u}_0\tilde{u}_{-t} \bullet 1 &= \cos\frac{t}{2} - i\sin\frac{t}{2}\ \alpha_j\alpha_k \\
u_t\tilde{u}_0\tilde{u}_{-t} \bullet 1 &= -\cos t\ \alpha_k - i\sin t\ \alpha_j \\
u_0u_t\tilde{u}_0\tilde{u}_{-t} \bullet 1 &= \cos t\ - i\sin t\ \alpha_j\alpha_k.
\end{aligned} \tag{7.72}$$

Thus $\exp_*$ maps $\mathfrak{m}$ onto the $\binom{d}{2}$-dimensional complex subspace spanned by elements of the form $\alpha_j\alpha_k = \alpha_j \wedge \alpha_k$, where $1 \le j < k \le d$. The action of $SO(2d)$ preserves the Hermitian metric, and so preserves the sphere $S^{2n+1}(r)$.

The shape operators of M

We have seen that the tangent space is spanned by vectors of the form α_I where I has cardinality 2. Thus the normal space is spanned by the remaining α_I, namely those for which $|I| \ge 4$ is even. Choosing ξ to be one of these normals, we can compute A_ξ as follows: let

$$\xi_t = u_0u_t\tilde{u}_0\tilde{u}_{-t} \bullet \xi.$$

Then $-A_\xi X$ is the normal component of $d\xi/dt$ evaluated at $t = 0$.

Rather than computing ξ_t completely, we can shorten the computation by differentiating in $\mathcal{C}\ell(V_{\mathbb{C}})$. The desired result can be obtained by applying

$$-u_0u_0\tilde{u}_0\tilde{u}_0' + u_0u_0'\tilde{u}_0\tilde{u}_0 = -\tilde{u}_0\tilde{u}_0' + u_0u_0'. \tag{7.73}$$

In case $X = G_{jk}^-$,

$$\begin{aligned} u_0u_0' &= -\frac{1}{4}(\alpha_k + \bar{\alpha}_k)(\alpha_j + \bar{\alpha}_j) \\ \tilde{u}_0\tilde{u}_0' &= \frac{1}{4}(\alpha_k - \bar{\alpha}_k)(\alpha_j - \bar{\alpha}_j), \end{aligned} \tag{7.74}$$

so that

$$\begin{aligned} -\tilde{u}_0\tilde{u}_0' + u_0u_0' &= -\frac{1}{4}\big((\alpha_k + \bar{\alpha}_k)(\alpha_j + \bar{\alpha}_j) + (\alpha_k - \bar{\alpha}_k)(\alpha_j - \bar{\alpha}_j)\big) \\ &= -\frac{1}{2}(\alpha_k\alpha_j + \bar{\alpha}_k\bar{\alpha}_j) = \frac{1}{2}(\alpha_j\alpha_k + \bar{\alpha}_j\bar{\alpha}_k). \end{aligned} \tag{7.75}$$

Thus

$$\left.\frac{d\xi}{dt}\right|_{t=0} = \frac{1}{2}(\alpha_j\alpha_k + \bar{\alpha}_j\bar{\alpha}_k) \bullet \xi. \tag{7.76}$$

Similarly, if $X = H_{jk}^-$, then

$$\begin{aligned} u_0u_0' &= -\frac{i}{4}(\alpha_k - \bar{\alpha}_k)(\alpha_j + \bar{\alpha}_j) \\ \tilde{u}_0\tilde{u}_0' &= -\frac{i}{4}(\alpha_j - \bar{\alpha}_j)(\alpha_k + \bar{\alpha}_k), \end{aligned} \tag{7.77}$$

so that

$$\begin{aligned} -\tilde{u}_0\tilde{u}_0' + u_0u_0' &= -\frac{i}{4}\big((\alpha_k - \bar{\alpha}_k)(\alpha_j + \bar{\alpha}_j) - (\alpha_j - \bar{\alpha}_j)(\alpha_k + \bar{\alpha}_k)\big) \\ &= -\frac{i}{2}(\alpha_k\alpha_j + \bar{\alpha}_j\bar{\alpha}_k) = \frac{i}{2}(\alpha_j\alpha_k - \bar{\alpha}_j\bar{\alpha}_k). \end{aligned} \tag{7.78}$$

Shape operator computation for d = 5

Let us choose $\xi = \alpha_1\alpha_2\alpha_3\alpha_4$ and set $j = 1, k = 2$. Then

$$\left.\frac{d\xi}{dt}\right|_{t=0} = \frac{1}{2}(\alpha_1\alpha_2 + \bar{\alpha}_1\bar{\alpha}_2) \bullet \alpha_1\alpha_2\alpha_3\alpha_4 = \alpha_3\alpha_4, \tag{7.79}$$

which corresponds to the tangent vector $-\frac{1}{r}G_{\overline{34}}$. Thus we can compute

$$A_\xi G_{\overline{12}} = \frac{1}{r} G_{\overline{34}}$$

or equivalently,

$$A_\xi(\alpha_1\alpha_2) = \frac{1}{r}(\alpha_3\alpha_4).$$

Continuing in a similar fashion, we get

$$\begin{aligned} A_\xi(\alpha_3\alpha_4) &= \frac{1}{r}(\alpha_1\alpha_2), \\ A_\xi(\alpha_1\alpha_3) &= \frac{1}{r}(\alpha_4\alpha_2), \quad A_\xi(\alpha_4\alpha_2) = \frac{1}{r}(\alpha_1\alpha_3), \\ A_\xi(\alpha_1\alpha_4) &= \frac{1}{r}(\alpha_2\alpha_3), \quad A_\xi(\alpha_2\alpha_3) = \frac{1}{r}(\alpha_1\alpha_4), \end{aligned} \tag{7.80}$$

so that $\alpha_1\alpha_2 + \alpha_3\alpha_4$, $\alpha_1\alpha_3 + \alpha_4\alpha_2$, and $\alpha_1\alpha_4 + \alpha_2\alpha_3$ span a 3-dimensional space of eigenvectors V^+ with eigenvalue $\frac{1}{r}$, while $\alpha_1\alpha_2 - \alpha_3\alpha_4$, $\alpha_1\alpha_3 - \alpha_4\alpha_2$, and $\alpha_1\alpha_4 - \alpha_2\alpha_3$ span a 3-dimensional space of eigenvectors V^- with eigenvalue $-\frac{1}{r}$.

It is easy to check using $H_{\overline{jk}}$ (or merely invoking the fact that the complex structure of $\mathbf{C}^{n+1}$ anticommutes with A_ξ) that iV^- and iV^+ are spaces of eigenvectors with respective eigenvalues $\frac{1}{r}$ and $-\frac{1}{r}$. Finally, $A_\xi(\alpha_j\alpha_5)$ and $A_\xi(i\alpha_j\alpha_5)$ vanish for all j. Thus, the eigenvalues of A_ξ are $0, 1/r, -1/r$, with respective multiplicities $8, 6, 6$.

We shall see that when $d = 5$, tubes over the submanifold $SO(2d)/U(d)$ are Hopf hypersurfaces with constant principal curvatures (see Theorem 8.26 on page 444). For $d \geq 6$, the analogous tubes are Hopf (of course) but do not have constant principal curvatures. When $d = 2$, we have $n + 1 = 2^{d-1} = 2$ so $n = 1$. Also, $m = 1$, so M is just $\mathbf{CP}^1$. Similarly, when $d = 3$, we have $n + 1 = 2^{3-1} = 4$ so that $n = 3$. Also, $m = 3$, so M is just $\mathbf{CP}^3$. Finally, when $d = 4$, we have $n + 1 = 2^{4-1} = 8$ so that $n = 7$. Also, $m = 6$. In this case, M is the complex quadric Q^6 in $\mathbf{CP}^7$.

Chapter 8
Hopf Hypersurfaces

In this chapter, we give the classification of Hopf hypersurfaces with constant principal curvatures due to Kimura [270] in $\mathbf{CP}^n$ (see Theorem 8.13) and Berndt [27] in $\mathbf{CH}^n$ (see Theorem 8.12). These classifications state that such a hypersurface is an open subset of a hypersurface on Takagi's list for $\mathbf{CP}^n$, and on Montiel's list for $\mathbf{CH}^n$.

We begin in Section 8.1 by proving a generalization of Cartan's formula to hypersurfaces of complex space forms due to Berndt [27]. Then in Section 8.2, we derive general formulas for the shape operators of parallel hypersurfaces and focal submanifolds of Hopf hypersurfaces with constant principal curvatures.

In Section 8.3, we prove Berndt's [27] classification of Hopf hypersurfaces with constant principal curvatures in $\mathbf{CH}^n$, which is based primarily on the Cartan's formula mentioned above. Kimura's [270] classification of Hopf hypersurfaces with constant principal curvatures in $\mathbf{CP}^n$ is proven in Section 8.4. This is significantly more complicated than the $\mathbf{CH}^n$ case, and it involves the analysis of the special complex submanifolds given in Sections 7.2–7.5.

In Section 8.5, we study several characterizations of the hypersurfaces on the lists of Takagi and Montiel based on conditions on their shape operators, curvature tensors or Ricci tensors. Complete proofs are given for many of the results and we have tried to demonstrate the typical kind of arguments used in the original papers. For example, we have presented Kon and Loo's classification of η-parallel hypersurfaces (Theorem 8.128) in detail. This involves extensive analysis under the non-Hopf assumption for which we introduce a formal framework at the beginning of the section. We have omitted similar lengthy arguments in discussing Theorems 8.98 and 8.112. In the latter two cases, the purpose is to prove that the given hypothesis implies that the hypersurface is Hopf, while in the former case one proves that it is either Hopf or ruled.

Section 8.6 deals with important examples of non-Hopf hypersurfaces in $\mathbf{CH}^n$ with special curvature properties. In Section 8.7, we discuss various generalizations of the notion of an isoparametric hypersurface to complex space forms. These

T.E. Cecil, P.J. Ryan, *Geometry of Hypersurfaces*, Springer Monographs in Mathematics, DOI 10.1007/978-1-4939-3246-7_8

include several conditions that are equivalent for hypersurfaces of real space forms, but are different for hypersurfaces of complex space forms. Finally, in Section 8.8 we list some open problems.

8.1 Cartan's Formula for Hopf Hypersurfaces

Let M be a submanifold of a Riemannian manifold $\tilde{M}$. Let v be a smooth section of the normal bundle NM and suppose that λ and μ are distinct eigenvalues of A_v with constant multiplicity and corresponding eigendistributions T_λ and T_μ. If $X \in T_\lambda$ and $Y \in T_\mu$, then one easily verifies that

$$\langle(\nabla_Z A_v)X, Y\rangle = (\lambda - \mu)\langle\nabla_Z X, Y\rangle$$

for all vectors Z tangent to M.

Now consider the special case where $\tilde{M}$ is a complex space form and M is a Hopf hypersurface. Observe that $\mathcal{W}^\perp$ is A-invariant and we denote the spectrum (i.e., set of eigenvalues) of A restricted to $\mathcal{W}^\perp$ by $\sigma(\mathcal{W}^\perp)$.

Lemma 8.1. *Let M be a Hopf hypersurface with constant principal curvatures in* $\mathbf{CP}^n$ *or* $\mathbf{CH}^n$. *For all λ, μ in $\sigma(\mathcal{W}^\perp)$, we have*

- $\nabla_X Y + \lambda\langle\varphi X, Y\rangle W \in T_\lambda$ *for all X, Y in T_λ*
- $\nabla_X Y \perp T_\lambda$ *if $X \in T_\lambda$, $Y \in T_\mu$, $\lambda \neq \mu$.*

Proof. It is easy to check that the first expression is orthogonal to W. Now take any $\mu \in \sigma(\mathcal{W}^\perp)$ with $\mu \neq \lambda$ and choose any $Z \in T_\mu$. By the Codazzi equation,

$$\begin{aligned} 0 &= \langle\big((\nabla_X A)Z - (\nabla_Z A)X\big), Y\rangle \\ &= (\mu - \lambda)\langle\nabla_X Z, Y\rangle - \langle\big(\nabla_Z(\lambda X) - A\nabla_Z X\big), Y\rangle \\ &= (\mu - \lambda)\langle\nabla_X Z, Y\rangle - (Z\lambda)\langle X, Y\rangle - \lambda\langle\nabla_Z X, Y\rangle + \lambda\langle\nabla_Z X, Y\rangle \\ &= (\lambda - \mu)\langle\nabla_X Y, Z\rangle - (Z\lambda)\langle X, Y\rangle. \end{aligned}$$

Since $Z\lambda = 0$ we have $\langle\nabla_X Y, Z\rangle = 0$ and thus, $\nabla_X Y + \lambda\langle\varphi X, Y\rangle W \in T_\lambda$. Note that the second assertion follows from the first since $\langle\nabla_X Z, Y\rangle = -\langle\nabla_X Y, Z\rangle$. □

From the first assertion in the lemma above, we immediately get the following corollary.

Corollary 8.2. *Under the hypothesis of Lemma 8.1, T_λ is integrable if and only if $\lambda = 0$ or $\varphi T_\lambda \subset T_\lambda^\perp$.*

The next lemma gives a formula similar to Cartan's formula for isoparametric hypersurfaces in real space forms (see Lemma 3.10 on page 93).

Lemma 8.3. *Let M be a Hopf hypersurface with constant principal curvatures in $\mathbf{CP}^n$ or $\mathbf{CH}^n$. Let $X \in \mathcal{W}^\perp$ be a unit principal vector at a point p with associated principal curvature λ. For any principal orthonormal basis $\{e_i\}_{i=1}^{2n-2}$ of $\mathcal{W}^\perp$ satisfying $Ae_i = \mu_i e_i$, we have*

$$\sum_{\substack{i=1 \\ \mu_i \neq \lambda}}^{2n-2} k_i \frac{\lambda \mu_i + c}{\lambda - \mu_i} = 0$$

where $k_i = 1 + 2\langle \varphi X, e_i \rangle^2$.

Proof. **Proof outline**
Let $Y \in \mathcal{W}^\perp$ be a second unit principal vector at p with corresponding principal curvature $\mu \neq \lambda$. Extend X and Y to be principal vector fields near p. Our proof will be broken down into several steps, as follows:

1. Using the Codazzi equation we show that

$$\langle (\nabla_{[X,Y]} A)X, Y \rangle = (\lambda - \mu)\big(\langle \nabla_X Y, \nabla_Y X \rangle + c\langle \varphi X, Y \rangle^2\big). \tag{8.1}$$

2. Using the Gauss equation, show that

$$\langle R(X, Y)Y, X \rangle = \lambda\mu + c(1 + 3\langle \varphi X, Y \rangle^2). \tag{8.2}$$

3. Using the definition of the curvature tensor, show that

$$\langle R(X, Y)Y, X \rangle = \langle \nabla_X Y, \nabla_Y X \rangle + \frac{1}{\lambda - \mu} \langle (\nabla_{[X,Y]} A)X, Y \rangle. \tag{8.3}$$

4. Using the Codazzi equation, show that for a unit principal vector $Z \in \mathcal{W}^\perp$ corresponding to a principal curvature ν not equal to λ or μ, we have

$$(\lambda - \nu)(\mu - \nu)\langle \nabla_X Y, Z \rangle \langle \nabla_Y X, Z \rangle = \langle (\nabla_Z A)X, Y \rangle^2. \tag{8.4}$$

5. Express $\langle \nabla_X Y, \nabla_Y X \rangle$ in terms of the orthonormal principal basis as follows:

$$\langle \nabla_X Y, \nabla_Y X \rangle = \sum_{\mu_i \neq \lambda, \mu} \langle \nabla_X Y, e_i \rangle \langle \nabla_Y X, e_i \rangle - \lambda\mu \langle \varphi X, Y \rangle^2. \tag{8.5}$$

6. Use the results of the previous steps to show that

$$2 \sum_{\mu_i \neq \lambda, \mu} \frac{\langle (\nabla_{e_i} A)X, Y \rangle^2}{(\lambda - \mu_i)(\mu - \mu_i)} = (\lambda\mu + c)(1 + 2\langle \varphi X, Y \rangle^2). \tag{8.6}$$

Now for any j with $\mu_j \neq \lambda$, we have (setting $Y = e_j$ in equation (8.6)),

$$\frac{\lambda\mu_j + c}{\lambda - \mu_j}(1 + 2\langle \varphi X, e_j\rangle^2) = 2 \sum_{\mu_i \neq \lambda, \mu_j} \frac{\langle (\nabla_{e_i} A)e_j, X\rangle^2}{(\lambda - \mu_i)(\lambda - \mu_j)(\mu_j - \mu_i)}. \tag{8.7}$$

Summing this over all j for which $\mu_j \neq \lambda$, we have

$$\sum_{\mu_j \neq \lambda} \frac{\lambda\mu_j + c}{\lambda - \mu_j}(1 + 2\langle \varphi X, e_j\rangle^2) = 2 \sum_{\substack{i,j \\ \mu_i \neq \mu_j \\ \mu_i, \mu_j \neq \lambda}} \frac{\langle (\nabla_{e_i} A)e_j, X\rangle^2}{(\lambda - \mu_i)(\lambda - \mu_j)(\mu_j - \mu_i)}. \tag{8.8}$$

Since the summand on the right side of equation (8.8) is skew-symmetric in $\{i,j\}$, the value of the sum is 0, and so the sum on the left is 0.

Proof details

1. First note that $\langle \nabla_X Y, W\rangle = -\langle Y, \nabla_X W\rangle = -\langle Y, \varphi AX\rangle = -\lambda\langle \varphi X, Y\rangle$. Similarly,

$$\langle \nabla_Y X, W\rangle = -\mu\langle \varphi Y, X\rangle = \mu\langle \varphi X, Y\rangle.$$

Thus, $\langle [X, Y], W\rangle = -(\lambda + \mu)\langle \varphi X, Y\rangle$. Next we compute

$$\begin{aligned}
\langle (\nabla_{[X,Y]} A)X, Y\rangle &= \langle (\nabla_X A)[X, Y], Y\rangle + c\langle [X, Y], W\rangle\langle \varphi X, Y\rangle \\
&= \langle [X, Y], (\nabla_X A)Y\rangle + c\langle [X, Y], W\rangle\langle \varphi X, Y\rangle \\
&= \langle [X, Y], (\nabla_Y A)X\rangle + c\langle [X, Y], 2\langle X, \varphi Y\rangle W\rangle \\
&\quad + c\langle [X, Y], W\rangle\langle \varphi X, Y\rangle \\
&= \langle [X, Y], (\nabla_Y A)X\rangle - c\langle [X, Y], W\rangle\langle \varphi X, Y\rangle \\
&= \langle [X, Y], (\nabla_Y A)X\rangle + c(\lambda + \mu)\langle \varphi X, Y\rangle^2.
\end{aligned} \tag{8.9}$$

But now,

$$\langle \nabla_X Y, (\nabla_Y A)X\rangle = \langle (\lambda I - A)\nabla_Y X, \nabla_X Y\rangle,$$

while

$$\begin{aligned}
\langle \nabla_Y X, (\nabla_Y A)X\rangle &= \langle \nabla_Y X, (\nabla_X A)Y\rangle + 2c\langle \varphi X, Y\rangle\langle \nabla_Y X, W\rangle \\
&= \langle (\mu I - A)\nabla_Y X, \nabla_X Y\rangle + 2c\mu\langle \varphi X, Y\rangle^2.
\end{aligned}$$

Thus

$$\langle [X, Y], (\nabla_Y A)X \rangle = (\lambda - \mu)\langle \nabla_X Y, \nabla_Y X \rangle - 2c\mu \langle \varphi X, Y \rangle^2. \tag{8.10}$$

On substituting in equation (8.9) we obtain equation (8.1) as desired.

2. This is immediate from the Gauss equation.
3. First note that $\langle \nabla_Y Y, X \rangle = 0$ by Lemma 8.1, and so

$$\langle \nabla_X \nabla_Y Y, X \rangle = -\langle \nabla_Y Y, \nabla_X X \rangle,$$

which vanishes, again by Lemma 8.1. Similarly, $\langle \nabla_X Y, X \rangle = 0$, so that

$$\langle \nabla_Y \nabla_X Y, X \rangle = -\langle \nabla_X Y, \nabla_Y X \rangle.$$

Finally, $\langle (\nabla_{[X,Y]} A)X, Y \rangle = (\lambda - \mu)\langle \nabla_{[X,Y]} X, Y \rangle$. We then compute

$$\langle \nabla_X \nabla_Y Y - \nabla_Y \nabla_X Y - \nabla_{[X,Y]})Y, X \rangle = \langle \nabla_X Y, \nabla_Y X \rangle + \frac{1}{\lambda - \mu}\langle (\nabla_{[X,Y]} A)X, Y \rangle,$$

which gives equation (8.3).

4. We compute

$$\langle (\nabla_Z A)X, Y \rangle = \langle (\nabla_X A)Z, Y \rangle = \langle Z, (\nabla_X A)Y \rangle = (\mu - \nu)\langle Z, \nabla_X Y \rangle.$$

The same calculation with X and Y interchanged gives

$$\langle (\nabla_Z A)X, Y \rangle = (\lambda - \nu)\langle Z, \nabla_Y X \rangle.$$

Multiplying these two equations together gives equation (8.4).

5. To establish equation (8.5), note that $-\lambda\mu \langle \varphi X, Y \rangle^2$ is just

$$\langle \nabla_X Y, W \rangle \langle \nabla_Y X, W \rangle.$$

Thus, we need only observe that the terms omitted from the full summation of the $\langle \nabla_X Y, e_i \rangle \langle \nabla_Y X, e_i \rangle$, (i.e., those i for which $\mu_i = \lambda$ or $\mu_i = \mu$) actually vanish and make no contribution to the sum. This is easily checked using Lemma 8.1.

6. Combine equations (8.1), (8.2), and (8.3) to get

$$2\langle \nabla_X Y, \nabla_Y X \rangle = \lambda\mu + c + 2c\langle \varphi X, Y \rangle^2.$$

Using this and the result of equation (8.4) in equation (8.5), we get equation (8.6).

□

Recalling Theorem 6.17 on page 357 and the accompanying Corollary 6.18, we can state the following lemma. In fact, it is true at each point, even without the assumption of constant principal curvatures.

Lemma 8.4. *Let M be a Hopf hypersurface with constant principal curvatures in* $\mathbf{CP}^n$ *or* $\mathbf{CH}^n$ *such that* $\alpha^2 + 4c \neq 0$*. Then for every* $\lambda \in \sigma(\mathcal{W}^\perp)$*, there is a unique* $\mu \in \sigma(\mathcal{W}^\perp)$ *such that the following equivalent conditions are satisfied:*

1. $\varphi T_\lambda = T_\mu$
2. $\lambda\mu = \frac{(\lambda+\mu)}{2}\,\alpha + c$
3. $(\lambda - \frac{\alpha}{2})(\mu - \frac{\alpha}{2}) = \frac{\alpha^2}{4} + c.$

In case $\mu = \lambda$*, we have* $\lambda^2 = \alpha\lambda + c$*, so there are at most two principal curvatures with this property.*

We now look at the possibility excluded above, namely $\alpha^2+4c = 0$. Suppose that $\lambda \neq \frac{\alpha}{2}$ is a principal curvature with a corresponding principal vector $X \in \mathcal{W}^\perp$. Then, as before, φX will be principal with principal curvature μ satisfying conditions (2) and (3) of Lemma 8.4. Thus $\mu = \frac{\alpha}{2}$ and we have at least one principal curvature with value equal to $\alpha/2$. Now apply Lemma 8.3 with $\lambda = \frac{\alpha}{2}$. Then

$$\lambda\mu_i + c = \frac{\alpha}{2}(\mu_i - \frac{\alpha}{2})$$

so that the summation in Lemma 8.3 reduces to

$$-(\sum k_i)\frac{\alpha}{2} = 0,$$

a contradiction. Thus, we have the proved the following theorem.

Theorem 8.5. *Let M be a Hopf hypersurface with constant principal curvatures in* $\mathbf{CH}^n$ *such that* $\alpha^2+4c = 0$*. Then* $\alpha/2$ *is a principal curvature of multiplicity* $2n-2$*.*

Theorem 8.6. *Let M be a Hopf hypersurface with constant principal curvatures in* $\mathbf{CP}^n$ *or* $\mathbf{CH}^n$ *such that* $\alpha^2 + 4c \neq 0$*. Then, for every* $\lambda \in \sigma(\mathcal{W}^\perp)$ *such that, in the sense of Lemma 8.4,* $\mu \neq \lambda$*, we have*

$$\sum_{\substack{\nu\in\sigma(\mathcal{W}^\perp)\\ \nu\neq\lambda}} m_\nu \frac{\lambda\nu + c}{\lambda - \nu} + 2\,\frac{\lambda\mu + c}{\lambda - \mu} = 0,$$

where m_ν *is the multiplicity of* ν *on* $\mathcal{W}^\perp$*. If* $\mu = \lambda$*, the extra* $\lambda - \mu$ *term does not occur, i.e.,*

$$\sum_{\substack{\nu\in\sigma(\mathcal{W}^\perp)\\ \nu\neq\lambda}} m_\nu \frac{\lambda\nu + c}{\lambda - \nu} = 0.$$

Proof. First consider the case where $\mu = \lambda$. Then each summand in Lemma 8.3 has $k_i = 1$. Collecting terms with the same value of μ_i yields

$$\sum_{\substack{\nu \in \sigma(\mathcal{W}^\perp) \\ \nu \neq \lambda}} m_\nu \frac{\lambda\nu + c}{\lambda - \nu} = 0.$$

Now suppose that μ is distinct from λ. We may choose the basis $\{e_i\}$ to contain X and φX. For $\mu_i \neq \lambda, \mu$, we still have $k_i = 1$. There are m_μ summands with $\mu_i = \mu$. One of these will have $k_i = 1 + 2\langle \varphi X, \varphi X \rangle = 3$; the rest will have $k_i = 1$. Thus we need to add the additional term

$$2\,\frac{\lambda\mu + c}{\lambda - \mu},$$

as shown, to the summation involving the multiplicities in order to agree with the formula from Lemma 8.3. □

Of course, Theorem 8.6 also holds (vacuously) when $\alpha^2 + 4c = 0$. The two formulas given in the theorem may be regarded as analogous to the famous "Cartan's Formula" for hypersurfaces with constant principal curvatures in real space forms (see Lemma 3.10 on page 93). One applies when $\lambda^2 = \alpha\lambda + c$ and the other when $\lambda^2 \neq \alpha\lambda + c$.

Corollary 8.7. *Let M be a Hopf hypersurface with constant principal curvatures in* $\mathbf{CH}^n$. *Then M has at most 3 distinct principal curvatures. In particular, if λ and ν are distinct elements of $\sigma(\mathcal{W}^\perp)$, then $\lambda\nu + c = 0$.*

Proof. Changing the sign of the unit normal ξ if necessary, we may assume that at least one element of $\sigma(\mathcal{W}^\perp)$ is positive. Arrange the positive elements of $\sigma(\mathcal{W}^\perp)$ in an ascending list $\{\lambda_i\}_{i=1}^k$.

We can choose an element λ so that there are no other elements between λ and $-c/\lambda$ as follows. If $\lambda_1^2 + c > 0$, choose $\lambda = \lambda_1$. If $\lambda_k^2 + c \leq 0$, choose $\lambda = \lambda_k$. Otherwise, there is a unique i such that $\lambda_{i-1}^2 + c \leq 0$, but $\lambda_i^2 + c > 0$. If

$$\lambda_{i-1} \leq -c/\lambda_{i-1} < \lambda_i,$$

choose $\lambda = \lambda_{i-1}$. If $\lambda_{i-1} \leq -c/\lambda_i \leq \lambda_i$, choose $\lambda = \lambda_i$. It is easy to check that the only other arrangement, $-c/\lambda_i < \lambda_{i-1} < \lambda_i \leq -c/\lambda_{i-1}$, yields a contradiction, and that the choice of λ has the required property.

Now one can check directly that for any $\nu \in \sigma(\mathcal{W}^\perp)$, $\nu \neq \lambda$, we have

$$\frac{\lambda\nu + c}{\lambda - \nu} \leq 0,$$

whether ν lies to the left or right of the range determined by λ and $-c/\lambda$. As a consequence, every term in the relevant equation from Theorem 8.6 vanishes. Thus, there can be at most two distinct elements λ and $-c/\lambda$ in $\sigma(\mathcal{W}^\perp)$. □

8.2 Parallel Hypersurfaces and Focal Submanifolds

In this section, we assume that M is a Hopf hypersurface with constant principal curvatures in $\mathbf{CP}^n$ or $\mathbf{CH}^n$. Let $\theta = (p, \xi)$. Since we will keep the same θ throughout, we abbreviate $(\gamma_\theta)_t$ as γ_t. Then along the geodesic γ_t, we have for all v tangent to $\tilde{M}$ at γ_t,

$$\tilde{R}(v, \overrightarrow{\gamma_t})\overrightarrow{\gamma_t} = c(v + 3\langle v, J\overrightarrow{\gamma_t}\rangle J\overrightarrow{\gamma_t} - \langle v, \overrightarrow{\gamma_t}\rangle\overrightarrow{\gamma_t}).$$

Take a principal vector $X \in \mathcal{W}^\perp$ with $AX = \lambda X$. Let B_t be the parallel vector field along γ_t satisfying $B_0 = X$. Also along γ_t, define

$$X_t = (\cos u - r\lambda \sin u)B_t \text{ for } \mathbf{CP}^n$$
$$X_t = (\cosh u - r\lambda \sinh u)B_t \text{ for } \mathbf{CH}^n, \tag{8.11}$$

where $u = t/r$. It is easy to check that $X_t'' = -cX_t = -\tilde{R}(X_t, \overrightarrow{\gamma_t})\overrightarrow{\gamma_t}$ so that X_t is a Jacobi field with initial conditions $(X, -\lambda X)$. Similarly, if B_t is constructed by parallel translation of W along γ_t, it turns out that $\tilde{R}(B_t, \overrightarrow{\gamma_t})\overrightarrow{\gamma_t} = 4cB_t$ so that we construct a Jacobi field by setting

$$W_t = (\cos 2u - \frac{r\alpha}{2} \sin 2u)B_t \text{ for } \mathbf{CP}^n,$$
$$W_t = (\cosh 2u - \frac{r\alpha}{2} \sinh 2u)B_t \text{ for } \mathbf{CH}^n, \tag{8.12}$$

where again $u = t/r$.

Note that these Jacobi fields span the tangent space to the tube M_t at γ_t. Because M is a hypersurface (codimension 1 in $\tilde{M}$), the second type of Jacobi field needed to span the tangent space in the general submanifold case does not occur.

It follows from the proof of Lemma 6.35 on page 383 that $\tilde{\nabla}_{X_t}\overrightarrow{\gamma_t} = X_t'$. Thus, the shape operator of the tube M_t (using $\overrightarrow{\gamma_t}$ as unit normal) satisfies $-AX_t = X_t'$. For each principal vector $X \in \mathcal{W}^\perp$ at $p \in M$ with corresponding principal curvature λ, X_t is a principal vector for M_t satisfying

$$AX_t = -X_t' = \frac{1}{r}\frac{\sin u + r\lambda \cos u}{\cos u - r\lambda \sin u}X_t = \frac{1}{r}\cot(\theta - u)X_t$$

in the $\mathbf{CP}^n$ case, where we have written λ as $\frac{1}{r}\cot\theta$. Similarly, in the $\mathbf{CH}^n$ case, we have

$$AX_t = -X_t' = \frac{1}{r}\frac{\sinh u - r\lambda \cosh u}{\cosh u - r\lambda \sinh u}X_t.$$

If $|\lambda| > 1$, we can write $\lambda = \frac{1}{r}\coth\theta$, and

$$AX_t = \frac{1}{r}\coth(\theta - u)X_t.$$

On the other hand, when $|\lambda| < 1$, we have

$$AX_t = \frac{1}{r}\tanh(\theta - u)X_t,$$

where $\lambda = \frac{1}{r}\tanh\theta$. Finally, if $\lambda = \pm\frac{1}{r}$, we have

$$AX_t = \pm\frac{1}{r}X_t.$$

The calculations are similar for the tangent direction arising from W. We have

$$\tilde{\nabla}_{W_t}\overrightarrow{\gamma_t} = W_t'$$

which yields

$$AW_t = \frac{2}{r}\cot 2(\theta - u)\; W_t.$$

for $\mathbf{CP}^n$, where $\alpha = \frac{2}{r}\cot 2\theta$, and

$$AW_t = \frac{2}{r}\coth 2(\theta - u)W_t$$

for $\mathbf{CH}^n$, where $\alpha = \frac{2}{r}\coth 2\theta$. In the remaining cases, $|\alpha| \le \frac{2}{r}$, we have

$$AW_t = \frac{2}{r}\tanh 2(\theta - u)W_t$$

with $\alpha = \frac{2}{r}\tanh 2\theta$, and

$$AW_t = \pm\frac{2}{r}W_t$$

with $\alpha = \pm\frac{2}{r}$.

Remark 8.8. Nothing in this section (so far) depends on the principal curvatures being constant.

Thus we have thus proved:

Theorem 8.9. *Let M be a Hopf hypersurface in $\mathbf{CP}^n$ or $\mathbf{CH}^n$. Then all parallel hypersurfaces M_t are also Hopf hypersurfaces whose Hopf principal curvature $\alpha(t)$ varies with t as shown in the previous paragraph. Further, the pointwise behavior of the other principal curvatures and principal spaces is preserved as indicated there.*

Corollary 8.10. *Let M be a Hopf hypersurface in $\mathbf{CP}^n$ or $\mathbf{CH}^n$ with constant principal curvatures. Then each parallel hypersurface M_t is also a Hopf hypersurface with constant principal curvatures whose values vary with t as indicated above.*

In the general case, focal points may occur at different values of t for different $p \in M$. However, for a given value of t, the B_t are linearly independent. Therefore, focal points will occur for precisely those values of t for which $r\lambda = \cot u$ or $r\alpha = 2\cot 2u$ (respectively, $r\lambda = \coth u$ or $r\alpha = 2\coth 2u$) in the $\mathbf{CP}^n$ and $\mathbf{CH}^n$ cases. Subject to certain genericity restrictions, we will locally have focal submanifolds whose principal spaces and principal curvatures can be determined by the same calculations.

When M has constant principal curvatures, these difficulties will not arise. Each M_t will either be a hypersurface or will consist entirely of focal points. In the latter case, M_t is a focal submanifold whose shape operator may be described by adapting the calculations we have done above.

Lemma 8.11. *Let M be a Hopf hypersurface in $\mathbf{CP}^n$ or $\mathbf{CH}^n$. Assume that $(\Pi \circ \Phi)(t)$ has constant rank ρ which is less than the dimension of M. Then M_t is a ρ-dimensional submanifold of $\tilde{M}$. Let $\phi_t(x) = \gamma_{(x,\,\xi_x)t}$ for $x \in M$, where ξ is the unit normal field. Then $\phi_t : M \to M_t$ is a submersion.*

- *For any $p \in M$ and any principal vector $X \in T_pM$ not in the kernel of $(\phi_t)_*$, the vector $\eta = \overrightarrow{\gamma_{(p,\,\xi_p)_t}}$ is a unit normal to the focal submanifold M_t and $A_\eta X_t = -X_t'$.*
- *For $q \in M_t$, $V = \phi_t^{-1}(q)$ is a ρ-dimensional submanifold of M whose tangent space at any point is a principal subspace coinciding with the kernel of $(\phi_t)_*$. The map*

$$\eta : V \to T_qM_t$$

defined by $\eta(x) = \overrightarrow{\gamma_{(x,\,\xi_x)_t}}$ satisfies $\eta_* X = X_t'$. In particular, η is a local diffeomorphism of V with the sphere of unit normals to M_t at q.

Shape operator of the focal submanifold

Assume now that M is a Hopf hypersurface and M_t is a focal submanifold. For a particular $p \in M$, and $q = \gamma_{(p,\,\xi_p)_t} \in M_t$, we have $\eta = \overrightarrow{\gamma_{(p,\xi_p)_t}}$, a unit normal to M_t at q. We distinguish two conditions that cause q to be a focal point:

1. The Jacobi field W_t arising from the Hopf principal curvature α vanishes at t.
2. W_t does not vanish, but for some principal vector $X \in \mathcal{W}^\perp$, the Jacobi field X_t vanishes at t.

Case 1 for $\mathbf{CP}^n$*:* $\frac{r\alpha}{2} = \cot 2u$

Let $\lambda_0 = \frac{1}{r}\cot u$. If $\lambda_0 \in \sigma(\mathcal{W}^\perp)$, then the codimension of M_t is one more than the multiplicity of λ_0 (as an eigenvalue of A restricted to $\mathcal{W}^\perp$). Otherwise, the codimension is 1. In either case, for each $\lambda \in \sigma(\mathcal{W}^\perp)$, where $\lambda \neq \lambda_0$, there is a corresponding eigenvalue $\tilde{\lambda}$ of A_η such that if $AX = \lambda X$, then $A_\eta X_t = \tilde{\lambda} X_t$. If $\lambda = \frac{1}{r}\cot\theta$, then $\tilde{\lambda} = \frac{1}{r}\cot(\theta - u)$.

Case 2 for $\mathbf{CP}^n$*:* $\frac{r\alpha}{2} \neq \cot 2u$

Again, let $\lambda_0 = \frac{1}{r}\cot u \in \sigma(\mathcal{W}^\perp)$. For every $\lambda \in \sigma(\mathcal{W}^\perp)$, where $\lambda \neq \lambda_0$, there is a corresponding eigenvalue $\tilde{\lambda}$ of A_η such that if $AX = \lambda X$, then $A_\eta X_t = \tilde{\lambda} X_t$. If $\lambda = \frac{1}{r}\cot\theta$, then $\tilde{\lambda} = \frac{1}{r}\cot(\theta - u)$. Further, if $\alpha = \frac{2}{r}\cot 2\theta$, then $A_\eta W_t = \frac{2}{r}\cot 2(\theta - u) W_t$.

Case 1 for $\mathbf{CH}^n$*:* $\frac{r\alpha}{2} = \coth 2u$

Let $\lambda_0 = \frac{1}{r}\coth u \in \sigma(\mathcal{W}^\perp)$. If $\lambda_0 \in \sigma(\mathcal{W}^\perp)$, then the codimension of M_t is one more than the multiplicity of λ_0 (as an eigenvalue of A restricted to $\mathcal{W}^\perp$). Otherwise, the codimension is 1. In either case, for each $\lambda \in \sigma(\mathcal{W}^\perp)$, where $\lambda \neq \lambda_0$, there is a corresponding eigenvalue $\tilde{\lambda}$ of A_η such that if $AX = \lambda X$, then $A_\eta X(t) = \tilde{\lambda} X_t$. Furthermore,

- if $\lambda = \frac{1}{r}\coth\theta$, then $\tilde{\lambda} = \frac{1}{r}\coth(\theta - u)$;
- if $\lambda = \frac{1}{r}\tanh\theta$, then $\tilde{\lambda} = \frac{1}{r}\tanh(\theta - u)$;
- if $\lambda = \pm\frac{1}{r}$, then $\tilde{\lambda} = \pm\frac{1}{r}$, respectively.

Case 2 for $\mathbf{CH}^n$*:* $\frac{r\alpha}{2} \neq \coth 2u$

Again, let $\lambda_0 = \frac{1}{r}\coth u \in \sigma(\mathcal{W}^\perp)$. For each $\lambda \in \sigma(\mathcal{W}^\perp)$, where $\lambda \neq \lambda_0$, there is a corresponding eigenvalue $\tilde{\lambda}$ of A_η such that if $AX = \lambda X$, then $A_\eta X_t = \tilde{\lambda} X_t$.

- If $\lambda = \frac{1}{r}\coth\theta$, then $\tilde{\lambda} = \frac{1}{r}\coth(\theta - u)$;
- If $\lambda = \frac{1}{r}\tanh\theta$, then $\tilde{\lambda} = \frac{1}{r}\tanh(\theta - u)$;

- If $\lambda = \pm\frac{1}{r}$, then $\tilde{\lambda} = \pm\frac{1}{r}$, respectively.

Also,

- if $\frac{r\alpha}{2} = \coth 2\theta$, then $A_\eta W_t = \frac{2}{r}\coth 2(\theta - u)W_t$;
- if $\frac{r\alpha}{2} = \tanh 2\theta$, then $A_\eta W_t = \frac{2}{r}\tanh 2(\theta - u)W_t$;
- if $\frac{r\alpha}{2} = \pm 1$, then $A_\eta W_t = \pm\frac{2}{r}W_t$, respectively.

8.3 Berndt's Classification in $\mathbf{CH}^n$

Berndt [27] classified the Hopf hypersurfaces with constant principal curvatures in $\mathbf{CH}^n$ as follows.

Theorem 8.12. *Let M be a Hopf hypersurface in $\mathbf{CH}^n$, where $n \geq 2$, having constant principal curvatures. Then M is an open subset of a hypersurface in Montiel's list, i.e., a hypersurface of Type A_0, Type A_1, Type A_2 or Type B. In particular, the number g of distinct principal curvatures is 2 or 3.*

Proof. Suppose first that $\alpha^2 + 4c \neq 0$. By Corollary 8.7, $\sigma(\mathcal{W}^\perp)$ has either 1 or 2 elements. Without loss of generality, we can assume that these elements are positive. If $\sigma(\mathcal{W}^\perp)$ consists of a single number λ, we may choose $u > 0$ such that either $r\lambda = \coth u$ or $r\lambda = \tanh u$. Because $\lambda^2 = \alpha\lambda + c$, we get $r\alpha = 2\coth 2u$ in either case. Thus M lies on a Type A_1 hypersurface. Otherwise, $\sigma(\mathcal{W}^\perp)$ has two elements which we may write as $\frac{1}{r}\coth u$ and $\frac{1}{r}\tanh u$. Substituting in the second equation of Lemma 8.4, we get $\alpha = \frac{2}{r}\tanh 2u$. Note that $\frac{1}{r}\coth u$ and $\frac{1}{r}\tanh u$ are principal curvatures of multiplicity $n-1$ whose principal spaces are interchanged by φ. As we have seen in Section 6.8, M lies on a tube over the totally real focal submanifold $\mathbf{RH}^n$.

If $\alpha^2 + 4c = 0$, we have already derived complete information on the shape operator (Theorem 8.5, page 426). There are no focal points and M lies on a horosphere. For a verification of this last fact, we refer to Berndt's argument [27, pp. 140–141]. □

8.4 Kimura's Classification in $\mathbf{CP}^n$

Kimura [270] classified the Hopf hypersurfaces with constant principal curvatures in $\mathbf{CP}^n$ as follows.

Theorem 8.13. *Let M be a Hopf hypersurface in $\mathbf{CP}^n$, where $n \geq 2$, having constant principal curvatures. Then M is an open subset of a homogeneous hypersurface (i.e., a member of Takagi's list). In particular, the number g of distinct principal curvatures is 2, 3, or 5.*

Proof. Let $\bar{M} = \pi^{-1}M$ be the lift of M to the sphere $S^{2n+1}(r)$ with unit normal ξ^L and shape operator $\bar{A}$. Then

$$\bar{A}\,\frac{iz}{r} = \frac{1}{r}\,W^L$$
$$\bar{A}W^L = \alpha W^L + \frac{1}{r}\,\frac{iz}{r}$$
$$\bar{A}X^L = \lambda X^L \tag{8.13}$$

for each principal vector $X \in \mathcal{W}^\perp$, where λ is the corresponding principal curvature.

Write $\alpha = \frac{2}{r}\cot 2\theta$ where $0 < \theta < \frac{\pi}{2}$. The eigenvalues of the matrix

$$\begin{bmatrix} 0 & \frac{1}{r} \\ \frac{1}{r} & \alpha \end{bmatrix}$$

are $\frac{1}{r}\cot\theta$ and $\frac{1}{r}\cot\,(\theta + \frac{\pi}{2}) = -\frac{1}{r}\tan\theta$. Now $\bar{M}$ is an open subset of an isoparametric hypersurface of $S^{2n+1}(r)$ having $\bar{g} = 1, 2, 3, 4$, or 6 distinct principal curvatures, according to Münzner's result (Theorem 3.49 on page 136).

Since we have already found two distinct principal curvatures for $\bar{M}$, we know that $\bar{g} \neq 1$. Also, if $\bar{g}$ were equal to 3, then according to Theorem 3.26 on page 108, the three distinct principal curvatures could be written $\frac{1}{r}\cot\phi$, $\frac{1}{r}\cot(\phi + \frac{\pi}{3})$, and $\frac{1}{r}\cot(\phi + \frac{2\pi}{3})$, where $0 < \phi < \frac{\pi}{3}$. Since the cotangent function is strictly decreasing on $[0, \pi]$, it is impossible for any two of these three numbers to be of the form $\cot\,\theta$ and $\cot\,(\theta + \frac{\pi}{2})$. Thus $\bar{g}$ cannot be equal to 3.

Again, according to Münzner [382] and Abresch [2] (see Remark 3.51, page 136), $\bar{g} = 6$ implies that $n = 3$ or $n = 6$. By Theorem 3.26 on page 108, there is a number $\phi \in (0, \frac{\pi}{6})$ such that the 6 distinct principal curvatures of $\bar{M}$ are given by

$$\frac{1}{r}\,\cot(\phi + \frac{k\pi}{6}), \quad 0 \leq k \leq 5.$$

Then there are three possibilities: $\theta = \phi$, $\theta = \phi + \frac{\pi}{6}$, and $\theta = \phi + \frac{\pi}{3}$, and no matter which of these possibilities holds, we can choose notation so that the principal curvatures of $\bar{M}$ are

$$\lambda_1 = \frac{1}{r}\cot\theta, \quad \lambda_2 = \frac{1}{r}\cot(\theta + \frac{\pi}{6}), \quad \lambda_3 = \frac{1}{r}\cot(\theta + \frac{\pi}{3}),$$
$$\lambda_4 = \frac{1}{r}\cot(\theta + \frac{\pi}{2}), \quad \lambda_5 = \frac{1}{r}\cot(\theta + \frac{2\pi}{3}), \quad \lambda_6 = \frac{1}{r}\cot(\theta + \frac{5\pi}{6}). \tag{8.14}$$

Note that λ_1 and λ_4 are the eigenvalues of the matrix

$$\begin{bmatrix} 0 & \frac{1}{r} \\ \frac{1}{r} & \alpha \end{bmatrix}.$$

If $\bar{g} = 6$ and $n = 6$, the principal curvatures are $(\alpha, \lambda_1, \lambda_2, \lambda_3, \lambda_4, \lambda_5, \lambda_6)$ with respective multiplicities $(1, 1, 2, 2, 1, 2, 2)$. By Lemma 8.4, the principal spaces of λ_1 and λ_4 are φ-invariant, since both of these principal curvatures satisfy $\lambda^2 = \alpha\lambda + c$. This contradicts the fact that each of these principal curvatures has multiplicity 1. We conclude that n cannot be 6.

If $\bar{g} = 6$ and $n = 3$, the principal curvatures of M are $(\alpha, \lambda_2, \lambda_3, \lambda_5, \lambda_6)$, all of multiplicity 1. We look at the focal set. First note that each parallel hypersurface M_u has the same principal curvature configuration as M, substituting $\theta - u$ for θ in the expressions for α and the various λ_i in equation (8.14). Consider the focal submanifold M_u where $u = \theta + \frac{\pi}{6}$. The eigenvalues of the shape operator A_η are obtained by substituting $\theta - u = -\frac{\pi}{6}$ in the formulas for M_u, omitting, of course, λ_2. This gives the eigenvalues as $(\alpha, \lambda_3, \lambda_5, \lambda_6) = \frac{1}{r\sqrt{3}}(-2, 3, 0, -1)$, each of multiplicity 1. Since this configuration of eigenvalues is not invariant under multiplication by -1, we have a contradiction to the fact that A_η and $A_{-\eta}$ have the same eigenvalues (see Lemma 6.27 on page 374). We conclude that n cannot be equal to 3. Thus the possibility that $\bar{g} = 6$ has been eliminated.

We have established that $\bar{g}$ is 2 or 4. If $\bar{g} = 2$, then we can write the two distinct principal curvatures of $\bar{M}$ as $\lambda_1 = \frac{1}{r}\cot\theta$ of multiplicity m_1 and $\lambda_2 = \frac{1}{r}\cot(\theta + \frac{\pi}{2})$ of multiplicity m_2. Then $\alpha = \frac{2}{r}\cot 2\theta$ is a principal curvature of M of multiplicity 1. If one of the multiplicities (say m_2) is equal to 1, then $g = 2$ and the only other principal curvature for M is λ_1 with multiplicity $m_1 - 1 = 2n - 2$. If both multiplicities are greater than 1, then $g = 3$ and both λ_1 and λ_2 are principal curvatures of respective multiplicities $m_1 - 1$ and $m_2 - 1$. Both principal subspaces of $\mathcal{W}^\perp$ are φ-invariant.

Next we deal with the case $\bar{g} = 4$. By a similar argument to that used when $\bar{g} = 6$, we can write the four distinct principal curvatures of $\bar{M}$ as

$$\lambda_1 = \frac{1}{r}\cot\theta, \quad \lambda_2 = \frac{1}{r}\cot(\theta + \frac{\pi}{4}),$$
$$\lambda_3 = \frac{1}{r}\cot(\theta + \frac{\pi}{2}), \quad \lambda_4 = \frac{1}{r}\cot(\theta + \frac{3\pi}{4}), \tag{8.15}$$

with respective multiplicities (m_1, m_2, m_1, m_2). If $m_1 = 1$, then $g = 3$ and the principal curvatures of M are $(\alpha, \lambda_2, \lambda_4)$ with respective multiplicities given by $(1, n-1, n-1)$. Otherwise, $g = 5$ and the principal curvatures are $(\alpha, \lambda_1, \lambda_2, \lambda_3, \lambda_4)$ with multiplicities $(1, m_1 - 1, m_2, m_1 - 1, m_2)$. When $g = 2$ or $g = 3$, M is an open subset of a Type A or Type B hypersurface from Takagi's list. □

The case $g = 5$

When $g = 5$, we need a significant amount of additional work to describe the focal set (see the following several pages), and this will complete the proof.

First note that the principal curvatures of the parallel hypersurfaces M_u are given by substituting $\theta - u$ for θ in equation (8.15) and in the equation $\alpha = \frac{2}{r}\cot 2\theta$. When $g = 5$ and $u = \theta$, M_u is a (complex, minimal) focal submanifold whose shape operator (in all directions, see Lemma 6.27, page 374) has eigenvalues

$$\lambda_2 = \frac{1}{r}\cot\frac{\pi}{4} = \frac{1}{r}, \quad \lambda_4 = \frac{1}{r}\cot\frac{3\pi}{4} = -\frac{1}{r},$$

each of multiplicity m_2, and $\lambda_3 = \frac{1}{r}\cot\frac{\pi}{2} = 0$, of multiplicity $m_1 - 1$. Since $m_1 + m_2 = n$, the focal submanifold M_θ therefore has dimension $m_1 - 1 + 2m_2$.

Principal curvatures for the g = 5 case

These hypersurfaces are tubes over their focal sets which are complex submanifolds. We know enough about the shape operators of the focal submanifolds to calculate the principal curvatures of the tubes. Making use of the results of Subsection 6.7.1, we can say that for a tube of radius ru,

- the Hopf principal curvature $\alpha = \frac{2}{r}\cot 2u$;
- there is a principal curvature $\lambda_2 = -\frac{1}{r}\cot(\frac{\pi}{4} - u)$ of multiplicity m_2;
- there is a principal curvature $\lambda_4 = -\frac{1}{r}\cot(\frac{3\pi}{4} - u)$ of multiplicity m_2;
- there is a principal curvature $\lambda_3 = -\frac{1}{r}\cot(\frac{\pi}{2} - u)$ of multiplicity $m_1 - 1$;
- since the codimension of the focal submanifold is $m_1 + 1$, the tube construction gives a principal curvature $\frac{1}{r}\cot u$ of multiplicity $m_1 - 1$. We may therefore write $\lambda_1 = -\frac{1}{r}\cot(-u)$ as our final principal curvature to fit with the notation of the other λ_i.

We will see from the construction of the possible focal submanifolds in Chapter 7 that the multiplicities are as follows:

- Type C: $m_1 - 1 = n - 3$, $m_2 = 2$;
- Type D: $m_1 - 1 = 4$, $m_2 = 4$;
- Type E: $m_1 - 1 = 8$, $m_2 = 6$;

Note that the five (distinct, constant) principal curvatures are the same for all three types. Also, the values of the principal curvature $\{\lambda_1, \lambda_3\}$ are those occurring in the Type A case while $\{\lambda_2, \lambda_4\}$ are principal curvatures for Type B hypersurfaces, when considered as a tubes over the complex quadric. Thus, we can also state

Proposition 8.14. *The principal curvatures of the hypersurfaces in* $\mathbf{CP}^n$ *of types C, D, and E have the following properties*

- $\lambda_1\lambda_3 + c = 0$; $\lambda_2\lambda_4 + c = 0$;
- $\lambda_1 + \lambda_3 = \alpha$; $(\lambda_2 + \lambda_4)\alpha + 4c = 0$;
- *the principal spaces of* λ_1 *and* λ_3 *are* φ*-invariant;*
- *the principal spaces of* λ_2 *and* λ_4 *are interchanged by* φ*.*

Finally, we remark that hypersurfaces of types C, D, or E cannot be pseudo-Einstein. We can see this from the fact that the pseudo-Einstein condition would require that (by analysis similar to that done in Section 6.5) $\alpha = \lambda_1 + \lambda_3 = \mathbf{m} = \lambda_2 + \lambda_4$ so that $\alpha^2 = (\lambda_2 + \lambda_4)\alpha = -4c$, which is a contradiction.

Certain Kähler submanifolds

We now study a class of Kähler submanifolds that includes the focal submanifold M_θ. We first define a (positive semi-definite) inner product on the space of normal vectors at any point by $G(u, v) = \operatorname{trace} A_u A_v$. Note that the identities $G(Ju, Jv) = G(u, v)$ and $G(u, Ju) = 0$ hold since $A_{Ju} = JA_u = -A_u J$. The focal submanifold M_θ arising from the $g = 5$ case satisfies the hypothesis of the following slightly more general lemma.

Lemma 8.15. *Let M be a Kähler submanifold of $\mathbf{CP}^n$ or $\mathbf{CH}^n$ with the property that there is a nonzero constant λ such that $A_\zeta^3 = \lambda^2 A_\zeta$ for every unit normal ζ. Let ξ be a particular unit normal such that A_ξ is nonzero and has a nontrivial nullspace T_0. If η is any unit normal such that*

$$\operatorname{trace} A_\xi A_\eta = \operatorname{trace} A_{J\xi} A_\eta = 0,$$

then $A_\eta T_0 \subset T_0^\perp$ and $A_\eta T_0^\perp \subset T_0$.

Proof. Let X be a unit eigenvector of A_ξ and write $A_\xi X = aX$. Then

$$A_\xi^3 X = a^3 X = \lambda^2 a X,$$

so that $a(a^2 - \lambda^2) = 0$ and $a \in \{0, \lambda, -\lambda\}$. Now $A_\xi J = -JA_\xi$ so $A_\xi JX = -aJX$. Thus, for each eigenspace T_a, we have $T_{-a} = JT_a$. Let

$$\ell = \dim T_\lambda = \dim T_{-\lambda},$$

so that T_0 has dimension $2(m - \ell)$, where m is the complex dimension of M.

We first show that $A_\eta T_0 \subset T_0^\perp$. First note that the statement is trivially true if $A_\eta = 0$. Thus, we may assume that $G(\eta, \eta) > 0$. For $X \in T_0$, we write $A_\eta X$ as the sum $u + v + w$ of vectors in T_λ, $T_{-\lambda}$, T_0, respectively. We need to show that $w = 0$.

First observe that $\xi + \eta$ is nonzero, otherwise $G(\xi, \eta) = 0$ would be violated. Denoting $|\xi + \eta|$ by κ, we have

$$A_{\xi+\eta}^3 X = \kappa^3 A_{\frac{\xi+\eta}{\kappa}}^3 X = \kappa^3 \lambda^2 A_{\frac{\xi+\eta}{\kappa}} X = \kappa^2 \lambda^2 A_{\xi+\eta} X = \lambda^2 \kappa^2 A_\eta X. \tag{8.16}$$

On the other hand,

$$\begin{aligned} A_{\xi+\eta}^3 X &= (A_\xi + A_\eta)^3 X = A_\xi^2 (u + v + w) + (A_\xi A_\eta + A_\eta A_\xi) A_\eta X + \lambda^2 A_\eta X \\ &= \lambda^2 (u + v) + \lambda^2 (u + v + w) + (A_\xi A_\eta + A_\eta A_\xi) A_\eta X. \end{aligned} \tag{8.17}$$

Comparing the two expressions for $A^3_{\xi+\eta}X$, we have

$$\lambda^2|\xi + \eta|^2(u + v + w) = \lambda^2(2u + 2v + w) + (A_\xi A_\eta + A_\eta A_\xi)A_\eta X. \tag{8.18}$$

If we replace η by $-\eta$, we get

$$-\lambda^2|\xi - \eta|^2(u + v + w) = -\lambda^2(2u + 2v + w) + (A_\xi A_\eta + A_\eta A_\xi)A_\eta X, \tag{8.19}$$

so that $(|\xi+\eta|^2+|\xi-\eta|^2)(u+v+w) = 2(2u+2v+w)$. Since $|\xi+\eta|^2+|\xi-\eta|^2 = 4$, we get $4w = 2w$, i.e., $w = 0$ as required. We have shown that $A_\eta T_0 \subset T_0^\perp$.

We now choose $X \in T_\lambda$ and again write $A_\eta X = u + v + w$. Following similar steps as before, we get

$$A^3_{\xi+\eta}X = \kappa^2\lambda^2 A_{\xi+\eta}X = \lambda^2\kappa^2(\lambda X + A_\eta X) = \lambda^2\kappa^2(\lambda X + (u + v + w)), \tag{8.20}$$

$$\begin{aligned}(A_\xi + A_\eta)^3 X = \lambda^3 X + \lambda^2(u + v) + \lambda A_\xi(u + v + w) + (A_\xi A_\eta + A_\eta A_\xi)A_\eta X \\ +\lambda^2(u + v + w) + \lambda A^2_\eta X + \lambda^2(u + v + w).\end{aligned} \tag{8.21}$$

Thus,

$$\begin{aligned}\lambda^2|\xi + \eta|^2(\lambda X + (u + v + w)) = \lambda^3 X + \lambda^2(u + v) + \lambda^2(u - v) \\ + \lambda A^2_\eta X + (A_\xi A_\eta + A_\eta A_\xi)A_\eta X + 2\lambda^2(u + v + w).\end{aligned} \tag{8.22}$$

so that

$$\begin{aligned}(|\xi + \eta|^2 - |\xi - \eta|^2)\lambda^3 X + \lambda^2(|\xi + \eta|^2 + |\xi - \eta|^2)(u + v + w) \\ = 2\lambda^2(u + v) + 2\lambda^2(u - v) + 4\lambda^2(u + v + w).\end{aligned} \tag{8.23}$$

Upon simplification, this yields

$$u = \nu\lambda X, \tag{8.24}$$

where $\nu = \langle\xi, \eta\rangle$. Similarly, if we take $X \in T_{-\lambda}$, and write $A_\eta X = u+v+w$, we get

$$v = -\nu\lambda X. \tag{8.25}$$

Let $\{e_i\}_{i=1}^{\ell}$ be an orthonormal basis for T_λ. Since $G(\xi, \eta) = 0$, we have

$$\begin{aligned}0 &= \sum\langle A_\eta A_\xi e_i, e_i\rangle + \sum\langle A_\eta A_\xi Je_i, Je_i\rangle \\ &= \lambda\Big(\sum(\langle A_\eta e_i, e_i\rangle - \langle A_\eta Je_i, Je_i\rangle)\Big) \\ &= \lambda\ell(\nu\lambda - (-\nu\lambda)) = 2\lambda^2\ell\nu,\end{aligned} \tag{8.26}$$

from which we get $\nu = 0$. Thus $A_\eta T_\lambda \subset T_0 + T_{-\lambda}$ and $A_\eta T_{-\lambda} \subset T_0 + T_\lambda$. Since $G(\xi, J\eta) = 0$, (this was part of the original hypothesis), we may apply the same argument to $J\eta$ to obtain $A_{J\eta} T_\lambda \subset T_0 + T_{-\lambda}$ and $A_{J\eta} T_{-\lambda} \subset T_0 + T_\lambda$. However, $A_{J\eta} = JA_\eta$, T_0 is J-invariant, and J interchanges T_λ and $T_{-\lambda}$. Therefore, $A_\eta T_\lambda \subset (T_0 + T_\lambda) \cap (T_0 + T_{-\lambda}) = T_0$. Similarly, $A_\eta T_{-\lambda} \subset T_0$. We conclude that $A_\eta T_0^\perp \subset T_0$. □

Corollary 8.16. *Under the conditions of Lemma 8.15,* $(A_\eta A_\xi)^2 = 0$.

Proof. Clearly, $(A_\eta A_\xi)^2 X = 0$ for any $X \in T_0$. Now consider $X \in T_0^\perp$. Then $Y = A_\xi X \in T_0^\perp$. Thus $A_\eta Y \in T_0$ so that $A_\xi A_\eta Y = 0$. Thus

$$(A_\eta A_\xi)^2 X = A_\eta (A_\xi A_\eta A_\xi X) = A_\eta (A_\xi A_\eta Y) = 0.$$

This completes the proof. □

Lemma 8.17. *Let M be a Kähler submanifold of* $\mathbf{CP}^n$ *or* $\mathbf{CH}^n$ *with the property that there is a nonzero constant λ such that $A_\zeta^3 = \lambda^2 A_\zeta$ for every unit normal ζ. If ξ and η are orthogonal unit normals (i.e., $\langle \xi, \eta \rangle = 0$), then they are orthogonal with respect to the inner product G.*

Proof. Let γ_t be a curve in the sphere of unit normals at a particular point p such that $\gamma_0 = \xi$ and $\overrightarrow{\gamma_0} = \eta$. For definiteness, set $\gamma_t = \cos t\ \xi + \sin t\ \eta$ so that we can write explicitly $A_{\gamma_t} = \cos t\ A_\xi + \sin t\ A_\eta$. From our hypothesis, trace $A_{\gamma_t}^2$ is a constant function of t (equal to $2\ell\lambda^2$ in the notation of Lemma 8.15). Thus

$$\begin{aligned} 0 &= \left.\frac{d}{dt}\right|_{t=0} \text{trace } A_{\gamma_t}^2 \qquad (8.27)\\ &= \left.\frac{d}{dt}\right|_{t=0} \text{trace } (\cos t\ A_\xi + \sin t\ A_\eta)^2 \\ &= \text{trace } (A_\xi A_\eta + A_\eta A_\xi) = 2 \text{ trace } (A_\xi A_\eta) = 2\, G(\xi, \eta). \end{aligned}$$

□

Lemma 8.18. *Under the conditions of Lemma 8.17, we have*

$$2 \text{ trace } (A_\xi^2 A_\eta^2) = \text{trace } A_\xi^4.$$

Proof. Apply the same technique as in Lemma 8.17, computing the second derivative of $A_{\gamma_t}^4$ at $t = 0$. This yields

$$\begin{aligned} \text{trace } A_\xi^4 &= \text{trace } A_\xi^2 A_\eta^2 + \text{trace } (A_\xi A_\eta)^2 + \text{trace } (A_\eta A_\xi^2 A_\eta) \\ &= 2 \text{ trace } A_\xi^2 A_\eta^2 + \text{trace } (A_\xi A_\eta)^2. \qquad (8.28) \end{aligned}$$

In view of Corollary 8.16, this completes the proof. □

Focal submanifolds for $g = 5$

Theorem 7.1 simplifies further when the hypotheses of Lemmas 8.15 and 8.17 are satisfied. Specifically, we have the following.

Proposition 8.19. *Let M be a Kähler submanifold of a complex space form $\tilde{M}$ of constant holomorphic curvature $4c$. Assume that there is a nonzero constant λ such that $A_\zeta^3 = \lambda^2 A_\zeta$ for every unit normal ζ. If $\{\xi_\alpha\}_{\alpha=1}^{2p}$ is an orthonormal basis for the normal space with corresponding shape operators $\{A_\alpha\}$, then*

$$\frac{1}{2}\Delta|\sigma|^2 = |\nabla'\sigma|^2 + 2(m+2)c|\sigma|^2 - \Sigma(\text{ trace } A_\alpha^2)^2 - 2 \text{ trace } (\Sigma A_\alpha^2)^2.$$

Here m is the complex dimension of M.

Proof. This is a direct consequence of Lemma 8.17. □

Theorem 8.20. *Let M be a Kähler submanifold of a complex space form $\tilde{M}$ ($\mathbf{CP}^n$ or $\mathbf{CH}^n$) of complex dimension m. Suppose M has the property that there is a nonzero constant λ such that the eigenvalues of A_ζ are λ, $-\lambda$ and 0 for every unit normal ζ at every point. Then*

1. *$\tilde{M} = \mathbf{CP}^n$, i.e., no such submanifold of $\mathbf{CH}^n$ exists;*
2. *M is a parallel submanifold of $\tilde{M}$, i.e., $\nabla'\sigma = 0$;*
3. *$\lambda = \pm\frac{1}{r}$ and $\ell = 1 + 2m - n$, where ℓ is the multiplicity of λ.*

Since $1 \leq \ell \leq m-1$, we have $m + 2 \leq n \leq 2m$.

Proof. We first remark that if $\mathbf{CH}^n$ admitted such a submanifold, then almost all tubes over it would be Hopf hypersurfaces with 4 or 5 distinct constant principal curvatures (as computed from equations (6.37) and (6.38)), violating Theorem 8.12. Further, even in $\mathbf{CP}^n$, such submanifolds cannot occur with complex codimension $p = 1$ since, according to the corresponding calculation (6.36) for tubes in $\mathbf{CP}^n$, they would lead to Hopf hypersurfaces with 4 distinct constant principal curvatures. Equation (6.36) combined with the material earlier in this section on $g = 5$ gives $\lambda = \pm\frac{1}{r}\cot\frac{\pi}{4} = \pm\frac{1}{r}$ so that $\lambda^2 = \frac{1}{r^2} = c$. Now the hypotheses of our theorem imply that $A_\zeta^3 = \lambda^2 A_\zeta$. Thus the simplified version of Theorem 7.1 (i.e., Proposition 8.19) applies. For substitution in the Simons' type formula, we compute

$$|\sigma|^2 = 4\ell\lambda^2 p,$$

and

$$\Sigma(\text{ trace } A_\alpha^2)^2 = 8p\ell^2\lambda^4.$$

Also,

$$2 \text{ trace } (\Sigma A_\alpha^2)^2 = 8p(p+1)\ell\lambda^4.$$

This last assertion requires some justification. First, we simplify by noting that $(JA_\alpha)^2 = A_\alpha^2$, so that in computing ΣA_α^2, we may restrict our summation to $1 \leq \alpha \leq p$ and multiply the result by 2. We now compute the square of this sum. Note that β also runs from 1 to p but omits the $\beta = \alpha$ term.

$$(\Sigma A_\alpha^2)^2 = \Sigma A_\alpha^4 + \Sigma A_\alpha^2 A_\beta^2 \tag{8.29}$$

$$\begin{aligned} \text{trace } (\Sigma A_\alpha^2)^2 &= \text{trace } \Sigma A_\alpha^4 + \Sigma \text{ trace } A_\alpha^2 A_\beta^2 \\ &= \text{trace } \Sigma A_\alpha^4 + \frac{p-1}{2} \Sigma \text{ trace } A_\alpha^4 \\ &= \frac{p+1}{2} \text{ trace } \Sigma A_\alpha^4 = \frac{p+1}{2} 2p\ell\lambda^4 = p(p+1)\ell\lambda^4, \end{aligned} \tag{8.30}$$

where we have used Lemma 8.18 to evaluate the β summation.

Noting that $\Delta|\sigma|^2 = 0$, we get $|\nabla'\sigma|^2 = 8\ell pc^2(\ell + p + 1 - m - 2)$. On the other hand, recalling the tube construction (see equation (6.36) and the discussion before Proposition 8.14), we see that tubes over M have two (constant) principal curvatures of multiplicity ℓ, one of multiplicity $2(m-\ell)$, one of multiplicity $2(p-1)$ in addition to the Hopf principal curvature. Thus $p - 1$ and $m - \ell$ must be equal from which it follows that $\ell + p + 1 - m - 2 = 0$ and $\nabla'\sigma = 0$, thus M is a parallel submanifold of $\tilde{M}$. Finally, since $m + p = n$, we have $\ell = m + 1 - p = 1 + m - (n - m) = 1 + 2m - n$ as required. □

Note that in terms of the corresponding isoparametric hypersurfaces in the sphere, $2(p - 1) = 2(m - \ell) = m_1 - 1$ and $\ell = m_2$ which is consistent with the fact that $m_1 + m_2 = n$.

Refining Theorem 8.20

We know that a Hopf hypersurface in $\mathbf{CP}^n$ with $g = 5$ distinct constant principal curvatures has a focal submanifold satisfying Theorem 8.20. In particular this focal submanifold is parallel. Parallel submanifolds of $\mathbf{CP}^n$ have been classified by Nakagawa and Takagi [391] (see also [33, p. 260]). Our parallel submanifolds will be open subsets of elements of their list (below) of complete parallel submanifolds. In order to complete the proof of Theorem 8.13, we need to determine which of these submanifolds actually satisfy the hypothesis of Theorem 8.20.

Nakagawa and Takagi's list of parallel submanifolds of $\mathbf{CP}^n$

- $\mathbf{CP}^m$ as a totally geodesic projective subspace;
- a complex quadric Q^m in a totally geodesic $\mathbf{CP}^{m+1}$ where $m+1 \leq n$;
- the Veronese embedding of $\mathbf{CP}^m$;
- the Segre embedding of $\mathbf{CP}^h \times \mathbf{CP}^k$;
- the Plücker embedding of the Grassmannian $G_{2,3}(\mathbf{C})$;
- the half-spin embedding of $SO(10)/U(5)$;
- the first canonical embedding of $E_6/(T \times Spin(10))$.

We have previously discussed most of these submanifolds. Now we will see how they fit in with Theorem 8.20.

The first two possibilities, the totally geodesic $\mathbf{CP}^m$ and the complex quadric Q^{n-1} have already been discussed and their tubes have g equal to 2 or 3. The higher codimension quadrics do not satisfy the conditions of Theorem 8.20 since there will be normal directions for which the shape operator vanishes. For $\mathbf{CP}^m$, all shape operators are zero while all shape operators of $Q^{n-1} \subset \mathbf{CP}^n$ have only two distinct eigenvalues (see Smyth [487]). Further, $Q^m \subset \mathbf{CP}^{m+1}$, where $m+1 < n$, does not qualify since a unit normal to $\mathbf{CP}^{m+1}$ at a point of Q^m will also be a unit normal to Q^m and therefore will have vanishing shape operator. None of these submanifolds satisfy Theorem 8.20.

The Veronese embedding

Theorem 8.21. *The Veronese embedding of* $\mathbf{CP}^m$ *does not satisfy the conditions of Theorem 8.20. This holds for all* $m \geq 1$.

Proof. The Ricci tensor of a complex space form of complex dimension k and constant holomorphic curvature 4κ is $2(k+1)\kappa I$. We look at the Gauss equation (7.11). First, the trace of the map $\{X \mapsto R(X,Y)Z - \tilde{R}(X,Y)Z\}$ is $-(m+1)c^2\langle Y, Z\rangle$ in our case (since the respective holomorphic curvatures are $2c^2$ and $4c^2$). However, by the Gauss equation, this must be the same as the trace of the map from X to

$$A_{\sigma(Y,Z)}X - A_{\sigma(X,Z)}Y.$$

Using a convenient orthonormal basis $\{\xi_\alpha\}$ for the normal space, we see that the displayed equation is the sum of the $2p$ terms

$$\langle A_\alpha Y, Z\rangle A_\alpha X - \langle A_\alpha X, Z\rangle A_\alpha Y,$$

where A_α denotes the shape operator corresponding to ξ_α. The trace of the map from X to this α-th term is $-\langle A_\alpha^2 Y, Z\rangle$, since A_α has zero trace. Thus

$$\Sigma A_\alpha^2 = (m+1)c^2 I.$$

Now, in the case of Theorem 8.20, the trace of each term on the left side is $2\ell\lambda^2$ so that the trace of the left side is $4p\ell\lambda^2$. The trace of the right side is $2m(m+1)c^2$. Also from Theorem 8.20, $\lambda^2 = c^2$ in our case. Finally, we have $p = n - m$ and $\ell = 2m - n + 1$. Since

$$n = \frac{m^2 + 3m}{2},$$

we get

$$2m(m+1) = 4(n-m)(2m-n+1) = 2m(m+1)(2m-n+1),$$

which gives $n = 2m$ and hence $m = 1$. However, this implies $\ell = m$, which contradicts the final assertion in Theorem 8.20. □

The Segre embedding

Although the case $h = k = 1$ does not qualify because it is a quadric Q^2, it can also be ruled out by the inequalities occurring in the proof of the following theorem which eliminates many possibilities for (h, k), based only on dimensional considerations.

Theorem 8.22. *The Segre embedding of* $\mathbf{CP}^h \times \mathbf{CP}^k$*, where* $h \leq k$*, does not satisfy the conditions of Theorem 8.20 for* $h \geq 3$ *or for* $h = 2$*,* $k \geq 3$*. (Eventually we will eliminate the possibility that* $h = k = 2$*, as well.)*

Proof. From Theorem 8.20, we have

$$h + k + 2 \leq h + k + hk \leq 2(h+k),$$

which simplifies to

$$2 \leq hk \leq h + k.$$

This shows that $k \geq 2$ and $k(h-1) \leq h \leq k$ from which we see that $h \leq 2$. If $h = 2$ however, we get $2k \leq 2 + k$ in which case $k = 2$ is the only possibility. □

If $h = k = 2$, then $m = 4$ and $n = 8$ so that $\ell = 1$. This contradicts equation (7.58) which implies that $\ell = 4$.

On the other hand, if $h = 1$, we have, in the notation of Theorem 8.20, $n = 2m - 1$ and $\ell = 1 + 2m - n = 2$. Recall that all shape operators have rank 2ℓ.

Thus, we can state the following:

Theorem 8.23. *The Segre embedding of* $\mathbf{CP}^h \times \mathbf{CP}^k$ *into* $\mathbf{CP}^n$*, where* $h \leq k$*, satisfies the conditions of Theorem 8.20 if and only if* $h = 1$ *and* $k \geq 2$*. Note that* $n = 2k + 1 = 2m - 1$.

Proof. Our previous discussion identifies those choices of (h, k) that cannot occur. In addition, our calculation shows that $\mathbf{CP}^1 \times \mathbf{CP}^k$ does, in fact, satisfy the conditions of Theorem 8.20 when $n = 2k + 1 \geq 5$. From the results of Subsection 6.7.1, tubes over such Segre embeddings are Hopf hypersurfaces with 5 distinct principal curvatures having multiplicities $(1,\ n-3,\ 2,\ n-3,\ 2)$. These are the Type *C* hypersurfaces, see [399, pp. 261–262]. □

The Plücker embedding

We first prove two lemmas that, taken together, restrict the values of h and k for which $G_{h,k}(\mathbf{C})$ can satisfy the conditions of Theorem 8.20.

Lemma 8.24. *For positive integers* h, k*, let* $m = hk$ *and* $n = \binom{h+k}{h} - 1$*. If* $n \geq 2m$*, then increasing* h *or* k *will make* $n > 2m$.

Proof. Suppose that

$$\frac{(h+k)!}{h!\,k!} \geq 1 + 2hk. \tag{8.31}$$

Then

$$\frac{(h+k+1)!}{(h+1)!\,k!} = \frac{h+1+k}{h+1}\,\frac{(h+k)!}{h!k!} \geq \frac{h+1+k}{h+1}\,(1+2hk). \tag{8.32}$$

It is routine to verify that

$$\frac{h+1+k}{h+1}\,(1+2hk) > 1 + 2(h+1)k \tag{8.33}$$

if and only if $2h(k-1) > 1$. Since $k > 1$, because of equation (8.31), we have the desired inequality. □

Based on the results of this lemma, we make the following four observations. Here we assume that $h \leq k$.

1. If $h = 1$, then $m = k$ and $n = k$. This violates $m + 2 \leq n$.
2. If $h = 2$ and $k = 5$, then $m = 10$ and $n = 20$ so that $n = 2m$. Thus every choice with $h = 2$ and $k \geq 6$ will violate $n \leq 2m$.
3. If $h = k = 3$, then $m = 9$ and $n = 19$ so that $n > 2m$. Thus if $h \geq 3$ and $k \geq 3$, we will have a violation of $n \leq 2m$.
4. If $h = k = 2$, then $m = 4$ and $n = 5$. This violates $m + 2 \leq n$.

Thus we have the following.

Lemma 8.25. *For positive integers $h \leq k$, let $m = hk$ and $n = \binom{h+k}{h} - 1$. Then the inequality $m + 2 \leq n \leq 2m$ requires that $\{h, k\}$ be $\{2, 3\}$, $\{2, 4\}$ or $\{2, 5\}$.*

The Plücker embedding satisfies the conditions of Theorem 8.20 if and only if $k = 3$.

Theorem 8.26. *Let M be a Kähler submanifold of $\mathbf{CP}^n$ of complex dimension m. Suppose M has the property that that the eigenvalues of A_ζ are $\frac{1}{r}$, $-\frac{1}{r}$ and 0 for every unit normal ζ at every point. Then M is congruent to an open subset of one of the following submanifolds*

1. $\mathbf{CP}^1 \times CP^{m-1}$ *(Segre embedding) with $2m = n + 1 \geq 6$.*
2. $SU(5)/S(U(2) \times U(3))$, $(G_{2,3}(\mathbf{C})$, *the Plücker embedding) so that $m = 6$ and $n = 9$.*
3. $SO(10)/U(5)$ *(half-spin embedding), so that $m = 10$ and $n = 15$.*

Proof. The first three items in Nakagawa and Takagi's list have been ruled out by the discussion immediately following the list. The Segre and Plücker embeddings have been dealt with in Theorems 8.23 and 7.11, respectively. The shape operator for $SO(10)/U(5)$ is computed in Section 7.5 and satisfies Theorem 8.20.

Finally, we consider $E_6/(Spin(10) \times T^1)$. Since E_6 has dimension 78 and $Spin(10) \times T^1$ has dimension 46, we get $2m = 32$ and hence $m = 16$. We also have $n = 26$. If Theorem 8.20 is to be satisfied, we must have $\ell = 7$. However, this requires the existence of an isoparametric hypersurface in $S^{2n+1}(r)$ with $\bar{g} = 4$, $m_1 = 19$ and $m_2 = 7$. As we can see from Section 3.9, no such hypersurface exists. □

This completes the proof of Kimura's [270] classification of Hopf hypersurfaces in $\mathbf{CP}^n$ (Theorem 8.13).

8.5 Characterization Theorems

The Hopf hypersurfaces with constant principal curvatures can be classified in terms of the properties of different geometric structures: shape operators, curvature tensors, Ricci tensors, and combinations thereof. Since these properties typically do not include the Hopf assumption, the first step (and sometimes the major step) in a classification theorem is often a proof that the condition being examined implies the Hopf condition. This requires us to do a significant amount of analysis under the non-Hopf assumption. As a by-product, we sometimes discover new interesting non-Hopf examples, especially in the case $n = 2$. We begin this section by introducing appropriate tools for this study.

8.5.1 Framework for studying non-Hopf hypersurfaces

Our study of Hopf hypersurfaces is facilitated by the breakdown of the tangent space into $\mathcal{W}$ and $\mathcal{W}^{\perp}$ components. We now introduce a standard setup for the non-Hopf case. Suppose that $\beta = |AW - \alpha W|$ is nonvanishing on a hypersurface M. Let U be the unit vector field satisfying $AW - \alpha W = \beta U$. Then $(W, U, \varphi U)$ is an orthonormal triple. In case $n = 2$, we call this the "standard non-Hopf frame," and there are functions λ, μ and ν such that the matrix of A with respect to this frame is

$$\begin{bmatrix} \alpha & \beta & 0 \\ \beta & \lambda & \mu \\ 0 & \mu & \nu \end{bmatrix}. \tag{8.34}$$

When $n \geq 3$, we define a *standard non-Hopf frame* to be an orthonormal frame whose first three elements are W, U and φU. In this case, the upper-left 3×3 submatrix of the matrix of A is of the form (8.34). The remaining items in the first column are zero.

When using this "standard non-Hopf" setup, we reserve the symbols U, β, λ, μ and ν for the purposes indicated.

Let $\mathcal{H}$ be the smallest A-invariant subspace of the tangent space containing the structure vector W. If $\mathcal{H}$ is an integrable distribution of dimension k, we say that the hypersurface is *k-Hopf*. Of course, Hopf and 1-Hopf are synonymous. If M is k-Hopf for some $k \geq 2$, then the matrix in equation (8.34) will be simplified. Another special class are the ruled hypersurfaces. A hypersurface M^{2n-1} in $\mathbf{CP}^n$ or $\mathbf{CH}^n$ is said to be *ruled* if its shape operator satisfies $A\mathcal{W}^{\perp} \subset \mathcal{W}$.

An aside on ruled hypersurfaces

Our definition is in terms of the shape operator and this serves our purposes best. However, a few remarks about the terminology are in order. We begin with a proposition.

Proposition 8.27. *For a real hypersurface M^{2n-1} in $\mathbf{CP}^n$ or $\mathbf{CH}^n$, the following are equivalent:*

- *M is ruled;*
- *The holomorphic distribution $\mathcal{W}^{\perp}$ is integrable and its leaves are totally geodesic.*

Proof. Let X and Y be vector fields in $\mathcal{W}^{\perp}$. Then

$$\begin{aligned} \langle [X, Y], W \rangle &= \langle \nabla_X Y, W \rangle - \langle \nabla_Y X, W \rangle \\ &= -\langle Y, \varphi AX \rangle + \langle X, \varphi AY \rangle. \end{aligned}$$

Suppose that $A\mathcal{W}^{\perp} \subseteq \mathcal{W}$. Then $\varphi AX = \varphi AY = 0$, so that $[X, Y] \in \mathcal{W}^{\perp}$. Furthermore, on any leaf L of $W^{\perp}$, the tangential component $(\nabla_X Y)^0$ of $\widetilde{\nabla}_X Y$ satisfies

$$\begin{aligned}\widetilde{\nabla}_X Y &= \nabla_X Y + \langle AX, Y\rangle\xi \\ &= (\nabla_X Y)^0 + \langle \nabla_X Y, W\rangle W + \langle AX, Y\rangle\xi \\ &= (\nabla_X Y)^0 - \langle Y, \varphi AX\rangle W + \langle AX, Y\rangle\xi \\ &= (\nabla_X Y)^0. \end{aligned} \tag{8.35}$$

Thus L is totally geodesic.

Conversely, let us assume that $\mathcal{W}^{\perp}$ is integrable with totally geodesic leaves. At any point p of M, let X and Y be vectors in $\mathcal{W}_p^{\perp}$ and extend them to local vector fields tangent to the leaf L through p. Then equation (8.35) holds locally. Because L is totally geodesic, we have $-\langle Y, \varphi AX\rangle W + \langle AX, Y\rangle\xi = 0$. In particular, $\langle Y, \varphi AX\rangle = 0$ at p. This shows that $A\mathcal{W}^{\perp} \subseteq \mathcal{W}$. □

This proposition means that M is foliated by leaves that are open subsets of $\mathbf{CP}^{n-1}$ (resp. $\mathbf{CH}^{n-1}$). These are the "rulings." In the $\mathbf{CH}^n$ case, it is possible for all rulings to be complete copies of $\mathbf{CH}^{n-1}$ and for M to be a complete hypersurface. However, this cannot happen in $\mathbf{CP}^n$.

Locally, ruled hypersurfaces can be constructed from a regular curve x_t as the union over t of the totally geodesic copies of $\mathbf{CP}^{n-1}$ (resp. $\mathbf{CH}^{n-1}$) which pass through x_t and are orthogonal at x_t to the holomorphic 2-plane spanned by $\{\overrightarrow{x_t}, J\overrightarrow{x_t}\}$. This is how they were first introduced by Kimura [271]. Of course, one may have to restrict to a small neighborhood of the curve to avoid singularities.

Ruled hypersurfaces cannot be Hopf (see Remark 8.44). In many of our classification theorems, the condition in question will imply that the hypersurface is either Hopf or ruled. Although we have many results characterizing specific Hopf hypersurfaces, the class of ruled hypersurfaces is quite broad and the problem of identifying and characterizing the "simplest" ruled hypersurfaces is yet unresolved.

In the standard non-Hopf setup, the only nonzero elements of the matrix of the shape operator of a ruled hypersurface are β and possibly α. In addition, a ruled hypersurface can have points where $\beta = 0$ and there may be no standard non-Hopf frame in a neighborhood of such points.

One simple type of ruled hypersurfaces are the *bisectors*. They play a significant role in construction of Dirichlet fundamental domains (see Goldman [176]). Just as tubes are the simplest hypersurfaces from our point of view, for the questions in which Goldman is primarily interested, the bisectors are the simplest hypersurfaces, providing a substitute for totally geodesic hypersurfaces which cannot occur in $\mathbf{CP}^n$ or $\mathbf{CH}^n$. Gorodski and Gusevskii [179] have also studied bisectors in the context of complete minimal hypersurfaces.

In our brief discussion of bisectors, we limit ourselves to the $\mathbf{CH}^n$ case. For any two points P_+ and P_-, the bisector they determine is the set of points equidistant from P_+ and P_-. This generalizes, for example, the "perpendicular bisector" of a line segment in the Euclidean plane. In real space forms, bisectors are totally geodesic hypersurfaces.

Returning to the $\mathbf{CH}^n$ case, let $P = x_0$ be the midpoint of the geodesic segment x_t joining P_+ and P_- and let γ_t be the geodesic through P in the direction $J\overrightarrow{x_0}$. For each t, let M_t be the unique $\mathbf{CH}^{n-1}$ through γ_t spanned by the orthogonal complement of the span of $\{\overrightarrow{\gamma_t}, J\overrightarrow{\gamma_t}\}$. Then the bisector is the union of these M_t. The geodesic γ_t is called the *spine* of the bisector.

For a point not on the spine, the standard non-Hopf setup holds and the upper-left 3×3 portion of the shape operator matrix is

$$\begin{bmatrix} 0 & \beta & 0 \\ \beta & 0 & 0 \\ 0 & 0 & 0 \end{bmatrix} \tag{8.36}$$

where β is essentially the distance from the spine. For points on the spine, the shape operator vanishes identically. The principal curvatures cannot be defined smoothly on any open set containing a point of the spine.

A final example of ruled hypersurfaces in $\mathbf{CH}^n$ is the minimal homogeneous Lohnherr hypersurface discussed in Section 8.6. This is also a good candidate for the designation "simplest" as it has constant principal curvatures – two nonzero principal curvatures of multiplicity 1 and a zero principal curvature of multiplicity $2n - 3$.

Although ruled hypersurfaces were not discussed in [399], a brief list of properties occurs there under the "Additional Topics" heading. These should be understood to be typical or generic for ruled hypersurfaces, but as our present discussion shows, there are exceptions. In particular, the Lohnherr hypersurface violates property (iii) in the list and points on the spine of a bisector violate properties (ii) and (iv).

Identities for the standard non-Hopf setup

Lemma 8.28. *In the standard non-Hopf setup, we have the following identities.*

1. $\langle(\nabla_X A)W, W\rangle = \langle(\nabla_W A)X, W\rangle = \langle(\nabla_W A)W, X\rangle$;
2. $(\nabla_W A)W = \operatorname{grad}\alpha + 2\beta A\varphi U$;
3. $\varphi\nabla_X(\beta\varphi U) = \beta\langle X, A\varphi U\rangle W - (\nabla_X A)W - (A - \alpha)\varphi AX + (X\alpha)W$;
4. $\nabla_X(\beta\varphi U) + \langle A^2 W, X\rangle W = \varphi(\nabla_X A)W + (\varphi A)^2 X + \alpha AX$;
5. $\nabla_W(\beta\varphi U) + (\alpha^2 + \beta^2)W = 3\beta\varphi A\varphi U + \alpha AW + \varphi \operatorname{grad}\alpha$.

Proof. 1. This is an easy consequence of the Codazzi equation and the symmetry of $\nabla_W A$.

2. Since $\beta U = AW - \alpha W$, we have

$$\beta\nabla_X U + (X\beta)U = (\nabla_X A)W + A\varphi AX - (X\alpha)W - \alpha\varphi AX. \tag{8.37}$$

Thus

$$\begin{aligned}
\beta\langle\nabla_X U, W\rangle &= \langle(\nabla_X A)W, W\rangle + \langle(A-\alpha)W, \varphi AX\rangle - X\alpha \\
-\beta\langle U, \varphi AX\rangle &= \langle(\nabla_X A)W, W\rangle + \beta\langle U, \varphi AX\rangle - X\alpha \\
\beta\langle A\varphi U, X\rangle &= \langle(\nabla_W A)W, X\rangle - \beta\langle A\varphi U, X\rangle - X\alpha
\end{aligned} \tag{8.38}$$

from which the result is immediate.

3.

$$\begin{aligned}
\varphi\nabla_X(\beta\varphi U) &= \varphi\nabla_X(\varphi AW) \\
&= \varphi(\nabla_X\varphi)AW + \varphi^2\nabla_X(AW) \\
&= \varphi(\langle AW, W\rangle AX - \langle AX, AW\rangle W) - \nabla_X(AW) \\
&\quad +\langle\nabla_X(AW), W\rangle W \\
&= \alpha\varphi AX - ((\nabla_X A)W + A\varphi AX) + (X\alpha)W - \langle AW, \varphi AX\rangle W \\
&= (\alpha - A)\varphi AX - (\nabla_X A)W + (X\alpha)W + \langle A\varphi AW, X\rangle W.
\end{aligned} \tag{8.39}$$

The result is now clear since $\varphi AW = \beta\varphi U$.

4. If we apply φ to the left side of (8.39), we get

$$-\nabla_X(\beta\varphi U) + \langle\nabla_X(\beta\varphi U), W\rangle W.$$

This second term is equal to $-\langle\beta\varphi U, \varphi AX\rangle W = -\beta\langle AU, X\rangle W$, since $\varphi^2 U = -U$. On the other hand, applying φ to the right side yields two nonzero terms, $-\varphi(\nabla_X A)W$ and $-\varphi(A-\alpha)\varphi AX = -(\varphi A)^2 X + \alpha\varphi^2 AX$. In order to reach our conclusion, we need only note that

$$\begin{aligned}
\alpha\varphi^2 AX &= -\alpha AX + \alpha\langle AX, W\rangle W \\
&= -\alpha AX + \langle A^2 W, X\rangle W - \beta\langle AU, X\rangle W.
\end{aligned}$$

5. We compute

$$\begin{aligned}
\varphi(\nabla_W A)W + (\varphi A)^2 W &= \varphi\,\mathrm{grad}\,\alpha + 2\beta\varphi A\varphi U + \varphi A\beta\varphi U \\
&= \varphi\,\mathrm{grad}\,\alpha + 3\beta\varphi A\varphi U.
\end{aligned}$$

Since $\langle A^2 W, W\rangle = \langle AW, AW\rangle = \alpha^2 + \beta^2$, this completes the proof. □

8.5.2 *Classifications involving shape operators*

Restrictions on the number of principal curvatures

We first discuss classification in terms of the number g of distinct principal curvatures. We have seen that totally umbilic hypersurfaces are impossible (see Theorem 6.6 on page 349). Thus every hypersurface has at least one point (and hence an open set) where $g \geq 2$. The hypersurfaces with $g \leq 2$ have been classified as follows:

Theorem 8.29. *Let M^{2n-1}, where $n \geq 3$, be a real hypersurface in $\mathbf{CP}^n$ or $\mathbf{CH}^n$ with $g \leq 2$ distinct principal curvatures at each point. Then M is a Hopf hypersurface with constant principal curvatures. Specifically, M is an open subset of a Type A_0 or Type A_1 hypersurface, or a Type B hypersurface in $\mathbf{CH}^n$ with radius $\frac{r}{2}\log(2+\sqrt{3})$.*

As a first step in proving this theorem, we show that M is Hopf. Since non-Hopf hypersurfaces with $g = 2$ do occur in $\mathbf{CP}^2$ and $\mathbf{CH}^2$ (see Remark 8.31), we phrase our lemma as follows:

Lemma 8.30. *Let M^{2n-1}, where $n \geq 2$, be a real hypersurface in $\mathbf{CP}^n$ or $\mathbf{CH}^n$ for which the standard non-Hopf setup holds. If $g \leq 2$ everywhere, then $n = 2$.*

Proof. Since M cannot consist entirely of umbilic points, we work in an open set where there are two distinct principal curvatures κ_1 and κ_2 with corresponding principal spaces T_1 and T_2. For convenience, we arrange that κ_1 has larger (or equal) multiplicity compared with κ_2. Then there are unique nonvanishing vector fields $X_1 \in T_1$ and $X_2 \in T_2$ such that

$$W = X_1 + X_2 \tag{8.40}$$

and so

$$AW = \kappa_1 X_1 + \kappa_2 X_2. \tag{8.41}$$

Note that $\{X_1, X_2\}$, $\{AW, W\}$, and $\{W, U\}$ have the same span, which is 2-dimensional, and thus this span is A-invariant. In (8.34), we must have $\mu = 0$. Now for any Y_1 and Z_1 in $X_1^{\perp} \cap T_1$, the Codazzi equation gives

$$(Y_1\kappa_1)Z_1 - (Z_1\kappa_1)Y_1 + (\kappa_1 - A)[Y_1, Z_1] = 2c\langle Y_1, \varphi Z_1\rangle W. \tag{8.42}$$

This shows that $\langle Y_1, \varphi Z_1\rangle = 0$. Otherwise, taking the inner product with X_1 would yield an immediate contradiction. Since the terms on the left side are in T_1 and T_2 respectively, each equals zero.

If $n \geq 3$, we may arrange that Y_1 and Z_1 are linearly independent, and thus conclude that $V\kappa_1 = 0$ for all $V \in X_1^{\perp} \cap T_1$. Now, the Codazzi equation for X_1 and Y_1 reduces to

$$(X_1\kappa_1)Y_1 + (\kappa_1 - A)[X_1, Y_1] = c(\langle X_1, W\rangle \varphi Y_1 + 2\langle X_1, \varphi Y_1\rangle W). \tag{8.43}$$

Taking the inner product with X_1 gives $3\langle X_1, \varphi Y_1\rangle = 0$. Thus φY_1 is not only orthogonal to $X_1^\perp \cap T_1$, but also to X_1. In addition,

$$\langle \varphi Y_1, X_2\rangle = \langle \varphi Y_1, W\rangle - \langle \varphi Y_1, X_1\rangle = 0,$$

and hence φY_1 lies in $X_2^\perp \cap T_2$.

Finally, φ maps $X_1^\perp \cap T_1$ into $X_2^\perp \cap T_2$ injectively since both spaces lie in $\mathcal{W}^\perp$. We conclude that the dimension of $X_2^\perp \cap T_2$ is at least 2. Thus, we can repeat the argument, reversing the roles of T_1 and T_2 and conclude that they must have the same dimension. This is a contradiction since M has odd dimension $2n - 1$.

Having shown that the assumption $n \geq 3$ leads to a contradiction, we conclude that $n = 2$. □

To complete the proof of Theorem 8.29, we need only show that if a Hopf hypersurface has $g \leq 2$ distinct principal curvatures, these principal curvatures must be constant. Of course, the constant α is a principal curvature. If $\lambda \neq \frac{\alpha}{2}$ is an eigenvalue of the restriction of A to $\mathcal{W}^\perp$ at some point, then we have, in the notation of Theorem 6.17,

$$\lambda\mu = \frac{\lambda + \mu}{2}\alpha + c. \tag{8.44}$$

If λ and μ are distinct then, extending by continuity, one of them, say λ, coincides with α in a neighborhood $\mathcal{U}$. Then, in $\mathcal{U}$, we have

$$\alpha\mu = \frac{\alpha + \mu}{2}\alpha + c. \tag{8.45}$$

Thus $\alpha \neq 0$ and $\mu = (\alpha^2 + 2c)/\alpha$ is also constant. By Theorems 8.12 and 8.13, $\mathcal{U}$ is an open subset of one of the hypersurfaces on Montiel's or Takagi's lists. The only one with the correct multiplicities (see Section 6.4) is the Type B hypersurface in $\mathbf{CH}^n$ for which $\coth u = 2 \tanh 2u$, i.e., u has the value indicated in the theorem statement. By continuity of the principal curvatures, the set of points of M for which they agree with the (constant) values that they have on $\mathcal{U}$ is closed (as well as being open). Since connectedness is assumed, this set is all of M. We conclude that M is indeed an open subset of the indicated Type B hypersurface.

On the other hand, if $\lambda = \mu$ on an open set, then $\lambda^2 = \alpha\lambda + c$ and λ is constant by the quadratic formula. In this case, M is an open subset of a Type A_1 hypersurface. The final possibility is that α and $\frac{\alpha}{2}$ are the only principal curvatures. In this case, M is an open subset of a Type A_0 hypersurface in $\mathbf{CH}^n$. This completes our proof of Theorem 8.29.

Remark 8.31. The condition that $n \geq 3$ is essential for Theorem 8.29. When $n = 2$ there is, in addition to the expected Hopf examples, a 2-parameter family of non-Hopf hypersurfaces with $g = 2$ principal curvatures and these principal curvatures are nonconstant. The proof of Lemma 8.30 (in case $n = 2$) shows that φU is a principal vector corresponding to the principal curvature ν. Also, since the span of $\{W, U\}$ is not a principal space, ν must coincide with one of the principal curvatures κ_1 or κ_2. This means that

$$\nu^2 - (\alpha + \lambda)\nu + (\lambda\alpha - \beta^2) = 0. \tag{8.46}$$

This observation is the starting point for the recent classification of non-Hopf hypersurfaces with $g = 2$ in $\mathbf{CP}^2$ and $\mathbf{CH}^2$ by Ivey and Ryan [225]. The construction and proof make extensive use of exterior differential systems and the new examples occur as solutions to a system of ordinary differential equations. This classification has also been carried out independently by Díaz-Ramos, Domínguez-Vázquez and Vidal-Castiñeira [132] using the notion of polar actions.

The classification problem for hypersurfaces with $g = 3$ principal curvatures is open. For Hopf hypersurfaces, we have the following result due to Böning [46].

Theorem 8.32. *Let M^{2n-1}, where $n \geq 3$, be a Hopf hypersurface in $\mathbf{CP}^n$ or $\mathbf{CH}^n$ with $g \leq 3$ distinct principal curvatures at each point. Then M has constant principal curvatures, provided that $\alpha^2 + 4c \neq 0$. Specifically, M is an open subset of a hypersurface of Type A or Type B.*

Proof. Since Theorem 8.29 covers the case $g \leq 2$, we will work in an open set where $g = 3$. Using the notation of Theorem 6.17, $\mathcal{W}^\perp$ is the direct sum of even-dimensional φ-invariant subspaces, each determined by two (possibly equal) principal curvatures λ and μ satisfying

$$\lambda\mu = \frac{\lambda + \mu}{2}\alpha + c. \tag{8.47}$$

In our case, the number of such subspaces cannot be 3 since this would require three distinct solutions to the quadratic equation $t^2 - \alpha t - c = 0$. Suppose now that there is a point where there are two such subspaces. Then, at least one of them must correspond to an "equal pair". The other must also have the equal-pair property, since otherwise, one of the principal curvatures, say μ, would have to coincide with α and using (8.47), we could express λ in terms of α. This situation would hold locally and all three distinct principal curvatures would be locally constant. However, none of the examples in the Takagi-Montiel list have this particular principal curvature configuration. The upshot is that both subspaces are principal spaces corresponding to locally constant principal curvatures, as occurs for the Type A_2 hypersurfaces.

The remaining possibility is that (at every point) there is only one such subspace and it is the sum of two principal spaces T_λ and T_μ, each of dimension $n-1$. Again, this situation holds locally. Since $n \geq 3$, we may apply the Codazzi equation to an orthonormal pair $\{X, Y\}$ of principal vectors in T_λ to obtain

$$(X\lambda)Y - (Y\lambda)X + (\lambda - A)[X, Y] = 2c\langle X, \varphi Y\rangle W. \tag{8.48}$$

Taking the inner product with Y yields $X\lambda = 0$. In other words, $Z\lambda = 0$ for all $Z \in T_\lambda$. Similarly, $Z\mu = 0$ for all $Z \in T_\mu$. On the other hand, differentiating (8.47) with respect to $Z \in T_\mu$ gives $(\mu - \frac{\alpha}{2})Z\lambda = 0$ so that $Z\lambda = 0$ (and similarly $Z\mu = 0$) for all $Z \in \mathcal{W}^\perp$. Finally, for a unit vector $Z \in T_\lambda$, we apply the Codazzi equation to the pair $\{Z, W\}$ to get

$$(\alpha - A)\nabla_Z W - (W\lambda)Z - (\lambda - A)\nabla_W Z + c\varphi Z = 0. \tag{8.49}$$

Since the first term reduces to a scalar multiple of φZ, taking the inner product with Z yields $W\lambda = 0$. Similarly, $W\mu = 0$ and thus λ and μ are constants. Thus the principal curvature configuration is that of the Type B hypersurfaces.

Having shown that the principal curvatures are locally constant, the usual connectedness argument completes our proof. □

Remark 8.33. The existence of Hopf hypersurfaces in $\mathbf{CH}^n$ with three distinct principal curvatures, not all constant, remains open. One can show that if such a hypersurface exists, then two of the three principal curvatures must be constants equal to $\alpha = 2/r$ and $\frac{\alpha}{2} = 1/r$.

Remark 8.34. Again, the specification $n \geq 3$ is necessary. There are examples of Hopf hypersurfaces in $\mathbf{CP}^2$ with $g = 3$ distinct principal curvatures $(0, \lambda, \mu)$ where λ and μ are nonconstant. In fact, these examples are pseudo-Einstein (see Kim and Ryan [260]).

On the other hand, we can ask about non-Hopf hypersurfaces with $g = 3$ principal curvatures, where the principal curvatures are assumed to be constant. It turns out that there exists a family of non-Hopf homogeneous hypersurfaces in $\mathbf{CH}^2$ (and, in fact, in $\mathbf{CH}^n$ for $n \geq 2$), having $g = 3$ distinct constant principal curvatures. These are the orbits of the Berndt group or *Berndt orbits*, see Berndt [30] and Kim et al. [257, 261]. We will discuss them and their generalizations in Section 8.6.

However, the situation is simple for $g = 2$.

Theorem 8.35. *Let M^{2n-1}, where $n \geq 2$, be a real hypersurface in $\mathbf{CP}^n$ or $\mathbf{CH}^n$ with $g = 2$ distinct principal curvatures at each point. Assume that these principal curvatures are constant. Then M is a Hopf hypersurface. Thus, M is an open subset of a Type A_0 or Type A_1 hypersurface, or a Type B hypersurface in $\mathbf{CH}^n$ with radius $\frac{r}{2}\log(2+\sqrt{3})$.*

Proof. In view of Theorem 8.29, we need only consider the case $n = 2$. Assume that M is not Hopf and consider the standard non-Hopf setup. Proceed as in the proof of Lemma 8.30 until we reach the point there $\mu = 0$. Note that κ_1 has multiplicity 2 and $\nu = \kappa_1$, while κ_2 has multiplicity 1. Now let $Z = \varphi U$. Applying the Codazzi equation to the pair (X_1, Z), we have

$$(X_1\kappa_1)Z - (Z\kappa_1)X_1 + (\kappa_1 - A)[X_1, Z] = 2c\langle X_1, \varphi Z\rangle W. \tag{8.50}$$

Now take the inner product of this equation with X_1. The left side vanishes and the right side is

$$2c\langle X_1, -U\rangle\langle W, X_1\rangle, \tag{8.51}$$

which cannot vanish. This contradiction shows that M must be Hopf. □

The question of hypersurfaces with constant principal curvatures has been settled for the case $g = 3$. See Takagi [508] and Berndt and Díaz-Ramos [34].

Theorem 8.36. *Let M^{2n-1}, where $n \geq 2$, be a real hypersurface in $\mathbf{CP}^n$ or $\mathbf{CH}^n$ with $g = 3$ distinct principal curvatures at each point. Assume that these principal curvatures are constant. Then either*

- *M is a Hopf hypersurface and hence is an open subset of a Type A_2 or Type B hypersurface, or*
- *M is non-Hopf and is an open subset of a Berndt orbit in $\mathbf{CH}^n$, or*
- *M is non-Hopf and is an open subset of a tube of radius $\frac{r}{2}\log(2+\sqrt{3})$ around a homogeneous ruled submanifold $F_{k,\theta}$ of $\mathbf{CH}^n$. We will discuss such submanifolds in conjunction with the Berndt-Tamaru classification of homogeneous hypersurfaces in $\mathbf{CH}^n$ (see Theorem 8.148).*

Díaz-Ramos, Domínguez-Vázquez et al. have embarked on a project to investigate hypersurfaces with constant principal curvatures for higher values of g and for weakened versions of the Hopf condition (see Section 8.6). The general approach is to attempt to classify k-Hopf hypersurfaces with constant principal curvatures according to the parameters (k, g). The theorems of Berndt and Kimura correspond to $k = 1$ and the possible values of g are 2, 3, and 5, the latter occurring in $\mathbf{CP}^n$ but not in $\mathbf{CH}^n$. See Section 8.6.

Algebraic conditions on the shape operator

As we have seen in Theorem 6.15, it is impossible for the shape operator A to vanish identically. In fact, we have $|\nabla A|^2 \geq 4c^2(n-1)$. Hypersurfaces for which this inequality becomes an equality form a familiar class, as we see in the next theorem.

Theorem 8.37. *Let M^{2n-1}, where $n \geq 2$, be a real hypersurface in $\mathbf{CP}^n$ or $\mathbf{CH}^n$. The following are equivalent:*

1. *M is an open subset of a Type A hypersurface;*
2. *$\varphi A = A\varphi$;*
3. *$|\nabla A|^2 = 4c^2(n-1)$;*
4. *$(\nabla_X A)Y + c(\langle\varphi X, Y\rangle W + \langle Y, W\rangle\varphi X) = 0$ for all X and Y tangent to M;*
5. *The cyclic sum $\langle(\nabla_X A)Y, Z\rangle + \langle(\nabla_Y A)Z, X\rangle + \langle(\nabla_Z A)X, Y\rangle$ over every triple (X, Y, Z) of tangent vectors vanishes.*

Work along these lines was initiated by Y. Maeda [350], Okumura [408], and Ki [238]. For a detailed proof of this theorem along with relevant references, the reader may consult Section 4 of Niebergall and Ryan [399]. This theorem allows us to characterize the compact Type A hypersurfaces by an inequality involving the shape operator as follows (recall that $\mathbf{m} = \text{trace } A$).

Theorem 8.38. *Let M^{2n-1}, where $n \geq 2$, be a compact hypersurface in $\mathbf{CP}^n$ or $\mathbf{CH}^n$. Then the following are equivalent:*

- $|A|^2 \leq 2(n-1)c + \mathbf{m}\alpha$;
- $|A|^2 = 2(n-1)c + \mathbf{m}\alpha$;
- *M is a Type A_2 hypersurface in $\mathbf{CP}^n$ or a geodesic sphere in $\mathbf{CP}^n$ or $\mathbf{CH}^n$.*

Proof. We recall Yano's formula [559]

$$\operatorname{div}(\nabla_X X - (\operatorname{div} X)X) = \langle SX, X\rangle + \frac{1}{2}|\mathcal{L}_X g|^2 - |\nabla X|^2 - (\operatorname{div} X)^2 \tag{8.52}$$

valid for any vector field X on a Riemannian manifold. (In order to display the formula, we needed a letter to denote the Riemannian metric and have used g temporarily for this purpose.) The proof of Yano's formula is straightforward, so we leave it as an exercise for the reader. A proof can also be found (in the "tensor notation" prevalent at the time) in Yano's paper pp. 39-40. We now set $X = W$ and compute

$$\begin{aligned}
\langle SW, W\rangle &= 2(n-1)c + \mathbf{m}\alpha - |AW|^2 \\
|\mathcal{L}_W g|^2 &= |\varphi A - A\varphi|^2 \\
|\nabla W|^2 &= |\varphi A|^2 = |A|^2 - |AW|^2 \\
\operatorname{div} W &= \text{trace } \varphi A = 0.
\end{aligned}$$

If M is orientable, we can integrate (8.52) with $X = W$ over M and apply Green's theorem ([283] vol. I, p. 281) to get

$$0 = \int \left(2(n-1)c + \mathbf{m}\alpha - |A|^2 + \frac{1}{2}|\varphi A - A\varphi|\right) dv \tag{8.53}$$

where dv is the standard volume element. Our hypothesis on A now implies that $|A|^2 = 2(n-1)c + \mathbf{m}\alpha$ and $\varphi A = A\varphi$. By Theorem 8.37, M is an open subset of a Type A hypersurface. Being compact, M is actually a Type A hypersurface and hence one of the examples listed. The orientability assumption is now seen to be superfluous. If M were not orientable, we could apply the same argument to the (compact orientable) twofold cover $\hat{M}$. Since $|\varphi A - A\varphi|$ and $2(n-1)c + \mathbf{m}\alpha - |A|^2$ agree at corresponding points of M and $\hat{M}$, we find that M itself is one of the listed examples.

Conversely, it is straightforward to check that $|A|^2 = 2(n-1)c + \mathbf{m}\alpha$ for all hypersurfaces of Type A. This completes the proof of our theorem. □

Remark 8.39. A proof of Theorem 8.38 in the $\mathbf{CP}^2$ case has been given by Deshmukh and Al-Gwaiz [129]. See also [128].

A tensor field T of type $(1,1)$ is said to be *cyclic parallel* if the cyclic sum of $\langle(\nabla_X T)Y, Z\rangle$ over every triple (X, Y, Z) of tangent vectors vanishes. Thus, Condition *5* of Theorem 8.37 could be expressed as "A is cyclic parallel." Condition *2* is equivalent to parallelism of the shape operator A' of the lifted hypersurface $M' = \pi^{-1}M$. See Lemma 4.2 of [399]. Ki and H.-J. Kim [240] derive another equivalent condition for Theorem 8.37, namely $\nabla C = 0$ where $C(X, Y, Z)$ is the cyclic sum used in Condition *5*. They call this condition "covariantly cyclic constant." See also the related work of J.J. Kim and Pyo [265].

Ghosh [173] proved the equivalence to Condition *2* of two weaker conditions. To describe these results, we need the following terminology. A tensor field T is said to be *semiparallel* if $R(X, Y) \cdot T = 0$ for all tangent vectors X and Y. The expression on the left is defined by

$$R(X, Y) \cdot T = \nabla_X \nabla_Y T - \nabla_Y \nabla_X T - \nabla_{[X,Y]} T. \tag{8.54}$$

Note that covariant differentiation, and hence $R(X, Y)$, acts on the algebra of tensor fields as a derivation that commutes with contractions. If T is of type (r, s), we can write $R \cdot T$ for the corresponding tensor field of type $(r, s+2)$. For example, if T is of type $(1, 1)$, then

$$(R \cdot T)(X, Y, Z) = (R(X, Y) \cdot T)Z.$$

A tensor field T is *recurrent* if there is a 1-form ω such that $\nabla_X T = \omega(X)T$ for all tangent vectors X. Using routine tensor algebra it is straightforward to verify that if T is a $(1, 1)$ tensor field satisfying the definition of recurrence, then

$$(R(X, Y) \cdot T)Z = ((\nabla_X \omega)Y - (\nabla_Y \omega)X)TZ \tag{8.55}$$

for all tangent vectors X, Y and Z

Ghosh's theorem can be stated as follows.

Theorem 8.40. *Let M^{2n-1}, where $n \geq 2$, be a real hypersurface in $\mathbf{CP}^n$ or $\mathbf{CH}^n$. Let $T = \varphi A - A\varphi$. The following are equivalent:*

1. *$T = 0$;*
2. *T is recurrent;*
3. *T is semiparallel.*

Proof. We break the proof down into two parts, first showing that Condition *3* implies Condition *1* and then showing that Condition *2* also implies Condition *1*. Since Condition *1* implies Conditions *2* and *3* trivially, this will be sufficient.

- Condition *3* implies Condition *1*:
  Suppose that T is semiparallel. Then, in particular,

$$0 = \langle [R(X,Y),T]W, W\rangle = 2\langle R(X,Y)TW, W\rangle. \tag{8.56}$$

Note that $TW = \varphi AW$ and thus $\langle TW, W\rangle = \langle TW, AW\rangle = 0$. From the Gauss equation, we get

$$\langle (AX \wedge AY)W, TW\rangle + c\langle (X \wedge Y)W, TW\rangle = 0$$

since $(\varphi X \wedge \varphi Y)W = \langle X, \varphi Y\rangle \varphi W = 0$. Thus

$$\begin{aligned}\langle AY, W\rangle\langle AX, TW\rangle &+ c\langle Y, W\rangle\langle X, TW\rangle \\ &= \langle AX, W\rangle\langle AY, TW\rangle + c\langle X, W\rangle\langle Y, TW\rangle\end{aligned} \tag{8.57}$$

for all tangent vectors X and Y. Setting $Y = TW$, we have

$$\langle AX, W\rangle\langle ATW, TW\rangle + c\langle X, W\rangle|TW|^2 = 0. \tag{8.58}$$

Setting $X = \varphi ATW$, we get $-\langle ATW, TW\rangle^2 = 0$, which allows us to conclude that $\langle ATW, TW\rangle = 0$. Substituting back in (8.58) yields

$$c\langle X, W\rangle|TW|^2 = 0 \tag{8.59}$$

which, upon setting $X = W$, gives $TW = \varphi AW = 0$. Thus M is Hopf. Consider now

$$0 = [R(X,W),T]W = -TR(X,W)W$$

which, by the Gauss equation, gives

$$T(\alpha A + c)X = 0. \tag{8.60}$$

for all X. Now $\mathcal{W}^{\perp}$ has an orthonormal basis consisting of $n-1$ pairs of principal vectors $(X, \varphi X)$ with corresponding principal curvatures (λ, μ) (see Theorem 6.17). Then $TX = (\varphi A - A\varphi)X = (\lambda - \mu)\varphi X$ and $T\varphi X = (\lambda - \mu)X$. If $T \neq 0$, there is a choice of X for which $\lambda \neq \mu$. But then, by (8.60) we have $\alpha\mu + c = 0$. The same reasoning applied to φX yields $\alpha\lambda + c = 0$. Since $\alpha \neq 0$ and $\lambda \neq \mu$, we have a contradiction and are forced to conclude that $T = 0$. In fact, M is an open subset of a Type A hypersurface by Theorem 8.37.

- Condition *2* implies Condition *1*:
  Since T is symmetric, it has real eigenvalues. We choose a point of M where the maximum number of eigenvalues are distinct. This insures that the distinct eigenvalues are smooth functions with smooth eigenspaces in a neighborhood $\mathcal{U}$

of this point. Assuming that $T \neq 0$, we know that at least one eigenvalue κ is nonzero and we can choose a smooth unit vector field Z in $\mathcal{U}$ such that $AX = \kappa X$. Now $\omega(X)\kappa = \omega(X)\langle TZ, Z\rangle = \langle(\nabla_Z T)Z, Z\rangle = X\kappa$ where we have used the fact that Z, being a unit vector, is orthogonal to its covariant derivative. Thus $X\kappa = \omega(X)\kappa$ and we have

$$d\kappa = \kappa\omega$$

on $\mathcal{U}$. By the Poincaré lemma,

$$0 = d^2\kappa = d\kappa \wedge \omega + \kappa d\omega = \kappa(\omega \wedge \omega + d\omega) = \kappa d\omega.$$

But $d\omega(X, Y) = (\nabla_X \omega)Y - (\nabla_Y \omega)X$ which shows that $R(X, Y) \cdot T = 0$ and T is semiparallel on $\mathcal{U}$. By the earlier argument "Condition *3* implies Condition *1*," we get $T = 0$ on $\mathcal{U}$, a contradiction. We conclude that $T = 0$ at all points of M. □

Following Maeda and Naitoh [348] we define a tensor field T of type $(1, 1)$ on a real hypersurface in $\mathbf{CP}^n$ or $\mathbf{CH}^n$ to be *strongly φ-invariant* if

$$\langle T\varphi X, \varphi Y\rangle = \langle TX, Y\rangle,$$

for all pairs (X, Y) of tangent vectors. It is easy to check that this is equivalent to $\varphi T\varphi = -T$.

We have the following theorem regarding the case where the shape operator is strongly φ-invariant.

Theorem 8.41. *Let M^{2n-1}, where $n \geq 2$, be a real hypersurface in $\mathbf{CP}^n$ or $\mathbf{CH}^n$. The following two conditions are equivalent:*

1. *the shape operator A is strongly φ-invariant;*
2. *M is an open subset of a Type A hypersurface with $\alpha = 0$.*

In particular, these conditions imply that

- *$\varphi A = A\varphi$;*
- *the ambient space is $\mathbf{CP}^n$;*
- *M lies on a tube of radius $\frac{\pi}{4}r$ over a totally geodesic $\mathbf{CP}^k$ for some k satisfying $1 \leq k \leq n-1$.*

Proof. Suppose that Condition 2 is satisfied. Note that $\alpha = 0$ rules out the possibility that the ambient space is $\mathbf{CH}^n$. For Type A hypersurfaces in $\mathbf{CP}^n$, we have $\alpha = \frac{2}{r}\cot 2u$, so that $u = \frac{\pi}{4}$ and M must be an open subset of a tube over $\mathbf{CP}^k$ as described in the statement of the theorem. Also, since $\varphi A = A\varphi$, we have $\varphi A\varphi X = A\varphi^2 X = -AX$ for all $X \in \mathcal{W}^\perp$ and $0 = AW = -\varphi A\varphi W$. Thus A is strongly φ-invariant.

Conversely, if $\varphi A\varphi = -A$, then $AW = 0$. Thus M is a Hopf hypersurface with $\alpha = 0$. Furthermore,

$$(\varphi A - A\varphi)\varphi X = -AX - A\varphi^2 X = 0,$$

for all $X \in \mathcal{W}^\perp$. Thus $\varphi A - A\varphi$ annihilates $\mathcal{W}^\perp$, as well as (trivially) $\mathcal{W}$, so that Condition 2 holds by Theorem 8.37, and the proof is complete. □

A hypersurface (in any ambient space) is said to be *semiparallel* if the shape operator A is itself semiparallel. Note that this is a weaker condition than $\nabla A = 0$. Of course, the latter cannot occur for a hypersurface in $\mathbf{CP}^n$ or $\mathbf{CH}^n$. It turns out that even semiparallelism is too strong a condition. We can state the following theorem.

Theorem 8.42. *Let M^{2n-1}, where $n \geq 2$, be a real hypersurface in $\mathbf{CP}^n$ or $\mathbf{CH}^n$. Then $R \cdot A$ cannot vanish identically.*

This was first proved for $\mathbf{CP}^n$ by S. Maeda [346] for $n \geq 3$. Later, Niebergall and Ryan [400] dealt with the case $n = 2$. Finally, Ortega [414] provided a proof for $\mathbf{CH}^n$, $n \geq 3$.

Lobos and Ortega studied two more general conditions and their work is embodied in Theorems 8.51 and 8.60 below. Theorem 8.42 follows from their results (see remarks following Proposition 8.58 below), and it is basically their approach that we use in our exposition.

The next few results provide further characterizations of the Type A_0 and Type A_1 hypersurfaces. The condition (8.61) analyzed in the first proposition was introduced by Matsuyama [354].

Proposition 8.43. *Let M^{2n-1}, where $n \geq 3$, be a real hypersurface in $\mathbf{CP}^n$ or $\mathbf{CH}^n$. Then its curvature tensor satisfies*

$$\big(R(AX, Y) - A \circ R(X, Y)\big)\mathcal{W}^\perp \subset \mathcal{W} \tag{8.61}$$

for all X and Y in $\mathcal{W}^\perp$ if and only if M is

- *a ruled real hypersurface, or*
- *an open subset of a Type A_0 or Type A_1 hypersurface.*

Remark 8.44. It is easy to check that a ruled real hypersurface cannot be Hopf. In fact, if it were, we would have $A\mathcal{W}^\perp = 0$, violating Theorem 6.17 (page 357). Thus Matsuyama's condition (8.61) characterizes ruled real hypersurfaces among all non-Hopf hypersurfaces, as well as characterizing the Type A_0 and A_1 hypersurfaces among all Hopf hypersurfaces.

So far, we have considered algebraic conditions relating tangent vectors "in general." We now begin to weaken such conditions by requiring them only on the holomorphic subspace $\mathcal{W}^\perp$. We begin with "umbilicity."

Lemma 8.45. *Let M^{2n-1}, where $n \geq 3$, be a real hypersurface in* $\mathbf{CP}^n$ *or* $\mathbf{CH}^n$*. Suppose that there is a function κ such that $\langle AX, Y\rangle = \kappa\langle X, Y\rangle$ for all X and Y in $\mathcal{W}^\perp$. Then κ is constant and M is*

- *a ruled real hypersurface if $\kappa = 0$;*
- *an open subset of a Type A_0 or Type A_1 hypersurface if $\kappa \neq 0$.*

Proof. If κ is identically zero, then AX is orthogonal to $\mathcal{W}^\perp$ for all $X \in \mathcal{W}^\perp$ and M must be ruled. Now suppose that there is a point where $\kappa \neq 0$. Assume that $AW \neq \alpha W$ there and use the standard non-Hopf setup in a neighborhood. Since $n \geq 3$, we can choose a unit tangent vector field Z orthogonal to the span of $\{W, U, \varphi U\}$. Note that $\langle AZ, W\rangle = 0$ so that $AZ = \kappa Z$. Also, $AU = \beta W + \kappa U$ and $A\varphi U = \kappa\varphi U$ (i.e., $\lambda = \nu = \kappa$ and $\mu = 0$). From the Codazzi equation (6.8), we have (since Z is also orthogonal to φU),

$$0 = (\nabla_U A)Z - (\nabla_Z A)U = (U\kappa)Z - (Z\kappa)U - (Z\beta)W - \beta\nabla_Z W + (\kappa - A)[U, Z].$$

Taking inner product with φZ and using the fact that $(\kappa - A)\varphi Z = 0$, the only term remaining gives

$$0 = \beta\langle\nabla_Z W, \varphi Z\rangle = \beta\langle\varphi AZ, \varphi Z\rangle = \beta\kappa.$$

This contradicts the fact that β and κ are nonzero. We must conclude that every point where $\kappa \neq 0$ has a Hopf neighborhood in which $AX = \kappa X$ for all $X \in \mathcal{W}^\perp$. Since this neighborhood has $g \leq 2$ distinct principal curvatures, it must be an open subset of a hypersurface of Type A_0 or A_1. (It is clear that the special Type B hypersurfaces with $g = 2$ do not qualify, since, in that case, both principal curvatures have principal vectors in $\mathcal{W}^\perp$). In particular, κ must be constant. By a standard connectedness argument, κ is constant on all of M and the conclusion follows. □

Remark 8.46. Given the condition $\langle AX, Y\rangle = \kappa\langle X, Y\rangle$, one need not assume that the function κ is continuous. It is automatically smooth, since it is related to $\mathbf{m} = \operatorname{trace} A$ by $\kappa = (\mathbf{m} - \alpha)/(2(n-1))$.

The following lemma essentially completes the proof of Proposition 8.43.

Lemma 8.47. *A hypersurface in* $\mathbf{CP}^n$ *or* $\mathbf{CH}^n$*, where $n \geq 3$, satisfying Matsuyama's condition (8.61) also satisfies*

$$\langle AX, Y\rangle = \frac{\mathbf{m} - \alpha}{2(n-1)}\langle X, Y\rangle$$

for all X and Y in $\mathcal{W}^\perp$.

Proof. Using the Gauss equation, the left side of Matsuyama's condition may be written as the sum of four terms:

- $A^2X \wedge AY - A \circ (AX \wedge AY)$;
- $c(AX \wedge Y - A \circ (X \wedge Y))$;
- $c(\varphi AX \wedge \varphi Y - A \circ (\varphi X \wedge \varphi Y))$;
- $2c(\langle AX, \varphi Y\rangle\, \varphi - \langle X, \varphi Y\rangle\, A\varphi)$.

Choose an orthonormal basis $\{e_j\}$ of $\mathcal{W}^\perp$. Set $X = e_j$, apply our formula to φe_j and sum over all j. Routine calculation (using the fact that φA and φA^2 have trace zero) shows that the first two terms each sum to zero. The third term yields

$$c\big((2n-3)A\varphi Y + \varphi AY - (\mathbf{m} - \alpha)\varphi Y\big),$$

while the fourth term reduces to $2c\langle A\varphi Y, W\rangle W$. Thus, for all $Z \in \mathcal{W}^\perp$, we have

$$(2n-3)\langle A\varphi Y, Z\rangle + \langle \varphi AY, Z\rangle - (\mathbf{m} - \alpha)\langle \varphi Y, Z\rangle = 0. \tag{8.62}$$

Interchanging the roles of Y and Z and using the symmetry and skew-symmetry properties of our operators, we get

$$-(2n-3)\langle \varphi AY, Z\rangle - \langle A\varphi Y, Z\rangle + (\mathbf{m} - \alpha)\langle \varphi Y, Z\rangle = 0.$$

Adding these two equations yields

$$2(n-2)\langle (\varphi A - A\varphi)Y, Z\rangle = 0.$$

Since $n \geq 3$, we have $\langle (\varphi A - A\varphi)Y, Z\rangle = 0$ for all Y and Z in $\mathcal{W}^\perp$. In particular, equation (8.62) becomes

$$2(n-1)\langle A\varphi Y, Z\rangle = (\mathbf{m} - \alpha)\langle \varphi Y, Z\rangle.$$

Since any pair of vectors in $\mathcal{W}^\perp$ can be written in the form $(\varphi Y, Z)$, we have the desired conclusion. □

Remark 8.48. One should also verify that every Type A_0, every Type A_1, and every ruled hypersurface satisfies the Matsuyama condition. Except for the ruled hypersurfaces, this is trivial since, if $A = \lambda I$ on $\mathcal{W}^\perp$, the condition reduces to

$$\langle (\lambda - A)R(X, Y)Z_1, Z_2\rangle = 0,$$

for every quadruple (X, Y, Z_1, Z_2) in $\mathcal{W}^\perp$. For ruled hypersurfaces, look at the breakdown of the condition into four terms, apply each term to Z_1, and take the inner product with Z_2. Repeatedly using the fact that $A\mathcal{W}^\perp \subset \mathcal{W}$, we find that every term vanishes.

The preceding proof also establishes a technical criterion for Matsuyama's condition. We state this as a corollary since we will also make use of it in the next proof.

Corollary 8.49. *A hypersurface in* $\mathbf{CP}^n$ *or* $\mathbf{CH}^n$*, where* $n \geq 3$*, satisfies Matsuyama's condition (8.61) if and only if*

$$(2n-3)\langle A\varphi Y, Z\rangle + \langle \varphi AY, Z\rangle - (\mathbf{m} - \alpha)\langle \varphi Y, Z\rangle = 0, \tag{8.63}$$

for all Y *and* Z *in* $\mathcal{W}^{\perp}$.

Remark 8.50. It is easy to check that condition (8.63) holds automatically when $n = 2$. In this case, Matsuyama's condition must be analyzed using a different method.

We now discuss two more criteria that are equivalent to Matsuyama's condition for $n \geq 3$. These were established by Ortega [414]. We combine them with Proposition 8.43 to obtain the following.

Theorem 8.51. *For a hypersurface* M *in* $\mathbf{CP}^n$ *or* $\mathbf{CH}^n$*, where* $n \geq 3$*, the following are equivalent:*

1. *M is a ruled real hypersurface or an open subset of a hypersurface of Type A_0 or Type A_1;*
2. $(R(X,Y)\cdot A)Z + (R(Y,Z)\cdot A)X + (R(Z,X)\cdot A)Y = 0$ *for all* X, Y, *and* Z *in* $\mathcal{W}^{\perp}$;
3. $(R(X,Y)\cdot A)Z + (R(Y,Z)\cdot A)X + (R(Z,X)\cdot A)Y \in \mathcal{W}$ *for all* X, Y, *and* Z *in* $\mathcal{W}^{\perp}$;
4. $\big(R(AX,Y) - A \circ R(X,Y)\big)\mathcal{W}^{\perp} \subset \mathcal{W}$ *for all* X *and* Y *in* $\mathcal{W}^{\perp}$.

Proof. In view of the first Bianchi identity we can write the cyclic sum of $(R(X,Y)\cdot A)Z$ as

$$R(X,Y)AZ + R(Y,Z)AX + R(Z,X)AY. \tag{8.64}$$

Using the Gauss equation, this can expressed as the sum of four terms:

- $(AX \wedge AY)AZ + (AY \wedge AZ)AX + (AZ \wedge AX)AY$;
- $c\big((X \wedge Y)AZ + (Y \wedge Z)AX + (Z \wedge X)AY\big)$;
- $c\big((\varphi X \wedge \varphi Y)AZ + (\varphi Y \wedge \varphi Z)AX + (\varphi Z \wedge \varphi X)AY\big)$;
- $2c\big(\langle X, \varphi Y\rangle \varphi AZ + \langle Y, \varphi Z\rangle \varphi AX + \langle Z, \varphi X\rangle \varphi AY\big)$.

Clearly, the first and second terms evaluate to zero. The third term reduces to

$$c\big(\langle(\varphi A + A\varphi)Y, Z\rangle \varphi X + \langle(\varphi A + A\varphi)Z, X\rangle \varphi Y + \langle(\varphi A + A\varphi)X, Y\rangle \varphi Z\big).$$

If $Z = \varphi Y$, this becomes

$$\begin{aligned} c\big(-(\langle \varphi A\varphi Y, Y\rangle - \langle AY, Y\rangle)\varphi X - (\langle AY, X\rangle - \langle \varphi A\varphi X, Y\rangle)\varphi Y \\ - \langle(\varphi A + A\varphi)X, Y\rangle Y\big), \end{aligned}$$

and the fourth term becomes

$$2c\big(-\langle\varphi X, Y\rangle\varphi A\varphi Y - \langle Y, Y\rangle\varphi AX + \langle Y, X\rangle\varphi AY\big).$$

Choose an orthonormal basis $\{e_j\}$ for $\mathcal{W}^\perp$, as in Lemma 8.47. Set $Y = e_j$, and sum over all j to obtain

$$2c\big((\mathbf{m} - \alpha)\varphi X - (\varphi A + A\varphi)X + \langle AW, \varphi X\rangle W\big),$$

and

$$-4c(n-2)\varphi AX.$$

Addition of these expressions yields

$$2c\big((\mathbf{m} - \alpha)\varphi X - (2n-3)\varphi AX - A\varphi X + \langle AW, \varphi X\rangle W\big).$$

If we take the inner product of this with any $Z \in \mathcal{W}^\perp$, and apply Corollary 8.49, we see that Condition *3* in our theorem implies Condition *4*, which is equivalent to Condition *1* by Proposition 8.43. Of course, Condition *2* implies Condition *3* trivially.

To complete the proof, we need only observe that Type A_0 and Type A_1 hypersurfaces, as well as ruled hypersurfaces, satisfy Condition *2*. For the Type A_0 and Type A_1 hypersurfaces, every vector that occurs is an eigenvector of A with the same eigenvalue. It follows that the third and fourth terms of (8.64) are negatives of each other. For ruled hypersurfaces, the third and fourth terms of (8.64) vanish, since $\varphi AX = 0$ for every $X \in \mathcal{W}^\perp$. □

As a corollary, we obtain the following result which was proved earlier by Gotoh [182] in the $\mathbf{CP}^n$ case.

Corollary 8.52. *For a hypersurface M^{2n-1} in $\mathbf{CP}^n$ or $\mathbf{CH}^n$, where $n \geq 3$, the following are equivalent:*

- $(R(X, Y) \cdot A)Z = 0$ *for all X, Y, and Z in $\mathcal{W}^\perp$;*
- *M is an open subset of a hypersurface of Type A_0 or Type A_1.*

Proof. For a hypersurface of Type A_0 or Type A_1, there is a constant λ such that A coincides with λI on $\mathcal{W}^\perp$. Using the Gauss equation, it is straightforward to check that $(R(X, Y) \cdot A)Z$ vanishes for all X, Y, and Z in $\mathcal{W}^\perp$.

For the converse, in view of Theorem 8.51, it is sufficient to show that ruled hypersurfaces do not satisfy our curvature condition. This can be verified by using the Gauss equation to show that, in terms of the standard non-Hopf setup, ruled hypersurfaces satisfy $(R(U, \varphi U) \cdot A)\varphi U = -2cAU \neq 0$. We leave the details to the reader. □

Remark 8.53. In fact, the same argument shows that the condition

$$(R(X,Y)\cdot A)Z = 0$$

can be replaced by $(R(X,Y)\cdot A)Z \in \mathcal{W}$.

Proposition 8.43, Lemma 8.45, Lemma 8.47, Theorem 8.51, and Corollary 8.52 do not hold for $n = 2$. We will now consider what happens in this case.

In the standard non-Hopf setup, the Gauss equation gives

$$\begin{aligned} R(U,\varphi U) &= (\lambda\nu - \mu^2 + 4c)U\wedge\varphi U + \beta\nu W\wedge\varphi U + \beta\mu W\wedge U;\\ R(W,U) &= (\lambda\alpha - \beta^2 + c)W\wedge U + \mu\alpha W\wedge\varphi U + \beta\mu U\wedge\varphi U;\\ R(W,\varphi U) &= (\nu\alpha + c)W\wedge\varphi U + \mu\alpha W\wedge U + \beta\nu U\wedge\varphi U. \end{aligned} \tag{8.65}$$

Then, a straightforward calculation yields

$$\begin{aligned} R(AU,\varphi U)\varphi U \;-\; AR(U,\varphi U)\varphi U &= -\mu q\varphi U \mod \mathcal{W};\\ R(AU,\varphi U)U \;-\; AR(U,\varphi U)U &= -\mu\big((\lambda-\nu)q + \beta^2\nu\big)\varphi U\\ &\quad +\mu(q-\beta^2)U \mod \mathcal{W};\\ (R(U,\varphi U)\cdot A)U &= -\big((\lambda-\nu)q + \beta^2\nu\big)\varphi U + 2\mu(q-\beta^2)U \mod \mathcal{W};\\ (R(U,\varphi U)\cdot A)\varphi U &= -\big((\lambda-\nu)q + \beta^2\nu\big)U - 2\mu q\varphi U \mod \mathcal{W} \end{aligned} \tag{8.66}$$

where we have used q as a temporary abbreviation for $\lambda\nu - \mu^2 + 4c$. This provides a basis for the following theorem in the non-Hopf case.

Theorem 8.54. *Let M^3 be a real hypersurface in $\mathbf{CP}^2$ or $\mathbf{CH}^2$ for which the standard non-Hopf setup is valid. Then*

1. *There is a function κ such that $\langle AX,Y\rangle = \kappa\langle X,Y\rangle$ for all X and Y in $\mathcal{W}^\perp$ if and only if $\mu = 0$ and $\lambda = \nu$. Although all ruled hypersurfaces satisfy this condition, there also exist non-ruled examples.*
2. *The following are equivalent:*
    - $\big(R(AX,Y) - A\circ R(X,Y)\big)\mathcal{W}^\perp \subseteq \mathcal{W}$ *for all X and Y in $\mathcal{W}^\perp$;*
    - $(R(X,Y)\cdot A)\mathcal{W}^\perp \subseteq \mathcal{W}$ *for all X and Y in $\mathcal{W}^\perp$;*
    - $\mu = 0$ *and* $(\lambda-\nu)(\lambda\nu + 4c) + \beta^2\nu = 0$.

    Although all ruled hypersurfaces satisfy these conditions, there also exist non-ruled examples.
3. *$(R(X,Y)\cdot A)Z$ cannot vanish for all X, Y and Z in $\mathcal{W}^\perp$.*

Proof. 1. If a function κ exists as indicated, then $\mu = \langle AU,\varphi U\rangle = 0$ whereas $\lambda = \langle AU,U\rangle = \kappa = \langle A\varphi U,\varphi U\rangle = \nu$. Conversely, suppose that $\mu = 0$ and $\lambda = \nu$. Then we can set $\kappa = \lambda = \nu$ and easily verify the desired identity.

2. If $\big(R(AX,Y) - A \circ R(X,Y)\big)\mathcal{W}^\perp \subseteq \mathcal{W}$ for all X and Y in $\mathcal{W}^\perp$, then the first two equations of (8.66) give $\mu q = 0$ and $\mu(q-\beta^2) = 0$ so that we have $\mu = 0$. Now consider

$$\begin{aligned} R(A\varphi U, U)\varphi U - AR(\varphi U, U)\varphi U &= -(\nu - A)(qU + \beta\nu W) \\ &= \big((\lambda - \nu)q + \beta^2\nu\big)U \mod \mathcal{W} \end{aligned}$$

where we have used the fact that $\mu = 0$. Thus, we have $(\lambda - \nu)q + \beta^2\nu = 0$.

If $(R(X,Y)\cdot A)\mathcal{W}^\perp \subseteq \mathcal{W}$ for all X and Y in $\mathcal{W}^\perp$, then the last two equations of (8.66) give $\mu q = 0$ and $\mu(q - \beta^2) = 0$ so that $\mu = 0$ as before. In this case, the conclusion that $(\lambda - \nu)q + \beta^2\nu = 0$ is evident without further calculation.

Conversely, if $\mu = 0$ and $(\lambda-\nu)q+\beta^2\nu = 0$, the right side of every equation in (8.66) vanishes. This establishes the second condition. For the first condition, we also need to check the cases involving $R(AU,U)$ and $R(A\varphi U,U)$. This is routine.

3. Assume that $(R(X,Y)\cdot A)Z = 0$ for all X, Y, and Z in $\mathcal{W}^\perp$. Using $\mu = 0$, we compute

$$\begin{aligned} (R(U,\varphi U)\cdot A)W &= (\alpha - A)R(U,\varphi U)W + \beta R(U,\varphi U)U \\ &= -\beta\nu(\alpha - A)\varphi U - \beta q\varphi U = \beta(\nu(\nu-\alpha) - q)\varphi U. \end{aligned}$$

Taking the inner product with φU and using the symmetry of $R(U,\varphi U)\cdot A$, we can conclude that $\nu(\nu - \alpha) = q$. However, it is shown in Ivey (personal communication, 2015) that this implies $\beta = 0$, a contradiction.

The existence of non-ruled examples follows from Theorem 20 of Ivey and Ryan [224]. There it is proved, using the theory of exterior differential systems, that a non-Hopf hypersurface with $\mu = 0$ exists with α, β, λ and ν satisfying any desired algebraic condition and having prescribed initial values. □

Remark 8.55. It is easy to check that for hypersurfaces in $\mathbf{CP}^2$ and $\mathbf{CH}^2$, Conditions *2* and *3* of Theorem 8.51 are automatically satisfied. Theorem 8.54 shows that Conditions *1* and *4* of Theorem 8.51 are inequivalent when $n = 2$.

We now look at the analogue of Theorem 8.54 for Hopf hypersurfaces.

Theorem 8.56. *For a Hopf hypersurface M^3 in $\mathbf{CP}^2$ or $\mathbf{CH}^2$, the following are equivalent:*

(i) *M is an open subset of a hypersurface of Type A_0 or Type A_1;*
(ii) *There is a function κ such that $\langle AX,Y\rangle = \kappa\langle X,Y\rangle$ for all X and Y in $\mathcal{W}^\perp$;*
(iii) *$\big(R(AX,Y) - A \circ R(X,Y)\big)\mathcal{W}^\perp \subset \mathcal{W}$ for all X and Y in $\mathcal{W}^\perp$;*
(iv) *$(R(X,Y)\cdot A)\mathcal{W}^\perp \subseteq \mathcal{W}$ for all X and Y in $\mathcal{W}^\perp$.*

In case (ii), *we have $\kappa = (\mathbf{m} - \alpha)/2$ and in case* (iv), *$(R(X,Y)\cdot A)Z = 0$ for all X, Y, and Z in $\mathcal{W}^\perp$.*

The proof of Theorem 8.56 is easy and we provide only a brief sketch. The curvature information given in (8.65) is still valid in the Hopf case if we set $\mu = \beta = 0$ and choose U to be any unit principal vector in $\mathcal{W}^{\perp}$. The critical condition is $(\lambda - \nu)(\lambda\nu + 4c) = 0$. However, an assumption that $\lambda \neq \nu$ will lead to a contradiction. In fact, $\lambda\nu + 4c = 0$ allows us to show that λ and ν are constant. This implies, in turn, that $\lambda\nu + c = 0$.

A hypersurface is said to be *cyclic-semiparallel* if

$$(R(X,Y) \cdot A)Z + (R(Y,Z) \cdot A)X + (R(Z,X) \cdot A)Y = 0$$

for all tangent vectors X, Y and Z.

Extending Condition 2 of Theorem 8.51 to $\mathcal{W}$, we obtain the following corollary (see Kimura and Maeda [277] and Choe [118]).

Corollary 8.57. *For a hypersurface M in $\mathbf{CP}^n$ or $\mathbf{CH}^n$, where $n \geq 3$, the following are equivalent:*

- *M is an open subset of a hypersurface of Type A_0 or Type A_1;*
- *M is cyclic-semiparallel.*

A hypersurface in $\mathbf{CP}^2$ or $\mathbf{CH}^2$ is cyclic-semiparallel if and only if it is Hopf.

Proof. We revisit our analysis of equation (8.64) from the proof of Theorem 8.51. The first and second terms still evaluate to zero. Suppose now that X and Y are in $\mathcal{W}^{\perp}$ but that $Z = W$. Then the third term reduces to

$$c\big(\langle A\varphi Y, W\rangle \varphi X + \langle \varphi AW, X\rangle \varphi Y\big),$$

while the fourth term reduces to

$$2c\langle X, \varphi Y\rangle \varphi AW.$$

If M is ruled, we may choose (at a point where $AW \neq \alpha W$), a unit vector $X \in \mathcal{W}^{\perp}$ such that $AW = \alpha W + \beta X$ where $\beta \neq 0$. Let $Y = \varphi X$. Then the third and fourth terms of (8.64) sum to

$$c(-\beta Y - 2\beta Y) = -3c\beta Y \neq 0.$$

This shows that ruled real hypersurfaces are not cyclic-semiparallel. We now check that Type A_0 and Type A_1 hypersurfaces are cyclic-semiparallel. As before, the third and fourth terms cancel for X, Y, and Z in $\mathcal{W}^{\perp}$. For X and Y in $\mathcal{W}^{\perp}$ and $Z = W$, both of these terms vanish. This proves our claim.

Having shown that Type A_0 and Type A_1 hypersurfaces are cyclic-semiparallel, it remains to remark that every cyclic-semiparallel hypersurface satisfies Condition 2 of Theorem 8.51 and hence is either ruled or an open subset of a Type A_0 or Type A_1 hypersurface. Ruled hypersurfaces having been excluded, the first part of our corollary is proved.

To prove the second assertion, we need to show that every cyclic parallel hypersurface in $\mathbf{CP}^2$ or $\mathbf{CH}^2$ is Hopf. We can do this by using the standard non-Hopf setup and considering

$$R(U, \varphi U)AW + R(\varphi U, W)AU + R(W, U)A\varphi U = 0.$$

Now take the inner product of this equation with φU and evaluate using the Gauss equation. We get

$$\beta(\nu(\alpha + \lambda) - \mu^2 + 4c) - \beta(\nu(\alpha + \lambda) + c) + \beta\mu^2 = 0, \tag{8.67}$$

which simplifies to $3c\beta = 0$, a contradiction which proves that cyclic parallel hypersurfaces are Hopf. On the other hand, it is now trivial to check, using an orthonormal principal basis of the form $(W, U, \varphi U)$, that every Hopf hypersurface satisfies the cyclic-semiparallelism condition. □

Although a hypersurface in $\mathbf{CP}^n$ or $\mathbf{CH}^n$ cannot be semiparallel, we have seen that a slightly weakened version (cyclic-semiparallelism) can be realized. We now turn to another way of weakening the semiparallelism condition. A hypersurface is said to be *pseudoparallel* if there is some function κ such that

$$(R(X, Y) - \kappa(X \wedge Y)) \cdot A = 0,$$

for all tangent vectors X and Y. A brief explanation of the notation is in order. The linear operator $X \wedge Y$ on tangent vectors extends naturally to the algebra of tensor fields as a derivation that annihilates scalar functions and commutes with contractions (just as $R(X, Y)$ does, see (8.54) on page 455). Thus,

$$((X \wedge Y) \cdot A)Z = (X \wedge Y)(AZ) - A((X \wedge Y)Z).$$

Note that semiparallelism is a special case of pseudoparallelism in which κ is identically zero.

We first observe the following.

Proposition 8.58. *A hypersurface in the Takagi/Montiel lists is pseudoparallel if and only if it is of Type A_0 or Type A_1. In the notation of Theorem 6.17, the value of κ is $\lambda^2 = \lambda\alpha + c \neq 0$.*

We leave this for the reader to check. However, the converse is also true, namely, every pseudoparallel hypersurface must be an open subset of a Type A_0 or Type A_1 hypersurface (see Theorem 8.60 below). In view of Proposition 8.58, Theorem 8.42 is a consequence of Theorem 8.60.

We begin by showing that a pseudoparallel hypersurface must be Hopf.

Lemma 8.59. *Let M^{2n-1}, where $n \geq 2$, be a pseudoparallel real hypersurface in* $\mathbf{CP}^n$ *or* $\mathbf{CH}^n$. *Then M is a Hopf hypersurface.*

Proof. The condition that M is pseudoparallel means that for all tangent vectors X and Y, we have

$$[R(X,Y) - \kappa(X \wedge Y), A] = 0,$$

which, in view of the Gauss equation means

$$\big[\big((c-\kappa)X \wedge Y + c(\varphi X \wedge \varphi Y + 2\langle X, \varphi Y\rangle \varphi) + AX \wedge AY\big), A\big] = 0. \tag{8.68}$$

We contract this formula as follows. Take an orthonormal basis $\{e_j\}$ for the tangent space. Set $Y = e_j$ and apply the formula to e_j, then sum over j. By a routine calculation,

- the $AX \wedge AY$ term yields $(\text{trace } A^2)AX - \mathbf{m}\, A^2X$;
- the $X \wedge Y$ term yields $(c-\kappa)(\mathbf{m}X - (2n-1)AX)$;
- the $\varphi X \wedge \varphi Y$ term yields $c(A\varphi^2X - \varphi A \varphi X)$;
- the $\langle X, \varphi Y\rangle \varphi$ term yields $2c(A\varphi^2X - \varphi A \varphi X)$.

Since A, A^2, and $\varphi A \varphi$ are symmetric operators, $A\varphi^2$ is also symmetric. This means that

$$A\varphi^2 = (A\varphi^2)^T = (-\varphi)^2 A = \varphi^2 A,$$

and hence

$$0 = A\varphi^2 W = \varphi^2 AW = -AW + \langle AW, W\rangle W = -(AW - \alpha W).$$

Thus M is Hopf. □

We are now ready to classify the pseudoparallel hypersurfaces. This result is due to Lobos and Ortega [334].

Theorem 8.60. *Let M^{2n-1}, where $n \geq 2$, be a real hypersurface in $\mathbf{CP}^n$ or $\mathbf{CH}^n$. Then M is pseudoparallel if and only if it is an open subset of a Type A_0 or Type A_1 hypersurface.*

Proof. Assuming that M is pseudoparallel, and therefore Hopf, we consider an arbitrary point of M and pairs (X, Y), where $Y = \varphi X$, belonging to an orthonormal basis for $\mathcal{W}^\perp$ at that point, as given by Corollary 6.18. Our calculation is completely pointwise – we do not need to consider smoothness of principal curvatures or of the function κ.

Applying the pseudoparallelism condition (8.68) to X, the $AX \wedge AY$ term yields $-\lambda\mu(\lambda - \mu)Y$. The remaining three terms simplify to $[(2c - \kappa)X \wedge Y - 2c\varphi, A]$ which, when applied to X, yields $-(4c - \kappa)(\lambda - \mu)Y$. The net result is that

$$(\lambda\mu + 4c - \kappa)(\lambda - \mu) = 0.$$

Assume now that $\lambda \neq \mu$, so that $\kappa = \lambda\mu + 4c$. If we consider

$$[R(W,X) - \kappa(W \wedge X), A]X = 0,$$

the analogous calculation gives $(\lambda-\alpha)(\lambda\alpha+c-\kappa) = 0$. Similarly, using Y instead of X, we get $(\mu - \alpha)(\mu\alpha + c - \kappa) = 0$. If neither λ nor μ is equal to α, we must have $\alpha = 0$ and hence $\kappa = c$. Also, from Theorem 6.17,

$$\lambda\mu = \frac{\lambda+\mu}{2}\alpha + c = c,$$

which contradicts $\kappa = \lambda\mu + 4c$.

Without loss of generality, we can thus assume that $\lambda = \alpha$. However, this gives $\lambda\mu + c - \kappa = 0$, again contradicting $\kappa = \lambda\mu + 4c$. We are forced to conclude that $\lambda = \mu$. Thus all principal curvatures on $\mathcal{W}^\perp$ satisfy the quadratic equation $t^2 = \alpha t + c$.

We now show that there can be only one such principal curvature. To see this, suppose that there were two, λ and ν. From the quadratic equation, $\lambda\nu = -c$ and $\lambda + \nu = \alpha$. Consider

$$[R(X,Z) - \kappa(X \wedge Z), A]Z = 0,$$

where $(X, \varphi X)$ and $(Z, \varphi Z)$ are basis pairs (as before) corresponding to λ and ν, respectively. By a similar calculation to those done earlier, we get $(\lambda - \nu)(\lambda\nu + c - \kappa) = 0$ which implies that $\kappa = 0$. On the other hand, as we have seen earlier, by considering $[R(W,X) - \kappa(W \wedge X), A]$, we have

$$(\lambda - \alpha)(\lambda\alpha + c - \kappa) = 0,$$

i.e., $0 = \nu(\lambda\alpha - \lambda\nu) = \nu\lambda^2$, a contradiction.

Since we have shown that $AX = \lambda X$ for all $X \in \mathcal{W}^\perp$ and λ is a constant (expressible in terms of α through the quadratic equation), we can conclude that M is an open subset of a hypersurface of Type A_0 or Type A_1. □

8.5.3 *In terms of curvature and Ricci tensors*

In the preceding subsection, we have dealt with classification of hypersurfaces in terms of its shape operator. We now move on to classification in terms of the intrinsic geometry of the hypersurface as expressed by its curvature tensor. The familiar simple conditions such as constant sectional curvature cannot be realized. In fact, there are no locally symmetric hypersurfaces (i.e., with $\nabla R = 0$). This condition may be weakened to $R \cdot R = 0$ (semisymmetric) and further to $R \cdot S = 0$ (Ricci-semisymmetric, also called *Ryan*) or to $\nabla S = 0$ (Ricci-parallel).

We will begin by classifying the pseudo-Einstein hypersurfaces. The first step is to show that such hypersurfaces are Hopf.

Lemma 8.61. *Let M^{2n-1}, where $n \geq 3$, be a real hypersurface in $\mathbf{CP}^n$ or $\mathbf{CH}^n$. Suppose that there is a scalar function ρ such that $SX = \rho X$ for all $X \in \mathcal{W}^{\perp}$. Then M is a Hopf hypersurface.*

Proof. Assume that $AW \neq \alpha W$ at some point and let $\mathcal{U}$ be an open neighborhood where the standard non-Hopf setup holds. Since $\mathcal{W}^{\perp}$ is S-invariant, W must also be an eigenvector and we may write $SW = (\rho + \sigma)W$ for some scalar function σ. Using (8.34) we compute

$$(\mathbf{m}A - A^2)W = (\mathbf{m}\alpha - \alpha^2)W + (\mathbf{m} - \alpha)\beta U - \beta AU.$$

Since W is an eigenvector of S, $(\mathbf{m}A - A^2)W$ must be a scalar multiple of W and hence AU lies in the span of $\{W, U\}$. Thus $\mu = 0$ and $AU = \beta W + \lambda U$ so that

$$(\mathbf{m}A - A^2)W = (\mathbf{m}\alpha - \alpha^2 - \beta^2)W + (\mathbf{m} - \alpha - \lambda)\beta U.$$

We conclude that

$$\mathbf{m} = \alpha + \lambda \tag{8.69}$$

and

$$(\mathbf{m}A - A^2)W = (\alpha\lambda - \beta^2)W.$$

Further, we can compute

$$(\mathbf{m}A - A^2)U = (\alpha\lambda - \beta^2)U$$

so that $\sigma = -3c$. Because all vectors in $\mathcal{W}^{\perp}$ share the same eigenvalue ρ with U, we can conclude that

$$\mathbf{m}A - A^2 = (\alpha\lambda - \beta^2)I \tag{8.70}$$

on the whole tangent space, which proves that $\mathcal{U}$ has at most two distinct principal curvatures. By Lemma 8.30, we have a contradiction. □

Remark 8.62. Lemma 8.61 is also true when $n = 2$. However, this requires some extra work. Our conditions lead, in view of Remark 8.31, to $\nu = \lambda\alpha - \beta^2 = 0$ and ultimately, using Lemma 6.8 of [399], to a contradiction. For details, see the proof of Proposition 2.13 in Kim and Ryan [260].

We are now ready to classify the pseudo-Einstein hypersurfaces for $n \geq 3$.

Theorem 8.63. *Let M^{2n-1}, where $n \geq 3$, be a pseudo-Einstein hypersurface in $\mathbf{CP}^n$ or $\mathbf{CH}^n$. Then M is a Hopf hypersurface with constant principal curvatures. Specifically, it is an open subset of one of the following hypersurfaces: (See Section 6.5.)*

- *Type* A_0 – *a horosphere in* $\mathbf{CH}^n$*;*
- *Type* A_1 – *a geodesic sphere in* $\mathbf{CP}^n$ *or* $\mathbf{CH}^n$ *or a tube over a totally geodesic* $\mathbf{CH}^{n-1}$ *in* $\mathbf{CH}^n$*;*
- *Type* A_2 – *a tube of radius ru over a totally geodesic* $\mathbf{CP}^k$ *in* $\mathbf{CP}^n$ *where* $1 \le k \le n-2$, $0 < u < \frac{\pi}{2}$, *and* $\cot^2 u = \frac{k}{\ell}$, *where* $k + \ell = n - 1$*;*
- *Type B – a tube of radius ru over a complex quadric* Q^{n-1} *in* $\mathbf{CP}^n$ *where* $0 < u < \frac{\pi}{4}$ *and* $\cot^2 2u = n - 2$.

Proof. By Lemma 8.61, M is a pseudo-Einstein Hopf hypersurface. Then

$$(\mathbf{m}A - A^2)X = (\rho - (2n-1)c)X \tag{8.71}$$

for all $X \in \mathcal{W}^\perp$. In particular, if λ and μ are principal curvatures corresponding to principal vectors in $\mathcal{W}^\perp$, we have

$$(\lambda - \mu)(\mathbf{m} - (\lambda + \mu)) = 0. \tag{8.72}$$

For convenience, assume that $\alpha^2 + 4c \neq 0$ so that $\frac{\alpha}{2}$ does not occur as a principal curvature. Consider now one particular point p in M and one particular choice of X and λ. Then there is unique number μ such that $A\varphi X = \mu\varphi X$ at p. If $\lambda \neq \mu$, then λ, μ, and α are the only principal curvatures at p and $\mathcal{W}^\perp$ splits into complementary subspaces of dimension $n-1$ consisting of eigenvectors of λ and μ, respectively. We may now choose an open neighborhood $\mathcal{U}$ on which the largest and smallest eigenvalues of A (restricted to $\mathcal{W}^\perp$) remain distinct. Then λ and μ and the corresponding eigenspaces extend smoothly to $\mathcal{U}$ and

$$\mathbf{m} = (n-1)(\lambda + \mu) + \alpha = \lambda + \mu, \tag{8.73}$$

so that $\lambda + \mu$ is a constant multiple of α. On the other hand, Theorem 6.17 gives

$$\lambda\mu = \frac{\lambda + \mu}{2}\alpha + c, \tag{8.74}$$

which allows us to express $\lambda - \mu$ in terms of α. The net result is that both λ and μ must be constant and hence $\mathcal{U}$ must be an open subset of a hypersurface from the Takagi/Montiel lists. A quick examination of these lists shows that the only possibility is the Type B hypersurface mentioned in the theorem statement. By a standard continuity and connectedness argument, M is an open subset of this same Type B hypersurface.

Returning now to our original choice of λ and μ, the alternative possibility is that $\lambda = \mu$, i.e., $\lambda^2 = \lambda\alpha + c$. If there is a second eigenvalue ν at p, then it must satisfy the same quadratic equation. The tangent space splits into two φ-invariant subspaces consisting of eigenvectors corresponding to eigenvalues λ and ν, respectively, with respective (even) dimensions 2ℓ and $2k$ (say). A similar argument shows that the same setup holds in a neighborhood with λ and ν as distinct constant principal curvatures. This neighborhood, and by the connectedness argument M itself, is an open subset of the Type A_2 hypersurface mentioned in the theorem statement.

We have already classified the Hopf hypersurfaces with $g \leq 2$ distinct principal curvatures (Theorem 8.29). Thus we can exclude the possibility that $AX = \lambda X$ for all $X \in \mathcal{W}^{\perp}$ at all $p \in M$ as long as we include the Type A_1 hypersurfaces in our theorem statement.

Finally, if $\alpha^2 + 4c = 0$, suppose that $\lambda \neq \frac{\alpha}{2}$ could be chosen. For all eigenvectors X corresponding to λ, we will get $A\varphi X = \frac{\alpha}{2}\varphi X$ and $\mathbf{m} = \lambda + \frac{\alpha}{2}$. The set of all such pairs $(X, \varphi X)$ spans a 2ℓ-dimensional space, where ℓ is the multiplicity of λ. Its complementary subspace in $\mathcal{W}^{\perp}$ is spanned by eigenvectors corresponding to the eigenvalue $\frac{\alpha}{2}$, and thus $\mathbf{m} = \alpha + \ell\lambda + (2n - 2 - \ell)\frac{\alpha}{2}$. This allows us to express λ as a constant multiple of α. By the same argument as used earlier, M must be an open subset of a Hopf hypersurface with constant principal curvatures. However, from Berndt's classification (Theorem 8.12), there is no hypersurface admitting such a principal curvature configuration. The Type A_0 hypersurface in the theorem statement is the only possibility satisfying $\alpha^2 + 4c = 0$. This completes the proof of our theorem. □

The analogue of Theorem 8.63 for $n = 2$ was proved by Kim and Ryan [260].

Theorem 8.64. *Let M be a pseudo-Einstein hypersurface in* $\mathbf{CP}^2$ *or* $\mathbf{CH}^2$. *Then M is one of the following:*

- *an open subset of a Type A_0 or Type A_1 hypersurface, or;*
- *a Hopf hypersurface with $\alpha = 0$ and nonconstant principal curvatures λ and ν satisfying $\lambda\nu = c$.*

As we have seen, there are no totally umbilic hypersurfaces. We say that the hypersurface is *η-umbilical* if there is a scalar function λ such that $AX = \lambda X$ for all $X \in \mathcal{W}^{\perp}$. The weakening of totally umbilic to η-umbilical is analogous to the weakening of Einstein to pseudo-Einstein. It also anticipates the notions of η-parallelism and η-recurrence, to be discussed in later sections. The terminology arises from the fact that the letter η has traditionally been used for the contact 1-form satisfying $\eta(X) = \langle X, W \rangle$ (so that $\mathcal{W}^{\perp}$ is the nullspace of η).

Clearly, for an η-umbilical hypersurface M, the holomorphic distribution $\mathcal{W}^{\perp}$ is A-invariant. Thus η-umbilical hypersurfaces are Hopf. Furthermore, using Theorem 6.17, we see that

$$\lambda^2 = \alpha\lambda + c, \tag{8.75}$$

so that λ is constant. Thus M is a Hopf hypersurface with constant principal curvatures. Using the classification theorem of Kimura and Berndt and inspecting the lists of Takagi and Montiel, we have the following.

Theorem 8.65. *Let M^{2n-1}, where $n \geq 2$, be a real hypersurface in* $\mathbf{CP}^n$ *or* $\mathbf{CH}^n$. *Then M is η-umbilical if and only if it an open subset of a hypersurface of Type A_1 or Type A_0.*

These hypersurfaces can also be characterized by properties of the curvature tensor, as follows:

Theorem 8.66. *Let M^{2n-1}, where $n \geq 2$, be a real hypersurface in $\mathbf{CP}^n$ or $\mathbf{CH}^n$. Then there is a real function κ such that, for all tangent vectors X and Y,*

$$(R(X,Y) - \kappa(X \wedge Y))W = 0 \tag{8.76}$$

if and only if M is

1. *an open subset of a hypersurface of Type A_1 or Type A_0, or*
2. *a Hopf hypersurface with Hopf principal curvature $\alpha = 0$.*

Proof. The first step is to show that any hypersurface satisfying (8.76) must be Hopf. We assume this for the moment and defer proof until the next lemma, Lemma 8.67. We now show that any η-umbilical hypersurface (Alternative 1) satisfies (8.76) with $\kappa = \lambda\alpha + c$. In fact, by the Gauss equation,

$$R(X,Y) = (\lambda^2 + c)(X \wedge Y) + c(\varphi X \wedge \varphi Y + 2\langle X, \varphi Y\rangle\varphi), \tag{8.77}$$

for all X and Y in $\mathcal{W}^\perp$. Thus $R(X,Y) - \kappa(X \wedge Y)$ annihilates W no matter what value of κ is chosen. On the other hand, if $X \in \mathcal{W}^\perp$ but $Y = W$,

$$R(X,Y) = (\lambda\alpha + c)(X \wedge Y). \tag{8.78}$$

This establishes our assertion.

On the other hand, if M is a Hopf hypersurface with $\alpha = 0$ (Alternative 2), then $R(X,W) = c(X \wedge W)$ while $R(X,Y)W = (X \wedge Y)W = 0$ for all principal vectors X and Y in $\mathcal{W}^\perp$. Thus every Hopf hypersurface with $\alpha = 0$ satisfies (8.76) with $\kappa = c$.

To complete our proof, it is sufficient to show that a Hopf hypersurface satisfying (8.76) must fall under Alternative 1 or Alternative 2. If $X \in \mathcal{W}^\perp$ is a principal vector satisfying $AX = \lambda X$, then by setting $Y = W$ in (8.76), we get $\kappa = \lambda\alpha + c$. Thus, unless $\alpha = 0$ (Alternative 2), two distinct values of λ cannot occur and M is η-umbilical (Alternative 1). □

Lemma 8.67. *Let M^{2n-1}, where $n \geq 2$, be a real hypersurface in $\mathbf{CP}^n$ or $\mathbf{CH}^n$. If there is a real function κ such that for all tangent vectors X and Y,*

$$(R(X,Y) - \kappa(X \wedge Y))W = 0, \tag{8.79}$$

then M is a Hopf hypersurface.

Proof. We assume the standard non-Hopf setup and will derive a contradiction. Setting $Y = U$ in (8.76) yields $(AX \wedge AU)W = 0$ for any X orthogonal to both W and U. This reduces to

$$\beta AX = 0. \tag{8.80}$$

In particular, taking $X = \varphi U$, this shows that $\mu = \nu = 0$. Also,

$$0 = R(X, W)W - \kappa(X \wedge W)W = (AX \wedge AW)W + (c - \kappa)(X \wedge W)W = (c - \kappa)X, \tag{8.81}$$

so that $\kappa = c$, a constant. Finally, we have

$$0 = R(U, W)W - \kappa(U \wedge W)W = (AU \wedge AW)W + (c - \kappa)(U \wedge W)W \tag{8.82}$$

which reduces to

$$\alpha AU = \beta AW. \tag{8.83}$$

Thus the first two columns of A are linearly dependent and A has rank 1. This contradicts the fact that a hypersurface with fewer than three distinct principal curvatures must be Hopf (Lemma 8.30) unless $n = 2$. For $n = 2$, using (8.34), we compute the matrix of the Ricci tensor

$$\begin{bmatrix} 2c + \alpha(\lambda + \nu) - \beta^2 & \nu\beta & -\mu\beta \\ \nu\beta & 5c + \lambda(\nu + \alpha) - \beta^2 - \mu^2 & \mu\alpha \\ -\mu\beta & \mu\alpha & 5c + \nu(\lambda + \alpha) - \mu^2 \end{bmatrix}. \tag{8.84}$$

Since $\mu = \nu = \alpha\lambda - \beta^2 = 0$, we see that M is pseudo-Einstein. However, pseudo-Einstein hypersurfaces must be Hopf (See Lemma 8.61 and Remark 8.62), so we have a contradiction. An explicit proof that there are no hypersurfaces with

$$S = \begin{bmatrix} 2c & 0 & 0 \\ 0 & 5c & 0 \\ 0 & 0 & 5c \end{bmatrix}, \tag{8.85}$$

may be found in [260, p. 108]. □

Remark 8.68. The condition addressed in Theorem 8.66 was introduced by Cho and Ki [111, 114]. However, their proof that M is Hopf is incomplete and the possibility of Alternative 2 occurring in $\mathbf{CH}^n$ is not recognized. See also Remark 8.126.

We now proceed to discuss further classification theorems. The $\nabla S = 0$ condition can be weakened in several ways:

A Riemannian manifold is said to be

- *cyclic-Ryan* if the cyclic sum of $(R(X, Y) \cdot S)Z$ over every triple (X, Y, Z) of tangent vectors vanishes. Note that since S is of type $(1, 1)$,

$$(R(X, Y) \cdot S)Z = R(X, Y)(SZ);$$

- of *harmonic curvature* if the Ricci tensor S is a Codazzi tensor, i.e., $(\nabla_X S)Y = (\nabla_Y S)X$ for every pair (X, Y) of tangent vectors.

We now summarize results on these conditions and others introduced earlier in this section.

Theorem 8.69. *Let M^{2n-1}, where $n \geq 2$, be a real hypersurface in* $\mathbf{CP}^n$ *or* $\mathbf{CH}^n$. *Then*

- *M is not Ricci-parallel;*
- *M is not Ricci-semisymmetric except perhaps if $n = 2$;*
- *M does not have harmonic curvature if $n \geq 3$;*
- *If $n = 2$, M is automatically cyclic-Ryan;*
- *If $n \geq 3$, M is cyclic-Ryan if and only if it is a pseudo-Einstein hypersurface in the Takagi-Montiel lists (see Theorem 8.63).*

For a further discussion of these results, we refer the reader to Theorems 6.20, 6.28, 6.29, and 6.30 of [399]. Some complete proofs are given there along with references to the original papers where they appeared. The fact that M cannot be Ricci-parallel when $n \geq 3$ was first proved by Ki [239] and a proof is included in [399]. The result for $n = 2$ is more recent, and can be found in Ryan [470]. It was also proved independently by U.K. Kim [269].

Remark 8.70. We know that there are no Hopf hypersurfaces in $\mathbf{CP}^2$ or $\mathbf{CH}^2$ that are Ricci-semisymmetric (see Theorem 6.29 in [399]). The existence of non-Hopf Ricci-semisymmetric hypersurfaces is an open question. However, there is a weaker condition that can be satisfied. A hypersurface is said to be *pseudo-Ryan* if $(R(X,Y)\cdot S)\mathcal{W}^\perp \subseteq \mathcal{W}$ for all X and Y in $\mathcal{W}^\perp$. In [224], Ivey and Ryan show how to construct a broad class of non-Hopf pseudo-Ryan hypersurfaces in $\mathbf{CP}^2$ and $\mathbf{CH}^2$. The Hopf pseudo-Ryan hypersurfaces in $\mathbf{CP}^n$ and $\mathbf{CH}^n$ coincide with the pseudo-Einstein ones for all $n \geq 2$. This is part of Theorem 6.30 of [399] for $n \geq 3$ and can be checked directly for $n = 2$ using Theorem 8.64. We state it as Theorem 8.71 below.

The existence question for hypersurfaces in $\mathbf{CP}^2$ and $\mathbf{CH}^2$ with harmonic curvature also seems to be open.

Theorem 8.71. *Let M^{2n-1}, where $n \geq 2$, be a Hopf hypersurface in* $\mathbf{CP}^n$ *or* $\mathbf{CH}^n$. *Then M is pseudo-Ryan if and only if it is pseudo-Einstein.*

The tensor fields $R \cdot A$ and $R \cdot S$ have a particularly simple form for Type A_0 and Type A_1 hypersurfaces. The following result is due to Kimura and Maeda [277] for $\mathbf{CP}^n$ and Choe [118] for $\mathbf{CH}^n$.

Theorem 8.72. *Let M^{2n-1}, where $n \geq 2$, be a real hypersurface in* $\mathbf{CP}^n$ *or* $\mathbf{CH}^n$. *The following are equivalent.*

- *M is an open subset of a Type A_0 or Type A_1 hypersurface;*
- *There is a function κ such that*

$$(R(X,Y)\cdot A)Z = \kappa\big(\langle Z,W\rangle(X\wedge Y)W + \langle (X\wedge Y)W, Z\rangle W\big). \tag{8.86}$$

The function κ is constant and equal to $-\lambda c$. Further, for $n \geq 3$,

- *There is a function κ such that*

$$(R(X,Y)\cdot S)Z = \kappa\big(\langle Z,W\rangle(X\wedge Y)W + \langle (X\wedge Y)W, Z\rangle W\big) \tag{8.87}$$

if and only if M is an open subset of a hypersurface of Type A_0 or Type A_1 or (in the case of $\mathbf{CP}^n$ where n is odd) a Type A_2 hypersurface with $\alpha = 0$ and principal curvatures $\pm\frac{1}{r}$ of equal multiplicities. The function κ is constant and equal to $-2nc\lambda^2$ ($-2c^2$ for Type A_2).

Finally, for $n = 2$,

- *There is a function κ satisfying* (8.87) *if and only if M is a Hopf hypersurface with $\alpha = 0$ and nonconstant principal curvatures (and $\kappa = -4c^2$) or M is an open subset of a hypersurface of Type A_0 or Type A_1 (and $\kappa = -4c\lambda^2$).*

Remark 8.73. In Theorem 8.72, λ denotes the principal curvature of multiplicity $2n-2$. Note that all hypersurfaces occurring in this theorem are pseudo-Einstein, but only one special pseudo-Einstein hypersurface of Type A_2 satisfies (8.87).

Pseudosymmetry Conditions

A Riemannian manifold is said to be *pseudosymmetric*, respectively, *Ricci-pseudosymmetric*, if there is some function κ such that

$$(R(X,Y) - \kappa(X\wedge Y))\cdot R = 0, \text{ respectively, } (R(X,Y) - \kappa(X\wedge Y))\cdot S = 0,$$

for all tangent vectors X and Y. (Recall notation from Proposition 8.58 on page 466.) Note that the case $\kappa = 0$ would correspond to semisymmetry (resp., Ricci-semisymmetry). Pseudosymmetry and related conditions were introduced and have been studied extensively by R. Deszcz [130].

Remark 8.74. Inoguchi [219] uses the terms "Ricci pseudo-parallel" and "with pseudo-parallel Ricci operator" instead of "Ricci-pseudosymmetric." The latter term is more widely used in the literature.

To deal with pseudosymmetry more efficiently, we introduce the following abbreviation:

$$Q(Z_1, Z_2) = R(Z_1, Z_2) - \kappa(Z_1 \wedge Z_2)$$

for tangent vectors Z_1 and Z_2. Since S is a contraction of R, it is easy to see that every pseudosymmetric manifold is Ricci-pseudosymmetric.

On the other hand, we observe that the cyclic sum of $(X\wedge Y)SZ$ vanishes for all X, Y, and Z. Thus every Ricci-pseudosymmetric manifold is cyclic-Ryan. Applying Theorem 8.69, we get the first part of the following classification theorem:

Theorem 8.75. *Let M^{2n-1}, where $n \geq 3$, be a Ricci-pseudosymmetric hypersurface in* $\mathbf{CP}^n$ *or* $\mathbf{CH}^n$. *Then M is a pseudo-Einstein Hopf hypersurface with constant principal curvatures. In fact, M must be an open subset of*

- *a Type A_0 or Type A_1 hypersurface (with $\kappa = \lambda\alpha + c$), or*
- *a Type A_2 hypersurface having principal curvatures $(0, 1/r, -1/r)$ with respective multiplicities $(1, 2k, 2k)$, where $k \geq 1$ and $n = 2k + 1$. The hypersurface lies on a tube of radius $\frac{\pi}{4}r$ about* $\mathbf{CP}^k$. *The value of κ is c (which is the same as $\lambda\alpha + c$ in this case).*

Proof. We need only consult the list of pseudo-Einstein hypersurfaces and check which ones are Ricci-pseudosymmetric. First take a unit vector X in $\mathcal{W}^\perp$ such that $AX = \lambda X$. Then

$$[Q(X, W), S]W = (\lambda\alpha + c - \kappa)(\rho_\alpha - \rho_\lambda)X.$$

Thus, Ricci-pseudosymmetry requires that $\kappa = \lambda\alpha + c$. If there is a second principal curvature μ, we must also have $\kappa = \mu\alpha + c$, and hence $\alpha = 0$. The Type B pseudo-Einstein hypersurfaces are not Ricci-pseudosymmetric since they do not have $\alpha = 0$. A Type A_2 Ricci-pseudosymmetric hypersurface must have $\alpha = 0$ and be pseudo-Einstein. Thus, it lies in $\mathbf{CP}^n$ and has $u = \frac{\pi}{4}$ and $k = \ell$. This eliminates all pseudo-Einstein hypersurfaces not occurring in the theorem statement. Conversely, it is routine to check that all Type A_0 and A_1 hypersurfaces and the special Type A_2 hypersurface occurring in the theorem statement are in fact Ricci-pseudosymmetric. This completes the proof. □

Theorem 8.75 was essentially proved by I.-B. Kim, H.J. Park, and H. Song [263]. However, their proof contained an error that caused them to exclude the Type A_1 hypersurfaces from their list. Subsequently, the error was corrected by Inoguchi [219].

When $n = 2$, the cyclic-Ryan condition offers no restriction. We must check for Ricci-pseudosymmetry directly. We first consider Hopf hypersurfaces.

Theorem 8.76. *Let M be a Hopf hypersurface in* $\mathbf{CP}^2$ *or* $\mathbf{CH}^2$. *Then M is Ricci-pseudosymmetric if and only if it is a pseudo-Einstein hypersurface with Hopf principal curvature $\alpha = 0$, or an open subset of*

- *a Type A_0 or Type A_1 hypersurface, or*
- *a Type B hypersurface in* $\mathbf{CP}^2$ *which is a tube of radius $\frac{\pi}{6}r$ over* $\mathbf{RP}^2$.

Proof. At an arbitrary point $p \in M$, we can choose an orthonormal basis $(W, X, Y = \varphi X)$ for T_pM such that $AW = \alpha W$, $AX = \lambda X$ and $AY = \mu Y$ for suitable scalars λ and μ. Then

$$Q(X, W) = (\lambda\alpha + c - \kappa)X \wedge W,$$
$$Q(Y, W) = (\mu\alpha + c - \kappa)Y \wedge W,$$

$$Q(X,Y) = (\lambda\mu + 4c - \kappa)X \wedge Y.$$

The Ricci tensor satisfies

$$SW = \rho_\alpha W = (2c + (\lambda + \mu)\alpha)W,$$
$$SX = \rho_\lambda X = (5c + (\lambda\mu + \lambda\alpha))X,$$
$$SY = \rho_\mu Y = (5c + (\lambda\mu + \mu\alpha))Y.$$

Then we have

$$[Q(X,W),S]W = (\lambda\alpha + c - \kappa)(\rho_\alpha - \rho_\lambda)X, \tag{8.88}$$

$$[Q(Y,W),S]W = (\mu\alpha + c - \kappa)(\rho_\alpha - \rho_\mu)Y, \tag{8.89}$$

$$\begin{aligned}[Q(X,Y),S]Y &= (\lambda\mu + 4c - \kappa)(\rho_\mu - \rho_\lambda)X \\ &= -\alpha(\lambda\mu + 4c - \kappa)(\lambda - \mu).\end{aligned} \tag{8.90}$$

Now suppose that M is Ricci-pseudosymmetric. Consider the possibility that $\alpha(\lambda - \mu) \neq 0$. From equation (8.90), we get $\kappa = \lambda\mu + 4c$. Also, $\lambda\alpha + c - \kappa$ and $\mu\alpha + c - \kappa$ cannot both be nonzero. Without loss of generality, we can assume that $\lambda\alpha + c - \kappa \neq 0$ so that $\rho_\alpha = \rho_\lambda$, which simplifies to $3c + (\lambda - \alpha)\mu = 0$. Also, $\mu\alpha + c = \kappa = \lambda\mu + 4c$, but this reduces to the same condition. Using Theorem 6.17 to express μ in terms of λ and α, we find that λ must satisfy the quadratic equation

$$\alpha t^2 - (\alpha^2 - 8c)t - 5c\alpha = 0.$$

By a similar calculation, μ must satisfy

$$\alpha t^2 - (\alpha^2 + 8c)t + 3c\alpha = 0.$$

This allows us to express λ and μ in terms of α. The analysis we have just completed extends to a neighborhood of p. Since α is constant, this neighborhood is a Hopf hypersurface with constant principal curvatures (α, λ, μ) where $\lambda \neq \mu$. By a simple continuity and connectedness argument, M is an open subset of a Type B hypersurface. One can check directly that for a Type B hypersurface with $\alpha \neq 0$, the Ricci-pseudosymmetry condition implies that $c > 0$, and

$$\alpha = \frac{2\sqrt{3}}{r}, \quad \lambda = -\frac{\sqrt{3}}{r}, \quad \mu = \frac{\sqrt{3}}{3r}.$$

Thus, $\kappa = \lambda\mu + 4c = 3c$, $\rho_\alpha = \rho_\lambda$ and $\mu\alpha + c = \kappa$. The Type B hypersurface must be a tube of radius $\frac{\pi}{6}r$ over $\mathbf{RP}^2$ in $\mathbf{CP}^2$.

Conversely, based on this discussion, it is easy to check that for this particular hypersurface, every expression of the form $[Q(Z_1, Z_2), S]Z_3$ vanishes, so that the hypersurface is indeed Ricci-pseudosymmetric.

Now consider the alternative scenario, namely $\alpha(\lambda - \mu) = 0$ everywhere. If $\alpha \neq 0$, then $\lambda = \mu$ everywhere, and M is an open subset of a Type A_1 or Type A_0 hypersurface. Also, note that such hypersurfaces are Ricci-pseudosymmetric with $\kappa = \lambda\alpha + c$. Finally, if $\alpha = 0$, we have $\lambda\mu = c$ by Theorem 6.17, so that $\rho_\lambda = \rho_\mu = 6c \neq 2c = \rho_\alpha$. Thus, the Ricci-pseudosymmetry condition (8.90) requires that $\kappa = c$, and M must be a pseudo-Einstein hypersurface. Since every pseudo-Einstein hypersurface in $\mathbf{CP}^2$ or $\mathbf{CH}^2$ is Hopf (see Theorem 8.64), we can now verify that those with $\alpha = 0$ must be Ricci-pseudosymmetric. This completes the proof of our theorem. □

Remark 8.77. For a Hopf hypersurface M^3 in $\mathbf{CP}^2$ or $\mathbf{CH}^2$, there are just two possibilities. Either there is a point where $\alpha(\lambda - \mu) \neq 0$, or every point satisfies $\alpha(\lambda - \mu) = 0$. In the former case, as we have seen, the hypersurface is Ricci-pseudosymmetric if and only if it is an open subset of a certain Type B hypersurface in $\mathbf{CP}^2$. For this hypersurface, we have $\kappa = 3c$. The second case splits in two. If $\alpha \neq 0$, then M is an open subset of a Type A_1 or Type A_0 hypersurface, and M is Ricci-pseudosymmetric with $\kappa = \lambda\alpha + c = \lambda^2$. The remaining possibility is that M everywhere matches the pointwise characterization of pseudo-Einstein hypersurfaces with $\alpha = 0$ (see [260], p.109).

Theorem 8.76 tells us that, except for the isolated Type B example, a Hopf hypersurface in $\mathbf{CP}^2$ or $\mathbf{CH}^2$ is Ricci-pseudosymmetric if and only if it is pseudo-Einstein. The classification of pseudo-Einstein hypersurfaces (see Theorem 8.64 and, for more detail, [260] and [222]) includes hypersurfaces with nonconstant principal curvatures, as well as the well-known Type A_0 and Type A_1 hypersurfaces. The existence question for non-Hopf Ricci-pseudosymmetric hypersurfaces still seems to be open.

We are now able to classify the pseudosymmetric hypersurfaces for $n \geq 3$ by refining Theorem 8.75.

Proposition 8.78. *For $n \geq 2$, all Type A_0 and Type A_1 hypersurfaces are pseudosymmetric, but Type A_2 hypersurfaces are not.*

Proof. Consider a Type A hypersurface and choose a unit principal vector $X \in \mathcal{W}^\perp$. Then for a suitable number λ, we have $AX = \lambda X$, $A\varphi X = \lambda\varphi X$ and $AW = \alpha W$, where $\lambda^2 = \lambda\alpha + c$. A necessary condition for pseudosymmetry is

$$[Q(X, W), R(X, \varphi X)] - R(Q(X, W)X, \varphi X) - R(X, Q(X, W)\varphi X) = 0. \tag{8.91}$$

By the Gauss equation, $R(X, W) = (\lambda\alpha + c)(X \wedge W)$. Also,

$$R(X, \varphi X) = (\lambda^2 + 2c)(X \wedge \varphi X) - 2c\varphi.$$

Similar statements hold for Q with respective coefficients $\lambda\alpha + c - \kappa$ and $\lambda^2 + 2c - \kappa$. It is easy to check that for any orthonormal triple (Z_1, Z_2, Z_3) of tangent vectors,

$$[Z_1 \wedge Z_2,\ Z_1 \wedge Z_3] = Z_3 \wedge Z_2$$

and

$$[\varphi, Z_1 \wedge Z_2] = Z_1 \wedge \varphi Z_2 + \varphi Z_1 \wedge Z_2.$$

Thus (8.91) implies that

$$(\lambda\alpha + c - \kappa)(\lambda^2 + 4c) = (\lambda\alpha + c - \kappa)(\lambda\alpha + c) \tag{8.92}$$

which reduces to $\kappa = \lambda\alpha + c = \lambda^2$.

Now suppose we have a pseudosymmetric hypersurface of Type A_2. By Theorem 8.75 we must have $\alpha = 0$, $\lambda = \frac{1}{r}$, $\mu = -\frac{1}{r}$ and hence $\kappa = c$. We show that $(Q(X, Y) \cdot R)(X, \varphi X) \neq 0$ where X and Y are unit principal vectors corresponding to respective principal curvatures λ and μ. To see this, note that

$$R(X, Y) = (\lambda\mu + c)X \wedge Y + c\,\varphi X \wedge \varphi Y,$$

so that

$$\begin{aligned}
[Q(X, Y), R(X, \varphi X)] &= -7c^2(\varphi X \wedge Y + X \wedge \varphi Y); \\
R(Q(X, Y)X, \varphi X) &= c^2\, X \wedge \varphi Y; \\
R(X, Q(X, Y)\varphi X) &= c^2\, \varphi X \wedge Y.
\end{aligned}$$

This contradiction shows that Type A_2 hypersurfaces cannot, in fact, be pseudosymmetric. On the other hand, it is straightforward, though tedious, to check that Type A_0 and Type A_1 hypersurfaces are pseudosymmetric. Specifically, if $(Z_i, \varphi Z_i)$ form an orthonormal basis for $\mathcal{W}^{\perp}$, we need to check that for all i, $Q(Z_i, \varphi Z_i) \cdot R$ sends all argument pairs of the form (Z_j, Z_k), $(Z_j, \varphi Z_j)$, (Z_j, W) to zero and, for all pairs of distinct indicies (i, j), $Q(Z_i, Z_j) \cdot R$ does the same. In this analysis, it is necessary to consider separately the cases of arguments involving distinct and coincident indices. □

Thus we have the following classification result for pseudosymmetric hypersurfaces.

Theorem 8.79. *Let M^{2n-1}, where $n \geq 3$ be a hypersurface in $\mathbf{CP}^n$ or $\mathbf{CH}^n$. Then M is pseudosymmetric if and only if it is an open subset of a hypersurface of Type A_0 or Type A_1. A Hopf hypersurface in $\mathbf{CP}^2$ or $\mathbf{CH}^2$ is pseudosymmetric if and only if it is Ricci-pseudosymmetric.*

Proof. Since every pseudosymmetric hypersurface is Ricci-pseudosymmetric, Theorem 8.75 together with Proposition 8.78 establishes our claim for $n \geq 3$ and for the Type A_0 and Type A_1 possibilities in case $n = 2$. It remains to check that the other hypersurfaces occurring in Theorem 8.76 are pseudosymmetric. For the Type B hypersurface, we again take $\kappa = 3c$ and note that $Q(X, \varphi X) = 0$. Then

$$[Q(X, W), R(X, \varphi X)] = 3c(\lambda\alpha - 2c)\,\varphi X \wedge W$$

while $R(Q(X,W)X,\varphi X) = (\lambda\alpha - 2c)(\mu\alpha + c)\,\varphi X \wedge W$ and $R(X, Q(X,W)\varphi X) = 0$. This is consistent with pseudosymmetry since the hypersurface in question satisfies $\mu\alpha = 2c$. By a similar calculation, we get $(Q(X,W)\cdot R)(\varphi X, W) = 0$. Trivially, $(Q(X,W)\cdot R)(X,W) = 0$. Since $Q(\varphi X, W) = (\mu\alpha + c - 3c)\,\varphi X \wedge W = 0$, we have shown that the Type B hypersurface is pseudosymmetric.

We now consider the pseudo-Einstein hypersurfaces with $\alpha = 0$. Take, as before, an orthonormal basis $(W, X, \varphi X)$ with $AX = \lambda X$ and $A\varphi X = \mu\varphi X$. Then $\lambda\mu = c$ and, since $n = 2$, we have $\varphi = \varphi X \wedge X$. Take $\kappa = c$. The only nonzero possibility for Q is $Q(X,\varphi X) = 5c\, X \wedge \varphi X$. Trivially, $(Q(X,\varphi X)\cdot R)(X,\varphi X)$ vanishes. In addition, $[Q(X,\varphi X), R(X,W)] = 4c^2\, W \wedge \varphi X$ while $R(Q(X,\varphi X)X, W) = 4cR(-\varphi X, W) = -4c^2\,\varphi X \wedge W$ and $R(X, Q(X,\varphi X)W) = 0$. This establishes pseudosymmetry for the pseudo-Einstein hypersurfaces with $\alpha = 0$ as required. □

Remark 8.80. The existence question for non-Hopf pseudosymmetric hypersurfaces in $\mathbf{CP}^2$ and $\mathbf{CH}^2$ seems to be open.

Special Forms of ∇S

If a hypersurface is pseudo-Einstein, the covariant derivative of the Ricci tensor has a special form,

$$(\nabla_X S)Y = \sigma(\langle \varphi AX, Y\rangle W + \langle Y, W\rangle \varphi AX) \tag{8.93}$$

for all tangent vectors X and Y, where σ is the constant satisfying

$$SX = \rho X + \sigma\langle X, W\rangle W.$$

We now ask about other possible hypersurfaces satisfying a similar condition. Of course Theorem 8.69 assures us that ∇S cannot vanish identically. It turns out that equation (8.93) characterizes the pseudo-Einstein hypersurfaces.

Theorem 8.81. *Let M^{2n-1}, where $n \geq 2$, be a real hypersurface in $\mathbf{CP}^n$ or $\mathbf{CH}^n$ satisfying the identity*

$$(\nabla_X S)Y = \kappa(\langle \varphi AX, Y\rangle W + \langle Y, W\rangle \varphi AX) \tag{8.94}$$

for some function κ which is not identically zero. Then M is pseudo-Einstein.

It is straightforward to deduce from (8.94) that

$$|\nabla S|^2 = 2\kappa^2(|A|^2 - |AW|^2). \tag{8.95}$$

Using the information given in Chapter 6, concerning the Takagi/Montiel lists, we can compute $|\nabla S|^2$ to get

Theorem 8.82. *For all $n \geq 2$, the pseudo-Einstein hypersurfaces in the Takagi-Montiel lists satisfy* (8.94) *with values of κ and $|\nabla S|^2$ as follows:*

- *Type A_0 in $\mathbf{CH}^n$: $\kappa = -2nc$; $|\nabla S|^2 = 16n^2(n-1)|c|^3$;*
- *Type A_1 in $\mathbf{CP}^n$: $\kappa = -2nc$; $|\nabla S|^2 = 16n^2(n-1)|c|^3 \cot^2 u$ (for geodesic sphere of radius ru);*
- *Type A_1 in $\mathbf{CH}^n$: $\kappa = -2nc$; $|\nabla S|^2 = 16n^2(n-1)|c|^3 \coth^2 u$ (for geodesic sphere of radius ru);*
- *Type A_1 in $\mathbf{CH}^n$: $\kappa = -2nc$; $|\nabla S|^2 = 16n^2(n-1)|c|^3 \tanh^2 u$ (for tube of radius ru over $\mathbf{CH}^{n-1}$);*
- *Type A_2 in $\mathbf{CP}^n$: $\kappa = -2c$; $|\nabla S|^2 = 16(n-1)|c|^3$;*
- *Type B in $\mathbf{CP}^n$: $\kappa = -2(2n-1)c$; $|\nabla S|^2 = 16n(n-1)\frac{(2n-1)^2}{n-2}|c|^3$.*

For $n = 2$, there is class of pseudo-Einstein hypersurfaces (see Theorem 8.64) not covered by Theorem 8.95. For these hypersurfaces, we have

Theorem 8.83. *For the pseudo-Einstein hypersurfaces in $\mathbf{CP}^2$ and $\mathbf{CH}^2$ that do not occur in the Takagi/Montiel lists,*

- *$\kappa = -4c$;*
- *$|\nabla S|^2 = 32c^2(\lambda^2 + \mu^2)$ where λ and μ are the nonconstant principal curvatures.*

Also, $\lambda\mu = c$ and $|\nabla S|^2$ is nonconstant.

There is an extensive discussion of material relevant to Theorem 8.81 in Section 6 of [399]. Here, we provide only the necessary additional argument for the proof.

Proof (of Theorem 8.81). Choose a point where $\kappa \neq 0$ and work in a neighborhood $\mathcal{U}$ of this point. By Lemmas 6.8 and 6.9 of [399], $\mathcal{U}$ is Hopf and $\mathbf{m}\alpha$ is constant there. In addition, the identity

$$(\kappa + 3c)\varphi AX = (A^2 - \mathbf{m}A + (\mathbf{m}\alpha - \alpha^2))\varphi AX \tag{8.96}$$

holds for all tangent vectors X. Note that either $\alpha = 0$ or $\mathbf{m}$ is constant. Every hypersurface has an open dense set (see Section 2.5) on which the principal curvatures have constant multiplicities. In the Hopf case, we can assert that on such a set, $\mathcal{W}^\perp$ consists locally of smooth φ-invariant distributions of the form T_λ or $T_\lambda + T_\mu$ where $T_\mu = \varphi T_\lambda$, as in Theorem 6.17. We can decrease the size of $\mathcal{U}$ if necessary so that this setup holds on $\mathcal{U}$. We consider the two cases separately.

Case I: $\alpha = 0$

For a pair (λ, μ) as in this setup, we have $\lambda\mu = c$. Thus neither λ nor μ vanishes. As a consequence of (8.96), we have

$$\kappa + 3c = \mu^2 - \mathbf{m}\mu = \lambda^2 - \mathbf{m}\lambda \tag{8.97}$$

so that $(\lambda - \mu)(\lambda + \mu - \mathbf{m}) = 0$. Note that if $\lambda = \mu$, then λ satisfies the quadratic equation $\lambda^2 = c$ and hence is constant. If $\lambda \neq \mu$, on the other hand, then $\lambda + \mu = \mathbf{m}$. Substituting in (8.96), we have $\kappa + 3c = \lambda^2 - (\lambda + \mu)\lambda = -c$. Therefore, we can rewrite (8.96) as

$$(A^2 - \mathbf{m}A + c)X = 0$$

for all $X \in \mathcal{W}^{\perp}$. If a principal curvature λ satisfying $\lambda^2 = c$ exists, we have $\mathbf{m}\lambda = 2c$. If not, then there is pair of principal curvatures (λ, μ) such that $\mathbf{m} = (n-1)(\lambda + \mu) = (n-1)\mathbf{m}$.

From this, there are only two possibilities,

1. $\mathbf{m}$ is constant and hence all principal curvatures are constant. $\mathcal{U}$ is an open subset of a Type A hypersurface with $\alpha = 0$. Type B, C, D, and E hypersurfaces do not have $\alpha = 0$.
2. $n = 2$ and M is a pseudo-Einstein hypersurface with nonconstant principal curvatures.

Case II: m is constant and $\alpha \neq 0$

Assume first that 0 is not a principal curvature. Then, using the same setup where $\lambda \neq \mu$ and $\lambda + \mu = \mathbf{m}$, we have $\lambda\mu = \frac{\mathbf{m}\alpha}{2} + c$ which is constant and consequently λ and μ are constants. Thus, all principal curvatures are constant and the local principal curvature data match that of a hypersurface on the Takagi/Montiel lists. If there is also a pair $(\tilde{\lambda}, \tilde{\mu})$ with $\tilde{\lambda} = \tilde{\mu}$, then

$$\tilde{\lambda}^2 - \mathbf{m}\tilde{\lambda} = (\kappa + 3c + \alpha^2 - \mathbf{m}\alpha) = \lambda^2 - \mathbf{m}\lambda$$

by (8.96), implying that $\lambda + \tilde{\lambda} = \mathbf{m} = \lambda + \mu$, a contradiction. This rules out Types C, D, and E. The only possible Type B match will be pseudo-Einstein, since $\lambda + \mu = \mathbf{m}$ will be satisfied (see Proposition 6.9). To see that the Type A_2 possibilities must also be pseudo-Einstein, apply the same argument involving (8.96) to the two φ-invariant principal subspaces. This results in the conclusion that the relevant principal curvatures satisfy $\lambda + \tilde{\lambda} = \mathbf{m}$, which is the pseudo-Einstein condition (again, see Proposition 6.9). Type A_0 and Type A_1 hypersurfaces are the only remaining possibilities.

The upshot is that either $\mathcal{U}$ is a pseudo-Einstein hypersurface with nonconstant principal curvatures or an open subset of a member of the Takagi/Montiel lists with principal curvatures of specific values and multiplicities. In the latter case, however, the set of points of M with these particular data is a closed set and our construction shows that it is also open. Therefore, it is all of M and M is an open subset of a particular pseudo-Einstein member of the Takagi/Montiel lists, as required.

Finally, we eliminate the possibility of 0 occurring as a principal curvature. Working again in $\mathcal{U}$, suppose that one principal curvature μ is identically zero. (If not, we get the desired conclusion by again reducing the size of $\mathcal{U}$.) Then we have a pair (λ, μ) with $\mu = 0$ and $\lambda = -2c/\alpha \neq 0$. Let X be a principal vector for λ. Then (8.96) gives $\kappa + 3c = \mathbf{m}\alpha - \alpha^2$. Thus (8.96) reduces to

$$(A^2 - \mathbf{m}A)\varphi AX = 0$$

and any principal curvatures other than λ, 0, and α must be equal to $\mathbf{m}$. This implies that all principal curvatures are constant which is a contradiction since, for the Takagi/Montiel lists, the Hopf principal curvature α is the only one that can be 0. □

An even simpler form of ∇S characterizes the Type A_0 and Type A_1 hypersurfaces. Relevant references are Kimura and Maeda [276], Taniguchi [522], and Choe [118]).

Theorem 8.84. *Let M^{2n-1}, where $n \geq 2$, be a real hypersurface in $\mathbf{CP}^n$ or $\mathbf{CH}^n$. Then M satisfies the identity*

$$(\nabla_X S)Y = \kappa(\langle \varphi X, Y\rangle W + \langle Y, W\rangle \varphi X), \tag{8.98}$$

for some nonzero constant κ if and only if M is an open subset of a Type A_0 or Type A_1 hypersurface.

The following appears in Ki and Suh [254].

Theorem 8.85. *Let M^{2n-1}, where $n \geq 3$, be a real hypersurface in $\mathbf{CP}^n$. Suppose that $\mathbf{m}$ and $\alpha = \langle AW, W\rangle$ are constant. Then M satisfies the identity*

$$\begin{aligned}(\nabla_X S)Y = &- c\,\mathbf{m}(\langle \varphi X, Y\rangle W + \langle Y, W\rangle \varphi X)\\ &+ c\,(\langle \varphi X, Y\rangle AW + \langle AY, W\rangle \varphi X) - 2c(\langle Y, W\rangle \varphi AX - \langle AX, \varphi Y\rangle W)\end{aligned}$$

if and only if M is an open subset of a hypersurface of Type A_1 or Type A_2.

Remark 8.86. This result is consistent with Theorem 8.84, as can be seen by setting $\kappa = -2n\lambda c$ for a Type A_1 hypersurface.

One can consider weaker conditions on ∇S and still obtain a strong conclusion. For example, Loo [337], considering only directions in $\mathcal{W}^\perp$, has proved the following:

Theorem 8.87. *Let M^{2n-1}, where $n \geq 3$, be a hypersurface in $\mathbf{CP}^n$ or $\mathbf{CH}^n$. If there is a constant κ such that*

- $(\nabla_X S)Y = \kappa\langle \varphi AX, Y\rangle W$ *for X and Y in $\mathcal{W}^\perp$, and*
- $[\varphi, S]\mathcal{W}^\perp \subset \mathcal{W}$,

then M is pseudo-Einstein.

Ikuta [213] has obtained the following result concerning the second derivative of the Ricci tensor.

Theorem 8.88. *Let M^{2n-1}, where $n \geq 3$, be a Hopf hypersurface in* $\mathbf{CP}^n$. *Then there is an even positive integer κ such that for all Z_1, Z_2, Z_3, and Z_4 in $\mathcal{W}^\perp$*

$$\begin{aligned}&\langle(\nabla^2 S)(Z_3; Z_2; Z_1), Z_4\rangle \\ &= -\kappa\big(\langle\varphi AZ_2, Z_3\rangle\langle\varphi AZ_1, Z_4\rangle + \langle\varphi AZ_1, Z_3\rangle\langle\varphi AZ_2, Z_4\rangle\big)\end{aligned}$$

if and only if M is pseudo-Einstein.

Kwon and Nakagawa [309, 310] studied the Ricci cyclic parallel condition.

Theorem 8.89. *Let M^{2n-1}, where $n \geq 2$, be a Hopf hypersurface in* $\mathbf{CP}^n$ *or* $\mathbf{CH}^n$. *Then the Ricci tensor of M is cyclic parallel if and only if M is an open subset of a hypersurface of Type A or Type B.*

Remark 8.90. A weaker version of Theorem 8.89 is stated as Theorem 6.21 in [399]. However, there is a typographical error there causing the Hopf condition to be omitted from the hypothesis. We will return to a discussion of Theorem 8.89 in the section on η-parallelism (see Theorem 8.136).

Commutativity conditions on the Ricci tensor

We have seen that the Type *A* hypersurfaces satisfy $[\varphi, A] = 0$. Clearly, this condition implies $[\varphi, S] = 0$. In fact, if we recall (6.12)

$$SX = (2n+1)cX - 3c\langle X, W\rangle W + \mathbf{m}AX - A^2X, \tag{8.99}$$

we note that φ automatically commutes with the first two terms of S, so that

$$[\varphi, S] = \mathbf{m}[\varphi, A] - [\varphi, A^2]. \tag{8.100}$$

We also have

Theorem 8.91. *Let M^{2n-1}, where $n \geq 2$, be a real hypersurface in* $\mathbf{CP}^n$ *or* $\mathbf{CH}^n$. *Then*

$$S \circ A = A \circ S$$

if and only if M is a Hopf hypersurface.

Proof. A commutes automatically with three of the four terms in the expression (8.99) for S and hence our condition is equivalent to

$$\langle X, W\rangle AW = \langle AX, W\rangle W \tag{8.101}$$

for all X. In particular, setting $X = W$, we have $AW = \alpha W$, the Hopf condition. On the other hand, if the Hopf condition is satisfied, then both sides of (8.101) vanish for $X \in \mathcal{W}^\perp$, and so we have $[S, A] = 0$. □

Clearly, all pseudo-Einstein hypersurfaces satisfy $\varphi \circ S = S \circ \varphi$. The following converse, due to Lim, Sohn, and Ahn [330] (see also [116]), holds for $n = 2$.

Theorem 8.92. *A hypersurface in* $\mathbf{CP}^2$ *or* $\mathbf{CH}^2$ *satisfies* $\varphi \circ S = S \circ \varphi$ *if and only if it is pseudo-Einstein.*

Proof. The fact that such a hypersurface is necessarily Hopf can be seen by looking at the Ricci tensor in the standard non-Hopf situation (8.84). First, $\varphi SW = S\varphi W = 0$ gives immediately that $\mu = \nu = 0$. Then, $\varphi SU = (5c + \lambda\alpha - \beta^2)\varphi U$ while $S\varphi U = 5c\varphi U$ implies that $\lambda\alpha - \beta^2 = 0$. This shows that a non-Hopf hypersurface satisfying $\varphi \circ S = S \circ \varphi$ is pseudo-Einstein when $n = 2$. In view of the classification of pseudo-Einstein hypersurfaces as discussed in the remark following Lemma 8.61, we conclude that all hypersurfaces in $\mathbf{CP}^2$ and $\mathbf{CH}^2$ satisfying $\varphi \circ S = S \circ \varphi$ are Hopf. The Ricci tensor now satisfies a modified version of (8.84), namely

$$\begin{bmatrix} 2c + \alpha\lambda & 0 & 0 \\ 0 & 5c + \lambda(\nu + \alpha) & 0 \\ 0 & 0 & 5c + \nu(\lambda + \alpha) \end{bmatrix}. \tag{8.102}$$

Using $\varphi \circ S = S \circ \varphi$, we get $\alpha(\nu - \lambda) = 0$ from which the pseudo-Einstein property is immediate. □

When $n \geq 3$, however, we cannot draw the conclusion that a hypersurface satisfying $\varphi \circ S = S \circ \varphi$ is Hopf. Instead, we shall pursue a discussion of Hopf hypersurfaces satisfying this condition.

Suppose that M^{2n-1}, where $n \geq 3$, is a Hopf hypersurface satisfying $\varphi \circ S = S \circ \varphi$. Choose a point p of M where the restriction of A to $\mathcal{W}^\perp$ has the maximum number of distinct eigenvalues. This ensures constant multiplicities and smoothness for the principal curvature functions nearby.

As in Theorems 8.32 and 8.81, $\mathcal{W}^\perp$ is the direct sum of even-dimensional φ-invariant subspaces, each determined by two (possibly equal) principal curvatures λ and μ satisfying

$$\lambda\mu = \frac{\lambda + \mu}{2}\alpha + c. \tag{8.103}$$

At most two of these subspaces have $\lambda = \mu$ since the quadratic equation $\lambda^2 - \alpha\lambda - c = 0$ must be satisfied. Let λ_1 and λ_2 be the roots of the quadratic equation and let $2k_1 \geq 2k_2$ (possibly 0) be their multiplicities as principal curvatures at p.

Thus for $X \in T_\lambda$, we use (8.100) to get

$$\varphi SX - S\varphi X = (\lambda - \mu)(\mathbf{m} - \lambda - \mu)\varphi X. \qquad (8.104)$$

When $\lambda \neq \mu$, we have $\lambda + \mu = \mathbf{m}$. Thus

$$\mathbf{m} = \alpha + 2k_1\lambda_1 + 2k_2\lambda_2 + k_3\mathbf{m} \qquad (8.105)$$

where $2k_3$ is the sum of the dimensions of the φ-invariant subspaces for which $\lambda \neq \mu$. Note that $k_1 + k_2 + k_3 = n - 1$. Because of how p has been chosen, this situation also holds locally with each λ_i and k_i constant. Assuming, for the moment, that $k_3 \neq 1$, we see that $\mathbf{m}$ is constant. Thus, in view of (8.103), we see that for each "unequal pair" $\{\lambda, \mu\}$, both $\lambda\mu$ and $\lambda + \mu = \mathbf{m}$, and hence λ and μ themselves, are constant. This shows that p has a neighborhood that is a Hopf hypersurface with constant principal curvatures and thus is an open subset of an element of the Takagi/Montiel list. The list and the corresponding values of the k_i are as follows. See discussion preceding Proposition 8.14.

- Type A_0 or Type A_1 with $k_1 = n - 1$
- Type A_2 with k_1 and k_2 positive, $k_1 + k_2 = n - 1$
- Type B with $k_3 = n - 1$
- Type C with $2k_1 = 2k_2 = n - 3$ and $k_3 = 2$
- Type D with $k_1 = k_2 = 2$ and $k_3 = 4$
- Type E with $k_1 = k_2 = 4$ and $k_3 = 6$.

As we have already observed, all Type A hypersurfaces satisfy $\varphi S = S\varphi$. Type B hypersurfaces satisfy $\mathbf{m} = \alpha + (n-1)(\lambda + \mu)$ and $(\lambda + \mu)\alpha + 4c = 0$. Since $\varphi S = S\varphi$ if and only if $\mathbf{m} = \lambda + \mu$ in this case, we can easily check that

- Type B hypersurfaces in $\mathbf{CH}^n$ do not satisfy $\varphi S = S\varphi$.
- A Type B hypersurface in $\mathbf{CP}^n$ satisfies $\varphi S = S\varphi$ if and only if it is pseudo-Einstein, i.e., as a tube over the complex quadric, its radius is ru where $\cot^2 2u = n - 2$. See Theorem 8.63.

Similarly, we may check that Type C, D, and E hypersurfaces satisfy $\varphi S = S\varphi$ if and only if their respective radii ru as tubes satisfy

- Type C: $\tan^2 2u = n - 2$;
- Type D: $\tan^2 2u = \frac{5}{3}$;
- Type E: $\tan^2 2u = \frac{9}{5}$.

In the notation of Proposition 8.14, the relevant condition is $\lambda_2 + \lambda_4 = \mathbf{m}$.

We now consider the possibility that $k_3 = 1$. Since $n \geq 3$, we have $k_1 > 0$. Our mean curvature calculation then yields

$$\alpha + 2k_1\lambda_1 + 2k_2\lambda_2 = 0. \qquad (8.106)$$

Recall that $\lambda_1\lambda_2 = -c$ and $\lambda_1 + \lambda_2 = \alpha$. Thus, if $c < 0$, then α, λ_1 and λ_2 all have the same sign. This contradicts (8.106). Applying the usual continuity and connectedness arguments, we have proved

Theorem 8.93. *Let M^{2n-1}, where $n \geq 3$, be a Hopf hypersurface in $\mathbf{CH}^n$. Then $\varphi \circ S = S \circ \varphi$ if and only if M is an open subset of a Type A hypersurface.*

Since the possibility that $k_3 = 1$ has not been eliminated in the $\mathbf{CP}^n$ case, we will state our conclusion as follows:

Theorem 8.94. *Let M^{2n-1}, where $n \geq 2$, be a hypersurface in $\mathbf{CP}^n$ that occurs in the Takagi list. Then $\varphi \circ S = S \circ \varphi$ if and only M it is*

- *Type A, or*
- *Type B with* $\cot^2 2u = n - 2$, *or*
- *Type C with* $\tan^2 2u = n - 2$, *or*
- *Type D with* $\tan^2 2u = \frac{5}{3}$, *or*
- *Type E with* $\tan^2 2u = \frac{9}{5}$.

We do know that M is a tube over a complex focal submanifold. Kimura [272] has stated the same result by listing some necessary conditions that this focal submanifold should satisfy. In particular, he states that certain tubes over certain complex curves will satisfy $\varphi S = S\varphi$ and have $k_3 = 1$. However, the problem of finding a complete classification is still open.

Remark 8.95. Our statement is a slightly more detailed version of Theorem 6.18 in [399]. Relevant references are Aiyama, Nakagawa, and Suh [8], Ki and Suh [253], and Kimura [270].

Variants on the condition $\varphi \circ S = S \circ \varphi$ are also considered in the literature. One such theorem is the following due to Ki and Suh [253] and Kim and Pyo [259].

Theorem 8.96. *Let M^{2n-1}, where $n \geq 3$, be a Hopf hypersurface in $\mathbf{CP}^n$ or $\mathbf{CH}^n$.*

1. *If there is a constant κ such that*

$$\varphi \circ S + S \circ \varphi = \kappa\varphi,$$

   then M is an open subset of a hypersurface of Type A or Type B.
2. *If there is a constant κ such that*

$$S \circ \varphi \circ S = \kappa\varphi,$$

   then M is an open subset of a hypersurface of Type A_1, Type A_0 or Type B.

For $n = 2$, the situation is slightly different.

Theorem 8.97. *Let M^3 be a Hopf hypersurface in* $\mathbf{CP}^2$ *or* $\mathbf{CH}^2$. *The following three conditions are equivalent:*

1. *There is a constant* κ *such that*

$$\varphi \circ S + S \circ \varphi = \kappa\varphi.$$

2. *There is a constant* κ *such that*

$$S \circ \varphi \circ S = \kappa\varphi,$$

3. *M is an open subset of a hypersurface of Type* A_1, *Type* A_0 *or Type B or M is a pseudo-Einstein hypersurface with* $\alpha = 0$ *and two nonconstant principal curvatures.*

Proof. In terms of the principal curvatures (α, λ, ν), the first condition reduces to

$$\big(5c + \nu(\alpha + \lambda)\big) + \big(5c + \lambda(\alpha + \nu)\big) = 10c + 2\lambda\nu + \alpha(\lambda + \nu) = \kappa.$$

Because of the relationship between λ and μ given by Theorem 6.17, we have $\alpha(\lambda + \nu) = 2(\lambda\nu - c)$ so that $\lambda\nu$ is constant. If $\alpha \neq 0$, then $\lambda + \nu$ is also constant and thus so are λ and ν individually. M is an open subset of a hypersurface of Type A_0, Type A_1, or Type B. Otherwise, $\alpha = 0$, and M is a pseudo-Einstein hypersurface with $\alpha = 0$ and nonconstant principal curvatures.

The second condition reduces to

$$\big(5c + \nu(\alpha + \lambda)\big)\big(5c + \lambda(\alpha + \nu)\big) = \kappa.$$

Upon expanding the left side and substituting for $\alpha(\nu + \lambda)$, we find that $\lambda\nu$ satisfies a quadratic equation with constant coefficients, and is therefore constant. Proceeding as before, we find that M is one of the hypersurfaces listed in *3*.

Conversely, if M is a hypersurface listed in *3*., one can verify directly that it satisfies the identities of *1*. and *2*. for suitable choices of the constant κ. □

Kimura and Maeda [278], and Kwon and Suh [311] considered the condition that the Ricci tensor S commutes with φA. As a result of their work and that of Ki and Suh [255], we have

Theorem 8.98. *Let* M^{2n-1}, *where* $n \geq 3$, *be a real hypersurface in* $\mathbf{CH}^n$. *If*

$$S \circ \varphi \circ A = \varphi \circ A \circ S \tag{8.107}$$

then M is an open subset of a hypersurface of Type A.

Remark 8.99. Since Type A hypersurfaces have the property that $\mathcal{W}^\perp$ consists of one or two φ-invariant principal subspaces, it is clear from (8.99) that they satisfy the identity $S \circ \varphi \circ A = \varphi \circ A \circ S$. This holds for all $n \geq 2$. The question arises: Does Theorem 8.98 extend to $n = 2$? In fact, it does not. The pseudo-Einstein hypersurfaces with nonconstant principal curvatures as discussed by Kim, Ivey, and Ryan in [260] and [222] are counterexamples. One can show, however, that these are the only possibilities (Ivey, personal communication, 2015).

For our discussion of Theorem 8.98, we temporarily denote by T the orthogonal projection on $\mathcal{W}$, i.e., $TX = \langle X, W\rangle W$. Clearly, T is a symmetric $(1, 1)$ tensor field. Then, using (8.99), we can write

$$[\varphi A, S] = A^2\varphi A - \varphi A^3 + \mathbf{m}(\varphi A^2 - A\varphi A) - 3c\varphi AT.$$

Thus $S \circ \varphi \circ A = \varphi \circ A \circ S$ if and only if

$$3c\varphi AT = A^2\varphi A - \varphi A^3 + \mathbf{m}(\varphi A^2 - A\varphi A). \tag{8.108}$$

Equivalently,

$$3cTA\varphi = A\varphi A^2 - A^3\varphi + \mathbf{m}(A^2\varphi - A\varphi A) \tag{8.109}$$

which is the negative of the transpose of (8.108). Multiplying (8.108) and (8.109) by A on the left and on the right, respectively, and adding, we get

$$3c(A\varphi AT + TA\varphi A) = 0.$$

Applying this to W yields

$$A\varphi AW = -\langle A\varphi AW, W\rangle W = -\langle \varphi AW, AW\rangle W = 0.$$

Thus, we have

$$A\varphi AW = 0. \tag{8.110}$$

Also, by applying (8.109) to W, we get

$$A\varphi A^2 W = 0. \tag{8.111}$$

Further, we can apply (8.108) to W to get $3c\varphi AW = -\varphi A^3 W + \mathbf{m}\varphi A^2 W$ which tells us that $(A^3 W - \mathbf{m}A^2 W + 3cAW) \in \mathcal{W}$. Thus,

$$A^4 W - \mathbf{m}A^3 W + 3cA^2 W \tag{8.112}$$

lies in the span of AW. On the other hand, if we apply (8.108) to AW, we get $3c\varphi A\alpha W = -(\varphi A^4 W - \mathbf{m}\varphi A^3 W)$ so that

$$A^4 W - \mathbf{m}A^3 W \tag{8.113}$$

lies in the span of $\{AW, W\}$. Comparing expressions (8.112) and (8.113), we conclude that A^2W also lies in the span of $\{AW, W\}$. We state these results as a lemma.

Lemma 8.100. *Let M^{2n-1}, where $n \geq 2$, be a real hypersurface in $\mathbf{CP}^n$ or $\mathbf{CH}^n$. If $S \circ \varphi \circ A = \varphi \circ A \circ S$, then*

1. *$A\varphi AW = 0$;*
2. *the span of $\{AW, W\}$ is A-invariant.*

Of course, this gives us no new information at points where $AW = \alpha W$, in particular for Hopf hypersurfaces. In fact, we can use the equations just derived to prove

Theorem 8.101. *For a Hopf hypersurface M^{2n-1}, where $n \geq 2$, in $\mathbf{CP}^n$ or $\mathbf{CH}^n$, S commutes with φA if and only if S commutes with φ.*

Proof. First of all, for any Hopf hypersurface, we have $[\varphi A, S]W = 0$ and $[\varphi, S]W = 0$. Thus we need consider only $\mathcal{W}^{\perp}$. As in our proof of Theorem 8.93, we consider basis pairs $\{X, \varphi X\}$ and note that, assuming that $[\varphi, S] = 0$, we have

$$[\varphi A, S]X = \lambda((A^2 - \mathbf{m}A)\varphi X - (\lambda^2 - \mathbf{m}\lambda)\varphi X)$$

But $(A^2 - \mathbf{m}A)\varphi X = (\mu^2 - \mathbf{m}\mu)\varphi X$ and thus

$$\lambda(\lambda - \mu)(\mathbf{m} - \lambda - \mu) = 0.$$

The same equation holds with the roles of λ and μ reversed. Since not both λ and μ can be zero, we have $(\lambda - \mu)(\mathbf{m} - \lambda - \mu) = 0$. With this, it is easy to check that $[\varphi, S]X = [\varphi, S]\varphi X = 0$ and $[\varphi, S]$ vanishes on all of $\mathcal{W}^{\perp}$.

Conversely, assuming that $[\varphi, S] = 0$ and using the same basis, we get $(\lambda - \mu)(\mathbf{m} - \lambda - \mu) = 0$ from (8.104). Applying the right side of (8.108) to X gives

$$\mu^2\lambda - \lambda^3 + \mathbf{m}(\lambda^2 - \mu\lambda) = \lambda((\mu^2 - \lambda^2) + \mathbf{m}(\lambda - \mu))$$

which is zero. A similar equation results from using φX. Consequently, (8.109) is satisfied on all of $\mathcal{W}^{\perp}$ and S commutes with φA. □

We now consider the non-Hopf situation. It is easy to check the following:

Lemma 8.102. *Let M^{2n-1}, where $n \geq 2$, be a real hypersurface in $\mathbf{CP}^n$ or $\mathbf{CH}^n$ satisfying $S \circ \varphi \circ A = \varphi \circ A \circ S$. Let p be a point where $AW \neq \alpha W$. Then, in terms of the standard non-Hopf setup in a neighborhood of p, the upper-left 3×3 submatrix of A (see (8.34)) takes the form*

$$\begin{bmatrix} \alpha & \beta & 0 \\ \beta & \lambda & 0 \\ 0 & 0 & 0 \end{bmatrix}, \tag{8.114}$$

i.e., $\mu = \nu = 0$. *In addition, the orthogonal complement of the span of* $\{W, U, \varphi U\}$ *is A-invariant and* φ*-invariant and has an orthonormal basis consisting of* $n-2$ *pairs* $\{X, \varphi X\}$ *satisfying* $(A^2 - \mathbf{m}A)X = \rho X$ *and* $(A^2 - \mathbf{m}A)\varphi X = \rho\varphi X$ *for a suitable function* ρ.

A major step in proving Theorem 8.98 is to show that the hypersurface in question is Hopf. Using the standard non-Hopf setup, a lengthy computation is necessary to reach a contradiction. We will not present this computation here, but the interested reader can fill in the details by following the argument in Ki and Suh [255].

8.5.4 The structure Jacobi operator R_W

At this point, we introduce the Jacobi operators which are $(1, 1)$ tensor fields derived from the curvature tensor, just as the Ricci tensor is.

For a vector field Z on a Riemannian manifold, the *Jacobi operator* R_Z is defined by

$$R_Z X = R(X, Z)Z \tag{8.115}$$

for all tangent vectors X. For a hypersurface in a complex space form with structure vector field $W = -J\xi$, the Jacobi operator R_W is called the *structure Jacobi operator*.

A related but distinct concept is that of normal Jacobi operator which we will introduce in the next chapter when discussing curvature-adapted hypersurfaces.

Before proceeding with the main business of this section, the classification of hypersurfaces in terms of the structure Jacobi operator, we deal briefly with a classification where the general Jacobi operators play a role.

A Riemannian manifold is said to be

- a *D'Atri space* if all its local geodesic symmetries are volume-preserving up to a sign;
- a *C-space* if the eigenvalues of the Jacobi operators are constant along corresponding geodesics.

Cho and Vanhecke [117] studied the D'Atri and C-space conditions. The following is their result.

Theorem 8.103. *Let* M^{2n-1}*, where* $n \geq 2$*, be a Hopf hypersurface in* $\mathbf{CP}^n$ *or* $\mathbf{CH}^n$*. The following are equivalent:*

- *M is an open subset of a Type A hypersurface;*
- *M is a D'Atri space;*
- *M is a C-space.*

Their proof relies on the fact that both D'Atri spaces and C-spaces have cyclic parallel Ricci tensor. By the theorem of Kwon and Nakagawa (Theorem 8.89), such hypersurfaces are locally of Type A or Type B. The Type B possibility can be eliminated. An alternative proof was given by Nagai [389]. The reader may find further background in papers of Vanhecke (with Willmore and Berndt, respectively) [39, 542], and in Nagai's survey [390].

The structure Jacobi operator, by definition, satisfies

$$R_W X = R(X, W)W. \tag{8.116}$$

A few of its basic properties can easily be checked using the Gauss equation and the first Bianchi identity.

Proposition 8.104. *The structure Jacobi operator R_W*

- *is symmetric, i.e., satisfies $\langle R_W X, Y\rangle = \langle X, R_W Y\rangle$;*
- *satisfies $R_W W = 0$;*
- *satisfies $R_W X = (\lambda\alpha + c)X$ if M is Hopf and $X \in \mathcal{W}^{\perp}$ is a principal vector corresponding to a principal curvature λ.*

Note that R_W takes on a particularly simple form when $\alpha = 0$. We can ask about other hypersurfaces sharing this property.

Theorem 8.105. *Let M^{2n-1}, where $n \geq 2$, be a Hopf hypersurface in $\mathbf{CP}^n$ or $\mathbf{CH}^n$. Suppose that there is a function κ such that*

$$R_W X = \kappa X \tag{8.117}$$

for all $X \in \mathcal{W}^{\perp}$. Then, κ is constant and M is

- *an open subset of a Type A_1 or Type A_0 hypersurface ($\kappa = \lambda^2 = \lambda\alpha + c$), or*
- *an open subset of a Type A_2 hypersurface in $\mathbf{CP}^n$ with radius $\frac{\pi}{4}r$ ($\kappa = c$ and $\alpha = 0$), or*
- *a Hopf hypersurface with $\alpha = 0$ and nonconstant principal curvatures (again, $\kappa = c$).*

Conversely, all hypersurfaces in this list satisfy the given hypothesis.

Proof. Suppose that M is a Hopf hypersurface satisfying equation (8.117). If $\alpha \neq 0$, then every principal vector $X \in \mathcal{W}^{\perp}$ satisfies $AX = \lambda X$ where $\lambda\alpha + c = \kappa$, i.e.,

$$\lambda = \frac{\kappa - c}{\alpha}.$$

Since the principal space of λ is φ-invariant, we have $\lambda^2 = \lambda\alpha + c = \kappa$. In particular, λ and κ are constants. Thus M is an open subset of a hypersurface of Type A_1 or A_0. It is also clear that these hypersurfaces satisfy (8.117). Suppose now that $\alpha = 0$. If M has constant principal curvatures, it must be an open subset of some hypersurface on the Takagi/Montiel list. However, the only list members for which $\alpha = 0$ is possible are the Type A_1 and Type A_2 hypersurfaces in $\mathbf{CP}^n$ with radius $\frac{\pi}{4}r$. The Type A_1 hypersurfaces have already been mentioned, so the only new possibility is the indicated Type A_2 hypersurface. Conversely, the condition $\alpha = 0$ guarantees that $R_W = cX$ for all principal vectors $X \in \mathcal{W}^{\perp}$. This completes the proof. □

Remark 8.106. It is easy to construct examples in $\mathbf{CP}^n$ that satisfy the third condition in Theorem 8.105. Take any complex submanifold that is not totally geodesic or part of a quadric. Construct the tube of radius $\frac{\pi}{4}r$ over a suitable open subset. This will give a Hopf hypersurface with $\alpha = 0$, as desired. In $\mathbf{CH}^n$, tubes will not suffice. We need to make a different construction (see Ivey and Ryan [222] and Ivey [221] for an indication of how to do this).

Remark 8.107. Characterization of non-Hopf hypersurfaces satisfying (8.117) is an open problem. It is not too difficult to show that there are none with α identically zero. For $n = 2$, however, it is not difficult to show that non-Hopf examples exist. In the standard non-Hopf setup, a hypersurface with $\alpha \neq 0$ satisfies (8.117) if and only if $\mu = 0$ and $\alpha(\lambda - \nu) = \beta^2$. Such hypersurfaces exist by Theorem 20 of [224] as explained in the proof of Theorem 8.54.

Clearly, the condition (8.117) implies that

$$R_W \circ \varphi = \varphi \circ R_W. \tag{8.118}$$

We now derive a useful characterization of this condition.

Lemma 8.108. *Let M^{2n-1}, where $n \geq 2$, be a real hypersurface in $\mathbf{CP}^n$ or $\mathbf{CH}^n$. Then, for all tangent vectors X,*

$$[R_W, \varphi]X = -\alpha(\varphi A - A\varphi)X + \langle \varphi AW, X\rangle AW + \langle AW, X\rangle \varphi AW. \tag{8.119}$$

Thus R_W commutes with φ if and only if

$$\alpha(\varphi A - A\varphi)X = \langle \varphi AW, X\rangle AW + \langle AW, X\rangle \varphi AW \tag{8.120}$$

for all X.

Proof. Using the Gauss equation, we have

$$\begin{aligned} R_W\varphi X - \varphi R_W X &= R(\varphi X, W)W - \varphi R(X, W)W \\ &= (A\varphi X \wedge AW)W - \varphi(AX \wedge AW)W \\ &\quad + c(\varphi X \wedge W)W - c\varphi(X \wedge W)W \\ &= \alpha A\varphi X + \langle X, \varphi AW\rangle AW \\ &\quad - \alpha\varphi AX + \langle AX, W\rangle \varphi AW \\ &\quad + c(\varphi X - \varphi X), \end{aligned}$$

which is essentially (8.119). □

Corollary 8.109. *Under the conditions of the preceding lemma,*

- *At a point where* $AW = \alpha W$, *we have*

$$[R_W, \varphi] = -\alpha(\varphi A - A\varphi).$$

In particular, this is the case globally for a Hopf hypersurface.

- *In a neighborhood of a point where* $AW \neq \alpha W$, *we have (in terms of the standard non-Hopf setup),*

$$[R_W, \varphi]X = -\alpha(\varphi A - A\varphi)X + \beta(\langle \varphi U, X\rangle AW + \langle AW, X\rangle \varphi U)$$

for all tangent vectors X.

The study of hypersurfaces in terms of the structure Jacobi operator was initiated by Cho and Ki [112, 113] who proved Theorem 8.105 for $\mathbf{CP}^n$, where $n \geq 3$. They then classified hypersurfaces in $\mathbf{CP}^n$ that satisfy

$$R_W \circ \varphi \circ A = A \circ \varphi \circ R_W.$$

Subsequently, Ki, Kim, and Lee in [241] obtained a classification (Theorem 8.112 below) that included the $\mathbf{CH}^n$ case. Our exposition of this theorem begins with an analysis of this condition.

Lemma 8.110. *Let* M^{2n-1}, *where* $n \geq 2$, *be a real hypersurface in* $\mathbf{CP}^n$ *or* $\mathbf{CH}^n$. *Then,*

- *For all tangent vectors X,*

$$(R_W\varphi A - A\varphi R_W)X = c(\varphi A - A\varphi)X + \langle \varphi AW, AX\rangle AW + \langle AW, X\rangle A\varphi AW. \tag{8.121}$$

- *At a point where* $AW = \alpha W$, *we have*

$$R_W\varphi A - A\varphi R_W = c[\varphi, A].$$

- *In a neighborhood of a point where* $AW \neq \alpha W$, *we have (in terms of the standard non-Hopf setup),*

$$(R_W\varphi A - A\varphi R_W)X = c(\varphi A - A\varphi)X + \beta(\langle \varphi U, AX\rangle AW + \langle AW, X\rangle A\varphi U).$$

- *Further, if* $A\varphi U = \nu\varphi U$, *then*

$$\begin{aligned}(R_W\varphi A - A\varphi R_W)X &= c(\varphi A - A\varphi)X + \beta\nu(\langle \varphi U, X\rangle AW + \langle AW, X\rangle \varphi U)\\ &= \nu[R_W, \varphi]X + (\alpha\nu + c)[\varphi, A]X.\end{aligned}$$

Proof. Again, from the Gauss equation, we have

$$\begin{aligned} R_W\varphi AX - A\varphi R_W X &= R(\varphi AX, W)W - A\varphi R(X,W)W \\ &= (A\varphi AX \wedge AW)W - A\varphi((AX \wedge AW)W) \\ &\quad + c(\varphi AX \wedge W)W - cA\varphi((X \wedge W)W) \\ &= \langle AX, \varphi AW\rangle AW + \langle AX, W\rangle A\varphi AW \\ &\quad + c(\varphi AX - A\varphi X), \end{aligned}$$

which is the first assertion. Since $\varphi AW = 0$ when $AW = \alpha W$, the second assertion follows. On the other hand, in the standard non-Hopf setup, $\varphi AW = \beta\varphi U$. Making this substitution, we get the third assertion. Finally, when $A\varphi U = \nu\varphi U$, the first part of the fourth assertion is immediate. Then, using Corollary 8.109 to substitute for $\beta(\langle\varphi U, X\rangle AW + \langle AW, X\rangle\varphi U)$, we get the second part of the fourth assertion. □

If we set $X = W$ in the first assertion, we see that

$$R_W\varphi AW = (\alpha A + c)\varphi AW.$$

In particular, for the standard non-Hopf setup, if $(R_W\varphi A - A\varphi R_W)W = 0$, then $(\alpha A + c)\varphi AW = 0$. Thus $\alpha \neq 0$ and

$$A\varphi U = \nu\varphi U,$$

where $\nu = -\frac{c}{\alpha}$. In view of the fourth assertion, we get

$$R_W \circ \varphi \circ A - A \circ \varphi \circ R_W = \nu(R_W \circ \varphi - \varphi \circ R_W). \tag{8.122}$$

This gives

Corollary 8.111. *Let M^{2n-1}, where $n \geq 2$, be a real hypersurface in $\mathbf{CP}^n$ or $\mathbf{CH}^n$. If $R_W \circ \varphi \circ A = A \circ \varphi \circ R_W$, then $R_W \circ \varphi = \varphi \circ R_W$.*

The major step needed to complete our characterization of the condition

$$R_W \circ \varphi \circ A = A \circ \varphi \circ R_W$$

is to show that it implies the Hopf condition. Assuming the standard non-Hopf setup, a lengthy computation is necessary to reach a contradiction. We will not present this computation here, but the interested reader can fill in the details by following the argument in [241]. Thus we get the following:

Theorem 8.112. *Let M^{2n-1}, where $n \geq 2$, be a real hypersurface in $\mathbf{CP}^n$ or $\mathbf{CH}^n$. Then*

$$R_W \circ \varphi \circ A = A \circ \varphi \circ R_W$$

if and only if M is an open subset of a Type A hypersurface.

Proof. We begin with the fact that a hypersurface satisfying $R_W \circ \varphi \circ A = A \circ \varphi \circ R_W$ must be Hopf. Then Lemma 8.110 shows that such a hypersurface must satisfy $\varphi \circ A = A \circ \varphi$ and hence by Theorem 8.37 must be an open subset of a Type A hypersurface. Conversely, Type A hypersurfaces satisfy $\varphi \circ A = A \circ \varphi$ and are Hopf. Therefore, they must satisfy $R_W \circ \varphi \circ A = A \circ \varphi \circ R_W$, again by Lemma 8.110. □

Lim and Sohn [329] considered the anti-commutative condition

$$R_W \circ \varphi \circ A + A \circ \varphi \circ R_W = 0$$

and prove that a hypersurface satisfying it must be Hopf. Although they conclude that such hypersurfaces are locally Type A, the appropriate conclusion is that this condition cannot be realized since it is easy to check that the Type A hypersurfaces do not, in fact, satisfy it.

Although R_W cannot be parallel, it can be cyclic parallel. The hypersurfaces for which R_W is cyclic parallel were classified by Ki and Kurihara [243] as follows:

Theorem 8.113. *Let M^{2n-1}, where $n \geq 3$, be a real hypersurface in $\mathbf{CP}^n$ or $\mathbf{CH}^n$. Then R_W is cyclic parallel if and only if M is an open subset of*

- *A Type A hypersurface, or*
- *A Type B hypersurface in $\mathbf{CP}^n$ whose radius ru as a tube over the complex quadric satisfies* $\tan^2 2u = 2$.

Further work

We conclude this section by listing a number of further results that have occurred in the literature recently. The first few assert non-existence of hypersurfaces satisfying particular conditions on R_W.

Theorem 8.114. *There are no real hypersurfaces in $\mathbf{CP}^n$ or $\mathbf{CH}^n$, where $n \geq 3$, satisfying any of the following conditions.*

- *$R_W \circ R_X = R_X \circ R_W$ for all tangent vectors X;*
- *R_W is parallel; i.e., $\nabla_X R_W = 0$ for all tangent vectors X;*
- *R_W is recurrent.*

Theorem 8.115. *There are no real hypersurfaces in $\mathbf{CP}^n$ or $\mathbf{CH}^n$, where $n \geq 2$, satisfying either of the following conditions*

- $\mathcal{L}_X R_W = 0$ *for all tangent vectors* X*;*
- R_W *is a Codazzi tensor, i.e., it satisfies* $(\nabla_X R_W)Y = (\nabla_Y R_W)X$ *for all tangent vectors* X *and* Y*.*

On the other hand, all Type A hypersurfaces satisfy $\mathcal{L}_W R_W = 0$ *and every hypersurface satisfying* $\mathcal{L}_W R_W = 0$ *is Hopf.*

These theorems include work of Pérez, Santos, and Suh [439, 440], Ivey and Ryan [223], and Theofanidis and Xenos [532].

Following Kaimakamis and Panagiotidou [233] we define a tensor field T to be *Lie recurrent* if there is a 1-form ω such that

$$\mathcal{L}_X T = \omega(X) T,$$

for all tangent vectors X. They obtained the following improvement of Theorem 8.115.

Theorem 8.116. *There are no real hypersurfaces in* $\mathbf{CP}^n$ *or* $\mathbf{CH}^n$*, where* $n \geq 2$*, whose shape operator is Lie recurrent.*

Theorem 8.117 deals with the situation where R_W commutes with both φ and the Ricci tensor and is due to Ki, Nagai, and Takagi [251]. Earlier versions of this result may be found in [242] and [250].

Theorem 8.117. *Let* M^{2n-1}*, where* $n \geq 2$*, be a real hypersurface in* $\mathbf{CP}^n$ *or* $\mathbf{CH}^n$ *satisfying*

$$R_W \circ \varphi = \varphi \circ R_W \text{ and } R_W \circ S = S \circ R_W.$$

Then M *is a Hopf hypersurface. Further, if* $\alpha \neq 0$*, then* M *is an open subset of a Type A hypersurface.*

Yet another way of way of weakening the $\nabla S = 0$ assumption has been studied by Ki and Nagai [248, 249] with the following result. See also Li and Ki [317].

Theorem 8.118. *Let* M^{2n-1}*, where* $n \geq 2$*, be a real hypersurface in* $\mathbf{CP}^n$ *satisfying* $\nabla_{AW-\alpha W} S = 0$.

- *If* $SW = \sigma W$ *for some constant* σ*, then* M *is Hopf.*
- *If* $SW = \sigma W$ *for some function* σ*, and* $|AW|^2 - \alpha^2$ *is constant, then* M *is Hopf.*
- *If* $R_W \circ S = S \circ R_W$ *and* $\langle SW, W \rangle$ *is constant, then* M *is Hopf.*
- *If* $R_W \circ A = A \circ R_W$ *and the mean curvature is constant, then* M *is Hopf.*

8.5.5 *W-parallelism and η-parallelism*

It is clear that the relationship between geometric structures and the holomorphic distribution $\mathcal{W}^\perp$ plays a large role in the study of hypersurfaces in complex space forms. The special properties of Hopf hypersurfaces constitute one example.

Classification in terms of the structure Jacobi operator R_W provides another. In this section, we take familiar geometric conditions, some of which are too strong to be realized in our context, and weaken them by applying them to $\mathcal{W}$ or $\mathcal{W}^\perp$ only.

W-parallelism

A tensor field T on a hypersurface M is said to be *W-parallel* if $\nabla_W T = 0$. Although the shape operator cannot be parallel, there are nontrivial examples for which it is W-parallel. Our first step in classifying such hypersurfaces is the following.

Proposition 8.119. *Let M^{2n-1}, where $n \geq 2$, be a real hypersurface in* $\mathbf{CP}^n$ *or* $\mathbf{CH}^n$ *with $\nabla_W A = 0$. Then M is a Hopf hypersurface.*

This is Proposition 7.2 in Niebergall and Ryan [399] where a detailed proof can be found. Using this, we get

Theorem 8.120. *Let M^{2n-1}, where $n \geq 2$, be a real hypersurface in* $\mathbf{CP}^n$ *or* $\mathbf{CH}^n$ *with $\nabla_W A = 0$. Then M is*

- *a Hopf hypersurface with $\alpha = 0$, or*
- *an open subset of a Type A hypersurface .*

Conversely, such hypersurfaces satisfy $\nabla_W A = 0$.

Proof. Assume that M satisfies $\nabla_W A = 0$. Then for any vector field X, the Codazzi equation gives $(\nabla_X A)W = -c\varphi X$ and so

$$\nabla_X(AW) = -c\varphi X + A\varphi AX. \tag{8.123}$$

Recalling that M is Hopf, let $X \in \mathcal{W}^\perp$ be a principal vector with corresponding principal curvature λ. In the notation of Theorem 6.17, we have

$$\alpha\lambda\varphi X = \nabla_X(\alpha W) = -c\varphi X + \lambda\mu\varphi X, \tag{8.124}$$

i.e., $\alpha\lambda = -c + \lambda\mu$. Since $\lambda\mu - c = \frac{\lambda+\mu}{2}\alpha$, this reduces to $\alpha(\lambda - \mu) = 0$. Unless $\alpha = 0$, we have $\lambda = \mu$ and hence T_λ is φ-invariant. Since the quadratic equation, $\lambda^2 - \alpha\lambda - c = 0$ is satisfied, there are only two possible values of λ and they are constant. Thus, M is an open subset of a Type A hypersurface. Conversely, if M is Hopf with $\alpha = 0$, then $(\nabla_W A)W = \nabla_W(AW) - A\varphi AW = 0$ while $(\nabla_W A)X = (\nabla_X A)W + c\varphi X$ for any principal vector $X \in \mathcal{W}^\perp$. Since $(\nabla_X A)W = \nabla_X(AW) - A\varphi AX$, we get

$$(\nabla_W A)X = (-\lambda\mu + c)\varphi X = -\frac{\alpha}{2}(\lambda + \mu) = 0$$

where λ and μ are the principal curvatures corresponding to X and φX, respectively. On the other hand, for a Type A hypersurface, the W-parallelism follows directly from Theorem 8.37. □

Theorem 8.120 was proved by Kimura and Maeda [275] for the $\mathbf{CP}^n$ case and has been generalized by Pyo and others [258, 454, 455]. In the same paper, Kimura and Maeda considered Hopf hypersurfaces in $\mathbf{CP}^n$ with W-parallel Ricci tensor under the assumption of constant mean curvature which was later removed by S. Maeda [347]. Although he considers only the $\mathbf{CP}^n$ case, his proof essentially applies in $\mathbf{CH}^n$ as well. It turns out that W-parallelism of the Ricci tensor is related to its commuting with φ which is discussed in Theorems 8.93 and 8.94.

Theorem 8.121. *Let M^{2n-1}, where $n \geq 2$, be a Hopf hypersurface in $\mathbf{CP}^n$ or $\mathbf{CH}^n$. Then $\nabla_W S = 0$ if and only if M satisfies $\alpha = 0$ or $\varphi S = S\varphi$.*

Proof. It is an easy consequence of the Codazzi equation (see Lemma 2.13 of [399]) that $W\mathbf{m} = 0$ for a Hopf hypersurface. From this and (6.12), we can deduce that every Hopf hypersurface satisfies $(\nabla_W S)W = 0$. Consider now any principal vector $X \in \mathcal{W}^\perp$ with corresponding principal curvature λ. Then

$$(\nabla_W S)X = (\mathbf{m} - A)(\nabla_W A)X - (\nabla_W A)AX$$

and the Codazzi equation gives $(\nabla_W A)X = (\nabla_X A)W + c\varphi X$. Thus

$$\begin{aligned}(\nabla_W S)X &= (\mathbf{m} - A - \lambda)((\nabla_X A)W + c\varphi X) \\ &= (\mathbf{m} - A - \lambda)(\lambda(\alpha - A)\varphi X + c\varphi X) \qquad (8.125) \\ &= (\mathbf{m} - \mu - \lambda)\frac{\lambda - \mu}{2}\alpha\varphi X \qquad (8.126)\end{aligned}$$

since $c - \lambda\mu = -\frac{\lambda+\mu}{2}\alpha$. Thus $(\nabla_W S)X = 0$ if and only if

$$\alpha(\lambda - \mu)(\lambda + \mu - \mathbf{m}) = 0.$$

In case $n = 2$, this is equivalent to $\alpha(\lambda - \mu) = 0$ which is precisely the condition for M to be pseudo-Einstein. The result then follows from Theorem 8.92. For $n \geq 3$, consider the discussion preceding Theorems 8.93 and 8.94. The fact that $\nabla_W S = 0$ implies $S\varphi = \varphi S$ is immediate from (8.104). Conversely, the same equation shows that if $S\varphi = \varphi S$, then $(\lambda - \mu)(\lambda + \mu - \mathbf{m}) = 0$. This completes the proof. □

Remark 8.122. Note that the condition $\nabla_W S = 0$ is weaker than $\nabla_W A = 0$ since, for $n \geq 3$, hypersurfaces of types *B-E* qualify, albeit for only one specific radius. We note that for the Type *B* case, those qualifying are precisely the pseudo-Einstein ones. The general classification problem for Hopf hypersurfaces satisfying $\nabla_W S = 0$ remains open as there is still the possibility of nonconstant λ and μ. We do know, however, that when such a $\{\lambda, \mu\}$ pair exists, all other principal curvatures will have φ-invariant principal spaces.

The Hopf assumption in Theorem 8.121 can be weakened to $SW = \sigma W$ for constant σ, with similar conclusions. See Kang and Ki [237]. Also see Ahn et al. [7] and Lee et al. [313] for related results.

Finally, we note that the W-parallelism condition can be realized even for the curvature tensor R (see Kimura and Maeda [275]).

We have seen in Theorem 8.114 that the structure Jacobi operator R_W cannot be parallel. However, W-parallelism is possible. We first look at Hopf hypersurfaces.

Theorem 8.123. *Let M^{2n-1}, where $n \geq 2$, be a Hopf hypersurface in $\mathbf{CP}^n$ or $\mathbf{CH}^n$. Then the structure Jacobi operator of M is W-parallel if and only if M is*

- *a Hopf hypersurface with $\alpha = 0$, or*
- *an open subset of a Type A hypersurface.*

Proof. Every hypersurface has an open dense set (see Section 2.5) on which the principal curvatures have constant multiplicities. In the Hopf case, we can assert that on such a set, $\mathcal{W}^{\perp}$ consists locally of smooth φ-invariant distributions of the form T_λ or $T_\lambda + T_\mu$ where $T_\mu = \varphi T_\lambda$, as in Theorem 6.17. For $X \in T_\lambda$,

$$\nabla_W(R_W X) = \nabla_W((\lambda\alpha + c)X) = \alpha(W\lambda)X + (\lambda\alpha + c)\nabla_W X$$

while

$$R_W \nabla_W X = (\alpha A + c)\nabla_W X$$

where we have used the Gauss equation and the fact that

$$\langle \nabla_W X, W \rangle = -\langle X, \nabla_W W \rangle = 0.$$

This gives

$$(\nabla_W R_W)X = \alpha\big((W\lambda)X + (\lambda - A)\nabla_W X\big). \tag{8.127}$$

It is trivial to check that $(\nabla_W R_W)W = 0$. If we apply the Codazzi equation to the pair (X, W), we get

$$\begin{aligned} \nabla_X(\alpha W) - A\varphi AX - \nabla_W(\lambda X) + A\nabla_W X &= -c\varphi X \qquad (8.128)\\ \lambda(\alpha - \mu)\varphi X - (W\lambda)X - (\lambda - A)\nabla_W X &= -c\varphi X. \end{aligned}$$

It is clear from (8.127) that every Hopf hypersurface with $\alpha = 0$ satisfies $\nabla_W R_W = 0$. For Type A hypersurfaces, we have $\lambda = \mu$ and $\lambda^2 = \lambda\alpha + c$. Thus (8.128) reduces to $(\lambda - A)\nabla_W X = 0$. Substituting in (8.127) yields $(\nabla_W R_W)X = 0$ as desired.

Conversely, equations (8.127) and (8.128) also show that every Hopf hypersurface satisfying $\nabla_W R_W = 0$ will have either $\alpha = 0$ or $\lambda(\alpha - \mu) + c = 0$. Substituting for $\lambda\mu$ from Theorem 6.17, we get $\alpha(\lambda - \mu) = 0$. Thus, unless $\alpha = 0$, we have $\varphi A = A\varphi$ on an open dense set of M, and hence globally. By Theorem 8.37, M is an open subset of a Type A hypersurface. □

Beginning with Cho and Ki, several authors have studied hypersurfaces satisfying $\nabla_W R_W = 0$. Various conditions have been found for such a hypersurface to be Hopf and therefore belong to the class characterized in Theorem 8.123.

Theorem 8.124. *Let M^{2n-1}, where $n \geq 2$, be a real hypersurface in* $\mathbf{CP}^n$ *or* $\mathbf{CH}^n$ *whose structure Jacobi operator is W-parallel. Each of the following conditions is sufficient for M to be Hopf:*

1. *R_W commutes with the shape operator A; (Cho and Ki [115])*
2. *R_W commutes with the Ricci tensor S; (Ki et al. [245, 247])*
3. *$\nabla_W S = 0$; (Ki, Pérez, Santos and Suh [252])*
4. *$R_W \circ \varphi \circ S = S \circ \varphi \circ R_W$; (Ki, Kurihara and Takagi [246])*

Conversely, all Type A hypersurfaces satisfy conditions 1–4. All Hopf hypersurfaces with $\alpha = 0$ satisfy conditions 1–3.

Proof. Proving sufficiency for the Hopf condition involves starting with the standard non-Hopf setup and deriving a contradiction. These proofs are rather long and somewhat similar (see also our discussion of Theorems 8.98 and 8.112). Because of space limitations, we will not present them here. It is easy to check conditions *1–4* for Type *A* hypersurfaces. It is also evident from Theorems 8.123 and 8.121 that all Hopf hypersurfaces with $\alpha = 0$ satisfy conditions *1–3*. □

The fourth condition in Theorem 8.124 entails an additional restriction.

Theorem 8.125. *Let M^{2n-1}, where $n \geq 2$, be a Hopf hypersurface in* $\mathbf{CP}^n$ *or* $\mathbf{CH}^n$ *whose structure Jacobi operator is W-parallel. Then $R_W \circ \varphi \circ S = S \circ \varphi \circ R_W$ if and only if*

- *M is an open subset of a Type A hypersurface, or*
- *$n = 2$, $\alpha = 0$ and M is a pseudo-Einstein hypersurface with nonconstant principal curvatures as described in Theorems 8.64 and 8.92, or*
- *M is a Hopf hypersurface in* $\mathbf{CP}^n$, *where $n \geq 3$, for which $\alpha = 0$, $\varphi S = S\varphi$ and not all principal curvatures are constant. See the discussion preceding Theorem 8.93 for more information.*

Remark 8.126. Corresponding theorems in the literature, for example [115], have been modified to deal with the condition $\alpha = 0$. Some authors have made the assertion, citing Berndt [27], that Hopf hypersurfaces in $\mathbf{CH}^n$ cannot have $\alpha = 0$. This is an incorrect reading of Berndt's result, which was obtained under the assumption of constant principal curvatures. In fact, as shown by Ivey and Ryan [222], α can take on all small values $0 \leq \alpha \leq \frac{2}{r}$ as well as the higher values that occur in the Montiel list, at least for $n = 2$. For $n \geq 3$, existence of such hypersurfaces is an open problem.

For further work along these lines, see [266, 268] and [267].

Remark 8.127. Note that Theorem 8.125 differs slightly from the version stated by Ki, Kurihara, and Takagi to take into account the $\alpha = 0$ possibilities. In particular, for $n = 2$, hypersurfaces with nonconstant principal curvatures can occur. We do not know whether any hypersurface satisfying the third condition in Theorem 8.125 actually exists. In terms of the analysis leading to Theorem 8.93, the hypersurface must contain an open set where $\lambda_1 = -\lambda_2 = \frac{1}{r}$ and $k_1 = k_2 = n - 2$ is even. Also, λ and μ are nonconstant principal curvatures of multiplicity 1 satisfying $\lambda\mu = c$.

The same authors also consider the related condition $R_W \circ \varphi \circ S = R_W \circ S \circ \varphi$ in [244, 308].

η-parallelism

The η-parallelism condition is somewhat more complicated and we will make separate definitions for each type of tensor field in which we are interested. Specifically, a tensor field T of type $(1, 1)$ is said to be *η-parallel* if

$$\langle(\nabla_X T)Y, Z\rangle = 0$$

for all X, Y, and Z in $\mathcal{W}^\perp$. Equivalently, for all $X \in \mathcal{W}^\perp$,

$$(\nabla_X T)\mathcal{W}^\perp \subset \mathcal{W}.$$

We also say that T is *cyclic η-parallel* if the cyclic sum of $\langle(\nabla_X T)Y, Z\rangle$ vanishes for all X, Y, and Z in $\mathcal{W}^\perp$. A hypersurface is said to be η-parallel (resp. cyclic η-parallel) if its shape operator is an η-parallel (resp. cyclic η-parallel) tensor field.

S.H. Kon and T.-H. Loo [292] obtained, for $n \geq 3$, the following remarkable characterization of η-parallel hypersurfaces.

Theorem 8.128. *Let M^{2n-1}, where $n \geq 3$, be a real hypersurface in* $\mathbf{CP}^n$ *or* $\mathbf{CH}^n$*. Then M is η-parallel if and only if it is*

- *a ruled hypersurface, or*
- *an open subset of a hypersurface of Type A or Type B.*

This completes (except for $n = 2$) a project begun by Kimura and Maeda [274] who classified the η-parallel Hopf hypersurfaces in $\mathbf{CP}^n$. Subsequently, Suh [504] dealt with the hyperbolic case. The result of Kimura, Maeda, and Suh is as follows:

Theorem 8.129. *Let M^{2n-1}, where $n \geq 2$, be a Hopf hypersurface in* $\mathbf{CP}^n$ *or* $\mathbf{CH}^n$*. Then M is η-parallel if and only if it is an open subset of a Type A or a Type B hypersurface.*

Theorem 8.129 was proved in detail in [399]. Following on the recent work of Kon and Loo [292, 293], we present here a self-contained proof of the complete result, Theorem 8.128, which we break down into a collection of theorems, lemmas, and propositions. The classification of η-parallel hypersurfaces in $\mathbf{CP}^2$ and $\mathbf{CH}^2$ remains an open problem.

It is easy to verify that every hypersurface of Type A or Type B is η-parallel. Because of the Codazzi equation and the fact that A is symmetric, the expression $\langle(\nabla_X A)Y, Z\rangle$ is symmetric in all three arguments used in the definition of η-parallelism. It is sufficient to check that this expression vanishes on principal basis vectors and since there are at most two distinct principal curvatures involved, we can assume that Y and Z correspond to the same principal curvature, say λ. Then

$$\langle(\nabla_X A)Y, Z\rangle = \langle(\lambda - A)\nabla_X Y, Z\rangle = \langle\nabla_X Y, (\lambda - A)Z\rangle = 0,$$

where we have used the fact that λ is constant. For ruled hypersurfaces, we have

Proposition 8.130. *Let M^{2n-1}, where $n \geq 2$, be a ruled hypersurface in $\mathbf{CP}^n$ or $\mathbf{CH}^n$. Then M is η-parallel.*

Proof. First note that for any scalar function ρ, and any vector fields X_1 and X_2 in $\mathcal{W}^\perp$,

$$\langle\nabla_{X_1}(\rho W), X_2\rangle = (X_1\rho)\langle W, X_2\rangle + \rho\langle\varphi A X_1, X_2\rangle = 0, \tag{8.129}$$

since $\varphi A X_1 = 0$. Now let X, Y, and Z be any vector fields in $\mathcal{W}^\perp$. Since AY and AZ are both scalar multiples of W, we get

$$\langle(\nabla_X A)Y, Z\rangle = \langle\nabla_X(AY), Z\rangle - \langle A\nabla_X Y, Z\rangle \tag{8.130}$$
$$= 0 - \langle\nabla_X Y, AZ\rangle = \langle Y, \nabla_X(\rho W)\rangle \tag{8.131}$$

for a suitable function ρ. Thus, $\langle(\nabla_X A)Y, Z\rangle = 0$ and M is η-parallel. □

To complement Theorem 8.129, we state our main result on the non-Hopf case. This follows from the exposition below and completes the proof of Theorem 8.128.

Theorem 8.131. *Let M^{2n-1}, where $n \geq 3$, be a non-Hopf η-parallel hypersurface in $\mathbf{CP}^n$ or $\mathbf{CH}^n$. Then M is ruled.*

Proof of Theorem 8.128

We develop, for η-parallel hypersurfaces, a useful formula relating the curvature operator to the shape operator and its W-derivative.

Lemma 8.132. *Let M^{2n-1}, where $n \geq 2$, be an η-parallel hypersurface in $\mathbf{CP}^n$ or $\mathbf{CH}^n$. Then, for all X, Y, U, V in $\mathcal{W}^\perp$,*

$$\begin{aligned}\langle[R(X,Y),A]\,U,V\rangle = \ & L(X,Y,U,V) + L(X,Y,V,U) \\ & -L(Y,X,U,V) - L(Y,X,V,U)\end{aligned} \tag{8.132}$$

where

$$L(X, Y, U, V) = \langle \varphi AX, U \rangle \langle (F - c\varphi)Y, V \rangle + \frac{1}{2} \langle \varphi AX, Y \rangle \langle FU, V \rangle$$

and $F = \nabla_W A$.

The proof of this lemma is obtained by applying ∇_X to the identity

$$\langle (\nabla_Y A)U, V \rangle = 0$$

to obtain an expression for $\nabla^2 A(;Y;X)$. Then, of course,

$$[R(X, Y), A]\, U = (R(X, Y) \cdot A)U = \nabla^2 A(;Y;X)U - \nabla^2 A(;X;Y)U,$$

for all X, Y, and U in $\mathcal{W}^{\perp}$.

We use the Codazzi equation liberally throughout to replace derivatives of A in $\mathcal{W}^{\perp}$ directions by $\nabla_W A = F$. The Codazzi equation also shows that F is symmetric. We leave the details to the reader.

Lemma 8.133. *Let* M^{2n-1}*, where* $n \geq 3$*, be an* η*-parallel hypersurface in* $\mathbf{CP}^n$ *or* $\mathbf{CH}^n$ *and let* $T_1 = \varphi^2 A \varphi^2$*,* $T_2 = \varphi A \varphi$*. Let* $p \in M$ *be a point where* $AW \neq \alpha W$*. Then* T_1 *and* T_2 *commute at* p*.*

Proof. Our proof is in two parts, in which we first show that

- T_1 and T_2 have a common eigenvector in $\mathcal{W}_p^{\perp}$, and then
- T_1 and T_2 commute at p.

Proof of the first assertion

We adopt the standard non-Hopf setup in a neighborhood of p. Let $Y \in \mathcal{W}_p^{\perp}$ be a unit eigenvector of T_1 with corresponding eigenvalue γ. Then, working pointwise at p, we have

$$AY = \gamma Y + \beta \langle Y, U \rangle W$$

from which it is straightforward to compute

$$A^2 Y = |AY|^2 Y + \beta^2 \langle U, Y \rangle (U - \langle U, Y \rangle Y) \mod \mathcal{W}. \tag{8.133}$$

To set up a proof by contradiction for the first assertion, we assume that T_1 and T_2 have no common eigenvectors in $\mathcal{W}_p^{\perp}$. In particular, Y is not an eigenvector of T_2. Then there is a unique nonzero vector Z orthogonal to W, Y, and φY such that

$$A\varphi Y = k\varphi Y + \varphi Z \mod \mathcal{W}$$

where $k = \langle A\varphi Y, \varphi Y\rangle$. In fact, $Z = -kY - \varphi A\varphi Y$. We carry out a number of steps, each proving one or more assertions, culminating in $Z = 0$, a contradiction. The outline is as follows:

1. $2\gamma(|AY|^2 - A^2)Y + \langle FY, Y\rangle(3A\varphi + \varphi A)Y = 0 \mod \mathcal{W}$;
2. $\langle FY, Y\rangle = 0$ and $\gamma \neq 0$ and thus $A^2Y = |AY|^2Y \mod \mathcal{W}$. From (8.133), either $Y = \pm U$ or Y is orthogonal to U. We can conclude that $\mathcal{W}^\perp$ has an orthonormal basis of eigenvectors of $\varphi^2A\varphi^2$ consisting of U and $2n-3$ vectors orthogonal to U which are also eigenvectors of A. Note that the span of $\{W, AW\}$ is A-invariant and has $\{W, U\}$ as an orthonormal basis. For the remaining steps in this list, we take $Y = U$, so that $\gamma = \lambda$, $k = \nu$ and $Z = -\nu U - \varphi A\varphi U$.
3. $\lambda\langle FZ, \varphi U\rangle = c|Z|^2$, $\langle FU, Z\rangle = 0$, $\langle FZ, Z\rangle = 0$ and $\langle \varphi AZ, Z\rangle = 0$;
4. $\langle F\varphi Z, \varphi Z\rangle = 0$ and $AZ = \tau Z$ where $\tau\langle FU, \varphi Z\rangle = c|Z|^2$;
5. $\langle F\varphi Z, \varphi U\rangle = 0$, $\langle F\varphi U, \varphi U\rangle = 0$ and $\lambda\langle FZ, \varphi U\rangle = c|Z|^2$;
6. $\tau(FU + c\varphi U) = cA\varphi U \mod \mathcal{W}$, and thus $\tau\langle FU, \varphi U\rangle = c(\nu - \tau)$;
7. $\lambda\langle FU, \varphi U\rangle = c(\lambda - \nu)$ and $2c|Z|^2 = \lambda\tau(\nu^2 + |Z|^2 - \nu\tau) + \lambda c(\nu - \tau)$;
8. $\nu(\lambda + \tau) = 2\lambda\tau$ and $(\lambda\tau + c)(\lambda - \tau) = -\tau\beta^2$;
9. $b\langle FU, \varphi Z\rangle + \langle FZ, \varphi Z\rangle = c|Z|^2$ where $b|Z^2| = \langle A\varphi Z, \varphi Z\rangle$;
10. $b\langle FZ, \varphi U\rangle - \langle FZ, \varphi Z\rangle = c|Z|^2$;
11. $\nu b = |Z|^2 - c$ and $b = \nu$. Thus $\nu^2 + c = |Z|^2$.
12. $2\lambda^2 - \lambda\tau + c = 0$.
13. $2\tau^2 - \lambda\tau + c = 0$.

We now show that Steps 1 through 13 imply a contradiction. To this end, first observe that the results of Steps 12 and 13 show that $\lambda = \pm\tau$. However, by Step 8, neither of these alternatives can occur.

Proof details for the first assertion

1. We apply (8.132) using arguments (X, Y, Y, Y) for $X \in \mathcal{W}^\perp$. Because the last two arguments are equal, our equation reduces to

$$\langle R(X, Y)AY, Y\rangle = L(X, Y, Y, Y) - L(Y, X, Y, Y).$$

It is straightforward to check that

$$L(X, Y, Y, Y) = \frac{3}{2}\langle \varphi AX, Y\rangle\langle FY, Y\rangle$$

$$L(Y, X, Y, Y) = -\frac{1}{2}\langle A\varphi X, Y\rangle\langle FY, Y\rangle$$

where we have used the fact that $\langle \varphi AY, Y\rangle = 0$. There is only one term of the Gauss equation that contributes to the curvature term, giving

$$\langle (AX \wedge AY)AY, Y\rangle = |AY|^2\langle AX, Y\rangle - \langle AX, AY\rangle\langle AY, Y\rangle.$$

Since $\langle AX, Y\rangle = \gamma\langle X, Y\rangle$ and $\langle AY, Y\rangle = \gamma\langle Y, Y\rangle = \gamma$, we have established the assertion of Step 1.

2. The fact that $\langle FY, Y\rangle = 0$ is clear from Step 1 if $\gamma = 0$ or $\langle Y, U\rangle = 0$. Suppose now that $\langle FY, Y\rangle \neq 0$ (and therefore $\gamma\langle Y, U\rangle \neq 0$). We obtain a contradiction by three applications of (8.132), as follows:

   - Look at

$$\langle R(X,Z)AY, Y\rangle = L(X,Z,Y,Y) - L(Z,X,Y,Y).$$

   for $X \in \mathcal{W}^{\perp}$. Since $Y, \varphi Y, AY, \varphi AY$ and A^2Y are all orthogonal to Z, the left side vanishes. On the other hand,

$$L(X,Z,Y,Y) = \langle \varphi AX, Y\rangle\langle (F - c\varphi)Z, Y\rangle + \frac{1}{2}\langle \varphi AX, Z\rangle\langle FY, Y\rangle$$

$$L(Z,X,Y,Y) = \langle \varphi AZ, Y\rangle\langle (F - c\varphi)X, Y\rangle + \frac{1}{2}\langle \varphi AZ, X\rangle\langle FY, Y\rangle.$$

   Since $\langle \varphi AZ, Y\rangle = 0$, our equation reduces to

$$\frac{1}{2}\langle FY, Y\rangle\langle (\varphi A + A\varphi)Z, X\rangle + \langle A\varphi Y, X\rangle\langle FZ, Y\rangle = 0 \tag{8.134}$$

   for all $X \in \mathcal{W}^{\perp}$. In particular, for $X = \varphi Y$, we have

$$(\varphi A + A\varphi)\varphi Y = \varphi A\varphi Y - AY = \langle \varphi A\varphi Y, Y\rangle Y - Z - \gamma Y \mod \mathcal{W}$$

   and the inner product with Z is just $-|Z|^2$. Thus,

$$\frac{1}{2}\langle FY, Y\rangle|Z|^2 + \langle FZ, Y\rangle\langle A\varphi Y, \varphi Y\rangle = 0$$

   which implies that both $\langle FZ, Y\rangle$ and $k = \langle A\varphi Y, \varphi Y\rangle$ are nonzero.

   - Now look at

$$\langle R(X,Y)AZ, Z\rangle = L(X,Y,Z,Z) - L(Y,X,Z,Z). \tag{8.135}$$

   It is easy to check that the only term from the Gauss equation that makes a contribution to the left side is

$$2c\langle X, \varphi Y\rangle\langle \varphi AZ, Z\rangle.$$

   Also,

$$L(X,Y,Z,Z) = \langle \varphi AX, Z\rangle\langle (F - c\varphi)Y, Z\rangle + \frac{1}{2}\langle \varphi AX, Y\rangle\langle FZ, Z\rangle$$

$$L(Y,X,Z,Z) = \langle \varphi AY, Z\rangle \langle (F - c\varphi)X, Z\rangle + \frac{1}{2}\langle \varphi AY, X\rangle \langle FZ, Z\rangle$$

so that (8.135) becomes

$$2c\langle X, \varphi Y\rangle \langle \varphi AZ, Z\rangle + \langle AX, \varphi Z\rangle \langle FY, Z\rangle + \frac{1}{2}\langle (A\varphi + \varphi A)Y, X\rangle \langle FZ, Z\rangle = 0. \tag{8.136}$$

Setting $X = Z$ and using the fact that $\langle FY, Z\rangle \neq 0$, we get $\langle \varphi AZ, Z\rangle = 0$. This simplifies (8.136) to

$$\langle FY, Z\rangle A\varphi Z + \frac{1}{2}\langle FZ, Z\rangle (A\varphi + \varphi A)Y = 0 \mod \mathcal{W}, \tag{8.137}$$

from which $A\varphi Z$ lies in the span of $\{\varphi Y, \varphi Z, W\}$. By (8.134), so does φAZ. Further, since $\langle \varphi AZ, \varphi Y\rangle = \langle AZ, Y\rangle = 0$, φAZ actually lies in the span of $\{\varphi Z, W\}$ and $\varphi^2 AZ$ is a scalar multiple of Z. Thus we can write

$$AZ = aW + bZ. \tag{8.138}$$

- Finally, look at

$$\langle R(X,Z)AZ, Z\rangle = L(X,Z,Z,Z) - L(Z,X,Z,Z). \tag{8.139}$$

Taking (8.138) into account, the curvature term reduces to

$$|AZ|^2\langle AX, Z\rangle - \langle AZ, AX\rangle \langle AZ, Z\rangle.$$

On the other hand,

$$L(X,Z,Z,Z) = \langle \varphi AX, Z\rangle \langle FZ, Z\rangle + \frac{1}{2}\langle \varphi AX, Z\rangle \langle FZ, Z\rangle$$

$$L(Z,X,Z,Z) = \langle \varphi AZ, Z\rangle \langle (F - c\varphi)X, Z\rangle + \frac{1}{2}\langle \varphi AZ, X\rangle \langle FZ, Z\rangle$$

Since $\langle \varphi AZ, Z\rangle = 0$, (8.139) reduces to

$$|AZ|^2\langle AX, Z\rangle - \langle AZ, AX\rangle \langle AZ, Z\rangle + \frac{1}{2}\langle FZ, Z\rangle \langle (3A\varphi Z + \varphi AZ), X\rangle = 0 \tag{8.140}$$

which simplifies to

$$a^2bZ - ab|Z|^2\beta U + \langle FZ, Z\rangle (3A\varphi Z + b\varphi Z) = 0 \mod \mathcal{W}. \tag{8.141}$$

Taking the inner product with Y gives $ab|Z|^2\beta\langle U, Y\rangle = 0$, i.e., $ab = 0$. Since $A\varphi Z \neq 0$, we have from (8.137) that $\langle FZ, Z\rangle \neq 0$. Thus (8.141) reduces to

$$3A\varphi Z + b\varphi Z = 0 \mod \mathcal{W}.$$

However, the inner product of this with φY yields $\langle A\varphi Z, \varphi Y\rangle = 0$ which contradicts the fact that $\langle A\varphi Y, \varphi Z\rangle = |\varphi Z|^2 = |Z^2|$. We are forced to conclude that $\langle FY, Y\rangle = 0$.

Now, considering (8.132) with arguments $(Z, Y, \varphi Y, \varphi Y)$, we will show that $\gamma \neq 0$. Start with

$$\langle R(Z, Y)A\varphi Y, \varphi Y\rangle = L(Z, Y, \varphi Y, \varphi Y) - L(Y, Z, \varphi Y, \varphi Y). \tag{8.142}$$

It is straightforward to check that the only nonzero part of the curvature term is

$$c\langle(\varphi Z \wedge \varphi Y)A\varphi Y, \varphi Y\rangle = -c|Z|^2.$$

On the other hand,

$$L(Z, Y, \varphi Y, \varphi Y) = \langle \varphi AZ, \varphi Y\rangle\langle(F - c\varphi)Y, \varphi Y\rangle + \frac{1}{2}\langle \varphi AZ, Y\rangle\langle F\varphi Y, \varphi Y\rangle$$

$$L(Y, Z, \varphi Y, \varphi Y) = \langle \varphi AY, \varphi Y\rangle\langle(F - c\varphi)Z, \varphi Y\rangle + \frac{1}{2}\langle \varphi AY, Z\rangle\langle F\varphi Y, \varphi Y\rangle.$$

Since $\langle \varphi AZ, \varphi Y\rangle = \langle AZ, Y\rangle$, $\langle \varphi AZ, Y\rangle = -\langle Z, A\varphi Y\rangle$ and $\langle \varphi AY, Z\rangle = \gamma\langle \varphi Y, Z\rangle$ all vanish, (8.142) reduces to

$$-c|Z|^2 = -\gamma\langle(F - c\varphi)Z, \varphi Y\rangle \tag{8.143}$$

from which it is clear that $\gamma \neq 0$ and $\gamma\langle FZ, \varphi Y\rangle = c|Z|^2$.

3. The first assertion has been established as a by-product of our calculation in Step 2. Now, revisiting our application of (8.132) with arguments (X, Z, U, U), we have

$$L(X, Z, U, U) = L(Z, X, U, U)$$

which reduces to

$$\langle \varphi AX, U\rangle\langle FU, Z\rangle = 0$$

since $\langle FU, U\rangle$ and $\langle \varphi AZ, U\rangle$ both vanish. As $A\varphi U \neq 0$, we have $\langle FZ, U\rangle = 0$. Continuing with (X, U, Z, Z) we get

$$2c\langle X, \varphi U\rangle\langle \varphi AZ, Z\rangle = \frac{1}{2}\langle FZ, Z\rangle(\langle \varphi AX, U\rangle - \langle \varphi AU, X\rangle).$$

Setting $X = \varphi Z$ gives $\langle \varphi Z, A\varphi U\rangle\langle FZ, Z\rangle = 0$. Thus $\langle FZ, Z\rangle = 0$ and, as a consequence, $\langle \varphi AZ, Z\rangle\varphi U = 0$, so that $\langle \varphi AZ, Z\rangle = 0$.

4. Using arguments $(X, U, \varphi Z, \varphi Z)$, we have

$$\langle R(X, U)A\varphi Z, \varphi Z\rangle = L(X, U, \varphi Z, \varphi Z) - L(U, X, \varphi Z, \varphi Z).$$

The only contribution from the curvature term is

$$c|Z|^2\langle Z, X\rangle.$$

To see this, we need to observe that U, AU, and A^2U are all orthogonal to φZ. On the other hand,

$$L(X, U, \varphi Z, \varphi Z) = \langle \varphi AX, \varphi Z\rangle\langle (F - c\varphi)U, \varphi Z\rangle + \frac{1}{2}\langle \varphi AX, U\rangle\langle F\varphi Z, \varphi Z\rangle$$

$$L(U, X, \varphi Z, \varphi Z) = \langle \varphi AU, \varphi Z\rangle\langle (F - c\varphi)X, \varphi Z\rangle + \frac{1}{2}\langle \varphi AU, X\rangle\langle F\varphi Z, \varphi Z\rangle.$$

Thus, we have

$$c|Z|^2 Z = \langle FU, \varphi Z\rangle AZ - \frac{1}{2}\langle F\varphi Z, \varphi Z\rangle(A\varphi U + \varphi AU). \quad (8.144)$$

Taking the inner product with φZ yields

$$0 = \langle F\varphi Z, \varphi Z\rangle(\langle A\varphi U, \varphi Z\rangle + \langle \varphi AU, \varphi Z\rangle) = \langle F\varphi Z, \varphi Z\rangle|Z|^2.$$

From this we get $\langle F\varphi Z, \varphi Z\rangle = 0$ and $c|Z|^2 Z = \langle FU, \varphi Z\rangle AZ$ which we rewrite as $AZ = \tau Z$.

5. Using arguments $(X, Z, \varphi Z, \varphi Z)$, we have

$$\langle R(X, Z)A\varphi Z, \varphi Z\rangle = L(X, Z, \varphi Z, \varphi Z) - L(Z, X, \varphi Z, \varphi Z).$$

The curvature term reduces to

$$c\langle A\varphi Z, \varphi Z\rangle\langle \varphi Z, \varphi X\rangle + c|Z|^2\langle \varphi A\varphi Z, X\rangle.$$

On the other hand, using the fact that $\langle F\varphi Z, \varphi Z\rangle = 0$, we get

$$L(X, Z, \varphi Z, \varphi Z) = \langle \varphi AX, \varphi Z\rangle\langle (F - c\varphi)Z, \varphi Z\rangle$$

$$L(Z, X, \varphi Z, \varphi Z) = \langle \varphi AZ, \varphi Z\rangle\langle (F - c\varphi)X, \varphi Z\rangle.$$

Thus

$$c(\langle A\varphi Z, \varphi Z\rangle Z + |Z|^2\varphi A\varphi Z) = \tau(\langle FZ, \varphi Z\rangle Z - |Z|^2 F\varphi Z) \mod \mathcal{W}. \quad (8.145)$$

Upon taking the inner product with φU we get $\langle F\varphi Z, \varphi U\rangle = 0$. Now using arguments $(X, Z, \varphi U, \varphi U)$, we have

$$\langle R(X,Z)A\varphi U, \varphi U\rangle = L(X,Z,\varphi U,\varphi U) - L(Z,X,\varphi U,\varphi U).$$

The curvature term reduces to $c|Z|^2\langle U,X\rangle$ and

$$\begin{aligned} L(X,Z,\varphi U,\varphi U) &= \langle FZ,\varphi U\rangle\langle AU,X\rangle - \frac{1}{2}\langle F\varphi U,\varphi U\rangle\langle A\varphi Z,X\rangle \\ L(Z,X,\varphi U,\varphi U) &= \frac{1}{2}\langle F\varphi U,\varphi U\rangle\langle \varphi AZ,X\rangle. \end{aligned}$$

Thus

$$c|Z|^2U = \langle FZ,\varphi U\rangle AU - \frac{1}{2}\langle F\varphi U,\varphi U\rangle(A\varphi Z + \varphi AZ) \mod \mathcal{W}. \tag{8.146}$$

Taking the inner product with φU and then with U, we get $\langle F\varphi U, \varphi U\rangle = 0$ and $\lambda\langle FZ, \varphi U\rangle = c|Z|^2$.

6. Use $(X, \varphi Z, U, U)$ to get $|Z|^2(FU + c\varphi U) = \langle FU, \varphi Z\rangle A\varphi U \mod \mathcal{W}$ which is essentially the required result.
7. For Step 7, we use arguments $(X, \varphi U, \varphi U, Z)$. Since the last two arguments are not the same, our application of (8.132) is more complicated. The curvature term is

$$\langle R(X,\varphi U)A\varphi U, Z\rangle - \langle R(X,\varphi U)\varphi U, AZ\rangle.$$

To evaluate this using the Gauss equation, we need

$$\begin{aligned} \langle (AX\wedge A\varphi U)A\varphi U,Z\rangle - \langle (AX\wedge A\varphi U)\varphi U,AZ\rangle &= \tau(|A\varphi U|^2 - \nu\tau)\langle Z,X\rangle \\ \langle (X\wedge \varphi U)A\varphi U,Z\rangle - \langle (X\wedge \varphi U)\varphi U,AZ\rangle &= (\nu - \tau)\langle Z,X\rangle \\ \langle (\varphi X\wedge \varphi^2 U)A\varphi U,Z\rangle - \langle (\varphi X\wedge \varphi^2 U)\varphi U,AZ\rangle &= 0 \\ 2\langle X,\varphi^2U\rangle(\langle \varphi A\varphi U,Z\rangle - \langle \varphi^2U,AZ\rangle) &= 2|Z|^2\langle U,X\rangle. \end{aligned} \tag{8.147}$$

For the right side of (8.132), we get

$$\begin{aligned} L(X,\varphi U,\varphi U,Z) &= \langle AX,U\rangle\langle F\varphi U,Z\rangle + \frac{1}{2}\langle F\varphi U,Z\rangle\langle AX,U\rangle = \frac{3}{2}c|Z|^2\langle U,X\rangle \\ L(X,\varphi U,Z,\varphi U) &= \frac{1}{2}c|Z|^2\langle U,X\rangle \qquad (8.148) \\ L(\varphi U,X,\varphi U,Z) &= \langle \varphi A\varphi U,\varphi U\rangle\langle (F - c\varphi)X,Z\rangle + \frac{1}{2}\langle \varphi A\varphi U,X\rangle\langle F\varphi U,Z\rangle \\ &= -\frac{1}{2}\langle \varphi U,FZ\rangle\langle (\nu U + Z),X\rangle \end{aligned}$$

$$L(\varphi U, X, Z, \varphi U) = \langle \varphi A \varphi U, Z\rangle\langle (F - c\varphi)X, \varphi U\rangle - \frac{1}{2}\langle \varphi U, FZ\rangle\langle (\nu U + Z), X\rangle.$$

Thus (8.132) reduces to

$$\begin{aligned}&\tau(\nu^2 + |Z|^2 - \nu\tau)Z + c(\nu - \tau)Z + 2c|Z|^2 U\\&= 2c|Z|^2 U + \langle FZ, \varphi U\rangle(\nu U + Z) + |Z|^2(F\varphi U - cU) \mod \mathcal{W}.\end{aligned}$$

from which the desired result follows.

8. The first result follows by equating the two expressions just obtained for $\langle FU, \varphi U\rangle$. For the second result, consider (8.132) with arguments (X, U, U, Z). It is straightforward to check that all four terms on the right side vanish. On the other hand, the only contributions to the curvature term are

$$\langle (AX \wedge AU)AU, Z\rangle - \langle (AX \wedge AU)U, AZ\rangle = (\tau|AU|^2 - \lambda\tau^2)\langle Z, X\rangle$$

and

$$c\langle (X \wedge U)AU, Z\rangle - c\langle (X \wedge U)U, AZ\rangle = c\langle (\lambda - \tau)Z, X\rangle.$$

from which we conclude

$$\tau(|AU|^2 - \lambda\tau)Z + c(\lambda - \tau)Z = 0 \mod \mathcal{W}$$

which simplifies to $(\lambda\tau + c)(\lambda - \tau) + \tau\beta^2 = 0$.

9. For Step 9, we use arguments $(Z, \varphi Z, \varphi Z, U)$. The left side of (8.132) is

$$\langle R(Z, \varphi Z)A\varphi Z, U\rangle - \langle R(Z, \varphi Z)\varphi Z, AU\rangle.$$

The only term from the Gauss equation that makes a nonzero contribution is the last one, and the left side of (8.132) reduces to $2c|Z|^4$. Evaluating the right side, we have

$$\begin{aligned}L(Z, \varphi Z, \varphi Z, U) &= \langle AZ, Z\rangle\langle F\varphi Z, U\rangle + \frac{1}{2}\langle F\varphi Z, U\rangle\langle AZ, Z\rangle\\&= \frac{3}{2}\tau|Z|^2\langle FU, \varphi Z\rangle\\L(Z, \varphi Z, U, \varphi Z) &= \frac{1}{2}\tau|Z|^2\langle FU, \varphi Z\rangle\\L(\varphi Z, Z, \varphi Z, U) &= \frac{1}{2}\langle \varphi A \varphi Z, Z\rangle\langle FU, \varphi Z\rangle = -\frac{1}{2}b|Z|^2\langle FU, \varphi Z\rangle\\L(\varphi Z, Z, U, \varphi Z) &= \langle \varphi A \varphi Z, U\rangle\langle (F - c\varphi)Z, \varphi Z\rangle - \frac{1}{2}b|Z|^2\langle FU, \varphi Z\rangle.\end{aligned}$$

This gives

$$(\langle FZ, \varphi Z\rangle + b\langle FU, \varphi Z\rangle + c|Z|^2)|Z|^2 \tag{8.149}$$

where we have used the result of Step 3 to substitute for τ. Setting this equal to the left side yields the desired identity.

10. Again, consider (8.132) using arguments $(X, \tilde{X}, Z, Z)$ where X and $\tilde{X}$ are arbitrary members of $\mathcal{W}^{\perp}$. The left side of (8.132) is

$$\langle R(X,\tilde{X})AZ, Z\rangle - \langle R(X,\tilde{X})Z, AZ\rangle = 0. \tag{8.150}$$

since $AZ = \tau Z$ and the curvature operator is skew-symmetric. Now, look at the right side of (8.132) for arguments $(\varphi U, \varphi Z, Z, Z)$. Since the third and fourth arguments are equal and $\langle FZ, Z\rangle = 0$, this right side reduces to

$$\langle \varphi A\varphi U, Z\rangle\langle (F - c\varphi)\varphi Z, Z\rangle - \langle \varphi A\varphi Z, Z\rangle\langle (F - c\varphi)\varphi U, Z\rangle$$

which we can evaluate as

$$-|Z|^2(\langle FZ, \varphi Z\rangle + c|Z|^2) + b|Z|^2\langle FZ, \varphi U\rangle.$$

Thus we have

$$\langle FZ, \varphi Z\rangle + c|Z|^2 = b\langle FZ, \varphi U\rangle. \tag{8.151}$$

11. Consider (8.132) with arguments $(\varphi U, \varphi Z, \varphi U, \varphi U)$. The two terms from the Gauss equation that contribute to the left side are

$$\langle (A\varphi U \wedge A\varphi Z)A\varphi U, \varphi U\rangle = \langle (A\varphi Z, A\varphi U\rangle\nu - |Z|^2(\nu^2 + |Z|^2)$$

and

$$c\langle (\varphi U \wedge \varphi Z)A\varphi U, \varphi U\rangle = c|Z|^2.$$

On the other hand, using the facts that $\langle F\varphi U, \varphi U\rangle = 0$ and $\langle A\varphi U, U\rangle = 0$, we find that $L(\varphi U, \varphi Z, \varphi U, \varphi U) = L(\varphi Z, \varphi U, \varphi U, \varphi U) = 0$. Since $\langle A\varphi Z, A\varphi U\rangle = \nu|Z|^2 + b|Z|^2$, we can simplify the left side to obtain

$$(\nu b - |Z|^2 + c)|Z|^2 = 0$$

from which the first statement is immediate. Also, adding the equations from Steps 9 and 10, we get

$$2\lambda\tau c|Z|^2 = b(c\tau|Z|^2 + \lambda c|Z|^2).$$

In other words $2\lambda\tau = b(\lambda + \tau)$. In view of Step 8, this gives $b = \nu$.

12. Consider (8.132) with arguments $(Z, \varphi Z, U, \varphi U)$. First note that $\langle AU, AZ\rangle = 0$, and it is straightforward to check that the only contribution from the Gauss equation is

$$-2c|Z|^2(\lambda - \nu).$$

Also,

$$L(Z, \varphi Z, U, \varphi U) = L(Z, \varphi Z, \varphi U, U) = \frac{1}{2}\tau|Z|^2\langle FU, \varphi U\rangle$$
$$L(\varphi Z, Z, U, \varphi U) = -|Z|^2\langle FZ, \varphi U\rangle - \frac{1}{2}b|Z|^2\langle FU, \varphi U\rangle$$
$$L(\varphi Z, Z, \varphi U, U) = -\frac{1}{2}b|Z|^2\langle FU, \varphi U\rangle.$$

Thus the right side of (8.132) is

$$\tau|Z|^2\langle FU, \varphi U\rangle + |Z|^2\langle FZ, \varphi U\rangle + b|Z|^2\langle FU, \varphi U\rangle$$

so that

$$2c(\lambda - \nu) + (\tau + b)\langle FU, \varphi U\rangle + \langle FZ, \varphi U\rangle = 0 \tag{8.152}$$

which, by Steps 5 and 6, gives

$$2\lambda\tau(\lambda - \nu) + \lambda(b + \tau)(\nu - \tau) + \tau|Z|^2 = 0. \tag{8.153}$$

Since we have just shown that $b = \nu$, and $|Z|^2 = \nu^2 + c$, we can rewrite this equation as

$$2\lambda\tau(\lambda - \nu) + (\lambda + \tau)\nu^2 - \lambda\tau^2 + \tau c = 0. \tag{8.154}$$

Applying $\nu(\lambda + \tau) = 2\lambda\tau$ from Step 8 allows us to simplify this equation to

$$2\lambda^2 - \lambda\tau + c = 0.$$

13. Consider (8.132) with arguments $(U, \varphi U, Z, \varphi Z)$. Using $AZ = \tau Z$ and the Gauss equation, we get

$$[R(U, \varphi U), A]Z = (\tau - A)R(U, \varphi U)Z = -2c(\tau - A)\varphi Z.$$

Thus the left side of (8.132) is

$$-2c\langle(\tau - A)\varphi Z, \varphi Z\rangle = -2c(\tau - \nu)|Z|^2.$$

Since $\langle \varphi AU, Z\rangle = \langle \varphi AU, \varphi Z\rangle = \langle \varphi A\varphi U, \varphi Z\rangle = 0$, we have

$$L(U, \varphi U, Z, \varphi Z) = \frac{1}{2}\langle \varphi AU, \varphi U\rangle\langle FZ, \varphi Z\rangle = \frac{1}{2}\lambda\langle FZ, \varphi Z\rangle;$$

$$L(U, \varphi U, \varphi Z, Z) = \frac{1}{2}\lambda\langle FZ, \varphi Z\rangle;$$

$$\begin{aligned} L(\varphi U, U, Z, \varphi Z) &= \langle \varphi A\varphi U, Z\rangle\langle (F - c\varphi)U, \varphi Z\rangle + \frac{1}{2}\langle \varphi A\varphi U, U\rangle\langle FZ, \varphi Z\rangle \\ &= -|Z|^2\langle FU, \varphi Z\rangle - \frac{1}{2}\nu\langle FZ, \varphi Z\rangle; \end{aligned}$$

$$L(\varphi U, U, \varphi Z, Z) = -\frac{1}{2}\nu\langle FZ, \varphi Z\rangle.$$

Substituting for $\langle FZ, \varphi Z\rangle$ from the result of Step 9 and for $\langle FU, \varphi Z\rangle$ from Step 4, we get

$$2c(\nu - \tau)c|Z|^2 = (\lambda + \nu)c|Z|^2 - \frac{\lambda\nu + \nu^2 - |Z|^2}{\tau}c|Z|^2$$

which, upon simplification using $\nu(\lambda + \tau) = 2\lambda\tau$ and $\nu^2 - |Z|^2 = c$ (Steps 8 and 11) gives

$$2\tau^2 - \lambda\tau + c = 0. \tag{8.155}$$

□

Proof of the second assertion

Proof. We first observe that if $v \in \mathcal{W}^\perp$ is an eigenvector of T_1, it is easy to check that v is also an eigenvector of T_2 if and only if φv is an eigenvector of T_1.

Take a unit vector $e_1 \in \mathcal{W}^\perp$ which is an eigenvector of both T_1 and T_2 and let $e_2 = \varphi e_1$. Then e_2 is a second eigenvector of T_1. Working now on the orthogonal complement of the span of $\{W, e_1, e_2\}$, choose another such pair. Continue until no further common eigenvectors exist. Let m be the number of pairs thus chosen. We know that $1 \le m \le n - 1$.

Suppose that $m < n - 1$. We complete the orthonormal basis by adding eigenvectors of T_1 that are not eigenvectors of T_2. Name the basis elements $\{e_i\}_1^{2n-2}$ so that $e_{2i} = \varphi e_{2i-1}$ for $1 \le i \le m$, and φe_i is not an eigenvector of T_1 for $2m + 1 \le i \le 2n - 2$. For $v \in \mathcal{W}^\perp$, $\varphi^2 A\varphi^2 v = \lambda v$ if and only if $\lambda v = -\varphi^2 Av = Av - \langle Av, W\rangle W$. Thus there are numbers λ_i (eigenvalues of T_1) such that

$$Ae_i - \langle Ae_i, W\rangle W = \lambda_i e_i, \tag{8.156}$$

for $1 \leq i \leq 2n-2$. However, $A\varphi e_i - \langle A\varphi e_i, W\rangle W$ is not a scalar multiple of φe_i when $i \geq 2m+1$.

We will obtain a contradiction by a series of steps that involve making particular choices of arguments in (8.132) and evaluating the left side using the Gauss equation. Let

$$\mathcal{B}_1 = \{e_{2i-1} \mid 1 \leq i \leq m\}$$
$$\mathcal{B}_2 = \{e_i \mid 2m+1 \leq i \leq 2n-2\}.$$

We make the following assertions:

(i) $\langle FY, \varphi Y\rangle = 0$ for all $Y \in \mathcal{B}_1$. Also, $\lambda_{2j-1} = \lambda_{2j}$ for all $j \leq m$.
(ii) $\langle FY, Y\rangle = 0$ for all $Y \in \mathcal{B}_2$. Also, all λ_j are nonzero for $1 \leq j \leq 2m$.
(iii) For $Y \in \mathcal{B}_2$, X and $\tilde{X}$ in $\mathcal{W}^\perp$,

$$\langle A\varphi Y, X\rangle\langle (F+c\varphi)Y, \tilde{X}\rangle = \langle A\varphi Y, \tilde{X}\rangle\langle (F+c\varphi)Y, X\rangle. \tag{8.157}$$

(iv) The λ_i are all equal. That is, there is a nonzero number γ such that $T_1X = \gamma X$ for all $X \in \mathcal{W}^\perp$.

Because of (iv), if $Y \in \mathcal{B}_2$, $T_1\varphi Y = \gamma\varphi Y$ contradicting the condition that φY is not an eigenvalue of T_1. We conclude that $m = n-1$ and that $\mathcal{W}^\perp$ has a basis consisting of $n-1$ pairs of the form $(X, \varphi X)$ which are common eigenvectors of T_1 and T_2. From this it is clear that T_1 and T_2 commute.

Proof (i):

Consider (8.132) with arguments $(X, \tilde{X}, Y, \varphi Y)$ with $X, \tilde{X} \in \mathcal{B}_2$, $Y \in \mathcal{B}_1$. There is only one nonzero term arising from the Gauss equation and we compute the left side of (8.132) as

$$2c\langle X, \varphi\tilde{X}\rangle\langle (\varphi A - A\varphi)Y, \varphi Y\rangle.$$

Using the fact that AY and $A\varphi Y$ are orthogonal to X and $\tilde{X}$, it is easy to deduce that

$$L(X, \tilde{X}, Y, \varphi Y) = L(X, \tilde{X}, \varphi Y, Y) = \frac{1}{2}\langle \varphi AX, \tilde{X}\rangle\langle FY, \varphi Y\rangle$$
$$L(\tilde{X}, X, Y, \varphi Y) = L(\tilde{X}, X, \varphi Y, Y) = \frac{1}{2}\langle \varphi A\tilde{X}, X\rangle\langle FY, \varphi Y\rangle$$

and so we get

$$\langle (\varphi A + A\varphi)X, \tilde{X}\rangle\langle FY, \varphi Y\rangle = 2c(\lambda_{2j-1} - \lambda_{2j})\langle X, \varphi\tilde{X}\rangle,$$

for a suitable value of $j \leq m$. If $\langle FY, \varphi Y\rangle \neq 0$, we can express $A\varphi X$ as a linear combination of W and φX which contradicts the fact that φX is not an eigenvector of T_2. (For this, we need to observe that $A\varphi X$ is orthogonal to the span of $\mathcal{B}_1$ and $\varphi\mathcal{B}_1$.) From $\langle FY, \varphi Y\rangle = 0$, it follows that $\lambda_{2j-1} = \lambda_{2j}$.

Proof (ii):

Consider (8.132) with arguments $(X, \varphi X, Y, \tilde{Y})$ where $X \in \mathcal{B}_1$ and $Y, \tilde{Y} \in \mathcal{B}_2$. Let γ be the eigenvalue of T_1 associated with X. Then

$$L(X, \varphi X, Y, \tilde{Y}) = \frac{1}{2}\gamma \, \langle FY, \tilde{Y}\rangle$$
$$L(\varphi X, X, Y, \tilde{Y}) = -\frac{1}{2}\gamma \, \langle FY, \tilde{Y}\rangle$$

so that

$$\langle [R(X, \varphi X), A]Y, \tilde{Y}\rangle = 2\gamma \, \langle FY, \tilde{Y}\rangle.$$

Once again, the only contribution of the Gauss equation is

$$-\, 2c\langle(\varphi A - A\varphi)Y, \tilde{Y}\rangle$$

and thus

$$\gamma \, \langle FY, \tilde{Y}\rangle = -c\langle(\varphi A - A\varphi)Y, \tilde{Y}\rangle. \tag{8.158}$$

If γ were zero, we could obtain a contradiction by expressing $A\varphi Y$ as a linear combination of W and φY. (Here we need to observe that $A\varphi Y$ is orthogonal to the span of $\mathcal{B}_1$.) Now set $\tilde{Y} = Y$ in (8.158). Since $\gamma \neq 0$ and $\langle AY, \varphi Y\rangle = 0$, we get $\langle FY, Y\rangle = 0$ as desired.

Proof (iii):

Recalling the analysis of (8.132) with arguments (X, Y, Y, Y) in the first part of the proof and using the fact that $\langle FY, Y\rangle = 0$, we get

$$A^2 Y = |AY|^2 Y \mod \mathcal{W}.$$

Then

$$\langle (AX \wedge A\tilde{X})AY, Y\rangle = \gamma |AY|^2 (\langle Y, \tilde{X}\rangle\langle Y, X\rangle - \langle Y, X\rangle\langle Y, \tilde{X}\rangle) = 0$$

and the other terms in the Gauss equation for $\langle R(X, \tilde{X})AY, Y\rangle$ vanish for more obvious reasons.

Since $\langle FY, Y\rangle = 0$, we have

$$L(X, \tilde{X}, Y, Y) = \langle \varphi AX, Y\rangle\langle (F - c\varphi)\tilde{X}, Y\rangle$$
$$L(\tilde{X}, X, Y, Y) = \langle \varphi A\tilde{X}, Y\rangle\langle (F - c\varphi)X, Y\rangle.$$

Equality of these two expressions is equivalent to (iii).

Proof (iv):

It follows from (iii) that there is a number κ such that

$$(F + c\varphi)Y - \kappa A\varphi Y = 0 \mod \mathcal{W}, \tag{8.159}$$

where we have used the fact that $A\varphi Y \neq 0$. Now take any $X \in \mathcal{B}_1$ and let γ be associated eigenvalue of T_1. Then (8.158) implies that

$$\gamma FY - cA\varphi Y + c\varphi AY \tag{8.160}$$

is orthogonal to the span of $\mathcal{B}_2$. Using (8.159) to replace FY, the same can be said of

$$(\gamma\kappa - c)A\varphi Y - c(\gamma - \gamma')\varphi Y \tag{8.161}$$

where γ' is the eigenvalue of T_1 associated with Y. It is easy to check that this expression is orthogonal to the span of $\mathcal{B}_1$ as well and thus is actually a multiple of W. We deduce that $\gamma\kappa - c = 0$ and hence $\gamma = \gamma'$. Otherwise, we could express $A\varphi Y$ as a linear combination of W and φY which is a contradiction. Since X and Y were chosen to be arbitrary elements of $\mathcal{B}_1$ and $\mathcal{B}_2$ respectively, we have completed our proof of (iv). □

This establishes Lemma 8.133.

Proposition 8.134. *Let M^{2n-1}, where $n \geq 3$, be an η-parallel hypersurface in $\mathbf{CP}^n$ or $\mathbf{CH}^n$. Then either*

1. *$\varphi(\varphi A - A\varphi)\varphi = 0$, i.e., $(\varphi A - A\varphi)\mathcal{W}^\perp \subseteq \mathcal{W}$, or*
2. *$\varphi(\varphi A + A\varphi - \kappa\varphi)\varphi = 0$, i.e., $(\varphi A + A\varphi - \kappa\varphi)\mathcal{W}^\perp \subseteq \mathcal{W}$, where $(n-1)\kappa = \mathbf{m} - \alpha$.*

Proof. Using the results of Lemma 8.133, we can choose an orthonormal basis for $\mathcal{W}^\perp$ consisting of $n-1$ orthonormal φ-invariant pairs. For any two such pairs $(X, \varphi X)$ and $(\tilde{X}, \varphi\tilde{X})$, there are suitable coefficients that allow us to write

$$\begin{aligned} AX &= \lambda X + tW; \\ A\varphi X &= \mu\varphi X + sW; \\ A\tilde{X} &= \tilde{\lambda}\tilde{X} + \tilde{t}W; \\ A\varphi\tilde{X} &= \tilde{\mu}\varphi\tilde{X} + \tilde{s}W. \end{aligned} \tag{8.162}$$

If it happens that $\lambda = \mu$ for all choices of X, then Condition *1* holds, i.e., $\varphi(\varphi A - A\varphi)\varphi = 0$. Otherwise, there is a choice of X with $\lambda \neq \mu$. To show that Condition 2 holds in this case, we first apply (8.132) with arguments $(\varphi X, \tilde{X}, \varphi X, \tilde{X})$ to get

$$\mu\tilde{\lambda}^2 - (\mu^2 + s^2 - c)\tilde{\lambda} + \mu(\tilde{t}^2 - c) = 0. \tag{8.163}$$

The calculation is similar to those done in the previous lemma and we omit the details. Using φX rather than X in the same equation yields

$$\lambda\tilde{\lambda}^2 - (\lambda^2 + s^2 - c)\tilde{\lambda} + \lambda(\tilde{t}^2 - c) = 0. \tag{8.164}$$

Multiplying (8.163) and (8.164) by λ and μ, respectively, and subtracting, we get

$$\tilde{\lambda}\big((\lambda - \mu)(\lambda\mu + c) - (\lambda s^2 - \mu t^2)\big) = 0. \tag{8.165}$$

Next apply (8.132) with arguments $(X, \varphi X, X, \varphi X)$ to obtain

$$(\lambda - \mu)(\lambda\mu + 5c) + 2\langle FX, \varphi X\rangle(\lambda + \mu) = \lambda s^2 - \mu t^2, \tag{8.166}$$

and then apply (8.132) with arguments $(\tilde{X}, \varphi\tilde{X}, X, \varphi X)$ to get

$$(\tilde{\lambda} + \tilde{\mu})\langle FX, \varphi X\rangle = -2c(\lambda - \mu). \tag{8.167}$$

Note that $\langle FX, \varphi X\rangle \neq 0$. Set

$$\kappa = \frac{-2c(\lambda - \mu)}{\langle FX, \varphi X\rangle}.$$

Because of (8.167) we may, without loss of generality, assume that $\tilde{\lambda} \neq 0$ (replacing $\tilde{X}$ by $\varphi\tilde{X}$ if necessary). Then we have

$$\lambda s^2 - \mu t^2 = (\lambda - \mu)(\lambda\mu + c) \tag{8.168}$$

from (8.165) and we can rewrite (8.166) as

$$4c(\lambda - \mu) + 2\langle FX, \varphi X\rangle(\lambda + \mu) = 0. \tag{8.169}$$

Thus

$$\lambda + \mu = \tilde{\lambda} + \tilde{\mu} = \kappa. \tag{8.170}$$

Having made particular choices of X and $\tilde{X}$, we now observe that $n - 3$ additional choices of $\tilde{X}$ can be made to complete a basis, and that (8.170) holds for all these choices since (8.167) does not depend on $\tilde{\lambda}$ being nonzero. Clearly $(n-1)\kappa = \mathbf{m} - \alpha$, and it is easy to verify that $(\varphi A + A\varphi - \kappa\varphi)\mathcal{W}^{\perp} \subseteq \mathcal{W}$, which establishes Condition 2.

So far, we have worked under the assumption that $AW \neq \alpha W$ at p. However, if p has a Hopf neighborhood, the same argument works with $s = t = \tilde{s} = \tilde{t} = 0$. Further, if p has no Hopf neighborhood but $AW = \alpha W$ there, then p is in the closure of the zero set of either $\varphi(\varphi A - A\varphi)\varphi$ or $\varphi(\varphi A + A\varphi - \kappa\varphi)\varphi$ and therefore one of the desired conditions holds at p. □

Using Proposition 8.134 we can now finish the proof of Theorem 8.128. Let M_0 be a nonempty connected open subset of M on which $\varphi(\varphi A - A\varphi)\varphi \neq 0$. Then the condition

$$\varphi(\varphi A + A\varphi - \kappa\varphi)\varphi = 0 \tag{8.171}$$

holds on M_0. We first show that M_0 is Hopf. Assume not and use the standard non-Hopf setup. Differentiating the left side of (8.171) with respect to X, applying the result to Y and taking the inner product with Z (where X, Y, and Z are in $\mathcal{W}^\perp$), we obtain

$$\begin{aligned} -(X\kappa)\langle\varphi Y, Z\rangle &= \beta\big(\langle U, Z\rangle\langle AX, Y\rangle - \langle U, Y\rangle\langle AX, Z\rangle\big) \\ &\quad -\beta\big(\langle\varphi U, Z\rangle\langle AX, \varphi Y\rangle - \langle\varphi U, Y\rangle\langle AX, \varphi Z\rangle\big). \end{aligned} \tag{8.172}$$

Setting $Z = \varphi Y$ in (8.172), where Y is nonzero and orthogonal to U and φU , we get $X\kappa = 0$ and thus (8.172) reduces to

$$\langle U, Z\rangle\langle AX, Y\rangle - \langle U, Y\rangle\langle AX, Z\rangle = \langle\varphi U, Z\rangle\langle AX, \varphi Y\rangle - \langle\varphi U, Y\rangle\langle AX, \varphi Z\rangle \tag{8.173}$$

for all X, Y, and Z in $\mathcal{W}^\perp$. Now suppose that $X \in \mathcal{W}^\perp$ is a unit vector such that $AX = \gamma X \mod \mathcal{W}$ for some $\gamma \neq 0$. Then choosing Y as above and $Z = U$ in (8.173), we get $\gamma\langle X, Y\rangle = 0$. In other words, $AX = 0 \mod \mathcal{W}$ for all $X \in \mathcal{W}^\perp$ orthogonal to U and φU. On the other hand, if we choose $Z = U$ and $Y = \varphi U$, we get $\langle AX, \varphi U\rangle = -\langle AX, \varphi U\rangle$ for all $X \in \mathcal{W}^\perp$. In particular, $\mu = \langle AU, \varphi U\rangle = 0$ and $\nu = \langle A\varphi U, \varphi U\rangle = 0$ so that $\mathbf{m} - \alpha = \lambda$ where λ and μ are the parameters used in the standard non-Hopf setup (and are not to be confused with those used in Proposition 8.134). We also have $0 = (\varphi A + A\varphi - \kappa\varphi)U = (\lambda - \kappa)\varphi U$ so that $\lambda = \kappa$. Thus $(n-1)\kappa = \kappa$ which implies that $\lambda = \kappa = 0$ since $n \geq 3$. We have shown that $A\mathcal{W}^\perp \subseteq \mathcal{W}$ and hence $\varphi(\varphi A - A\varphi)\varphi = 0$, a contradiction. We conclude that M_0 is Hopf.

Furthermore, M_0 is an open subset of a Type B hypersurface. To see this, we revisit the proof of Proposition 8.134 following the same argument with $s = t = \tilde{s} = \tilde{t} = 0$ up to equation (8.168). Thus, for each pair $(X, \varphi X)$, we have $AX = \lambda X$ and $A\varphi X = \mu\varphi X$ where

$$(\lambda\mu + c)(\lambda - \mu) = 0.$$

We also have $\lambda + \mu = \kappa$. Thus, if $\lambda \neq \mu$, then

$$-c = \lambda\mu = \frac{\lambda + \mu}{2}\alpha + c = \frac{\kappa\alpha}{2} + c$$

so that κ is a nonzero constant. Consequently, both λ and μ are nonzero constants and are the same two constants for every choice of X for which the principal curvatures corresponding to X and φX are distinct. On the other hand, if $\lambda = \mu$ for a choice of X, then $\lambda^2 - \alpha\lambda - c = 0$ so that such principal curvatures are also constant. In the standard classification theorems for Hopf hypersurfaces with constant principal curvatures (see Theorems 8.13 and 8.12) the only hypersurfaces that match our criteria are of Type B, since those of Type A already satisfy $\varphi(\varphi A - A\varphi)\varphi = 0$, while for those of types C, D, and E, the values of $\lambda + \mu$ are not all the same. For Type B, however, there are two principal curvatures, each of multiplicity $n-1$ and it is easy to check that

$$\varphi A + A\varphi - \kappa\varphi = 0.$$

The standard continuity and connectedness argument shows (since M_0 is nonempty) that $M_0 = M$. Thus M is an open subset of a Type B hypersurface.

The alternative is that $\varphi(\varphi A - A\varphi)\varphi = 0$ everywhere on M. If M has an open connected subset $\mathcal{U}$ that is Hopf, then $\varphi A = A\varphi$ there and $\mathcal{U}$ is an open subset of a Type A hypersurface by Theorem 8.37. Since the set of points of M matching the corresponding principal curvature data is closed, we conclude that M itself is an open subset of a Type A hypersurface. Suppose now that there is no such $\mathcal{U}$. We use the standard non-Hopf setup. Differentiating

$$\varphi(\varphi A - A\varphi)\varphi = 0 \tag{8.174}$$

with respect to X, applying the result to Y and taking the inner product with Z , we obtain

$$\begin{aligned}&\langle U, Z\rangle\langle AX, Y\rangle + \langle U, Y\rangle\langle AX, Z\rangle\\ &+\langle\varphi U, Z\rangle\langle AX, \varphi Y\rangle + \langle\varphi U, Y\rangle\langle AX, \varphi Z\rangle = 0\end{aligned} \tag{8.175}$$

for all X, Y, and Z in $\mathcal{W}^\perp$. Recall that $\mathcal{W}^\perp$ has an orthonormal basis consisting of $n-1$ pairs $(X, \varphi X)$ with respect to which the shape operator A can be expressed as in (8.162). Then

$$\begin{aligned}0 = \varphi(\varphi A - A\varphi)\varphi X &= \varphi(\varphi\mu\varphi X + \lambda X)\\ &= (\lambda - \mu)\varphi X\end{aligned} \tag{8.176}$$

so that $\lambda = \mu$ in (8.162). Now take $Y = \varphi X$ in (8.175) to get

$$\lambda\big(\langle U, \varphi X\rangle\langle X, Z\rangle - \langle\varphi U, Z\rangle + \langle\varphi U, \varphi X\rangle\langle X, \varphi Z\rangle\big) = 0. \tag{8.177}$$

This shows that either $\lambda = 0$ or φU lies in the span of X and φX. However, the latter is impossible, since it would imply the contradictory condition

$$\varphi U = -\langle\varphi U, X\rangle X - \langle U, X\rangle\varphi X.$$

Thus $\lambda = \mu = 0$ and $A\mathcal{W}^\perp \subseteq \mathcal{W}$ on a dense open subset of M and hence on M itself. We conclude that M is ruled.

This completes our proof of Theorem 8.128.

Other results on η-parallelism

Although the hypothesis of Theorem 8.128 requires $n \geq 3$, we have a partial result that covers $n = 2$.

Theorem 8.135. *Let M^3 be an η-parallel hypersurface in $\mathbf{CP}^2$ or $\mathbf{CH}^2$. If*

$$(\varphi \circ A - A \circ \varphi)\mathcal{W}^\perp \subset W$$

then either M is ruled or M is an open subset of a Type A hypersurface.

Returning to the Hopf case, we have Kwon and Nakagawa's result [310] about the Ricci tensor.

Theorem 8.136. *Let M^{2n-1}, where $n \geq 3$, be a Hopf hypersurface in $\mathbf{CP}^n$ or $\mathbf{CH}^n$. If the Ricci tensor is cyclic η-parallel, then M is an open subset of a Type A or a Type B hypersurface.*

A complete proof is given in [399] presented as Theorem 6.23. This proof also provides detail relevant to the proof of Theorem 8.89. For another version of this result, see Suh [504].

We define a $(1,3)$ tensor field T to be *η-parallel* if

$$((\nabla_Z T)(X,Y))\mathcal{W}^\perp \subset \mathcal{W}$$

for Z, X and Y in $\mathcal{W}^\perp$. Baikoussis, Lyu, and Suh [17] have the following result.

Theorem 8.137. *Let M^{2n-1}, where $n \geq 3$, be a Hopf hypersurface in $\mathbf{CP}^n$ or $\mathbf{CH}^n$. If the curvature tensor is η-parallel, then M is an open subset of a Type A or a Type B hypersurface.*

Relaxing the Hopf condition somewhat, Sohn [488] has proved the following.

Theorem 8.138. *Let M^{2n-1}, where $n \geq 3$, be a real hypersurface in $\mathbf{CP}^n$ or $\mathbf{CH}^n$ with η-parallel Ricci tensor. If $[S,\varphi]\mathcal{W}^\perp \subset \mathcal{W}$, then M is an open subset of a hypersurface of Type A or Type B.*

It is also possible to weaken the η-parallelism condition, while introducing a more restrictive algebraic condition. I.-B. Kim, K.H. Kim, and W.H. Sohn [262] have the following result (see Mayuko Kon [290] for $n = 2$).

Theorem 8.139. *Let M^{2n-1}, where $n \geq 3$, be a real hypersurface in $\mathbf{CP}^n$ or $\mathbf{CH}^n$. Assume that*

$$(\varphi \circ A - A \circ \varphi)\mathcal{W}^\perp \subset \mathcal{W}. \tag{8.178}$$

- *If M is cyclic η-parallel, then it is either ruled or an open subset of a Type A hypersurface;*

- *If the Ricci tensor S is cyclic η-parallel, then M is an open subset of a Type A hypersurface.*

8.5.6 Recurrence conditions

We recall that a tensor field T is *recurrent* if there is a 1-form ω such that $\nabla_X T = \omega(X)T$ for all tangent vectors X. The field T is *birecurrent* if the same sort of relationship holds for second covariant derivatives, namely (see Section 7.1 for notation)

$$\nabla^2 T(;Y;X) = h(X,Y)\,T,$$

for some tensor field h of type $(0,2)$.

As we have seen, there are many conditions that can be realized for hypersurfaces of real space forms (totally umbilic, Einstein, semisymmetric, etc.), but are not possible in $\mathbf{CP}^n$ or $\mathbf{CH}^n$. Continuing this theme, we have the following.

Theorem 8.140. *Let M^{2n-1}, where $n \geq 2$, be a real hypersurface in $\mathbf{CP}^n$ or $\mathbf{CH}^n$. Then,*

- *the shape operator cannot be birecurrent or recurrent;*
- *if $n \geq 3$, the Ricci tensor cannot be birecurrent or recurrent.*

The relationship between birecurrence and recurrence is clarified in the following algebraic lemma.

Lemma 8.141. *Let T be a tensor field of type $(1,1)$ on a Riemannian manifold. Then,*

- *if T is recurrent, it is also birecurrent;*
- *if T is symmetric, birecurrent, and nonzero, then T is semiparallel.*

Proof. Suppose that $(\nabla_Y T)Z = \omega(Y)TZ$ for all tangent vectors Y and Z. Then, by a few lines of calculation, we get

$$\begin{aligned}\nabla^2 T(;Y;X)Z &= (\nabla_X\nabla_Y T - \nabla_{\nabla_X Y}T)Z\\ &= ((\nabla_X\omega)Y)TZ + \omega(X)\omega(Y)TZ,\end{aligned}$$

for all tangent vectors X, Y and Z. If we set

$$h(X,Y) = (\nabla_X\omega)Y + \omega(X)\omega(Y),$$

we have $\nabla^2 T(;Y;X) = h(X,Y)T$ as required. This proves that recurrence implies birecurrence.

Now, suppose that T is symmetric, birecurrent, and nonzero. Consider T^2 and note that for all tangent vectors Y,

$$\nabla_Y T^2 = \nabla_Y T \circ T + T \circ \nabla_Y T.$$

Thus

$$\begin{aligned}\nabla^2 T^2(;Y;X) =& \nabla_X \nabla_Y T^2 - \nabla_{\nabla_X Y} T^2 \\ =& (\nabla_X \nabla_Y T) \circ T + \nabla_Y T \circ \nabla_X T + \nabla_X T \circ \nabla_Y T + T \circ \nabla_X \nabla_Y T \\ & - \nabla_{\nabla_X Y} T \circ T - T \circ \nabla_{\nabla_X Y} T \\ =& \nabla^2 T(;Y;X) \circ T + T \circ \nabla^2 T(;Y;X) \\ & + \nabla_Y T \circ \nabla_X T + \nabla_X T \circ \nabla_Y T.\end{aligned}$$

Since T is birecurrent, we have

$$\nabla^2 T^2(;Y;X) = 2h(X,Y)T^2 + \nabla_Y T \circ \nabla_X T + \nabla_X T \circ \nabla_Y T.$$

Taking the trace commutes with covariant differentiation. Thus

$$\begin{aligned}\nabla^2(\text{trace } T^2)(;Y;X) =& \text{ trace } \nabla^2 T^2(;Y;X) \\ =& 2h(X,Y) \text{ trace } T^2 + 2 \text{ trace } (\nabla_X T \circ \nabla_Y T).\end{aligned}$$

Since the left side is symmetric in X and Y and trace $T^2 = |T^2| \neq 0$, we must have $h(X,Y) = h(Y,X)$. This means that

$$R(X,Y) \cdot T = \nabla^2 T(;Y;X) - \nabla^2 T(;X;Y) = 0,$$

for all tangent vectors X and Y. In other words, T is semiparallel. □

Applying Lemma 8.141 to the shape operator, and invoking Theorem 8.42, we see that the first claim in Theorem 8.140 has been established. (Note that the possible vanishing of A is not an issue since every hypersurface has at least one point where $A \neq 0$. Birecurrence in a neighborhood of this point yields a contradiction.)

Similarly, every hypersurface has a point (and hence a neighborhood) where $S \neq 0$. Applying the same lemma to S in this neighborhood, we see that birecurrence of S implies that S is semiparallel, i.e., M is Ricci-semisymmetric. Since this contradicts Theorem 8.69, we have proved the second claim of Theorem 8.140.

η-recurrence

We have seen in Subsection 8.5.6 that the shape operator of a real hypersurface cannot be recurrent. However, there are hypersurfaces that satisfy this condition restricted to $\mathcal{W}^\perp$. Specifically, a $(1,1)$ tensor field T is said to be *η-recurrent* if there is a 1-form ω such that

$$(\nabla_X T - \omega(X)T)\mathcal{W}^\perp \subset \mathcal{W} \tag{8.179}$$

for all $X \in \mathcal{W}^\perp$.

We have the following result due to Hamada, Lyu, and Suh [188, 343].

Theorem 8.142. *Let M^{2n-1}, where $n \geq 3$, be a Hopf hypersurface in* $\mathbf{CP}^n$ *or* $\mathbf{CH}^n$ *with η-recurrent shape operator. Then M is an open subset of a hypersurface of Type A or Type B.*

Theorem 8.143. *Let M^{2n-1}, where $n \geq 3$, be a Hopf hypersurface in* $\mathbf{CP}^n$ *or* $\mathbf{CH}^n$ *with η-recurrent Ricci tensor. Then M is an open subset of a hypersurface of Type A or Type B.*

This is due to Kon and Loo [291]. The result was proved earlier by Baikoussis, Lyu, and Suh [18] under the assumption of constant mean curvature.

Nakajima [392] and Nagai [388] have obtained some classification results for hypersurfaces of Types A and B in terms of the behavior of $\nabla^2 A$ on vectors in $\mathcal{W}^\perp$.

8.6 Non-Hopf Possibilities in $\mathbf{CH}^n$

As we have seen, the homogeneous hypersurfaces in $\mathbf{CP}^n$ are Hopf and have constant principal curvatures. After Hopf hypersurfaces with constant principal curvatures in $\mathbf{CH}^n$ had been classified, the question of homogeneous hypersurfaces remained.

Almost a decade elapsed before M. Lohnherr discovered an example of a homogeneous non-Hopf hypersurface in $\mathbf{CH}^n$. Since this hypersurface is homogeneous, it has constant principal curvatures. In fact, it is also complete and ruled. The number of distinct principal curvatures is $g = 3$ and, in terms of the standard non-Hopf setup, the upper-left 3×3 submatrix of the shape operator is

$$\frac{1}{r}\begin{bmatrix} 0 & 1 & 0 \\ 1 & 0 & 0 \\ 0 & 0 & 0 \end{bmatrix}, \tag{8.180}$$

and the rest of the shape operator matrix is zero.

Remark 8.144. The original construction of Lohnherr's example may be found in his thesis [335] or in [336]. In fact, we now know (see Theorem 2 of [36]) that Lohnherr's hypersurface is characterized by these principal curvature data.

Berndt [30] realized this hypersurface as one of a family of homogeneous hypersurfaces M_σ each of which is an orbit of a cohomogeneity-1 action on $\mathbf{CH}^n$ by an appropriate subgroup (the *Berndt subgroup*, see [257]) of the isometry group of $\mathbf{CH}^n$. Again, each M_σ is non-Hopf and complete. However only M_0 is ruled and only M_0 is minimal. The upper-left 3×3 submatrix of the shape operator can be written

$$\frac{1}{r}\begin{bmatrix} 3\sigma - \sigma^3 & \tau & 0 \\ \tau & \sigma^3 & 0 \\ 0 & 0 & \sigma \end{bmatrix}, \tag{8.181}$$

where $\tau = (1-\sigma^2)^{\frac{3}{2}}$. The rest of the matrix is zero, except that all the remaining diagonal entries are the same as the $(3,3)$ entry. Thus the principal curvatures are ν of multiplicity $2n-3$ and

$$\frac{3}{2}\nu \pm \frac{1}{r}\sqrt{1-\frac{3}{4}r^2\nu^2}, \tag{8.182}$$

each of multiplicity 1, where $\nu = \sigma/r$. Then $\mathbf{m} = 2n\nu$, which explains our statement about minimality. The form of the matrix also shows that M_σ is ruled if and only if $\sigma = 0$. It also turns out that the Berndt orbits M_σ form a parallel family of hypersurfaces for $-1 < \sigma < 1$. They are parametrized by σ as σ ranges through the interval $(-1, 1)$.

In the notation of Section 6.7, this family can also be regarded as a family of parallel hypersurfaces $\{M_u\}$ where u ranges through the real numbers. The hypersurface M_0 is the Lohnherr example and M_u is the parallel hypersurface at (signed) distance ru from it. Using the general techniques for computing shape operators of tubes (see Theorem 6.36), we find that the shape operator for M_u takes the form

$$\frac{1}{r}\begin{bmatrix} \tanh^3 u - 3\tanh u & \operatorname{sech}^3 u & 0 \\ \operatorname{sech}^3 u & -\tanh^3 u & 0 \\ 0 & 0 & -\tanh u \end{bmatrix}, \tag{8.183}$$

with the extension to higher dimensions as in (8.181). In the notation of the previous paragraph, we have $\sigma = -\tanh u$. Our calculation shows that this family of parallel hypersurfaces has no focal points.

The Berndt orbits form a new class of model hypersurfaces in $\mathbf{CH}^n$. For $n = 2$, their discovery completes the catalogue of hypersurfaces with constant principal curvatures. Specifically, Berndt and Díaz-Ramos [35] have proved the following:

Theorem 8.145. *Let M^3 be a non-Hopf hypersurface in $\mathbf{CH}^2$ with constant principal curvatures. Then M is an open subset of a Berndt orbit.*

The Lohnherr hypersurface may be generalized to a "ruled minimal submanifold" F_k of dimension $2n-k$ with "rulings" which are complex hyperbolic spaces $\mathbf{CH}^{n-k}$, totally geodesic in $\mathbf{CH}^n$. This can be done for $2 \le k \le n-1$. Tubes around such submanifolds are homogeneous hypersurfaces that typically have $g = 4$ constant principal curvatures. However, (reminiscent of the Type B hypersurfaces), there is one particular radius for which g reduces to 3. We have

Theorem 8.146. *Let M^{2n-1}, where $n \geq 3$, be a non-Hopf hypersurface in* $\mathbf{CH}^n$ *with constant principal curvatures. If the number of distinct principal curvatures is $g = 3$, then M is either*

- *a Berndt orbit, or*
- *a tube of radius $\frac{r}{2}\log(2+\sqrt{3})$ around a ruled minimal submanifold F_k where $2 \leq k \leq n-1$.*

The classification of non-Hopf hypersurfaces in $\mathbf{CH}^n$ and $\mathbf{CP}^n$ with constant principal curvatures is still an active area of research (see, for example, Díaz-Ramos and Domínguez-Vázquez [131]).

The classification of non-Hopf homogeneous hypersurfaces in $\mathbf{CH}^n$ has been completed by Berndt and Tamaru [38]. As a first step in this classification, we can state

Theorem 8.147. *Let M^{2n-1}, where $n \geq 2$, be a homogeneous non-Hopf hypersurface in* $\mathbf{CH}^n$ *with no focal points. Then M belongs to the family of Berndt orbits.*

The remaining non-Hopf homogeneous hypersurfaces break down into two families of tubes over their focal submanifolds. The first is the set of tubes over the submanifolds F_k.

Also, for each real number θ with $0 < \theta < \frac{\pi}{2}$, and each integer k satisfying $0 < 2k < n$, there is a submanifold $F_{k,\theta}$ of dimension $2(n-k)$ that arises as the orbit under the action of a certain closed subgroup of the isometry group of $\mathbf{CH}^n$.

We can now state the Berndt-Tamaru classification.

Theorem 8.148. *Let M^{2n-1}, where $n \geq 2$, be a homogeneous non-Hopf hypersurface in* $\mathbf{CH}^n$ *with at least one focal point. Then M is either*

- *a tube over some F_k, or*
- *a tube over some $F_{k,\theta}$.*

Conversely, all such tubes are homogeneous.

The submanifolds F_k and $F_{k,\theta}$ were constructed by Berndt and Brück [32] and are called the *Berndt-Brück submanifolds* (see, for example, [131]). They are homogeneous, minimal and "ruled." We do not formally define "ruled" in this context but the notion is a natural generalization from the hypersurface case. Here, the real codimension is k and the holomorphic subbundle is a foliation with complex dimension $n-k$ and totally geodesic leaves isometric to $\mathbf{CH}^{n-k}$.

8.7 Isoparametric Hypersurfaces in $\mathbf{CP}^n$ and $\mathbf{CH}^n$

We adopt the following definition:

A hypersurface M in a Riemannian manifold is *isoparametric* if M and its nearby parallel hypersurfaces have constant mean curvature. This is equivalent to our definition for hypersurfaces in real space forms given in Section 3.1 (see

Theorem 3.6). In our case, however, not all isoparametric hypersurfaces have constant principal curvatures. In fact, for $\mathbf{CP}^n$, we have the following theorem due to Qi-Ming Wang [547].

Theorem 8.149. *For hypersurfaces M^{2n-1} in $\mathbf{CP}^n$ where $n \geq 2$, any two of the following properties imply the third.*

(1) *M has constant principal curvatures;*
(2) *M is Hopf;*
(3) *M is isoparametric.*

On the other hand, the non-Hopf homogeneous hypersurfaces in $\mathbf{CH}^n$ satisfy (1) and (3) but not (2). However, as we have also seen, (1) and (2) imply (3).

Although there are isoparametric hypersurfaces in $\mathbf{CP}^n$ that do not have constant principal curvatures, every known example of a hypersurface with constant principal curvatures is in fact Hopf. We also know from Theorems 8.29 and 8.36 that if there is a non-Hopf hypersurface in $\mathbf{CP}^n$ with constant principal curvatures, it must satisfy $g \geq 4$ (and hence $n \geq 3$). Even though isoparametric hypersurfaces in complex space forms offer a wider variety of principal curvature configurations than can occur in the real case, it turns out that there is a close relationship between the two contexts, as follows.

Lemma 8.150. *Let M^{2n-1}, where $n \geq 2$, be a real hypersurface in $\mathbf{CP}^n$ (resp. $\mathbf{CH}^n$). Then M is isoparametric if and only if $\pi^{-1}M$ is an isoparametric hypersurface of $S^{2n+1}(r)$ (resp. $\mathbf{H}$)*

Here, we take the definition of isoparametric hypersurface in the anti-de Sitter space $\mathbf{H}$ to be the more general one – all nearby parallel hypersurfaces have constant mean curvature. In this case, an isoparametric hypersurface must have constant principal curvatures and Cartan's formula (3.13) (page 91) applies. However, the hypersurfaces in question are Lorentz and the shape operator, which is self-adjoint with respect to the Lorentz metric, may not be diagonalizable and its eigenvalues can be complex. In this context, we require for validity of Cartan's formula that λ_i be a real principal curvature whose algebraic and geometric multiplicities coincide. There is no restriction on the other principal curvatures λ_j in the summation.

In spite of this complication, the trace of the shape operator of $\pi^{-1}M$ is equal to $\mathbf{m}$, the trace of the shape operator of M, for all hypersurfaces M in $\mathbf{CP}^n$ or $\mathbf{CH}^n$. This follows immediately from Lemma 6.12.

Using the Cartan formula, one can deduce the following about the principal curvatures of a Lorentzian isoparametric hypersurface.

Theorem 8.151. *Let M^{2n} be a Lorentzian isoparametric hypersurface in the anti-de Sitter space $\mathbf{H}$. Then the number g' of distinct principal curvatures will be 1, 2, 3, or 4. Further,*

- *If there is a complex principal curvature ν, then ν and its complex conjugate $\bar{\nu}$ are principal curvatures of multiplicity 1. There are two subcases:*

  (i) *ν is purely imaginary, $g' = 3$, and the third principal curvature is zero;*

(ii) $g' = 4$ *and the real principal curvatures* λ *and* μ *satisfy* $\lambda\mu + c = 0$*;*

- *If all principal curvatures are real, there are two possibilities:*

  (i) $g' = 2$ *and the principal curvatures* λ *and* μ *satisfy* $\lambda\mu + c = 0$*;*
  (ii) $g' = 1$.

This theorem is due to Xiao [557] who made use of earlier work by Hahn [187] and Magid [351]. A detailed proof of the theorem was presented by Domínguez-Vázquez in his thesis [133, pp. 69–73]. He goes on to prove the following pointwise result.

Theorem 8.152. *Let* M^{2n-1}*, where* $n \geq 2$*, be an isoparametric hypersurface in* $\mathbf{CH}^n$*. Then at any point* $p \in M$*,*

- *The number g of distinct principal curvatures is 2, 3, 4, or 5;*
- *The dimension h of* $\mathcal{H}$ *(the smallest A-invariant subspace containing W) is 1, 2, or 3;*
- *Neither g nor h need be constant.*

The second assertion implies that W can be written as the sum of three or fewer principal vectors corresponding to distinct principal curvatures.

An analysis of the proof gives the following analogue of Theorem 8.149 for $\mathbf{CH}^n$.

Theorem 8.153. *Let* M^{2n-1} *be a Hopf hypersurface in* $\mathbf{CH}^n$ *where* $n \geq 2$*. Then M is isoparametric if and only if its principal curvatures are constant.*

For $\mathbf{CP}^n$, Xiao [558], relying in part on the work of Park [427], studied the possible principal curvature configurations. In particular, it follows from his work that

Theorem 8.154. *Let* M^{2n-1}*, where* $n \geq 2$*, be an isoparametric hypersurface in* $\mathbf{CP}^n$*. Suppose that the isoparametric hypersurface* $\pi^{-1}M$ *in* $S^{2n+1}(r)$ *has* g' *distinct principal curvatures with multiplicities* $m_1 \leq m_2$*. Then*

(i) g' *is* 2, 4, *or* 6*;*
(ii) *If* $g' = 2$*, then* m_1 *and* m_2 *are odd;*
(iii) *If* $g' = 4$*, then either* $m_1 = 1$ *or* $m_1 m_2$ *is even;*
(iv) *If* $g' = 6$*, then* $m_1 = m_2 = 1$ *so that* $n = 3$.

In fact, if $g' = 2$, we can conclude that M is a familiar Hopf hypersurface with 2 or 3 constant principal curvatures. Neither principal space of $\pi^{-1}M$ is horizontal and M is an open subset of a Type A hypersurface.

As seen in Theorem 8.13 on page 432, hypersurfaces of types B, C, D, and E all correspond to $g' = 4$. In each case, exactly two principal subspaces for $\pi^{-1}M$ are horizontal. In fact, it is a consequence of Xiao's work that this property characterizes the Hopf hypersurfaces with constant principal curvatures among all isoparametric hypersurfaces in $\mathbf{CP}^n$.

We now look at non-Hopf isoparametric hypersurfaces with $g' = 4$. First note that $h \leq 3$ at all points. Consequently

$$h \leq g \leq h + 4 \leq 7.$$

To see this we use Lemma 8.156. Note that $h + 1$ of the 4 distinct principal curvatures of $\pi^{-1}M$ will have non-horizontal principal spaces and for those that have dimension greater than one, the corresponding principal curvatures will also be principal curvatures for M. For example, if $h = 3$ and

$$m_2 \geq m_1 \geq 2,$$

then M has 7 distinct principal curvatures.

If $g' = 6$, then (as we have seen in the proof of Theorem 8.13), M cannot be among the Hopf hypersurfaces with constant principal curvatures. In other words, M cannot be an open subset of a homogeneous hypersurface.

When $g' = 6$, we can also show, again with the help of Lemma 8.156, that $h = 2$ cannot hold on any open set. This is because $h = 2$ implies that three of the six principal curvatures must have horizontal principal spaces that project to three 1-dimensional principal subspaces with the same three numbers as principal curvatures. By considering the restriction of the characteristic polynomial of the shape operator of $\pi^{-1}M$ to the complementary 3-dimensional subspace, we can show that the two principal curvatures corresponding to the principal subspaces required to represent W must be constant. Thus M is a non-Hopf hypersurface with constant principal curvatures and $h = 2$. This contradicts the results of [131].

However, our argument has not eliminated the possible existence of points where $h = 2$, or even $h = 1$. In any case, no matter what the value of h, we have for $g' = 6$,

$$\max(h, 5 - h) \leq g \leq 5.$$

The fact that $5 - h \leq g$ follows by a similar argument to the one used for $h = 2$. There are $5-h$ principal curvatures of $\pi^{-1}M$ with horizontal principal spaces. Their projections lie in distinct principal spaces of M with the same $5 - h$ numbers as principal curvatures. Summarizing, we have

Theorem 8.155. *Let M^{2n-1}, where $n \geq 2$, be an isoparametric hypersurface in $\mathbf{CP}^n$. Then at any point $p \in M$,*

- *The number g of distinct principal curvatures satisfies $g \leq 7$;*
- *$\mathcal{H}$ (the smallest A-invariant subspace containing W) is of dimension $h \leq 5$.*

Neither g nor h need be constant.

The following algebraic lemma has been useful in our analysis.

Lemma 8.156. *Let T be an $(m + 1) \times (m + 1)$ real symmetric matrix. Assume that $T\epsilon_i = \alpha_i\epsilon_i + \mu_i\epsilon_{m+1}$ for $1 \leq i \leq m$ and that $\langle T\epsilon_{m+1}, \epsilon_{m+1}\rangle = 0$. Assume further that the α_i are distinct and that the μ_i are all nonzero. Then*

- *T has $m + 1$ distinct eigenvalues;*
- *No α_i is an eigenvalue;*

- *Every eigenvector has a nonzero ϵ_{m+1} component.*

As a consequence of this lemma, the h principal curvatures of M corresponding to the principal spaces required to express W determine $h+1$ distinct principal curvatures of $\pi^{-1}M$ and $h+1$ non-horizontal mutually orthogonal principal vectors of $\pi^{-1}M$. The $(2n-1-h)$-dimensional orthogonal complement of their span is horizontal. As a direct sum of (subspaces of) horizontal principal spaces, it projects to a direct sum of subspaces of principal spaces of M. However, we need to determine which of the corresponding principal curvatures are distinct from the h principal curvatures already postulated. As a consequence of Xiao's work, we can state:

Theorem 8.157. *Let M^{2n-1}, where $n \geq 2$, be a non-Hopf isoparametric hypersurface in $\mathbf{CP}^n$. Suppose that the isoparametric hypersurface $\pi^{-1}M$ in $S^{2n+1}(r)$ has g' distinct principal curvatures. Then, either $g' = 4$ or $g' = 6$.*

- *If $g' = 4$, then $g = 7$ and $h = 3$ generically with $g = 5$ and $h = 1$ on a lower-dimensional subset;*
- *If $g' = 6$, then $n = 3$. Generically $g = 5$ but g drops to 4 on a lower-dimensional subset. The values assumed by h are 5, 3 and 1.*

A deeper analysis of the situation by Domínguez-Vázquez [134] yields an almost complete classification of isoparametric hypersurfaces in $\mathbf{CP}^n$, except for $n = 15$. For further generalizations of the notion of isoparametric hypersurfaces, see J. Ge, Z.-Z. Tang and W. Yan [171]. For results on isoparametric submanifolds of higher codimension in $\mathbf{CP}^n$, see Domínguez-Vázquez [134]. See also the notion of equifocal submanifolds in symmetric spaces due to Terng and Thorbergsson [530, 531] which is discussed in Subsection 3.8.6.

8.8 Open Problems

In this section, we summarize the problems raised in this chapter that still appear to be open. Since the literature these topics is vast and many researchers are currently working in the area, some of the problems may have been already resolved. In any case, we present the following list:

1. Theorem 8.32 classifies the Hopf hypersurfaces in $\mathbf{CH}^n$ where $n \geq 3$ having $g \leq 3$ principal curvatures under the assumption that $\alpha^2 + 4c \neq 0$. Do there exist Hopf hypersurfaces with $\alpha^2 + 4c = 0$ other than open subsets of a horosphere? See Remark 8.33.
2. Classify hypersurfaces in $\mathbf{CP}^n$ and $\mathbf{CH}^n$, where $n \geq 3$, having $g = 3$ distinct principal curvatures. This is an open-ended problem. It has been settled in case of constant principal curvatures (Theorem 8.36). Also, ruled hypersurfaces have $g = 3$. What other interesting families of hypersurfaces are possible?

3. Find appropriate criteria for classifying hypersurfaces in $\mathbf{CP}^n$ and $\mathbf{CH}^n$ with $g \geq 4$ constant principal curvatures. See discussion following Theorem 8.36.
4. Do there exist non-Hopf Ricci-semisymmetric hypersurfaces in $\mathbf{CP}^2$ or $\mathbf{CH}^2$? See Remark 8.70.
5. Do there exist hypersurfaces in $\mathbf{CP}^2$ or $\mathbf{CH}^2$ with harmonic curvature, i.e., whose Ricci tensor is a Codazzi tensor?
6. Do there exist non-Hopf pseudosymmetric hypersurfaces (or even Ricci-pseudosymmetric hypersurfaces) in $\mathbf{CP}^2$ or $\mathbf{CH}^2$? See Remarks 8.77 and 8.80.
7. For $n \geq 3$, complete the classification of Hopf hypersurfaces in $\mathbf{CP}^n$ satisfying $\varphi \circ S = S \circ \varphi$ by giving a more precise description of those that do not have constant principal curvatures. See discussion following Theorem 8.94.
8. For $n \geq 3$, find and characterize an interesting family of non-Hopf hypersurfaces in $\mathbf{CP}^n$ and/or $\mathbf{CH}^n$ satisfying $\varphi \circ S = S \circ \varphi$.
9. Classify non-Hopf hypersurfaces in $\mathbf{CP}^n$ and $\mathbf{CH}^n$, where $n \geq 2$, satisfying $R_W X = \kappa X$ for some function κ and all $X \in \mathcal{W}^\perp$. See Theorem 8.105 and the subsequent remarks.
10. Classify Hopf hypersurfaces $\mathbf{CP}^n$ and $\mathbf{CH}^n$, where $n \geq 3$, with W-parallel Ricci tensor. For $\mathbf{CP}^n$, this is related to the $\varphi \circ S = S \circ \varphi$ condition mentioned in point 7 above (see Theorem 8.121). However, as that theorem indicates, this problem also relates to the condition $\alpha = 0$ which is not well understood in the $\mathbf{CH}^n$ case (see Remarks 8.106 and 8.126).
11. Classify η-parallel hypersurfaces in $\mathbf{CP}^2$ and $\mathbf{CH}^2$. See Theorems 8.128 and 8.129 and subsequent remarks.

Chapter 9
Hypersurfaces in Quaternionic Space Forms

In 1986, Martinez and Pérez [353] began the study of real hypersurfaces in quaternionic space forms, and in 1991, Berndt [28] found a list of standard examples of real hypersurfaces in quaternionic space forms with constant principal curvatures, leading to further research in this area. These examples and classification results are described in this section.

We begin with the construction of the standard models of the quaternionic space forms, quaternionic projective space $\mathbf{HP}^n$ and quaternionic hyperbolic space $\mathbf{HH}^n$.

9.1 Quaternionic Projective Space

For $z = (z_0, \ldots, z_n)$, $w = (w_0, \ldots, w_n)$ in $\mathbf{H}^{n+1}$, write

$$F(z, w) = \sum_{k=0}^{n} z_k \bar{w}_k,$$

and let $\langle z, w\rangle = \Re F(z, w)$, the real part of $F(z, w)$. The sphere $S^{4n+3}(r)$ of radius r is defined by

$$S^{4n+3}(r) = \{z \in \mathbf{H}^{n+1} \mid \langle z, z\rangle = r^2\}.$$

We may identify $\mathbf{H}^{n+1}$ with $\mathbf{R}^{4n+4}$, defining $u, v \in \mathbf{R}^{4n+4}$ by

$$\begin{aligned} z_\ell &= u_{4\ell} + u_{4\ell+1}\,\mathrm{i} + u_{4\ell+2}\,\mathrm{j} + u_{4\ell+3}\,\mathrm{k} \\ w_\ell &= v_{4\ell} + v_{4\ell+1}\,\mathrm{i} + v_{4\ell+2}\,\mathrm{j} + v_{4\ell+3}\,\mathrm{k}, \end{aligned} \tag{9.1}$$

for $0 \leq \ell \leq n$. Then

T.E. Cecil, P.J. Ryan, *Geometry of Hypersurfaces*, Springer Monographs in Mathematics, DOI 10.1007/978-1-4939-3246-7_9

$$\langle z, w \rangle = \langle u, v \rangle = \sum_{\ell=0}^{4n+3} u_\ell v_\ell$$

is the usual inner product on $\mathbf{R}^{4n+4}$. We will use $\langle z, w\rangle$ and $\langle u, v\rangle$ interchangeably. When desired, we can work exclusively in real terms by introducing the operators J_1, J_2, and J_3 for (left) multiplication by the quaternionic units i, j, and k. Note that for $z \in S^{4n+3}(r)$,

$$T_z S^{4n+3}(r) = \{w \in \mathbf{R}^{4n+4} \mid \langle z, w\rangle = 0\}.$$

The restriction of $\langle\ ,\ \rangle$ to $S^{4n+3}(r)$ is a Riemannian metric whose Levi-Civita connection $\tilde{\nabla}$ satisfies

$$D_X Y = \tilde{\nabla}_X Y - \langle X, Y\rangle \frac{z}{r^2}$$

for X, Y tangent to $S^{4n+3}(r)$ at z, where D is the Levi-Civita connection of $\mathbf{R}^{4n+4}$. The usual calculations of the Gauss equation show that the curvature tensor $\tilde{R}$ of $\tilde{\nabla}$ satisfies

$$\tilde{R}(X, Y) = \frac{1}{r^2} X \wedge Y. \tag{9.2}$$

Let $\mathcal{V}$ be the span of $\{J_1 z, J_2 z, J_3 z\}$ and write down the orthogonal decomposition into so-called vertical and horizontal subspaces,

$$T_z S^{4n+3}(r) = \mathcal{V} \oplus \mathcal{V}^\perp.$$

The quaternionic projective space $\mathbf{HP}^n$ is the set of 1-dimensional subspaces of $\mathbf{H}^{n+1}$, considered as a left vector space over $\mathbf{H}$. The canonical projection is

$$\pi : S^{4n+3}(r) \to \mathbf{HP}^n,$$

with fiber S^3, the group of unit quaternions.

9.2 Quaternionic Hyperbolic Space

Next, we introduce the quaternionic hyperbolic space $\mathbf{HH}^n$. The construction is parallel to that of $\mathbf{HP}^n$ with some important differences. For z, w in $\mathbf{H}^{n+1}$, write

$$F(z, w) = -z_0 \bar{w}_0 + \sum_{k=1}^{n} z_k \bar{w}_k,$$

and let $\langle z, w\rangle = \Re F(z, w)$. Using the same identification of $\mathbf{H}^{n+1}$ with $\mathbf{R}^{4n+4}$ we get

$$\langle z, w\rangle = \langle u, v\rangle = -\sum_{\ell=0}^{3} u_\ell v_\ell + \sum_{\ell=4}^{4n+3} u_\ell v_\ell.$$

Set

$$\mathbf{H} = \{z \in \mathbf{H}^{n+1} \mid \langle z, z\rangle = -r^2\}.$$

The restriction of $\langle\ ,\ \rangle$ to $\mathbf{H}$ makes it into the pseudohyperbolic space $H_3^{4n+3}(r)$, a semi-Riemannian space form of constant curvature $-\frac{1}{r^2}$ and index 3 (see [412, p. 110]). Its tangent space is given by

$$T_z\mathbf{H} = \{w \in \mathbf{H}^{n+1} \mid \langle z, w\rangle = 0\},$$

and its Levi-Civita connection $\tilde{\nabla}$ satisfies

$$D_X Y = \tilde{\nabla}_X Y + \langle X, Y\rangle \frac{z}{r^2}$$

at z. The Gauss equation takes the form

$$\tilde{R}(X, Y) = -\frac{1}{r^2} X \wedge Y. \tag{9.3}$$

Again we get an orthogonal decomposition

$$T_z\mathbf{H} = \mathcal{V} \oplus \mathcal{V}^\perp.$$

Denote by $\mathbf{HH}^n$ the image of $\mathbf{H}$ by the canonical projection π to complex projective space,

$$\pi : \mathbf{H} \to \mathbf{HH}^n \subset \mathbf{HP}^n.$$

Thus, topologically, $\mathbf{HH}^n$ is an open subset of $\mathbf{HP}^n$. However, as Riemannian manifolds, they have quite different structures.

9.3 Quaternionic Space Forms

From here on we make a uniform exposition covering both $\mathbf{HP}^n$ and $\mathbf{HH}^n$. When convenient, we make use of the letter ϵ to distinguish the two cases. It will serve as the sign of the constant quaternionic curvature $4c = 4\epsilon/r^2$. For example, (9.2) and (9.3) could be written as

$$\tilde{R}(X, Y) = \frac{\epsilon}{r^2} X \wedge Y.$$

We also use $\tilde{M}$ to stand for either $\mathbf{HP}^n$ or $\mathbf{HH}^n$ and $\tilde{M}'$ for $S^{4n+3}(r)$ or $\mathbf{H}$.

Note that $\pi_* \mathcal{V} = 0$ but that π_* is an isomorphism on $\mathcal{V}^\perp$. Let z be any point of $\tilde{M}'$. For $X \in T_{\pi z}\tilde{M}$, let X^L be the vector in $\mathcal{V}_z^\perp$ that projects to X. Then X^L is called the *horizontal lift* of X to z. Define a Riemannian metric on $\tilde{M}$ by $\langle X, Y\rangle = \langle X^L, Y^L\rangle$. It is well defined since the metric on $\tilde{M}'$ is invariant by the fiber S^3. Since $\mathcal{V}^\perp$ is invariant by J_1, J_2, and J_3, the J_i determine $(1, 1)$ tensor fields on $\tilde{M}$ which we (by abuse of notation) also denote by J_1, J_2, and J_3. The reader can easily distinguish by context. Specifically, we define for $X \in T_z\tilde{M}$

$$J_iX = \pi_*(J_iX^L).$$

It is easy to check that the J_i are complex structures on $T_z\tilde{M}$ and that $\langle\ ,\ \rangle$ is Hermitian with respect to each of them. Further, they satisfy the identities

$$\begin{aligned} J_i \circ J_{i+1} &= J_{i+2} \text{ (indices mod 3)} \\ J_i \circ J_j &= -J_j \circ J_i \end{aligned} \tag{9.4}$$

for $1 \le i \ne j \le 3$. The Levi-Civita connection $\tilde{\nabla}$ on $\tilde{M}$ satisfies

$$\tilde{\nabla}_X Y = \pi_*(\tilde{\nabla}_{X^L} Y^L).$$

We also note that on $\tilde{M}'$

$$\tilde{\nabla}_{X^L} V = \tilde{\nabla}_V X^L = J_i X^L = (J_i X)^L \tag{9.5}$$

for $V = J_i z \in \mathcal{V}$, while

$$\tilde{\nabla}_{\mathcal{V}} \mathcal{V} \subset \mathcal{V}.$$

See [412] for background on Riemannian submersions.

The curvature tensor of $\tilde{M}$ follows from the relationship between the respective Levi-Civita connections on $\tilde{M}$ and $\tilde{M}'$.

Theorem 9.1. *The curvature tensor $\tilde{R}$ of $\tilde{M}$ satisfies*

$$\tilde{R}(X, Y)Z = \frac{\epsilon}{r^2}\left(X \wedge Y + \sum_{i=1}^{3} (J_iX \wedge J_iY + 2\langle X, J_iY\rangle J_i)\right)Z. \tag{9.6}$$

Denote by $\mathcal{J}$ the span of $\{J_1, J_2, J_3\}$. $\mathcal{J}$ is a rank 3 sub-bundle of the bundle of $(1, 1)$ tensor fields on $\tilde{M}$ and gives $\tilde{M}$ the structure of a *quaternionic Kähler manifold* (see [28]). In particular, it follows from (9.6) that every quaternionic 2-plane (i.e., one with a basis of the form $\{X, JX\}$ where $J \in \mathcal{J}$) has sectional curvature $4\epsilon/r^2$ which we can write as $4c$. Such a space is said to have constant quaternionic sectional curvature.

We recall that in the case of a Kähler manifold (for example, a complex space form), the complex structure J is parallel, i.e., $\tilde{\nabla} J = 0$. In the case of a quaternionic Kähler manifold, it is the quaternionic Kähler structure $\mathcal{J}$ that is parallel. Specifically, for quaternionic space forms, we have

Proposition 9.2. *On a quaternionic space form $\tilde{M}$, there are 1-forms q_1, q_2, and q_3 such that for all tangent vectors X,*

$$\tilde{\nabla}_X J_i = q_{i+2}(X)J_{i+1} - q_{i+1}(X)J_{i+2} \tag{9.7}$$

(with indices mod 3*) for* $1 \leq i \leq 3$.

Remark 9.3. The result of this proposition is really a property of the quaternionic Kähler structure $\mathcal{J}$ in the sense that it holds for any local basis of $\mathcal{J}$ consisting of complex structures satisfying (9.4) on each tangent space.

9.4 Tubes over Submanifolds

Let M be a submanifold of a quaternionic space form $\tilde{M}$ of constant quaternionic sectional curvature $4c = \frac{4\epsilon}{r^2}$. For $\theta = (p, v)$ in the unit normal bundle, let X be an eigenvector of A_v corresponding to an eigenvalue λ. Let γ_t be the (normal) geodesic determined by θ. Let B_t be the parallel vector field along γ_t with $B_0 = X$. We are interested in the shape operators of tubes over M.

Lemma 9.4. *Assume that X is orthogonal to the span of $\{J_1 v, J_2 v, J_3 v\}$.*

- *If $c > 0$, then $X_t = (\cos u - r\lambda \sin u)B_t$ is a Jacobi field along γ_t with $X_0 = X$ and $X_0' = -\lambda X$;*
- *If $c < 0$, then $X_t = (\cosh u - r\lambda \sinh u)B_t$ is a Jacobi field along γ_t with $X_0 = X$ and $X_0' = -\lambda X$,*

where $u = t/r$.

Proof. First it is easy to check that

$$\tilde{\nabla}_t^2 X_t = X_t'' = -cX_t.$$

We need to compute $\tilde{R}(X_t, \vec{\gamma_t})\vec{\gamma_t}$. We consider the terms of (9.6) separately. First, we see that

$$\frac{\epsilon}{r^2}(X_t \wedge U_t)U_t = \frac{\epsilon}{r^2}(\langle U_t, U_t\rangle X_t - \langle X_t, U_t\rangle\, U_t) = cX_t, \tag{9.8}$$

where U_t is the unit vector $\overrightarrow{\gamma_t}$. Since X_t is a scalar multiple of B_t, $\langle B_t, U_t\rangle$ is constant along γ_t (being the inner product of two vector fields that are parallel along γ_t), and $\langle B_0, U_0\rangle = \langle X, v\rangle = 0$. Now,

$$\begin{aligned}\big(J_iX_t \wedge J_iU_t + 2\langle X, J_iU_t\rangle J_i\big)U_t &= -\langle J_iX_t, U_t\rangle J_iU_t + 2\langle X, J_iU_t\rangle J_iU_t \\ &= 3\langle X_t, J_iU_t\rangle J_iU_t. \end{aligned} \tag{9.9}$$

We claim that

$$\sum_{i=1}^{3} \langle X_t, J_iU_t\rangle J_iU_t = 0. \tag{9.10}$$

To see this, we show that

$$\sum_{i=1}^{3} \tilde{\nabla}_t(\langle B_t, J_iU_t\rangle J_iU_t) = 0, \tag{9.11}$$

and use the fact that X_t is a scalar multiple of B_t. Differentiation of B_t and U_t yields zero, but differentiation of J_i requires substitution from (9.7). One can check that the coefficient of each J_iU_t in the result consists of 4 terms which cancel in pairs. This completes the proof. □

Lemma 9.5. *Assume that X lies in the span of $\{J_1v, J_2v, J_3v\}$.*

- *If $c > 0$, then $X_t = (\cos 2u - \frac{r}{2}\lambda \sin 2u)B_t$ is a Jacobi field along γ_t with $X_0 = X$ and $X_0' = -\lambda X$.*
- *If $c < 0$, then $X_t = (\cosh 2u - \frac{r}{2}\lambda \sinh 2u)B_t$ is a Jacobi field along γ_t with $X_0 = X$ and $X_0' = -\lambda X$,*

where $u = t/r$.

Proof. The proof is similar to that of the previous lemma. First of all, we get

$$\tilde{\nabla}_t^2 X_t = X_t'' = -4cX_t.$$

The first term in the curvature expression is unchanged from (9.8). Furthermore, (9.11) still holds. This means that $\sum_{i=1}^{3}\langle B_t, J_iU_t\rangle J_iU_t$ is parallel along γ_t. Its initial value is $\sum_{i=1}^{3}\langle X, J_iv\rangle J_iv$. However, this is just X since $\{J_iv\}$ an orthonormal triple. Thus the summation,

$$\sum_{i=1}^{3}\langle B_t, J_iU_t\rangle J_iU_t$$

coincides with B_t and hence

$$\sum_{i=1}^{3} \langle X_t, J_i U_t \rangle J_i U_t = X_t.$$

Consequently,

$$\tilde{R}(X_t, \vec{\gamma_t})\vec{\gamma_t} = cX_t + 3cX_t = 4cX_t, \tag{9.12}$$

and the Jacobi equation is satisfied. □

To complete the tangent space to the tube, we need to look at normals orthogonal to v.

Lemma 9.6. *Assume that (p, w) in N_1M with $\langle w, v \rangle = \langle w, J_i v \rangle = 0$ for $1 \leq i \leq 3$. Let W_t be parallel along γ_t with $W_0 = w$.*

- *If $c > 0$, then $Y_t = (\sin u) W_t$ is a Jacobi field along γ_t with $Y_0 = 0$ and $Y_0' = \frac{w}{r}$.*
- *If $c < 0$, then $Y_t = (\sinh u) W_t$ is a Jacobi field along γ_t with $Y_0 = 0$ and $Y_0' = \frac{w}{r}$,*

where $u = t/r$.

Proof. Again, it is easy to check that $Y_t'' = -cY_t$ and that the first term of the curvature expression is cY_t. The expression corresponding to (9.11) is again zero. However, this time $\sum_{i=1}^{3} \langle W_t, J_i U_t \rangle J_i U_t$ vanishes since $W_0 = w$ is orthogonal to all of the $J_i v$. Hence, Y_t is a Jacobi field satisfying $Y_0 = 0$ and $Y_0' = \frac{w}{r}$ as required. □

Lemma 9.7. *Assume that (p, w) in N_1M with $\langle w, v \rangle = 0$ but*

$$w \in \operatorname{Span}\{J_1 v, J_2 v, J_3 v\}.$$

Let W_t be parallel along γ_t with $W_0 = w$.

- *If $c > 0$, then $Y_t = \frac{1}{2} \sin 2u \; W_t$ is a Jacobi field along γ_t with $Y_0 = 0$ and $Y_0' = \frac{w}{r}$,*
- *If $c < 0$, then $Y_t = \frac{1}{2} \sinh 2u \; W_t$ is a Jacobi field along γ_t with $Y_0 = 0$ and $Y_0' = \frac{w}{r}$,*

where $u = t/r$.

Proof. Following the same procedures as in earlier cases, we find that $Y_t'' = -4cY_t$ and that $\sum_{i=1}^{3} \langle W_t, J_i U_t \rangle J_i U_t = W_t$ so that $\tilde{R}(Y_t, \vec{\gamma_t})\vec{\gamma_t} = 4cY_t$, and the Jacobi equation is satisfied. □

9.5 Real Hypersurfaces

Now take any space $\tilde{M}$ of constant quaternionic curvature $4c$ with real dimension $4n$ and Levi-Civita connection

$\tilde{\nabla}$. For an immersed manifold $f : M^{4n-1} \to \tilde{M}$, the Levi-Civita connection ∇ of the induced metric and the shape operator A of the immersion are characterized respectively by

$$\tilde{\nabla}_X Y = \nabla_X Y + \langle AX, Y\rangle \xi,$$

and

$$\tilde{\nabla}_X \xi = -AX,$$

where ξ is the local choice of unit normal. We omit mention of the immersion f for brevity of notation. Then analogous with the situation for complex space forms, we have the unit vector fields $W_i = -J_i\xi$ on M. Also, we get the $(1, 1)$ tensor fields φ_i on M by projection of the J_i. Specifically, for all tangent vectors X, we define

$$\varphi_i X = J_i X - \langle J_i X, \xi\rangle \xi = J_i X - \langle X, W_i\rangle \xi.$$

Let $\mathcal{W}$ be the span of $\{W_1, W_2, W_3\}$. Then each φ_i preserves $\mathcal{W}$ and $\mathcal{W}^\perp$. Specifically, we have

Lemma 9.8. *The W_i and the φ_i satisfy the following identities (with indices* mod 3*)*

$$\begin{aligned} \varphi_i W_i &= 0 \\ \varphi_i W_{i+1} &= W_{i+2}; \quad \varphi_i W_{i+2} = -W_{i+1} \\ \varphi_i X &= J_i X \text{ for } X \in W_i^\perp \\ \varphi_i^2 X &= -X \text{ for } X \in W_i^\perp. \end{aligned} \tag{9.13}$$

The relationship between $\tilde{\nabla}$ and ∇ gives rise to the Gauss and Codazzi equations for the hypersurface

$$R(X, Y) = AX \wedge AY + c\big(X \wedge Y + \sum_{i=1}^{3}(\varphi_i X \wedge \varphi_i Y + 2\langle X, \varphi_i Y\rangle \varphi_i)\big) \tag{9.14}$$

$$(\nabla_X A)Y - (\nabla_Y A)X = c\big(\sum_{i=1}^{3}(\langle X, W_i\rangle \varphi_i Y - \langle Y, W_i\rangle \varphi_i X + 2\langle X, \varphi_i Y\rangle W_i)\big). \tag{9.15}$$

From the Gauss equation, we can compute the Ricci tensor S which is given by

$$SX = (4n + 7)cX - 3c\sum_{i=1}^{3}\langle X, W_i\rangle W_i + \mathbf{m}AX - A^2 X \tag{9.16}$$

where $\mathbf{m}$ is the trace of A.

A hypersurface is said to be *pseudo-Einstein* if the Ricci tensor acts as a multiple of the identity on both $\mathcal{W}$ and $\mathcal{W}^{\perp}$. Specifically, M is pseudo-Einstein if there exist functions ρ and σ such that

$$SX = \rho X + \sigma \sum_{i=1}^{3} \langle X, W_i \rangle W_i,$$

for all tangent vectors X. Although it is traditional to require that ρ and σ be constant, we will not do this since (as we shall see below) it follows from the classification, at least when $n \geq 3$. In fact, it is easy to see that even the smoothness of ρ and σ need not be assumed. Note that if σ is identically zero, we have the familiar Einstein condition. In contrast to the situation in complex projective space, Einstein hypersurfaces can occur, specifically there is one particular radius for which the geodesic sphere in quaternionic projective space is Einstein (see next section).

9.6 Examples in Quaternionic Projective Space

Let M be a totally geodesic $\mathbf{HP}^k$ in $\mathbf{HP}^n$. For $0 < u < \frac{\pi}{2}$, the tube of radius ru over M is a hypersurface in $\mathbf{HP}^n$.

Type A_2

If $1 \leq k \leq n-2$, then we have the following principal curvatures:

- $\frac{2}{r} \cot 2u$ of multiplicity 3;
- $\frac{1}{r} \cot u$ of multiplicity 4ℓ;
- $-\frac{1}{r} \tan u$ of multiplicity $4k$.

where $k + \ell = n - 1$.

Type A_1

If $k = n - 1$, we have the following principal curvatures:

- $\frac{2}{r} \cot 2u$ of multiplicity 3;
- $-\frac{1}{r} \tan u$ of multiplicity $4k = 4n - 4$.

The Type A_1 hypersurfaces are tubes over quaternionic projective hyperplanes. They are also geodesic spheres. For example, the geodesic sphere centered at πe_0 with radius $r(\frac{\pi}{2} - u)$ coincides with the tube of radius ru over the totally geodesic $\mathbf{HP}^{n-1}$ given by $\pi\{z \mid z_0 = 0\}$. In fact, if we abuse notation slightly and set $k = 0$ in the prescription for Type A_2 hypersurfaces, we get

- $\frac{2}{r} \cot 2u$ of multiplicity 3;
- $\frac{1}{r} \cot u$ of multiplicity $4n - 4$.

which, upon substitution of $\frac{\pi}{2} - u$, would give the configuration of principal curvatures derived for the Type A_1 case (with a change of sign).

Type B

Let M be a totally geodesic $\mathbf{CP}^n$ in $\mathbf{HP}^n$. This can be obtained, for example, by taking any $J \in \mathcal{J}$ and taking

$$M = \pi(\{x + Jy \mid x, y \in \mathbf{R}^{n+1}, |x|^2 + |y|^2 = r^2\}) \subset \mathbf{HP}^n. \tag{9.17}$$

(In particular, we could pick $J = J_1$). For $0 < u < \frac{\pi}{4}$, the tube of radius ru over M is a hypersurface in $\mathbf{HP}^n$. Its principal curvatures are as follows:

- $\alpha = \frac{2}{r} \cot 2u$ of multiplicity 1;
- $\beta = -\frac{2}{r} \tan 2u$ of multiplicity 2;
- $\lambda = \frac{1}{r} \cot u$ of multiplicity $2(n-1)$;
- $\mu = -\frac{1}{r} \tan u$ of multiplicity $2(n-1)$.

9.7 Examples in Quaternionic Hyperbolic Space

Type A_0

These are the horospheres in $\mathbf{HH}^n$. Our description follows closely that used for $\mathbf{CH}^n$. For $t > 0$ consider the hypersurface

$$M' = \{z \in \mathbf{H}^{n+1} | \langle z, z\rangle = -r^2, |z_0 - z_1|^2 = t\}$$

in $\mathbf{H}$. Then the corresponding horosphere M is $\pi M'$. For $z \in M'$, let ξ' be the unit normal at z with corresponding shape operator A'. For any $J \in \mathcal{J}$ (normalized so that $J^2 = -I$), let $V = \frac{1}{r} Jz$ and $U = -J\xi'$. Then $A'V = \frac{1}{r} U$ and $A'U = -\frac{1}{r} V$. Further, $A'X = \frac{1}{r} X$ for X orthogonal to all such U and V. On projection down to $\mathbf{HH}^n$, the $\pi_* V$ all vanish and the $\pi_* U$ span the 3-dimensional subspace $\mathcal{W}$. The principal curvatures of the horosphere M are

- $\frac{2}{r}$ with multiplicity 3;
- $\frac{1}{r}$ with multiplicity $4n - 4$.

The respective principal spaces are $\mathcal{W}$ and $\mathcal{W}^{\perp}$.

Let M be a totally geodesic $\mathbf{HH}^k$ in $\mathbf{HH}^n$. For $u > 0$ the tube of radius ru over M is a hypersurface in $\mathbf{HH}^n$.

Type A_2

If $1 \le k \le n-2$, then we have the following principal curvatures:

- $\frac{2}{r} \coth 2u$ of multiplicity 3;
- $\frac{1}{r} \coth u$ of multiplicity 4ℓ;
- $\frac{1}{r} \tanh u$ of multiplicity $4k$.

where $k + \ell = n - 1$.

Type A_1

If $k = n - 1$, we have the following principal curvatures:

- $\frac{2}{r}\coth 2u$ of multiplicity 3;
- $\frac{1}{r}\tanh u$ of multiplicity $4k = 4n - 4$.

The Type A_1 hypersurfaces are tubes over quaternionic projective hyperplanes. Unlike the case in $\mathbf{HP}^n$, the geodesic spheres form a distinct class of hypersurfaces, formally corresponding to the $k = 0$ case. The principal curvatures of a geodesic sphere of radius ru are:

- $\frac{2}{r}\coth 2u$ of multiplicity 3;
- $\frac{1}{r}\coth u$ of multiplicity $4n - 4$.

Type B

Let M be a totally geodesic $\mathbf{CH}^n$ in $\mathbf{HH}^n$. In a similar fashion to the $\mathbf{HP}^n$ case, we take

$$M = \pi(\{x + Jy \mid x, y \in \mathbf{R}^{n+1}, \langle x + Jy, x + Jy\rangle = -r^2\}) \subset \mathbf{HH}^n. \tag{9.18}$$

(In particular, we could pick $J = J_1$). For $u > 0$, the tube of radius ru over M is a hypersurface in $\mathbf{HH}^n$. Its principal curvatures are as follows:

- $\alpha = \frac{2}{r}\coth 2u$ of multiplicity 1;
- $\beta = \frac{2}{r}\tanh 2u$ of multiplicity 2;
- $\lambda = \frac{1}{r}\coth u$ of multiplicity $2(n-1)$;
- $\mu = \frac{1}{r}\tanh u$ of multiplicity $2(n-1)$.

Proposition 9.9.
1. *Type A_1 hypersurfaces are pseudo-Einstein with*

$$SX = \rho_\lambda X + (\rho_\alpha - \rho_\lambda)\sum_{i=1}^{3}\langle X, W_i\rangle W_i,$$

where

$$\begin{aligned}\rho_\lambda &= (4n+7)c + (4n-5)\lambda^2 + 3\lambda\alpha = 4(n+1)c + 2(2n-1)\lambda^2\\ \rho_\alpha &= (4n+4)c + (4n-4)\lambda\alpha + 2\alpha^2 = 8c + 4(n-1)\lambda^2 + 2\alpha^2\end{aligned} \tag{9.19}$$

and λ is the principal curvature of multiplicity $4n - 4$.

2. *Type A_0 hypersurfaces are pseudo-Einstein with*

$$SX = \rho_\lambda X + (\rho_\alpha - \rho_\lambda)\sum_{i=1}^{3}\langle X, W_i\rangle W_i,$$

where

$$\begin{aligned} \rho_\lambda &= -6\lambda^2 \\ \rho_\alpha &= 4(n-1)\lambda^2 \end{aligned} \tag{9.20}$$

and λ *is the principal curvature of multiplicity* $4n-4$. *(Note that these values are limits as* $u \to \infty$ *of the respective values for the corresponding Type* A_1 *hypersurfaces.)*

3. *A Type* A_1 *hypersurface in* $\mathbf{HP}^n$ *(considered as a tube of radius ru over* $\mathbf{HP}^{n-1}$*) is Einstein precisely when* $\cot^2 u = 2n$. *A geodesic sphere of radius ru is Einstein precisely when* $\tan^2 u = 2n$. *No Type* A_1 *hypersurface in* $\mathbf{HH}^n$ *is Einstein.*
4. *The principal vectors of Type* A_2 *hypersurfaces are also eigenvectors of the Ricci tensor S with corresponding eigenvalues*

$$\begin{aligned} \rho_\lambda &= 4(\ell+2)c + 2(2\ell+1)\lambda^2 \\ \rho_\mu &= 4(k+2)c + 2(2k+1)\mu^2 \\ \rho_\alpha &= 4c + 2\big((2\ell+1)\lambda^2 + (2k+1)\mu^2\big) \end{aligned} \tag{9.21}$$

5. *No Type* A_2 *hypersurface is Einstein. A Type* A_2 *hypersurface in* $\mathbf{HP}^n$ *is pseudo-Einstein if and only if*

$$\cot^2 u = \frac{2k+1}{2\ell+1}.$$

In this case, $\rho_\lambda = \rho_\mu = 2(2n+3)c$ *while* $\rho_\alpha = 4(n+1)c$. *No Type* A_2 *hypersurface in* $\mathbf{HH}^n$ *is pseudo-Einstein.*

The situation for Type B hypersurfaces is more complicated since A has two distinct eigenspaces in $\mathcal{W}$ and two in $\mathcal{W}^\perp$. Of course, the Ricci tensor S is a multiple of the identity on each of these eigenspaces and the corresponding multipliers can be computed as

$$\begin{aligned} \rho_\alpha &= 2(n-1)(2c+\alpha^2) \\ \rho_\beta &= -4(n-2)c + \beta^2 \\ \rho_\lambda &= 2((n+4)c + (n-1)\lambda^2 + \beta\lambda) \\ \rho_\mu &= 2((n+4)c + (n-1)\mu^2 + \beta\mu). \end{aligned} \tag{9.22}$$

Lemma 9.10.

1. *A Type B hypersurface in* $\mathbf{HP}^n$ *satisfies* $\rho_\alpha = \rho_\beta$ *if and only if* $\tan^2 2u = 2(n-1)$. *There are no Type B hypersurfaces in* $\mathbf{HH}^n$ *for which* $\rho_\alpha = \rho_\beta$*;*
2. *A Type B hypersurface in* $\mathbf{HP}^n$ *satisfies* $\rho_\lambda = \rho_\mu$ *if and only if* $\tan^2 2u = n-1$. *There are no Type B hypersurfaces in* $\mathbf{HH}^n$ *for which* $\rho_\lambda = \rho_\mu$.

Proof. From our formula for the Ricci tensor of a hypersurface, we note that $\rho_\alpha = \rho_\beta$ if and only if $\mathbf{m}\alpha - \alpha^2 = \mathbf{m}\beta - \beta^2$. Also,

$$\mathbf{m} = \alpha + 2\beta + 2(n-1)(\lambda + \mu) = \alpha + 2\beta + 2(n-1)\alpha,$$

since $\lambda + \mu = \alpha$. Using $\alpha \neq \beta$, we can see that $\rho_\alpha = \rho_\beta$ if and only if $\mathbf{m} = \alpha + \beta$. This is equivalent to

$$\alpha^2 = \frac{2c}{n-1},$$

since $\alpha\beta = -4c$. We conclude that c is positive and $\tan^2 2u = 2(n-1)$. The condition that $\rho_\lambda = \rho_\mu$ can be treated similarly. However, in this case, we have $\mathbf{m} = \lambda + \mu = \alpha$, which leads to the condition

$$\alpha^2 = \frac{4c}{n-1}.$$

Again c is positive. In this case, we get $\tan^2 2u = n - 1$. This completes the proof. □

9.8 Curvature-adapted Hypersurfaces

Let M be a submanifold of a Riemannian manifold $\tilde{M}$. For a unit normal vector ξ at a point $p \in M$, the normal Jacobi operator

$$R_\xi : T_pM \to T_p\tilde{M}$$

is defined by $R_\xi X = \tilde{R}(X, \xi)\xi$. M is said to be *curvature-adapted* if for all $p \in M$,

$$R_\xi(T_pM) \subset T_pM \text{ and } R_\xi \circ A_\xi = A_\xi \circ R_\xi \tag{9.23}$$

for all unit normals ξ at p. Note that $R_\xi(T_pM) \subset T_pM$ holds automatically for a hypersurface. We can check the following for familiar ambient spaces.

Proposition 9.11.

- *Every hypersurface in a real space form is curvature-adapted;*
- *A hypersurface in a nonflat complex space form is curvature-adapted if and only if it is Hopf;*
- *A hypersurface in a nonflat quaternionic space form is curvature-adapted if and only if its shape operator A satisfies $A\mathcal{W} \subset \mathcal{W}$.*

Proof. For a real space form of constant curvature c, we have

$$R_\xi X = \tilde{R}(X, \xi)\xi = c(X \wedge \xi)\xi = cX \tag{9.24}$$

so that R_ξ acts as a scalar multiple of the identity and thus commutes with A automatically. For a complex space form of constant holomorphic curvature $4c$, we have, for any tangent vector X,

$$\begin{aligned}\tilde{R}(X,\xi)\xi &= c\big((X \wedge \xi) + JX \wedge J\xi + 2\langle X, J\xi\rangle J)\big)\xi \\ &= c(X + 3\langle X, W\rangle W)\end{aligned} \tag{9.25}$$

from which it is clear that $R_\xi X$ is tangent. Further, $R_\xi X$ commutes with A if and only if

$$\langle X, W\rangle AW = \langle X, AW\rangle W. \tag{9.26}$$

By setting $X = W$, we see that M is Hopf if $R_\xi X$ commutes with A. On the other hand, if M is Hopf, then the same equation is satisfied, both for $X = W$ and for $X \in \mathcal{W}^\perp$.

Finally we look at the quaternionic case. From (9.6), we see that the expression for R_ξ is just a more complicated version of (9.25). Specifically, for a tangent vector X,

$$\begin{aligned}\tilde{R}(X,\xi)\xi &= c\big((X \wedge \xi) + \sum_{i=1}^{3}(J_iX \wedge J_i\xi + 2\langle X, J_i\xi\rangle J_i)\big)\xi \\ &= c(X + 3\sum_{i=1}^{3}\langle X, W_i\rangle W_i).\end{aligned} \tag{9.27}$$

Again, it is evident that $R_\xi X$ is tangent and that R_ξ commutes with A if and only if

$$\sum_{i=1}^{3}\langle X, W_i\rangle AW_i = \sum_{i=1}^{3}\langle X, AW_i\rangle W_i. \tag{9.28}$$

for all tangent vectors X. Setting $X = W_j$ in (9.28) for each $j \in \{1,2,3\}$ in turn, we see that $AW_j = \sum\langle AW_i, W_j\rangle W_i$ for each j and hence that $A\mathcal{W} \subset \mathcal{W}$. Conversely, if $\mathcal{W}$ is A-invariant, then (9.28) is satisfied for each individual W_j and hence $[R_\xi, A]W_j = 0$. Also, if $X \in \mathcal{W}^\perp$, then (9.28) is satisfied and we can conclude that R_ξ and A commute.

This completes our proof for all three cases. □

For curvature-adapted hypersurfaces in nonflat quaternionic space forms, we have the following analogue of our fundamental formula for Hopf hypersurfaces (see Theorem 6.17 or Lemma 2.2 of [399]).

Lemma 9.12. *Let M be a curvature-adapted hypersurface in $\mathbf{HP}^n$ or $\mathbf{HH}^n$. Choose $i \in \{1,2,3\}$ and suppose that $AW_i = \alpha W_i$. Then*

$$A\varphi_i AX = \frac{\alpha}{2}(A\varphi_i + \varphi_i A)X + c\varphi_i X$$

for all $X \in \mathcal{W}^\perp$.

This and other useful material may be found in Adachi and Maeda [3, 4].

Proposition 9.13. *All Type A hypersurfaces in nonflat quaternionic space forms satisfy $A \circ \varphi_i = \varphi_i \circ A$ for $i \in \{1,2,3\}$.*

In fact, this property characterizes the Type A hypersurfaces as follows:

Theorem 9.14. *Let M^{4n-1}, where $n \geq 2$, be a hypersurface in $\mathbf{HP}^n$ or $\mathbf{HH}^n$. Then $A \circ \varphi_i = \varphi_i \circ A$ for $1 \leq i \leq 3$ if and only if M is an open subset of a hypersurface of Type A.*

We also have for hypersurfaces in quaternionic projective space

Theorem 9.15. *Let M^{4n-1}, where $n \geq 2$, be a hypersurface in $\mathbf{HP}^n$. Then M is curvature-adapted if and only if it is an open subset of a hypersurface of Type A or Type B.*

In the hyperbolic case, however, we need an additional assumption.

Theorem 9.16. *Let M^{4n-1}, where $n \geq 2$, be a hypersurface in $\mathbf{HH}^n$ with constant principal curvatures. Then M is curvature-adapted if and only if it is an open subset of a hypersurface of Type A or Type B.*

Thus we see that "curvature-adapted" property is a rather strong one for hypersurfaces of quaternionic space forms, standing in contrast to the situation for complex space forms where there many nontrivial examples of Hopf hypersurfaces. We note that all known examples of curvature-adapted hypersurfaces in $\mathbf{HH}^n$ have constant principal curvatures.

Theorems 9.15 and 9.16 were proved by Berndt [28]. Martinez and Pérez had previously proved Theorem 9.15 under the assumption of constant principal curvatures. As in the case of complex space forms, the proofs involve a study of parallel hypersurfaces, tubes and focal submanifolds.

Just as in the case of complex space forms, it is impossible for the shape operator A to vanish identically. In fact, we have $|\nabla A|^2 \geq 24c^2(n-1)$. Further, we get the following analogue of Theorem 8.37.

Theorem 9.17. *Let M^{4n-1}, where $n \geq 2$, be a real hypersurface in $\mathbf{HP}^n$ or $\mathbf{HH}^n$. The following are equivalent:*

1. *M is an open subset of a Type A hypersurface;*
2. *$\varphi_i A = A\varphi_i$ for all $i \in \{1,2,3\}$;*
3. *$|\nabla A|^2 = 24c^2(n-1)$;*
4. *$(\nabla_X A)Y + c\sum_{i=1}^{3}(\langle\varphi_i X, Y\rangle W_i + \langle Y, W_i\rangle\varphi_i X) = 0$ for all X and Y tangent to M.*

These results are due to Pak [424] and Lyu, Pérez and Suh [342]. Pérez [430] has also proved for the case of $\mathbf{HP}^n$ that a hypersurface M is cyclic parallel if and only if it satisfies the conditions of Theorem 9.17. Further, Ki, Suh and Pérez [256] have shown (again, for $\mathbf{HP}^n$) that having conditions *2.* and *4.* of Theorem 9.17 both holding on $\mathcal{W}^\perp$ is also equivalent. Note that on $\mathcal{W}^\perp$, Condition *4.* takes the simpler form $(\nabla_X A)Y + c\sum_{i=1}^3 \langle \varphi_i X, Y\rangle W_i = 0$.

The Type A and Type B hypersurfaces in $\mathbf{HP}^n$ may also be characterized by properties of the (tangent) Jacobi operators. Recalling the definition of the Jacobi operators from equation (8.115), we have

Theorem 9.18. *Let M^{4n-1}, where $n \geq 3$, be a hypersurface in $\mathbf{HP}^n$. Then all Jacobi operators of M commute if and only if M is curvature-adapted.*

Remark 9.19. This is proved in Ortega, Pérez, and Suh [417]. The following questions still seem to be open:

- Is this theorem true for $n = 2$?
- Is this theorem true for hypersurfaces in $\mathbf{HH}^n$?

The analogue of Theorem 8.29 for the quaternionic case is the following due to Martinez, Pérez and Ortega [353, 415].

Theorem 9.20. *Let M^{4n-1}, where $n \geq 3$, be a hypersurface in $\mathbf{HP}^n$ or $\mathbf{HH}^n$. Then M has $g \leq 2$ principal curvatures at each point if and only if it is an open subset of a hypersurface of Type A_1 or Type A_0.*

The $n = 2$ case of this theorem is an open problem.

The situation for homogeneous hypersurfaces in the quaternionic space forms is similar to the complex case. For the quaternionic projective space, we have the following results of Iwata [226] and D'Atri [126].

Theorem 9.21. *Let M^{4n-1}, where $n \geq 2$, be a homogeneous hypersurface in $\mathbf{HP}^n$. Then M is a hypersurface of Type A or Type B. Conversely, all Type A and Type B hypersurfaces are homogeneous.*

For the quaternionic hyperbolic space, the classification problem is still open. See Berndt [31] for a discussion of this and related questions. We do know that there are examples other than the Type A and B hypersurfaces. First we define the notion of ruled hypersurfaces in the quaternionic context.

A hypersurface in $\mathbf{HP}^n$ (resp. $\mathbf{HH}^n$) is said to be *ruled* if $\mathcal{W}^\perp$ is a foliation with totally geodesic leaves locally congruent to $\mathbf{HP}^{n-1}$ (resp. $\mathbf{HH}^{n-1}$). As in the case of complex space forms, a hypersurface is ruled if and only if $A\mathcal{W}^\perp \subseteq \mathcal{W}$.

A family of homogeneous ruled real hypersurface in $\mathbf{HH}^n$ was constructed by Adachi, Maeda and Udagawa [6]. These hypersurfaces are not curvature-adapted but have constant principal curvatures. The construction is motivated by that of Lohnherr (see Section 8.6 and [335, 336]).

9.9 Einstein, Pseudo-Einstein, and Related Conditions

As we have seen in section 9.7, geodesic spheres of radius ru in $\mathbf{HP}^n$ are Einstein when $\tan^2 u = 2n$. Also, all Type A_1 and Type A_0 hypersurfaces in the nonflat quaternionic space forms are pseudo-Einstein. The only Type A_2 hypersurfaces that are pseudo-Einstein are those in $\mathbf{HP}^n$ for which $\cot^2 u = \frac{2k+1}{2\ell+1}$. Martinez, Ortega, Pérez and Santos [353, 415, 435] showed that for $n \geq 3$, these are essentially all the pseudo-Einstein hypersurfaces. Specifically, we can state

Theorem 9.22. *Let M^{4n-1}, where $n \geq 3$, be a pseudo-Einstein hypersurface in $\mathbf{HP}^n$ or $\mathbf{HH}^n$. Then M is an open subset of a Type A_1 or Type A_0 hypersurface or a Type A_2 hypersurface in $\mathbf{HP}^n$ for which $\cot^2 u = \frac{2k+1}{2\ell+1}$.*

The classification question for pseudo-Einstein hypersurfaces remains open for $n = 2$. The same authors also introduced a slightly different condition, "almost-Einstein." This means that there exist functions ρ and σ such that

$$SX = \rho X + \sigma \sum_{i=1}^{3} \langle AX, W_i \rangle W_i$$

for all tangent vectors X. For $n \geq 3$, every almost-Einstein hypersurface is pseudo-Einstein. Thus the almost-Einstein hypersurfaces can be classified by checking the standard examples. However, for $n = 2$, the almost-Einstein condition is sufficiently strong to allow the same classification. The result is as follows:

Theorem 9.23. *Let M^{4n-1}, where $n \geq 2$, be an almost-Einstein hypersurface in $\mathbf{HP}^n$ or $\mathbf{HH}^n$. Then M is an open subset of a Type A_1 or Type A_0 hypersurface or a Type A_2 hypersurface in $\mathbf{HP}^n$ for which $\cot^2 u = \frac{2k+1}{2\ell+1}$.*

We see that for a Type B hypersurface in $\mathbf{HP}^n$, one particular radius gives $\rho_\alpha = \rho_\beta$ and a different radius gives $\rho_\lambda = \rho_\mu$. In neither case is the pseudo-Einstein condition satisfied. This example motivated Ortega and Pérez [416] to consider the condition that S acts as a multiple of the identity on $\mathcal{W}^\perp$. They called this condition "D-Einstein". Their result is as follows:

Theorem 9.24. *Let M^{4n-1}, where $n \geq 3$, be a real hypersurface in $\mathbf{HP}^n$ or $\mathbf{HH}^n$. Suppose that there is a function ρ such that $SX = \rho X$ for all $X \in \mathcal{W}^\perp$. Then either M is pseudo-Einstein or is an open subset of a Type B hypersurface in $\mathbf{HP}^n$ satisfying $\tan^2 2u = n - 1$.*

Of course, this gives a specific list of Type A and B hypersurfaces that are characterized by the D-Einstein condition.

Further classification theorems and problems

There are no hypersurfaces in $\mathbf{HP}^n$ or $\mathbf{HH}^n$ with parallel shape operator. In fact, there are none satisfying the semiparallelism condition $R \cdot A = 0$. However, Pérez [432] has classified the hypersurfaces in $\mathbf{HP}^n$, where $n \geq 2$, satisfying $\nabla_{\mathcal{W}} A = 0$. Weakening the parallelism condition in another way, he has classified the cyclic parallel hypersurfaces, (again for $\mathbf{HP}^n$, where $n \geq 2$), namely those for which the cyclic sum of $\langle (\nabla_X A)Y, Z \rangle$ vanishes [430].

There are no locally symmetric hypersurfaces in $\mathbf{HP}^n$ or $\mathbf{HH}^n$. In fact, there are none satisfying the semisymmetry condition $R \cdot R = 0$. Pérez and Suh [441] have classified the hypersurfaces in $\mathbf{HP}^n$, where $n \geq 3$, satisfying $\nabla_{\mathcal{W}} R = 0$. In addition, Pérez and his collaborators have studied various conditions on the Ricci tensor, mostly in the context of $\mathbf{HP}^n$ where $n \geq 3$. In particular, they have

Theorem 9.25. *Let M^{4n-1}, where $n \geq 2$, be a real hypersurface in $\mathbf{HP}^n$. Then following are equivalent:*

1. *M is Einstein;*
2. *M is Ricci-parallel (i.e., $\nabla S = 0$) [433];*
3. *M has harmonic curvature (i.e., S is a Codazzi tensor) and A acts on $\mathcal{W}$ as a scalar multiple of the identity [437];*
4. *M is cyclic-Ryan (i.e., $R(X,Y)SZ + R(Y,Z)SX + R(Z,X)SY = 0$ for all triples (X,Y,Z) of tangent vectors) [431, 436].*

For other results, see [383, 434, 438] and [314]. Several results pertain only to $\mathbf{HP}^n$, where $n \geq 3$. There are still many questions unresolved for $\mathbf{HH}^n$ and/or $n = 2$.

9.10 Open Problems

Although our treatment of the quaternionic ambient spaces has been much less detailed than that of the complex case, we list here a few problems that appear to be open for hypersurfaces in the quaternionic space forms.

1. Theorem 9.18 characterizes hypersurfaces in $\mathbf{HP}^n$, where $n \geq 3$, whose Jacobi operators commute. Does this theorem extend to $n = 2$? Does it extend to the quaternionic hyperbolic space $\mathbf{HH}^n$?
2. Theorem 9.20 classifies hypersurfaces in $\mathbf{HP}^n$ and $\mathbf{HH}^n$, where $n \geq 3$, having $g \leq 2$ distinct principal curvatures. Does this theorem extend to $n = 2$?
3. Theorem 9.21 classifies the homogeneous hypersurfaces in $\mathbf{HP}^n$. What is the analogous classification theorem for $\mathbf{HH}^n$?
4. Classify the pseudo-Einstein hypersurfaces in $\mathbf{HP}^2$ and $\mathbf{HH}^2$. See Theorem 9.22. See also remarks following Theorem 9.25.

9.11 Further Research

All the material we have discussed in this book falls under the heading of "Hypersurfaces of symmetric spaces." Our ambient spaces have progressed in complexity, beginning with the real space forms, then moving to the complex space forms, and finally the quaternionic space forms. Important classification criteria have been constancy of principal curvatures, the curvature-adapted/Hopf property, and special identities involving geometric objects such as the shape operator, the curvature tensor and other tensors derived from it, principally the Ricci tensor and the structure Jacobi operator. A key unifying property of many of the most important classes of hypersurfaces is that they are tubes over their focal submanifolds , or at least share many of the algebraic properties of such tubes.

The most active area of study over the past decade has been that of hypersurfaces in the next most complicated ambient spaces, the complex two-plane Grassmannians (see Section 7.4). This topic was introduced by Berndt and Suh [29, 37] and approximately 100 papers have appeared in which many of the problems discussed in Chapters 8 and 9 have been studied in this new context. Unfortunately, limitations of time and space do not permit us to discuss these results in the current volume.

Appendix A
Summary of Notation

The following is a list of notations that are used frequently in the book and whose meaning is usually assumed to be known. Some symbols are used in more than one way.

1. Let **R**, **C**, **H**, **O** denote the real, complex, quaternion, and Cayley numbers, respectively.

$\mathbf{Z}$	ring of integers
$\mathbf{Z}_2$	field of integers modulo 2
$\mathbf{R}^n$	vector space of n-tuples of real numbers $(x_1, \ldots, x_n)$
$\mathbf{R}^n_k$	$\mathbf{R}^n$ endowed with an indefinite metric of signature $(k, n-k)$
$\mathbf{C}^n$	vector space of n-tuples of complex numbers $(z_1, \ldots, z_n)$
$\mathbf{H}^n$	vector space of n-tuples of quaternions $(z_1, \ldots, z_n)$
$\langle x, y \rangle$	nondegenerate bilinear form, signature depends on context
$\lvert x \rvert$	length of a vector x
S^n	unit sphere in $\mathbf{R}^{n+1}$
H^n	n-dimensional hyperbolic space with constant curvature -1
D^n	n-dimensional unit disk in $\mathbf{R}^n$
$\mathbf{RP}^n$	n-dimensional real projective space
$\mathbf{CP}^n$	n-dimensional complex projective space
$\mathbf{CH}^n$	n-dimensional complex hyperbolic space
$\mathbf{HP}^n$	n-dimensional quaternionic projective space
$\mathbf{HH}^n$	n-dimensional quaternionic hyperbolic space
$\mathbf{OP}^2$	Cayley projective plane
Q^{n+1}	Lie quadric in $\mathbf{RP}^{n+2}$
Λ^{2n-1}	manifold of projective lines on Q^{n+1}
$GL(n)$	general linear group
$PGL(n)$	group of projective transformations of $\mathbf{RP}^n$
$O(n)$	orthogonal group for the standard metric on $\mathbf{R}^n$

T.E. Cecil, P.J. Ryan, *Geometry of Hypersurfaces*, Springer Monographs in Mathematics, DOI 10.1007/978-1-4939-3246-7

$SO(n)$	special orthogonal group
$O(n-k,k)$	orthogonal group of $\mathbf{R}^n_k$
$U(n)$	unitary group
$SU(n)$	special unitary group
$M(n,\mathbf{F})$	space of all $n \times n$ matrices over a field $\mathbf{F}$
$H(n,\mathbf{F})$	space of all $n \times n$ Hermitian matrices over $\mathbf{F}$
$U(n,\mathbf{F})$	space of all $n \times n$ unitary matrices over $\mathbf{F}$
$\text{grad}\, F$	gradient vector field of a function F
$\text{grad}^E F$	gradient vector field of a function $F : \mathbf{R}^n \to \mathbf{R}$
$\text{grad}^S F$	gradient of the restriction of F to the unit sphere S^{n-1}
ΔF	Laplacian of a function F
$\Delta^E F$	Laplacian of $F : \mathbf{R}^n \to \mathbf{R}$
$\Delta^S F$	Laplacian of the restriction of F to the unit sphere S^{n-1}
$\Delta_1 F$	first Beltrami differential parameter $\Delta_1 F = \lvert\text{grad}\, F\rvert^2$
$\Delta_2 F$	second Beltrami differential parameter $\Delta_2 F = \Delta F$
$\text{div}\, X$	divergence of a vector field X
G	Lie group G
$\mathfrak{g}$	Lie algebra of Lie group G
C_m	Clifford algebra generated by 1 and $e_1, \ldots, e_m$
$V_{m+1,2}$	Stiefel manifold of orthogonal 2-frames in $\mathbf{R}^{m+1}$
$V_2(C_{m-1})$	Clifford–Stiefel manifold of Clifford orthogonal 2-frames

2. Critical point theory for a smooth function $\phi : M \to \mathbf{R}$ on a smooth n-dimensional manifold M.

$M_r(\phi)$	$\{x \in M \mid \phi(x) \le r\}$
$M_r^-(\phi)$	$\{x \in M \mid \phi(x) < r\}$
$M_r^+(\phi)$	$\{x \in M \mid \phi(x) > r\}$
$H_k(M,\mathbf{F})$	k-th homology group of M over the field $\mathbf{F}$
$H^k(M,\mathbf{F})$	k-th cohomology group of M over the field $\mathbf{F}$
$\beta_k(M,\mathbf{F})$	$\dim_{\mathbf{F}} H_k(M,\mathbf{F})$, the k-th $\mathbf{F}$-Betti number of M
$\beta(M,\mathbf{F})$	the sum of the $\mathbf{F}$-Betti numbers of M
$\beta_k(\phi,r,\mathbf{F})$	$\dim_{\mathbf{F}} H_k(M_r(\phi),\mathbf{F})$
$\mu_k(\phi,r)$	number of critical points of ϕ of index k on $M_r(\phi)$
$\mu_k(\phi)$	number of critical points of ϕ of index k on M
$\mu(\phi)$	number of critical points of ϕ on compact manifold M
l_p	linear height function $l_p : \mathbf{R}^n \to \mathbf{R}$, $l_p(q) = \langle p,q\rangle$
L_p	distance function $L_p : \mathbf{R}^n \to \mathbf{R}$, $L_p(q) = \lvert p-q\rvert^2$
$\chi(M)$	Euler characteristic of M
$[X,Y]$	Lie bracket of two vector fields X and Y on M

3. For a manifold M immersed in a Riemannian manifold $(\tilde{M}, g)$.

$g(X,Y)$	Riemannian metric on $\tilde{M}$
f	immersion $f : M \to \tilde{M}$
f_t	parallel hypersurface or tube at distance t from f
M_t	parallel hypersurface at distance t from M
T_xM	tangent space to M at x
$T_x^{\perp}M$	normal space to M at x
TM	tangent bundle of M
T_1M	unit tangent bundle of M
NM	normal bundle of M
BM	unit normal bundle of M
E	normal exponential map $E : NM \to \tilde{M}$
$\tilde{\nabla}_X$	covariant derivative for Levi-Civita connection $\tilde{\nabla}$ of $\tilde{M}$
∇_X	covariant derivative for induced metric on M
D_X	covariant derivative in Euclidean space $\mathbf{R}^n$
$\overrightarrow{\gamma_t}$	velocity vector of curve γ_t
ξ	unit normal vector field on M
$\nabla^{\perp}$	normal connection
$\sigma(X,Y)$	second fundamental form of M
A_ξ	shape operator for normal vector ξ
A	shape operator for an oriented hypersurface
A_t	shape operator of tube f_t or parallel hypersurface M_t
$\lambda(x)$	principal curvature of a hypersurface M at a point x
$T_\lambda(x)$	eigenspace for a principal curvature λ at x
T_λ	principal distribution on M determined by λ
$T_\lambda^{\perp}$	orthogonal complement of T_λ
M/T_λ	space of leaves of principal foliation T_λ
f_λ	focal map determined by principal curvature function λ
K_λ	curvature sphere determined by principal curvature λ
$R(X,Y)Z$	Riemann curvature tensor of M
$\tilde{R}(X,Y)Z$	Riemann curvature tensor of $\tilde{M}$

4. For a real hypersurface M in complex projective space $\mathbf{CP}^n$ or complex hyperbolic space $\mathbf{CH}^n$ (Chapters 6–8).

J	complex structure of $\mathbf{CP}^n$ or $\mathbf{CH}^n$
φ	projection of complex structure J to M
$S^{2n+1}(r)$	$(2n+1)$-dimensional sphere of radius r in $\mathbf{C}^{n+1} = \mathbf{R}^{2n+2}$
$H_1^{2n+1}(r)$	anti-de Sitter space of constant curvature $-1/r^2$
$\mathbf{H}$	anti-de Sitter space $H_1^{2n+1}(r)$
π	projection $\pi : S^{2n+1}(r) \to \mathbf{CP}^n$ or $\pi : H_1^{2n+1}(r) \to \mathbf{CH}^n$

$\mathcal{V}$	span of $\{Jz\}$ at $z \in S^{2n+1}(r)$ or $H_1^{2n+1}(r)$
$\mathcal{V}^\perp$	orthogonal complement of $\mathcal{V}$
X^L	horizontal lift of X to $z \in S^{2n+1}(r)$ or $H_1^{2n+1}(r)$
ξ	unit normal vector field on M
W	$W = -J\xi$, the structure vector field on M
$\mathcal{W}$	1-dimensional distribution spanned by W
$\mathcal{W}^\perp$	holomorphic distribution orthogonal to $\mathcal{W}$
A	shape operator of M
$\mathbf{m}$	trace A
$\sigma(X, Y)$	second fundamental form of submanifold
$\sigma(\mathcal{W}^\perp)$	spectrum of A restricted to $\mathcal{W}^\perp$
S	Ricci tensor S of M
Q^{n-1}	complex quadric in $\mathbf{CP}^n$
$\mathbf{RP}^n$	totally geodesic real projective space in $\mathbf{CP}^n$
$\mathbf{RH}^n$	totally geodesic real hyperbolic space in $\mathbf{CH}^n$
α	$\langle AW, W\rangle$: Hopf principal curvature in Hopf case
β, λ, μ, ν	entries in matrix of A in standard non-Hopf setup
NM	normal bundle of M
BM	unit normal bundle of M
E	normal exponential map $E : NM \to \mathbf{CP}^n$ (or $\mathbf{CH}^n$)
M_μ	parallel hypersurface of M
A_μ	shape operator of M_μ
TM	tangent bundle of M
SM	unit tangent bundle of M
$E_{j\ell}$	matrix with 1 in the (j, ℓ) position and 0 elsewhere
$F_{j\ell}$	matrix $E_{j\ell} - E_{\ell j}$
$G_{h,k}(\mathbf{C})$	complex Grassmann manifold of complex h-planes in $\mathbf{C}^{h+k}$
$\mathcal{C}\ell(V)$	Clifford algebra generated by a vector space V
R_W	structure Jacobi operator $R_W X = R(X, W)W$

5. For a real hypersurface M in quaternionic projective space $\mathbf{HP}^n$ or quaternionic hyperbolic space $\mathbf{HH}^n$ (Chapter 9).

J_1	operator obtained by left multiplication by quaternion i
J_2	operator obtained by left multiplication by quaternion j
J_3	operator obtained by left multiplication by quaternion k
φ_i	projection of J_i to M for $i = 1, 2, 3$
$S^{4n+3}(r)$	$(4n + 3)$-dimensional sphere of radius r in $\mathbf{H}^{n+1} = \mathbf{R}^{4n+4}$
$H_3^{4n+3}(r)$	space form of constant curvature $-1/r^2$ and index 3
$\mathbf{H}$	space form $H_3^{4n+3}(r)$
π	projection $\pi : S^{4n+3}(r) \to \mathbf{HP}^n$ or $\pi : H_3^{4n+3}(r) \to \mathbf{HH}^n$
$\mathcal{V}$	span of $\{J_1z, J_2z, J_3z\}$ at $z \in S^{4n+3}(r)$ or $H_3^{4n+3}(r)$
$\mathcal{V}^\perp$	orthogonal complement of $\mathcal{V}$

$\mathcal{J}$	span of $\{J_1, J_2, J_3\}$ on $\mathbf{HP}^n$ or $\mathbf{HH}^n$
X^L	horizontal lift of X to $z \in S^{4n+3}(r)$ or $H_3^{4n+3}(r)$
ξ	unit normal vector field on M
W_i	$W_i = -J_i\xi$, unit vector field on M for $i = 1, 2, 3$
$\mathcal{W}$	3-dimensional distribution spanned by $\{W_1, W_2, W_3\}$
$\mathcal{W}^\perp$	distribution orthogonal to $\mathcal{W}$
A	shape operator of M
$\mathbf{m}$	trace A
S	Ricci tensor S of M

References

1. N. Abe, K. Hasegawa, A generalization of Cartan's identities for isoparametric hypersurfaces and its applications. Results Math. **52**, 197–210 (2008)
2. U. Abresch, Isoparametric hypersurfaces with four or six distinct principal curvatures. Math. Ann. **264**, 283–302 (1983)
3. T. Adachi, S. Maeda, Curvature-adapted real hypersurfaces in quaternionic space forms. Kodai Math. J. **24**, 98–119 (2001)
4. T. Adachi, S. Maeda, Quaternionic distribution of curvature-adapted real hypersurfaces in a quaternionic hyperbolic space. J. Geom. **75**, 1–14 (2002)
5. T. Adachi, S. Maeda, Isoparametric hypersurfaces with less than four principal curvatures in a sphere. Colloq. Math. **105**, 143–148 (2006)
6. T. Adachi, S. Maeda, S. Udagawa, Ruled real hypersurfaces in a nonflat quaternionic space form. Monatsh. Math. **145**, 179–190 (2005)
7. S.S. Ahn, S.-G. Han, N.-G. Kim, S.-B. Lee, Real hypersurfaces with ξ-parallel Ricci tensor in a complex space form. Commun. Korean Math. Soc. **13**, 825–838 (1998)
8. R. Aiyama, H. Nakagawa, Y.J. Suh, A characterization of real hypersurfaces of type C, D and E of a complex projective space. J. Korean Math. Soc. **27**, 47–67 (1990)
9. M. Alexandrino, Singular Riemannian foliations with sections. Ill. J. Math. **48**, 1163–1182 (2004)
10. L. Alías, A. Brasil, Jr., O. Perdomo, A characterization of quadric constant mean curvature hypersurfaces of spheres. J. Geom. Anal. **18**, 687–703 (2008)
11. S. de Almeida, A. Brasil, Dupin hypersurfaces with constant scalar curvature. 10th School on Differential Geometry (Portuguese) (Belo Horizonte, 1998). Mat. Contemp. **17**, 29–44 (1999)
12. S. de Almeida, F. Brito, Closed 3-dimensional hypersurfaces with constant mean curvature and constant scalar curvature. Duke Math. J. **61**, 195–206 (1990)
13. J.C. Álvarez Paiva, Contact topology, taut immersions, and Hilbert's fourth problem, in *Differential and Symplectic Topology of Knots and Curves*. American Mathematical Society Translations Series 2, vol. 190 (American Mathematical Society, Providence, 1999), pp. 1–21
14. J.C. Álvarez Paiva, Tautness is invariant under Lie sphere transformations. Preprint (2001). see: http://www.math.poly.edu/$\sim$alvarez/pdfs/invariance.pdf
15. E. Artin, *Geometric Algebra* (Wiley-Interscience, New York, 1957)
16. M.F. Atiyah, R. Bott, A. Shapiro, Clifford modules. Topology **3**, 3–38 (1964)
17. C. Baikousssis, S.M. Lyu, Y.J. Suh, Real hypersurfaces in complex space forms with η-parallel curvature tensor. Bull. Greek Math. Soc. **40**, 49–55 (1998)
18. C. Baikousssis, S.M Lyu, Y.J. Suh, Real hypersurfaces in complex space forms with η-recurrent Ricci tensor. Math. J. Toyama Univ. **23**, 41–61 (2000)

T.E. Cecil, P.J. Ryan, *Geometry of Hypersurfaces*, Springer Monographs in Mathematics, DOI 10.1007/978-1-4939-3246-7

19. T. Banchoff, Tightly embedded 2-dimensional polyhedral manifolds. Am. J. Math. **87**, 462–472 (1965)
20. T. Banchoff, The spherical two-piece property and tight surfaces in spheres. J. Differ. Geom. **4**, 193–205 (1970)
21. T. Banchoff, The two-piece property and tight n-manifolds-with-boundary in E^n. Trans. Am. Math. Soc. **161**, 259–267 (1971)
22. T. Banchoff, Tight polyhedral Klein bottles, projective planes and Moebius bands. Math. Ann. **207**, 233–243 (1974)
23. T. Banchoff, W. Kühnel, Equilibrium triangulations of the complex projective plane. Geom. Dedicata **44**, 313–333 (1992)
24. T. Banchoff, W. Kühnel, Tight submanifolds, smooth and polyhedral, in *Tight and Taut Submanifolds*, MSRI Publications, vol. 32 (Cambridge University Press, Cambridge, 1997), pp. 51–118
25. T. Banchoff, W. Kühnel, Tight polyhedral models of isoparametric families, and PL-taut submanifolds. Adv. Geom. **7**, 613–629 (2007)
26. A. Bartoszek, P. Walczak, S. Walczak, Dupin cyclides osculating surfaces. Bull. Braz. Math. Soc. (N.S.) **45**, 179–195 (2014)
27. J. Berndt, Real hypersurfaces with constant principal curvatures in complex hyperbolic space. J. Reine Angew. Math. **395**, 132–141 (1989)
28. J. Berndt, Real hypersurfaces in quaternionic space forms. J. Reine Angew. Math. **419**, 9–26 (1991)
29. J. Berndt, Riemannian geometry of complex two-plane Grassmannians. Rend. Sem. Mat. Univ. Politec. Torino **55**, 19–83 (1997)
30. J. Berndt, Homogeneous hypersurfaces in hyperbolic spaces. Math. Z. **229**, 589–600 (1998)
31. J. Berndt, A note on hypersurfaces in symmetric spaces, in *Proceedings of the Fourteenth International Workshop on Differential Geometry*, vol. 14 (2010), 31–41
32. J. Berndt, M. Brück, Cohomogeneity one actions on hyperbolic spaces. J. Reine Angew. Math. **541**, 209–235 (2001)
33. J. Berndt, S. Console, C. Olmos, *Submanifolds and Holonomy*. Chapman and Hall/CRC Research Notes in Mathematics Series, vol. 434 (Chapman and Hall, Boca Raton, 2003)
34. J. Berndt, J.C. Díaz-Ramos, Real hypersurfaces with constant principal curvatures in complex hyperbolic spaces. J. Lond. Math. Soc. (2) **74**, 778–798 (2006)
35. J. Berndt, J.C. Díaz-Ramos, Real hypersurfaces with constant principal curvatures in the complex hyperbolic plane. Proc. Am. Math. Soc. **135**, 3349–3357 (2007)
36. J. Berndt, J.C. Díaz-Ramos, Homogeneous hypersurfaces in complex hyperbolic spaces. Geom. Dedicata **138**, 129–150 (2009)
37. J. Berndt, Y.J. Suh, Real hypersurfaces in complex two-plane Grassmannians. Monatsh. Math. **127**, 1–14 (1999)
38. J. Berndt, H. Tamaru, Cohomogeneity one actions on noncompact symmetric spaces of rank one. Trans. Am. Math. Soc. **359**, 3425–3438 (2007)
39. J. Berndt, L. Vanhecke, Two natural generalizations of locally symmetric spaces. Differ. Geom. Appl. **2**, 57–80 (1992)
40. L. Biliotti, C. Gorodski, Polar actions on compact rank one symmetric spaces are taut. Math. Z. **255**, 335–342 (2007)
41. D.E. Blair, *Contact Manifolds in Riemannian Geometry*. Lecture Notes in Mathematics, vol. 509 (Springer, Berlin, 1976)
42. W. Blaschke, *Vorlesungen über Differentialgeometrie und geometrische Grundlagen von Einsteins Relativitätstheorie*, vol. 3 (Springer, Berlin, 1929)
43. A. Bobenko, E. Huhnen-Venedey, Curvature line parametrized surfaces and orthogonal coordinate systems; discretization with Dupin cyclides. Geom. Dedicata **159**, 207–237 (2012)
44. G. Bol, *Projektive Differentialgeometrie*, vol. 3 (Vandenhoeck and Ruprecht, Göttingen, 1967)
45. J. Bolton, Transnormal systems. Q. J. Math. Oxford (2) **24**, 385–395 (1973)

46. R. Böning, Curvature surfaces of Hopf hypersurfaces in complex space forms. Manuscripta Math. **87**, 449–458 (1995)
47. A. Borel, F. Hirzebruch, Characteristic classes and homogeneous spaces I. Am. J. Math. **80**, 458–538 (1958)
48. K. Borsuk, Sur la courbure totales des courbes. Ann. de la Soc. Math. Pol. **20**, 251–265 (1947)
49. R. Bott, H. Samelson, Applications of the theory of Morse to symmetric spaces. Am. J. Math. **80**, 964–1029 (1958). Corrections in **83**, 207–208 (1961)
50. A. Brasil, Jr., A. Gervasio, O. Palmas, Complete hypersurfaces with constant scalar curvature in spheres. Monatsh. Math. **161**, 369–380 (2010)
51. S. Buyske, Lie sphere transformations and the focal sets of certain taut immersions, Trans. Am. Math. Soc. **311**, 117–133 (1989)
52. E. Cartan, *Théorie des groupes finis et continus et la géométrie différentielle traitées par la méthode du repère mobile* (Gauthiers-Villars, Paris, 1937)
53. E. Cartan, Familles de surfaces isoparamétriques dans les espaces à courbure constante. Annali di Mat. **17**, 177–191 (1938)
54. E. Cartan, Sur des familles remarquables d'hypersurfaces isoparamétriques dans les espaces sphériques. Math. Z. **45**, 335–367 (1939)
55. E. Cartan, *Sur quelque familles remarquables d'hypersurfaces* (C.R. Congrès Math. Liège, 1939), pp. 30–41
56. E. Cartan, Sur des familles d'hypersurfaces isoparamétriques des espaces sphériques à 5 et à 9 dimensions. Revista Univ. Tucuman, Serie A **1**, 5–22 (1940)
57. E. Cartan, *Leçons sur la géométrie des espaces de Riemann* (Gauthiers-Villars, Paris, 1946)
58. E. Cartan, *The Theory of Spinors* (Hermann, Paris, 1966). Reprinted by Dover, New York, 1981
59. S. Carter, N. Mansour, A. West, Cylindrically taut immersions. Math. Ann. **261**, 133–139 (1982)
60. S. Carter, Z. Şentürk, The space of immersions parallel to a given immersion. J. Lond. Math. Soc. **50**, 404–416 (1994)
61. S. Carter, A. West, Tight and taut immersions. Proc. Lond. Math. Soc. **25**, 701–720 (1972)
62. S. Carter, A. West, Totally focal embeddings. J. Differ. Geom. **13**, 251–261 (1978)
63. S. Carter, A. West, Totally focal embeddings: special cases. J. Differ. Geom. **16**, 685–697 (1981)
64. S. Carter, A. West, A characterisation of isoparametric hypersurfaces in spheres. J. Lond. Math. Soc. **26**, 183–192 (1982)
65. S. Carter, A. West, Convexity and cylindrical two-piece properties. Ill. J. Math. **29**, 39–50 (1985)
66. S. Carter, A. West, Isoparametric systems and transnormality. Proc. Lond. Math. Soc. **51**, 520–542 (1985)
67. S. Carter, A. West, Generalized Cartan polynomials. J. Lond. Math. Soc. **32**, 305–316 (1985)
68. S. Carter, A. West, Isoparametric and totally focal submanifolds. Proc. Lond. Math. Soc. **60**, 609–624 (1990)
69. A. Cayley, On the cyclide. Q. J. Pure Appl. Math. **12** (1873), 148–165. See also Collected Papers, vol. 9, pp. 64–78
70. T.E. Cecil, A characterization of metric spheres in hyperbolic space by Morse theory. Tôhoku Math. J. **26**, 341–351 (1974)
71. T.E. Cecil, Geometric applications of critical point theory to submanifolds of complex space. Nagoya Math. J. **55**, 5–31 (1974)
72. T.E. Cecil, Taut immersions of non-compact surfaces into a Euclidean 3-space. J. Differ. Geom. **11**, 451–459 (1976)
73. T.E. Cecil, On the Lie curvatures of Dupin hypersurfaces. Kodai Math. J. **13**, 143–153 (1990)
74. T.E. Cecil, Lie sphere geometry and Dupin submanifolds, in *Geometry and Topology of Submanifolds III* (World Scientific, River Edge, 1991), pp. 90–107
75. T.E. Cecil, Focal points and support functions in affine differential geometry. Geom. Dedicata **50**, 291–300 (1994)

76. T.E. Cecil, Taut and Dupin submanifolds, in *Tight and Taut Submanifolds*, MSRI Publications, vol. 32 (Cambridge University Press, Cambridge, 1997), pp. 135–180
77. T.E. Cecil, *Lie Sphere Geometry*, 2nd edn. (Springer, New York, 2008)
78. T.E. Cecil, Isoparametric and Dupin hypersurfaces. SIGMA: Symmetry, Integrability Geom. Methods Appl. **4**, Paper 062, 28 pp. (2008)
79. T.E. Cecil, S.-S. Chern, Tautness and Lie sphere geometry. Math. Ann. **278**, 381–399 (1987)
80. T.E. Cecil, S.-S. Chern, Dupin submanifolds in Lie sphere geometry, in *Differential Geometry and Topology, Proceedings Tianjin 1986–1987*. Lecture Notes in Mathematics, vol. 1369 (Springer, Berlin/New York, 1989), pp. 1–48
81. T.E. Cecil, Q.-S. Chi, G.R. Jensen, Isoparametric hypersurfaces with four principal curvatures. Ann. Math. **166**, 1–76 (2007)
82. T.E. Cecil, Q.-S. Chi, G.R. Jensen, Dupin hypersurfaces with four principal curvatures II. Geom. Dedicata **128**, 55–95 (2007)
83. T.E. Cecil, Q.-S. Chi, G.R. Jensen, Classifications of Dupin hypersurfaces, in *Pure and Applied Differential Geometry, PADGE, 2007*, ed. by F. Dillen, I. Van de Woestyne (Shaker Verlag, Aachen, 2007), pp. 48–56
84. T.E. Cecil, Q.-S. Chi, G.R. Jensen, On Kuiper's question whether taut submanifolds are algebraic. Pac. J. Math. **234**, 229–248 (2008)
85. T.E. Cecil, G.R. Jensen, Dupin hypersurfaces with three principal curvatures. Invent. Math. **132**, 121–178 (1998)
86. T.E. Cecil, G.R. Jensen, Dupin hypersurfaces with four principal curvatures. Geom. Dedicata **79**, 1–49 (2000)
87. T.E. Cecil, M. Magid, L. Vrancken, An affine characterization of the Veronese surface. Geom. Dedicata **57**, 55–71 (1995)
88. T.E. Cecil, P.J. Ryan, Focal sets of submanifolds. Pac. J. Math. **78**, 27–39 (1978)
89. T.E. Cecil, P.J. Ryan, Focal sets, taut embeddings and the cyclides of Dupin. Math. Ann. **236**, 177–190 (1978)
90. T.E. Cecil, P.J. Ryan, Distance functions and umbilical submanifolds of hyperbolic space. Nagoya Math. J. **74**, 67–75 (1979)
91. T.E. Cecil, P.J. Ryan, Tight and taut immersions into hyperbolic space. J. Lond. Math. Soc. **19**, 561–572 (1979)
92. T.E. Cecil, P.J. Ryan, Conformal geometry and the cyclides of Dupin. Can. J. Math. **32**, 767–782 (1980)
93. T.E. Cecil, P.J. Ryan, *Tight Spherical Embeddings*. Lecture Notes in Mathematics, vol. 838 (Springer, Berlin/New York, 1981), pp. 94–104
94. T.E. Cecil, P.J. Ryan, Focal sets and real hypersurfaces in complex projective space. Trans. Am. Math. Soc. **269**, 481–499 (1982)
95. T.E. Cecil, P.J. Ryan, *Tight and Taut Immersions of Manifolds*. Research Notes in Mathematics, vol. 107 (Pitman, London, 1985)
96. T.E. Cecil, P.J. Ryan, The principal curvatures of the monkey saddle. Am. Math. Mon. **93**, 380–382 (1986)
97. M. Cezana, K. Tenenblat, A characterization of Laguerre isoparametric hypersurfaces of the Euclidian space. Monatsh. Math. **175**, 187–194 (2014)
98. B.-Y. Chen, *Geometry of Submanifolds* (Marcel Dekker, New York, 1973)
99. C.S. Chen, On tight isometric immersions of codimension two. Am. J. Math. **94**, 974–990 (1972)
100. C.S. Chen, More on tight isometric immersions of codimension two. Proc. Am. Math. Soc. **40**, 545–553 (1973)
101. C.S. Chen, Tight embedding and projective transformation. Am. J. Math. **101**, 1083–1102 (1979)
102. S.-S. Chern, Curves and surfaces in Euclidean space, in *Studies in Global Geometry and Analysis*. MAA Studies in Mathematics, vol. 4 (1967), 16–56, Prentice-Hall, Englewood Cliffs, NJ

103. S.-S. Chern, R.K. Lashof, On the total curvature of immersed manifolds I. Am. J. Math. **79**, 306–318 (1957)
104. S.-S. Chern, R.K. Lashof, On the total curvature of immersed manifolds II. Mich. Math. J. **5**, 5–12 (1958)
105. Q.-S. Chi, Isoparametric hypersurfaces with four principal curvatures revisited. Nagoya Math. J. **193**, 129–154 (2009)
106. Q.-S. Chi, A note on the paper "Isoparametric hypersurfaces with four principal curvatures", Hongyou Wu Memorial Volume. Pac. J. Appl. Math. **3**, 127–134 (2011)
107. Q.-S. Chi, Isoparametric hypersurfaces with four principal curvatures II. Nagoya Math. J. **204**, 1–18 (2011)
108. Q.-S. Chi, A new look at Condition A. Osaka J. Math. **49**, 133–166 (2012)
109. Q.-S. Chi, Isoparametric hypersurfaces with four principal curvatures III. J. Differ. Geom. **94**, 469–504 (2013)
110. Q.-S. Chi, Taut submanifolds are algebraic. Preprint (2014) [arXiv:1102.1704v3 [math.DG]]
111. J.T. Cho, Real hypersurfaces of a complex hyperbolic space satisfying a pointwise nullity condition. Indian J. Pure Appl. Math. **31**, 265–276 (2000)
112. J.T. Cho, U.-H. Ki, Jacobi operators on real hypersurfaces of a complex projective space. Tsukuba J. Math. **22**, 145–156 (1998)
113. J.T. Cho, U.-H. Ki, Real hypersurfaces of a complex projective space in terms of the Jacobi operators. Acta Math. Hungar. **80**, 155–167 (1998)
114. J.T. Cho, U.-H. Ki, Real hypersurfaces of a complex projective space satisfying a pointwise nullity condition. Tsukuba J. Math. **23**, 279–291 (1999)
115. J.T. Cho, U.-H. Ki, Real hypersurfaces in complex space forms with Reeb flow symmetric structure Jacobi operator. Can. Math. Bull. **51**, 359–371 (2008)
116. J.T. Cho, M. Kimura, Ricci solitons of compact real hypersurfaces in Kähler manifolds. Math. Nachr. **284**, 1385–1393 (2011)
117. J.T. Cho, L. Vanhecke, Hopf hypersurfaces of D'Atri- and C-type in a complex space form. Rend. Mat. Appl. (7) **18**(1998), 601–613 (1999)
118. Y.-W. Choe, Characterizations of certain real hypersurfaces of a complex space form. Nihonkai Math. J. **6**, 97–114 (1995)
119. U. Christ, Homogeneity of equifocal submanifolds. J. Differ. Geom. **62**, 1–15 (2002)
120. L. Conlon, *Differentiable Manifolds: A First Course*, 1st edn. (Birkhäuser, Boston, 1993)
121. S. Console, C. Olmos, Clifford systems, algebraically constant second fundamental form and isoparametric hypersurfaces. Manuscripta Math. **97**, 335–342 (1998)
122. J. Dadok, Polar actions induced by actions of compact Lie groups. Trans. Am. Math. Soc. **288**, 125–137 (1985)
123. J. Dadok, V. Kac, Polar representations. J. Algebra **92**, 504–524 (1985)
124. M. Dajczer, L. Florit, R. Tojeiro, Reducibility of Dupin submanifolds. Ill. J. Math. **49**, 759–791 (2005)
125. G. Darboux, *Leçons sur la théorie générale des surfaces*, 2nd edn. (Gauthiers-Villars, Paris, 1941)
126. J.E. D'Atri, Certain isoparametric families of hypersurfaces in symmetric spaces. J. Differ. Geom. **14**, 21–40 (1979)
127. W. Degen, Generalized cyclides for use in CAGD, in *Computer-Aided Surface Geometry and Design (Bath, 1990)* The Institute of Mathematics and its Applications Conference Series New Series, vol. 48 (Oxford University Press, New York, 1994), pp. 349–363
128. S. Deshmukh, Real hypersurfaces of a complex space form. Proc. Indian Acad. Sci. (Math. Sci.) **122**, 629–634 (2012)
129. S. Deshmukh, M.A. Al-Gwaiz, Hypersurfaces of the two-dimensional complex projective space. Nihonkai Math. J. **3**, 1–7 (1992)
130. R. Deszcz, On pseudosymmetric spaces. Bull. Soc. Math. Belg. Sér. A **44**(1), 1–34 (1992)
131. J.C. Díaz-Ramos, M. Domínguez-Vázquez, Non-Hopf real hypersurfaces with constant principal curvatures in complex space forms. Indiana Univ. Math. J. **60**, 859–882 (2011)

132. J.C. Díaz-Ramos, M. Domínguez-Vázquez, C. Vidal-Castiñeira, Real hypersurfaces with two principal curvatures in complex projective and hyperbolic planes. Preprint (2013)
133. M. Domínguez-Vázquez, *Isoparametric Foliations and Polar Actions on Complex Space Forms*. Publ. Dto. Geom. Topol., vol. 126 (Universidade de Santiago de Compostela, 2013)
134. M. Domínguez-Vázquez, Isoparametric foliations on complex projective spaces. Trans. Am. Math. Soc. doi: 10.1090/S0002-9947-2014-06415-5
135. J. Dorfmeister, E. Neher, An algebraic approach to isoparametric hypersurfaces I, II. Tôhoku Math. J. **35**, 187–224, 225–247 (1983)
136. J. Dorfmeister, E. Neher, Isoparametric triple systems of algebra type. Osaka J. Math. **20**, 145–175 (1983)
137. J. Dorfmeister, E. Neher, Isoparametric triple systems of FKM-type I. Abh. Math. Sem. Hamburg **53**, 191–216 (1983)
138. J. Dorfmeister, E. Neher, Isoparametric triple systems of FKM-type II. Manuscripta Math. **43**, 13–44 (1983)
139. J. Dorfmeister, E. Neher, Isoparametric hypersurfaces, case $g = 6, m = 1$. Commun. Algebra **13**, 2299–2368 (1985)
140. J. Dorfmeister, E. Neher, Isoparametric triple systems with special Z-structure. Algebras Groups Geom. **7**, 21–94 (1990)
141. L. Druoton, L. Fuchs, L. Garnier, R. Langevin, The non-degenerate Dupin cyclides in the space of spheres using geometric algebra. Adv. Appl. Clifford Algebr. **24**, 515–532 (2014)
142. L. Druoton, R. Langevin, L. Garnier, Blending canal surfaces along given circles using Dupin cyclides. Int. J. Comput. Math. **91**, 641–660 (2014)
143. C. Dupin, *Applications de géométrie et de méchanique: à la marine, aux points et chausées, etc.* (Bachelier, Paris, 1822)
144. J. Eells, N. Kuiper, Manifolds which are like projective planes. Publ. Math. I.H.E.S. **14**, 128–222 (1962)
145. S. Eilenberg, N. Steenrod, *Foundations of Algebraic Topology* (Princeton University Press, Princeton, 1952)
146. L. Eisenhart, *A Treatise on the Differential Geometry of Curves and Surfaces* (Ginn, Boston, 1909)
147. J.-H. Eschenburg, V. Schroeder, Tits distance of Hadamard manifolds and isoparametric hypersurfaces. Geom. Dedicata **40**, 97–101 (1991)
148. H. Ewert, Über isoparametrische Hyperflächen in Räumen konstanter Krümmung. Diplomarbeit, Universität Köln (1994)
149. F. Fang, Multiplicities of principal curvatures of isoparametric hypersurfaces. Preprint, Max Planck Institut für Mathematik (1996)
150. F. Fang, On the topology of isoparametric hypersurfaces with four distinct principal curvatures. Proc. Am. Math. Soc. **127**, 259–264 (1999)
151. F. Fang, Topology of Dupin hypersurfaces with six principal curvatures. Math. Z. **231**, 533–555 (1999)
152. F. Fang, Equifocal hypersurfaces in symmetric spaces. Chin. Ann. Math. Ser. B **21**, 473–478 (2000)
153. I. Fary, Sur la courbure totale d'une courbe gauche faisant un noeud. Bull. Soc. Math. France **77**, 128–138 (1949)
154. H. Federer, *Geometric Measure Theory*, Die Grundlehren der mathematischen Wissenschaften, vol. 153 (Springer, New York, 1969)
155. W. Fenchel, Über die Krümmung und Windung geschlossener Raumkurven. Math. Ann. **101**, 238–252 (1929)
156. E.V. Ferapontov, Dupin hypersurfaces and integrable hamiltonian systems of hydrodynamic type, which do not possess Riemann invariants. Differ. Geom. Appl. **5**, 121–152 (1995)
157. E.V. Ferapontov, Isoparametric hypersurfaces in spheres, integrable nondiagonalizable systems of hydrodynamic type, and *N*-wave systems. Differ. Geom. Appl. **5**, 335–369 (1995)
158. M.L. Ferro, L.A. Rodrigues, K. Tenenblat, On a class of Dupin hypersurfaces in $\mathbf{R}^5$ with nonconstant Lie curvature. Geom. Dedicata **169**, 301–321 (2014)

159. D. Ferus, Notes on isoparametric hypersurfaces. in *Escola de geometria differencial* (Univ. Estadual de Campinas, Campinas, 1980)
160. D. Ferus, H. Karcher, H.-F. Münzner, Cliffordalgebren und neue isoparametrische Hyperflächen. Math. Z. **177**, 479–502 (1981) (see also an English translation by T.E. Cecil (2011) [arXiv:1112.2780v1 [mathDG]])
161. K. Fladt, A. Baur, *Analytische Geometrie spezieller Flächen und Raumkurven* (Friedr. Vieweg and Sohn, Braunschweig, 1975)
162. H. Freudenthal, Oktaven, Ausnahmegruppen und Oktavengeometrie. Geom. Dedicata **19**, 7–63 (1985)
163. G. Friedel, Les états mésomorphes de la matière. Ann. de Physique **18**, 273–474 (1923)
164. S. Fujii, Homogeneous isoparametric hypersurfaces in spheres with four distinct principal curvatures and moment maps. Tôhoku Math. J. **62**, 191–213 (2010)
165. S. Fujii, H. Tamaru, Moment maps and isoparametric hypersurfaces in spheres—an introduction, in *Proceedings of the 15th International Workshop on Differential Geometry and the 4th KNUGRG-OCAMI Differential Geometry Workshop [Volume 15]* (Natl. Inst. Math. Sci. (NIMS), Taejon, 2011), pp. 19–27
166. L. Garnier, H. Barki, S. Foufou, L. Puec, Computation of Yvon-Villarceau circles on Dupin cyclides and construction of circular edge right triangles on tori and Dupin cyclides. Comput. Math. Appl. **68**(12, part A), 1689–1709 (2014)
167. J. Ge, H. Ma, Anisotropic isoparametric hypersurfaces in Euclidean spaces. Ann. Global Anal. Geom. **41**, 347–355 (2012)
168. J. Ge, Z.-Z. Tang, Chern conjecture and isoparametric hypersurfaces, in *Differential Geometry*. Advanced Lectures in Mathematics (ALM), vol. 22 (International Press, Somerville, 2012), pp. 49–60
169. J. Ge, Z.-Z. Tang, Isoparametric functions and exotic spheres. J. Reine Angew. Math. **683**, 161–180 (2013)
170. J. Ge, Z.-Z. Tang, Geometry of isoparametric hypersurfaces in Riemannian manifolds. Asian J. Math. **18**, 117–125 (2014)
171. J. Ge, Z.-Z. Tang, W. Yan, A filtration for isoparametric hypersurfaces in Riemannian manifolds. J. Math. Soc. Japan. **67**, 1179–1212 (2015)
172. L. Geatti, C. Gorodski, Polar orthogonal representations of real reductive algebraic groups. J. Algebra **320**, 3036–3061 (2008)
173. A. Ghosh, Certain results of real hypersurfaces in a complex space form. Glasg. Math. J. **54**, 1–8 (2012)
174. A. Giunta, C. Sanchez, Projective spaces in the algebraic sets of planar normal sections of homogeneous isoparametric submanifolds. Rev. Un. Mat. Argent. **55**, 139–154 (2014)
175. A. Goetz, *Introduction to Differential Geometry* (Addison-Wesley, Reading, 1970)
176. W.M. Goldman, *Complex Hyperbolic Geometry* (Oxford University Press, New York, 1999)
177. C. Gorodski, Taut submanifolds. Rev. Un. Mat. Argent. **47**, 61–72 (2006)
178. C. Gorodski, Taut representations of compact simple Lie groups. Ill. J. Math. **52**, 121–143 (2008)
179. C. Gorodski, N. Gusevskii, Complete minimal hypersurfaces in complex hyperbolic space. Manuscripta Math. **103**, 221–240 (2000)
180. C. Gorodski, G. Thorbergsson, Cycles of Bott-Samelson type for taut representations. Ann. Global Anal. Geom. **21**, 287–302 (2002)
181. C. Gorodski, G. Thorbergsson, The classification of taut irreducible representations. J. Reine Angew. Math. **555**, 187–235 (2003)
182. T. Gotoh, Geodesic hyperspheres in complex projective space. Tsukuba J. Math. **18**, 207–215 (1994)
183. M. Greenberg, *Lectures on Algebraic Topology* (Benjamin, New York, 1967)
184. K. Grove, S. Halperin, Dupin hypersurfaces, group actions, and the double mapping cylinder. J. Differ. Geom. **26**, 429–459 (1987)
185. K. Grove, S. Halperin, Elliptic isometries, condition (C) and proper maps. Arch. Math. (Basel) **56**, 288–299 (1991)

186. L.C. Grove, C.T. Benson, *Finite Reflection Groups*. Graduate Texts in Mathematics, 2nd edn., vol. 99 (Springer, New York, 1985)
187. J. Hahn, Isoparametric hypersurfaces in the pseudo-Riemannian space forms. Math. Z. **187**, 195–208 (1984)
188. T. Hamada, On real hypersurfaces of a complex projective space with η-recurrent second fundamental tensor. Nihonkai Math. J. **6**, 153–163 (1995)
189. C. Harle, Isoparametric families of submanifolds. Bol. Soc. Brasil Mat. **13**, 35–48 (1982)
190. T. Hawkins, Line geometry, differential equations and the birth of Lie's theory of groups, in *The History of Modern Mathematics*, vol. I (Academic Press, San Diego, 1989), pp. 275–327
191. J. Hebda, Manifolds admitting taut hyperspheres. Pac. J. Math. **97**, 119–124 (1981)
192. J. Hebda, Blaschke manifolds with taut geodesics. Duke Math. J. **48**, 85–91 (1981)
193. J. Hebda, Some new tight embeddings which cannot be made taut. Geom. Dedicata **17**, 49–60 (1984)
194. J. Hebda, The possible cohomology of certain types of taut submanifolds. Nagoya Math. J. **111**, 85–97 (1988)
195. E. Heintze, X. Liu, Homogeneity of infinite dimensional isoparametric submanifolds. Ann. Math. **149**, 149–181 (1999)
196. E. Heintze, C. Olmos, G. Thorbergsson, Submanifolds with constant principal curvatures and normal holonomy groups. Int. J. Math. **2**, 167–175 (1991)
197. G. Henry, J. Petean, Isoparametric hypersurfaces and metrics of constant scalar curvature. Asian J. Math. **18**, 53–67 (2014)
198. U. Hertrich-Jeromin, *Introduction to Möbius Differential Geometry*. London Mathematical Society Lecture Note Series, vol. 300 (Cambridge University Press, Cambridge, 2003)
199. D. Hilbert, S. Cohn-Vossen, *Geometry and the Imagination* (Chelsea, New York, 1952)
200. M. Hirsch, *Differential Topology* (Springer, New York/Heidelberg/Berlin, 1976)
201. H. Hopf, Systeme symmetrischer Bilinearformen und euklidsche Modelle der projektiven Räume. Vierteljahrschrift der Naturforschenden Gesellschaft in Zürich **85**, 165–177 (1940)
202. W.-Y. Hsiang, H.B. Lawson, Jr., Minimal submanifolds of low cohomogeneity. J. Differ. Geom. **5**, 1–38 (1971)
203. W.-Y. Hsiang, R. Palais, C.-L. Terng, The topology of isoparametric submanifolds. J. Differ. Geom. **27**, 423–460 (1988)
204. Z. Hu, D. Li, Möbius isoparametric hypersurfaces with three distinct principal curvatures. Pac. J. Math. **232**, 289–311 (2007)
205. Z. Hu, H.-Z. Li, Classification of Möbius isoparametric hypersurfaces in S^4. Nagoya Math. J. **179**, 147–162 (2005)
206. Z. Hu, H.-Z. Li, C.-P. Wang, Classification of Möbius isoparametric hypersurfaces in S^5. Monatsh. Math. **151**, 201–222 (2007)
207. Z. Hu, X.X. Li, S. Zhai, On the Blaschke isoparametric hypersurfaces in the unit sphere with three distinct Blaschke eigenvalues. Sci. China Math. **54**, 2171–2194 (2011)
208. Z. Hu, X. Tian, On Möbius form and Möbius isoparametric hypersurfaces. Acta Math. Sin. **25**, 2077–2092 (2009)
209. Z. Hu, S. Zhai, Classification of Möbius isoparametric hypersurfaces in the unit six-sphere. Tôhoku Math. J. **60**, 499–526 (2008)
210. Z. Hu, S. Zhai, Möbius isoparametric hypersurfaces with three distinct principal curvatures, II. Pac. J. Math. **249**, 343–370 (2011)
211. A. Hurwitz, Über die Komposition der quadratischen Formen. Math. Ann. **88**, 1–25 (1923)
212. D. Husemöller, *Fibre Bundles*. Graduate Texts in Mathematics, 2nd edn., vol. 20 (Springer, New York, 1975)
213. K. Ikuta, Real hypersurface of a complex projective space. J. Korean Math. Soc. **36**, 725–736 (1999)
214. S. Immervoll, Smooth generalized quadrangles and isoparametric hypersurfaces of Clifford type. Forum Math. **14**, 877–899 (2002)
215. S. Immervoll, Isoparametric hypersurfaces and smooth generalized quadrangles. J. Reine Angew. Math. **554**, 1–17 (2003)

216. S. Immervoll, A characterization of isoparametric hypersurfaces of Clifford type. Beiträge Algebra Geom. **45**, 697–702 (2004)
217. S. Immervoll, The geometry of isoparametric hypersurfaces with four distinct principal curvatures in spheres. Adv. Geom. **5**, 293–300 (2005)
218. S. Immervoll, On the classification of isoparametric hypersurfaces with four distinct principal curvatures in spheres. Ann. Math. **168**, 1011–1024 (2008)
219. J.-I. Inoguchi, Real hypersurfaces in complex space forms with pseudo-parallel Ricci operator. Differ. Geom. Dyn. Syst. **14**, 69–89 (2012)
220. T.A. Ivey, Surfaces with orthogonal families of circles. Proc. Am. Math. Soc. **123**, 865–872 (1995)
221. T.A. Ivey, A d'Alembert formula for Hopf hypersurfaces. Results Math. **60**, 293–309 (2011)
222. T.A. Ivey, P.J. Ryan, Hopf hypersurfaces of small Hopf principal curvature in $\mathbf{CH}^2$. Geom. Dedicata **141**, 147–161 (2009)
223. T.A. Ivey, P.J. Ryan, The structure Jacobi operator for hypersurfaces in $\mathbf{CP}^2$ and $\mathbf{CH}^2$. Results Math. **56**, 473–488 (2009)
224. T.A. Ivey, P.J. Ryan, The *-Ricci tensor for hypersurfaces in $\mathbf{CP}^n$ and $\mathbf{CH}^n$. Tokyo J. Math. **34**, 445–471 (2011)
225. T.A. Ivey, P.J. Ryan, Hypersurfaces in $\mathbf{CP}^2$ and $\mathbf{CH}^2$ with two distinct principal curvatures. Glasg. Math. J. doi: 10.1017/S0017089515000105
226. K. Iwata, Classification of compact transformation groups on cohomology quaternion projective spaces with codimension one orbits. Osaka J. Math. **15**, 475–508 (1978)
227. S. Izumiya, D. Pei, M.C. Romero Fuster, M. Takahashi, The horospherical geometry of submanifolds in hyperbolic space. J. Lond. Math. Soc. **71**, 779–800 (2005)
228. S. Izumiya, D. Pei, T. Sano, Singularities of hyperbolic Gauss maps. Proc. Lond. Math. Soc. **86**, 485–512 (2003)
229. G.R. Jensen, Dupin hypersurfaces in Lie sphere geometry (2014) [arXiv:1405.5198v1 [math.DG]]
230. G.R. Jensen, E. Musso, L. Nicolodi, Surfaces in classical geometries by moving frames (in preparation)
231. X. Jia, Role of moving planes and moving spheres following Dupin cyclides. Comput. Aided Geom. Design **31**, 168–181 (2014)
232. P.J. Kahn, Codimension-one imbedded spheres. Invent. Math. **10**, 44–56 (1970)
233. G. Kaimakamis, K. Panagiotidou, Real hypersurfaces in a non-flat complex space form with Lie recurrent structure Jacobi operator. Bull. Korean Math. Soc. **50**, 2089–2101 (2013)
234. N. Kamran, K. Tenenblat, Laplace transformation in higher dimensions. Duke Math. J. **84**, 237–266 (1996)
235. J. Kaneko, Wave equation and Dupin hypersurface. Mem. Fac. Sci. Kyushu Univ. Ser. A **40**, 51–55 (1986)
236. J. Kaneko, Wave equation and isoparametric hypersurfaces, in *Geometry of Manifolds (Matsumoto, 1988)*. Perspect. Math. vol. 8 (Academic, Boston, 1989), pp. 165–180
237. E.-H. Kang, U.-H. Ki, Real hypersurfaces satisfying $\nabla_\xi S = 0$ of a complex space form. Bull. Korean Math. Soc. **35**, 819–835 (1998)
238. U.-H. Ki, Cyclic-parallel real hypersurfaces of a complex space form. Tsukuba J. Math. **12**, 259–268 (1988)
239. U.-H. Ki, Real hypersurfaces with parallel Ricci tensor of a complex space form. Tsukuba J. Math. **13**, 73–81 (1989)
240. U.-H. Ki, H.-J. Kim, A note on real hypersurfaces of a complex space form. Bull. Korean Math. Soc. **26**, 69–74 (1989)
241. U.-H. Ki, H.-J. Kim, A.-A. Lee, The Jacobi operator of real hypersurfaces in a complex space form. Commun. Korean Math. Soc. **13**, 545–560 (1998)
242. U.-H. Ki, S.J. Kim, S.-B. Lee, The structure Jacobi operator on real hypersurfaces in a nonflat complex space form. Bull. Korean Math. Soc. **42**, 337–358 (2005)
243. U.-H. Ki, H. Kurihara, Real hypersurfaces with cyclic-parallel structure Jacobi operators in a nonflat complex space form. Bull. Aust. Math. Soc. **81**, 260–273 (2010)

244. U.-H. Ki, H. Kurihara, Jacobi operators along the structure flow on real hypersurfaces in a nonflat complex space form II. Bull. Korean Math. Soc. **48**, 1315–1327 (2011)
245. U.-H. Ki, H. Kurihara, S. Nagai, R. Takagi, Characterizations of real hypersurfaces of type (A) in a complex space form in terms of the structure Jacobi operator. Toyama Math. J. **32**, 5–23 (2009)
246. U.-H. Ki, H. Kurihara, R. Takagi, Jacobi operators along the structure flow on real hypersurfaces in a nonflat complex space form. Tsukuba J. Math. **33**, 39–56 (2009)
247. U.-H. Ki, H. Liu, Some characterizations of real hypersurfaces of type (A) in a nonflat complex space form. Bull. Korean Math. Soc. **44**, 157–172 (2007)
248. U.-H. Ki, S. Nagai, Real hypersurfaces of a nonflat complex space form in terms of the Ricci tensor. Tsukuba J. Math. **29**, 511–532 (2005)
249. U.-H. Ki, S. Nagai, The Ricci tensor and structure Jacobi operator of real hypersurfaces in a complex projective space. J. Geom. **94**, 123–142 (2009)
250. U.-H. Ki, S. Nagai, R. Takagi, Structure Jacobi operator of real hypersurfaces with constant scalar curvature in a nonflat complex space form. Tokyo J. Math. **30**, 441–454 (2010)
251. U.-H. Ki, S. Nagai, R. Takagi, The structure vector field and structure Jacobi operator of real hypersurfaces in nonflat complex space forms. Geom. Dedicata **149**, 161–176 (2010)
252. U.-H. Ki, J.D. Pérez, F.G. Santos, Y.J. Suh, Real hypersurfaces in complex space forms with ξ-parallel Ricci tensor and structure Jacobi operator. J. Korean Math. Soc. **44**, 307–326 (2007)
253. U.-H. Ki, Y.J. Suh, On real hypersurfaces of a complex space form. Math. J. Okayama Univ. **32**, 207–221 (1990)
254. U.-H. Ki, Y.J. Suh, Characterizations of some real hypersurfaces in $P_n\mathbf{C}$ in terms of Ricci tensor. Nihonkai Math. J. **3**, 133–162 (1992)
255. U.-H. Ki, Y.J. Suh, Real hypersurfaces in complex hyperbolic space with commuting Ricci tensor. Kyungpook Math. J. **48**, 433–442 (2008)
256. U.-H. Ki, Y.J. Suh, J.D. Pérez, Real hypersurfaces of type A in quaternionic projective space. Int. J. Math. Math. Sci. **20**, 115–122 (1997)
257. H.S. Kim, I.-B. Kim, R. Takagi, Extrinsically homogeneous real hypersurfaces with three principal curvatures in $\mathbf{H}_n(\mathbb{C})$. Osaka J. Math. **41**, 853–863 (2004)
258. H.S. Kim, J.-H. Kim, Y.-S. Pyo, Certain characterizations of real hypersurfaces of Type A in a complex space form. Tsukuba J. Math. **23**, 369–381 (1999)
259. H.S. Kim, Y.-S. Pyo (Y.S. Pho), Certain real hypersurfaces of a complex space form II. Nihonkai Math. J. **8**, 155–170 (1997)
260. H.S. Kim, P.J. Ryan, A classification of pseudo-Einstein real hypersurfaces in $\mathbf{CP}^2$. Differ. Geom. Appl. **26**, 106–112 (2008)
261. I.-B. Kim, On Berndt subgroups and extrinsically homogeneous real hypersurfaces in complex hyperbolic spaces, in *Proceedings of the Sixth International Workshop on Differential Geometry (Taegu, 2001)* (Kyungpook National University, Taegu, 2002), pp. 45–59
262. I.-B. Kim, K.H. Kim, W.H. Sohn, Characterization of real hypersurfaces in a complex space form. Can. Math. Bull. **50**, 97–104 (2007)
263. I.-B. Kim, H.J. Park, H. Song, Ricci-pseudo-symmetric real hypersurfaces in complex space forms. Nihonkai Math. J. **18**, 1–9 (2007)
264. I.-B. Kim, T. Takahashi, Isoparametric hypersurfaces in a space form and metric connections. Tsukuba J. Math. **21**, 15–28 (1997) (erratum 185–185)
265. J.J. Kim, Y.-S. Pyo, Real hypersurfaces with parallely cyclic condition of a complex space form. Bull. Korean Math. Soc. **28**, 1–10 (1991)
266. N.-G. Kim, Characterization of real hypersurfaces of type A in a nonflat complex space form whose structure Jacobi operator is ξ-parallel. Honam Math. J. **31**, 185–201 (2009)
267. N.-G. Kim, U.-H. Ki, ξ-parallel structure Jacobi operators of real hypersurfaces in nonflat complex space form. Honam Math. J. **28**, 573–589 (2006)
268. N.-G. Kim, U.-H. Ki, H. Kurihara, Characterizations of real hypersurfaces of type A in a complex space form used by the ξ-parallel structure Jacobi operator. Honam Math. J. **30**, 535–550 (2008)

269. U.K. Kim, Nonexistence of Ricci-parallel real hypersurfaces in $P_2\mathbb{C}$ or $H_2\mathbb{C}$. Bull. Korean Math. Soc. **41**, 699–708 (2004)
270. M. Kimura, Real hypersurfaces and complex submanifolds in complex projective space. Trans. Am. Math. Soc. **296**, 137–149 (1986)
271. M. Kimura, Sectional curvature of holomorphic planes on a real hypersurface in $P^n(\mathbf{C})$. Math. Ann. **276**, 487–497 (1987)
272. M. Kimura, Some real hypersurfaces of a complex projective space. Saitama Math. J. **5**, 1–5 (1987). Correction in **10**, 33–34 (1992)
273. M. Kimura, Totally umbilic hypersurfaces and isoparametric hypersurfaces in space forms, in *Contemporary Aspects of Complex Analysis, Differential Geometry and Mathematical Physics* (World Scientific, Hackensack, 2005), pp. 149–157
274. M. Kimura, S. Maeda, On real hypersurfaces of a complex projective space. Math Z. **202**, 299–311 (1989)
275. M. Kimura, S. Maeda, On real hypersurfaces of a complex projective space II. Tsukuba J. Math. **15**, 547–561 (1991)
276. M. Kimura, S. Maeda, Characterizations of geodesic hyperspheres in a complex projective space in terms of Ricci tensors. Yokohama Math. J. **40**, 35–43 (1992)
277. M. Kimura, S. Maeda, On real hypersurfaces of a complex projective space III. Hokkaido Math. J. **22**, 63–78 (1993)
278. M. Kimura, S. Maeda, Lie derivatives on real hypersurfaces in a complex projective space. Czechoslov. Math. J. **45**(120), 135–148 (1995)
279. M. Kimura, S. Maeda, Geometric meaning of isoparametric hypersurfaces in a real space form. Can. Math. Bull. **43**, 74–78 (2000)
280. Y. Kitagawa, Flat tori in the 3-dimensional sphere. Sugaku Expositions **21**, 133–146 (2008)
281. F. Klein, *Vorlesungen über höhere Geometrie* (Springer, Berlin, 1926). Reprinted by Chelsea, New York, 1957
282. N. Knarr, L. Kramer, Projective planes and isoparametric hypersurfaces. Geom. Dedicata **58**, 193–202 (1995)
283. S. Kobayashi, K. Nomizu, *Foundations of Differential Geometry*, vols. I, II (Wiley-Interscience, New York, 1963, 1969)
284. N. Koike, The quadratic slice theorem and the equiaffine tube theorem for equiaffine Dupin hypersurfaces. Results Math. **47**, 69–92 (2005)
285. N. Koike, Blaschke Dupin hypersurfaces and equiaffine tubes. Results Math. **48**, 97–108 (2005)
286. N. Koike, A Cartan type identity for isoparametric hypersurfaces in symmetric spaces. Tôhoku Math. J. (2) **66**, 435–454 (2014)
287. N. Koike, Collapse of the mean curvature flow for isoparametric submanifolds in non-compact symmetric spaces. Kodai Math. J. **37**, 355–382 (2014)
288. N. Koike, Homogeneity of infinite dimensional anti-Kaehler isoparametric submanifolds. Tokyo J. Math. **37**, 159–178 (2014)
289. Masahiro Kon, Pseudo-Einstein real hypersurfaces in complex space forms. J. Differ. Geom. **14**, 339–354 (1979)
290. Mayuko Kon, 3-dimesional real hypersurfaces and Ricci operator, Differ. Geom. Dyn. Syst. **16**, 189–202 (2014)
291. S.H. Kon, T.-H. Loo, On Hopf hypersurfaces in a non-flat complex space form with η-recurrent Ricci tensor. Kodai Math. J. **33**, 240–250 (2010)
292. S.H. Kon, T.-H. Loo, Real hypersurfaces in a complex space form with η-parallel shape operator. Math. Z. **269**, 47–58 (2011)
293. S.H. Kon, T.-H. Loo, On characterizations of real hypersurfaces in a complex space form with η-parallel shape operator. Can. Math. Bull. **55**, 114–126 (2012)
294. M. Kotani, The first eigenvalue of homogeneous minimal hypersurfaces in a unit sphere $S^{n+1}(1)$. Tôhoku Math. J. **37**, 523–532 (1985)
295. L. Kramer, Compact ovoids in quadrangles II: the classical quadrangles. Geom. Dedicata **79**, 179–188 (2000)

296. L. Kramer, Compact ovoids in quadrangles III: Clifford algebras and isoparametric hypersurfaces. Geom. Dedicata **79**, 321–339 (2000)
297. L. Kramer, *Homogeneous Spaces, Tits Buildings, and Isoparametric Hypersurfaces*. Memoirs - American Mathematical Society, vol. 752 (American Mathematical Society, Providence, 2002)
298. L. Kramer, H. Van Maldeghem, Compact ovoids in quadrangles I: geometric constructions. Geom. Dedicata **78**, 279–300 (1999)
299. W. Kühnel, T. Banchoff, The 9-vertex complex projective plane. Math. Intell. **5**, 11–22 (1983)
300. N. Kuiper, *Immersions with Minimal Total Absolute Curvature*. Coll. de géométrie diff., Centre Belg. de Recherches Math., Bruxelles (1958), pp. 75–88
301. N. Kuiper, On convex maps. Nieuw Archief voor Wisk. **10**, 147–164 (1962)
302. N. Kuiper, Minimal total absolute curvature for immersions. Invent. Math. **10**, 209–238 (1970)
303. N. Kuiper, Tight embeddings and maps. Submanifolds of geometrical class three in E^n, in *The Chern Symposium 1979 (Proceedings of International Symposium, Berkeley, 1979)* (Springer, Berlin/New York, 1980), pp. 97–145
304. N. Kuiper, Taut sets in three space are very special. Topology **23**, 323–336 (1984)
305. N. Kuiper, Geometry in total absolute curvature theory, in *Perspectives in Mathematics*, ed. by Jäger, W. et al. (Birkhäuser, Basel, 1984), pp. 377–392
306. N. Kuiper, W. Meeks, The total curvature of a knotted torus. J. Differ. Geom. **26**, 371–384 (1987)
307. N. Kuiper, W. Pohl, Tight topological embeddings of the real projective plane in E^5. Invent. Math. **42**, 177–199 (1977)
308. H. Kurihara, The structure Jacobi operator for real hypersurfaces in the complex projective plane and the complex hyperbolic plane. Tsukuba J. Math. **35**, 53–66 (2011)
309. J.-H. Kwon, H. Nakagawa, Real hypersurfaces with cyclic-parallel Ricci tensor of a complex projective space. Hokkaido Math. J. **17**, 355–371 (1988)
310. J.-H. Kwon, H. Nakagawa, Real hypersurfaces with cyclic η-parallel Ricci tensor of a complex space form. Yokohama Math. J. **37**, 45–55 (1989)
311. J.-H. Kwon, Y.J. Suh, A new characterization of homogeneous real hypersurfaces in complex space forms. Nihonkai Math. J. **9**, 77–90 (1998)
312. G. Lancaster, Canonical metrics for certain conformally Euclidean spaces of dimension three and codimension one. Duke Math. J. **40**, 1–8 (1973)
313. S.-B. Lee, S.-G. Han, N.-G. Kim, S.S. Ahn, On real hypersurfaces of a complex space form in terms of the Ricci tensor. Commun. Korean Math. Soc. **11**, 757–766 (1996)
314. S.H. Lee, J.D. Pérez, Y.J. Suh, A characterization of Einstein real hypersurfaces in quaternionic projective space. Tsukuba J. Math. **22**, 165–178 (1998)
315. T. Levi-Civita, Famiglie di superficie isoparametrische nell'ordinario spacio euclideo. Atti. Accad. naz. Lincei Rend. Cl. Sci. Fis. Mat. Natur. **26**, 355–362 (1937)
316. C. Li, J. Wang, The classification of isoparametric surfaces in S^3_1. Kobe J. Math. **22**, 1–12 (2005)
317. C.J. Li, U.-H. Ki, Structure eigenvectors of the Ricci tensor in a real hypersurface of a complex projective space. Kyungpook Math. J. **46**, 463–476 (2006)
318. H.-Z. Li, H.-L. Liu, C.-P. Wang, G.-S. Zhao, Möbius isoparametric hypersurfaces in S^{n+1} with two distinct principal curvatures. Acta Math. Sin. (Engl. Ser.) **18**, 437–446 (2002)
319. T.Z. Li, H.-F. Sun, Laguerre isoparametric hypersurfaces in $\mathbf{R}^4$. Acta Math. Sin. **28**, 1179–1186 (2012)
320. T.Z. Li, C.-P. Wang, Laguerre geometry of hypersurfaces in $\mathbf{R}^n$. Manuscripta Math. **122**, 73–95 (2007)
321. T.Z. Li, C.-P. Wang, A note on Blaschke isoparametric hypersurfaces. Int. J. Math. **25**(12), 1450117 (9 p) (2014)
322. X.X. Li, Y.J. Peng, Classification of Blaschke isoparametric hypersurfaces with three distinct Blaschke eigenvalues. Results Math. **58**, 145–172 (2010)

323. X.X. Li, F.Y. Zhang, On the Blaschke isoparametric hypersurfaces in the unit sphere. Acta Math. Sin. **25**, 657–678 (2009)
324. Z.-Q. Li, Lorentzian isoparametric hypersurfaces in the Lorentz sphere S_1^{n+1}, in *Recent Developments in Geometry and Analysis*. Advanced Lectures in Mathematics (ALM), vol. 23 (International Press, Somerville, 2012), pp. 267–328
325. Z.-Q. Li, X.-H. Xie, Space-like isoparametric hypersurfaces in Lorentzian space forms. Translated from the Chinese original [J. Nanchang Univ. Natur. Sci. Ed. **28**, 113–117 (2004)]. Front. Math. China **1**. 130–137 (2006)
326. S. Lie, Über Komplexe, inbesondere Linien- und Kugelkomplexe, mit Anwendung auf der Theorie der partieller Differentialgleichungen. Math. Ann. **5** (1872), 145–208, 209–256 (Ges. Abh. 2, 1–121)
327. S. Lie, G. Scheffers, *Geometrie der Berührungstransformationen* (Teubner, Leipzig, 1896). Reprinted by Chelsea, New York, 1977
328. R. Lilienthal, Besondere Flächen, *Encyklopädie der Mathematischen Wissenschaften*, vol. III (Teubner, Leipzig, 1902–1927), pp. 269–354
329. D.H. Lim, W.H. Sohn, Real hypersurfaces in a complex space form with non-commuting operators. Differ. Geom. Appl. **30**, 622–630 (2012)
330. D.H. Lim, W.H. Sohn, S.-S. Ahn, The property of real hypersurfaces in 2-dimensional complex space form with Ricci operator. Turk. J. Math. **38**, 920–923 (2014)
331. J. Liouville, Note au sujet de l'article précedént. J. de Math. Pure et Appl. (1) **12**, 265–290 (1847)
332. J.A. Little, Manifolds with planar geodesics. J. Differ. Geom. **11**, 265–285 (1976)
333. J.A. Little, W.F. Pohl, On tight immersions of maximal codimension. Invent. Math. **13**, 179–204 (1971)
334. G. Lobos, M. Ortega, Pseudo-parallel real hypersurfaces in complex space forms. Bull. Korean Math. Soc. **41**, 609–618 (2004)
335. M. Lohnherr, *On Ruled Real Hypersurfaces of Complex Space Forms*. Ph.D. thesis, University of Cologne, 1998
336. M. Lohnherr, H. Reckziegel, On ruled real hypersurfaces in complex space forms. Geom. Dedicata **74**, 267–286 (1999)
337. T.-H. Loo, Holomorphic distribution on pseudo-Einstein real hypersurfaces of a complex space form. Kyungpook Math. J. **43**, 355–361 (2003)
338. A. Lytchak, Notes on the Jacobi equation. Differ. Geom. Appl. **27**, 329–334 (2009)
339. A. Lytchak, Geometric resolution of singular Riemannian foliations. Geom. Dedicata **149**, 379–395 (2010)
340. A. Lytchak, G. Thorbergsson, Variationally complete actions on nonnegatively curved manifolds. Ill. J. Math. **51**, 605–615 (2007)
341. A. Lytchak, G. Thorbergsson, Curvature explosion in quotients and applications. J. Differ. Geom. **85**, 117–139 (2010)
342. S.M. Lyu, J.D. Pérez, Y.J. Suh, On real hypersurfaces in a quaternionic hyperbolic space in terms of the derivative of the second fundamental tensor. Acta Math. Hungar. **97**, 145–172 (2002)
343. S.M. Lyu, Y.J. Suh, Real hypersurfaces in complex hyperbolic space with η-recurrent second fundamental tensor. Nihonkai Math. J. **8**, 19–27 (1997)
344. H. Ma, Y. Ohnita, On Lagrangian submanifolds in complex hyperquadrics and isoparametric hypersurfaces in spheres. Math. Z. **261**, 749–785 (2009)
345. H. Ma, Y. Ohnita, Hamiltonian stability of the Gauss images of homogeneous isoparametric hypersurfaces I. J. Differ. Geom. **97**, 275–348 (2014)
346. S. Maeda, Real hypersurfaces of complex projective spaces. Math. Ann. **263**, 473–478 (1983)
347. S. Maeda, Ricci tensors of real hypersurfaces in a complex projective space. Proc. Am. Math. Soc. **122**, 1229–1235 (1994)
348. S. Maeda, H. Naitoh, Real hypersurfaces with φ-invariant shape operator in a complex projective space. Glasg. Math. J. **53**, 347–358 (2011)

349. S. Maeda, H. Tanabe, Characterizations of isoparametric hypersurfaces in a sphere. Mem. Fac. Sci. Eng. Shimane Univ. Ser. B Math. Sci. **45**, 41–48 (2012)
350. Y. Maeda, On real hypersurfaces of a complex projective space. J. Math. Soc. Jpn. **28**, 529–540 (1976)
351. M. Magid, Lorentzian isoparametric hypersurfaces. Pac. J. Math. **118**, 165–197 (1985)
352. A.-L. Mare, Cohomology of isoparametric hypersurfaces in Hilbert space. Geom. Dedicata **85**, 21–43 (2001)
353. A. Martinez, J.D. Pérez, Real hypersurfaces in quaternionic projective space. Ann. Mat. Pura Appl. (4) **145**, 355–384 (1986)
354. Y. Matsuyama, A characterization of real hypersurfaces of complex projective space III. Yokohama Math. J. **46**, 119–126 (1999)
355. C.R.F. Maunder, *Algebraic Topology* (Van Nostrand Reinhold, London, 1970)
356. J.C. Maxwell, On the cyclide. Q. J. Pure Appl. Math. **34** (1867). See also Collected Works, vol. 2, pp. 144–159
357. J. Milnor, On the total curvature of knots. Ann. Math. **52**, 248–257 (1950)
358. J. Milnor, On the total curvature of space curves. Math. Scand. **1**, 289–296 (1953)
359. J. Milnor, *Morse Theory*. Annals of Mathematics Studies, vol. 51 (Princeton U. Press, Princeton, 1963)
360. J. Milnor, *Topology from the Differentiable Viewpoint* (University of Virginia Press, Charlottesville, 1965)
361. R. Miyaoka, Minimal hypersurfaces in the space form with three principal curvatures. Math. Z. **170**, 137–151 (1980)
362. R. Miyaoka, Complete hypersurfaces in the space form with three principal curvatures. Math. Z. **179**, 345–354 (1982)
363. R. Miyaoka, Compact Dupin hypersurfaces with three principal curvatures. Math. Z. **187**, 433–452 (1984)
364. R. Miyaoka, Taut embeddings and Dupin hypersurfaces, in *Differential Geometry of Submanifolds, Proceedings of Conference Kyoto, Japan, 1984*. Lecture Notes in Mathematics, vol. 1090 (Springer, Berlin/New York, 1984), pp. 15–23
365. R. Miyaoka, Dupin hypersurfaces and a Lie invariant. Kodai Math. J. **12**, 228–256 (1989)
366. R. Miyaoka, Dupin hypersurfaces with six principal curvatures. Kodai Math. J. **12**, 308–315 (1989)
367. R. Miyaoka, The linear isotropy group of $G_2/SO(4)$, the Hopf fibering and isoparametric hypersurfaces. Osaka J. Math. **30**, 179–202 (1993)
368. R. Miyaoka, Re-examination of isoparametric hypersurfaces. Sugaku **53**, 18–33 (2001)
369. R. Miyaoka, Classification of isoparametric hypersurfaces with four principal curvatures by T.E. Cecil, Q.S. Chi, G.R. Jensen. Sugaku **58**, 225–238 (2006)
370. R. Miyaoka, The Dorfmeister–Neher theorem on isoparametric hypersurfaces. Osaka J. Math. **46**, 675–715 (2009)
371. R. Miyaoka, Geometry of G_2 orbits and isoparametric hypersurfaces. Nagoya Math. J. **203**, 175–189 (2011)
372. R. Miyaoka, Isoparametric hypersurfaces and the moment map, in *Proceedings of the 16th International Workshop on Differential Geometry and the 5th KNUGRG-OCAMI Differential Geometry Workshop [Volume 16]* (Natl. Inst. Math. Sci. (NIMS), Taejon, 2012), pp. 11–23
373. R. Miyaoka, Isoparametric hypersurfaces with $(g, m) = (6, 2)$. Ann. Math. **177**, 53–110, 315–337 (2013)
374. R. Miyaoka, Errata on isoparametric hypersurfaces with $(g, m) = (6, 2)$, Mathematical Institute, Tôhoku University (2013). Available at http://www.math.tohoku.ac.jp/people/miyaoka/publication-ev3.html. To appear in Ann. Math.
375. R. Miyaoka, Transnormal functions and transnormal systems, in *Proceedings of the 17th International Workshop on Differential Geometry and the 7th KNUGRG-OCAMI Differential Geometry Workshop [Volume 17]* (Natl. Inst. Math. Sci. (NIMS), Taejon, 2013), pp. 13–20
376. R. Miyaoka, Transnormal functions on a Riemannian manifold. Differ. Geom. Appl. **31**, 130–139 (2013)

377. R. Miyaoka, T. Ozawa, Construction of taut embeddings and Cecil-Ryan conjecture, in *Geometry of Manifolds*, ed. by Shiohama, K., Perspect. Math. vol. 8 (Academic, New York, 1989), pp. 181–189
378. S. Montiel, Real hypersurfaces of a complex hyperbolic space. J. Math. Soc. Jpn. **37**, 515–535 (1985)
379. M. Morse, S. Cairns, *Critical Point Theory in Global Analysis and Differential Topology* (Academic, New York, 1969)
380. S. Mullen, Isoparametric systems on symmetric spaces, in *Geometry and Topology of Submanifolds VI*, ed. by Dillen, F. et al. (World Scientific, River Edge, 1994), pp. 152–154
381. H.-F. Münzner, Isoparametrische Hyperflächen in Sphären I. Math. Ann. **251**, 57–71 (1980)
382. H.-F. Münzner, Isoparametrische Hyperflächen in Sphären II. Math. Ann. **256**, 215–232 (1981)
383. T. Murphy, Real hypersurfaces of complex and quaternionic hyperbolic spaces. Adv. Geom. **14**, 1–10 (2014)
384. E. Musso, L. Nicolodi, A variational problem for surfaces in Laguerre geometry. Trans. Am. Math. Soc. **348**, 4321–4337 (1996)
385. E. Musso, L. Nicolodi, Laguerre geometry for surfaces with plane lines of curvature. Abh. Math. Sem. Univ. Hamburg **69**, 123–138 (1999)
386. H. Muto, The first eigenvalue of the Laplacian of an isoparametric hypersurface in a unit sphere. Math. Z. **197**, 531–549 (1988)
387. H. Muto, Y. Ohnita, H. Urakawa, Homogeneous minimal hypersurfaces in the unit sphere and the first eigenvalue of the Laplacian. Tôhoku Math. J. **36**, 253–267 (1984)
388. S. Nagai, Some characterization of a real hypersurface of type (B) in complex space forms. Saitama Math. J. **19**, 37–46 (2001)
389. S. Nagai, Some applications of homogeneous structures on Hopf hypersurfaces in a complex space form. Math. J. Toyama Univ. **25**, 77–86 (2002)
390. S. Nagai, On homogeneous structures of real hypersurfaces in non-flat complex space forms, in *Proceedings of the Eighth International Workshop on Differential Geometry (Taegu 2003)* (Kyungpook National University, Taegu, 2004), pp. 63–72
391. H. Nakagawa, R. Takagi, On locally symmetric Kaehler submanifolds in a complex projective space. J. Math. Soc. Jpn. **28**, 638–667 (1976)
392. S. Nakajima, Real hypersurfaces of a complex projective space. Tsukuba J. Math. **23**, 235–243 (1999)
393. R. Niebergall, Dupin hypersurfaces in $\mathbf{R}^5$ I. Geom. Dedicata **40**, 1–22 (1991)
394. R. Niebergall, Dupin hypersurfaces in $\mathbf{R}^5$ II. Geom. Dedicata **41**, 5–38 (1992)
395. R. Niebergall, P.J. Ryan, Isoparametric hypersurfaces – the affine case, in *Geometry and Topology of Submanifolds V* (World Scientific, River Edge, 1993), pp. 201–214
396. R. Niebergall, P.J. Ryan, Affine isoparametric hypersurfaces. Math. Z. **217**, 479–485 (1994)
397. R. Niebergall, P.J. Ryan, Focal sets in affine geometry, in *Geometry and Topology of Submanifolds VI* (World Scientific, River Edge, 1994), pp. 155–164
398. R. Niebergall, P.J. Ryan, Affine Dupin surfaces. Trans. Am. Math. Soc. **348**, 1093–1117 (1996)
399. R. Niebergall, P.J. Ryan, Real hypersurfaces in complex space forms, in *Tight and Taut Submanifolds*, MSRI Publications, vol. 32 (Cambridge University Press, Cambridge, 1997), pp. 233–305
400. R. Niebergall, P.J. Ryan, Semi-parallel and semi-symmetric real hypersurfaces in complex space forms. Kyungpook Math. J. **38**, 227–234 (1998)
401. K. Nomizu, *Fundamentals of Linear Algebra* (McGraw-Hill, New York, 1966)
402. K. Nomizu, Characteristic roots and vectors of a differentiable family of symmetric matrices. Linear Multilinear Algebra **2**, 159–162 (1973)
403. K. Nomizu, Some results in E. Cartan's theory of isoparametric families of hypersurfaces. Bull. Am. Math. Soc. **79**, 1184–1188 (1973)

404. K. Nomizu, Élie Cartan's work on isoparametric families of hypersurfaces, in *Proceedings of Symposia in Pure Mathematics*, vol. 27, Part 1 (American Mathematical Society, Providence, 1975), pp. 191–200
405. K. Nomizu, L. Rodriguez, Umbilical submanifolds and Morse functions. Nagoya Math. J. **48**, 197–201 (1972)
406. K. Nomizu, B. Smyth, A formula of Simons' type and hypersurfaces with constant mean curvature. J. Differ. Geom. **3**, 367–377 (1969)
407. Y. Ohnita, Geometry of Lagrangian submanifolds and isoparametric hypersurfaces, in *Proceedings of the 14th International Workshop on Differential Geometry and the 3rd KNUGRG-OCAMI Differential Geometry Workshop [Volume 14]* (Natl. Inst. Math. Sci. (NIMS), Taejon, 2010), pp. 43–67
408. M. Okumura, On some real hypersurfaces of a complex projective space. Trans. Am. Math. Soc. **212**, 355–364 (1975)
409. C. Olmos, Isoparametric submanifolds and their homogeneous structure. J. Differ. Geom. **38**, 225–234 (1993)
410. C. Olmos, Homogeneous submanifolds of higher rank and parallel mean curvature. J. Differ. Geom. **39**, 605–627 (1994)
411. B. O'Neill, The fundamental equations of a submersion. Michigan Math. J. **13**, 459–469 (1966)
412. B. O'Neill, *Semi-Riemannian Geometry* (Academic, New York, 1983)
413. M. Ooguri, On Cartan's identities of equiaffine isoparametric hypersurfaces. Results Math. **46**, 79–90 (2004)
414. M. Ortega, Classifications of real hypersurfaces in complex space forms by means of curvature conditions. Bull. Belg. Math. Soc. Simon Stevin **9**, 351–360 (2002)
415. M. Ortega, J.D. Pérez, On the Ricci tensor of a real hypersurface of quaternionic hyperbolic space. Manuscripta Math. **93**, 49–57 (1997)
416. M. Ortega, J.D. Pérez, D-Einstein real hypersurfaces in quaternionic space forms. Ann. Mat. Pura Appl. **178**, 33–44 (2000)
417. M. Ortega, J.D. Pérez, Y.J. Suh, Real hypersurfaces in quaternionic projective spaces with commuting tangent Jacobi operators. Glasg. Math. J. **45**, 79–89 (2003)
418. H. Osborn, *Vector Bundles*, vol. 1 (Academic, New York, 1982)
419. T. Otsuki, Minimal hypersurfaces in a Riemannian manifold of constant curvature. Am. J. Math. **92**, 145–173 (1970)
420. T. Otsuki, Minimal hypersurfaces with three principal curvatures in S^{n+1}. Kodai Math. J. **1**, 1–29 (1978)
421. T. Ozawa, On the critical sets of distance functions to a taut submanifold. Math. Ann. **276**, 91–96 (1986)
422. H. Ozeki, M. Takeuchi, On some types of isoparametric hypersurfaces in spheres I. Tôhoku Math. J. **27**, 515–559 (1975)
423. H. Ozeki, M. Takeuchi, On some types of isoparametric hypersurfaces in spheres II. Tôhoku Math. J. **28**, 7–55 (1976)
424. J.S. Pak, Real hypersurfaces in quaternionic Kaehlerian manifolds with constant Q-sectional curvature. Kodai Math. Sem. Rep. **29**, 22–61 (1977)
425. R. Palais, *A Global Formulation of the Lie Theory of Transformation Groups*. Memoirs - American Mathematical Society, vol. 22 (American Mathematical Society, Providence, 1957)
426. R. Palais, C.-L. Terng, *Critical Point Theory and Submanifold Geometry*. Lecture Notes in Mathematics, vol. 1353 (Springer, Berlin/New York, 1988)
427. K.S. Park, Isoparametric families on projective spaces. Math. Ann. **284**, 503–513 (1989)
428. G.P. Paternain, *Geodesic Flows*. Progress in Mathematics, vol. 180 (Birkhäuser Boston, Inc., Boston, 1999)
429. C.-K. Peng, Z. Hou, A remark on the isoparametric polynomials of degree 6, in *Differential Geometry and Topology, Proceedings Tianjin 1986–1987*, ed. by Jiang, B. et al., Lecture Notes in Mathematics, vol. 1369 (Springer, Berlin/New York, 1989), pp. 222–224

430. J.D. Pérez, Cyclic-parallel real hypersurfaces of quaternionic projective space. Tsukuba J. Math. **17**, 189–191 (1993)
431. J.D. Pérez, On certain real hypersurfaces of quaternionic projective space II. Algebras Groups Geom. **10**, 13–24 (1993)
432. J.D. Pérez, Real hypersurfaces of quaternionic projective space satisfying $\nabla_{U_i}A = 0$. J. Geom. **49**, 166–177 (1994)
433. J.D. Pérez, On the Ricci tensor of real hypersurfaces of quaternionic projective space. Int. J. Math. Math. Sci. **19**, 193–197 (1996)
434. J.D. Pérez, A characterization of almost-Einstein real hypersurfaces of quaternionic projective space. Tsukuba J. Math. **21**, 207–212 (1997)
435. J.D. Pérez, F.G. Santos, On pseudo-Einstein real hypersurfaces of the quaternionic projective space. Kyungpook Math. J. **25**, 15–28 (1985)
436. J.D. Pérez, F.G. Santos, On certain real hypersurfaces of quaternionic projective space. Int. J. Math. Math. Sci. **14**, 205–207 (1991)
437. J.D. Pérez, F.G. Santos, On real hypersurfaces with harmonic curvature of a quaternionic projective space. J. Geom. **40**, 165–169 (1991)
438. J.D. Pérez, F.G. Santos, Cyclic-parallel Ricci tensor on real hypersurfaces of a quaternionic projective space. Rev. Roumaine Math. Pures Appl. **38**, 37–40 (1993)
439. J.D. Pérez, F.G. Santos, Y.J. Suh, Real hypersurfaces in nonflat complex space forms with commuting structure Jacobi operator. Houston J. Math. **33**, 1005–1009 (2007)
440. J.D. Pérez, F.G. Santos, Y.J. Suh, Real hypersurfaces in complex projective space whose structure Jacobi operator is of Codazzi type. Can. Math. Bull. **50**, 347–355 (2007)
441. J.D. Pérez, Y.J. Suh, Real hypersurfaces of quaternionic projective space satisfying $\nabla_{U_i}R = 0$. Differ. Geom. Appl. **7**, 211–217 (1997)
442. U. Pinkall, *Dupin'sche Hyperflächen*. Dissertation, University of Freiburg, 1981
443. U. Pinkall, W-Kurven in der ebenen Lie-Geometrie I, II. Elemente der Mathematik **39**, 28–33, 67–78 (1984)
444. U. Pinkall, Letter to T.E. Cecil, December 5 (1984)
445. U. Pinkall, Dupin'sche Hyperflächen in E^4. Manuscr. Math. **51**, 89–119 (1985)
446. U. Pinkall, Dupin hypersurfaces. Math. Ann. **270**, 427–440 (1985)
447. U. Pinkall, Curvature properties of taut submanifolds. Geom. Dedicata **20**, 79–83 (1986)
448. U. Pinkall, G. Thorbergsson, Deformations of Dupin hypersurfaces. Proc. Am. Math. Soc. **107**, 1037–1043 (1989)
449. U. Pinkall, G. Thorbergsson, Taut 3-manifolds. Topology **28**, 389–401 (1989)
450. U. Pinkall, G. Thorbergsson, Examples of infinite dimensional isoparametric submanifolds. Math. Z. **205**, 279–286 (1990)
451. M.J. Pratt, Cyclides in computer aided geometric design. Comput. Aided Geom. Design **7**, 221–242 (1990)
452. M.J. Pratt, Cyclides in computer aided geometric design II. Comput. Aided Geom. Design **12**, 131–152 (1995)
453. A. Pressley, G. Segal, *Loop Groups* (Oxford University Press, New York, 1986)
454. Y.-S. Pyo, On real hypersurfaces in a complex space form (I). Tsukuba J. Math. **18**, 483–492 (1994)
455. Y.-S. Pyo, On real hypersurfaces in a complex space form (II). Commun. Korean Math. Soc. **9**, 369–383 (1994)
456. C. Qian, Z.-Z. Tang, W. Yan, New examples of Willmore submanifolds in the unit sphere via isoparametric functions II. Ann. Glob. Anal. Geom. **43**, 47–62 (2013)
457. H. Reckziegel, Krümmungsflächen von isometrischen Immersionen in Räume konstanter Krümmung. Math. Ann. **223**, 169–181 (1976)
458. H. Reckziegel, On the eigenvalues of the shape operator of an isometric immersion into a space of constant curvature. Math. Ann. **243**, 71–82 (1979)
459. H. Reckziegel, Completeness of curvature surfaces of an isometric immersion. J. Differ. Geom. **14**, 7–20 (1979)

460. C.M. Riveros, Dupin hypersurfaces with four principal curvatures in $\mathbf{R}^5$ with principal coordinates. Rev. Mat. Complut. **23**, 341–354 (2010)
461. C.M. Riveros, A characterization of Dupin hypersurfaces in $\mathbf{R}^4$. Bull. Belg. Math. Soc. Simon Stevin **20**, 145–154 (2013)
462. C.M. Riveros, L.A. Rodrigues, K. Tenenblat, On Dupin hypersurfaces with constant Möbius curvature. Pac. J. Math. **236**, 89–103 (2008)
463. C.M. Riveros, K. Tenenblat, On four dimensional Dupin hypersurfaces in Euclidean space. An. Acad. Brasil Ciênc. **75**, 1–7 (2003)
464. C.M. Riveros, K. Tenenblat, Dupin hypersurfaces in $\mathbf{R}^5$. Can. J. Math. **57**, 1291–1313 (2005)
465. L.A. Rodrigues, K. Tenenblat, A characterization of Moebius isoparametric hypersurfaces of the sphere. Monatsh. Math. **158**, 321–327 (2009)
466. D. Rowe, The early geometrical works of Sophus Lie and Felix Klein, in *The History of Modern Mathematics*, vol. I (Academic, San Diego, 1989), pp. 209–273
467. D. Ruberman, Null-homotopic embedded spheres of codimension one, in *Tight and Taut Submanifolds (Berkeley, CA, 1994)*. Math. Sci. Res. Inst. Publ., vol. 32 (Cambridge University Press, Cambridge, 1997), pp. 229–232
468. P.J. Ryan, Homogeneity and some curvature conditions for hypersurfaces. Tôhoku Math. J. **21**, 363–388 (1969)
469. P.J. Ryan, Hypersurfaces with parallel Ricci tensor. Osaka J. Math. **8**, 251–259 (1971)
470. P.J. Ryan, Intrinsic properties of real hypersurfaces in complex space forms, in *Geometry and Topology of Submanifolds X, Beijing-Berlin 1999* (World Scientific, River Edge, 2000), pp. 266–273
471. H. Samelson, Orientability of hypersurfaces in $\mathbf{R}^n$. Proc. Am. Math. Soc. **22**, 301–302 (1969)
472. P. Samuel, *Projective Geometry* (Springer, Berlin, 1988)
473. C. Sanchez, Triality and the normal sections of Cartan's isoparametric hypersurfaces. Rev. Un. Mat. Argent. **52**, 73–88 (2011)
474. M. Scherfner, S. Weiss, Towards a proof of the Chern conjecture for isoparametric hypersurfaces in spheres, 33. Süddeutsches Kolloquium über Differentialgeometrie, 1–13, Institut für Diskrete Mathematik und Geometrie. Technische Universität Wien, Vienna (2008)
475. M. Scherfner, S. Weiss, S.-T. Yau, A review of the Chern conjecture for isoparametric hypersurfaces in spheres, in *Advances in Geometric Analysis*. Advanced Lectures in Mathematics (ALM), vol. 21 (International Press, Somerville, 2012), pp. 175–187
476. R. Schoen, S.-T. Yau, Lectures on differential geometry, in *Conference Proceedings and Lecture Notes in Geometry and Topology, I* (International Press, Cambridge, 1994)
477. J.A. Schouten, Über die konforme Abildung n-dimenionaler Mannigfaltigkeiter mit quadratisher Maßbestimmung auf eine Mannigfaltigkeit mit euklidischer Maßbestimmung. Math. Z. **11**, 58–88 (1921)
478. M. Schrott, B. Odehnal, Ortho-circles of Dupin cyclides. J. Geom. Graph. **10**, 73–98 (2006)
479. B. Segre, Una proprietá caratteristica de tre sistemi ∞^1 di superficie. Atti. Accad. Sci. Torino **59**, 666–671 (1924)
480. B. Segre, Famiglie di ipersuperficie isoparametrische negli spazi euclidei ad un qualunque numero di demesioni. Atti. Accad. naz, Lincei Rend. Cl. Sci. Fis. Mat. Natur. **27**, 203–207 (1938)
481. V. Shklover, Schiffer problem and isoparametric hypersurfaces. Rev. Mat. Iberoam. **16**, 529–569 (2000)
482. S. Shu, Y. Li, Laguerre characterizations of hypersurfaces in $\mathbf{R}^n$. Bull. Korean Math. Soc. **50**, 1781–1797 (2013)
483. S. Shu, B. Su, Para-Blaschke isoparametric hypersurfaces in a unit sphere $S^{n+1}(1)$. Glasg. Math. J. **54**, 579–597 (2012)
484. A. Siffert, Classification of isoparametric hypersurface in spheres with $(g, m) = (6, 1)$. Proc. Amer. Math. Soc. doi: http://dx.doi.org/10.1090/proc/12924
485. J. Simons, Minimal varieties in Riemannian manifolds. Ann. Math. **88**, 62–105 (1968)
486. D. Singley, Smoothness theorems for the principal curvatures and principal vectors of a hypersurface. Rocky Mt. J. Math. **5**, 135–144 (1975)

487. B. Smyth, Differential geometry of complex hypersurfaces. Ann. Math. **85**, 246 –266 (1967)
488. W.H. Sohn, Characterizations of real hypersurfaces of complex space forms in terms of Ricci operators. Bull. Korean Math. Soc. **44**, 195–202 (2007)
489. B. Solomon, The harmonic analysis of cubic isoparametric minimal hypersurfaces I: dimensions 3 and 6. Am. J. Math. **112**, 157–203 (1990)
490. B. Solomon, The harmonic analysis of cubic isoparametric minimal hypersurfaces II: dimensions 12 and 24. Am. J. Math. **112**, 205–241 (1990)
491. B. Solomon, Quartic isoparametric hypersurfaces and quadratic forms. Math. Ann. **293**, 387–398 (1992)
492. C. Somigliana, Sulle relazione fra il principio di Huygens e l'ottica geometrica. Atti. Accad. Sci. Torino **54**, 974–979 (1918–1919). Also in: Memorie Scelte, 434–439
493. Y.P. Song, Laguerre isoparametric hypersurfaces in $\mathbf{R}^n$ with two distinct non-zero principal curvatures. Acta Math. Sin. **30**, 169–180 (2014)
494. Y.P. Song, C.-P. Wang, Laguerre minimal surfaces in $\mathbf{R}^3$. Acta Math. Sin. **24**, 1861–1870 (2008)
495. M. Spivak, *A Comprehensive Introduction to Differential Geometry*, 3rd edn., vols. 1–5 (Publish or Perish, Houston, 1999)
496. Y.L. Srinivas, D. Dutta, Blending and joining using cyclides. ASME Trans. J. Mech. Design **116**, 1034–1041 (1994)
497. Y.L. Srinivas, D. Dutta, An intuitive procedure for constructing complex objects using cyclides. Comput. Aided Design **26**, 327–335 (1994)
498. Y.L. Srinivas, D. Dutta, Cyclides in geometric modeling: computational tools for an algorithmic infrastructure. ASME Trans. J. Mech. Design **117**, 363–373 (1995)
499. Y.L. Srinivas, D. Dutta, Rational parametrization of parabolic cyclides. Comput. Aided Geom. Design **12**, 551–566 (1995)
500. S. Sternberg, *Lectures on Differential Geometry*, 2nd edn. (Prentice-Hall, Englewood Cliffs, 1964; Chelsea, New York, 1983)
501. J. Stoker, *Differential Geometry* (Wiley, New York, 1969)
502. S. Stolz, Multiplicities of Dupin hypersurfaces. Invent. Math. **138**, 253–279 (1999)
503. W. Strübing, Isoparametric submanifolds. Geom. Dedicata **20**, 367–387 (1986)
504. Y.J. Suh, On real hypersurfaces of a complex space form with η-parallel Ricci tensor. Tsukuba J. Math. **14**, 27–37 (1990)
505. S.-S. Tai, Minimum embeddings of compact symmetric spaces of rank one. J. Differ. Geom. **2**, 55–66 (1968)
506. H. Takagi, A condition for isoparametric hypersurfaces of S^n to be homogeneous. Tôhoku Math. J. (2) **37**, 241–250 (1985)
507. R. Takagi, On homogeneous real hypersurfaces in a complex projective space. Osaka J. Math. **10**, 495–506 (1973)
508. R. Takagi, Real hypersurfaces in a complex projective space with constant principal curvatures I. J. Math. Soc. Jpn. **27**, 43–53 (1975)
509. R. Takagi, Real hypersurfaces in a complex projective space with constant principal curvatures II. J. Math. Soc. Jpn. **27**, 507–516 (1975)
510. R. Takagi, A class of hypersurfaces with constant principal curvatures in a sphere. J. Differ. Geom. **11**, 225–233 (1976)
511. R. Takagi, T. Takahashi, On the principal curvatures of homogeneous hypersurfaces in a sphere, in *Differential Geometry in honor of K. Yano* (Kinokuniya, Tokyo, 1972), pp. 469–481
512. M. Takeuchi, Proper Dupin hypersurfaces generated by symmetric submanifolds. Osaka Math. J. **28**, 153–161 (1991)
513. M. Takeuchi, S. Kobayashi, Minimal imbeddings of R-spaces. J. Differ. Geom. **2**, 203–215 (1968)
514. Z.-Z. Tang, Isoparametric hypersurfaces with four distinct principal curvatures. Chin. Sci. Bull. **36**, 1237–1240 (1991)

515. Z.-Z. Tang, Multiplicities of equifocal hypersurfaces in symmetric spaces. Asian J. Math. **2**, 181–214 (1998)
516. Z.-Z. Tang, Y. Xie, W. Yan, Schoen-Yau-Gromov-Lawson theory and isoparametric foliations. Commun. Anal. Geom. **20**, 989–1018 (2012)
517. Z.-Z. Tang, Y. Xie, W. Yan, Isoparametric foliation and Yau conjecture on the first eigenvalue, II. J. Funct. Anal. **266**, 6174–6199 (2014)
518. Z.-Z. Tang, W. Yan, New examples of Willmore submanifolds in the unit sphere via isoparametric functions. Ann. Glob. Anal. Geom. **42**, 403–410 (2012)
519. Z.-Z. Tang, W. Yan, Critical sets of eigenfunctions and Yau conjecture (2012) [arXiv:1203.2089v1 [math.DG]]
520. Z.-Z. Tang, W. Yan, Isoparametric foliation and Yau conjecture on the first eigenvalue. J. Differ. Geom. **94**, 521–540 (2013)
521. Z.-Z. Tang, W. Yan, Isoparametric foliation and a problem of Besse on generalization of Einstein condition (2013) [arXiv:1307.3807v2 [math.DG]]
522. T. Taniguchi, Characterizations of real hypersurfaces of a complex hyperbolic space in terms of Ricci tensor and holomorphic distribution. Tsukuba J. Math. **18**, 469–482 (1994)
523. Y. Tashiro, S. Tachibana, On Fubinian and *C*-Fubinian manifolds. Kodai Math. Sem. Rep. **15**, 176–183 (1963)
524. L. Taylor, Taut codimension one spheres of odd order, in *Geometry and Topology: Aarhus (1998)*. Contemporary Mathematics, vol. 258 (American Mathematical Society, Providence, 2000), pp. 369–375
525. C.-L. Terng, Isoparametric submanifolds and their Coxeter groups. J. Differ. Geom. **21**, 79–107 (1985)
526. C.-L. Terng, Convexity theorem for isoparametric submanifolds. Invent. Math. **85**, 487–492 (1986)
527. C.-L. Terng, Submanifolds with flat normal bundle. Math. Ann. **277**, 95–111 (1987)
528. C.-L. Terng, Proper Fredholm submanifolds of Hilbert space. J. Differ. Geom. **29**, 9–47 (1989)
529. C.-L. Terng, Recent progress in submanifold geometry, in *Proceedings of Symposia in Pure Mathematics*, vol. 54, Part 1 (American Mathematical Society, Providence, 1993), pp. 439–484
530. C.-L. Terng, G. Thorbergsson, Submanifold geometry in symmetric spaces. J. Differ. Geom. **42**, 665–718 (1995)
531. C.-L. Terng, G. Thorbergsson, Taut immersions into complete Riemannian manifolds, in *Tight and Taut Submanifolds*, MSRI Publications, vol. 32 (Cambridge University Press, Cambridge, 1997), pp. 181–228
532. T. Theofanidis, P.J. Xenos, Non-existence of real hypersurfaces equipped with Jacobi structure operator of Codazzi type. Int. J. Pure Appl. Math. **65**, 31–40 (2010)
533. G. Thorbergsson, Dupin hypersurfaces. Bull. Lond. Math. Soc. **15**, 493–498 (1983)
534. G. Thorbergsson, Highly connected taut submanifolds. Math. Ann. **265**, 399–405 (1983)
535. G. Thorbergsson, Tight immersions of highly connected manifolds. Comment. Math. Helv. **61**, 102–121 (1986)
536. G. Thorbergsson, Homogeneous spaces without taut embeddings. Duke Math. J. **57**, 347–355 (1988)
537. G. Thorbergsson, Isoparametric foliations and their buildings. Ann. Math. **133**, 429–446 (1991)
538. G. Thorbergsson, A survey on isoparametric hypersurfaces and their generalizations, in *Handbook of Differential Geometry*, vol. I (North-Holland, Amsterdam, 2000), pp. 963–995
539. G. Thorbergsson, Singular Riemannian foliations and isoparametric submanifolds. Milan J. Math. **78**, 355–370 (2010)
540. D. Töben, Parallel focal structure and singular Riemannian foliations, Trans. Am. Math. Soc. **358**, 1677–1704 (2006)
541. B.L. Van der Waerden, *Modern Algebra* (Ungar, New York, 1949)
542. L. Vanhecke, T.J. Willmore, Interaction of tubes and spheres. Math. Ann. **263**, 31–42 (1983)

543. L. Verhóczki, Isoparametric submanifolds of general Riemannian manifolds, in *Differential Geometry and Its Applications (Eger, 1989)*, ed. by Szenthe, J., Tamássy, L., Colloquium Mathematical Society János Bolyai, vol. 56 (North-Holland, Amsterdam, 1992), pp. 691–705
544. C.-P. Wang, Surfaces in Möbius geometry. Nagoya Math. J. **125**, 53–72 (1992)
545. C.-P. Wang, Möbius geometry for hypersurfaces in S^4. Nagoya Math. J. **139**, 1–20 (1995)
546. C.-P. Wang, Möbius geometry of submanifolds in S^n. Manuscripta Math. **96**, 517–534 (1998)
547. Q.-M. Wang, Isoparametric hypersurfaces in complex projective spaces, in *Symposium on Differential Geometry and Differential Equations (Beijing, 1980)*, vol. 3 (Science Press, Beijing, 1982), pp. 1509–1523
548. Q.-M. Wang, Isoparametric maps of Riemannian manifolds and their applications, in *Advances in Science of China, Mathematics*, ed. by Gu, C.H., Wang, Y., vol. 2 (Wiley-Interscience, New York, 1986), pp. 79–103
549. Q.-M. Wang, Isoparametric functions on Riemannian manifolds I. Math. Ann. **277**, 639–646 (1987)
550. Q.-M. Wang, On the topology of Clifford isoparametric hypersurfaces. J. Differ. Geom. **27**, 55–66 (1988)
551. X.M. Wang, Dupin hypersurfaces with constant mean curvatures, in *Differential Geometry (Shanghai, 1991)* (World Scientific, River Edge, 1993), pp. 247–253
552. B. Wegner, Morse theory for distance functions to affine subspaces of Euclidean spaces, in *Proceedings of the Conference on Differential Geometry and Its Applications, Part 1* (Charles University, Prague, 1984), pp. 165–168
553. A. West, Isoparametric systems, in *Geometry and Topology of Submanifolds I* (World Scientific, River Edge, 1989), pp. 220–230
554. S. Wiesendorf, Taut submanifolds and foliations. J. Differ. Geom. **96**, 457–505 (2014)
555. B. Wu, Isoparametric submanifolds of hyperbolic spaces. Trans. Am. Math. Soc. **331**, 609–626 (1992)
556. B. Wu, A finiteness theorem for isoparametric hypersurfaces. Geom. Dedicata **50**, 247–250 (1994)
557. L. Xiao, Lorentzian isoparametric hypersurfaces in H_1^{n+1}. Pac. J. Math. **189**, 377–397 (1999)
558. L. Xiao, Principal curvatures of isoparametric hypersurfaces in $\mathbf{C}P^n$. Trans. Am. Math. Soc. **352**. 4487–4499 (2000)
559. K. Yano, On harmonic and Killing vector fields. Ann. Math. **55**, 38–45 (1952)
560. K. Yano, M. Kon, Generic submanifolds. Ann. Mat. Pura Appl. **123**, 59–92 (1980)
561. S.-T. Yau, *Seminar on Differential Geometry, Problem Section*. Annals of Mathematics Studies, vol. 102 (Princeton University Press, 1982)
562. S.-T. Yau, Open problems in geometry, in *Proceedings of Symposia in Pure Mathematics*, vol. 54, Part 1 (American Mathematical Society, Providence, 1993), pp. 1–28
563. Q. Zhao, Isoparametric submanifolds of hyperbolic spaces. Chin. J. Contemp. Math. **14**, 339–346 (1993)

Index

T.E. Cecil, P.J. Ryan, *Geometry of Hypersurfaces*, Springer Monographs in Mathematics, DOI 10.1007/978-1-4939-3246-7

I

J

K

U

V

W

X

Y

Z

CPI Antony Rowe
Chippenham, UK
[illegible]